ORIGIN, DIAGENESIS, AND PETROPHYSICS OF CLAY MINERALS IN SANDSTONES

Edited by

David W. Houseknecht, Department of Geological Sciences, University of Missouri, Columbia, Missouri 65211
and
Edward D. Pittman, Department of Geosciences, University of Tulsa, Tulsa, Oklahoma 74104

Copyright 1992 by
SEPM Special Publication No. 47

Barbara H. Lidz, Editor of Special Publications

Tulsa, Oklahoma, U.S.A. *April, 1992*

CONTENTS

CLAY MINERALS IN SANDSTONES: HISTORICAL PERSPECTIVE AND CURRENT TRENDS . . . . . *David W. Houseknecht and Edward D. Pittman*   1

ISOTOPIC DATING OF DIAGENETIC ILLITES IN RESERVOIR SANDSTONES: INFLUENCE OF THE INVESTIGATOR EFFECT *Norbert Clauer, Joshua D. Cocker, and Sam Chaudhuri*   5

CONTROLS ON POREWATER EVOLUTION DURING SANDSTONE DIAGENESIS, WESTERN CANADA SEDIMENTARY BASIN: AN OXYGEN ISOTOPE PERSPECTIVE . . . . *Fred J. Longstaffe, Barbara J. Tilley, Avner Ayalon, and Catherine A. Connolly*   13

K-AR DATING OF ILLITE DIAGENESIS IN THE MIDDLE ORDOVICIAN ST. PETER SANDSTONE, CENTRAL MICHIGAN BASIN, USA: IMPLICATIONS FOR THERMAL HISTORY . . . . *David A. Barnes, Jean-Pierre Girard, and James L. Aronson*   35

ILLITIC-CLAY FORMATION DURING EXPERIMENTAL DIAGENESIS OF ARKOSES . . . . *Wuu-Liang Huang*   49

CLAY MINERALS IN NORTH SEA SANDSTONES . . . . *Knut Bjørlykke and Per. Aagaard*   65

AUTHIGENIC CLAYS, DIAGENETIC SEQUENCES AND CONCEPTUAL DIAGENETIC MODELS IN CONTRASTING BASIN-MARGIN AND BASIN-CENTER NORTH SEA JURASSIC SANDSTONES AND MUDSTONES . . . *Stuart D. Burley and Joe H. S. MacQuaker*   81

VOLUMETRIC RELATIONS BETWEEN DISSOLVED PLAGIOCLASE AND KAOLINITE IN SANDSTONES: IMPLICATIONS FOR ALUMINUM MASS TRANSFER IN THE SAN JOAQUIN BASIN, CALIFORNIA . . . . *Michael J. Hayes and James R. Boles*   111

AUTHIGENIC MINERALOGY OF SANDSTONES INTERCALATED WITH ORGANIC-RICH MUDSTONES: INTEGRATING DIAGENESIS AND BURIAL HISTORY OF THE MESAVERDE GROUP, PICEANCE BASIN, NW COLORADO . . . . *Laura J. Crossey and Daniel Larsen*   125

CLAY MINERALOGY OF AN INTERBEDDED SANDSTONE, DOLOMITE, AND ANHYDRITE: THE PERMIAN YATES FORMATION, WINKLER COUNTY, TEXAS . . . . *J. S. Janks, M. R. Yusas, and C. M. Hall*   145

INFILTRATED MATERIALS IN CRETACEOUS VOLCANOGENIC SANDSTONES, SAN JORGE BASIN, ARGENTINA . . . . *Thomas L. Dunn*   159

CLAY MINERALOGY, SPORE COLORATION AND DIAGENESIS IN MIDDLE MIOCENE SEDIMENTS OF THE NIGER DELTA . . . . *Francisca E. Oboh*   175

ORIGIN AND DIAGENESIS OF CLAY MINERALS IN THE OLIGOCENE SESPE FORMATION, VENTURA BASIN . . . . *Lori A. Hathon and David W. Houseknecht*   185

DEPOSITIONAL, INFILTRATED AND AUTHIGENIC CLAYS IN FLUVIAL SANDSTONES OF THE JURASSIC SERGI FORMATION, RECÔNCAVO BASIN, NORTHEASTERN BRAZIL . . . . *Marco A. S. Moraes and Luiz F. De Ros*   197

INHERITED GRAIN-RIMMING CLAYS IN SANDSTONES FROM EOLIAN AND SHELF ENVIRONMENTS: THEIR ORIGIN AND CONTROL ON RESERVOIR PROPERTIES . . . . *Michael D. Wilson*   209

CLAY MINERALS IN ATOKAN DEEP-WATER SANDSTONE FACIES, ARKOMA BASIN: ORIGINS AND INFLUENCE ON DIAGENESIS AND RESERVOIR QUALITY . . . . *David W. Houseknecht and Louis M. Ross, Jr.*   227

CLAY COATS: OCCURRENCE AND RELEVANCE TO PRESERVATION OF POROSITY IN SANDSTONES . . . . *Edward D. Pittman, Richard E. Larese, and Milton T. Heald*   241

INFLUENCE OF AUTHIGENIC CLAY MINERALS ON PERMEABILITY . . . . *James J. Howard*   257

FORMATION CLAYS: ARE THEY REALLY A PROBLEM IN PRODUCTION? . . . . *George E. King*   265

COLOR PLATES   273

SUBJECT INDEX   279

ACKNOWLEDGMENTS

We extend our thanks to the following, whose incisive reviews made this volume possible: Joanne Ajdukiewicz, Jim Aronson, Jim Boles, Roger Burtner, Matthew Blauch, Laura Crossey, Tom Dunn, Shirley Dutton, Eric Eslinger, Steve Franks, Milton Heald, Jim Howard, Ian Hutcheon, Jonathan Janks, Rick Lahann, Dick Larese, Mingchou Lee, Paul Lundegard, Earle McBride, Kitty Milliken, Marco Moraes, Paul Nadeau, Franca Oboh, Nadia Pallatt, Dennis Prezbindowski, Bob Siebert, David Simon, Sharon Stonecipher, Lori Summa, Tony Walton, Joann Welton, and Mike Wilson.

This volume would not have made it off the editors' desks if not for the efforts of Marsha Bulen and Debbie Kelly (University of Missouri), who helped tremendously with editorial chores. Eric Mosley (University of Missouri) transformed an editor's vague concept and some SEM photomicrographs into the cover logo.

Barbara Lidz, SEPM Special Publications Editor, thoroughly and efficiently polished all manuscripts, and unequivocally demonstrated that her editorial comb has finer teeth than those of the co-editors.

We thank Exxon and Mobil for financial contributions to the Clay Minerals in Sandstones symposium, from which this Special Publication grew.

Dave Houseknecht and Ed Pittman, editors

CLAY MINERALS IN SANDSTONES: HISTORICAL PERSPECTIVE AND CURRENT TRENDS

DAVID W. HOUSEKNECHT
Department of Geological Sciences, University of Missouri, Columbia, Missouri 65211
AND
EDWARD D. PITTMAN
Department of Geosciences, University of Tulsa, Tulsa, Oklahoma 74104

INTRODUCTION

This volume grew out of a symposium held at the 27th Annual Meeting of the Clay Minerals Society in Columbia, Missouri, during October, 1990. The symposium was designed to present a current synthesis of research devoted to the origin, diagenesis, and petrophysics of clay minerals in sandstones. Appropriately, this meeting of the Clay Minerals Society was held in honor of Walter D. Keller in his 90th year. Keller's pioneering work in sedimentary petrology, geochemistry, and clay mineralogy, including the application of scanning electron microscopy (e.g., Borst and Keller, 1969), has contributed much to our understanding of the origin of clay minerals in diverse rock types, including sandstones.

This introductory paper presents a brief historical perspective of research that has molded our understanding of clay minerals in sandstones, and provides perspective for this collection of papers, which represents current trends in research activities.

HISTORICAL PERSPECTIVE

Perceptions of clay minerals in sandstones have evolved over the past 50 years with advances in our understanding of sedimentological and geochemical processes, and with improved instrument capabilities. An early, comprehensive evaluation of clay minerals in a sandstone and their effects on reservoir quality was Krynine's (1940) "Petrology and Genesis of the Third Bradford Sand." Although limited to transmitted light observations, Krynine clearly described clay morphology and distribution within pore spaces, and reasoned that the presence of clay minerals influenced permeability and wettability of the reservoir. Despite petrographic descriptions that implied an authigenic origin of clay minerals in certain sandstones (e.g., Krynine, 1940; Quaide, 1956; Carrigy and Mellon, 1964), most geologists considered clay minerals in sandstones to have a detrital origin until approximately 20 years ago (e.g., Dott, 1964). This perspective resulted from the facts that research into the origin of clay minerals in sandstones was mostly restricted to those sandstones that display relatively large volumes of "matrix," and because the primary tools available to attack the problem were transmitted light microscopy and x-ray diffractometry.

During the 1960s, petrographers recognized that clay minerals in sandstones may have diverse origins. For example, Curray (1960) demonstrated clay emplacement by bioturbation. Cummins (1962) discussed the origin of "matrix" resulting from chemical alteration of labile lithic fragments, a concept verified experimentally by Hawkins and Whetten (1969). Petrologists also realized that some "matrix" originated by compaction of ductile lithic fragments, a process described by Dickinson (1970), who termed the product "pseudomatrix."

Early application of the scanning electron microscope during the late 1960s led to recognition of authigenic clay minerals in sandstones other than "graywackes." Published examples started to appear in the early 1970s, including papers that recognized the effects of authigenic clays on sandstone-reservoir quality (e.g., Sarkisyan, 1971, 1972; Gaida and others, 1973; Stalder, 1973). More recently, widespread application of scanning electron microscopy has opened a new window of opportunity to sandstone petrographers and, as a result, it has been demonstrated that most sandstones contain clay minerals, in at least small volumes, and that a large proportion of those clay minerals appears to be of authigenic origin (e.g., Wilson and Pittman, 1977).

The visual impact of euhedral clay crystals depicted in SEM photomicrographs induced a proliferation of published examples, and authigenic clay minerals are now well documented in the geologic literature. The emphasis on authigenic clays during the 1970s and 1980s overshadowed contributions regarding clay minerals of other origins, although pioneering work by Crone (1975) and Walker and others (1978) demonstrated emplacement of detrital clay into sand by infiltration, and Folk (1976) documented the formation of soil-derived cutans. Similar processes previously had been described by soil scientists (e.g., Brewer, 1964), although these contributions were not widely recognized by geologists.

Petrographers have increasingly realized that clay minerals of various origins are commonly present in sandstones, and criteria are gradually emerging for their recognition. For example, within the past few years, a growing number of studies has documented the importance of detrital clay emplaced into sand by infiltration of muddy, depositional water (e.g., Matlack and others, 1989; Moraes and De Ros, 1990). Papers in this volume continue that trend by presenting criteria by which infiltrated clays (Dunn; Moraes and De Ros) and "inherited" clay rims (Wilson) can be recognized.

Recently, geochemical analyses have been widely applied to problems of clay-mineral genesis in sandstones. Oxygen-isotopic analyses of clay minerals, first applied to shales (e.g., Yeh and Savin, 1977), were initially used in sandstones during the early 1980s (e.g., Longstaffe, 1984) and are currently being widely applied (c.f., several papers in this volume). Similarly, isotopic age-dating techniques, first applied to clay minerals in sandstones during the 1970s (e.g., Lee and Brookins, 1978; Sommer, 1978), have been

---

Origin, Diagenesis, and Petrophysics of Clay Minerals in Sandstones, SEPM Special Publication No. 47
Copyright © 1992, SEPM (Society for Sedimentary Geology), ISBN 0-918985-95-1

refined in recent years (e.g., Lee and others, 1985; Liewig and others, 1987) and are currently being widely applied (c.f., several papers in this volume). Integration of petrographic and geochemical techniques is becoming an increasingly necessary approach to deciphering diagenetic histories in many basins (e.g., papers in this volume by Longstaffe and others, Bjørlykke and Aagaard, and Burley and McQuaker).

Clay minerals in sandstone reservoirs commonly reduce porosity and permeability, influence wire-line log response,

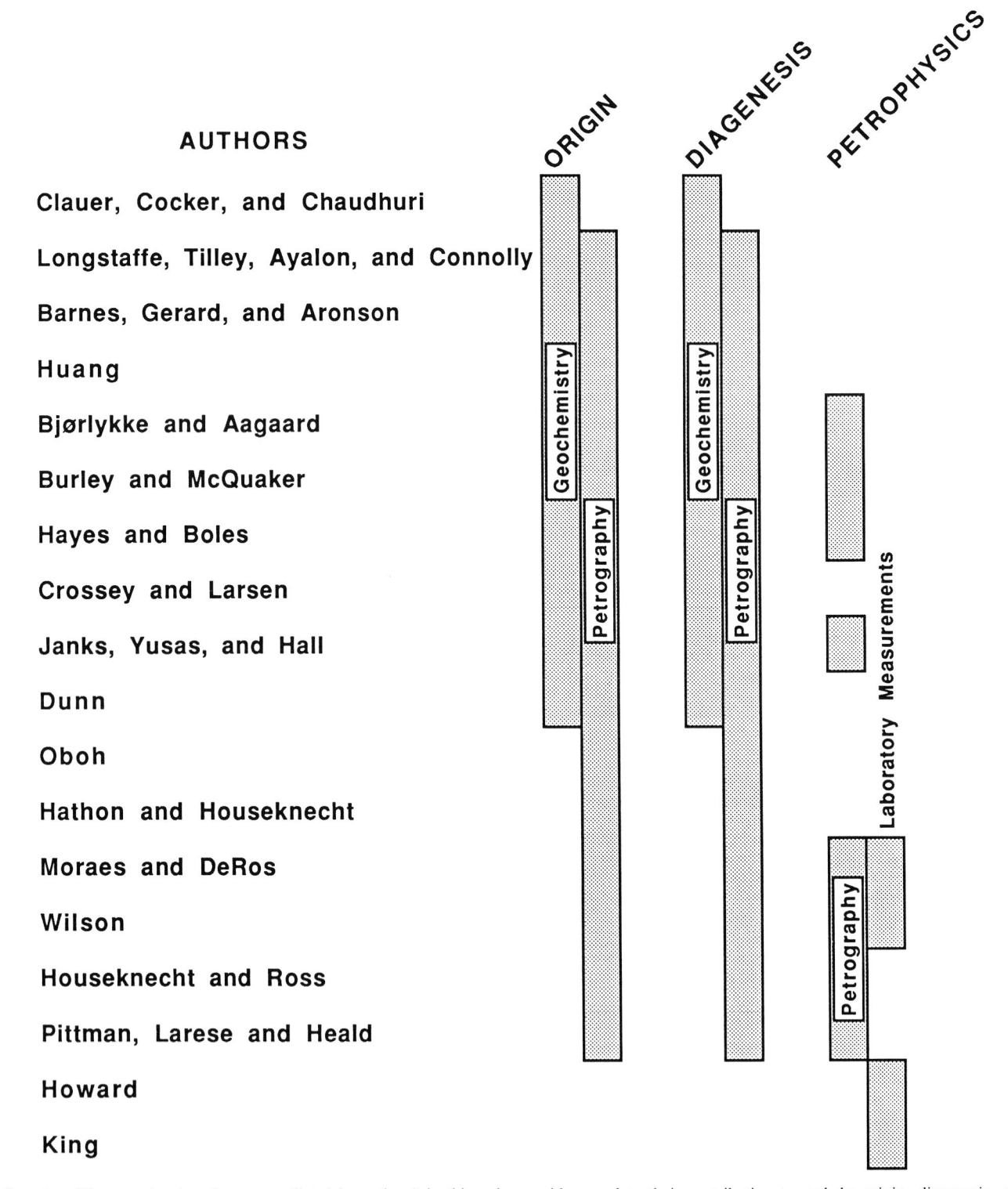

FIG. 1.—Diagram showing the papers (listed by authors) in this volume with regard to their contribution toward the origin, diagenesis, and petrophysics of clay minerals. Most of the papers apply a multidisciplinary approach to solving problems relating to clay minerals in sandstones.

and pose formation-damage potential (see Pittman, 1989, for a review). Early papers on formation damage in the engineering literature were concerned with problems related to perforating of reservoir rocks (e.g., Allen and Worzel, 1956; Krueger, 1956). The first papers dealing with formation damage related specifically to clay minerals in sandstones appeared in the early 1960s (e.g., Jones and Neil, 1960; Jones, 1964). Papers on formation damage due to clay minerals did not appear in the geological literature until the 1970s (e.g., Almon, 1977). King's contribution to this volume provides an engineer's viewpoint on the evaluation of formation damage that differs from the approach favored by geologists.

CURRENT TRENDS

There is a definite trend toward use of an integrated approach to solve problems related to diagenesis of sandstones. Integration of burial history, thermal maturity, and petrography has been profitably pursued in recent years. Commonly, geochemical and petrographic or petrographic and petrophysical techniques are combined to study problems. Some workers have used experimental petrology and petrography to attack problems related to clay minerals in sandstones (e.g., Huang; Pittman and others, this volume).

Figure 1 shows the contributions (by authors) to this volume on clay minerals in sandstones. The vertical bars on the figure indicate whether the papers deal with the petrographic or geochemical aspects of origin and diagenesis, or with the petrographic or laboratory measurements approach to petrophysics. Note that some papers deal only with petrography or with laboratory-oriented petrophysics. Other papers combine petrography and petrophysics or petrography and geochemistry. One paper is concerned only with geochemistry.

In summary, it has become clear that clay minerals in sandstones have diverse origins and that integration of increasingly sophisocated petrographic, geochemical, and petrophysical techniques provides the potential for successfully determining their origin, diagenesis, and effects on reservoir quality. Papers in this volume contribute to an understanding of the origin, diagenesis, and petrophysics of clay minerals in sandstones, and illustrate the multifaceted approach required to solve complex academic and applied problems related to clay minerals in sandstones.

REFERENCES

ALLEN, T. O., AND WORZEL, H. C., 1956, Productivity method of evaluating gun perforating: American Petroleum Institute, Drilling and Production Practice 1956, p. 112–125.

ALMON, W. R., 1977, Sandstone diagenesis in stimulating design factor: Oil and Gas Journal, June 13, p. 56–59.

BORST, R. L., AND KELLER, W. D., 1969, Scanning electron micrographs of API reference clay minerals and other selected samples, in Heller, L., ed., Proceedings, International Clay Conference, Tokyo, Israel University Press, Jerusalem, p. 871–901.

BREWER. R., 1964, Fabric and Mineral Analysis of Soils: New York, John Wiley and Sons, 470 p.

CARRIGY, M. A., AND MELLON, G. B., 1964, Authigenic clay mineral cements in Cretaceous and Tertiary sandstones of Alberta: Journal of Sedimentary Petrology, v. 34, p. 461–472.

CRONE, A. J., 1975, Laboratory and field studies of mechanically-infiltrated matrix clay in arid fluvial sediments: Unpublished Ph.D. Dissertation, University of Colorado, Boulder, 162 p.

CUMMINS, W. A., 1962, The greywacke problem: Liverpool Manchester Geological Journal, v. 3, p. 51–72.

CURRAY, J. R., 1960, Sediments and history of Holocene transgression, continental shelf, northwest Gulf of Mexico, in Shepard, F. P., Phleger, F. B., and Van Andel, T. H., eds., Recent Sediments, Northwest Gulf of Mexico: Tulsa, American Association of Petroleum Geologists, p. 221–266.

DICKINSON, W. R., 1970, Interpreting detrital modes of graywacke and arkose: Journal of Sedimentary Petrology, v. 40, p. 695–707.

DOTT, R. H., Jr., 1964, Wacke, graywacke and matrix—what approach to immature sandstone classification?: Journal of Sedimentary Petrology, v. 34, p. 625–632.

FOLK, R. L., 1976, Reddening of desert sands, Simpson Desert, Northern Territory, Australia: Journal of Sedimentary Petrology, v. 46, p. 604–615.

GAIDA, K. H., RUHL, W., AND ZIMMERLEE, W., 1973, Rasterelektronenmikroskopische untersuchen des porenraumes von sandstein: Erdoel Erdgas Zeitschrift, v. 89, p. 336–343.

HAWKINS, J. W., AND WHETTEN, J. T., 1969, Graywacke matrix: hydrothermal reactions with Columbia River sediments: Science, v. 166, p. 868–870.

JONES, F. O., Jr., 1964, Influence of chemical composition of water on clay blocking of permeability: Journal of Petroleum Technology, April, p. 441–445.

JONES, F. O., Jr., AND NEIL, J. D., 1960, The effect of clay blocking and low permeability on formation testing: Society of Petroleum Engineers 35th Annual Fall Meeting, Denver, Colorado, October 2–5, 1960, Preprint 1515-G, 12 p.

KRUEGER, R. F., 1956, Joint bullet and jet perforation tests: American Petroleum Institute, Drilling and Production Practice 1956, p. 126–140.

KRYNINE, P. D., 1940, Petrology and Genesis of the Third Bradford Sand: Pennsylvania State University Mineral Industry Experimental Station Bulletin, v. 29, 134 p.

LEE, M., ARONSON, J. L., AND SAVIN, S. M., 1985, K/Ar dating of time of gas emplacement in Rotliegendes Sandstone, Netherlands: American Association of Petroleum Geologists Bulletin, v. 69, p. 1381–1385.

LEE, M. J., AND BROOKINS, D. G., 1978, Rubidium-strontium minimum ages of sedimentation, uranium mineralization, and provenance, Morrison Formation (Upper Jurassic), Grants Mineral Belt, New Mexico: American Association of Petroleum Geologists Bulletin, v. 62, p. 1673–1683.

LIEWIG, N., CLAUER, N., AND SOMMER, F., 1987, Rb-Sr and K-Ar dating of clay diagenesis in Jurassic sandstone oil reservoir, North Sea: American Association of Petroleum Geologists Bulletin, v. 71, p. 1467–1474.

LONGSTAFFE, F. J., 1984, The role of meteoric water in diagenesis of shallow sandstones: stable isotope studies of the Milk River aquifer and gas pool, southeastern Alberta, in McDonald, D. A., and Surdam, R. C., eds., Clastic Diagenesis: American Association of Petroleum Geologists Memoir 37, p. 81–98.

MATLACK, K. S., HOUSEKNECHT, D. W., AND APPLIN, K. R., 1989, Emplacement of clay into sand by infiltration: Journal of Sedimentary Petrology, v. 59, p. 77–87.

MORAES, M. A. S., AND DE ROS, L. F., 1990, Infiltrated clays in fluvial Jurassic sandstones of Recôncavo Basin, northeastern Brazil: Journal of Sedimentary Petrology, v. 60, p. 809–819.

PITTMAN, E. D., 1989, Problems related to clay minerals in reservoir sandstones, in Mason, J. F., and Dickey, P. A., eds., Oil Field Development Techniques: American Association of Petroleum Geologists Studies in Geology, no. 28, p. 237–244.

QUAIDE, W. L.,1956, Petrography and clay mineralogy of Pliocene sedimentary rocks from the Ventura Basin, California: Unpublished Ph.D. Dissertation, University of California, Berkley, 81 p.

SARKISYAN, S. G., 1971, Application of the scanning electron microscope in the investigation of oil and gas reservoir rocks: Journal of Sedimentary Petrology, v. 41, p. 289–292.

SARKISYAN, S. G., 1972, Origin of authigenic clay minerals and their significance in petroleum geology: Sedimentology, v. 7, p. 1–22.

SOMMER, F., 1978, Diagenesis of Jurassic sandstones in the Viking graben: Journal of Geological Society of London, v. 135, p. 63–68.

STALDER, P. J., 1973, Influence of crytallographic habit and aggregate structure of authigenic clay minerals on sandstone permeability: Geologie en Mijnbouw, v. 52, p. 217–220.

WALKER, T. R., WAUGH, B., AND CRONE, A. J., 1978, Diagenesis in first-cycle desert alluvium of Cenozoic age, southwestern United States and northwestern Mexico: Geological Society of America Bulletin, v. 89, p. 19–32.

WILSON, M. D., AND PITTMAN, E. D., 1977, Authigenic clays in sandstones: recognition and influence on reservoir properties and paleoenvironmental analysis: Journal of Sedimentary Petrology, v. 47, p. 3–31.

YEH, H., AND SAVIN, S. M., 1977, Mechanism of burial metamorphism of argillaceous sediments: 3. O-isotope evidence: Geological Society of America Bulletin, v. 88, p. 1321–1330.

ISOTOPIC DATING OF DIAGENETIC ILLITES IN RESERVOIR SANDSTONES: INFLUENCE OF THE INVESTIGATOR EFFECT

NORBERT CLAUER
Centre de Géochimie de la Surface, 1 rue Blessig, 67084-Strasbourg, France
JOSHUA D. COCKER
Mobil Exploration and Production, P.O. Box 5444, Denver, Colorado 80217, USA
AND
SAM CHAUDHURI
Department of Geology, Kansas State University, Manhattan, Kansas 66506, USA

ABSTRACT: Authigenic illite generated during diagenetic events occurring in clastic reservoirs is datable by isotopic methods. Questionable analytical procedures adopted by the investigators can lead to doubtful meanings of the data. To avoid these ambiguities, precautionary steps should be systematically followed in selection, preparation, and characterization of the illite-enriched clay material. These steps include specific techniques, such as gentle disaggregation of the rocks instead of routine crushing, and extensive use of electron microscopy (TEM and SEM).

Potential exists for $^{40}Ar/^{39}Ar$-age determinations of these illites, which would avoid the time-consuming separation of the clay materials by size fractionation, but would not avoid its careful characterization. Basic tests for the systematics have still to be determined before routine application of this technique can be expected and investigator factor controlled.

INTRODUCTION

Many recent petrographic and mineralogic studies have emphasized the occurrence of authigenic clay minerals in oil-bearing sandstone reservoirs (e.g., Johns and Shimoyama, 1972; Lambert-Aikhionbare and Shaw, 1982; McHardy and others, 1982; Kantorowicz, 1984; Ahn and Peacor, 1985; Hurst, 1985) and geothermal-sandstone reservoirs (e.g., Muffler and White, 1969; McDowell and Elders, 1980; Liewig and others, 1984; Yau and others, 1987). The authigenic clay minerals in sandstone reservoirs result from combined effects of sedimentologic and tectonic conditions and varied diagenetic processes.

Interest in isotopic dating of diagenetic events occurring in sandstone reservoirs has greatly increased in the last two decades, with increasing discoveries of hydrocarbons in these rocks because geologically meaningful ages can provide crucial information about hydrocarbon accumulation. Authigenic-illite minerals, which are often formed during such events as a result of interactions between the migrating fluids and their host rocks, are theoretically datable by isotopic methods as shown in several recent publications (Clauer and others, 1982; Aronson and Burtner, 1983; Liewig and others, 1984; Lee and others, 1985; Malley and others, 1985; Glasmann and others, 1986; Liewig and others, 1987a and 1987b; Glasmann and others, 1989). However, the inherent complexity in the crystallo-chemical evolution of the authigenic clay minerals has contributed much confusion and speculation about the interpretation of isotopic dates relative to the time of their authigenesis in reservoirs within a basin or among different sub-basins.

Much of the present efforts of dating sedimentary materials has evolved from a suggestion of Bonhomme and others (1966), who first proposed doing isotopic age analyses of mineralogically well-defined authigenic-clay samples to understand the diagenetic history of sedimentary rocks containing these mineral phases. Many recent studies have helped refine this approach into a generalized application of isotopic dating of sedimentary rocks, which allows differentiation of a depositional event from post-depositional evolution (Clauer, 1979, 1982a; Bonhomme, 1982). This approach emphasizes the need for consideration of the following criteria to help depict the geologic context of the rocks under study: (1) field observations; (2) mineralogical determination and crystallographic characterization of the clay fractions; and (3) geochemical analyses of the carbonates associated with the studied silicic-rock samples. The two latter criteria are essential for isotopic analysis of authigenic-clay materials from reservoir sandstones. However, they do not safeguard against the "investigator effect," which can contribute to a confusing and even conflicting interpretation of isotopic data. Indeed, much confusion about isotopic dating can be avoided, or at least reduced, by taking careful steps to isolate the innate "geologic factors" from the "investigator factor." These steps include the preparation and separation of selected components and the detailed characterization of the different clay-size fractions through X-ray diffraction and optical and electron microscopies (transmission and scanning) prior to isotopic determination.

Any authigenic mineral that is potentially datable by the Rb-Sr and K-Ar isotopic methods, for instance, has to be isotopically homogeneous at t = O, which is either the time of precipitation of the mineral or of the recrystallization of its precursor. Isotopically homogeneous minerals incorporate, during their formation, Sr and Ar of the same isotopic composition. This isotopic homogeneity is not influenced by the mineral Rb/Sr or K/Ar ratios, which can be variable. This means that *all* clay mineral assemblages are *not* systematically suitable for dating purposes because some of the clay might be of detrital origin or only have been partly recrystallized during diagenetic episodes. Therefore, it is evident that a clear knowledge of the genesis of the clay material studied is needed. This knowledge might be deduced from evidence gathered by complementary X-ray and electron microscopic techniques mentioned above.

The aim of this paper is to compare published isotopic data on authigenic illitic materials separated from North Sea sandstone-reservoir rocks of similar stratigraphic ages and with similar diagenetic evolutions. The purposes of this comparison are to focus on the possible influence of the investigators on discrepancies among published K-Ar and/

or Rb-Sr data of similar materials, and to demonstrate the need for using careful procedures to ensure interpretable isotopic data of these materials.

EVALUATION OF AVAILABLE ILLITE DATA

Hamilton and others (1987) reported K-Ar ages of diagenetic illite from the Lower Rannoch Formation of the Brent Group in the North Sea. The values are scattered between 59 and 46 Ma; the younger ages being close to the age suggested by Liewig and others (1987a) for the formation of illite in equivalent strata from the Alwyn field. The mineralogic purity of the analyzed separates was checked only by XRD, which usually cannot detect the presence of discrete contaminants occurring in amounts less than about 3 to 4%. Furthermore, the interpretation was partly obscured by the scatter of the K-Ar values, with large overlaps in the analytical uncertainties.

Hogg and others (1987) reported illite K-Ar dating from the south Alwyn oil field in the North Sea. They described three morphologies of illite particles and found an overall range for the K-Ar values scattered between 45.5 and 60.3 Ma. As in the study of Hamilton and others (1987), the youngest apparent ages were obtained for the shallower samples. The authors of both studies believed that illitization may have been initiated as soon as the sediments reached specific temperature during progressive burial, and that it may have begun at relatively shallow depths. Such an interpretation, however, remains questionable as meaningful K-Ar ages are generally preserved in K-bearing authigenic-clay minerals only as long as they do not reach the "Ar diffusion temperature," which is probably close to their crystallization temperature (Liewig and others, 1987b). Therefore, it is difficult to assume that the authigenic-clay minerals of the rocks in question were able to evolve in closed systems during further burial, accompanied by an increase in the temperature. Lee and others (1985), for instance, obtained a decrease of the K-Ar values with depth, which is more consistent with a resetting of the K-Ar system of the deeper illites with increased temperature. One possibility for authigenic-illite fractions remaining a closed K-Ar system at greater depths is that they crystallized at shallow depth during a diagenetic event that induced abnormal physical (high temperature) and chemical (K-enriched fluids) conditions (Mossmann and others, pers. commun., 1991).

Jourdan and others (1987) claimed that the diagenetic history of the Brent Formation of the North Sea is long and complex. Based on K-Ar data of clay assemblages, which were not described in the publication, this evolution has been divided into four episodes, ranging from the end of the Lower Cretaceous (100 Ma) to the Eocene (50–40 Ma). Some of the data used by the authors are those of Liewig and others (1987a), who showed that the scatter in the ages reflects the occurrence of contaminating detrital K-feldspars. Although the steps of the proposed diagenetic history could have occurred, their ages, based on the K-Ar dates of the illite material, are highly questionable for lack of independent evidence on the degree of purity of the analyzed fractions.

Lee and others (1985) suggested, on the basis of scattered K-Ar dates on illitic-clay fractions, that the diagenetic formation of these authigenic minerals in the Rotliegendes Sandstone gas reservoirs of the Netherlands could have extended over long periods of time. If so, the formation of illite in the shallow gas or oil zones might have terminated earlier than that in the underlying water zones. Duration of the diagenetic events occurring in oil-bearing sandstone reservoirs probably can be fairly estimated by comparing the K-Ar ages of the illitic materials formed in the oil zones with those formed in the water zones. The argument of Lee and others (1985) is not supported by K-Ar data of Cocker and others (1988), who found only a limited scatter in the K-Ar values for illitic size fractions <0.4 µm across the oil-water zone of a Brent oil field in the North Sea. Cocker and others (1988) used a gentle freezing-thawing disaggregation technique, which has proven, as will be shown later, to be a more appropriate technique for obtaining the innate grain sizes in the rock samples. The K-Ar values of the illite materials, which were collected in the water zone and extracted using the gentle freezing-thawing technique, were found to be only about 2 Ma younger than their counterparts from the oil-bearing zone. These preliminary results suggest that oil entrapment might not extend over long periods, as suggested by Lee and others (1985) for the Rotliegendes Formation. Of course, the situation that existed during oil emplacement in the Brent Formation could have been different from the one during gas emplacement in the Rotliegendes Formation.

Unlike Hamilton and others (1987), Hogg and others (1987) and Jourdan and others (1987), who set only limited controls on the separation of illite fractions from Brent sandstones, Liewig and others (1987a) followed rigorous procedures carefully controlled by X-ray diffraction and electron microscopy for separation of the authigenic-clay minerals from oil-bearing sandstone reservoirs of the same formation. Liewig and others (1987a) determined the K-Ar and Rb-Sr isotope dates of clay-size fractions separated using a classic crushing technique, with a jaw crusher and a ball mill, and a repetitive freezing-thawing technique for disaggregation of pieces of sandstone samples. One clay-size fraction (<0.6 µm) of each of three samples of sandstones (6–1, 6–2, and 6–3) collected at different intervals from the same core and gently disaggregated by the freezing-thawing technique was investigated for K-Ar isotope dates. As evidenced by XRD analyses and electron microscopic observations, the clay fractions consisted of 100% illite for one sample and about 90% illite and 10% dickite for the two others, and they all were free of any detrital component. The K-Ar dates of these three illite-rich fractions narrowly ranged between 35 and 44 Ma (Table 1), with an average at 41 Ma. On the other hand, the dates for the clay fractions separated from equivalent sandstones following conventional crushing varied widely between 50 and 129 Ma (Table 1). The presence of varied amounts of detrital feldspar in these clay fractions was considered to be responsible for the wide disparity in the K-Ar dates (Liewig and others, 1987a). This becomes evident in comparing the apparent K-Ar dates and the illite-to-feldspar ratios, which were obtained by XRD semi-quantitative determination us-

TABLE 1.—MINERALOGICAL AND ISOTOPIC DATA OF CLAY MATERIALS FROM BRENT SANDSTONES (AFTER LIEWIG AND OTHERS, 1987A).

Samples (1)		Depth (m)	K-spar/illite (2)	K-Ar age (3)	$^{87}Sr/^{86}Sr$ (4)	$^{87}Sr/^{86}Sr$ (5)
1-1	<2 μm	3456.5	n.d.	n.d.	0.71923	0.7183
1-2		3467.5	n.d.	n.d.	0.72621	0.7239
1-3		3467.7	n.d.	n.d.	0.72380	0.7215
2-1	<2 μm	3308.4	16.7	n.d.	0.71471	0.7140
2-2		3312.6	11.1	n.d.	0.71425	0.7137
2-3		3315.5	12.3	n.d.	0.71485	0.7142
2-4		3319.5	15.8	n.d.	0.70913	0.7091
2-5		3325.5	n.d.	n.d.	0.71516	0.7146
2-6		3327.1	n.d.	n.d.	0.71531	0.7147
2-7		3331.7	8.7	n.d.	0.71632	0.7152
2-8		3338.2	64.1	n.d.	0.71787	0.7169
2-9		3339.5	n.d.	n.d.	0.71739	0.7165
2-10		3340.5	n.d.	n.d.	0.71656	0.7158
3-1	<1 μm	3261.2	34.4	100±3	n.d.	n.d.
3-2		3273.6	22.8	129±6	n.d.	n.d.
4-1		3419.4	13.0	72±3	n.d.	n.d.
4-2		3424.0	9.9	52±3	n.d.	n.d.
5-1		3549.5	9.2	54±2	n.d.	n.d.
5-2		3565.6	7.9	50±3	n.d.	n.d.
2-5	<0.4 μm	3325.5	n.d.	n.d.	0.71451	0.7138
	0.4-1 μm		13.2	77±2	0.71522	0.7147
	1-2 μm		20.1	89±2	0.71468	0.7143
2-6	<0.4 μm	3327.1	8.5	n.d.	0.71705	0.7161
	0.4-1 μm		19.5	n.d.	0.71655	0.7159
	1-2 μm		24.9	79±2	0.71643	0.7158
2-7	<0.4 μm	3331.7	7.4	n.d.	0.71804	0.7169
	0.4-1 μm		10.3	58±2	0.71770	0.7169
	1-2 μm		11.4	70±2	0.71814	0.7173
6-1	<0.6 μm	3238.5	0	44±2	n.d.	n.d.
6-2		3239.7	0	44±2	n.d.	n.d.
6-3		3242.0	0	35±1	n.d.	n.d.

(1) for complete description see Liewig and others, 1987a
(2) authigenic-illite/detrital-feldspar ratios in 10^2
(3) ages in Ma±2σ
(4) measured $^{87}Sr/^{86}Sr$ ratios
(5) calculated initial $^{87}Sr/^{86}Sr$ ratios for an assumed age of 40 Ma

ing an internal standard (Table 1 and Fig. 1). This correlation could be obtained because most of the clay-size fractions analyzed consisted essentially of mixtures of only two distinct end-member components: an authigenic K-bearing illitic clay and a detrital feldspar. An extension of the correlation trend to pure illite gives an extrapolated age at about 40 Ma. This value is analytically identical to the average age of 41 Ma obtained directly on the pure illitic fractions separated after repetitive freezing-thawing processing of the rocks. The agreement in the results between the two independent means should not, however, be construed as support for the use of the conventional crushing technique in sample preparation and subsequently determining the age of illite by means of an extrapolation from a relation be-

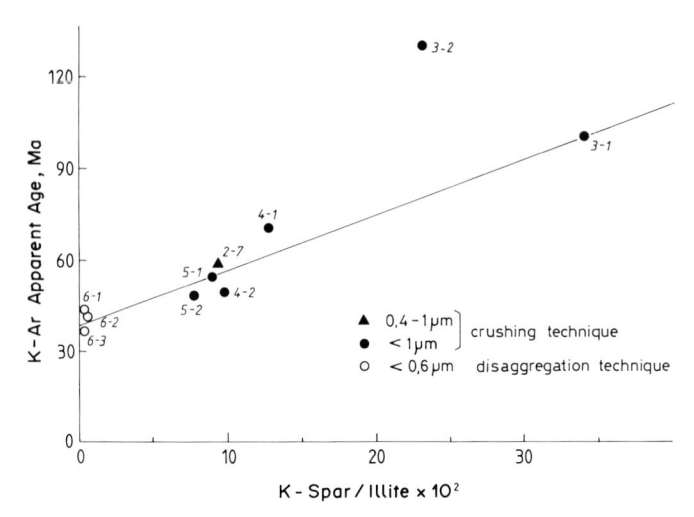

FIG. 1.—Relation between K-Ar apparent ages and K-feldspar/illite ratios of clay subfractions of Brent sandstones (after Liewig and others, 1987a).

tween the age and the authigenic/detrital ratio of the clay assemblages studied. Such correlation will, of course, fail if more than one influencing detrital mineral component is present in the fractions. The presence of any influential detritus will also prevent clear definition of the relative ages of morphologically different clay minerals, which could possibly form at distinctly different periods during the diagenetic history of a sandstone reservoir.

Reports in the literature are numerous about K-Ar isotopic dates on diagenetic illitic-clay fractions in sandstones, but accounts on dates by the Rb-Sr method are surprisingly scarce. Rb-Sr dating of diagenetic-clay fractions depends on a clear knowledge of the initial ^{87}Sr/^{86}Sr ratio of these minerals during their crystallization, which is that of the fluids from which they formed. If this ^{87}Sr/^{86}Sr ratio is well known, model ages for the clay materials can be calculated, but these values can be far from the true diagenetic ages due to faulty assumptions on the initial Sr-isotopic composition of the fluids that generated the clay minerals and, of course, due to incomplete purification of the clay materials. For these reasons, the Rb-Sr method is far less used for dating diagenetic-clay materials of sandstones. Rb-Sr data published by Liewig and others (1987a), on diagenetic-clay material of the Middle Jurassic Brent sandstones in the North Sea, give some perspective on the complexity in the interpretation of such isotopic dates. The only Rb-Sr results available from this study are those determined on the clay fractions separated following the classic crushing of the rocks. The data presented in Table 1 indicate that the fractions are moderately enriched in ^{87}Sr, as the ^{87}Sr/^{86}Sr ratios range from 0.71425 to 0.72621, except for an ankerite-enriched fraction (sample 2-4) whose ^{87}Sr/^{86}Sr ratio is 0.70913. The Sr-isotopic compositions and the Rb/Sr ratios are poorly correlated, so no meaningful isochron age could be developed. The model ages for the different clay fractions range widely from about 220 to about 350 Ma, which is considerably higher than the 50 to 130 Ma for the K-Ar apparent ages obtained on the same aliquots.

As in the case of the K-Ar ages, the Rb-Sr ages are somewhat higher for the coarser fractions (1-2 μm) than for their finer counterparts (<0.4 μm). This size-dependent age trend has been reported in many other studies, especially on clay fractions from shales and mudstones, and it has often been considered as a reflection of increased influence of detrital components in the larger size fractions. However, we have yet to find a logical reason why the model Rb-Sr ages are higher than the K-Ar ages for the same aliquots. As preferential diffusional loss of radiogenic ^{40}Ar out of K-bearing clay particles is no longer a convincing argument for the low K-Ar dates relative to the Rb-Sr values, a reason for this difference could be that the Rb-Sr model ages are calculated based on too low initial ^{87}Sr/^{86}Sr ratios for the fluids from which the clay material formed. Presence of low K-bearing components, such as detrital Ca-plagioclase or authigenic/detrital dolomitic carbonates in the clay fractions, produce lower initial ^{87}Sr/^{86}Sr of contaminated clay fractions, and therefore discordant Rb-Sr and K-Ar dates.

Taking the 40 Ma K-Ar age as being the true age of formation of the diagenetic illite in the Brent sandstone reservoirs of the North Sea, we have calculated the initial ^{87}Sr/^{86}Sr ratios for the different size fractions (Table 1). These ratios range from about 0.7136 to 0.7239, but the values are within much narrower limits if we consider the samples from the same drill hole. For example, the initial ^{87}Sr/^{86}Sr values for the 10 fractions taken within a 30-m interval from core 2, which was collected at a depth of 3308 to 3340 m, range within narrow limits between 0.7136 and 0.7169 (Table 1). Part of the variations in these Sr-isotopic compositions can be attributed to the presence of varied amounts of Rb-bearing detrital minerals carrying radiogenic ^{87}Sr, but limited-scale isotopic variations of fluids can also be a reason for the overall variations in the initial ^{87}Sr/^{86}Sr of the measured size fractions. Chaudhuri and others (pers. commun., 1991) recently demonstrated that the Sr-isotopic compositions of reservoir fluids can be spatially and temporarily varied in the same oil-field sandstone reservoir, but no clear explanation has been provided for such variations. Such difference in the initial ^{87}Sr/^{86}Sr of authigenic illite is suggested here by two trends between the Sr-isotopic compositions and the detrital feldspar-to-authigenic-illite ratios of the clay assemblages from Brent sandstones of Alwyn oil field (Fig. 2). By extrapolating these trends toward a pure illite phase in the mixtures, as already done for the K-Ar ages, at least two different Sr seem to have been incorporated by these minerals, one with an initial ^{87}Sr/^{86}Sr ratio at about 0.712, which was found for most of the fractions used for the correlation, and the other with about 0.709 (sample 2-7), which is close to the value measured for the ankerite-enriched sample 2-4.

Glasmann and others (1989) also carefully analyzed K-Ar ages of illitic-clay minerals in Brent sandstones in the Bergen High area of the North Sea. Different size fractions were separated after disaggregation of the samples by gentle crushing and mild sonic treatment. The different fractions were saturated with Mg to remove potentially exchangeable K from clay particles. Analyses of <0.1-μm illitic-clay fractions yielded K-Ar dates between 98 and 38 Ma. The

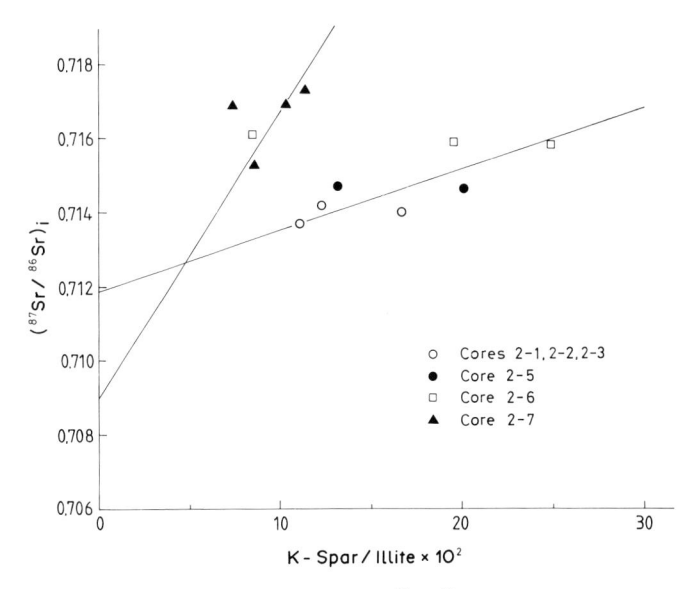

FIG. 2.—Relation between the initial ^{87}Sr/^{86}Sr ratios calculated from an assumed crystallization age of 40 Ma for the authigenic illites and the K-feldspar/illite ratios of clay subfractions of Brent sandstones.

authors also observed that a lath-shaped, <0.1-μm illite from a hydrocarbon-saturated zone yielded a K-Ar age of about 58 Ma, but similarly shaped illite of the transitional zone between water-saturated and hydrocarbon-saturated zones yielded an age of 38 Ma. The authors suggested that the difference between the two values is related to illitization at the upper level terminating soon after accumulation of the hydrocarbons. Different size fractions of clays from the transition zone gave K-Ar ages ranging from 38 to 83 Ma, which increased with increasing particle size. The authors acknowledged that the spread in the measured ages was due to increasing amounts of detrital materials in the coarser fractions. Nevertheless, the lowest reported age of 38 Ma agrees reasonably well with the average of 41 Ma obtained by Liewig and others (1987a) in Alwyn field and Cocker and others (1988) in Hutton field, both of whom followed the careful freeze-heat disaggregation procedure for separation of the clay materials. However, because of the wide spread of dates among the samples, the blocking of the diagenesis at 58 Ma in the zone where hydrocarbons accumulated remains questionable when compared to the 43 Ma obtained by Cocker and others (1988).

In addition to SEM and TEM observations for characterization of the different clay separates, it is helpful to date several size fractions of each sandstone sample. In many cases, SEM observations show that the *in situ* clay particles are in a 10-μm size range, whereas the K-Ar dating attempts give reliable ages only on particles smaller by one to two orders of magnitude. The reason for this change in size is mainly the occurrence of detrital contamination. However, dating slightly coarser fractions (for instance between 0.4 and 2 μm) free of detrital, non-micaceous silicate minerals would help determine the size limit for authigenic clay growth accompanied by complete reset of the radiogenic clock of the detrital micaceous particles. This is the case for all studied clay materials up to 2 μm, and for some up to 6 μm, separated from gas-bearing Rotliegendes sandstones of western Germany (Clauer, Cocker and Liewig, unpublished data).

CRITERIA FOR EVALUATING ILLITE SUITABLE FOR ISOTOPIC DATING

From examples presented earlier, what becomes clear is that rigorous examination of clay splits is needed for selection of the best suited materials for dating clay authigenesis in sandstone reservoirs. Various tools can be used to improve this selection.

Optical microscopy gives useful information about the clay distribution in the samples. SEM work is vital, providing information about the type(s) of clay minerals, their relations to other components (detrital or authigenic), the effects of interaction with brines, the alteration state of detrital grains, and the nature of diagenetic cements.

Critical-point drying helps in understanding the morphology of authigenic illites and, therefore, should be used when possible, that is when the drill cores are kept humid after recovery. Cocker (1986), for instance, showed that illite formed during a single diagenetic event usually presents one morphology in cores dried through the critical point, whereas it can be found with different morphologies in air-dried samples. Therefore, morphology variations of illites from air-dried cores do not represent arguments for long or multiple periods of growth, as Lee (1984) suggested.

X-ray diffraction is essential for characterization of the clay fractions. It is especially useful in the study of granulometric size fractions in order to separate minerals, such as kaolinite and/or chlorite from illite, or authigenic illite from detrital components. In addition, the crystallinity index and the polymorphic forms of illite can be obtained, as well as a more precise idea of the layering of the illite-smectite mixed layers using computer programs by Reynolds (1967, 1980) or Mossmann (1987, 1991).

Transmission electron microscopy can complement the mineralogical aspect and help to differentiate minerals by their shape and, therefore, act as an efficient tool for crystallographic characterization of illite and other clay phases of different origins.

Chemical and Sr-isotopic analyses of the fluids, when available, might give information about their origin, migration paths and interactions with host rocks (Chaudhuri and others, 1987), or about hydrologic aspects (Sunwall and Pushkar, 1979; Stueber and others, 1984, 1987). The Sr-isotopic composition of the leachates from clay particles may also provide useful information about the chemical environment during crystallization of the clays, as suggested by Clauer (1982b) and Clauer and others (1986).

THE POTENTIAL OF ^{40}Ar/^{39}Ar ILLITE DATING

What is obvious is that an investigator effect exists, and that isotopic dating of diagenetic events recorded in the authigenic-clay minerals of sandstones is only successful when extraction and purification are achieved properly. This most important step needs to be carefully controlled by XRD studies and TEM and SEM observations, and there should be no exception to these procedures. As these essential steps are more time consuming than difficult, interest is high in exploring alternative dating techniques that will avoid the time-consuming task of mineral separation. However, such methods will not eliminate the obligation of fully documenting the mineralogic and petrographic contexts of the material studied. Presently, the technique, that appears to have the best potential is the ^{40}Ar/^{39}Ar technique, with or without a laser probe.

The ^{40}Ar/^{39}Ar technique, a variation of the ^{40}K/^{39}Ar method, is in its nascent state for application to geologic materials and has yet to meet critical tests in studies of sedimentary and low-grade metamorphic rocks (Sigurgeirsson, 1962; Merrihue and Turner, 1966). Based on the production of ^{39}Ar by neutron activation of ^{39}K, this technique has the advantage that radioactive-parent and radiogenic-daughter isotopes can be measured simultaneously from a single isotopic determination of the Ar. This furthermore allows analysis by stepwise progressive heating, which leads to a separation of the low-temperature Ar (of potential authigenic components) from that released at higher temperatures (of potential detrital components).

Results of published ^{40}Ar/^{39}Ar-dating attempts on clay minerals are somewhat disappointing, but may not be definitive as further investigations should be made. Glauconies, for instance, seem to be not suitable for ^{40}Ar/^{39}Ar dating because of a systematic loss of ^{39}Ar due to recoil loss during or subsequent to irradiation. This behavior was mentioned by Yanase and others (1975) and Klay and Jessberger (1984) and documented by Brereton and others (1976) and Foland and others (1984), who showed that the recoil loss is not proportional to K content and crystallinity of the particles, or irradiation intensity of the reactor. As shown in many studies, glauconies crystallize at low temperatures (see discussion in Odin, 1982) and any irradiation in a reactor can, therefore, be expected to produce a recoil effect of ^{39}Ar, but also a temperature-induced diffusion of radiogenic ^{40}Ar. However, this behavior may not be generalized to all clay minerals. Burghele and others (1984), Hunziker and others (1981, 1986), Kligfield and others (1986), Reuter and Dallmeyer (1987a and 1987b) and Dallmeyer and others (1989) obtained a number of similar ages by both the ^{40}Ar/^{40}K and ^{40}Ar/^{39}Ar methods on the same aliquots of illitic fractions from very low- to low-grade metamorphic domains, whose particle sizes were in a range from 0.6 to 2 µm. Reuter and Dallmeyer (1989) documented by systematic studies the recoil effect and especially the redistribution of ^{39}Ar from high- to low-K phases such as chlorite or albite, which might be present in the illite fractions. This recoil and redistribution effect seems to occur preferentially in particles with large surface areas and poor grain edges. As grain size and grain edge quality increase in the middle to upper zone of very low-grade metamorphism, this effect tends to disappear. Therefore, the potential of the application of the ^{40}Ar/^{39}Ar technique to well-crystallized diagenetic illites from sandstone reservoirs will probably require larger particles, maybe in the 10-µm range. The effects of irradiation on such larger particles has to be controlled because these larger particles may have formed at temperatures close to those occurring in most reactors.

Systematic studies like those of Foland and others (1984), Hunziker and others (1986), and Reuter and Dallmeyer (1987a and b) should be pursued, especially on other types of micaceous particles, before any expectation of routine application of the ^{40}Ar/^{39}Ar method on sediments can be made. These systematic studies seriously should consider the factors that might be introduced by the investigators, adding confusion to interpretations of the results. The suggestion of Hearn and Sutter (1985), about the potential application of the ^{40}Ar/^{39}Ar technique on separates of feldspar grains representing mixtures of detrital cores and authigenic overgrowths, is also of interest. The accuracy of such an application depends on precise determination of the relative amounts of detrital and authigenic parts by cathodoluminescence in an SEM. However, a recent application by Girard and Onstott (1991) underscored some difficulties, such as Ar release from detrital and authigenic feldspars at a similar temperature range.

Analyses of individual in situ grains are also possible with a laser microprobe (Sutter and Hartung, 1984; Sutter and others, 1986). Some results of laser ^{40}Ar/^{39}Ar dating of micro-size illite associated with uranium deposits in northern Saskatchewan are already available (Bray and others, 1987). The ^{40}Ar/^{39}Ar data obtained with the laser were as scattered as those obtained with the conventional K-Ar technique after classic sample crushing. In both cases, other mineral components seem to have supplied some radiogenic ^{40}Ar. Sample crushing reduces the size of these minerals into the clay fraction, whereas the laser probe fuses components around clay particles in the samples. Bell (1985) has published a few data derived by laser-fusion technique on "alteration products" of basement samples close to U deposits of the southern Carswell structure, and which probably include mainly clays. As mentioned by Bell, such a technique analyzes clusters of grains, which consist of individual clay particles with sizes on the order of one to a few tens of micrometers, compared to the 200- to 300-µm diameter of the laser beam. Bell's values scattered between 500 and 1615 Ma, with most between 1175 and 1615 Ma. The size and probably the power of the laser beam might have induced gas releases from non-clay minerals around or within the melted "alteration products," some of which could be detrital. In summary, the presently available in situ ^{40}Ar/^{39}Ar data on clay minerals from U ore deposits are as much scattered as the corresponding K-Ar data from extracted particles for reasons which probably include incomplete characterization of the material, but also uncontrolled effects of the method.

To our best knowledge, no laser ^{40}Ar/^{39}Ar data on illites from oil-bearing sandstone reservoirs are presently available in the literature. Technical problems related to the type, size and power of the laser beams, the volume estimation of the spots melted by the beam, and the precise location of the laser shots in thin sections, which have been noted in the previously mentioned studies, need to be overcome.

CONCLUSION

K-Ar and Rb-Sr isotope dating of authigenic-illite-enriched clay fractions of oil-field and geothermal sandstone reservoirs depends on our ability to overcome problems of selection, preparation, purification and characterization of the authigenic clay minerals.

The dating attempts on authigenic illite extracted from oil-producing sandstones of different ages and from different localities emphasize the importance of understanding the dating and duration of any diagenetic event in a productive sedimentary basin. To achieve this, confusion related to the "investigator effect" should be essentially removed, which requires very careful sample preparation and clay separation followed by a verification of the purity of the fractions prior to isotopic analyses. These preparation steps should include XRD evaluation, optical and scanning electron microscopy, and examination of the morphology of the particles and determination of detrital minerals by transmission electron microscopy.

Potential exists for successful $^{40}Ar/^{39}Ar$ dating of clay separates and for *in situ* age determinations with laser probes. But for these techniques, basic systematics have still to be explored, even if preliminary attempts seem only faintly encouraging.

ACKNOWLEDGMENTS

We thank J.L. Aronson (Case Western Reserve University, Cleveland, Ohio) for a constructive and improving review of the manuscript, and D.W. Houseknecht (University of Missouri, Columbia, Missouri) for invitation to participate in this SEPM Special Publication.

REFERENCES

AHN, J. H., AND PEACOR, D. R., 1985, Transmission electron microscopic study of diagenetic chlorite in Gulf Coast argillaceous sediments: Clays and Clay Minerals, v. 33, p. 228–236.

ARONSON, J., AND BURTNER, R. L., 1983, K-Ar dating of illitic clays in Jurassic Nugget Sandstone and timing of petroleum migration in Wyoming Overthrust Belt (abs.): American Association of Petroleum Geologists Bulletin, v. 67, p. 414.

BELL, D., 1985, Geochronology of the Carswell area, northern Saskatchewan, *in* Lainé, R., Alonso, D., and Svab, M., eds., The Carswell Structure Uranium Deposits, Saskatchewan: Geological Association of Canada, Special Paper, v. 29, p. 33–46.

BONHOMME, M. G., 1982, The use of Rb-Sr and K-Ar methods as a stratigraphic tool applied to sedimentary rocks and minerals: Precambrian Research, v. 18, p. 5–25.

BONHOMME, M., LUCAS, J., AND MILLOT, G., 1966, Signification des déterminations isotopiques dans la géochronologie des sédiments: Actes du 151e Colloque International du CNRS, Nancy, 1965, p. 541–565.

BRAY, C. J., SPOONER, E. T. C., HALL, C. M., YORK, D., BILLS, T. M., AND KRUEGER, H. W., 1987, Laser probe $^{40}Ar/^{39}Ar$ and conventional K/Ar dating of illites associated with the McClean unconformity-related uranium deposits, north Saskatchewan, Canada: Canadian Journal of Earth Sciences, v. 24, p. 10–23.

BRERETON, N. R., HOOKER, P. T., AND MILLER, J. A., 1976, Some conventional potassium-argon and $^{40}Ar/^{39}Ar$ age studies on glauconite: Geological Magazine, v. 113, p. 329–340.

BURGHELE, A., ZIMMERMANN, T., CLAUER, N., AND KROENER, A., 1984, Interpretation of $^{40}Ar/^{39}Ar$ and K-Ar dating of fine clay mineral fractions in Precambrian sediments (abs.): Terra Cognita, v. 4, p. 130.

CHAUDHURI, S., BROEDEL, V., AND CLAUER, N., 1987, Strontium isotopic evolution of oil-field waters from carbonate reservoir rocks in Bindley field, central Kansas, USA: Geochimica et Cosmochimica Acta, p. 51, p. 45–53.

CLAUER, N., 1979, A new approach to Rb-Sr dating of sedimentary rocks, *in* Jaeger, E., and Hunziker, J. C., eds., Lectures in Isotope Geology: Berlin, Heidelberg, Springer-Verlag, p. 30–51.

CLAUER, N., 1982a, The rubidium-strontium method applied to sediments: certitudes and uncertainties, *in* Odin, G. S., ed., Numerical Dating in Stratigraphy: Chichester, J. Wiley and Sons Ltd., p. 245–276.

CLAUER, N., 1982b, Strontium isotopes of Tertiary phillipsites from the southern Pacific: timing of the geochemical evolution: Journal of Sedimentary Petrology, v. 52, p. 1003–1009.

CLAUER, N., CHAUDHURI, S., AND MASSEY, K. W., 1986, Relationship between clay minerals and environment by the Sr isotopic composition of their leachates: Annual Meeting of the Geological Society of America, Abstracts and Programs, p. 565–566.

CLAUER, N., SOMMER, F., AND LIEWIG, N., 1982, Difficulties in Rb-Sr and K-Ar dating of oil reservoirs (abs.): Fifth International Conference on Geochronology, Cosmochronology and Isotope Geology, Nikko National Park, Japan, p. 57.

COCKER, J. D., 1986, Authigenic illite morphology: appearances can be deceiving (abs.): American Association of Petroleum Geologists Bulletin, v. 70, p. 575.

COCKER, J. D., CLAUER, N., TSUI, T. F., AND SWARBICK, R. E., 1988, A diagenetic model for the Northwest Hutton field: Conference on Clay Mineral Diagenesis in Hydrocarbon Reservoirs and Shales, Cambridge, United Kingdom, 1 p.

DALLMEYER, R. D., REUTER, A., CLAUER, N., AND LIEWIG, N., 1989, Chronology of Caledonian tectonothermal activity within the Gaissa and Lakefjord nappe complexes (Lower Allochthon), Finmark, Norway, *in* Gayer, R. A., ed., The Caledonide Geology of Scandinavia: London, Graham and Trotman Ltd., p. 9–26.

FOLAND, K. A., LINDER, J. S., LASKOWSKI, T. E., AND GRANT, N. K., 1984, $^{40}Ar/^{39}Ar$ dating of glauconites: measured ^{39}Ar recoil loss from well-crystallized specimens: Chemical Geology, Isotope Geoscience Section, v. 2, p. 241–264.

GIRARD, J. P., AND ONSTOTT, T. C., 1991, Application of $^{40}Ar/^{39}Ar$ laser-probe and step-heating techniques to the dating of diagenetic K-feldspar overgrowths: Geochimica et Cosmochimica Acta, v. 52, p. 2207–2214.

GLASMANN, J. R., CLARK, R. A., AND LARTER, S. R., 1986, Episodic diagenesis on the Bergen High: implications of illite/smectite K/Ar dates (abs.): Terra Cognita, v. 6, p. 110.

GLASMANN, J. R., CLARK, R. A., LARTER, S. R., BRIEDIS, N. A., AND LUNDEGARD, P. D., 1989, Diagenesis and hydrocarbon accumulation, Brent Sandstone (Jurassic), Bergen High area, North Sea: American Association of Petroleum Geologists Bulletin, v. 73, p. 1341–1360.

HAMILTON, P. J., FALLICK, A. E., MACINTYRE, R. M., AND ELLIOT, S., 1987, Isotopic tracing of the provenance and diagenesis of lower Brent Group sands, North Sea, *in* Brooks, J., and Glennie, K., eds., Petroleum Geology of North-West Europe: London, Graham and Trotman Ltd., p. 939–949.

HEARN, P. P., JR., AND SUTTER, J. F., 1985, Authigenic potassium feldspar in Cambrian carbonates: evidence of Alleghanian brine migration: Science, v. 228, p. 1529–1531.

HOGG, A. J. C., PEIRSON, M. J., FALLICK, A. E., HAMILTON, P. J., AND MACINTYRE, R. M., 1987, Clay mineral and isotope evidence for controls on reservoir properties of Brent Group sandstones, British North Sea (abs.): Terra Cognita, v. 7, p. 342.

HUNZIKER, J. C., CLAUER, N., DALLMEYER, R. D., FREY, M., AND FRIEDRICHSEN, H., 1981, The evolution of illite to muscovite in low-grade metamorphism (abs.): Terra Cognita, v. 2, p. 70.

HUNZIKER, J. C., FREY, M., CLAUER, N., DALLMEYER, R. D., FRIEDRICHSEN, H., FLEHMIG, W., HOCHSTRASSER, K., ROGGWILLER, P., AND SCHWANDER, H., 1986, The evolution of illite to muscovite: mineralogical and isotopic data from the Glarus Alps, Switzerland: Contributions to Mineralogy and Petrology, v. 92, p. 157–180.

HURST, A., 1985, Diagenetic chlorite formation in some Mesozoic shales from the Sleipner area of the North Sea: Clay Minerals, v. 20, p. 69–79.

JOHNS, W. A., AND SHIMOYAMA, A., 1972, Clay minerals and petroleum forming reactions during burial and diagenesis: American Association of Petroleum Geologists Bulletin, v. 56, p. 2160–2167.

JOURDAN, A., THOMAS, M., BREVART, O., ROBSON, P., SOMMER, F., AND SULLIVAN, M., 1987, Diagenesis as the control of the Brent Sandstone reservoir properties in the greater Alwyn area (East Shetland Basin), *in* Brooks, J., and Glennie, K., eds., Petroleum Geology of Northwest Europe: London, Graham and Trotman Ltd., p. 951–961.

KANTOROWICZ, J. D., 1984, The nature, origin and distribution of authigenic clay minerals from Middle Jurassic Ravenscar and Brent Group sandstones: Clay Minerals, v. 19, p. 359–375.

KLAY, N., AND JESSBERGER, E. K., 1984, ^{40}Ar/^{39}Ar dating of glauconites (abs.): Terra Cognita, Special Issue, p. 22.

KLIGFIELD, R., HUNZIKER, J. C., DALLMEYER, R. D., AND SCHAMEL, S., 1986, Dating of deformation phases using K-Ar and ^{40}Ar/^{39}Ar techniques: results from Northern Apennines: Journal of Structural Geology, v. 8, p. 781–798.

LAMBERT-AIKHIONBARE, D. O., AND SHAW, H. F., 1982, Significance of clays in the petroleum geology of the Niger delta: Clay Minerals, v. 17, p. 91–104.

LEE, M., 1984, Diagenesis of the Permian Rotliegendes Sandstone, North Sea: K/Ar, O^{18}/O^{16}, and petrographic evidence: Unpublished Ph.D. Dissertation, Case Western Reserve University, Cleveland, Ohio, 346 p.

LEE, M., ARONSON, J. L., AND SAVIN, S. M., 1985, K-Ar dating of gas emplacement in Rotliegendes sandstones, Netherlands: American Association of Petroleum Geologists Bulletin, v. 69, p. 1381–1385.

LIEWIG, N., CLAUER, N., FRITZ, B., AND JEANNETTE, D., 1984, Sr and Ar evidences for brine-clay interactions in a geothermal reservoir: Annual Meeting of the Clay Mineral Society, Programs and Abstracts, p. 80.

LIEWIG, N., CLAUER, N., AND SOMMER, F., 1987a, Rb-Sr and K-Ar dating of clay diagenesis in Jurassic sandstone reservoirs, North Sea: American Association of Petroleum Geologists Bulletin, v. 71, p. 1467–1474.

LIEWIG, N., MOSSMANN, J. R., AND CLAUER, N., 1987b, Datation isotopique K-Ar d'argiles diagénétiques de réservoirs gréseux: Mise en évidence d'anomalies thermiques du Lias inférieur en Europe nord-occidentale: Comptes Rendus de l'Académie des Sciences, Paris, v. 304 II, p. 707–712.

MALLEY, P., CLAUER, N., JOURDAN, A., AND SOMMER, F., 1985, Fluid inclusion chronology and K-Ar dating of illite neoformation in sandstones from the Brent Formation of the North Sea (abs.): American Association of Petroleum Geologists, Research Conference on Radiogenic Isotopes and Evolution of Sedimentary Basins, New Orleans, p. 1.

MCDOWELL, S. D., AND ELDERS, W. E., 1980, Authigenic layer silicate minerals in borehole Elmore 1, Salton Sea geothermal field, California, USA: Contributions to Mineralogy and Petrology, v. 74, p. 293–310.

MCHARDY, W. J., WILSON, M. J., AND TAIT, J. M., 1982, Electron microscope and X-ray diffraction studies of filamentous illitic clay from sandstones of the Magnus field: Clay Minerals, v. 17, p. 23–40.

MERRIHUE, C. M., AND TURNER, G., 1966, Potassium-argon dating by activation with fast neutrons: Journal of Geophysical Research, v. 71, p. 2852–2857.

MOSSMANN, J. R., 1987, Les conditions physico-chimiques d'évolution de réservoirs gréseux. Approche pétrologique, minéralogique et isotopique. Application aux grès rhétiens du Bassin de Paris: Thèse de Doctorat, Université de Strasbourg, France, 124 p.

MOSSMANN, J. R., 1991, K-Ar dating of authigenic illite-smectite clay material: application to complex mixtures of mixed-layer assemblages: Clay Minerals, v. 26, p. 189–198.

MUFFLER, L. J. P., AND WHITE, D. E., 1969, Active metamorphism of Upper Cenozoic sediment in the Salton Sea geothermal field and the Salton Trough, southeastern California: Geological Society of America Bulletin, v. 80, p. 157–182.

ODIN, G. S., ed., 1982, Numerical Dating in Stratigraphy (2 vol.): Chichester, J. Wiley and Sons Ltd., 1040 p.

REUTER, A., AND DALLMEYER, R. D., 1987a, ^{40}Ar/^{39}Ar age spectra of whole-rock and constituent grain-size fractions from anchizonal slates: Chemical Geology, v. 66, p. 73–88.

REUTER, A., AND DALLMEYER, R. D., 1987b, ^{40}Ar/^{39}Ar dating of cleavage formation in tuffs during anchizonal metamorphism: Contributions to Mineralogy and Petrology, v. 97, p. 352–360.

REUTER, A., AND DALLMEYER, R. D., 1989, K-Ar and ^{40}Ar/^{39}Ar dating of cleavage formed during very low-grade metamorphism: a review, *in* Daly, J. S., Cliff, R. A., and Yardley, B. W. D., eds., Evolution of Metamorphic Belts: Geological Society of London Special Publication, Oxford, Blackwell, p. 161–172.

REYNOLDS, R. C., Jr., 1967, Interstratified clay system: calculation of the total one-dimensional diffraction function: American Mineralogist, v. 52, p. 661–672.

REYNOLDS, R. C., Jr., 1980, Interstratified clay minerals, *in* Brindley, C. W., and Brown, G., eds., Crystal Structures of Clay Minerals and Their X-ray Identification: Mineralogical Society of London, p. 249–303.

SIGURGEIRSSON, T., 1962, Dating recent basalt by potassium-argon method: Report of the Physical Laboratory, University of Iceland, 9 p.

STUEBER, A. M., PUSHKAR, P., AND HETHERINGTON, E. A., 1984, A strontium isotopic study of Smackover brines and associated solids, southern Arkansas: Geochimica et Cosmochimica Acta, v. 48, p. 1637–1650.

STUEBER, A. M., PUSHKAR, P., AND HETHERINGTON, E. A., 1987, A strontium isotopic study of formation waters from the Illinois Basin, U.S.A.: Applied Geochemistry, v. 2, p. 477–494.

SUNWALL, M. T., AND PUSHKAR, P., 1979, The isotopic composition of strontium in brines from petroleum fields of southeastern Ohio: Chemical Geology, v. 24, p. 189–197.

SUTTER, J. F., AND HARTUNG, J. B., 1984, Laser microprobe ^{40}Ar/^{39}Ar dating of mineral grains in situ: Scanning Electron Microscopy, IV, Chicago, SEM Inc., p. 1525–1529.

SUTTER, J. F., HARTUNG, J. B., AND KUNK, M., 1986, In situ ^{40}Ar/^{39}Ar ages of single mineral grains using a laser microprobe (abs.): Terra Cognita, v. 6, p. 149.

YANASE, Y., WAMPLER, J. M., AND DOOLEY, R. E., 1975, Recoil-induced loss of ^{39}Ar from glauconite and other minerals (abs.): Transactions of the American Geophysical Union, v. 56, p. 472.

YAU, Y. C., PEACOR, D. R., AND MCDOWELL, S. D., 1987, Smectite-to-illite reactions in Salton Sea shales: a transmission and analytical electron microscopy study: Journal of Sedimentary Petrology, v. 57, p. 335–342.

CONTROLS ON POREWATER EVOLUTION DURING SANDSTONE DIAGENESIS, WESTERN CANADA SEDIMENTARY BASIN: AN OXYGEN ISOTOPE PERSPECTIVE

FRED J. LONGSTAFFE
Department of Geology, University of Western Ontario, London, Ontario, Canada N6A 5B7
BARBARA J. TILLEY
Department of Geology, University of Alberta, Edmonton, Alberta, Canada T6G 2E3
AVNER AYALON
Geological Survey of Israel, 30 Malchei Israel Street, Jerusalem 95501, Israel
AND
CATHERINE A. CONNOLLY
Department of Geology, University of Alberta, Edmonton, Alberta, Canada T6G 2E3

ABSTRACT: The oxygen isotope compositions of diagenetic minerals from sandstones and conglomerates have been determined for a Permian to Upper Cretaceous section through the Alberta deep basin. These data have been used to reconstruct variations in the $\delta^{18}O$ values of pore water during diagenesis of intercalated sandstones and shales. The results confirm earlier observations on the oxygen isotope evolution of pore water in the western Canada sedimentary basin. First, meteoric water was abundant during early diagenesis, even in sediments deposited in shallow marine environments. Second, the maximum $\delta^{18}O$ values attained by pore waters during burial diagenesis were generally $<+3‰$, much lower than in other shale-dominated basins (e.g., Gulf Coast). However, pore waters in sandstones located adjacent to, or intercalated with, carbonates and shales of the underlying Paleozoic section had substantially higher $\delta^{18}O$ values ($+7$ to $+9‰$) at or near maximum burial. Third, late diagenesis was dominated by low-^{18}O meteoric waters.

Saturation of the sedimentary section by meteoric water during early diagenetic processes probably resulted from infiltration during subaerial exposure associated with sea level fluctuations of the inland sea. Influx of meteoric water probably continued episodically throughout burial diagenesis. Contribution of meteoric water during early diagenesis is reflected in the low maximum porewater $\delta^{18}O$ values that characterize the peak of burial diagenesis. These results also indicate that the smectite-to-illite reaction in Mesozoic shales did not dominate oxygen isotope evolution of sandstone pore waters in the western Canada sedimentary basin. Any increase in the $\delta^{18}O$ value of pore water during burial probably resulted from variable mixing between early diagenetic, meteoric-dominated pore waters and ^{18}O-rich pore waters that had equilibrated with the underlying Paleozoic carbonate platform. The $\delta^{18}O$ values of late diagenetic minerals and present pore waters reflect both this process and the subsequent, substantial influx of surface-derived meteoric water during post-Laramide uplift and erosion.

INTRODUCTION

Diagenetic minerals in coarse-grained clastic sedimentary rocks normally obtain oxygen isotope compositions in equilibrium with the formation water from which they crystallized (Longstaffe, 1983, 1989). Provided that the paragenetic sequence of crystallization can be deduced, the $\delta^{18}O$ values of these minerals can be used to reconstruct variations in formation-water composition (e.g., relative fractions of sea water, brines, meteoric water) throughout the diagenetic evolution of the sedimentary basin. Such studies of several Cretaceous sandstones and conglomerates in the western Canada sedimentary basin have demonstrated that: (1) depositional environment (continental, estuarine/deltaic, marine) had some influence on the oxygen isotope composition of early diagenetic-formation waters (-15 to $-1‰$), but that most of these waters have low $\delta^{18}O$ values; (2) porewater $\delta^{18}O$ values rose during burial diagenesis; and (3) uplift of the basin, concurrent with maximum burial, resulted in large-scale recharge of permeable units by low-^{18}O meteoric water and crystallization of late diagenetic minerals such as kaolinite, dickite, I/S and smectite (Longstaffe, 1986; Longstaffe and Ayalon, 1987; Ayalon and Longstaffe, 1988; Tilley and Longstaffe, 1989; Longstaffe and Ayalon, 1990).

In this paper, we focus on the scale of ^{18}O-enrichment of pore waters that occurred during burial diagenesis of clastic sedimentary rocks in the western Canada sedimentary basin, prior to post-Laramide recharge by meteoric water. Reported maximum $\delta^{18}O$ values of pore water during burial diagenesis are generally no higher than $+2$ to $+3‰$, and commonly are much lower (to $-6‰$) (Longstaffe and Ayalon, 1987; Ayalon and Longstaffe, 1988; Tilley and Longstaffe, 1989; Longstaffe and Ayalon, in prep.). Such values are lower than determined for other shale-dominated basins ($\delta^{18}O \approx +5$ to $+8‰$; e.g., Texas Gulf Coast, Great Valley Sequence), in which sandstone-porewater compositions have been buffered by the smectite-to-illite reaction in shales (Yeh and Savin, 1977; Suchecki and Land, 1983; Taylor, 1990).

The first objective of this paper is to present new data for a vertical section through the Alberta deep basin (Fig. 1) that was sampled specifically to evaluate the scale of porewater ^{18}O-enrichment during burial diagenesis. These data generally confirm earlier observations concerning the limited nature of this enrichment. The second objective is to evaluate the significance of this conclusion, including: (1) the extent of meteoric-water infiltration during burial diagenesis; (2) the contribution of the smectite-to-illite reaction in shales to sandstone pore waters; and (3) the contribution of formation water from Paleozoic carbonate rocks to formation waters in overlying sandstones.

POREWATER EVOLUTION IN THE ALBERTA DEEP BASIN

The Alberta deep basin is located immediately east of the Rocky Mountains in Alberta and British Columbia (Fig. 1), and forms the deepest portion of the southwestward-dipping, asymmetrical western Canada sedimentary basin. The area is of special economic interest because of the large gas reserves present throughout most of the Mesozoic section (Masters, 1984). In the deep basin, Precambrian basement

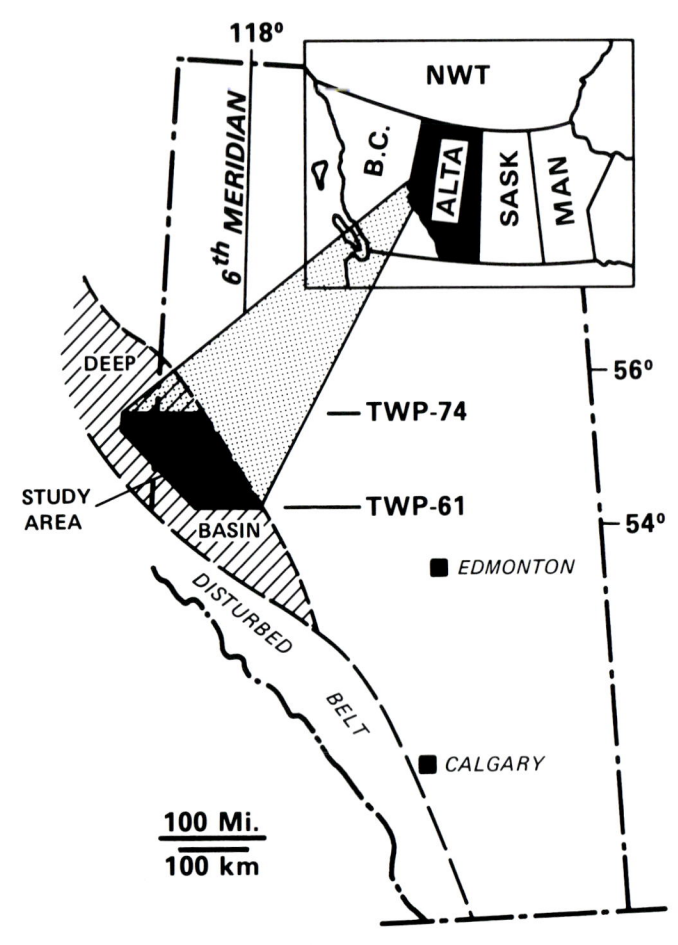

FIG. 1.—Location of deep basin study area, Alberta (modified from Tilley, 1988).

at temperatures corresponding to more typical geothermal gradients ($\approx 27°C/km$).

The regional applicability of these results is explored in this reconnaissance study of sandstones, which lie above and below the Falher Member (Spirit River Formation) and Cadomin Formation. The rocks examined in this study are predominantly very fine-, fine-, and medium-grained sandstones from the Upper Cretaceous Cardium Formation, the Lower Cretaceous Paddy and Cadotte Members of the Peace River Formation, the Lower Cretaceous Bluesky Formation, the Triassic Halfway Formation and the Permian Belloy Formation (Fig. 2). Permeabilities of these sandstones are significantly lower than in the conglomerates studied earlier from the Falher Member and Cadomin Formation (0.1–1 md versus 1–1,000 md; Cant and Ethier, 1984). Accordingly, the volume of pore water transmitted through the sandstones was probably significantly less than in the Falher Member or Cadomin Formation conglomerates.

Summary of Depositional Settings

Core locations are shown in Figure 3, subdivided according to the stratigraphic unit (Fig. 2) that has been sampled. Within the study area, the Cardium Formation is comprised of shelf to shoreline sedimentary rocks deposited in six major coarsening-upward sequences, separated by pebble beds overlying regional erosion surfaces (Plint and others, 1986; Plint and Walker, 1987). Both the Kakwa and Dismal Rat Members of the Cardium Formation have been sampled. The Kakwa Member is a very fine- to medium-grained shoreline sandstone, 6 to 18 m thick, capped by floodplain and lagoonal deposits (Plint and others, 1986; Plint and Walker, 1987). The Dismal Rat Member is a ≈ 10-m-thick, shallowing-upward shelf sequence of bioturbated silty and sandy mudstones containing thin stringers of pebbles; its upper boundary is defined by a regional erosional surface (Plint and others, 1986; Plint and Walker, 1987).

The Paddy Member of the Peace River Formation is a 16-m-thick sandstone, which was deposited as part of a northwest-southeast-trending tidal channel (Smith and others, 1984). The channel is surrounded by subtidal-bay deposits of interbedded, thin sandstones and shales. The Cadotte Member of the Peace River Formation is composed mostly of sandstones and conglomerates that comprise a wave-dominated sand delta (Smith and others, 1984). The Falher Member of the Spirit River Formation is composed of shoreline conglomerates, sandstones, shales and coal in the southern part of the study area, and sandstone in the northern part (Cant, 1983; Smith and others, 1984).

The Bluesky Formation is stratigraphically situated above the continental Gething Formation and below the marine Wilrich Member (Spirit River Formation) shale (Fig. 2). Its sedimentology has been described recently by Smith and others (1984) and O'Connell (1988). It is comprised of a barrier-bar/offshore-bar, regressive sandstone system represented by coarsening-upward sandstone. Lagoonal and bay sediments, consisting of siltstone, shale and coal, cap most of the barrier system. Low-energy bay sediments are predominant in the southeastern part of the study area, where the Bluesky Formation is represented by thin, highly bur-

is overlain by Paleozoic rocks, mostly carbonates, which are in turn overlain by Mesozoic clastic sedimentary rocks that reach thicknesses of $\approx 4,600$ m. The Mesozoic section consists mostly of shales and interlayered sandstones and conglomerates. A simplified stratigraphy of the Alberta deep basin for the interval of interest is given in Figure 2.

Using petrographic and isotopic data for sandstones and conglomerates from the Lower Cretaceous Falher Member (Spirit River Formation) of the Alberta deep basin (Fig. 2), Tilley and Longstaffe (1989) and Tilley and others (1989) showed that pore water evolved to a maximum $\delta^{18}O$ value of +3‰ during burial and became depleted in ^{18}O during uplift as a result of an influx of meteoric water. In contrast, formation waters that acquired $\delta^{18}O$ values as high as +7‰ by the time of maximum burial were indicated for the more deeply buried Cadomin Formation. Such values are much higher than typical for most other Cretaceous sandstones/conglomerates in the western Canada sedimentary basin. Tilley and others (1989) also suggested that hot waters were involved in diagenesis of Falher Member sandstones and conglomerates in the western part of the deep basin, where unexpectedly high geothermal gradients ($\approx 38°C/km$) were encountered. To the east, and in the Cadomin Formation in general, diagenetic minerals precipitated from pore waters

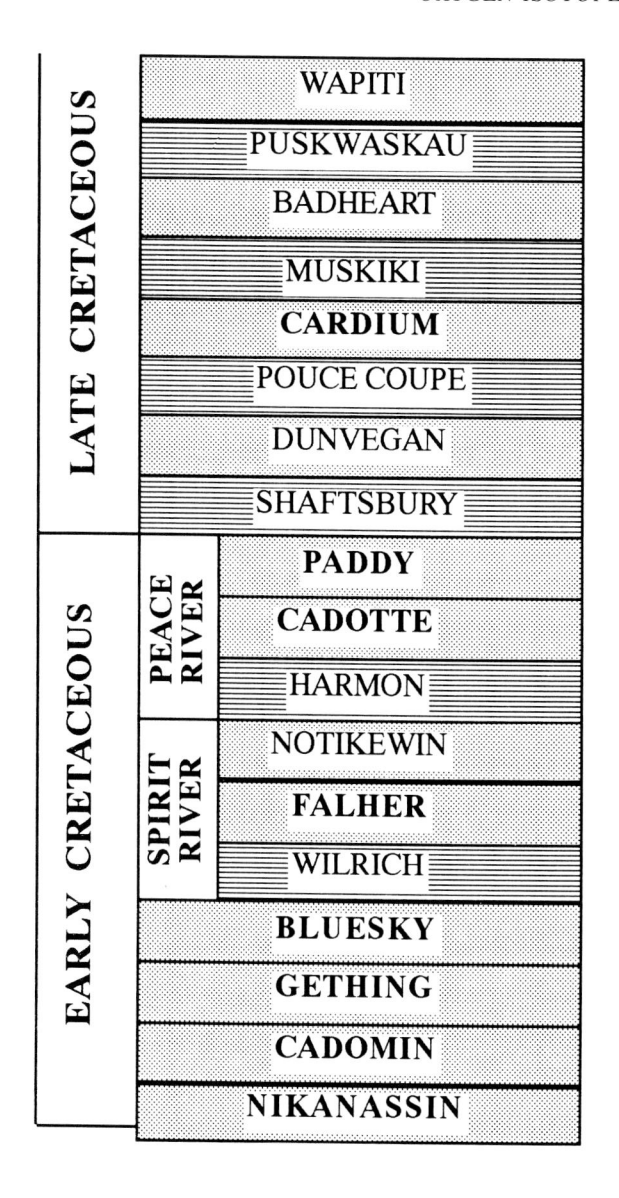

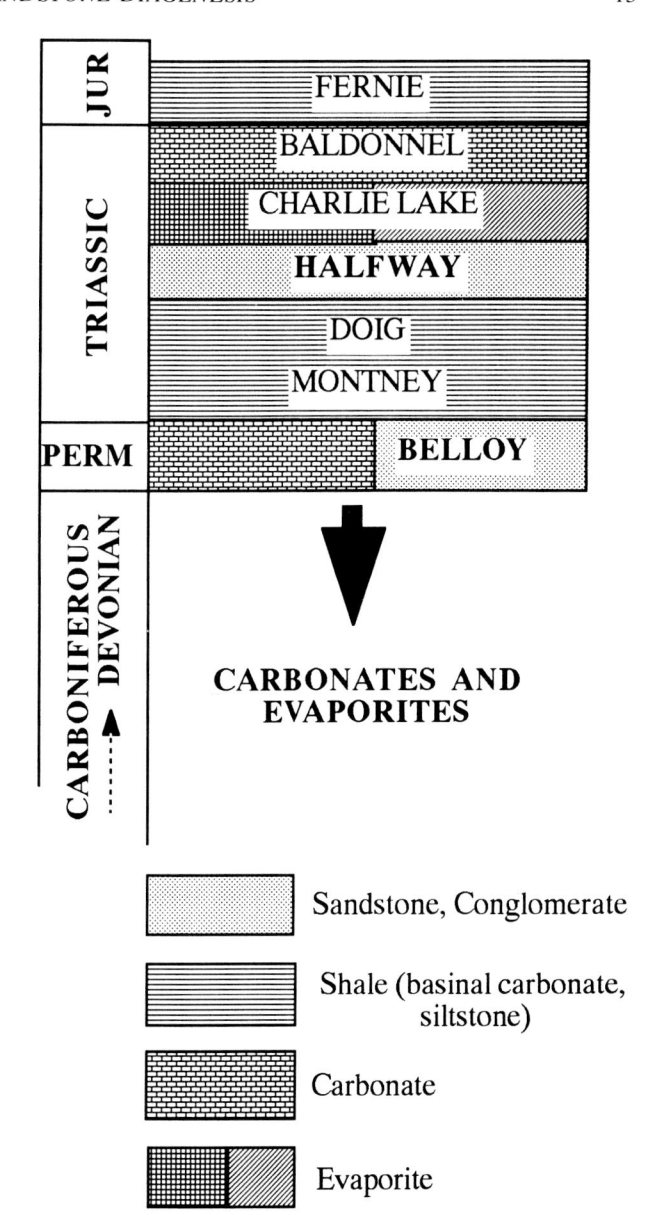

FIG. 2.—Generalized stratigraphic nomenclature of the Alberta deep basin. Unit thicknesses are arbitrary. Units listed in bold type are considered in this paper.

rowed sandstone, siltstone and shale containing thin molluscan shell banks.

The Cadomin Formation is a distal, gravel-dominated alluvial-braidplain deposit (Varley, 1982, 1984; Smith and others, 1984) ranging in thickness from 1 to 19 m. It unconformably overlies Upper Jurassic to Lower Cretaceous sandstones of the Nikanassin Formation and is overlain by continental coals and fine-grained sedimentary rocks of the Gething Formation.

The Triassic Halfway Formation and Permian Belloy Formation sandstones are surrounded by shale, limestone, dolomite and anhydrite. Cant (1986) has interpreted the Halfway Formation to be a complex of transgressive marine sands, barriers, tidal estuaries and coquina banks, which are capped by supratidal-sabkha deposits of the lower Charlie Lake Formation. Detailed sedimentological studies of the Belloy Formation sandstone are not available to us.

Analytical Methods

The sandstones have been examined using optical petrography, scanning electron microscopy, energy dispersive spectrometry, X-ray diffraction, and oxygen and carbon isotope geochemistry. The stable-isotope data are presented in the normal δ- notation relative to SMOW for oxygen (Craig, 1961) and PDB for carbon (Craig, 1957). An oxygen isotope CO_2-H_2O fractionation factor of 1.0412 at 25°C (Friedman and O'Neil, 1977) has been employed to calibrate the reference gas for the mass spectrometer. Oxygen was extracted from silicate minerals using the BrF_5 tech-

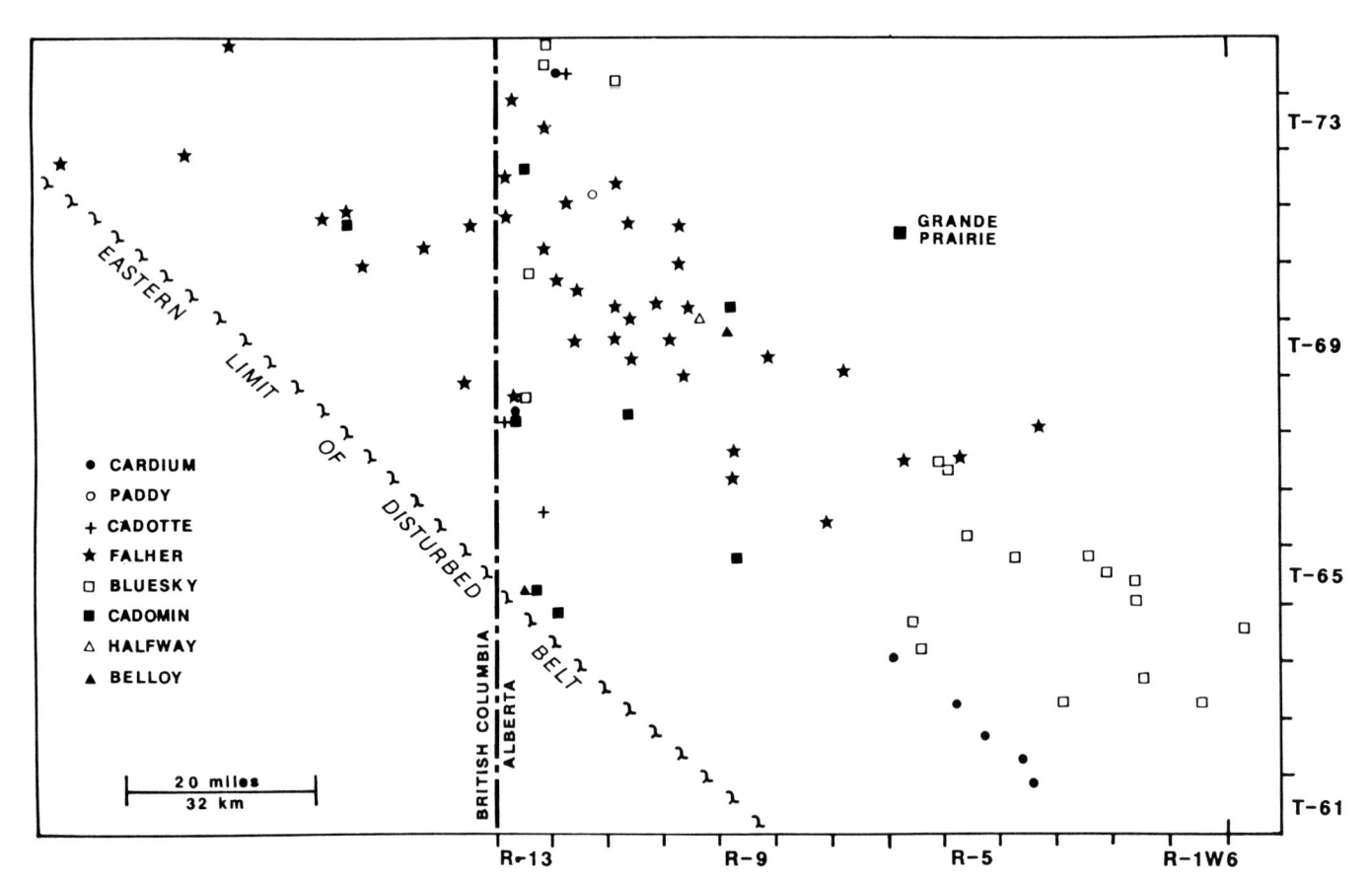

FIG. 3.—Location of drill-core samples from various units in the Alberta deep basin (modified after Tilley, 1988). Details of location and sample depth are given in Table 1. Samples from the Falher Member (Spirit River Formation) and the Cadomin Formation are described in Tilley and Longstaffe (1989), and are illustrated here for comparison.

nique of Clayton and Mayeda (1963). During these experiments, the average oxygen-isotope ratio for the NBS-28 silica standard was +9.6 ± 0.2‰. Carbonate minerals were prepared for isotopic analysis by reacting powdered samples in anhydrous H_3PO_4 at 25°C, following procedures described by Walters and others (1972), modified after McCrea (1950) and Epstein and others (1964). Average oxygen and carbon isotope ratios for calcite standard NBS-20 were +26.50 ± 0.02‰ and −1.17 ± 0.01‰, respectively. The following total carbonate CO_2-H_3PO_4 extracted CO_2 fractionation factors (α) at 25°C were used: 1.01025 for calcite and 1.01110 for dolomite (after Sharma and Clayton, 1965), and 1.01163 for siderite (Rosenbaum and Sheppard, 1986). The dolomite α was used in all calculations for ankerite. Using the α (25°C) for ankerite reported by Rosenbaum and Sheppard (1986) would lower the values reported here by about 0.66‰. These differences are insufficient to affect the sense of the arguments and conclusions that follow. The older fractionation factors were retained to maintain consistency with the results of Tilley and Longstaffe (1989). Details concerning the physical and chemical methods used to obtain minerals separates are given in Tilley and Longstaffe (1989).

Petrographic and Isotopic Results

In this reconnaissance study, too few samples were obtained to characterize fully the mineralogy and variability of each unit. However, certain generalizations are possible, following descriptions provided by Tilley (1988). Cardium Formation sandstones are composed mostly of quartz with lesser amounts of chert and siliceous-rock fragments; chert-rich conglomerates are also present. The Paddy Member sandstone is quartz rich, with minor quantities of rock fragments and trace amounts of mica and feldspar; rock fragments include chert, shale, siltstone and volcanic fragments. The Cadotte Member is also silica rich; chert is predominant in the conglomerates and lithic grains in the sandstones. The Bluesky Formation sandstones are composed of quartz rock fragments (mostly chert and sedimentary-rock fragments) and dolomite grains. Micas are minor and feldspar is rare. The amount of chert varies with grain size; medium- or fine-grained sandstones are dominated by chert. Glauconite is abundant in the uppermost Bluesky Formation sandstones, and is present in smaller amounts throughout the rest of the unit. The Triassic Halfway Formation and Permian Belloy Formation sandstones are quartz rich.

Cadotte Member.—

Tilley and Longstaffe (1989) showed that siderite from the Cadotte Member probably formed early in diagenesis from infiltrating meteoric to brackish pore waters (−9 ± 3‰). Quartz druse crystallized relatively late in the diagenetic sequence in vugs created by partial dissolution of siderite concretions and within horizontal fractures in Cadotte Member sandstones, primarily in the western portion of the study area. The similarity in oxygen isotope composition of vug-filling and fracture-filling quartz druse suggests that they formed under nearly identical conditions. Fluid inclusion data for the quartz druse indicate that temperatures as high as 176°C occurred during quartz crystallization (Tilley and others, 1989), and that porewater $\delta^{18}O$ values as high as +2‰ were achieved (Fig. 11). The oxygen isotope composition of pore-filling dickite from the same rocks is compatible with crystallization under such conditions (Fig. 11), but lower temperature formation from pore waters with lower $\delta^{18}O$ values is an equally valid interpretation.

Figure 11 also illustrates the possible range of porewater compositions needed to crystallize fracture-filling calcite and Fe-dolomite/ankerite in the Cadotte Member. Fracture-filling calcite is among the very last phases to have precipitated, and is in isotopic equilibrium at present formation temperatures with current pore waters (assuming pore waters in the Paddy and Cadotte Members have very similar $\delta^{18}O$ values). The oxygen isotope composition of the fracture-filling Fe-dolomite/ankerite is also compatible with a late diagenetic origin, perhaps at higher temperatures earlier in the erosional history of the unit.

Falher Member.—

Pore waters in Falher Member sandstones and conglomerates evolved from $\delta^{18}O$ values of about −10 ± 3‰ during early diagenesis to ≈ +2 to +3‰ during burial, even in the western part of the study area, where hot, silica-saturated fluids may have preferentially entered the conglomerates (Tilley and Longstaffe, 1989; Tilley and others, 1989). Formation waters subsequently acquired lower $\delta^{18}O$ values during the post-Laramide influx of meteoric water.

A few data, however, suggest that waters more ^{18}O rich than proposed by Tilley and Longstaffe (1989) may have locally penetrated the conglomerates and fractures of the Falher Member. Tilley and Longstaffe (1989) list two calcite cements with anomalously high $\delta^{18}O$ values (+23.4 to +23.6‰; $\delta^{13}C$ = +1.4 to +1.9‰). These are late-stage microcrystalline cements that overlie quartz druse in conglomerate pores. If this calcite crystallized under present conditions (65°C), pore waters significantly enriched in ^{18}O relative to regional Falher Member formation waters (+2 versus −6‰) are required. If the calcite crystallized at the maximum burial temperatures achieved in this locality (≈125°C using the Lopatin-Waples calculation), pore waters with $\delta^{18}O$ values as high as +8‰ are possible.

Fracture-filling calcite from the westernmost portion of the study area also has a much higher $\delta^{18}O$ value (+17.1‰; Table 1) than most other Falher Member calcite (≈ +11.5‰). Interpretation of this result, however, is equivocal in the absence of independent temperature data. Pore-water $\delta^{18}O$ values as high as +6‰ could be indicated if the fracture calcite precipitated at the maximum temperature (≈170°C) determined for fluid inclusions in pore-filling calcite from the western portion of the Falher Member. In contrast, if fracture calcite crystallized at present formation temperatures, a porewater $\delta^{18}O$ value ≈ −5‰ is indicated, similar to formation water from the overlying Paddy Member. The latter scenario seems more probable, as is discussed later.

Bluesky Formation.—

Most samples from the Bluesky Formation are located in the eastern portion of the study area, away from the supposed influence of hot formation waters to the west. Oxygen isotope data for these samples are illustrated in Figures 12A and B. Early diagenetic kaolinite has a high $\delta^{18}O$ value, typical of formation at low temperatures from fresh to brackish pore waters. Similar diagenetic kaolinite has been described from the Viking Formation in west-central Alberta (Longstaffe and Ayalon, 1987).

Illitic clays from glauconite pellets do not have oxygen isotope compositions consistent with crystallization from marine pore waters early in diagenesis. Glauconite formation at 20 to 30°C would require waters with $\delta^{18}O$ values of −10 to −8‰, characteristic of fresh to brackish waters (Fig. 12A). To form the glauconite from typical sea water (0‰) would require much higher temperatures (≈90°C). The best explanation for this anomaly is that the glauconite recrystallized and isotopically re-equilibrated during burial. Some evidence favoring such a possibility is provided by the glauconite's fibrous texture (Fig. 7H). Discrete diagenetic illite and glauconite have virtually identical $\delta^{18}O$ values (Table 1).

If glauconite recrystallization and illite formation occurred at the maximum burial temperatures suggested by vitrinite-reflectance data for this area (Lopatin-Waples method, 115°C), a maximum porewater $\delta^{18}O$ value of +2‰ is indicated (Figs. 12A,B). However, fluid-inclusion temperatures (>108°C) for diagenetic calcite crystallized after illite (Tilley and others, 1989) indicate that pore water at

FIG. 7.—(A) SEM photomicrograph of kaolinite (KAOL) filling pore space after quartz overgrowths. Bluesky Formation (11-24-67-6W6, 2125.2 m). (B) SEM photomicrograph of pore-filling illite. Cardium Formation (15-16-68-13W6, 1159.4 m). (C) SEM photomicrograph of illite (ILL) coating kaolinite (K). Cardium Formation (15-34-61-4W6, 1929.6 m). (D) SEM photomicrograph of pore-lining illite. Kaolinite (KAOL) formed after illite. Paddy Member of the Peace River Formation (7-11-72-12W6, 1684.4 m). (E) SEM photomicrograph of illite partially coating kaolinite (KAOL). Paddy Member of the Peace River Formation (7-11-72-12W6, 1684.4 m). (F) SEM photomicrograph of illite forming a bridge between authigenic quartz and pyrite. Paddy Member of the Peace River Formation (7-11-72-12W6, 1684.4 m). (G) SEM photomicrographs of illite (ILL) lining the pore wall and partially coating kaolinite (KAOL). Bluesky Formation (7-18-67-5W6, 2151.2 m). (H) SEM photomicrograph showing the recrystallized nature of illite on the surface of a glauconite pellet. Bluesky Formation (10-16-65-2W6, 2195.5 m).

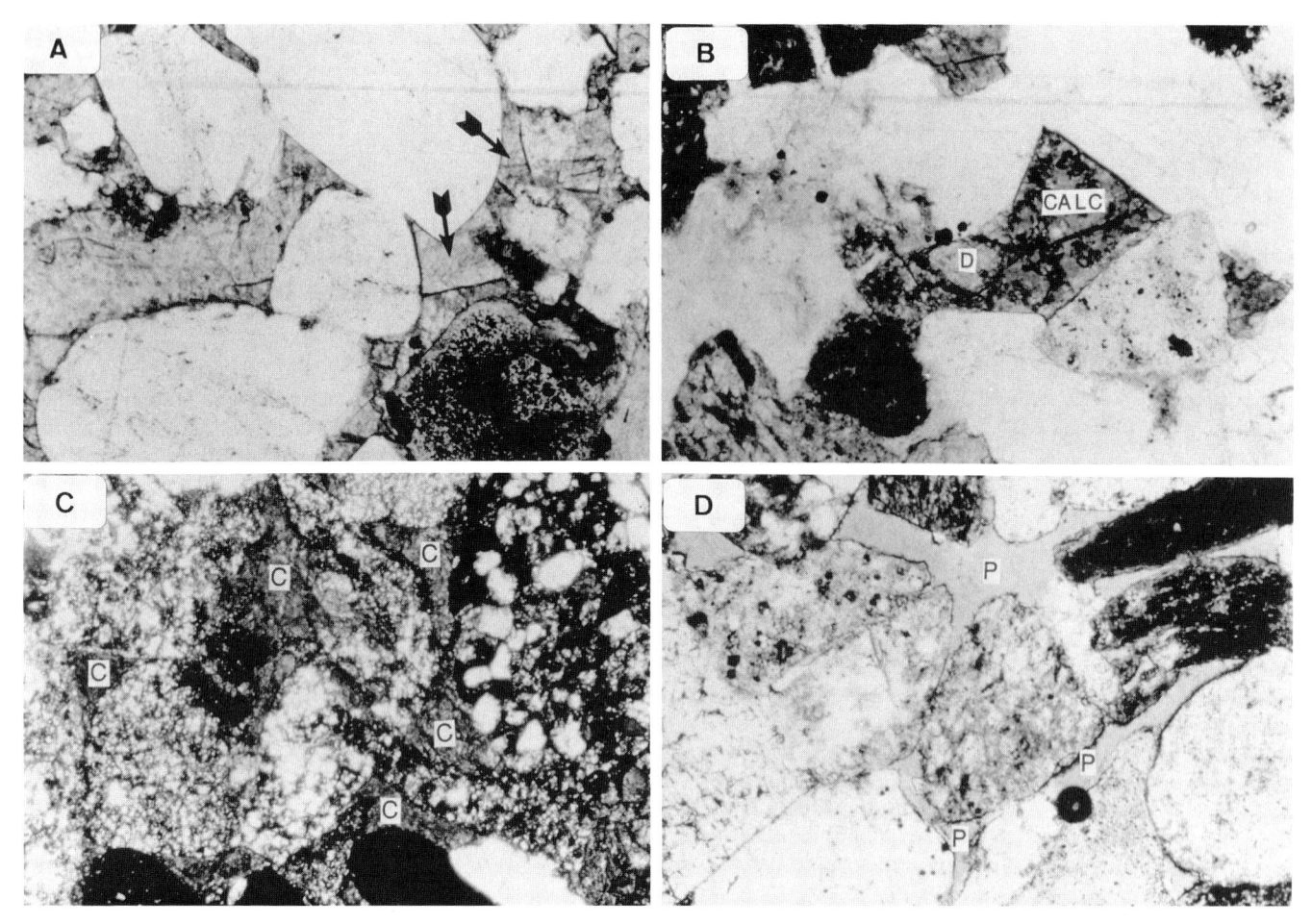

FIG. 8.—(A) Thin-section photomicrograph of ankerite (arrows) and calcite cement (adjacent to ankerite crystals). Plane polarized light. Paddy Member of the Peace River Formation (7-11-72-12W6, 1693.7 m). Field of view = 1.34 mm wide. (B) Thin-section photomicrograph of ankerite-rimmed dolomite (D) surrounded by calcite cement (CALC). The darker rim around the dolomite grain is ankerite. Plane polarized light. Bluesky Formation (11-20-64-26W5, 2206.6 m). Field of view = 1.34 mm wide. (C) Thin-section photomicrograph of late-stage, poikilotopic calcite cement (C). The calcite cement fills a partially dissolved chert grain (center). Crossed polarized light. Bluesky Formation (7-18-67-5W6, 2148.2 m). Field of view = 1.34 mm wide. (D) Thin-section photomicrograph of a sandstone sample adjacent to that observed in (C). Pore space (P) is completely open. Plane polarized light. Bluesky Formation (11-24-67-6W6, 2125.0 m). Field of view = 1.34 mm wide.

or shortly following maximum burial had even lower $\delta^{18}O$ values ($\approx -1‰$; Figs. 12A,B). Regardless of the model employed, pore waters with $\delta^{18}O$ values significantly higher than +2 to +3‰ seem not to be recorded in the diagenetic-mineral record. Finally, the influx of meteoric water that affected other Cretaceous sandstones and conglomerates, beginning at maximum burial and uplift (Hitchon, 1984), again best accounts for the lowering of $\delta^{18}O$ porewater values necessary to explain the paragenetic progression and isotopic compositions of illite, quartz, calcite, ankerite and kaolinite (Fig. 12B).

Similar arguments can be used to explain results for samples from the western part of the study area, where higher temperatures were reached during burial diagenesis. Minerals formed during burial diagenesis in that locality have systematically lower $\delta^{18}O$ values than in the eastern part of the Bluesky Formation (illite, +13.3 to +13.8‰ versus +14.3 to +15.1‰; late calcite, +12.0 to +13.6‰ versus +14.1 to +15.6‰; Table 1). By comparison, two $\delta^{18}O$ values for late diagenetic-kaolin group minerals, which formed following maximum burial, during the meteoric-water recharge event, show less variation (+10.9 in the east versus +11.4‰ in the west; Table 1).

The possibility of significantly higher maximum temperatures in the western portion of the Bluesky Formation, either because of migration of hot fluids through the most permeable units or because the vitrinite-reflectance equation of Barker and Pawlewicz (1986) is more appropriate (Tilley and others, 1989), makes calculation of maximum pore water $\delta^{18}O$ values more problematical. If temperatures as high as 190–200°C were achieved in this portion of the Bluesky Formation, porewater compositions between +3 ± 1 and +6.5 ± 0.5‰ can be calculated from calcite and illite data. At regional temperatures of $\approx 170°C$, typical of the im-

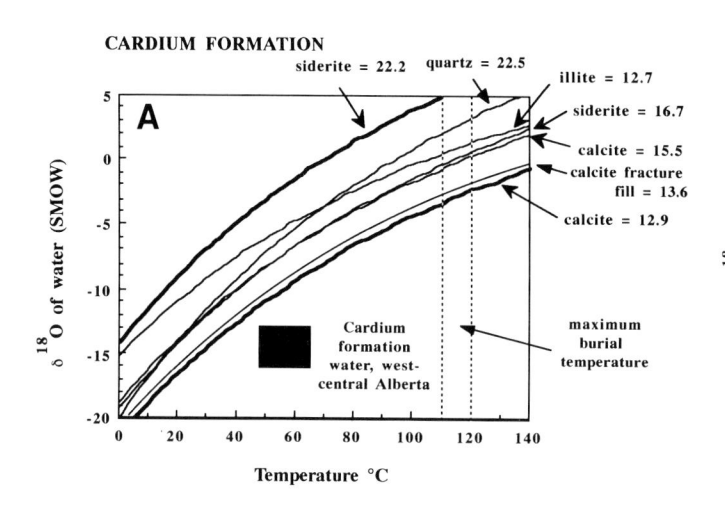

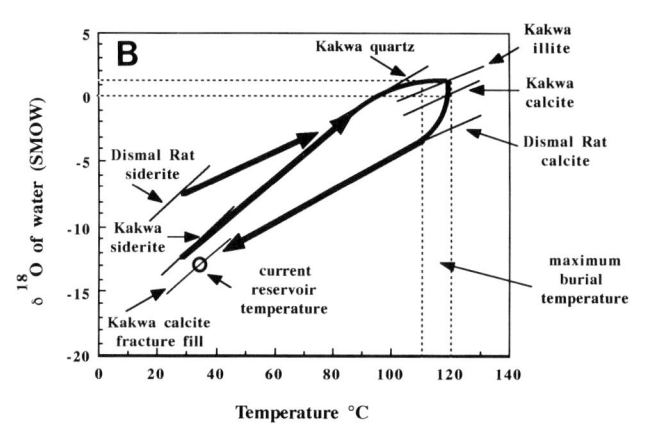

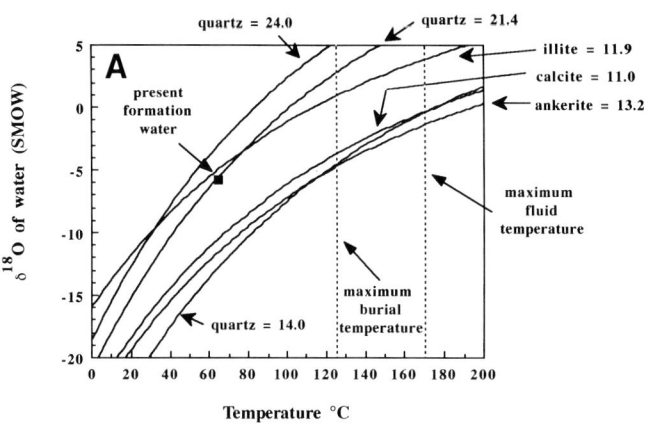

FIG. 9.—(A) Oxygen isotope composition of porewater versus crystallization temperature for diagenetic minerals from the Cardium Formation. Kakwa Member—light, solid curves; Dismal Rat Member—heavy, solid curves. The curves for each diagenetic mineral have been calculated using the mineral $\delta^{18}O$ value and appropriate mineral-water oxygen isotope fractionation equations: siderite-water, Becker and Clayton, 1976; quartz-water, Clayton and others, 1972 as modified by Friedman and O'Neil, 1977; calcite-water, O'Neil and others, 1969 as modified by Friedman and O'Neil, 1977; illite-water, after Yeh and Savin, 1977 as corrected by Savin and Lee, 1988. The field for formation waters from the Cardium Formation, west-central Alberta, is shown by the shaded box. (B) Schematic porewater-evolution paths (heavy arrows) for the Cardium Formation sandstone. Pore waters in the Dismal Rat and Kakwa Members follow separate pathways during early diagenesis. The pathway has been drawn assuming that illite crystallized at maximum burial temperatures; maximum porewater $\delta^{18}O$ values under such conditions are ≈ +1‰. Slightly lower maximum porewater $\delta^{18}O$ values (≈0‰) are indicated if calcite (Kakwa Member), rather than illite, formed at maximum burial temperatures. However, significantly lower porewater $\delta^{18}O$ values are probably associated with late diagenetic crystallization of calcite. For example, the open circle indicates that fracture-filling calcite (the latest phase) could precipitate from formation waters with $\delta^{18}O$ values ≈ −13‰ (within the measured range of pore waters in the Cardium Formation) at current subsurface temperatures (35°C).

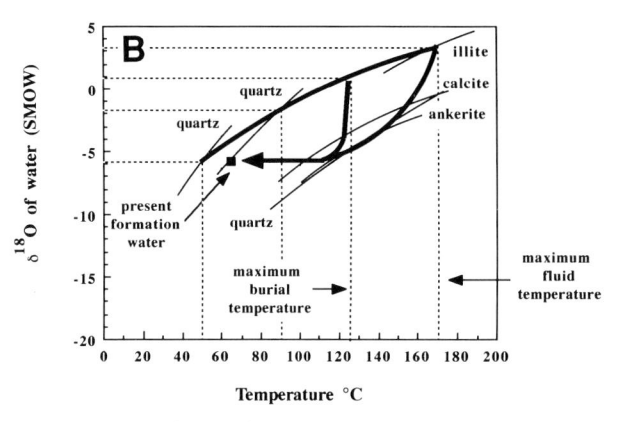

FIG. 10.—(A) Oxygen isotope composition of porewater versus crystallization temperature for diagenetic minerals from the Paddy Member of the Peace River Formation. The mineral-water oxygen isotope fractionation equations are as listed in Figure 9 plus ankerite-water from Dutton and Land (1985). The present formation-water composition is taken from Tilley and Longstaffe (1989). (B) Schematic porewater-evolution paths for Paddy Member (Peace River Formation) sandstones. The lower temperature path assumes that hot fluids from the Falher Member (Spirit River Formation) did not enter the Paddy Member sandstones. The higher temperature path assumes that hot fluids from below penetrated Paddy Member sandstones.

mediately overlying Wilrich Member shale at this depth (Connolly, 1989), these porewater values would be lower, +2 to +5‰.

Cadomin and Nikanassin Formations.—

Tilley and Longstaffe (1989) showed that infiltration of saline fluids from pre-Cretaceous units could have produced pore waters in the Cadomin Formation with $\delta^{18}O$ values as high as +4 to +7‰ during illite crystallization at or near maximum burial. Druse quartz formed at temperatures ≤150°C from pore waters with $\delta^{18}O$ values ≤ +3‰. Similar arguments best explain the single oxygen isotope composition of a quartz overgrowth from the immediately

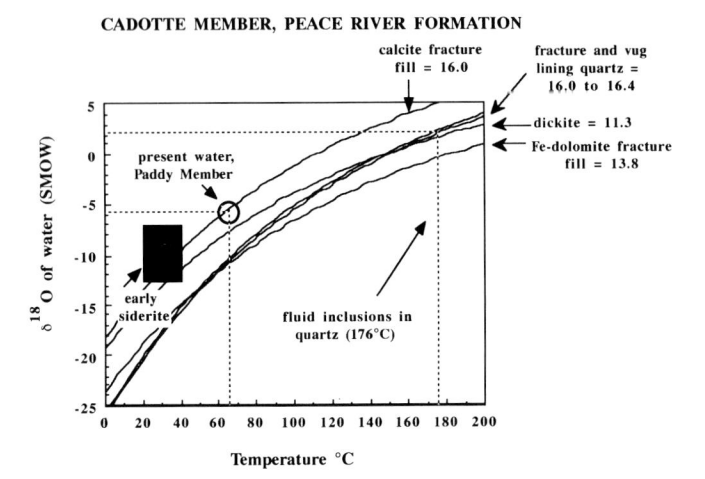

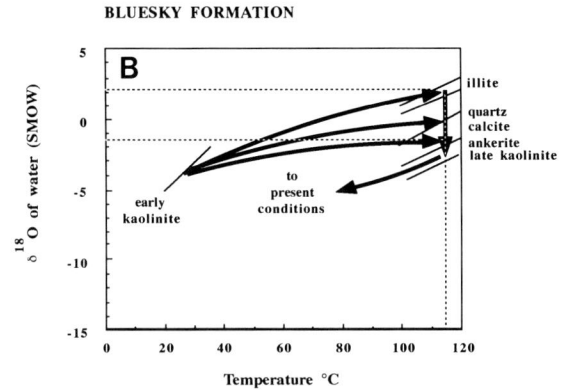

FIG. 11.—Oxygen isotope composition of porewater versus crystallization temperature for diagenetic minerals from the Cadotte Member of the Peace River Formation. The mineral-water oxygen isotope fractionation equations are as listed in previous figures plus kaolinite-water from Land and Dutton (1978). The shaded area for early siderite is taken from Tilley and Longstaffe (1989). The average temperature for crystallization of quartz druse, as determined from fluid inclusions by Tilley and others (1989), is also indicated. The present temperature and isotopic composition of formation water from the immediately overlying Paddy Member (Peace River Formation) sandstone is also shown (open circle).

underlying Nikanassin Formation sandstone. In the absence of specific temperature data, the high-$\delta^{18}O$ value of a late diagenetic Fe-dolomite/ankerite fracture fill (+20.5‰), which postdates diagenetic quartz in the Nikanassin Formation cannot be unequivocally interpreted. The fracture fill would be in oxygen isotope equilibrium with pore water of $\delta^{18}O \approx -2$ to 0‰ at current subsurface temperatures. Its composition fits the trend of higher $\delta^{18}O$ and lower $\delta^{13}C$ values with depth observed for fracture-filling carbonate minerals in the Lower Cretaceous, Triassic and Permian sandstones.

Halfway and Belloy Formations.—

The few oxygen isotope data for diagenetic quartz and pore-filling calcite from the Triassic Halfway Formation and Permian Belloy Formation sandstones are illustrated in Figure 13. Quartz precipitation was followed by crystallization of pore-filling calcite, which in turn is postdated by fracture-filling calcite. If the pore-filling calcite crystallized or recrystallized at or near maximum burial temperatures (145–150°C, Lopatin-Waples method), associated pore waters could have achieved $\delta^{18}O$ values as high as +7‰ in the Halfway Formation sandstone and +9‰ in the Belloy Formation sandstone. Even higher values (+9 to +11‰) are indicated if the vitrinite-reflectance equation of Barker and Pawlewicz (1986) is employed. Earlier formed quartz overgrowths would have precipitated from somewhat less ^{18}O-rich pore waters.

The very low-$\delta^{13}C$ values of the ^{18}O-rich, pore-filling calcite (−27.6 to −24.4‰) can be interpreted to support its crystallization near maximum burial temperatures. These compositions are typical of CO_2 released during thermally driven organic reactions active at such temperatures (e.g.,

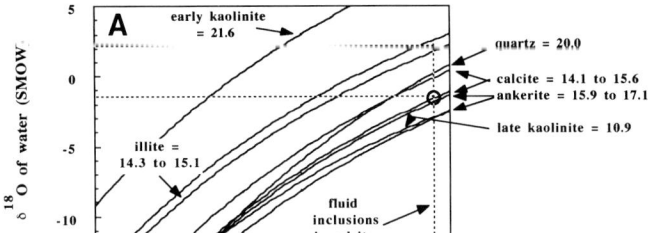

FIG. 12.—(A) Oxygen isotope composition of porewater versus crystallization temperature for diagenetic minerals from the Bluesky Formation sandstone. Mineral-water oxygen isotope fractionation equations are as listed in previous figures. Fluid inclusions in the calcite cement homogenize at ≈115°C (Tilley and others, 1989). The open circle represents the point on the porewater-evolution path fixed by stable-isotope and fluid-inclusion measurements. (B) Schematic porewater-evolution paths for Bluesky Formation sandstones. Different maximum porewater oxygen isotope compositions are possible, depending on whether illite, quartz or calcite formed at maximum burial temperatures.

decarboxylation of organic matter) (Irwin and others, 1977). In contrast, $\delta^{13}C$ values of low-^{18}O diagenetic calcites (non-fracture filling; $\delta^{18}O = +11.0$ to +15.6‰) from the overlying Cretaceous sandstones and conglomerates vary irregularly from −15.8 to +9.7‰ (Table 1 and Tilley and Longstaffe, 1989), consistent with variable mixing of CO_2 produced by near-surface and deep diagenetic processes.

Fracture-filling calcite in the Halfway Formation sandstone has a virtually identical oxygen isotope composition to the pore-filling calcite. However, the higher $\delta^{13}C$ value of the fracture calcite (−15.3‰) suggests that it probably crystallized from different formation waters. Crystallization at present subsurface temperatures (75–90°C) would require formation waters with $\delta^{18}O$ values of −0.5 to +1.3‰. These values are virtually identical to formation waters in nearby Triassic oil pools (−1.7 to +0.8‰; Hitchon and Friedman, 1969). The lower $\delta^{18}O$ value of the formation water is consistent with infiltration of meteoric water deep into the basin during late diagenesis. The $\delta^{13}C$ value of the

HALFWAY AND BELLOY FORMATIONS

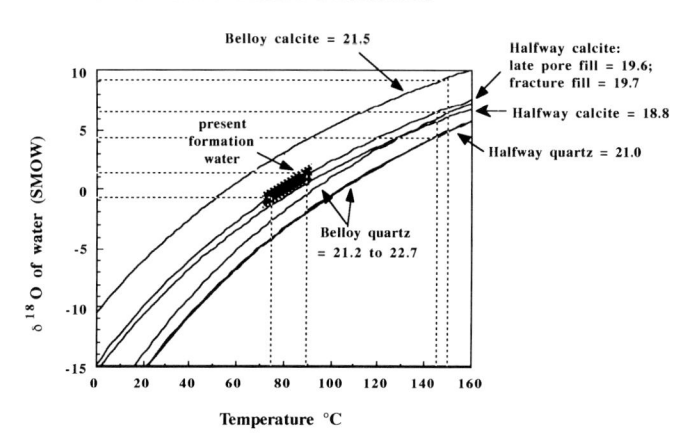

FIG. 13.—Oxygen-isotope composition of porewater versus crystallization temperature for diagenetic minerals from the Halfway Formation and Belloy Formation sandstones. Mineral-water oxygen isotope fractionation equations are as listed in previous figures. Porewater $\delta^{18}O$ values of +7 (Triassic Halfway Formation sandstone) and +9‰ (Permian Belloy Formation sandstone) are possible if late diagenetic calcite crystallized at maximum burial temperatures (Lopatin-Waples method) in these localities. However, lower values (+4 to +7‰) are indicated if quartz crystallized at maximum burial temperatures. The range of oxygen isotope compositions for nearby formation waters from Triassic oil fields is shown in the shaded area (Hitchon and Friedman, 1969). These data suggest that the latest stage, fracture-filling calcite may be in isotopic equilibrium with current formation waters at present subsurface temperatures.

fracture calcite may reflect mixing of inorganically derived CO_2 ($\approx$0‰) (carried by the meteoric water?) and CO_2 resulting from thermal reactions involving organic matter. That the carbon isotope compositions of carbonate fracture fills in overlying Jurassic to Lower Cretaceous rocks increase upward from −4.5 to +2.2‰ adds support to such a mixing model. Alternatively, all of the CO_2 could be provided by thermal decarboxylation-style reactions that had waned as reactants were exhausted and as the unit cooled during uplift. The $\delta^{13}C$ values of produced CO_2 could increase under such conditions; the most ^{13}C-depleted gas should have been preferentially released earlier in the reaction.

DISCUSSION

Comparison of Upper and Lower Cretaceous sandstones and conglomerates from the deep basin demonstrates a similarity in the major diagenetic processes that have affected these rocks. With increasing burial, early siderite or chlorite, kaolinite, quartz overgrowths and then illitic clays formed. Crystallization of quartz druse, dickite/late kaolinite, calcite and ankerite followed during uplift and erosion. Whereas burial depths, geographic position and depositional environment of each unit have varied, one important unifying control on diagenesis has been similarity in grain mineralogy, dominantly quartz and chert. In addition, an event common to all units, the post-Laramide influx of meteoric water, strongly influenced late diagenetic reactions (at least until the pores became saturated by methane). Meteoric water and its chemically evolved equivalents have been the transport agent responsible for much of the crystallization of quartz druse, dickite, calcite and ankerite in the conglomerates, and calcite, ankerite and quartz overgrowths in sandstones.

Porewater evolution during burial diagenesis of the Alberta deep basin, and indeed the entire western Canada sedimentary basin, prior to the meteoric-water-recharge event, was more complicated. First, meteoric water was common during early diagenesis. Second, the maximum oxygen isotope compositions acquired by pore waters during burial diagenesis seem, in most instances, to have been low relative to other shale-dominated basins. Third, ^{18}O-rich formation waters appear to have penetrated the clastic sedimentary section from underlying Paleozoic carbonate rocks. What are the implications of these observations?

Origin of Early Meteoric Water

Low-^{18}O pore water regularly appears in the early diagenetic-mineral record in the Alberta deep basin: (1) siderite formation in the Cardium Formation, Cadotte Member, Falher Member and Cadomin Formation sandstones and conglomerates; (2) chlorite formation in the Falher Member (Tilley and Longstaffe, 1989); and (3) early kaolinite crystallization in the Bluesky Formation sandstones. Meteoric water has also controlled early siderite formation in the Badheart Formation and equivalent sandstones, which overlie the Cardium Formation in the deep basin (McKay and others, 1989), and carbonate formation in mudstones and siltstones of the Harmon Member (Peace River Formation), which immediately underlies the conglomerates of the Cadotte Member (Bloch, 1990).

The importance of low-^{18}O meteoric water during early diagenetic-mineral formation has been noted elsewhere in the western Canada sedimentary basin: early chlorite and calcite, basal Belly River Formation sandstone, central Alberta (Longstaffe, 1983, 1986; Ayalon and Longstaffe, 1988); early siderite and kaolinite, Viking Formation, west-central Alberta (Longstaffe and Ayalon, 1987); early siderite and calcite, Cardium Formation, central Alberta (Machemer and Hutcheon, 1988); early berthierine, smectite and calcite, Clearwater Formation, northeastern Alberta (Longstaffe and others, 1989a,b,c; 1990) and early chlorite, Brazeau Group, west-central Alberta (Longstaffe and Ayalon, in prep.). A few of these units, like the sandstones of the Brazeau Group, were deposited in continental settings; hence, a preponderance of meteoric water during early diagenetic processes is not surprising. However, many of the units were deposited in marine settings. Even given a brackish inland sea, the ^{18}O-poor nature of many early diagenetic pore waters cannot be explained, except by infiltration of meteoric water early in diagenesis.

The source of the meteoric water, especially that associated with the widespread development of early diagenetic siderite, which is commonly distributed in thin beds adjacent to sequence boundaries, is probably related to emergent episodes and relative fluctuations in sea level (McKay and others, 1989; Wickert and others, 1989). Subsurface invasion of meteoric water during sea-level lowstands would strongly affect early diagenetic development of sediments originally deposited in marine settings.

An enormous volume of water is required to transport the ions needed to form diagenetic minerals in most sandstones (Land, 1984). Sufficient water generally cannot be derived by compaction and diagenesis of shales (Blatt, 1979; Land and Dutton, 1978; Land, 1984). Land (1984) concluded that unless the solubility of silica in formation waters is much higher than presently believed, or unless the supply of dissolved silica within the sandstones is much greater than presently accepted, large-scale cycling of connate waters, or meteoric waters, linking sandstones and shales within a basin, is essential.

The post-Laramide introduction of meteoric water into the western Canada sedimentary basin provided substantial transport capacity for the late diagenetic mineral reactions. We suggest here that repeated fluxes of meteoric water, directly linked to fluctuating sea level, also helped to attenuate the diagenetic "water shortage" during earlier burial diagenetic processes. Mechanisms such as diffusion (Lahann, 1980) and convection (Wood and Hewitt, 1982, 1984), which are potential agents for the transportation of solute from shales into sandstones, need not be invoked.

Scale of ^{18}O Enrichment during Burial Diagenesis

The generally limited ^{18}O-enrichment of pore waters during burial diagenesis of clastic sedimentary rocks in the western Canada sedimentary basin has been confirmed by this study. Porewater δ^{18}O values at maximum burial temperatures were generally no higher than +2 to +3‰ in the Cardium Formation, the Paddy Member, the Cadotte Member and the Bluesky Formation. Depending on the absolute timing of diagenetic-mineral formation, even lower values are likely in many of these units. Similar results have been reported for the Falher Member in the deep basin (Tilley and Longstaffe, 1989), and for other Cretaceous clastic sedimentary rocks in the western Canada sedimentary basin: Lower Cretaceous Viking Formation, +2‰ (Longstaffe and Ayalon, 1987), Lower Cretaceous Clearwater Formation, −3‰ (Longstaffe and others, 1989a); Upper Cretaceous basal Belly River Formation sandstone, 0‰ (Ayalon and Longstaffe, 1988), Upper Cretaceous Brazeau Group sandstones, −3‰ (Longstaffe and Ayalon, in prep.). In the Alberta deep basin, pore waters with higher δ^{18}O values appear only in the lowermost Lower Cretaceous rocks, and the Jurassic, Triassic and Permian units.

This observation has implications for the putative role of the smectite-to-illite reaction in the western Canada sedimentary basin. Transfer of solute (e.g., silica) from shales to sandstones can be an important control on sandstone diagenesis. For example, the smectite-to-illite reaction can increase the silica concentration of shale pore waters (Siever, 1962; Towe, 1962; Hower and others, 1976; Boles and Franks, 1979), the exact amount of silica enrichment depending upon the nature of the reaction (Hower, 1981). However, of greater importance is not how much silica is produced by the smectite-to-illite transformation, but whether any is actually expelled with shale pore waters into sandstones (Land, 1984). Diagenetic illitic clays in shales from the Texas Gulf Coast (δ^{18}O ≈ +16 to +19‰; Yeh and Savin, 1977) are in equilibrium at 130 to 170°C with pore waters of δ^{18}O ≈ +5 to +10‰. Porewater δ^{18}O values in associated sandstones are similar (Kharaka and Carothers, 1986; Taylor, 1990), suggesting that expulsion of shale pore waters into the sandstones has been a significant process.

Shale comprises at least 40 to 50% of the Mesozoic section in the western Canada sedimentary basin. However, the low maximum δ^{18}O values (< +3‰) of sandstone pore waters, during the period when the smectite-to-illite reaction in intercalated shales should have been most active, raises at least three serious questions concerning the relative contribution of shale-buffered pore waters to sandstone diagenesis in this system.

First, it may be that episodic infiltration of meteoric water resulting from fluctuating sea level has diluted/muted the shale-porewater isotopic signal during burial diagenesis. Second, the "shale-buffer smectite-illite" control on sandstone pore waters in these rocks may not be significant. Cretaceous shales in the western Canada sedimentary basin contain abundant illitic-clay minerals but are generally poor in smectitic clays, except where bentonite beds are abundant (Williams, 1960; Longstaffe, 1984; Dean, 1986; Longstaffe and Ayalon 1987; Ayalon and Longstaffe, 1988; Connolly, 1989). Illitic clay in many of these shales may be mostly of detrital origin. Conversion of smectite to illite in amounts sufficient to buffer shale pore waters to the appropriate oxygen isotope compositions may never have occurred. Third, the shale-pore water oxygen isotope signal resulting from the smectite-illite reaction may be different in some of these rocks. Only a few oxygen isotope data are available for <2-μm and <0.2-μm illite-rich separates from shales in the western Canada sedimentary basin (Table 2). Most values (+16 to +20‰) are similar to illite-rich separates from shales in the Texas Gulf Coast and the Great Valley sequence (Yeh and Savin, 1977; Suchecki and Land, 1983). Because the shales from western Canada and the United States have been subjected to approximately the same range of burial temperatures, similar shale porewater δ^{18}O values (≈ +5 to +9‰) would be expected, assuming a diagenetic origin at least for the finest, illite-rich fraction.

A few illite samples, however, have lower δ^{18}O values. For example, illite/smectite (<0.2 μm) from shales in-

TABLE 2.—δ^{18}O VALUES OF ILLITE-RICH CLAY FRACTIONS, CRETACEOUS SHALES, ALBERTA.

Unit	δ^{18}O ‰	Reference
Mudstone bed, Upper Cretaceous continental Brazeau sandstones, < 0.2 μm	+12.6	Longstaffe and Ayalon (1990)
Mudstone bed, Upper Cretaceous basal Belly River sandstone, <0.2 μm	+16.9	Ayalon and Longstaffe (1988)
Upper Cretaceous Lea Park/Pakowki shale,		
< 2 μm	+16.6 - +18.9	Longstaffe (1984)
<0.2 μm	+16.1 - +16.5	Longstaffe (1984)
Upper Cretaceous Colorado shale, < 2 μm	+16.2 - +16.5	Longstaffe (1983)
Shale beds in Lower Cretaceous		
Viking sandstones, < 2 μm	+13.5, +17.1	Longstaffe and Ayalon (1987)
<0.2 μm	+13.3, +14.8	Longstaffe and Ayalon (1990)
Lower Cretaceous Viking shales, < 2 μm (Joli Fou Formation equivalent)	+17.1 - +19.8	Longstaffe (1983)
Lower Cretaceous Clearwater shale, < 2 μm	+16.0 - +18.2	Longstaffe (unpubl. data)

terbedded with sandstones of the Brazeau Group has a $\delta^{18}O$ value of +12.6 ‰ (Table 2). The ordering of this illite/smectite, burial history reconstructions, and oxygen isotope quartz-I/S geothermometry all suggest a diagenetic origin at ≈80°C for this clay, implying a shale porewater composition of ≈ -3‰ during the smectite-to-illite reaction (Longstaffe and Ayalon, in prep.). These mudstones are intercalated with sandstones deposited in a continental setting. Low-^{18}O meteoric pore waters may have been trapped in the mud during deposition and/or early diagenesis. Although the smectite-to-illite conversion would result in ^{18}O enrichment of the diagenetic pore water, the resulting isotopic signal contributed to sandstone pore waters would be impossible to distinguish from evolved meteoric water of other origins. In essence, we really do not know enough about the stable isotopic compositions of shale pore waters.

Influence of Paleozoic Carbonates

It remains questionable that the smectite-to-illite conversion significantly influenced the oxygen isotope evolution of sandstone pore waters in the western Canada sedimentary basin. Nevertheless, some increase in the $\delta^{18}O$ value of pore water did occur during burial diagenesis, particularly near the base of the Mesozoic section. Did this enrichment result from reactions between formation waters and minerals in the pore system, or did it arise from mixing of two or more formation waters of different composition and origin? Within the sandstones and conglomerates, only carbonate grains and cements would be sufficiently susceptible to dissolution and reprecipitation reactions needed for oxygen isotope exchange with pore water. Although such processes undoubtedly occurred, the effective mineral/water ratios (for oxygen) would have been too low to produce the necessary increase in the $\delta^{18}O$ value of the pore water. Introduction of ^{18}O-rich pore waters from another source is a better explanation.

Hitchon and Friedman (1969) showed that mixing between evolved sea water and surface-derived meteoric water has largely controlled the isotopic composition of formation waters in the western Canada sedimentary basin. Furthermore, recent data indicate that formation waters derived from Paleozoic units have contributed to present sandstone pore water compositions, at least in the lower portions of the Mesozoic section. Connolly and others (1990a,b) demonstrated that formation waters from central Alberta can be subdivided into three hydrostratigraphic units on the basis of their chemical and isotopic compositions. The lowermost Group I waters dominate Paleozoic carbonate rocks, but also occur in deeper parts of some clastic units through to the lowermost Cretaceous. The chemistry of these waters has been modified by carbonate- and clay-mineral reactions, suggesting at least some interaction with intercalated shales. Group II waters are most common in Jurassic and lower Cretaceous clastic sedimentary rocks, and have been modified by leaching of feldspar and clay minerals. Both Groups I and II are composed of a brine (residual evaporated sea water) that has been subsequently diluted by meteoric water. By comparison, Group III waters occur mostly in Upper Cretaceous clastic sedimentary rocks and consist mostly of dilute meteoric waters that have become decoupled from the more saline, stratigraphically lower Groups I and II waters. Diffusional flow and density stratification have occurred between Group I and II formation waters, following their gradual isolation from the post-Laramide recharge of meteoric water.

These hydrostratigraphic zones are broadly reflected by an increase in the $\delta^{18}O$ values of formation waters from Group III (Upper Cretaceous units) to Group I (Paleozoic units) (Hitchon and Friedman, 1969; Connolly and others, 1990a). This relation holds true for the few data available for formation waters in the vicinity of the Alberta deep basin: Cretaceous, -10.8 to -5.7‰; Triassic, -1.7 to +0.8‰; Paleozoic, -1.2 to +7.2‰ (avg. ≈ +3‰) (Hitchon and Friedman, 1969). Moreover, the systematic variation in formation water $\delta^{18}O$ values that can be calculated from the composition of fracture-filling carbonates supports this pattern: Upper Cretaceous Cardium Formation, -13‰; Lower Cretaceous Cadotte Member, -6‰; Lower Cretaceous Falher Member, -5‰; Jurassic Nikanassin Formation, -1‰, and Triassic Halfway Formation, 0‰. We further speculate that mixing between Groups I and II can be patchy. For example, the anomalous ^{18}O-rich, late diagenetic calcite (+23.4 to +23.6‰) in conglomerates from the western portion of the Falher Member may represent a directional influx of formation water from the Paleozoic carbonate platform. These conglomerates have a previous history as a preferred conduit for externally derived diagenetic fluids (Tilley and Longstaffe, 1989). Present formation waters in Upper Devonian carbonates near this area have $\delta^{18}O$ values of +3.0 to +3.1‰. These values are similar to those that can be calculated for late-calcite crystallization in the Falher Member at current subsurface temperatures.

The data also indicate that formation waters from the carbonate-dominated Paleozoic sedimentary section infiltrated the Mesozoic clastic sedimentary section during burial diagenesis, prior to uplift and meteoric recharge in post-Laramide times. The greatest effects of such mixing should be observed in the lowermost Mesozoic sandstones and conglomerates, and diminish upward in the section. This is the observed pattern. Porewater ^{18}O-enrichment during burial diagenesis may have reached values ≥ +7‰ in the Triassic Halfway Formation sandstone, near the base of the Mesozoic clastic sedimentary section, and ≥ +9‰ in the Permian Belloy Formation sandstone, within the carbonate-dominated interval itself. Assuming equilibration with the average Paleozoic carbonate in the western Canada sedimentary basin (≈ +23‰; Hitchon and Friedman, 1969; Longstaffe, 1984; Connolly and others, 1990a), calculated formation water $\delta^{18}O$ values range from +2.5‰ at 100°C to +8.6‰ at 175°C. The susceptibility of carbonate, especially calcite, to oxygen-isotopic exchange with formation waters during dissolution and reprecipitation is well known (Clayton, 1959; Clayton and others, 1966). Moreover, the necessary effective water/rock ratio needed for such large-scale modification is easily attained in porous and permeable carbonate rocks. The large-scale dolomitization event(s) that have affected much of the Paleozoic section in the western Canada sedimentary basin demonstrate that such extensive water/rock interaction has occurred. The Paleozoic section also contains thick, fine-grained, carbonate- and clay-rich (mostly illite) intervals. However, without data concerning the diagenetic-mineral reactions that determined the present composition of these rocks, their role in the oxygen isotope evolution of formation waters in the Paleozoic section remains unclear.

CONCLUSIONS

The oxygen isotope compositions of diagenetic minerals from Mesozoic sandstones and conglomerates in the western Canada sedimentary basin, and in particular, the Alberta deep basin, have been used to deduce the evolution of formation waters during early diagenesis, burial diagenesis and late diagenesis (during uplift and erosion). Meteoric water has played a major role throughout the diagenetic evolution of these rocks. It has been the dominant source of water during late diagenetic processes, which are largely responsible for most porosity and permeability modification in these rocks. Meteoric water was also important during early diagenetic mineral formation, even in units originally deposited in marine environments. Episodic recharge of the sediments by meteoric water during early diagenetic processes probably occurred in response to fluctuating sea level during Cretaceous time.

The $\delta^{18}O$ values achieved by formation water at maximum burial depths and temperatures in the Mesozoic section are significantly lower than typical for other shale-dominated basins. This result suggests that the smectite-to-illite reaction in shales did not dominate the oxygen isotope evolution of pore waters during burial diagenesis. The modest increase in porewater oxygen isotope composition that did occur during burial can be explained by mixing of early diagenetic, meteoric-dominated pore waters and ^{18}O-rich pore waters from the underlying Paleozoic carbonate platform. The contribution of Paleozoic-derived formation waters was most pronounced in the lowermost Mesozoic sandstones and conglomerates. Current formation water oxygen isotope compositions reflect this mixing plus the further, commonly substantial contribution of surface-derived meteoric water during uplift and erosion of these sedimentary rocks.

ACKNOWLEDGMENTS

We gratefully acknowledge financial support from the Natural Sciences and Engineering Research Council of Canada (Grant A7387), Canadian Hunter Exploration Ltd., Petro-Canada Resources, Imperial Oil University Research Grant and the Alberta Research Council. Our appreciation is also extended to Mingchou Lee and Dennis Prezbindowski for their reviews of the manuscript.

REFERENCES

AYALON, A., AND LONGSTAFFE, F. J., 1988, Oxygen isotope studies of diagenesis and pore-water evolution in the western Canada sedimentary basin: evidence from the Upper Cretaceous basal Belly River sandstone, Alberta: Journal of Sedimentary Petrology, v. 58, p. 489–505.
BARKER, Ch. E., AND PAWLEWICZ, M. J., 1986, The correlation of vitrinite reflectance with maximum temperature in organic matter, in Buntebarth, G., and Stegena, L., eds., Paleogeothermics, Lecture Notes in Earth Sciences, v. 5: Berlin, Springer-Verlag, p. 79–93.
BECKER, R. H., AND CLAYTON, R. N., 1976, Oxygen isotope study of a Precambrian banded iron-formation, Hamersley Range, Western Australia: Geochimica et Cosmochimica Acta, v. 40, p. 1153–1165.
BLATT, H., 1979, Diagenetic processes in sandstones, in Scholle, P. A., and Schluger, P. R., eds., Aspects of Diagenesis: Society of Economic Paleontologists and Mineralogists Special Publication 26, p. 141–157.
BLOCH, J., 1990, Stable isotopic composition of authigenic carbonates from the Albian Harmon Member (Peace River Formation): evidence of early diagenetic processes: Bulletin of Canadian Petroleum Geology, v. 38, p. 39–52.
BOLES, J. R., AND FRANKS, S. G., 1979, Clay diagenesis in Wilcox sandstones of southwest Texas: implications of smectite diagenesis on sandstone cementation: Journal of Sedimentary Petrology, v. 49, p. 55–70.
CANT, D. J., 1983, Spirit River Formation—a stratigraphic-diagenetic gas trap in the deep basin of Alberta: American Association of Petroleum Geologists Bulletin, v. 67, p. 577–587.
CANT, D. J., 1986, Hydrocarbon trapping in the Halfway Formation (Triassic), Wembley Field, Alberta: Bulletin of Canadian Petroleum Geology, v. 34, p. 329–338.
CANT, D. J., AND ETHIER, V. G., 1984, Lithology-dependent diagenetic control of reservoir properties of conglomerates, Falher Member, Elmworth Field, Alberta: American Association of Petroleum Geologists Bulletin, v. 68, p. 1044–1054.
CLAYTON, R. N., 1959, Oxygen isotope fractionation in the system calcium carbonate-water: Journal of Chemical Physics, v. 30, p. 1246–1250.
CLAYTON, R. N., FRIEDMAN, I., GRAF, D. L., MAYEDA, T. K., MEENTS, W. F., AND SHIMP, N. F., 1966, The origin of saline formation waters, 1. Isotopic composition: Journal of Geophysical Research, v. 71, p. 3869–3882.
CLAYTON, R. N., AND MAYEDA, T. K., 1963, The use of bromine pentafluoride in the extraction of oxygen from oxides and silicates for isotopic analysis: Geochimica et Cosmochimica Acta, v. 27, p. 43–52.
CLAYTON, R. N., O'NEIL, J. R., AND MAYEDA, T. K., 1972, Oxygen isotope exchange between quartz and water: Journal of Geophysical Research, v. 77, p. 3057–3067.
CONNOLLY, C. A., 1989, Thermal history and diagenesis of the Wilrich Member shale, Spirit River Formation, northwest Alberta: Bulletin of Canadian Petroleum Geology, v. 37, p. 182–197.
CONNOLLY, C. A., WALTER, L. M., BAADSGAARD, H., AND LONGSTAFFE, F. J., 1990a, Origin and evolution of formation waters, Alberta Basin, western Canada sedimentary basin. II. Isotope systematics and water mixing: Applied Geochemistry, v. 5, p. 397–413.
CONNOLLY, C. A., WALTER, L. M., BAADSGAARD, H., AND LONGSTAFFE, F. J., 1990b, Origin and evolution of formation waters, Alberta Basin, western Canada sedimentary basin. I. Chemistry: Applied Geochemistry, v. 5, p. 375–395.
CRAIG, H., 1957, Isotopic standards for carbon and oxygen and correction factors for mass-spectrometric analysis of carbon dioxide: Geochimica et Cosmochimica Acta, v. 12, p. 133–149.
CRAIG, H., 1961, Standards for reporting concentrations of deuterium and oxygen-18 in natural waters: Science, v. 133, p. 1833–1834.
DEAN, M. E., 1986, Diagenesis of the Viking Formation, south-central Alberta: Unpublished M.S. Thesis, University of Alberta, Edmonton, Alberta, 195 p.
DUTTON, S. P., AND LAND, L. S., 1985, Meteoric burial diagenesis of Pennsylvanian arkosic sandstones, southwestern Anadarko Basin, Texas: American Association of Petroleum Geologists Bulletin, v. 69, p. 22–38.
EPSTEIN, S., GRAF, D. L., AND DEGENS, E. T., 1964, Oxygen isotope studies on the origin of dolomite, in Craig, H., Miller, S. L., and Wasserburg, G. J., eds., Isotopic and Cosmic Chemistry: North Holland Publishing Company, p. 169–180.
FRIEDMAN, I., AND O'NEIL, J. R., 1977, Compilation of stable isotope fractionation factors of geochemical interest, in Fleischer, M., ed., Data of Geochemistry (6th ed.): U. S. Geological Survey Professional Paper 440-KK, 12 p.
GIES, R. M., 1984, Case history for a major Alberta deep basin gas trap: the Cadomin Formation, in Masters, J. A., ed., Elmworth Case Study of a Deep Basin Gas Field: American Association of Petroleum Geologists Memoir 38, p. 115–140.
HITCHON, B., 1984, Geothermal gradients, hydrodynamics, and hydrocarbon occurrences, Alberta, Canada: American Association of Petroleum Geologists Bulletin, v. 68, p. 713–743.

HITCHON, B., AND FRIEDMAN, I., 1969, Geochemistry and origin of formation waters in the western Canada sedimentary basin – 1. Stable isotopes of hydrogen and oxygen: Geochimica et Cosmochimica Acta, v. 33, p. 1321–1349.

HOWER, J., 1981, Shale diagenesis, in Longstaffe, F. J., ed., Clays and the Resource Geologist: Mineralogical Association of Canada, Short Course, v. 7, p. 60–80.

HOWER, J., ESLINGER, E. V., HOWER, M. E., AND PERRY, E. A., 1976, Mechanisms of burial metamorphism of argillaceous sediment. 1. Mineralogical and chemical evidence: Geological Society of America Bulletin, v. 87, p. 725–737.

IRWIN, H., CURTIS, C., AND COLEMAN, M. L., 1977, Isotopic evidence for source of diagenetic carbonates formed during burial of organic-rich sediments: Nature, v. 269, p. 209–213.

JACKSON, M. L., 1979, Soil Chemical Analysis—Advanced Course (2nd ed.), 11th Printing: Published by the author, Madison, Wisconsin, U.S.A., 895 p.

KALKREUTH, W., AND MCMECHAN, M. E., 1984, Regional pattern of thermal maturation as determined from coal-rank studies, Rocky Mountain foothills and front ranges north of Grande Cache, Alberta—implications for petroleum exploration: Bulletin of Canadian Petroleum Geology, v. 32, p. 249–271.

KHARAKA, Y. K., AND CAROTHERS, W. W., 1986, Oxygen and hydrogen isotope geochemistry of deep basin brines, in Fritz, P., and Fontes, J. Ch., eds., Handbook of Environmental Isotope Geochemistry, v. 2: Amsterdam, Elsevier, p. 305–360.

LAHANN, R. W., 1980, Smectite diagenesis and sandstone cement: the effect of reaction temperature: Journal of Sedimentary Petrology, v. 50, p. 755–760.

LAND, L. S., 1984, Frio sandstone diagenesis, Texas Gulf Coast: a regional isotopic study, in McDonald, D. A., and Surdam, R. C., eds., Clastic Diagenesis: American Association of Petroleum Geologists Memoir 37, p. 47–62.

LAND, L. S., AND DUTTON, S. P., 1978, Cementation of a Pennsylvanian deltaic sandstone: isotopic data: Journal of Sedimentary Petrology, v. 48, p. 1167–1176.

LEE, M., AND SAVIN, S. M., 1985, Isolation of diagenetic overgrowths on quartz sand grains for oxygen isotopic analysis: Geochimica et Cosmochimica Acta, v. 49, p. 497–501.

LONGSTAFFE, F. J., 1983, Diagenesis, IV. Stable isotope studies of diagenesis in clastic rocks: Geoscience Canada, v. 10, p. 44–58.

LONGSTAFFE, F. J., 1984, The role of meteoric water in diagenesis of shallow sandstones: stable isotope studies of the Milk River aquifer and gas pool, in Surdam, R., and MacDonald, D., eds., Clastic Diagenesis: American Association of Petroleum Geologists Memoir 37, p. 81–98.

LONGSTAFFE, F. J., 1986, Oxygen-isotope studies of diagenesis in the basal Belly River sandstone, Pembina I-Pool, Alberta: Journal of Sedimentary Petrology, v. 56, p. 78–88.

LONGSTAFFE, F. J., 1989, Stable isotopes as tracers in clastic diagenesis, in Hutcheon, I. E., ed., Burial Diagenesis: Mineralogical Association of Canada Short Course, v. 15, p. 201–277.

LONGSTAFFE, F. J., AND AYALON, A., 1987, Oxygen-isotope studies of clastic diagenesis in the Lower Cretaceous Viking Formation, Alberta: implications for the role of meteoric water, in Marshall, J. D., ed., Diagenesis of Sedimentary Sequences: Geological Society of America Special Publication 36, p. 277–296.

LONGSTAFFE, F. J., AND AYALON, A., 1990, Hydrogen-isotope geochemistry of diagenetic clay minerals from Cretaceous sandstones, Alberta, Canada: evidence for exchange: Applied Geochemistry, v. 5, p. 657–668.

LONGSTAFFE, F. J., AYALON, A., RACKI, M. A., AND BIRD, G. W., 1989a, Natural diagenesis of Clearwater Formation reservoirs in the Cold Lake area, Alberta, Part II: Stable isotope studies of water/mineral interaction: Exploration Update '89, Calgary, Programs and Abstracts, p. 142.

LONGSTAFFE, F. J., AYALON, A., AND RACKI, M. A., 1989b, Natural diagenesis of Clearwater Formation reservoirs in the Cold Lake area, Alberta, Part I: Mineralogical studies: Exploration Update '89, Calgary, Programs and Abstracts, p. 130.

LONGSTAFFE, F. J., AYALON, A., AND RACKI, M. A., 1989c, Oxygen- and carbon-isotope studies of diagenesis in heavy oil deposits of the Clearwater Formation, northeastern Alberta: Geological Association of Canada, Programs with Abstracts, v. 14, p. A85.

LONGSTAFFE, F. J., RACKI, M. A., AYALON, A., WICKERT, L. M., WIGHTMAN, D. M., AND BIRD, G. W., 1990, Water-mineral-organic matter interactions during clastic diagenesis of Cretaceous heavy oil reservoirs, Cold Lake area, Alberta: American Association of Petroleum Geologists Bulletin, v. 74, p. 1306–1307.

LOPATIN, N. V., 1971, Temperature and geologic time as factors in coalification: Akademiya Nauk SSR Izvestiya, Seriya Geologicheskikh, no. 3, p. 95–106.

MACHEMER, S. D., AND HUTCHEON, I., 1988, Geochemistry of early carbonate cements in the Cardium Formation, central Alberta: Journal of Sedimentary Petrology, v. 58, p. 136–147.

MASTERS, J. A., 1984, Elmworth—Case Study of a Deep Basin Gas Field: American Association of Petroleum Geologists Memoir 38, 316 p.

MCCREA, J. M., 1950, On the isotopic chemistry of carbonates and a paleotemperature scale: Journal of Chemical Physics, v. 18, p. 849–857.

MCKAY, J. L., LONGSTAFFE, F. J., AND PLINT, A. G., 1989, Geochemical variations across marine-nonmarine transitions within the Bad Heart Formation, northwestern Alberta and northeastern British Columbia: Geological Association of Canada, Programs with Abstracts, v. 14, p. A14.

O'CONNELL, S. C., 1988, The distribution of Bluesky facies in the region overlying the Peace River Arch, northwestern Alberta, in James, D. P., and Leckie, D. A., eds., Sequences, Stratigraphy, Sedimentology: Surface and Subsurface: Canadian Society of Petroleum Geologists Memoir 15, p. 387–400.

O'NEIL, J. R., CLAYTON, R. N., AND MAYEDA, T. K., 1969, Oxygen isotope fractionation in divalent metal carbonates: Journal of Chemical Physics, v. 51, p. 5547–5558.

PLINT, A. G., AND WALKER, R. G., 1987, Cardium Formation 8. Facies and environments of the Cardium shoreline and coastal plain in the Kakwa field and adjacent areas, northwestern Alberta: Bulletin of Canadian Petroleum Geology, v. 35, p. 48–64.

PLINT, A. G., WALKER, R. G., AND BERGMAN, K. M., 1986, Cardium Formation 6. Stratigraphic framework of the Cardium in subsurface: Bulletin of Canadian Petroleum Geology, v. 34, p. 213–225.

ROSENBAUM, J., AND SHEPPARD, S. M. F., 1986, An isotopic study of siderites, dolomites and ankerites at high temperatures: Geochimica et Cosmochimica Acta, v. 50, p. 1147–1150.

SAVIN, S. M., AND LEE, M., 1988, Isotopic studies of phyllosilicates, in Bailey, S. W., ed., Hydrous Phyllosilicates (exclusive of micas): Reviews in Mineralogy, v. 19, p. 189–223.

SHARMA, T., AND CLAYTON, R. N., 1965, Measurement of O^{18}/O^{16} ratios of total oxygen of carbonates: Geochimica et Cosmochimica Acta, v. 29, p. 1347–1353.

SIEVER, R., 1962, Silica solubility, 0–200°C, and the diagenesis of siliceous sediments: Journal of Geology, v. 70, p. 127–150.

SMITH, D. G., ZORN, C. E., AND SNEIDER, R. M., 1984, The paleogeography of the Lower Cretaceous of western Alberta and northeastern British Columbia in and adjacent to the deep basin of the Elmworth area, in Masters, J. A., ed., Elmworth Case Study of a Deep Basin Gas Field: American Association of Petroleum Geologists Memoir 38, pp. 79–114.

SUCHECKI, R. K., AND LAND, L. S., 1983, Isotopic geochemistry of burial-metamorphosed volcanogenic sediments, Great Valley sequence, northern California: Geochimica et Cosmochimica Acta , v. 47, p. 1487–1499.

SYERS, J. K., CHAPMAN, S. L., JACKSON, M. L., REX, R. W., AND CLAYTON, R. N., 1968, Quartz isolation from rocks, sediments and soils for determination of oxygen isotope composition: Geochimica et Cosmochimica Acta, v. 32, p. 1022–1025.

TAYLOR, T. R., 1990, The influence of calcite dissolution on reservoir porosity in Miocene sandstones, Picaroon Field, offshore Texas Gulf Coast: Journal of Sedimentary Petrology, v. 60, p. 322–334.

TILLEY, B. J., 1988, Diagenesis and porewater evolution in Cretaceous sedimentary rocks of the Alberta deep basin: unpublished Ph.D. Dissertation, University of Alberta, Edmonton, Alberta, 205 p.

TILLEY, B. J., AND LONGSTAFFE, F. J., 1989, Diagenesis and isotopic evolution of porewaters in the Alberta deep basin: the Falher Member and Cadomin Formation: Geochimica et Cosmochimica Acta, v. 53, p. 2529–2546.

TILLEY, B. J., NESBITT, B. E., AND LONGSTAFFE, F. J., 1989, Thermal history of Alberta deep basin: comparative study of fluid inclusion and vitrinite reflectance data: American Association of Petroleum Geologists Bulletin, v. 73, p. 1206–1222.

Towe, K. M., 1962, Clay mineral diagenesis as a possible source of silica cement in sedimentary rocks: Journal of Sedimentary Petrology, v. 32, p. 26–28.

Varley, C. J., 1982, The sedimentology and diagenesis of the Cadomin Formation, Elmworth area, northwestern Alberta: unpublished M.S. Thesis, University of Calgary, Calgary, Alberta, 173 p.

Varley, C. J., 1984, Sedimentology and hydrocarbon distribution of the Lower Cretaceous Cadomin Formation, northwest Alberta, in Koster, E. H., and Steel, R. J., eds., Sedimentology of Gravels and Conglomerates: Canadian Society of Petroleum Geologists Memoir 10, p. 175–187.

Walters, L. J., Jr., Claypool, G. E., and Choquette, P. W., 1972, Reaction rates and $\delta^{18}O$ variation for the carbonate-phosphoric acid preparation method: Geochimica et Cosmochimica Acta, v. 36, p. 129–140.

Waples, D., 1980, Time and temperature in petroleum formation: application of Lopatin's method to petroleum exploration: American Association of Petroleum Geologists Bulletin, v. 64, p. 916–926.

Weiss, H. M., 1985, Geochemische und petrographische Untersuchungen am organischen Material kretazischer Sedimentgesteine aus dem Deep Basin, Westkanada: Unpublished Ph.D. Dissertation, Rheinsich-Westfälischen Technischen Hochschule, Aachen, Federal Republic of Germany, 261 p.

Wickert, L. M., Longstaffe, F. J., and Pemberton, S. G., 1989, A diagenetic investigation of sequence stratigraphy in the lower Cretaceous Clearwater Formation, Cold Lake oil sands, east-central Alberta: Geological Association of Canada, Programs with Abstracts, v. 14, p. A86.

Williams, G. D., 1960, The Mannville Group, central Alberta: Unpublished Ph.D. Dissertation, University of Alberta, Edmonton, Alberta, 160 p.

Wood, J. R., and Hewett, T. A., 1982, Fluid convection and mass transfer in porous sandstones—a theoretical approach: Geochimica et Cosmochimica Acta, v. 46, p. 1707–1713.

Wood, J. R., and Hewett, T. A., 1984, Reservoir diagenesis and convective fluid flow, in McDonald, D. A., and Surdam, R. C., eds., Clastic Diagenesis: American Association of Petroleum Geologists Memoir 37, p. 81–98.

Yeh, H-W., and Savin, S. M., 1977, Mechanism of burial metamorphism of argillaceous sediments: 3. O-isotope evidence: Geological Society of America Bulletin, v. 88, p. 1321–1330.

Youn, S. H., 1983, Depositional environments and their significance on diagenetic processes, Falher Member (Lower Cretaceous), Spirit River Formation, Elmworth area, Alberta: Unpublished report, University of Calgary, 317 p.

Zhu, J., 1981, Vitrinite reflectance measurements and geothermal gradients on the Grande Prairie, northwest Alberta plains, Canada: Unpublished report, University of Calgary, 79 p.

K-Ar DATING OF ILLITE DIAGENESIS IN THE MIDDLE ORDOVICIAN ST. PETER SANDSTONE, CENTRAL MICHIGAN BASIN, USA: IMPLICATIONS FOR THERMAL HISTORY

DAVID A. BARNES
Department of Geology, Western Michigan University, Kalamazoo, Michigan 49008
AND
JEAN-PIERRE GIRARD AND JAMES L. ARONSON
Department of Geological Sciences, Case Western Reserve University, Cleveland, Ohio 44106

ABSTRACT: The Middle Ordovician St. Peter Sandstone (dominantly quartzarenites and feldsarenites) occurs throughout the central Michigan Basin at depths ranging from 1.5 to 3.5 km and produces condensate and/or natural gas from several horizons in more than 36 fields. The primary mineralogy, textures and reservoir characteristics have been dramatically modified by a complex diagenesis. The main diagenetic features, generally observable throughout the basin, are, chronologically: early marine carbonate cement; quartz overgrowth and pressure solution; burial dolomite and anhydrite; dissolution of framework grains and early cements; and pervasive authigenic illite and chlorite cementation. This late clay cement commonly occurs in secondary porosity formed after the dolomite cement.

A K-Ar study of the fine-grained authigenic illite indicates that this regionally significant episode of clay cementation in the Michigan Basin occurred during Late Devonian-Mississippian times. Sixteen samples from 11 wells located mainly in the central part of the basin yielded consistent ages ranging from 367 Ma to 327 Ma with an average of 346 ± 11 Ma. Combined with burial-history reconstructions for the central basin, these ages indicate that illite formed at depths of approximately 3 km.

Fluid-inclusion temperatures for the dolomite and quartz cements associated with the diagenetic illite suggest temperatures of formation for the latter on the order of 150°C or higher. Combined with the depth estimate, these temperatures imply the existence of elevated geothermal gradients, i.e., 38°C/km or greater, in the central Michigan Basin at the time of illitization. Because illite and/or hydrocarbons typically fill secondary pores in the St. Peter Sandstone, the K-Ar age data also place constraints on the timing of development of secondary porosity and hydrocarbon emplacement.

INTRODUCTION

In the past decade, a growing number of diagenetic studies of reservoir sandstones has used petrographic and geochemical characterizations of diagenetic phases combined with the geodynamic history of basins to elucidate the timing and conditions of formation of cements as well as the temperature, nature and origin of the pore fluids involved in the cementation (see recent reviews by Longstaffe, 1989; Lundegard, 1989). Reservoir quality and hydrocarbon production are commonly influenced by sedimentary facies and diagenesis, the latter being often correlated to the former. Knowledge of factors controlling or influencing post-depositional transformations through time and space is of economic and practical importance. Several instances of integrated studies of diagenesis have yielded economically valuable information on the time of hydrocarbon emplacement in a reservoir (Lee and others, 1985), or on the origin and migration pathways of hydrocarbons (Glasmann and others, 1989a) and/or diagenetic fluids (Whitney and Northrop, 1987; Girard and others, 1989; Lee and others, 1989; Glasmann and others, 1989b) in a basin.

In recent years there has been increasing emphasis on the reconstruction of diagenetic histories of sandstones in terms of absolute geologic time (Hamilton and others, 1989; Lundegard, 1989). Diagenetic illite is a common cement in sandstone reservoirs, where its fibrous nature can greatly deteriorate permeability and interfere with petroleum production (Stalder, 1973). Illite is of particular importance because it is often a late diagenetic cement precipitated from fluids preceding or accompanying hydrocarbon emplacement in reservoirs (Glasmann and others, 1989b). Characterization of the timing and conditions of formation of diagenetic illite cement, therefore, can constrain the timing and mechanisms of hydrocarbon emplacement as well as the formation of other diagenetic cements (Sommer, 1975; Lee and others, 1985; Liewig and others, 1987; Girard and others, 1989; Glasmann and others, 1989b).

The Middle Ordovician St. Peter Sandstone produces gas, condensate, and lesser amounts of oil from several horizons throughout the central Michigan Basin. Estimated hydrocarbon reserves are on the order of 2.5 tcf of gas with associated condensate in some wells up to 200 bbl/day (Mazuchowski, pers. commun., 1990). Detailed petrographic studies (Barnes, 1988) indicate that the formation has experienced major modifications of primary mineralogy and porosity during diagenesis. These post-depositional transformations have strongly influenced hydrocarbon reservoir quality in the St. Peter in Michigan (Barnes and others, 1989). In this paper we present an extensive set of preliminary results of ongoing research on the diagenesis of the St. Peter Sandstone. Emphasis is placed on the K-Ar geochronology of authigenic illite and its integration with petrographic and fluid-inclusion data in an attempt to reconstruct the timing and the conditions of late diagenesis. Implications on the thermal history of the formation and the emplacement of hydrocarbons in the reservoir are discussed. Broader scale implications on the hydrodynamics and geodynamic evolution of the Michigan Basin must await further data.

GEOLOGICAL BACKGROUND

The Michigan Basin (Fig. 1) is a circular intracratonic basin that contains upward of 4,880 m (16,000 ft) of Paleozoic sedimentary strata and is surrounded by basement highs (arches). The Paleozoic sediments range in age from Cambrian through Pennsylvanian. They are predominantly alternating siliciclastic and dolomitic sediments in the Cambrian to Middle Ordovician, carbonates and evaporites in

---

Origin, Diagenesis, and Petrophysics of Clay Minerals in Sandstones, SEPM Special Publication No. 47
Copyright © 1992, SEPM (Society for Sedimentary Geology), ISBN 0-918985-95-1

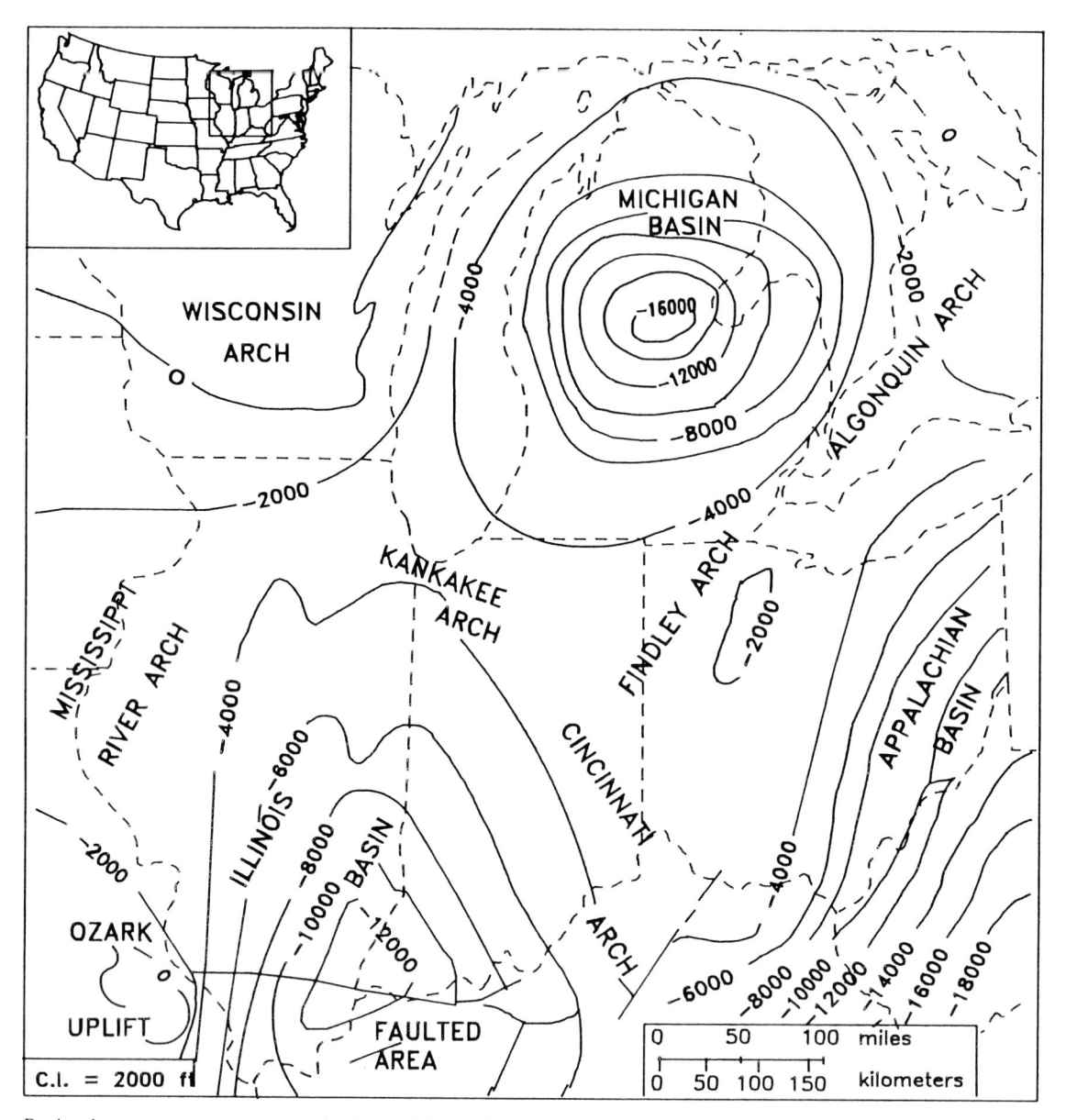

FIG. 1.—Regional structure contour map on the Precambrian surface in the Midwestern Basins and Arches geological province. Modified from Droste and Shaver (1983).

the Upper Ordovician to Middle Devonian and essentially clastics in the post-late Devonian part of the section. The stratigraphic succession is punctuated by interregional unconformities that define the well-known "Native American" Sequences of Sloss (1963). The renowned St. Peter Sandstone of the North American midcontinent is recognized as the basal formation of the Tippecanoe Sequence in the Michigan Basin (Fig. 2) and lies above the "Brazos shale" of the underlying Foster Formation and below the fine-grained argillaceous sediments of the Glenwood Formation.

The St. Peter Sandstone is nearly 366 m (1,200 ft) thick in the central part of the basin and thins drastically or is absent on the margins (Fig. 3). In the central Michigan Basin the formation consists predominantly of quartz sandstone interbedded with dolomitic sandstone, siltstone and argillaceous dolomicrite. The main lateral facies change within the formation is a significant enrichment in carbonate strata southward and eastward (Fisher and Barratt, 1985).

There still is some debate about the origin and nature of the subsidence that led to the thick accumulation of Paleozoic sediments in the Michigan Basin. Subsidence may have been initiated by an early Paleozoic, deep-seated thermal event with associated lithospheric stretching and subsequent thermal contraction (Nunn and others, 1984; Klein and Hsui, 1987). Associated fault-controlled mechanical subsidence may also have been a factor (Watso and Klein, 1989). Extrabasinal tectonic influences have also been suggested (Quinlan and Beaumont, 1984; Quinlan, 1987); however, a combination of mechanisms is most probable

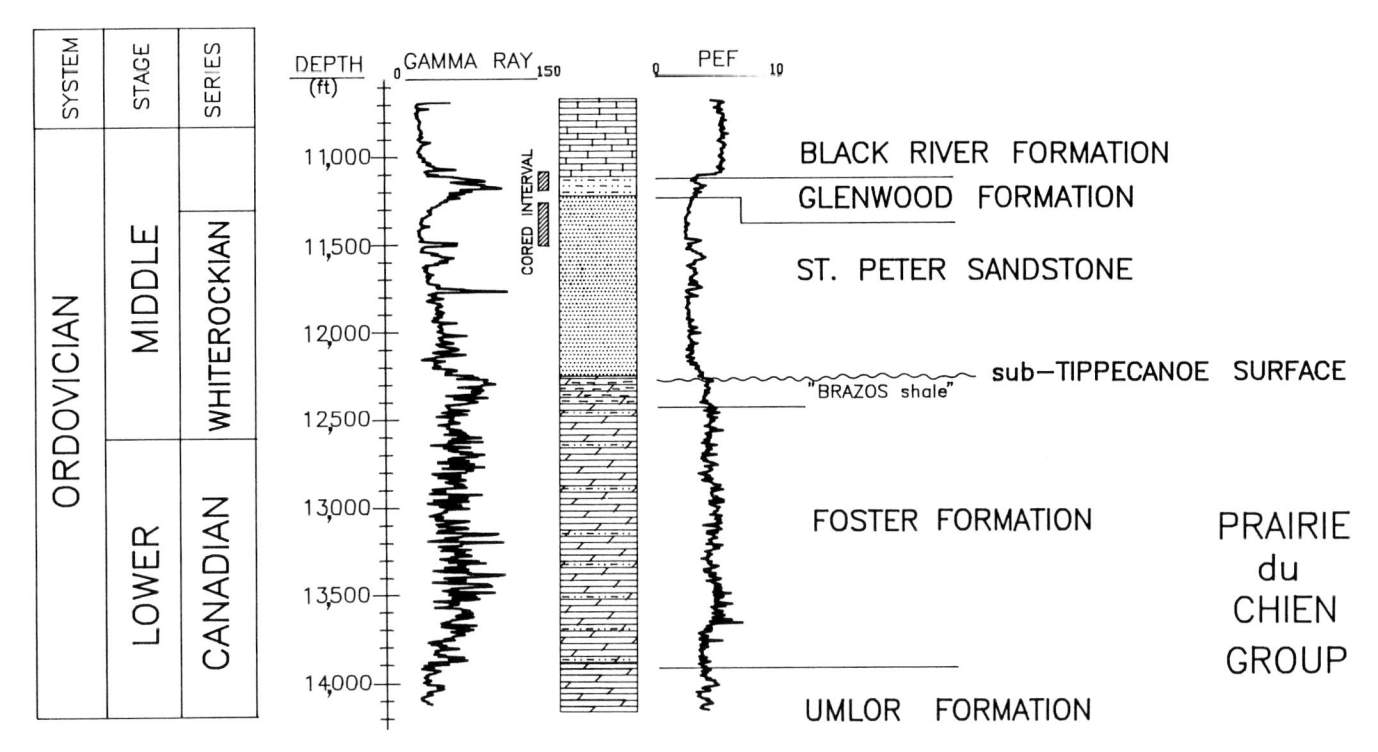

FIG. 2.—Reference well section from the Hunt Martin well in Gladwin County, Michigan, showing the stratigraphic terminology used in this paper. Left log track is gamma-ray log, right track is photoelectric effect (PEF) log. See Figure 6 for well location.

(Howell and van der Pluijm, 1990). Crude estimates of relative tectonic subsidence in the cratonic interior have been determined based on regional isopach maps (Sloss, 1988). During the early Paleozoic (Sauk and Tippecanoe Sequence time), the Basins and Arches province experienced subsidence rates no greater than 30 m/ma. This maximum rate of subsidence in the Michigan Basin is nearly an order of magnitude less than known rates of subsidence at passive continental margins of 250 m/ma (Pitman, 1978) and illustrates the slow but relatively persistent nature of the subsidence in the basinal area of the Basins and Arches province during the early Paleozoic.

SEDIMENTOLOGY AND DIAGENESIS

The sedimentology and diagenesis of the St. Peter Sandstone in Michigan are reported elsewhere in detail (Barnes and others, in prep.) and will only be summarized here. Four sedimentologic lithofacies can be recognized in the reference well, Hunt Martin, Gladwin County (Fig. 4), located in the central basin. These lithofacies, which can be correlated from well to well throughout the basin, indicate an upward transgressive succession consisting of, from base to top: (1) as much as 250 m of sandstone deposited in intertidal and supratidal sandflat environments (locally associated with minor dolomitic lagoonal deposits) and shallow subtidal-shoreface environments (lithofacies 1 and 2); (2) 25 to 75 m of sandstones deposited in subtidal-shoreface to upper offshore environments (lithofacies 3); and (3) about 30 to 50 m of dolomitic and argillaceous sands deposited in a storm-dominated epeiric-sea shelf setting (lithofacies 4).

The primary composition and average grain size of the sandstones are closely related to their depositional environment. Coarse- to fine-grained quartz arenites mainly occur in high-energy littoral facies, whereas fined-grained feldspathic and carbonate-rich sands and silts occur predominantly in low-energy shelf facies (Fig. 5). The relations among framework grain composition, grain size, and depositional environment observed in the St. Peter in Michigan are in close agreement with those reported by Odom and others (1976) in several quartz-rich cratonic sandstones. Sandstones in the St. Peter, especially those deposited in lower energy depositional environments, may also contain substantial amounts of primary carbonate sediment. Extreme examples typically exhibit open grain-packing textures with a continuous carbonate matrix comprising up to 35 to 40% of the rock volume. In such samples, as well as in the rest of the formation, the carbonate is ubiquitously dolomite as a result of diagenesis (see later discussion).

Petrographic analysis of over 500 thin sections of the St. Peter Sandstone from 29 wells throughout the Michigan Basin (Fig. 6) documents the dramatic post-depositional modifications of the primary textures and mineralogy. Although the details of the diagenetic pathways followed by individual samples are correlated to their depositional facies and primary mineralogy, petrographic studies to date provide a reasonably consistent paragenetic sequence generally applicable throughout the basin. The generalized succession of well-characterized diagenetic events in the sandstones of the St. Peter in Michigan (Fig. 7) is as follows: early marine carbonate cementation, early (syndepositional?) dolomitization, development of quartz overgrowths, burial dolomite and anhydrite cementation, selective mineral

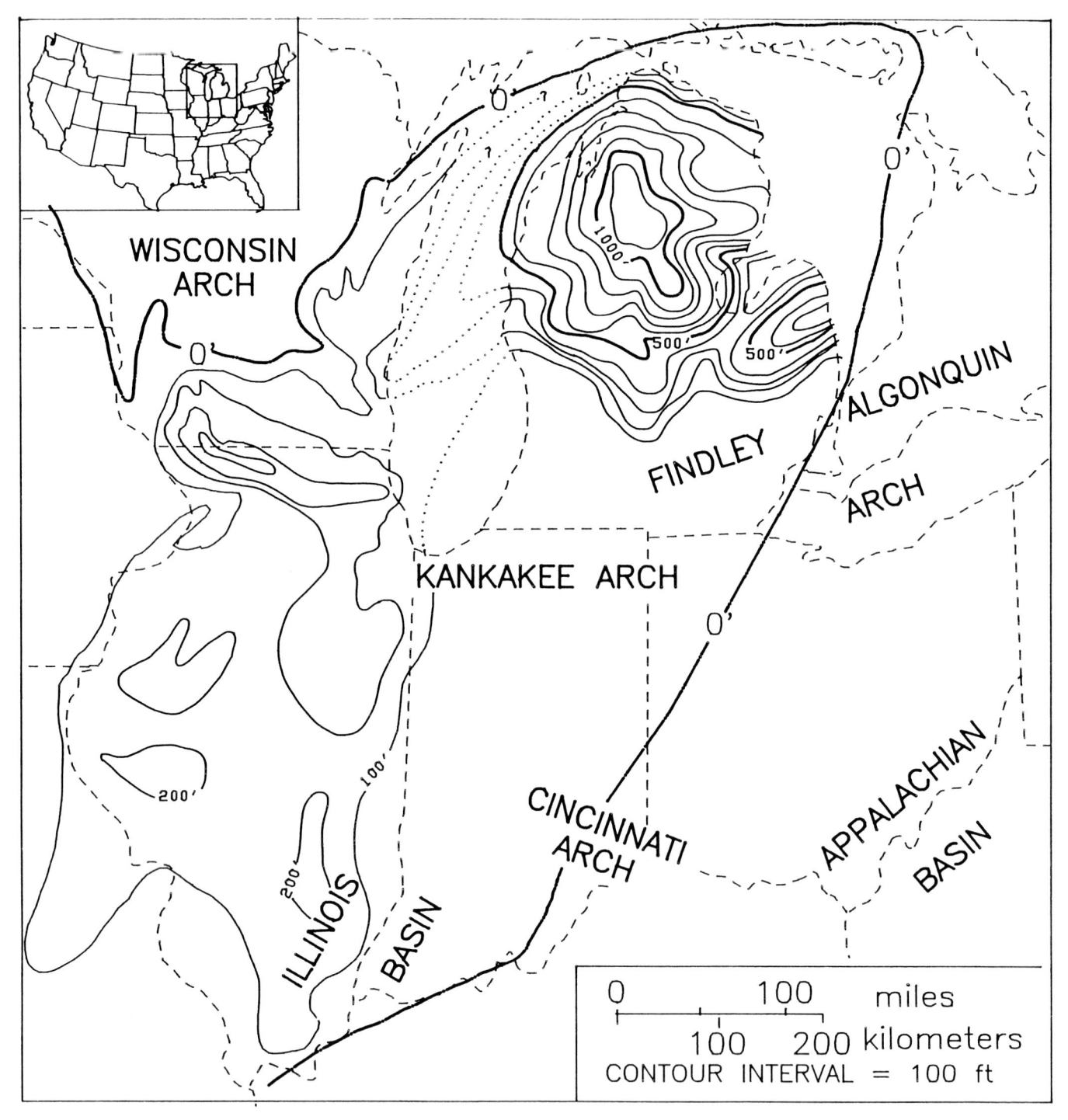

FIG. 3.—Regional isopach map of the St. Peter Sandstone in the Midwestern Basins and Arches geological province after Droste and Shaver (1983), Fisher and Barratt (1985), Mai and Dott (1985), and Lundgren (1991).

dissolution, illite (±chlorite) cementation and continued formation of quartz overgrowths. Minor K-feldspar overgrowths are also present but their time relation with other diagenetic phases is not known. Extensive chemical compaction, including stylolitization and quartz pressure solution, is also important in some parts of the formation.

The nature and/or the intensity of the diagenetic transformations observed in the St. Peter Sandstone are strongly correlated to the primary texture and detrital composition, i.e., the depositional lithofacies. The sandstones from the lower part of the St. Peter section (littoral facies) typically contain abundant quartz overgrowths (Fig. 8B), saddle dolomite, and anhydrite cements. Local dissolution of the latter cements is responsible for the development of some secondary porosity, commonly partially filled with minor amounts of late authigenic clays. In contrast, the sandstones

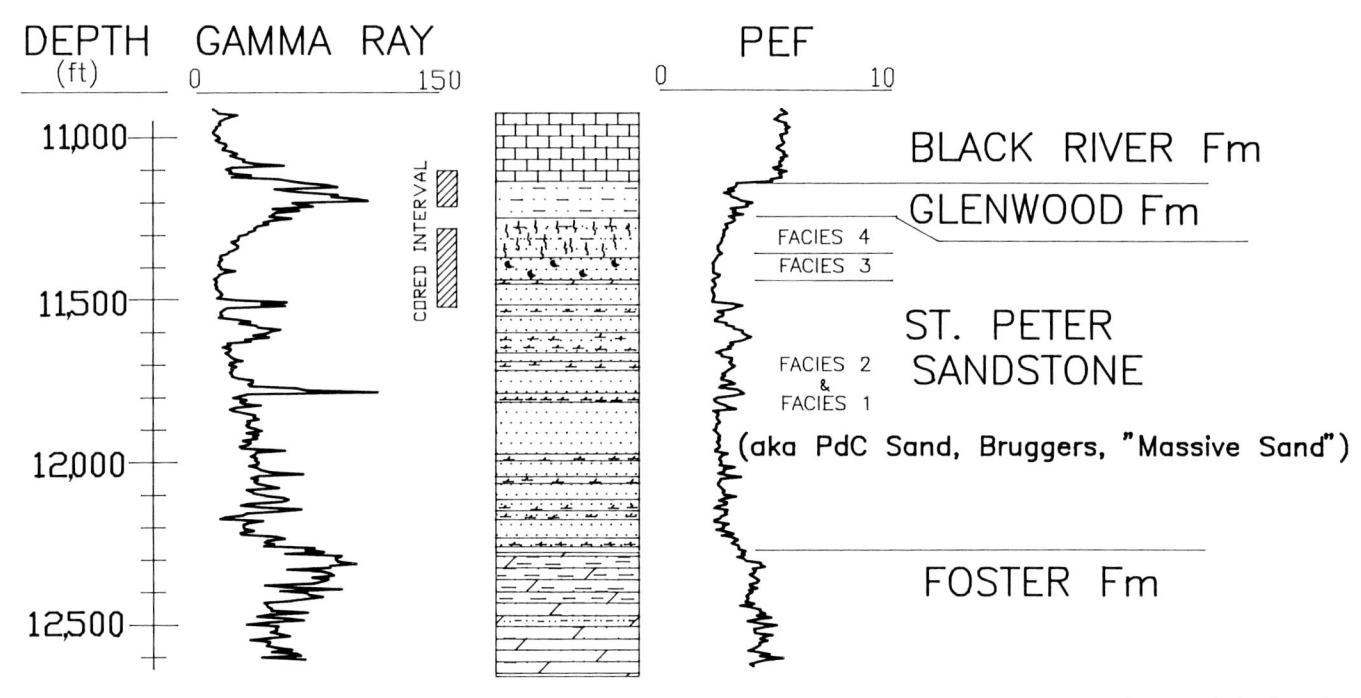

FIG. 4.—Detail of wireline log response (gamma-ray and photoelectric effect, PEF curves), interpreted lithology and sedimentary facies from the central Michigan Basin reference well, Hunt Martin, Gladwin County, Michigan. Available conventional core in this well is indicated.

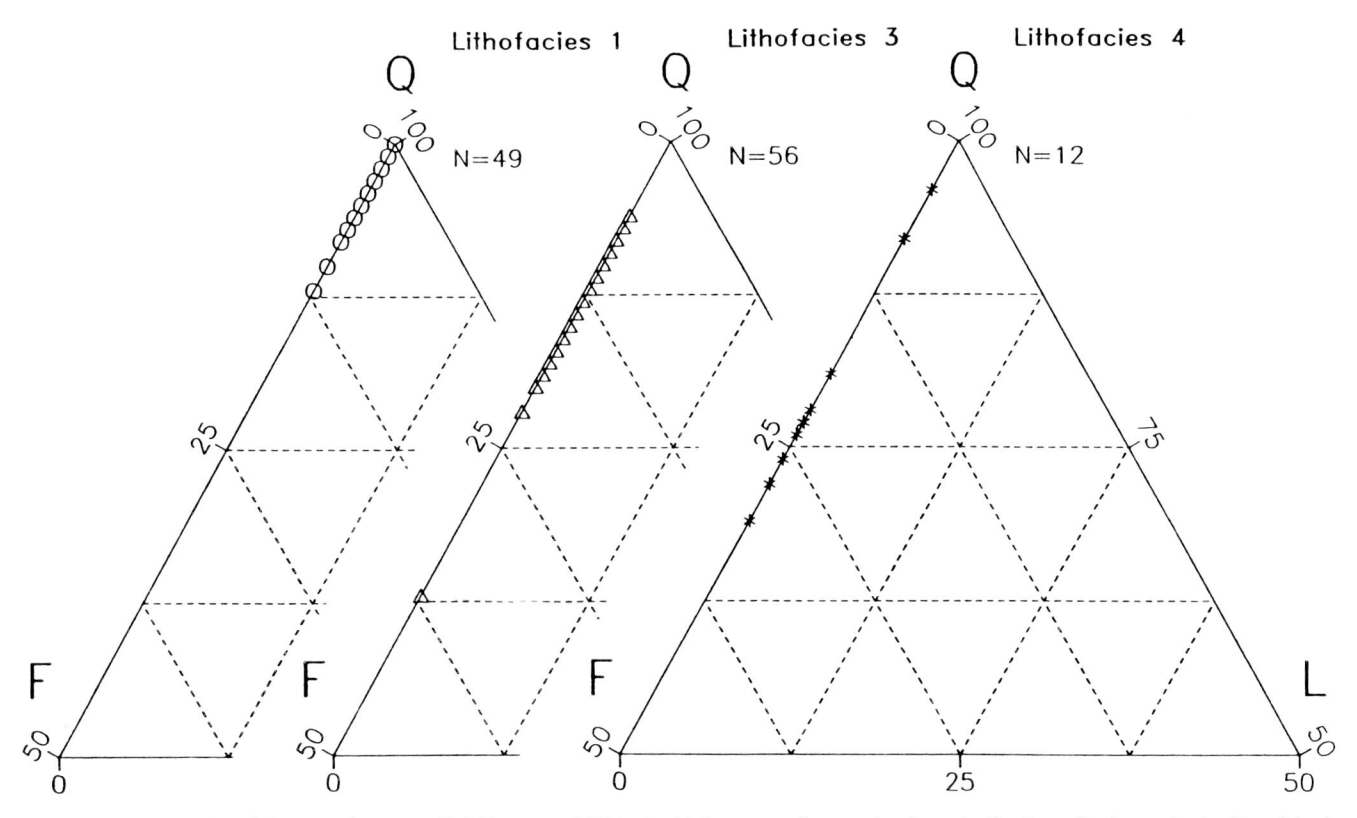

FIG. 5.—Ternary plot of framework quartz, K-feldspar, and lithic (rock) fragments for samples from the St. Peter Sandstone in the Hunt Martin well. Lithofacies are differentiated by symbol. Polycrystalline quartz and chert are included in the quartz field. Core from this well contains portions of all sandstone lithofacies in the St. Peter. Data from Lundgren (1991). Classification scheme after Folk and others (1970).

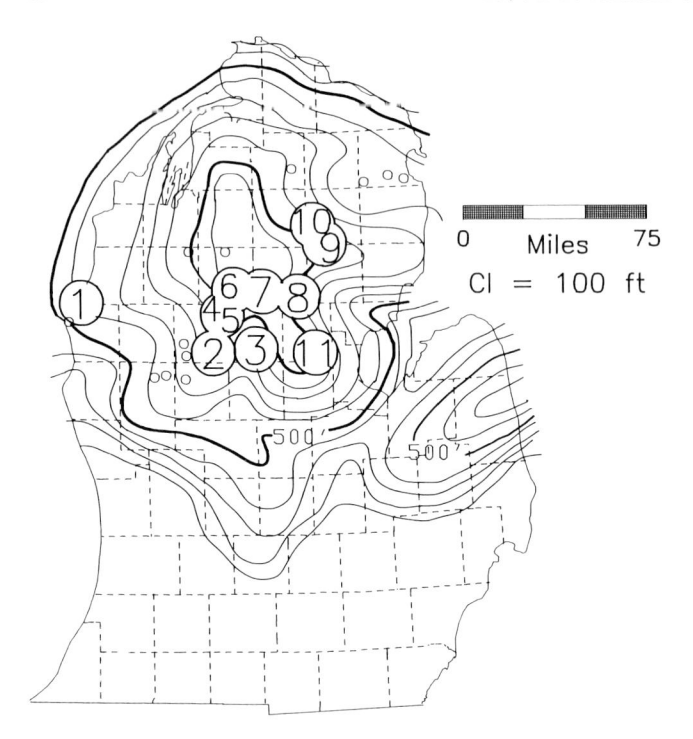

FIG. 6.—Locations of wells with core samples examined for this study. Well locations are superimposed on an isopach map of the St. Peter Sandstone showing thickening of the formation into the central Michigan Basin. Wells with samples analyzed for K-Ar ages of illite are indicated by numbered circles (numbers are consistent with Table 1) and are as follows: 1 = Miller Carnagle, Mason; 2 = JEM Freudenberg, Osceola; 3 = JEM Weingartz, Clare; 4 = Hunt Robinson, Osceola; 5 = JEM McCormick, Osceola; 6 = Patrick Gilde, Missauke; 7 = Federated Kitchenhoff, Roscommon; 8 = Sun Roseville, Roscommon; 9 = Sun Mentor, Oscoda; 10 = Sun Consumers Power, Oscoda; 11 = Hunt Martin, Gladwin.

from the upper part of the formation (shelf facies) are characterized by smaller amounts of quartz cement, minor remnants of intergranular dolomite and anhydrite cement, abundant secondary porosity, and abundant authigenic clay coatings and pore fillings (Figs. 8A and B). The distribution of the diagenetic cements does not seem to follow any distinctive pattern with depth of burial or geographic position in the basin other than the one described.

ORIGIN OF THE CLAYS IN THE ST. PETER

Clays show a variety of modes of occurrence in the St. Peter Sandstone in the Michigan Basin. Shale beds are rare, even in distal storm-shelf facies. Distinctly detrital clays are found in sandstones most commonly as thin, disrupted laminae and as rip-up clasts. Disseminated intergranular clay is also encountered and can be very abundant (i.e., in clay-rich sandstones from the upper part of the section) or almost absent as in most clean quartz arenites from the lower part of the section (Fig. 8A). Many petrographic observations point to the diagenetic origin of most of this clay. However, in the clay-rich samples from the lower energy depositional environments, it is possible that some or all of the intergranular clay was initially a primary detrital matrix. If so, petrographic evidence, as well as K-Ar ages, suggest it was extensively recrystallized during diagenesis (see later discussion).

Clay coatings on detrital grains and intergranular clay are very common in both the shoreface-upper offshore and offshore-shelf sandstone facies (lithofacies 3 and 4, Fig. 8A). Where abundant, especially in highly bioturbated sandstones (Plate 1A), the intergranular clay gives the sandstone a pronounced light green color visible to the naked eye. This is distinct from the dark brown color of the detrital-clay laminae observable in the rare occurrences of shaly sandstones and silts found at the top of the St. Peter formation (lithofacies 4, Plate 1B). The intergranular greenish clay is also commonly present in planar to low-angle cross-stratified beds of the subtidal-shoreface to upper offshore sandstones (lithofacies 3, Plate 1C). The occurrence of this clay in current-laminated sandstones is hydrodynamically inconsistent with a detrital origin for the clay.

A distinctive macroscopic feature of this intergranular clay is its interfingering contact, oblique to the bedding, with the dolomite cement (Plate 1D). In such samples, petrographic evidence (see later discussion) suggests that the clay postdates the dolomite and, therefore, is clearly of diagenetic origin. This common oblique-to-bedding, interdigitate geometry of the contact between authigenic clay and dolomite can be seen as an alteration front along which diagenetic clay is formed while dolomite is dissolved.

In thin section, this clay exhibits a similar greenish color in plane light and a low birefringence under crossed polars. It occurs most often as anhedral pore fillings and rarely as pore linings. Microtextures demonstrate the occurrence of the clay in secondary, intragranular pores after K-feldspar (Plate 1E) and in intergranular secondary pores after dolomite (Plate 1F, see plate caption for details). These microtextures are most common in shelf lithofacies 3 and 4. In subtidal shelf sandstones early carbonate "matrix," either dolomitized carbonate mud or dolomitized early marine carbonate cement, apparently inhibited subsequent development of quartz overgrowths. However, clays are present and also occur in secondary pores after the dolomite. In such instances, the diagenetic origin of the clay is unequivocal. Additional evidence for the authigenic origin of most of the clay was obtained by scanning electron microscopy (SEM). Most examples of clay coatings and pore fills examined under the SEM have poorly developed crystal habits. Pore-filling clay in particular commonly exhibits a massive, compact texture in which individual crystallites are not easily recognizable. However, delicate, euhedral crystal morphologies may be observed in places (Figs. 9A and B). These are undoubtedly of authigenic origin. Our samples were not critical-point dried and, therefore, delicate clay morphologies might have been irreversibly modified during drying (Pallatt, 1990). Although we cannot rule out the existence of detrital clay in our samples, SEM evidence, in spite of the oven-drying procedure, does demonstrate the presence of authigenic clays. If some or all of the massive pore-filling clay was detrital, SEM observations suggest that most likely it was recrystallized during burial and/or that significant diagenetic clay neoformed within it.

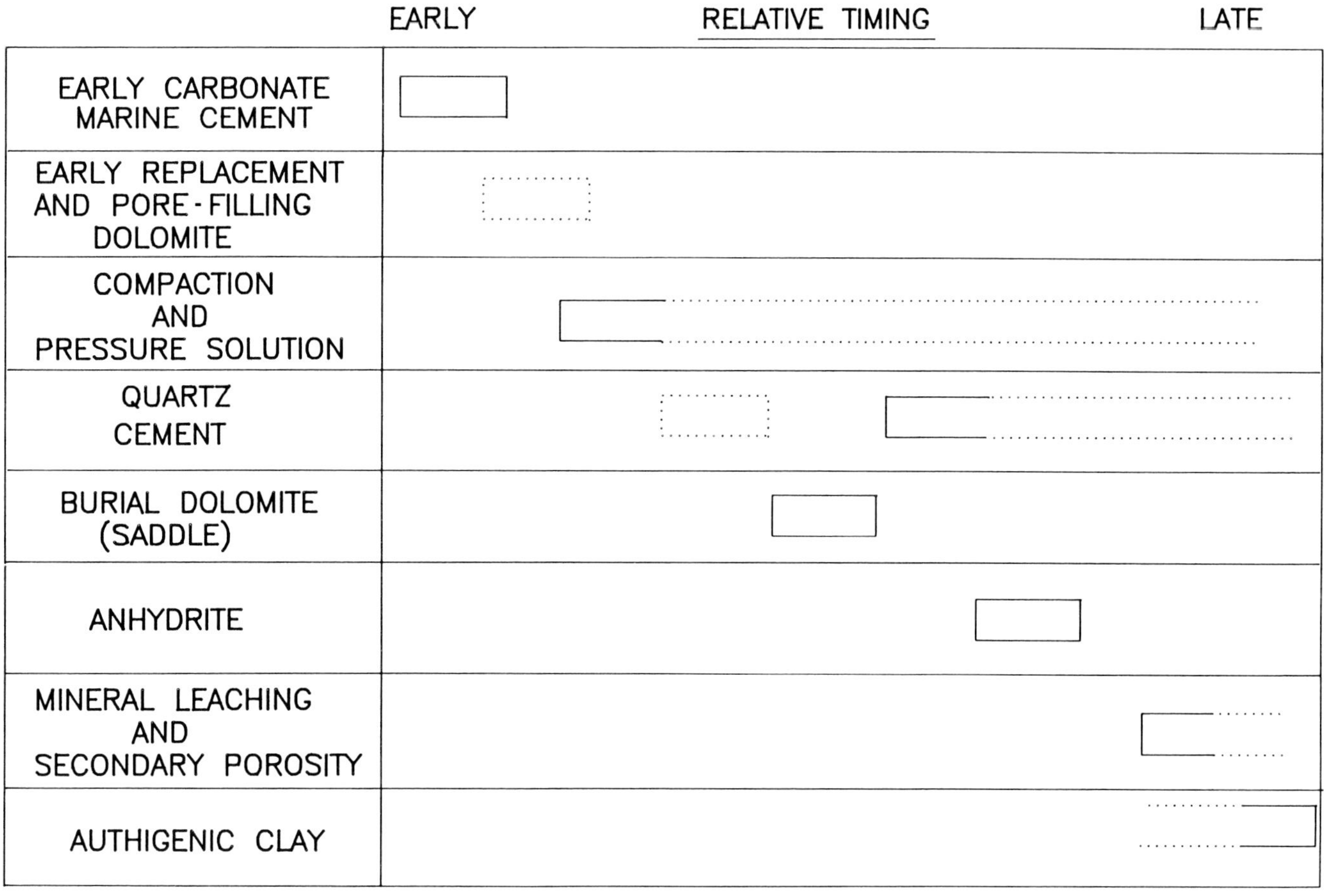

FIG. 7.—Generalized paragenetic relations in the St. Peter Sandstone in the central Michigan Basin. Few sandstones show all components of the sequence, although the relative timing of these diagenetic minerals is consistent in all samples observed. Dotted symbol represents poorly constrained paragenetic relation.

Considered together, macroscopic, thin-section and SEM observations suggest that most of the clay, and in particular the very fine (submicron) clay used for dating, is authigenic (i.e., neoformed in pores and/or recrystallized after detrital clay). This is consistent with the transmission electron microscopy (TEM) and K-Ar data obtained on clay separates (see later discussion).

X-ray diffraction (XRD) patterns and energy-dispersive analyses indicate that diagenetic-clay coatings and pore fills are consistently made of an intimate mixture of illite and chlorite in variable proportions throughout the formation. The non-uniform distribution of the authigenic clay (clay-rich versus clay-poor samples) may be related to the incongruent dissolution of precursor minerals in the formation including K-feldspar, dolomite, micas, and detrital clay. Where these precursor minerals were more common (in the subtidal-shelf sandstones), authigenic clay formation was enhanced. Availability of local sources of K, Si, Al, Mg, and Fe may have controlled the distribution of authigenic clay in the St. Peter, although external sources of these components cannot be excluded at this time.

ILLITE K-Ar GEOCHRONOLOGY

Petrographic evidence in many wells in the central Michigan Basin indicates that illitization followed the formation of substantial secondary porosity in the St. Peter Sandstone. This dissolution porosity also constitutes the main reservoir porosity and is now occupied by hydrocarbons in many wells. Therefore, it was of great interest to date the diagenetic illite in order to place constraints on the timing of the dissolution event and, possibly, the timing of the hydrocarbon migration in the St. Peter in Michigan. The K-Ar geochronology study of the illite clay cement was undertaken for that purpose.

Methods and Materials

After careful selection based on petrographic observations, 15 illitic sandstones and one shale were sampled from conventional cores of the St. Peter Sandstone in 11 different wells in central Michigan (Fig. 6, Table 1) in order to provide broad areal coverage. Composite samples, representing several to as much as 60 cm of core, were selected

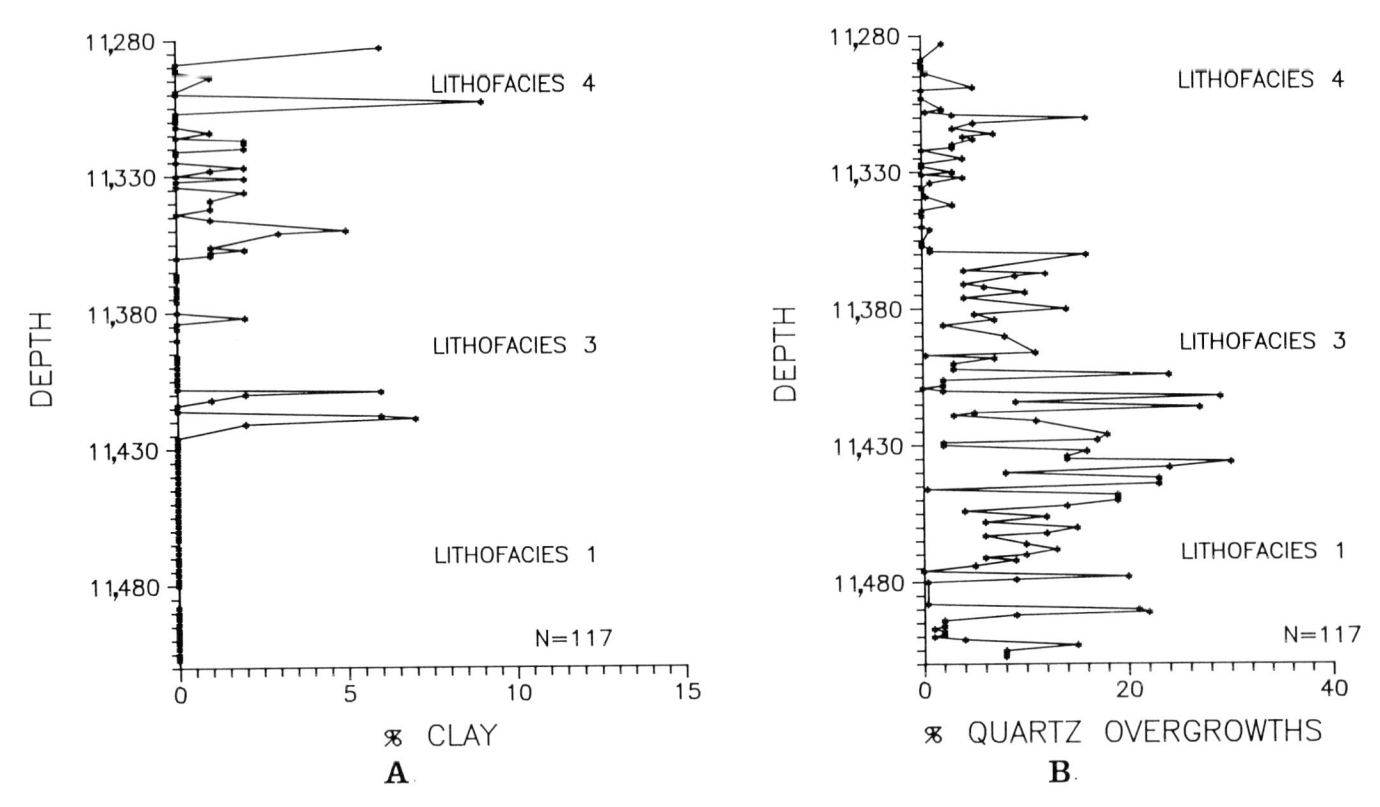

FIG. 8.—(A) Percent authigenic clay (by point count) versus depth in the Hunt Martin reference well. Clay correlates to lower energy depositional facies and increased proportions of non-quartz detrital grains. (B) Percent quartz overgrowths (by point count) versus depth in the Hunt Martin reference well. Quartz overgrowths predominate in high-energy, well-sorted facies with high proportions of detrital-quartz grains.

from both high- and low-energy facies. The sample set also included both clay-rich and clay-poor sandstones as well as one thin shale bed.

Individual samples, after being soaked in water overnight, were gently disaggregated with a mortar and pestle in order to preserve as much as possible the original grain size of detrital and diagenetic particles. No intensive hammering or ball-mill-type crushing technique was used. The fine, clay-rich fraction was recovered in suspension and ultrasonically agitated. Carbonates, organic matter and iron oxides were then removed using standard chemical treatments described by Jackson (1979). Finally, the clay separates were deflocculated and size separated by use of a high-speed centrifuge. The following series of size separates was produced for clay-rich samples; 2–1, 1–0.5, 0.5–0.2, 0.2–0.1, and <0.1 μm. For clay-poor and/or smaller samples, which did not yield enough clay to produce a complete suite, the finest size fraction available was the <0.2 or <0.15 μm. All size separates were dialyzed in deionized water for several days. XRD analysis indicated that most size separates, in particular coarser size fractions, contained chlorite in addition to illite (Table 1). However, because chlorite lacks potassium, its presence did not affect the K-Ar age determination of the illite component (Girard and others, 1989).

Potassium content for 70 to 150 mg of illite-chlorite separates was determined by flame photometry using a lithium (Li) internal standard. Argon measurements were made using an MS10 mass spectrometer equipped with an online, multiple-load extraction system and a bulb-pipetted ^{38}Ar tracer calibrated against the LP6 interlaboratory standard. K-Ar ages were calculated using the ^{40}Ar abundance and decay constants of Steiger and Jager (1977). Uncertainties in the ages were calculated using a method, modified after Cox and Dalrymple (1967), which takes into account errors affecting the sample weight, the ^{36}Ar/^{38}Ar and ^{40}Ar/^{38}Ar ratio measurements, the composition of the tracer, the sample inhomogeneity, and the uncertainty in the K$_2$0 analysis.

Reliability of Apparent K-Ar Ages

The suitability of authigenic illite as a reliable K-Ar clock is well established (Aronson and Lee, 1986; Hamilton and others, 1989; Lundegard, 1989). In such K-Ar investigations, considerable difficulty is often encountered in the separation of diagenetic illite from detrital illite and other potassium-bearing detrital phases (Liewig and others, 1987; Girard and others, 1989; Lee and others, 1989).

In order to minimize the risk of contamination of authigenic illite by detrital contaminants, samples were chosen on the basis of abundant diagenetic clay and minimal obvious detrital clay (e.g., rip-up clasts and mica flakes). However, many of the clay-rich samples investigated contained significant K-feldspar observable in thin section, in

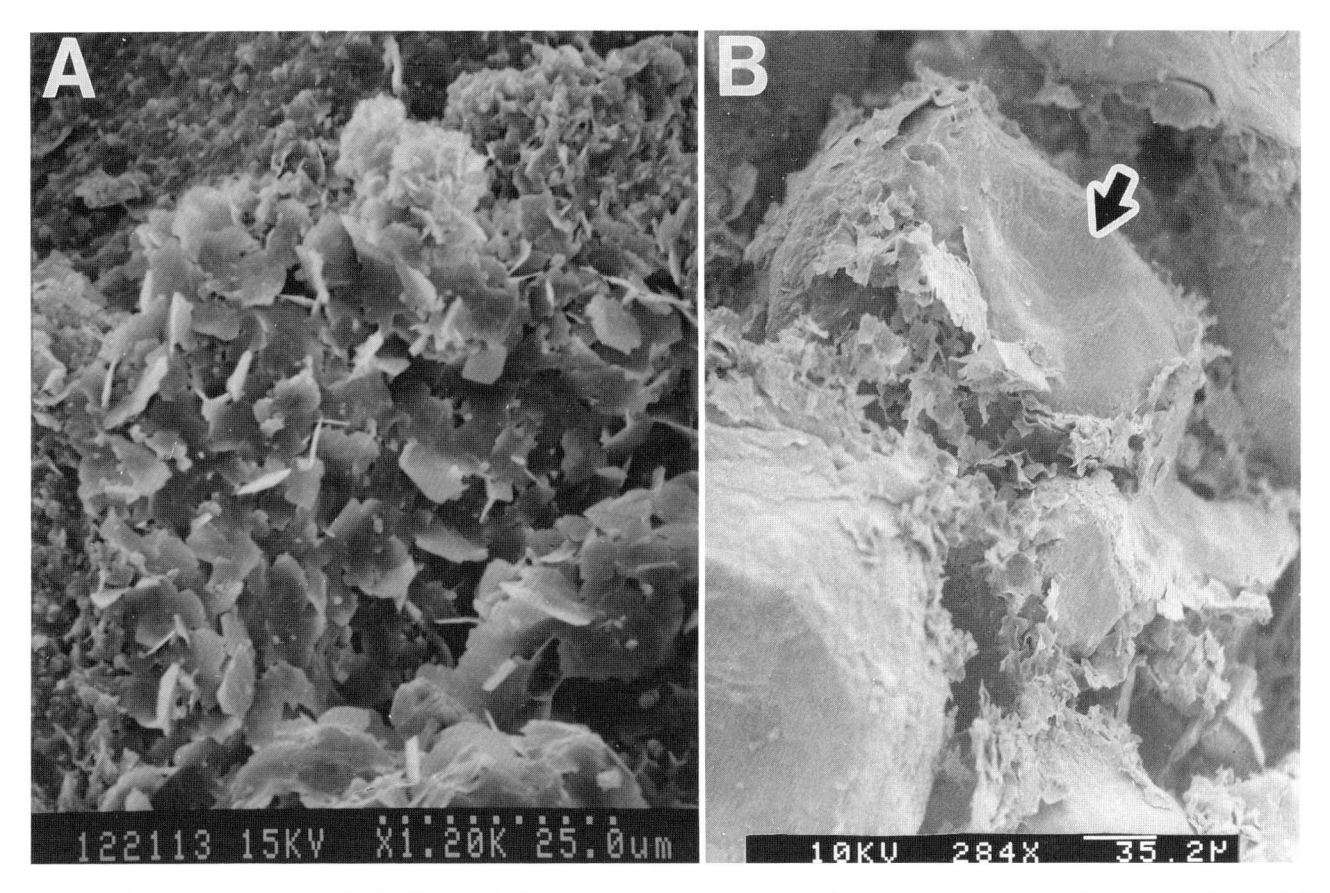

FIG. 9.—(A) SEM photomicrograph of delicate, euhedral, authigenic-clay grain coating in the St. Peter Sandstone. Energy dispersive and XRD analyses indicate that the clay in this sample is an intimate physical mixture of illite and chlorite. Hunt Martin, 3,446 m (11,299 ft). Illite age is 353 ± 6 Ma (see Table 1). Dotted scale bar is 25 μm. (B) SEM photomicrograph of intergranular illite adjacent to a quartz grain. Note pressure solution scar (arrow) where illite is absent. This texture suggests a post-compaction, authigenic origin of this clay. Miller Carnagle, 1,677 m (5,499 ft). Illite age is 367 ± 6 Ma (see Table 1). Scale bar is 35.2 μm.

the XRD patterns of the bulk clay fraction, and occasionally in the XRD pattern of the coarser clay-size separates (see samples, Hunt Martin 11292–96 and 11423 in Table 1). Furthermore, some samples may have contained detrital illite (see earlier discussion).

In all fine-clay separates (<0.2 μm) no contaminant other than chlorite was detectable by XRD (see Table 1). However, in order to verify that the apparent illite ages were not influenced by significant detrital contamination, we systematically analyzed the two finest size fractions of each sample. For three sandstone samples that appeared most prone to contamination and for the shale sample, several (3 to 5) successive size fractions were dated. When the apparent K-Ar ages of different size fractions asymptotically approached a constant value (within analytical uncertainty) as particle size decreased, we considered this age to be the true diagenetic age of the fine illite analyzed. Concordant ages for two different contaminated size fractions from the same sample would imply identical proportions of contaminants in each size fraction. This is believed to be highly unlikely.

Figure 10 is a plot of K-Ar apparent age versus particle size for three clay-rich sandstones and one shale. For each of the three sandstones, the fine-clay separates (smaller than 1.0 μm) yielded similar ages, ranging from 384 to 352 Ma (Table 1). The three finest size fractions in two of the sandstones and the two finest size fractions in the third sandstone gave identical ages within analytical uncertainty. This internal consistency (i.e., asymptotic trend to a constant age) is good evidence against any significant detrital contamination in the finest size fractions. In contrast, the fine-clay separates (<1.0 μm) from the shale showed a large spread of apparent ages ranging from 545 Ma (i.e., older than deposition) to 393 Ma (Table 1). Even though ultra-fine size separates (<0.1 μm) were not obtained from this sample, the variation of ages with decreasing particle size clearly did not trend toward a constant age. When compared to the sandstones, for any given size fraction, the clay separate from the shale always gave an older age. This indicated substantial contamination of the fine clay with detrital discrete illite and is to be expected in a shale where illite constitutes one of the main detrital phases.

Careful examination of the finer size fractions extracted from the sandstones using transmission electron microscopy (TEM) provided an additional check on their purity. Figure 11 is a TEM photograph of the <0.1-μm clay sep-

TABLE 1.—K-Ar ANALYTICAL DATA FOR THE CLAY SEPARATES FROM 15 SANDSTONES AND ONE SHALE IN THE ST. PETER SANDSTONE, MICHIGAN BASIN. REFER TO FIGURE 6 FOR WELL LOCATIONS.

Well	Sample Interval (ft)	Particle size (μm)	Mineral #	K_2O (%)	Ar* (%)	K-Ar age (Ma)
1	5,499 -500	0.5-0.2	I	8.4	98	377±6
		<0.2	I	8.7	98	367±6
2	9,677 - 92	0.2-0.15	I,ch	9.6	98	353±6
		<0.15	I	9.4	96	342±14
3	10,789 - 91	0.2-0.15	I,ch	9.4	98	354±6
		<0.15	I	9.2	97	345±6
4	10,393 - 401	0.2-0.15	I,ch	9.9	98	328±5
		<0.15	I,ch	9.7	98	327±6
5	9,839 - 49	0.2-0.15	I,ch	9.7	98	331±5
		<0.15	I	9.6	98	328±6
6	10,589 - 91	0.2-0.15	I	9.6	98	348±6
		<0.15	I	9.5	98	333±6
7	11,021 - 24	0.2-0.15	I	9.6	98	344±6
		<0.15	I	9.4	98	345±6
8	11,698 -701	1.0-0.5	I,ch	7.8	98	384±6
		0.5-0.2	I,ch	9.1	97	361±6
		0 2-0.1	I,ch	9.0	97	360±6
		<0.1	I	9.3	98	352±6
9	10,060 - 65	0.2-0.15	I	9.4	98	343±6
		<0.15	I	9.2	97	342±6
10	9,516 - 17	0.2-0.15	I	9.5	98	339±6
		<0.15	I	9.2	97	343±6
11	11,289 - 91	0.5-0.2	I,ch	8.1	98	377±6
		0.2-0.1	I,ch	8.8	97	361±7
		<0.1	I,ch	8.7	97	359±6
	11,292 - 96	0.5-0.2	I,ch,k,q	7.5	97	383±6
		0.2-0.1	I,ch	8.7	95	359±6
		<0.1	I,ch	8.6	97	353±6
	11,309 - 10	0.5-0.2	I,ch	4.9	98	345±6
		<0.2	I,ch	8.7	98	343±6
	11,380 - 81	0.5-0.2	I,ch	5.7	98	373±6
		<0.2	I,ch	9.3	98	346±6
	11,408 - 09	0.2-0.15	I,ch	9.5	94	347±6
		<0.15	I,ch	9.5	97	339±7
Shale sample						
11	11,423 - 24	5.0-2.0	I,ch,k,q	8.6	98	570±9
		2.0-1.0	I,ch,k	8.8	98	550±9
		1.0-0.5	I,ch	8.1	98	544±9
		0.5-0.25	I,ch	8.8	98	448±8
		<0.25	I,ch	8.6	97	393±7

\# = mineralogy as indicated by XRD patterns, I = illite, ch = chlorite, k = K-feldspar, q = quartz; Ar* = radiogenic Ar.

arate from a typical clay-rich sandstone. Particles exhibit euhedral morphologies, either as laths or plates, clearly authigenic and typical of diagenetic illite in sandstones (Hamilton and others, 1989; Glasmann and others, 1989b).

Illite K-Ar Ages

A complete summary of all K-Ar age determinations performed on our sample set is given in Table 1. All samples but two, Patrick Gilde 10,589–91 and Hunt Martin 11,380–81, satisfy the internal consistency (concordance of the ages of the two finest size fractions) requirement and are considered to yield the true ages of the fine diagenetic illite. The age (346 Ma) of the finest size fraction (<0.2 μm) of Hunt Martin 11,380–81 is identical to the ages of similar size fractions from other samples in the same well (Table 1) and is also probably valid. Although not strictly identical

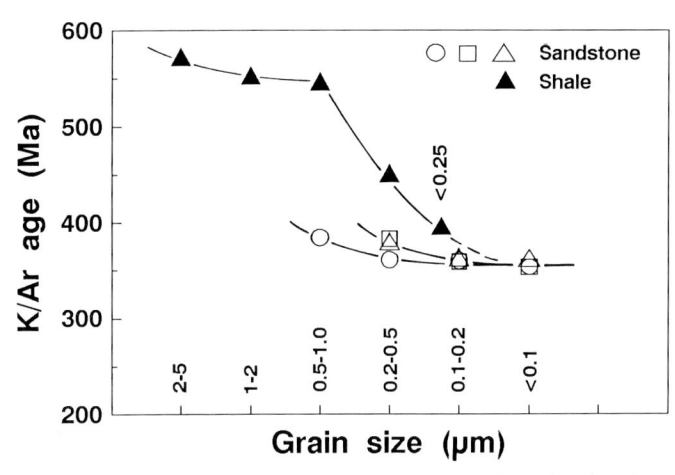

FIG. 10.—Plot of apparent K-Ar age versus particle size for three sandstones and one shale (Sun Roseville, 11,698–701 ft; Hunt Martin, 11,289–91, 11,292–96 and 11,423–24 ft; see Table 1 and Fig. 6). Symbol size approximates analytical uncertainty on K-Ar age (±6–9 Ma). Note the asymptotic trend toward a constant and concordant age for the finest size fractions of the sandstones, indicating a lack of detrital contamination (see text).

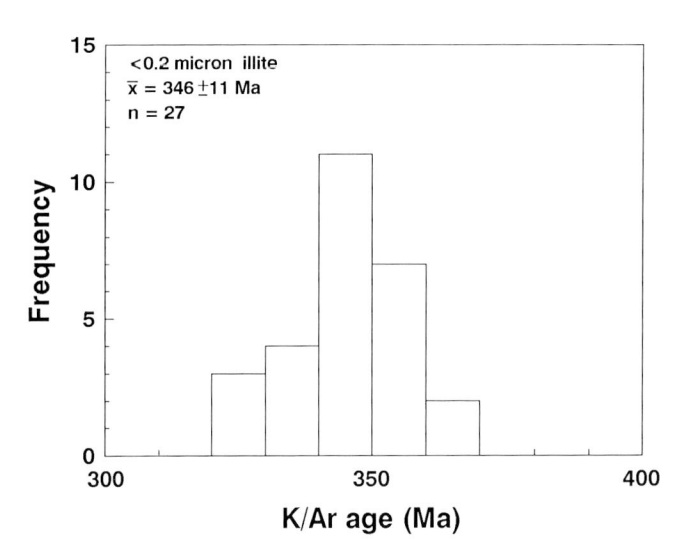

FIG. 12.—Histogram of apparent K-Ar ages of all <0.2-µm size separates (i.e., <0.1, <0.15, 0.1–0.2, 0.15–0.2, <0.2 µm, see Table 1) from 15 illitic sandstones in the St. Peter Sandstone, Michigan Basin. The distribution of ages is unimodal with a range of 327 Ma to 367 Ma and an average of 346 ± 11 Ma.

within their respective analytical uncertainties, the ages of two finest size fractions of sample Patrick Gilde 10,589–91 are close to one another and essentially similar to the ages yielded by the other samples. We consider them as valid diagenetic ages.

Figure 12 is a histogram of all ages corresponding to the clay material smaller than 0.2 µm in size, corresponding to the one or two finest size fractions of each sample that we demonstrated are purely diagenetic in nature. Ages range from 327 Ma to 367 Ma with a unimodal distribution and an average of 346 ± 11 Ma, indicating a major, regional episode of illitization during the Mississippian.

As discussed earlier, the ages of the finest clay separates from the shale sample (Table 1) do not agree with the ages for the sandstones because of detrital contamination (Fig. 10). However, the 393-Ma age of the <0.25-µm size fraction is over 100 Ma younger than depositional age and only slightly less than 50 Ma older than the diagenetic illite in the sandstones. This indicates that a substantial amount of diagenetic illite is present in the shale, either as a result of recrystallization of detrital illite or neoformation. The coarser size fractions (>1 µm) yielded ages in excess of 500 Ma (i.e., significantly older than deposition of the St. Peter), which places some constraints on the age of the source rocks.

The spread of ages observed for the diagenetic illite from the sandstones is 40 Ma. This is significantly larger than the analytical uncertainty and may reflect actual variations in the time of precipitation of the illite. Figure 13 is a plot of the age of the finest size fraction of each sample versus depth. Although the shallowest sample yielded the oldest age, no distinct trend of ages with depth is apparent at this time. Unfortunately, the majority of the samples are from a rather narrow depth range of 3,048 m (10,000 ft) to 457 m (1,500 ft) and our sample set may not be most appropriate to test a possible correlation between age and depth. It is interesting to note that the five samples from the Hunt Martin well show a distinct trend toward younger ages down depth. A similar trend has been described in the Permian Rotliegende Sandstone, North Sea (Lee and others, 1985, 1989), and was interpreted as reflecting the downward migration of the water-gas contact in the reservoir. In the Hunt Martin well the shallowest and the deepest samples dated are only 37 m (120 ft) apart and the difference in age is

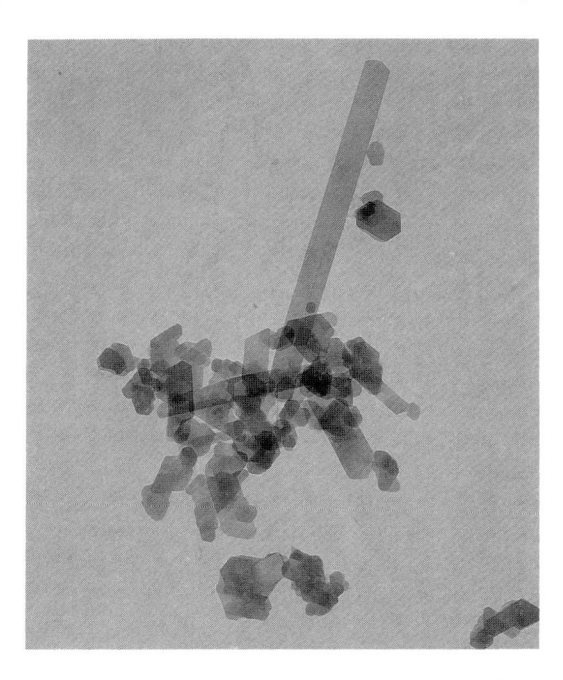

FIG. 11.—Transmission electron micrograph of the <0.1-µm clay separate from sample at 31,698–701 ft, Sun Roseville well, Roscommon. This separate contains only illite, which was dated at 352 ± 6 Ma (see Table 1). Euhedral lath and plate morphologies attest to the authigenic origin of the clay and the purity of the separate. Scale is given by the width of the longest lath, which is exactly 0.07 µm. Photograph by R. Glasmann, UNOCAL.

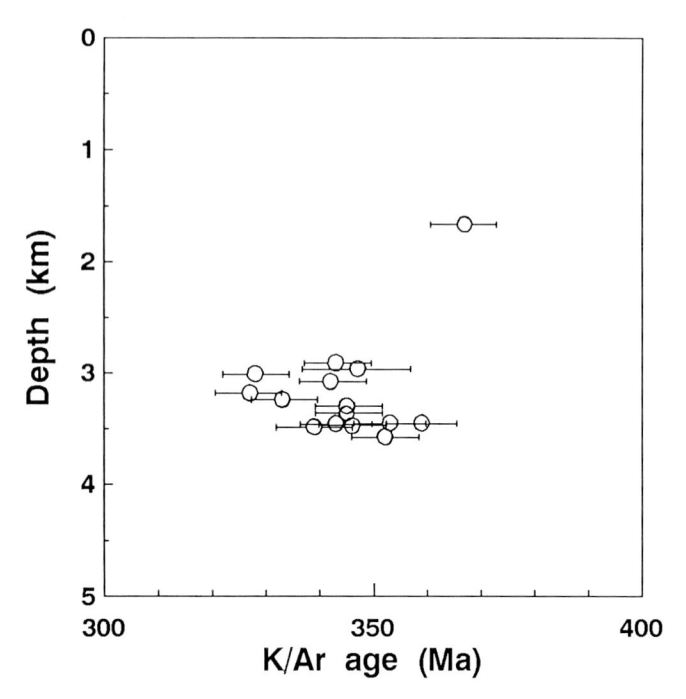

FIG. 13.—Plot of apparent K-Ar ages for the finest size fractions from 15 illitic sandstones in the St. Peter Sandstone versus present burial depth. Error bars represent uncertainties on ages as reported in Table 1.

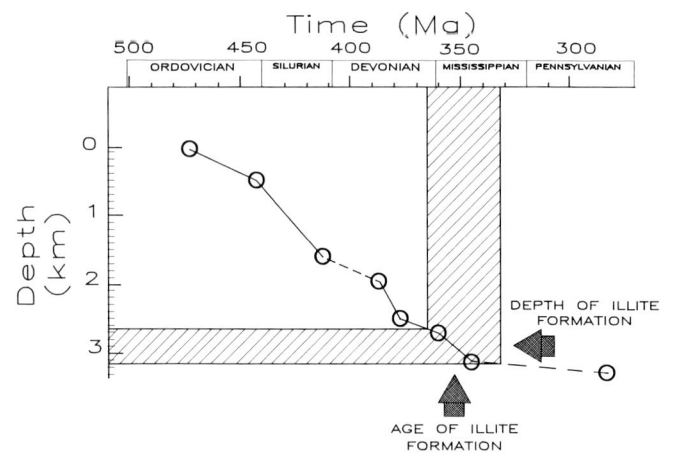

FIG. 14.—Burial-history curve (uncorrected for compaction) drawn for the upper part of the St. Peter Sandstone in the Hunt Martin well, Gladwin County, Michigan. The age of deposition of the Middle Ordovician St. Peter is estimated at about 470 Ma. The K-Ar ages of the authigenic illite in this well range from 339 ± 7 Ma to 359 ± 6 Ma (see Table 1). A total age range of 332 to 365 Ma, including analytical uncertainties, was used to determine the range of burial depth during illitization (2.7 to 3.4 km). The dashed portions of the curve indicate uncertainty concerning the extent of erosion associated with the Silurian-Devonian and late Mississippian unconformities (see text for discussion).

only slightly larger than the analytical uncertainty (respectively, 20 Ma and 12–14 Ma). We must, therefore, await further data to confirm the significance of this trend.

At this point, no definite trend to the geographic distribution of the ages is discernable from our data. However, if this areal variation is real, it might reflect local differences in the thermal regime or burial rates undergone by the sediments and/or in the migration pathways of the illitizing fluids.

IMPLICATIONS

Petrographic observations and K-Ar geochronology of authigenic illite in the St. Peter Sandstone indicate a major, regional-scale episode of illitization in the Late Devonian-Mississippian (367–327 Ma) in the Michigan Basin. Illitization resulted in both neoformation of clay in secondary pores as well as recrystallization of precursor detrital-clay minerals. Because illite commonly occupies economically significant secondary porosity, mineral dissolution must have preceded the illitization event. Hydrocarbons also occupy the dissolution porosity and must have migrated either contemporaneously with, or subsequent to, the precipitation of authigenic illite (Hamilton and others, 1989).

A crude reconstruction of the burial history of the uppermost, clay-rich portion of the St. Peter Sandstone in the Hunt Martin well is shown in Figure 14. Illite ages (339 to 359 Ma) and associated analytical uncertainties suggest that illitization occurred between 332 and 365 Ma in this well (Table 1). Combining this age bracket with the burial history (Fig. 14) yields paleodepths of illitization ranging from 2.7 to 3.2 km. This paleodepth range is a minimum estimate as no compensation for compaction of the sediments nor sediment removal at unconformities was applied to our burial curve. Decompaction would certainly not have a significant effect on the depth estimate as no major shale unit occurs in the pre-Middle Ordovician part of the stratigraphic section.

Three significant periods of erosion are recognized in the Michigan Basin: during the Late Silurian-Early Devonian; the Late Mississippian; and from Late Pennsylvanian to Jurassic time (Dorr and Eschman, 1971; Cercone, 1984). Cercone (1984) estimated a total removal of 1,000 m, most of which occurred in the Late Mississippian and Pennsylvanian and predominantly affected the edges of the basin. Only the Late Mississippian erosion could possibly have influenced the burial history of the St. Peter during the illitization event in the central basin (Fig. 14). Although it is difficult to estimate accurately how much erosion could have occurred in the central Michigan Basin at that time, it is very unlikely to have exceeded the amount of erosion documented at the margins of the basin. Maximum estimates, based on stratigraphic evidence, of late Mississippian erosional stripping along the Kankakee Arch in northern Indiana (Fig. 1) is 200 m (Gray, pers. commun., 1991). Therefore, taking all uncertainties into account, a maximum range of possible paleodepths of illitization in the St. Peter Sandstone in the central Michigan Basin is 2.7 to 3.4 km.

A preliminary study of two-phase aqueous-fluid inclusions in the saddle dolomite in the St. Peter Sandstone (Tsui, pers. commun., 1989) indicates homogenization temperatures ranging from 110° to 145°C. Fluid-inclusion freezing data indicate that dolomite-forming fluids were brines (17–24 weight percent equivalent NaCl). These homogenization temperatures represent minimum temperatures of formation for the dolomite immediately preceding illitization. If we assume that no significant decrease of the geothermal gradient occurred between the dolomitization and the illitization event, then the temperature of formation of the diagenetic illite must have been on the order of 150°C or higher. Combining this temperature of illitization with the paleodepths previously derived and assuming a surface temperature of 20°C yields paleogeothermal gradients of 38 to 48°C/km. These are in good agreement with the 35 to 45°C/km geothermal gradients inferred by Cercone (1984) for pre-Carboniferous times. Present geothermal gradients in the basin are estimated to be between 19°C/km (Vugrinovich, 1989) and 25°C/km (Cercone, 1984). Regardless of the magnitude of the uncertainties on our depth and temperature estimates for the diagenetic illite, our data indicate that the geothermal gradients during the Late Devonian-Mississippian in the central Michigan Basin must have been in excess of the present gradients.

For the sake of argument and in spite of the lack of evidence in support of the following, we consider the possibility that the fluid-inclusion temperatures obtained from the dolomite cement actually reflect maximum burial temperatures (as a result of re-equilibration). According to Cercone (1984), the maximum burial depth of the upper part of the St. Peter Sandstone in the central Michigan Basin could have been 4 km at the most at the end of the Paleozoic (Pennsylvanian-Permian). This maximum burial estimate corresponds to a minimum paleogeothermal gradient of 32.5°C/km, i.e., significantly higher than present gradients.

Additional fluid-inclusion data obtained from quartz overgrowths (Tsui, pers. commun., 1990) indicate a wide range of homogenization temperatures from 80 to 170°C, consistent with the petrographic evidence that quartz overgrowths formed over an extended period of time throughout the diagenesis of the St. Peter Sandstone. Bearing in mind that these temperatures are minimum formation temperatures, they support the previous conclusion that significantly hotter thermal regimes than today prevailed in the Michigan Basin in the Late Paleozoic.

The illitization event is interpreted to have occurred following a significant mineral dissolution event and prior to, or contemporaneous with, hydrocarbon migration in the St. Peter Sandstone in Michigan. Other workers have argued that the formation of secondary porosity in hydrocarbon reservoirs is closely related to the onset of thermal maturation in associated hydrocarbon source rocks and the generation of organic acids (Surdam and others, 1989).

No conclusive evidence currently exists in the published literature concerning the source rock for hydrocarbons of the St. Peter Sandstone in Michigan. Proprietary data, made available to the authors informally, suggest potential origin of hydrocarbons from algae-laminated, predominantly carbonate strata of the underlying Prairie du Chien Group (Foster Formation of Fig. 2). Ubiquitously low total organic-carbon content in the immediately subjacent "Brazos shale" indicates that these strata are unlikely hydrocarbon source rocks for the St. Peter. The close stratigraphic proximity of the Foster Formation and the St. Peter Sandstone suggests similar burial-thermal histories for these formations. The spatial and temporal relations of dissolution porosity, authigenic illite and hydrocarbon migration in the St. Peter Sandstone tentatively suggest a genetic relation between these processes. Although our data set does not presently constrain the main controls on illitization, further studies may reveal that possible areal variation in thermal regime in the Michigan Basin had strong influence on the timing of both organic and inorganic diagenesis in lower Paleozoic strata.

CONCLUSIONS

1. The Middle Ordovician St. Peter Sandstone in the Michigan Basin has experienced major modifications of primary mineralogy and textures as a result of a complex diagenesis. The most important diagenetic features are marine carbonate cement, quartz cement, burial dolomite and anhydrite, mineral dissolution, and the formation of authigenic illite and chlorite.

2. Diagenetic transformations were markedly influenced by depositional environments. Quartz overgrowth cement is more common in the coarser grained, more quartz-rich sandstones of paralic facies found in the lower part of the formation. Authigenic clay is more abundant in generally finer grained, more calcareous and feldspathic sandstones of the offshore shelf facies located primarily in the upper part of the formation.

3. K-Ar age determinations of the fine-grained authigenic illite in the St. Peter indicate a major, regional episode of illitization between 327 Ma to 367 Ma, during Late Devonian to Mississippian times.

4. Constraints placed by fluid-inclusion temperatures for the dolomite and quartz cements and burial-history reconstructions suggest that diagenetic illite formed between 2.7 and 3.4 km at temperatures of 150°C or higher in the central Michigan Basin. These estimates imply the existence of geothermal gradients significantly greater than present gradients at the time of illitization in the central basin.

ACKNOWLEDGMENTS

We express special thanks to M.C. Lee and T. Tsui, Mobil Research & Development, and C. Lundgren, Arco Exploration, for access to data on the St. Peter Sandstone and for constructive comments and suggestions. We thank R. Glasmann, Unocal Sciences and Technologies, for performing the TEM study; L.L. Gray, Indiana State Geological Survey, for stratigraphic data; Don Mazuchowski, Michigan Public Services Commission, for St. Peter Sandstone production data; and L. Abel (CWRU) for help with drafting. This work was supported by the American Chemical Society Petroleum Research Fund (Grants 22408-B2 to D. Barnes, Western Michigan University, and 23313-AC8 to S. Savin, CWRU). Additional support was provided by Arco Oil & Gas. This paper benefitted from reviews by L. Crossey and R. Siebert.

REFERENCES

ARONSON, J. L., AND LEE, M., 1986, K-Ar systematics of bentonite and shale in a contact metamorphic zone, Cerillos, New Mexico: Clays and Clay Minerals, v. 34, p. 483–487.

Barnes, D. A., 1988, Burial diagenesis in the St. Peter Sandstone, deep Michigan Basin: Geological Society of America, Abstracts with Programs, v. 20, p. 333.
Barnes, D. A., Turmelle, T. M., and Adam, R., 1989, Diagenetic controls on reservoir heterogeneity in the St. Peter Sandstone, Michigan Basin (abs.): American Association of Petroleum Geologists Bulletin, v. 73, p. 331–332.
Cercone, K. R., 1984, Thermal history of the Michigan basin: American Association of Petroleum Geologists Bulletin, v. 68, p. 130–136.
Cox, A., and Dalrymple, G. B., 1967, Statistical analysis of geomagnetic reversal data and the precision of Potassium-Argon dating: Journal of Geophysical Research, v. 72, p. 2603–2614.
Dorr, J. A., Jr., and Eschman, D. F., 1971, Geology of Michigan: University of Michigan Press, Ann Arbor, Michigan, 476 p.
Droste, J. B., and Shaver, R. H., 1983, Atlas of Early and Middle Paleozoic Paleogeography of the Southern Great Lakes Region: Indiana Geological Survey, Special Report 32, 32 p.
Fisher, J. H., and Barratt, M. W., 1985, Exploration in Ordovician of central Michigan Basin: American Association of Petroleum Geologists Bulletin, v. 69, p. 2065–2076.
Folk, R. L., Andrews, P. B., and Lewis, D. W., 1970, Detrital sedimentary rock classification for use in New Zealand: New Zealand Journal of Geology and Geophysics, v. 13, p. 937–68.
Girard, J. P., Savin, S. M., and Aronson, J. L., 1989, Diagenesis of the Lower Cretaceous arkoses of the Angola margin: petrologic, K-Ar dating and $^{18}O/^{16}O$ evidence: Journal of Sedimentary Petrology, v. 59, p. 519–538.
Glasmann, J. R., Clark, R. A., Larter, S., Briedis, N. A., and Lundegard, P. D., 1989a, Diagenesis and hydrocarbon accumulation, Brent Sandstone (Jurassic), Bergen High area, North Sea: American Association of Petroleum Geologists Bulletin, v. 73, p. 1341–1360.
Glasmann, J. R., Lundegard, P. D., Clark, R. A., Penny, B. K., and Collins, I. D., 1989b, Geochemical evidence for the history of diagenesis and fluid migration: Brent Sandstone, Heather field, North Sea: Clay Minerals, v. 24, p. 255–284.
Hamilton, P. J., Kelley, S., and Fallick, A. E., 1989, K-Ar dating of illite in hydrocarbon reservoirs: Clay Minerals, v. 24, p. 215–231.
Howell, P. D., and van der Pluijm, B. A., 1990, Early history of the Michigan Basin: subsidence and Appalachian tectonics: Geology, v. 18, p. 1195–1198.
Jackson, M. L., 1979, Soil Chemical Analysis—Advanced Course (2nd ed.), 11th Printing: Published by the author, Madison, Wisconsin, U.S.A., 895 p.
Klein, G. deV., and Hsui, A. T., 1987, Origin of cratonic basins: Geology, v. 15, p. 1094–98.
Lee, M., Aronson, J. L., and Savin, S. M., 1985, K-Ar dating of time of gas emplacement in Rotliegendes Sandstone, Netherlands: American Association of Petroleum Geologists Bulletin, v. 69, p. 1381–1385.
Lee, M., Aronson, J. L., and Savin, S. M., 1989, Timing and conditions of Permian Rotliegende Sandstone diagenesis, southern North Sea: K-Ar and oxygen isotopic data: American Association of Petroleum Geologists Bulletin, v. 73, p. 195–215.
Liewig, N., Clauer, N., and Sommer, F., 1987, Rb-Sr and K-Ar dating of clay diagenesis in Jurassic sandstone oil reservoir, North Sea: American Association of Petroleum Geologists Bulletin, v. 71, p. 1467–1474.
Longstaffe, F. J., 1989, Stable isotopes as tracers in clastic diagenesis, in Hutcheon, I. E., ed., Burial Diagenesis: Mineralogical Association of Canada Short Course, v. 15, p. 201–278.

Lundegard, P. D., 1989, Temporal reconstruction of sandstone diagenetic history, in Hutcheon, I. E., ed., Burial Diagenesis: Mineralogical Association of Canada Short Course, v. 15, p.161–200.
Lundgren, C. E., 1991, Diagenesis of the St. Peter Sandstone, Michigan Basin: Unpublished M.S. Thesis, Western Michigan University, Kalamazoo, Michigan, 117 p.
Mai, H., and Dott, R. H. Jr., 1985, A subsurface study of the St. Peter Sandstone in southern and eastern Wisconsin: Wisconsin Geological and Natural History Survey, Madison, Wisconsin, Report #47, 26 p.
Nunn, J. A., Sleep, N. H., and Moore, W. E., 1984, Thermal subsidence and generation of hydrocarbons in the Michigan Basin: American Association of Petroleum Geologists Bulletin, v. 68, p. 296–315.
Odom, I. E., Doe, T. W., and Dott, R. H., Jr., 1976, Nature of feldspar-grain size relations in some quartz-rich sandstones: Journal of Sedimentary Petrology, v. 46, p. 862–870.
Pallatt, N., 1990, Critical point drying applied to clay minerals in sandstones (abs.): Clay Minerals Society 27th Annual Meeting, Columbia, Missouri, p. 100.
Pitman, W. C., 1978, Relationship between eustasy and stratigraphic sequences of passive margins: Geological Society of America Bulletin, v. 89, p. 1389–1403.
Quinlan, G. M., 1987, Models of subsidence mechanisms in intracratonic basins and their applicability to North American examples, in Beaumont, C., and Tankard, A. J., eds., Sedimentary Basins and Basin-Forming Mechanisms: Canadian Society of Petroleum Geologists Memoir No. #12, p. 463–481.
Quinlan, G. M., and Beaumont, C., 1984, Appalachian thrusting, lithospheric flexure and the Paleozoic stratigraphy of the eastern interior of North America: Canadian Journal of Earth Sciences, v. 21, p. 973–96.
Sloss, L. L., 1963, Sequences in the cratonic interior of North America: Geological Society of America Bulletin, v. 74, p. 93–113.
Sloss, L. L., 1988, Tectonic evolution of the craton in Phanerozoic time, in Sloss, L. L., ed., Sedimentary Cover, North American Craton: U.S. Geological Society of America DNAG, The Geology of North America, v. D-2, p. 25–51.
Sommer, F., 1975, Histoire diagenetique d'une serie greseuse de mer du Nord. Datation de l'introduction des hydrocarbures: Revue de l'Institut Francais du Petrole, v. 30, p. 729–741.
Stalder, P. J., 1973, Influence of crystallographic habit and aggregate structure of authigenic clay minerals on sandstone permeability: Geologie en Mijnbouw, v. 52, p. 217–220.
Steiger R. H. and Jager, R. H., and Jager, E., 1977, Subcommission on geochronology: convention on the use of decay constant in geo- and cosmochronology: Earth and Planetary Science Letters, v. 36, p. 359–362.
Surdam, R. C., Crossey, L. J., Hagen, E. S., and Heasler, H. P., 1989, Organic-inorganic interactions and sandstone diagenesis: American Association of Petroleum Geologists Bulletin, v. 73, p. 1–32.
Vurgrinovich, R., 1989, Subsurface temperatures and surface heat flow in the Michigan Basin and their relationships to regional subsurface fluid movement: Marine and Petroleum Geology, v. 6, p. 60–70.
Watso, D. C., and Klein, G. deV., 1989, Origin of the Cambrian-Ordovician sedimentary cycles of Wisconsin using tectonic subsidence analysis: Geology, v. 17, p. 879–881.
Whitney, G., and Northrop, H. R., 1987, Diagenesis and fluid flow in the San Juan Basin, New Mexico − Regional zonation in the mineralogy and stable isotope composition of clay minerals in sandstones: American Journal of Science, v. 287, p. 353–382.

ILLITIC-CLAY FORMATION DURING EXPERIMENTAL DIAGENESIS OF ARKOSES

WUU-LIANG HUANG
Exxon Production Research Company, P.O. Box 2189, Houston, Texas 77252

ABSTRACT: Model arkoses containing K-feldspar (or Na-feldspar) + kaolinite + quartz (or silica glass or boehmite) were reacted in solutions of a variety of compositions in the KCl-NaCl-H$_2$O system, and in sea water at 200 to 350°C and 500–1,000 bars in rapid-quench, cold-seal pressure vessels. Experiments reveal that: (1) mixed-layer illite/smectite (I/S) with or without discrete illite are the major neoformed phases for all runs with low fluid/rock ratio; (2) the precipitation rate for neoformed clays is similar for near-neutral solutions of varying compositions, but the expandability of the I/S depends strongly on solution composition at the same temperature and pressure; (3) discrete illite appears only in solutions rich in K$^+$; (4) sea water dramatically retards the formation of illite layers in I/S; (5) illitic clay with fibrous habit grows more effectively in solutions with silica oversaturated with respect to quartz; illitic clays formed in initially alkaline and acidic solutions tend to be platy; and (6) illitic clays form in near-neutral solutions at much slower rates than those in alkaline solutions.

The experimental study verifies that the nonequilibrium mineral assemblage feldspar + kaolinite + quartz serves as a control for fluid composition, which favors precipitation of illitic clays in rock-dominant systems. The diagenetic products of a feldspar-bearing sandstone are determined primarily by the presence or absence of kaolinite and by the flow rate, and secondly by the initial fluid composition or temperature. The initial fluid composition becomes crucial only in the fluid-dominant system, whereas temperature is more important in controlling the kinetics than in shifting the thermodynamics of illitic-clay formation. The commonly observed illitization of kaolinite-bearing arkosic sandstones during burial diagenesis is attributed more to the reduction of the flow rate of the existing fluids than to the influx of new illitization fluids, or to the increase of temperature alone.

INTRODUCTION

The reservoir quality of sandstones relates closely to their diagenetic history (Wilson and Pittman, 1977; Galloway, 1979; Schmidt and McDonald, 1979). In particular, the dramatic reduction of permeability of arkosic-sandstone reservoirs has been attributed to the precipitation of authigenic fibrous (hairy) illite in pore spaces (Stalder, 1973; Sommer, 1978; Rossel, 1982; Seemann, 1982; Thomas, 1986). Our ability to predict and evaluate the quality of arkosic-sandstone reservoirs depends largely on understanding the major controls on pore-filling, fibrous illite. The objective of this study is to establish experimentally factors affecting authigenic-illite formation in arkosic or feldspathic sandstone, and to develop a model based on our experiments as well as on previous knowledge about the occurrence and geochemistry of authigenic illite.

Among the most common depth-related diagenetic sequences of arkosic sandstones reported are kaolinitization at shallow to moderate depths, and illitization of kaolinite at greater depth (Waugh, 1978; Hancock and Taylor, 1978; Seemann, 1979; Wilson, 1982; Bjørlykke and others, 1986; Dutta and Suttner, 1986). These observations suggest that illitization is controlled by some depth-related processes.

The geochemistry, particularly the ion-activity phase relations, of reservoir minerals have been used to predict the occurrence of authigenic clays (e.g., Hurst and Irwin, 1982). On the basis of calculated phase relations, Huang (1986) suggested that a nonequilibrium mineral assemblage, K-feldspar + kaolinite + quartz, may account for authigenic-illite formation during burial diagenesis. The three minerals together provide a chemical force to drive the fluid composition to a divariant stability field of illite over a wide range of reservoir conditions. The convergence of the activity of Si(OH)$_4$ in pore fluids toward the quartz-saturation level, and the counter driving of the activity ratio for K$^+$ and H$^+$ (i.e. a_{K^+}/a_{H^+}) into the stability field of illite by kaolinite and K-feldspar constitute the basis of the overall control of illite formation in arkosic sands. The importance of the reaction of forming illite from K-feldspar + kaolinite has also been suggested by Bjørlykke (1983) and Bjørkum and Gjelsvik (1988). Similarly, Rex (1965) suggested that the mineral assemblage K-feldspar + mica + kaolinite around the invariant point seems to approximate a geochemical convergence for the weathering of sandstone. The experimental results of Montoya and Hemley (1975) on the stability relations between K-feldspar and mica were used to construct activity-activity diagrams for the stability relations between illite and other reservoir minerals and served as the guide for the experiments in the present study.

METHODS

Starting Materials

The starting materials used in the experiments include eight three-phase mixtures containing various proportions of minerals (Tables 1 and 3) similar to those in arkosic sandstones. The mixture consisting of K-feldspar + kaolinite + quartz (model arkoses, mixture 1 and mixture 2 in Table 3) was used in most runs. The sources and characteristics of those starting minerals are listed in Table 1.

Solutions of various compositions were used in the experiments: distilled water; 0.01M KCl; 0.1M KCl; 1M KCl; 3M KCl; 0.48M NaCl; 1M NaCl; and sea water. As indicated in the experimental data in Table 3, some starting solutions were adjusted to specific pH values of 2 and 10 by using HCl or hydroxides; some solutions, which were prepared using water with dissolved silicic acid, contained approximately 145 or 215 ppm of Si in order to simulate subsurface silica-bearing water. The pH of distilled water used in some runs was 5.3 due to the dissolved CO$_2$ from air. Sea water used was Standard Sea Water (P99) from the Institute of Oceanographic Sciences, England, and contains 0.01M of KCl, 0.477M of NaCl and 0.00002M of Si with pH = 7.8.

TABLE 1.—SOLID STARTING MATERIALS USED IN EXPERIMENTS.

Minerals	Source	Characteristics
K-feldspar	National Bureau of Standards, No. 70a	powder < 10µ; NBS No. 70a contains 11.8 wt % K_2O, 2.5 wt % Na_2O, and .11 wt % CaO
Na-feldspar	British Chemical Standards, No. 375	powder < 10µ; BCS No. 375 contains 10.4 wt % Na_2O, .79 wt % K_2O, and .89 wt % CaO
Kaolinite	Synthesized at 300°C, 500 bars for 6 days	powder 1-2 µ
Kaolinite	Clay Minerals Society KGa-1	powder, surface area = 10.05 m^2/gm TiO_2 = 1.39% Fe_2O_3 = 0.13%
Boehmite	Synthesized at 250°C 500 bars for 5 to 7 days	
Quartz	Natural quartz crystal	powder 10 - 44 µ
Silica glass	Ultrapure silica glass kindly supplied by G. E. Lofgren, Johnson Space Center	powder < 10µ

Procedures

The hydrothermal experiments were carried out in rapid-quench, cold-seal pressure vessels using gold capsules as sample containers. Experiments were performed at temperatures ranging from 200°C to 350°C and pressures of 500 and 1,000 bars with run time of as many as 141 days. Detailed descriptions of the hydrothermal equipment and experimental procedures are presented in Huang and others (1986). The solution products of selected runs were analyzed for pH, Si, K and Na, whereas the solid products were routinely characterized by X-ray diffraction (XRD) and scanning electron microscope (SEM). The expandability of mixed-layer illite/smectite (hereinafter called I/S for illite-layer rich and S/I for smectitic-layer rich) was determined for the ethylene-glycolated samples using the method of Reynolds (1980). The rates of clay formation were semi-quantitatively estimated by comparing the XRD patterns of run products with those of standard mixtures. The uncertainty of the measurements of percent clay formed may be as high as ±15 weight percent.

TEM Characterization of Illitic Clays

Selected air-dried samples were examined with analytical electron microscope using SEM, transmission electron microscope (TEM), select area diffraction (SAD) and energy dispersive spectrum (EDS) modes. The equipment and analytical procedure are similar to those reported in Güven and Huang (in prep.). Chemical composition of the individual particles was determined by a KEVEX 8000 X-ray microanalyzer. A selected area of 0.5 μm^2 from a clay particle was scanned for 100 seconds at 100 kv potential. The net spectral intensities of an element were then converted to atomic ratios by the relation shown in Güven and Huang (in prep.). Table 3 summarizes the morphology, chemistry and polytype of the neoformed clays grown under a variety of experimental conditions.

Three morphologic textures of illitic clays are commonly found in the run products: (1) equal-dimensional or short lath-like platelets; (2) elongated lath-like platelets; and (3) thin crumbled films. The analytical data show that lath-like platelets are illite or I/S, whereas the crumbled films are

TABLE 2.—MORPHOLOGY, CHEMISTRY AND STRUCTURE FOR ILLITIC CLAYS IN RUN PRODUCTS.

Run no.*	Figure No.	TEM Morphology	Al/Si	K/Si	Na/Si	a (Å)	b (Å)	Polytypes	Claytypes
214	1a	long lath				5.17	8.93	1M	illite
214	1b	parallel platelets	0.39	0.16					I/S
288	1c	filaments	0.51	0.12					I/S
288	3d	crumpled films	0.56	0.10					I/S
221	1d	long laths	0.37			5.14	9.02	1M	pyrophyllite
221		equant lamella	0.69					$2M_1$	pyrophyllite
221		platelets	0.88	0.32	0.01				illite
221	3c	crumpled films	0.33	0.02	0.01				I/S
213	2a	platelet	0.65	0.19		5.19	9	$2M_1$	illite
213		thin platelet				5.21	8.92	1M	illite
215	2b	platelet				5.19	8.93	1M	illite
215	3b	crumpled films	0.6	0.15 - 0.20					I/S
239	2c	platelets	0.94	0.43				2M	illite (k-rich)
301	3a	crumpled films	0.32					1M	smectite
300		crumpled films	0.38-0.41						smectite

Note: * Run 239 used kaolinite and KOH solution at 200°C/6days.
Run 288 used synthetic illite glass and distilled water at 300°C/35 days.
Starting materials, solutions and experimental conditions of other runs in Table 3.
** with uncertainty ±0.001.

smectite or S/I. Similar observations have been found in illitic clays grown from a synthetic glass of illite composition (Güven and Huang, in prep.) as well as from samples from hydrothermal environments (Inoue and others, 1987). The elongated lath-like illitic platelets are similar to the authigenic fibrous illite typically found in arkosic sandstones (McHardy and others, 1982). Examples of these three types of illitic clays found in run products are shown, respectively, in Figures 1 to 3.

Elongated illitic laths were commonly grown in neutral solutions with a high concentration of dissolved silica. Examples are found in runs containing amorphous silica (glass). Figures 1A and 1B show the TEM micrographs of illitic laths that were grown from a mixture of K-feldspar + kaolinite + amorphous silica (glass) in KCl solution (run 214 in Table 3). The laths range in width from 0.1 to 0.2 μm and in length from 0.5 to 2.0 μm. The individual laths typically develop radially from a dense core with equal-dimensional illite platelets and silica spheres (Fig. 1A). The radiating growth texture has also been found for authigenic illite in sandstones (Güven and others, 1980; Glasmann and others, 1989). The unit cell dimensions and pattern obtained from SAD for these illitic laths indicate a dioctahedral-layer mica with 1M stacking sequence. The clay laths are sometimes arranged in a parallel set as shown in Figure 1B. The EDS analysis shows that these laths have a composition similar to illite but with significant potassium deficiency (Table 2). This K-deficiency suggests the presence of some smectite layers in the illitic clay.

Acicular- or lath-like illites have also been grown from synthetic glass of illite composition (Fig. 1C; run 288 in Table 3). The laths may form reticulated aggregates, which intersect at an angle of 60° (or 120°). The SAD pattern obtained from the lattice arrangement of illite laths shows six inner spots having an intensity distribution that is characteristic of 2M$_1$ mica polymorph. Similar lattice-like arrangements of illite laths were found in authigenic illite from a sandstone reservoir (Güven and others, 1980). EDS analysis of the laths shows that the Al/Si atomic ratio is close to that expected for an illite, but the K/Si ratio is signifi-

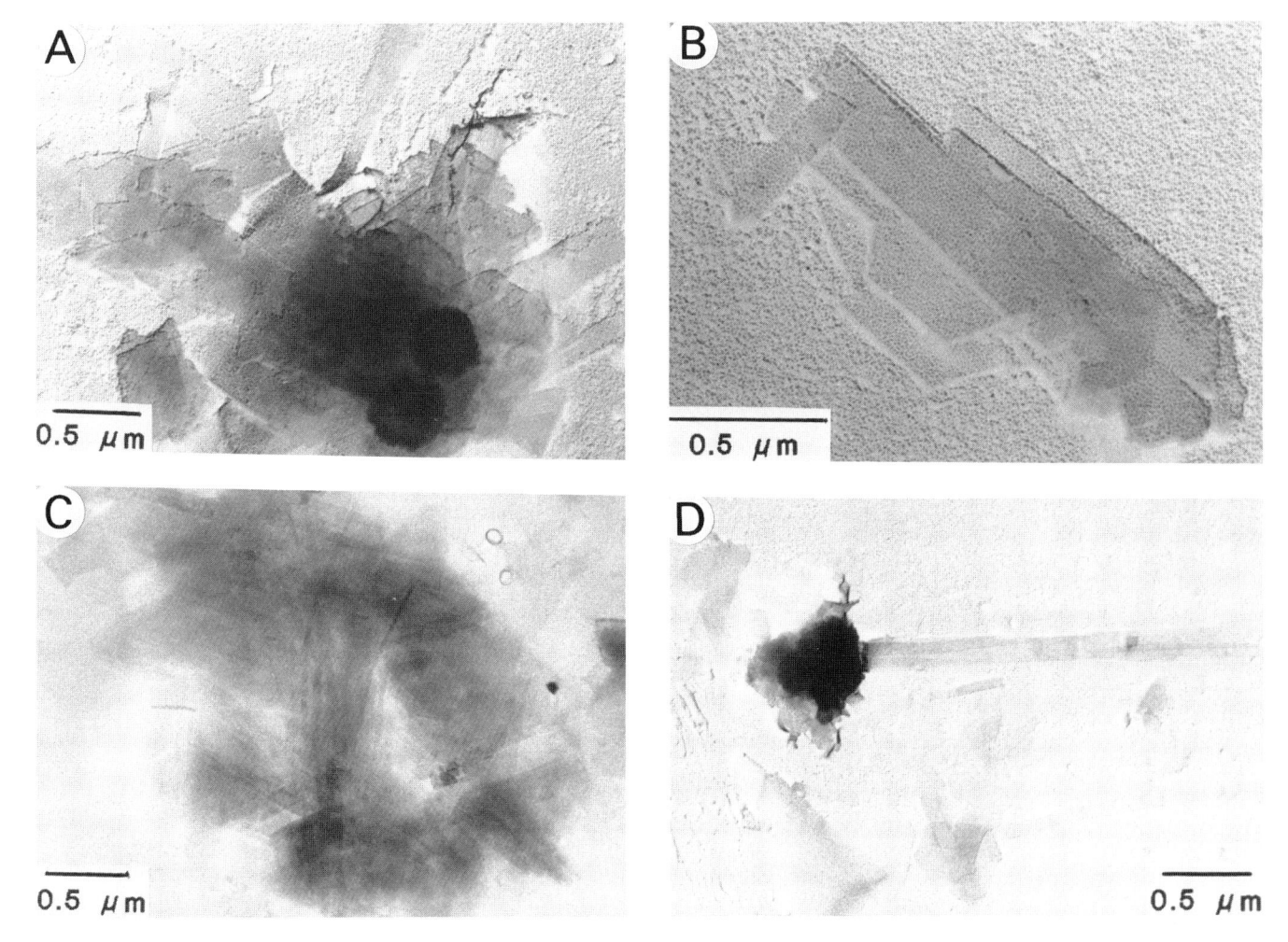

FIG. 1.—TEM photomicrographs showing laths of neoformed clays in run products (for detailed description see text). (A) Illite laths developed radially from a dense core of illite platelets (grown from K-feldspar + kaolinite + silica in KCl solution, run 214). (B) Illite laths arranged in a parallel set (run 214). (C) Illitic filaments arranged in a triangular set (grown from a synthetic glass with illite composition in water, run 288). (D) Pyrophyllite laths developed radially from a dense core (grown from K-feldspar + kaolinite + silica in KCl solution, run 221).

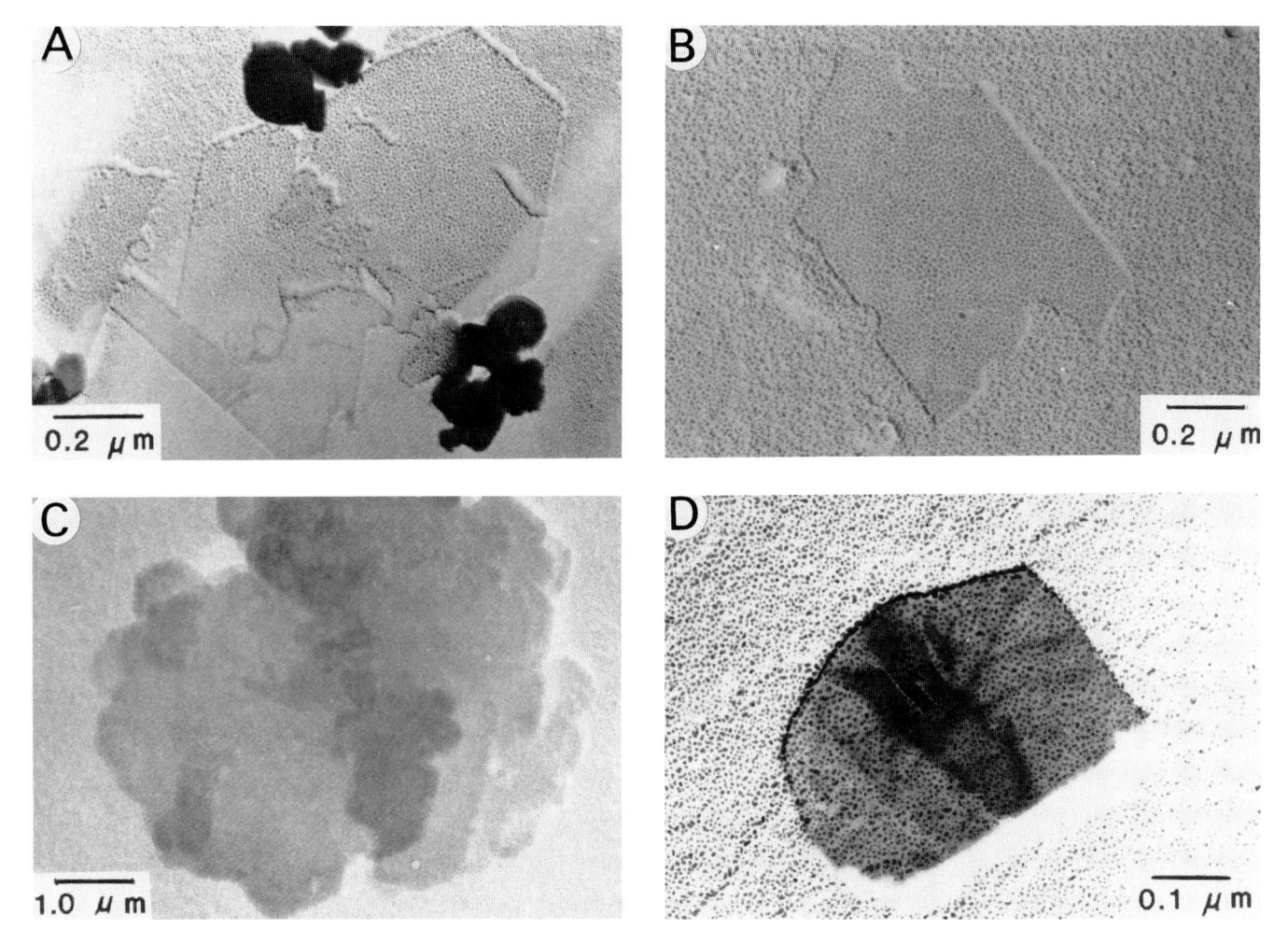

Fig. 2.—TEM photomicrographs showing equant platelets of neoformed illitic clays in run products. (A) Pseudohexagonal illite platelets with minor amounts of lath-like illite (grown from K-feldspar + kaolinite + boehmite in KCl solution, run 213); EDS analysis of the hexagonal platelets suggests an illite composition with some potassium deficiency (Table 2); SAD pattern obtained from the illite platelets indicates a $2M_1$ stacking sequence of a dioctahedral mica; however, some extremely thin films of similar illitic platelets from the same experiment show a 1M stacking sequence, suggesting that two polytypes of illite can coexist under the same chemical and physical conditions. (B) Illite platelet (grown from albite + kaolinite + boehmite in KCl solution); EDS analysis and the SAD pattern indicate a dioctahedral illite with 1M stacking sequence. (C) Illite platelets with hexagonal shape (grown from kaolinite in KOH solution); crystal mosaic is composed of minute pseudohexagonal platelets 0.1-μm size; EDS analysis indicates an illite composition with rather high K content (run 239 in Table 2); a spot SAD pattern similar to that for a single crystal indicates a perfect crystallographic arrangement of these minute platelets. (D) Illite platelet with anhedral shape (grown from albite in KCl + HCl solution with initial pH = 2 at 200°C, Huang and others, 1986); the EDS, SAD, and XRD of the sample indicates an illite with 1M stacking and rather high K content.

cantly lower than that of illite. These illitic clays exhibit some K-deficiency, which is also found in some authigenic fibrous illite in sandstones.

Figure 1D shows the TEM micrograph of elongated clay laths grown from a mixture of K-feldspar + kaolinite + silica glass in KCl solution (run 221 in Table 3). In addition to the elongated clay laths, well-developed platelets of illite and leaf-like smectite-rich clay have also been found in the same run product. The elongated clay laths range in width from 0.1 to 0.5 μm and in length from 0.5 to 5 μm. The radiating growth habit has also been found for the clay laths (Fig. 1D). EDS analysis of an individual lath (Al/Si = 0.37 and no detectable K, Table 2) is close to pyrophyllite. The SAD obtained from the individual long lath indicates a 1M stacking sequence for a dioctahedral-type layer. The clay laths also often reticulate by intersecting at 120°. The SAD obtained for the crossing laths, however, shows a pattern expected from a 2M stacking. It appears that the 2M stacking sequence develops as these laths reticulate at 120° (N. Güven, pers. commun., 1983). However, it is not clear whether the reticulation is by physical aggregation or by growth. These reticulated laths have an Al/Si intensity ratio (0.44) as expected for pyrophyllite.

Equant illitic platelets (Fig. 2) are grown typically from runs in the neutral solutions with a low concentration of dissolved silica, such as those using boehmite instead of quartz as a starting material, or from runs with the initial alkaline or acidic solutions. Figure 2 shows examples of

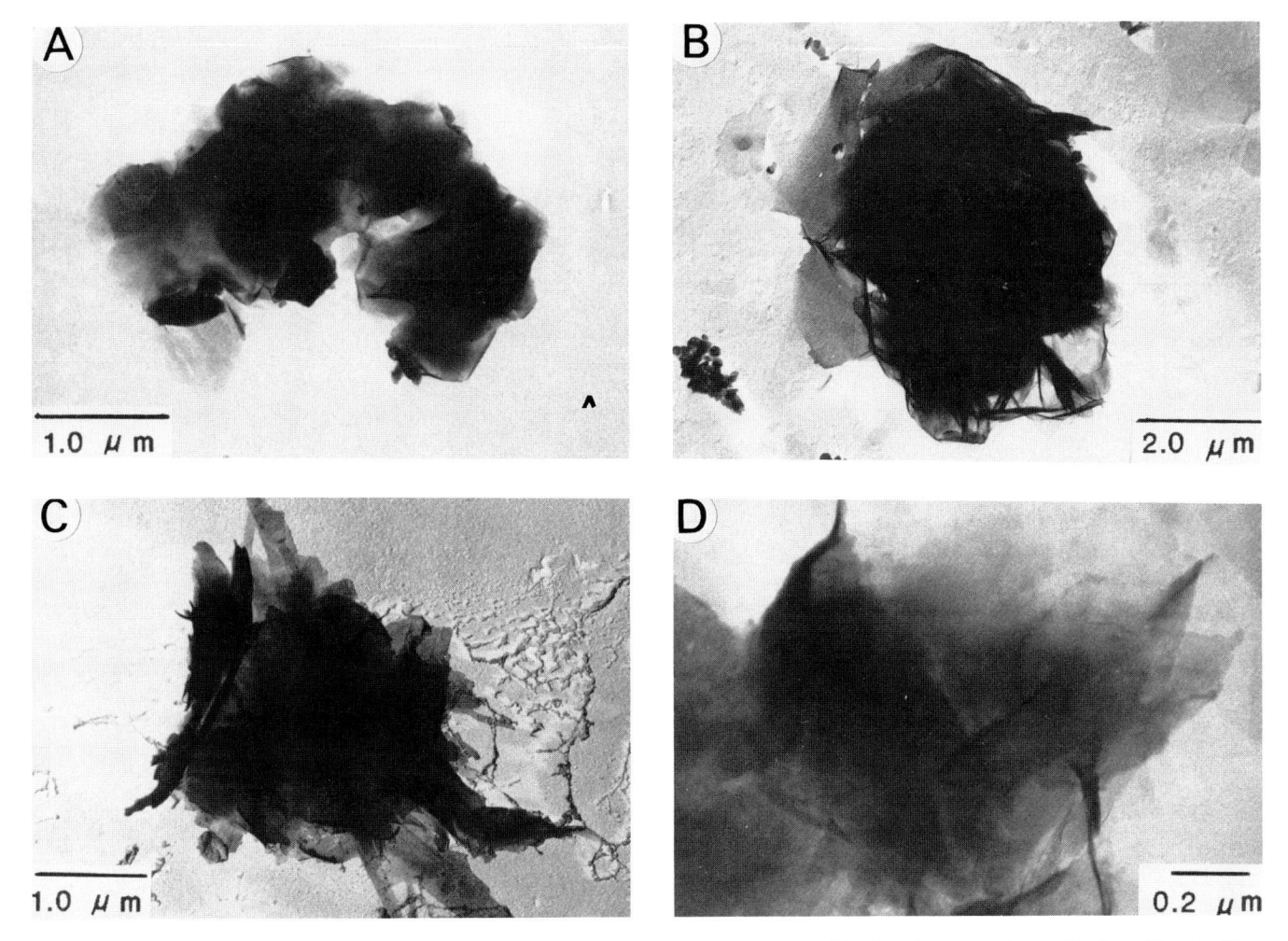

FIG. 3.—TEM photomicrographs showing the crumpled films of neoformed illitic or smectitic clays in run products: (A) Smectitic-clay films (grown from K-feldspar + kaolinite + quartz in NaCl solution, run 301); EDS analysis of the clay particles shows Al/Si atomic ratio (0.32) close to smectite with negligible K contents, suggesting smectite. (B) Smectite-rich S/I clay films with thin illitic platelets (grown from albite + kaolinite + quartz in KCl solution, run 215); the aggregates consist of crumpled thin films with rhombic platelets that appear to grow at the edges of the aggregates; EDS analysis of the aggregates reveals an average composition similar to a mixed-layer smectite/illite; the crumpled thin films may be smectite or smectite-rich S/I, whereas the rhombic platelets are more illite-rich. (C) Smectite-rich S/I thin films (with a large majority of illite and pyrophyllite grown from K-feldspar + kaolinite + silica in KCl solution); elongated crumpled thin films constitute only a small part of the neoformed clays in the run product, which are mainly illite and pyrophyllite (Fig. 1D); EDS data for Al/Si and K/Si ratios in these crumpled films are very similar to those observed from Na-smectite (in Wyoming bentonite). A small amount of potassium was also detected, suggesting some illitic layers in the smectite-rich clay or K-exchange ions. (D) S/I thin crumpled films (grown from a synthetic glass with illite composition in water, run 288). EDS analysis shows that the Al/Si intensity ratio of the aggregates is much higher than that for a pure smectite, whereas the K/Si ratio is lower than that for a pure illite. This suggests that the morphologically smectite-like clay is a mixed-layer smectite/illite.

equant illite platelets grown from a mixture of albite + kaolinite + boehmite in a KCl solution (Fig. 2B; run 215), from kaolinite in alkaline solutions (Fig. 2C; run 239 in Table 3; Huang and Otten, 1985), and from albite in an initial acidic solution (Huang and others, 1986). The detailed descriptions of these platy illites are shown in the captions of Figure 2.

Crumpled films of clays are mostly found in, but not restricted to, runs with a Na-rich solution. Figure 3 shows examples of thin crumpled films of clays formed from a variety of starting materials: a mixture of K-feldspar + kaolinite + quartz in 1M NaCl solution (run 301 in Table 3), a mixture of albite + kaolinite + boehmite in a KCl solution (run 215, Table 3), a mixture of K-feldspar + kaolinite + silica in KCl solution (run 221 in Table 3) and a synthetic glass with illite composition in distilled water (run 288 in Table 3). The characteristics of these clays are described in the captions of Figure 3.

EXPERIMENTAL RESULTS AND DISCUSSION

The experiments aim to: (1) confirm the role of the non-equilibrium mineral assemblage, feldspar + kaolinite + quartz, in controlling fluid composition for illite formation; (2) characterize the neoformed clays; (3) define the optimum conditions for the growth of fibrous illite; and 4) es-

TABLE 3.—SUMMARY OF EXPERIMENTAL DATA.

Run No.	Press. Bars	Time Days	Starting Solids	pH	Starting Si mM	Solutions Salt	F/R Ratio μl/mg	Quench pH	Si mM	Solutions K mM	Na mM	Neoformed Clay Type by XRD	% I in I/S & Ordering	Development of fibrous habit
200°C														
300	500	85	mx 2	2.0	5.16	1M NaCl	5	4.11	--	29.2	854	S (tr.)	--	--
301	500	85	mx 2	10.0	5.16	1M NaCl	5	4.73	--	8.3	915	I/S (tr.)	--	--
298	500	85	mx 4	2.0	5.16	1M KCl	5	4.38	--	847	10.0	I (tr.)	--	--
299	500	85	mx 4	10.0	5.16	1M KCl	5	4.53	--	803	8.4	I (tr.)	--	--
275°C														
212	500	32	mx 3	5.3	7.66	0.01M KCl	10	--	--	--	--	I/S	--	--
213	500	32	mx 3	9.8	7.66	0.01M KCl*	10	5.36	--	--	--	I/S	--	--
210	500	32	mx 5	5.3	7.66	0.01M KCL	10	7.12	--	--	--	I/S	--	--
211	500	32	mx 5	5.1	7.66	1M KCl	10	--	--	--	--	I/S + I	--	--
300°C														
293	1000	15	mx 2	2.0	5.16	1M NaCl	5	4.15	capsule	Leaking	237	I/S	20 R0	--
294	1000	15	mx 2	5.2	5.16	1M NaCl	5	4.27	--	8.6	238	I/S	30 R0	--
295	1000	25	mx 2	10.0	5.16	1M NaCl	5	5.2	--	9.1	--	I/S	20 R0	--
226	1000	45	mx 1	5.2	7.66	0.1M KCl	10	4.47	16.5	--	--	I/S + I	87 R0	poor
227	1000	45	mx 1	5.2	7.66	0.1M KCl	2	3.01	--	--	--	I/S + I	70 R1	poor
228	1000	45	mx 1	5.1	7.66	1M KCl	2	2.49	33.8	--	--	I/S + I	85 R?	poor
217	1000	45	mx 6	5.2	7.66	1M KCl	10	3.29	25.2	--	--	I/S (tr.)	--	fair
218	1000	45	mx 6	5.2	7.66	0.1M KCl	10	8.13	--	--	--	Kf overgrowth	--	--
219	1000	45	mx 6	8.2	7.66	1M KHCO3	10	5.63	--	--	--	I/S	50 R0	--
317	500	18	mx 2	5.5	0	Water	1	2.02	--	--	--	I/S + I (?)	90 R3	--
318	500	18	mx 2	5.7	0	3M KCl	1	2.46	--	49	426	I/S + I (?)	<10 R0	--
308	500	35	mx 2	2.0	0	Sea Water	5	2.65	--	46	421	I/S + I (?)	<10 R0	--
307	500	35	mx 2	8.0	0	Sea Water	5	2.61	--	--	--	I/S + I (tr.)	<10 R0	--
306	500	35	mx 2	10.0	0	Sea Water	5	4.11	--	28	511	I/S + I	30 R0	--
324	500	45	mx 2	8.1	0	0.48M NaCl	5	3.15	--	53	446	I/S + I	<10 R0	--
325	500	45	mx 2	8.1	7.66	Sea Water	2	--	--	--	--	I/S	50 R1	--
313	500	61	mx 2	6.6	7.66	Water	2	--	--	--	--	I/S	60 R1	--
312	500	61	mx 2	6.7	7.66	0.1M KCl	2	--	--	--	--	I/S + I	--	TEM scale
288	500	35	mx 7	5.3	0	Water	1	--	--	--	--			
350°C														
254	1000	45	mx 1	5.3	0	Water	2	--	--	--	--	I/S + I (?)	--	fair
225	1000	45	mx 1	5.2	7.66	0.1M KCl	2	--	--	--	--	I/S + I	90 R2	fair
215	1000	45	mx 5	5.2	7.66	0.1M KCl	10	3.42	--	21.3	91	I/S + I	75 R1	--
222	1000	45	mx 5	8.2	7.66	0.1M KHCO3	10	7.85	--	--	--	I/S (tr.)	--	fair
214	1000	45	mx 6	5.1	7.66	1M KCl	10	2.19	55.9	--	--	I/S + I	--	good
216	1000	45	mx 6	8.2	7.66	1M KHCO3	10	8.15	--	--	--	Kf overgrowth	--	good
255	1000	45	mx 6	5.3	7.66	Water	10	--	--	--	--	I/S (tr.)	--	good
256	1000	45	mx 6	5.3	7.66	0.1M KCl	2	2.36	--	--	--	I + I/S (tr.)	--	good
223	1000	45	mx 6	5.1	7.66	1M KCl	2	3.72	--	--	--	I/S + Py (tr.)	--	good
221	1000	45	mx 6	5.2	7.66	0.1M KCl	2	5.21	--	--	--	I/S + Py (tr.)	80 R1	fair
425	1000	141	mx 8		7.66	0.1M KCl	2					I/S		

Abbreviations see Table 1: Additional: S = smectite, I = discrete illite, I/S = mixed-layer smectite/illite, (tr.) = trace amounts, py = pyrophyllite, ? = possible, -- indicates no data.

Starting solid mixtures used in the experiments (wt.%).

MX1 = Ki (70%) + Ka (10%) + Qz (20%)
MX2 = Kf (40%) + Ka (37%) + Qz (23%)
MX3 = Kf (30%) + Ka (50%) + B (20%)
MX4 = Nf (40%) + Ka (37%) + Qz (22%)
MX5 + Nf (30%) + Ka (50%) + B (20%)
MX6 = Kf (70%) + Ka (10%) + Silica glass (20%)
MX7 = illite glass
MX8 = Kf (40%) + Kga (40%) + Qz (20%)
Kga is KGA-1 reference kaolinite

timate the precipitation rate of neoformed illitic clays. The temperatures of this study were chosen to be higher than those in diagenetic environments in order to have significant reactions within reasonable experimental time. The experimental results are listed in Table 3.

Role of K-feldspar + Kaolinite + Quartz in Illite Formation

Model arkoses containing various proportions of K-feldspar (or Na-feldspar), kaolinite, and quartz were reacted with distilled water, KCl or NaCl solutions and sea water with a pH of 2, 5.3 and 10, representing acidic, near-neutral, and alkaline pore fluids. The fluid/rock ratios used are less than 10 for simulating rock-dominant conditions. The final solutions may not be in equilibrium because of the presence of metastable relict phases (either kaolinite or feldspar) in the run products. However, the solution data provide information about the evolution pathway of solutions.

The activities of K^+, Na^+ and H^+ in starting and final solutions at run conditions were calculated based on the fluid-chemistry analyses of the quench solutions using the Gt geochemical program (C. Bethke, University of Illinois, Urbana, Illinois). The program uses equilibrium thermodynamics to calculate the distribution of an aqueous species in water that is in equilibrium with a solid. Figures 4 and 5 show the experimental data for initial (open symbols), the final solution compositions (solid symbols) and the possible evolution pathways (dashed lines with arrows) on the calculated reference phase diagrams from the Gt geochemical program. The diagrams show the stability fields of mus-

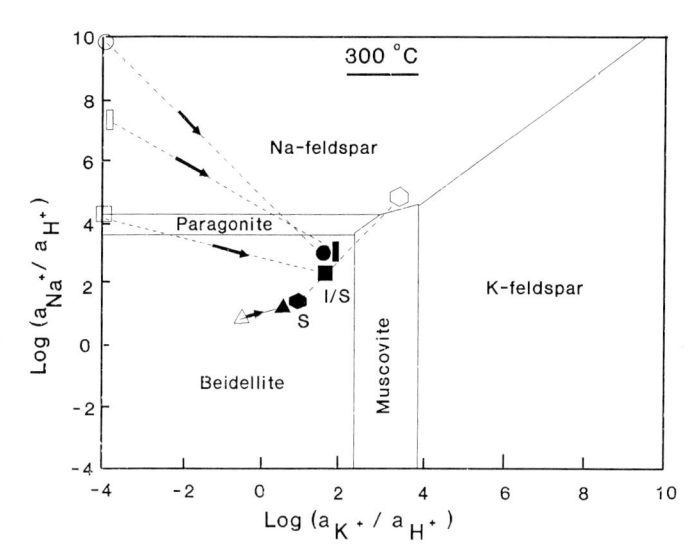

FIG. 5.—Experimental results similar to those shown in Figure 4 but at 300°C. Symbols: circle (in 1M NaCl, initial pH = 10, used K-feldspar, run 295), rectangle (in 0.48M NaCl, initial pH = 8.1, used K-feldspar, run 324), square (1M NaCl, initial pH = 5.2, used K-feldspar, run 294), triangle (in sea water, initial pH = 2, used K-feldspar, run 308) and hexagon (in sea water, initial pH = 8, used K-feldspar, run 307). Note that the solid products are mixed-layer illite/smectite (I/S), with or without discrete illite, and smectite (S), with trace of illite layers. The reference diagram was calculated for solution saturated with quartz.

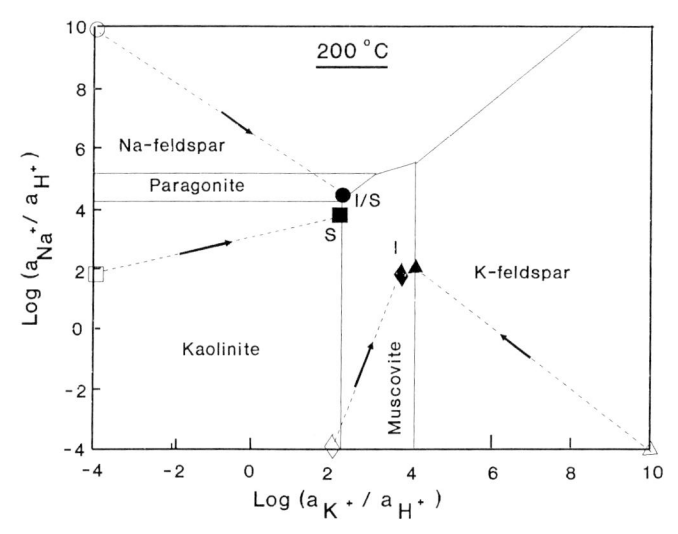

FIG. 4.—Experimental results showing the evolution of fluid composition when a mineral assemblage of feldspar + kaolinite + quartz was reacted with a variety of fluids at 200°C and 500 bars at fluid/rock ratio of 10 or less. The open and solid symbols indicate, respectively, the initial and final fluid compositions, whereas the dashed lines with arrow indicate the directions of the fluid composition change. Symbols are circle (in 1M NaCl, initial pH = 10, used K-feldspar, run 301), square (in 1M NaCl, initial pH = 2, used K-feldspar, run 300), diamond (in 1M KCl, initial pH = 2, used Na-feldspar, run 298), triangle (in 1M KCl, initial pH = 10, used Na-feldspar, run 299). Note that solid products are illite (I) or mixed-layer illite/smectite (I/S). The reference activity diagram was calculated for solutions saturated with respect to quartz.

covite (illite) and related minerals in term of activity ratios of Na^+/H^+ and K^+/H^+ of solutions that are saturated with quartz. Our solution data, when plotted on the diagrams, show that all initial solutions evolve toward specific fields on the diagrams.

Experimental data at 200°C show that the reactions of Na-feldspar + kaolinite + quartz with KCl solutions precipitate illite with final solution compositions near the calculated muscovite/K-feldspar boundary (Fig. 4), whereas the reaction of K-feldspar + kaolinite + quartz with NaCl solutions precipitates S/I or smectite with final solution compositions near the calculated paragonite/muscovite boundary (Fig. 4). Experimental data at 300°C show that the reaction of K-feldspar + kaolinite + quartz precipitates I/S and discrete illite in NaCl and KCl solutions and precipitates I/S (about 50% I) in distilled water. In both cases, the final solution compositions approach the calculated paragonite/muscovite/beidellite invariant point (Fig. 5). In contrast, the reaction precipitates smectite with less than 10% of illite layers and with or without a trace amount of discrete illite in sea water, and the final solution composition is in the calculated beidellite stability field (Fig. 5).

The analyses of both solutions and solid products consistently show that at low fluid/rock ratios (mass ratio = 10 or less), mixed-layer illite/smectite (I/S or S/I) with or without discrete illite or smectite is the dominant run product in spite of the variety of fluid compositions used (Table 3). This verifies the proposed concept that the arkosic-mineral assemblage feldspar + kaolinite + quartz can always lead to the formation of illitic clays over a temperature range of 200°C to 350°C, and at 500 or 1,000 bars. The experiments additionally confirm the presence of smectite layers in the illitic clays, even in the Na-depleted system, and that Na-feldspar (oligoclase) behaves similarly to K-feldspar for I/S formation as long as the solution contains K.

The confirmation has also been extended to lower temperatures, where the reaction rates are too slow to be studied in the laboratory, by using the EQ 3/6 geochemical model of Wolery (1979). In the modeling, a mineral assemblage containing K-feldspar, kaolinite and quartz (with different K-feldspar/kaolinite ratio) was added stepwise into a fixed volume of pore fluid and allowed to equilibrate with the fluid before proceeding with the next step. In each step, the rock added contains a constant mineral proportion that is the same as that in the initial assemblage. We are particularly interested in the cumulative amount of rock added when muscovite (illite) first appears (i.e., fluid begins to saturate with respect to muscovite). The fluid-to-rock mass ratio at this moment was defined as the fluid/rock ratio plotted in Figures 6 and 7. This is the maximum fluid/rock ratio for illite formation. This ratio depends on the temperature as well as the K-feldspar/kaolinite ratio in the starting mineral assemblage. With a fluid/rock ratio higher than the maximum value, no illite forms because the fluid chemistry does not effectively move into the illite field.

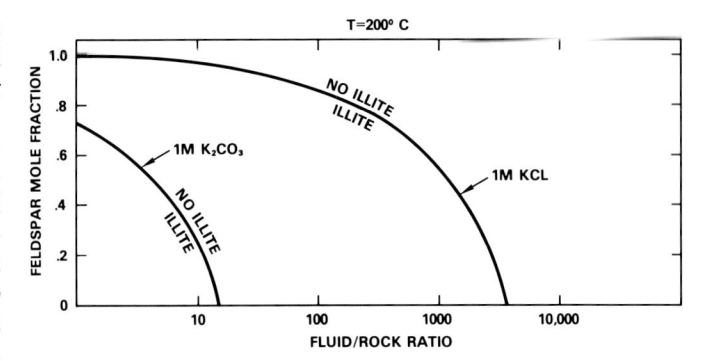

FIG. 7.—EQ3/6 modeling results showing that CO_3^{2-}-bearing fluid significantly reduces the illite field compared to the Cl^{-1}-bearing fluid. The maximum fluid/rock ratio required for illite formation depends on mole fraction of feldspar (relative to kaolinite) as well as carbonate-ion concentration.

Figure 6 shows an example of the calculated relations showing that illitization could occur at reservoir temperatures due to the decrease of the fluid/rock ratio during burial diagenesis.

Experiments were also conducted in $KHCO_3$ solutions in order to examine the effect of bicarbonate ions on illite formation (Table 3). In KCl solution (runs 217 and 214), the solid mixture, K-feldspar + kaolinite + quartz, reacts to form I/S and the final pH of solutions drifts to around 2.5. In contrast, in $KHCO_3$ solutions (runs 219 and 216), a similar reaction precipitates K-feldspar and the final pH buffered by the bicarbonate ions is around 8. This suggests that the bicarbonate buffer more effectively controls the activity ratio (a_{K+}/a_{H+}) of the fluid than the mineral assemblage in the experiments. However, EQ 3/6 modeling shows that as the fluid/rock ratio decreases to some level, the driving force of the mineral assemblage may surmount the bicarbonate buffer and lead to the precipitation of illite. Figure 7 shows that the stability field of illite (in terms of fluid/rock ratio and mole fraction of feldspar relative to kaolinite) significantly decreases if KCl fluid is replaced by K_2CO_3 fluid.

In some experiments, we used boehmite or silica glass instead of quartz in the starting solid mixtures to simulate fluid conditions with Si concentration undersaturated or supersaturated with respect to quartz. In both cases I/S was also a major neoformed phase; however, the amounts of I/S clays formed in a glass-bearing system are less than the amounts formed using quartz. In silica glass experiments, pyrophyllite or K-feldspar sometimes appears in the products (runs 221 and 223), suggesting that the solution composition may shift to the boundary between I/S and pyrophyllite (or feldspar) fields.

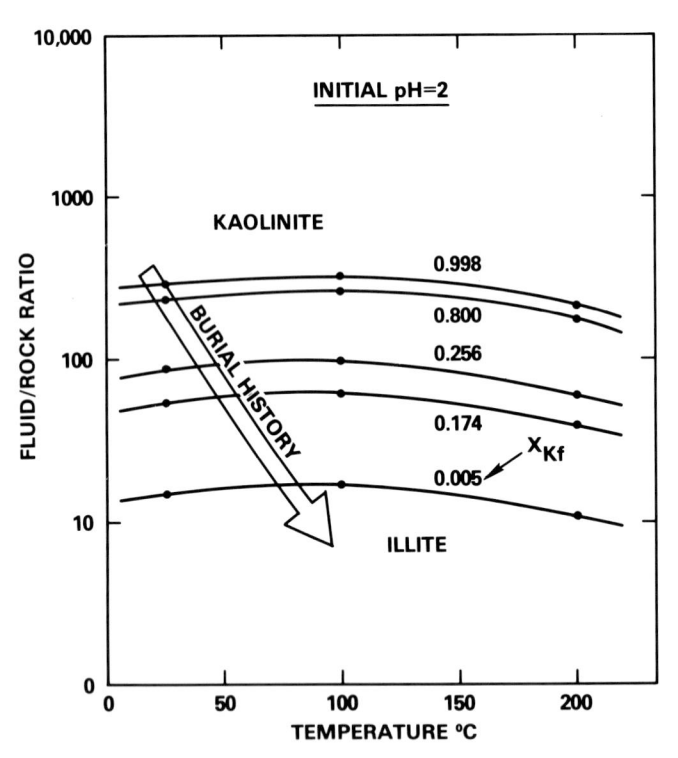

FIG. 6.—EQ3/6 modeling results showing the maximum fluid/rock ratio for illite formation as a function of temperatures and Kf/(Kf + Ka), the mole fraction of K-feldspar (relative to kaolinite), at an initial pH of 2 in 1M KCl fluids. The solutions evolve to near-neutral as the reaction proceeds. Diagenetically, the Kf/(Kf + Ka) ratio reflects the relative decomposition rate of K-feldspar and kaolinite. The arrows indicate the general trend during burial diagenesis: increase of temperatures and decrease of flow rate with increasing depth.

Expandabilities of Neoformed Illitic Clays

XRD analyses indicate that solid products from kaolinite + K-feldspar + quartz are mainly mixed-layer illite/smectite (I/S), with or without discrete illite, in Na-rich as well as Na-depleted systems. This observation is consistent with the thermodynamic calculation (Helgeson and others, 1969;

Garrels, 1984; Gt program) as well as the experimental observations (Sass and others, 1987) that suggest that smectite may be a stable phase in addition to illite or muscovite in the K_2O-Al_2O_3-SiO_2 system. Furthermore, the observations show that the expandability of I/S remains essentially constant with increasing time at the same temperatures. This is consistent with the recent field study of Meunier and Velde (1989), who showed that at constant temperature, pressure and bulk composition, I/S with a constant I/S ratio can coexist with discrete illite. In this study, all the expandable layers detected by XRD are reported as smectite layers. However, some of the smectite (expandable layers) in I/S detected by XRD in the run products may also result from interparticle diffraction (Nadeau and others, 1984). It is likely that the clays showing less than 50% expandability may actually consist of only illite. The distinction between these two types of expandable layer requires further TEM efforts.

The experimental results show that the amounts of illite layers in I/S precipitated in K-bearing fluids are proportional to the K concentration of initial solutions. For instance, runs 318, 312 and 313 with 3M, 0.1M KCl and distilled water, respectively, yield I/S with 90, 70 and 50% illite (Table 3). The observation that more illitic layers form from K^+-bearing solutions than from K-feldspar in distilled water suggests that the dissolution rate of K-feldspar, rather than illite precipitation, limits the rate for illite formation. In contrast, Na^+-bearing fluids retard illite but favor expandable-layer formation in I/S from kaolinite + K-feldspar + quartz. Experiments (runs 293, 294 and 324) in initial NaCl solutions yield only 20 to 30% illite in contrast to 50 to 90% illite in I/S in KCl-bearing fluids. The results also show that sea water more severely retards formation of illite layers in I/S from kaolinite + K-feldspar + quartz than a NaCl-bearing solution. An experiment (run 324) with 0.48M NaCl solution yields I/S with 30% illite layers, whereas similar experiments with sea water containing the same amount of NaCl produce less than 10% illite layers in smectite. The different amounts of illite formed in distilled water and sea water may be attributed to the degree of deviation of initial-solution composition from the K-feldspar/illite equilibrium, as suggested by Hutcheon (pers. commun., 1990). His calculations using a solubility speciation computer code indicates that the a_{K+}/a_{H+} ratio of sea water at 300°C is very close to the illite/K-feldspar-equilibrium boundary. Therefore, little illite would have to form to establish equilibrium between K-feldspar and illite. The fewer illite layers in the S/I precipitated from sea water may be also contributed to the presence of other cations, such as Na and Mg, which favor smectite formation (Roberson and Lahann, 1981).

Our results also show that the accompanying discrete illite formed only in those runs with KCl in the initial solutions or in sea water (containing 0.01M KCl). No discrete illite was detected by XRD in those runs without K^+ in the initial solutions, such as distilled-water or NaCl solutions, although the systems contain K in K-feldspar. This suggests that in order to form discrete illite, a minimum amount of potassium (e.g., 0.01M KCl in sea water) in the initial solutions is required.

Conditions for Fibrous-Illite Formation

The SEM characterization of illitic clays formed from model arkoses reveals that neoformed clay appears as a "cellular" aggregate of crumpled flakes with small amounts of fibers protruding from the edges of the clay flakes (Figs. 8A and 8D) and that the longer the run time, the more the illite fibers grow. This texture is similar to the authigenic illite or fibrous I/S commonly observed in arkosic sandstones (e.g., Pollastro, 1985; Keller and others, 1986). The presence of a smectite component in the clay may contribute to the formation of the "cellular" texture of the illite.

The observed growth texture suggests that the illitic-clay fibers were formed later in the course of the experiment than the clay platelets. The change of clay morphology is probably due to the drifting of solution chemistry during the course of the experiments toward a condition that favors the formation of a fibrous texture. We have searched for factors that may influence the growth of illitic-clay fibers by changing experimental parameters, such as the fluid/rock ratio and the concentration of H^+, K^+ or $Si(OH)_4$ in the solution. In the experiments where silica glass was used to enrich Si in solutions (Table 3), we were able to grow a much greater quantity of clay fibers (Figs. 1A, 8A and 9). XRD indicates that these fibers are mainly illite or illite/smectite; the TEM confirms the presence of illite laths (Fig. 1A). However, TEM analysis of one of the experimental products (run 221) shows that some of the clay laths contain predominantly Al and Si with only negligible amounts of K or other cations, an indication of the presence of pyrophyllite (Fig. 9B). The fibrous habits of the neoformed pyrophyllite may be inherited from fibrous illite by pseudomorphous substitution.

These experimental observations suggest that illitic-clay platelets and fibers can be grown from the model arkosic sandstone in near-neutral solutions, in contrast to the mostly platy illite grown rapidly from kaolinite in alkaline solution (Fig. 2C) and from feldspar in acidic solution (Fig. 2D). Illite laths, which were identified based on TEM, were also reported to have grown from synthetic illite gels in distilled water (Velde and Weir, 1979; Güven and Huang, in prep.) and from albite in neutral KCl solutions (Güven and others, 1982). These synthesis studies suggest that neutral solutions favor the formation of illite laths or fibers. Our results also show that the growth of illite fibers is more efficient in pore fluids with Si concentrations oversaturated with respect to quartz. The close association of fibrous illite with quartz overgrowth in arkosic sandstones (Glasmann and others, 1989) also indicates a high-silica concentration in the pore fluids during the growth of fibrous illite. This conclusion does not preclude the possibility that there could be other factors, as yet unidentified, that may significantly enhance the formation of illite fibers.

Recently, Chermak and Rimstidt (1990) reported the growth of hairy illite from kaolinite in KCl solution. The results are consistent with our conclusion that illite grows in near-neutral conditions. However, illite grown from their experiments manifests much more fibrous texture and a larger crystal size than those grown in the present study under

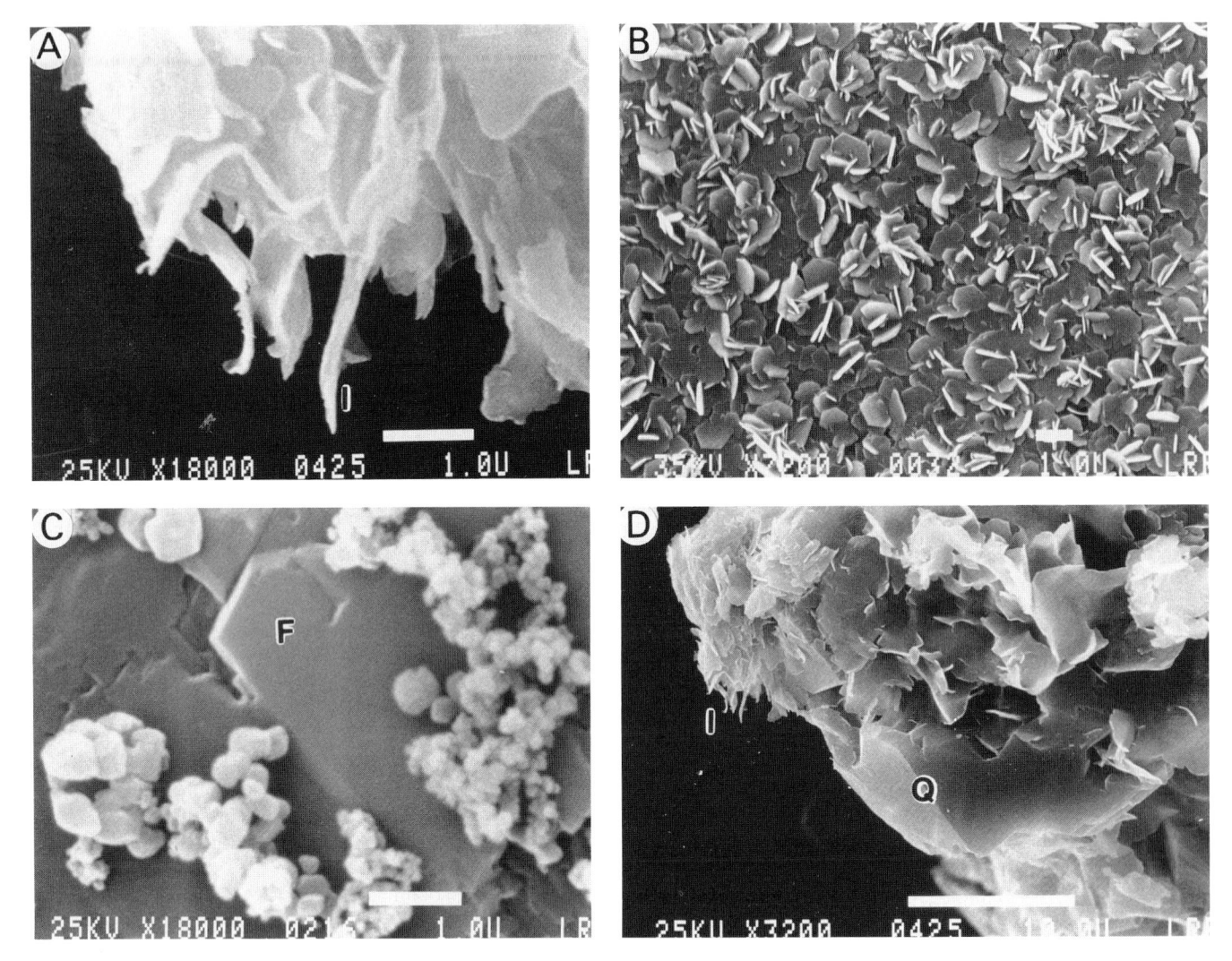

Fig. 8.—SEM micrographs of reservoir minerals experimentally grown from model arkosic sandstones. (A) Illitic fibers (indicated by I) grown from edges of clay flakes, which mainly are mixed-layer illite/smectite from a mixture of K-feldspar + kaolinite + quartz in 0.1M KCl fluid at 350°C and 1,000 bars for 141 days. (B) Pseudohexagonal kaolinite platelets grown on the surface of an albite grain in 1M KCl fluids (quench pH = 5) at 200°C and 500 bars for 53 days. (C) K-feldspar (F) overgrown on the mixture of K-feldspar + kaolinite + quartz in 1M K_2CO_3 fluid at 350°C and 1,000 bars for 45 days. (D) Neoformed quartz (Q) intergrown with illite (I) from a mixture of K-feldspar + kaolinite + quartz in water at 300°C and 500 bars for 61 days.

similar conditions. The major differences between these two studies are the volume of the free-vapor space in the sample container and the total pressure. Chermak and Rimstidt (1990) used a rigid titanium container with a large free-vapor space at vapor pressures, whereas the present study used deformable gold capsules with nearly no vapor space and the experiments were run at water pressure higher than vapor pressure. Further experiments are required to unveil and differentiate the crystal growth processes under these two laboratory conditions.

Rate of Illite/Smectite Formation in Arkosic Sandstones

The precipitation rate of illite has been measured semi-quantitatively at 300°C from runs with mixtures of K-feldspar + kaolinite + quartz. Our preliminary experimental results (Fig. 10) show that the total amounts of I/S clays formed in near-neutral solutions with different compositions are similar at the same experimental conditions, whereas the proportion of illite layers in neoformed I/S depends strongly on solution composition. The rate of illitic-clay formation estimated from this study is slightly lower than that reported by Chermak and Rimstidt (1990) for a similar reaction.

The precipitation rates of illitic clays in near-neutral solutions are much slower than those in alkaline KOH solutions (Huang and Otten, 1985) and perhaps slower than those in initial acidic solutions of pH 2 (Huang and others, 1986). Figure 10 compares the reaction time required to convert 50% of kaolinite to illitic clays in near-neutral solutions with those in alkaline solutions and with those for smectite-to-illite conversion in KCl solution. For each reaction, the percent of conversion as a function of time at constant tem-

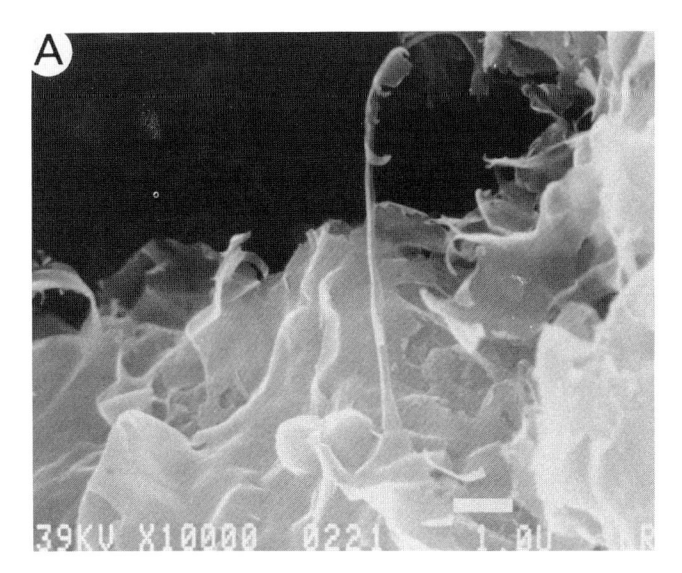

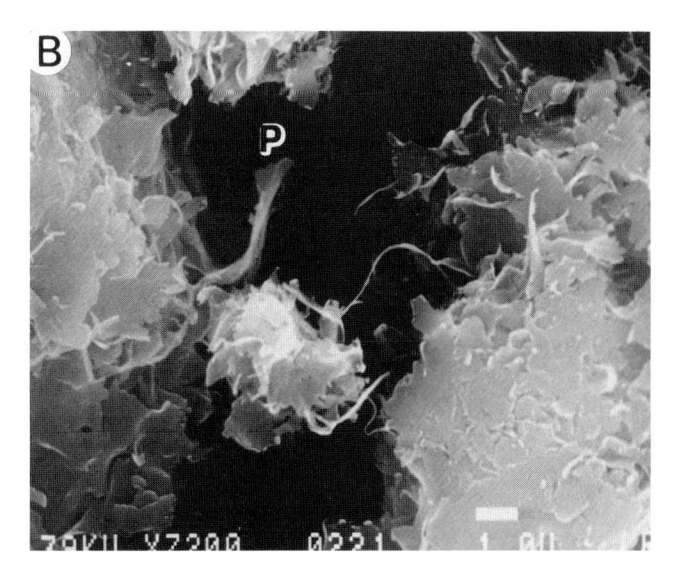

FIG. 9.—SEM photomicrographs showing a significant amount of clay fibers grown from a mixture containing K-feldspar + kaolinite + silica glass in 0.1M KCl solution (Run 221). The clay flakes and fibers are mainly illitic clays and pyrophyllite (indicated by P).

perature was first determined. Then the rate constant of the reaction at that temperature was calculated by fitting the data to an empirical kinetic model. Finally, the time required for 50% of conversion was calculated for each temperature and plotted as a single point on the diagram shown in Figure 10. The rate of kaolinite-to-illite conversion in KOH solution was experimentally measured at 225, 250, 300 and 350°C (Huang and Otten, 1985; Huang, in prep.). At each temperature, the mole percent of conversion was found to be a linear function of the square root of experimental time. The rate constants obtained from different temperatures are consistent with the Arrhenius equation with activation energy of 12.8 ± 1.5 kcal/mole. The conversion rate of smectite to illite was experimentally measured for Na-saturated SWy-1 reference clay, montmorillonite, at temperatures from 250 to 325°C in KCl solutions with [K$^+$] ranging from 0.1 to 6 M/l. The preliminary results show that the smectite-to-illite conversion rate significantly depends on [K$^+$]. The reaction times for 50% conversion with 200 and 10,000 ppm [K$^+$] at different temperatures are presented in Figure 10, whereas the detailed results will be published elsewhere (Huang and others, in prep.). Figure 10 shows that the conversion rate of K-feldspar + kaolinite + quartz to illitic clay is slightly faster than the smectite-to-illite conversion in a K$^+$ concentration commonly found in subsurface brine (200 ppm).

The results from the experimental kinetic studies of illitic-clay formation under a variety of conditions suggest that illite can be formed over a wide range of reaction rates depending on the precursor minerals as well as the fluid chemistry, particularly the pH. A further systematic study on the kinetics of illite formation from kaolinite as a function of solution pH may provide the important information for quantitatively predicting the occurrence of illite in arkosic sandstones.

MODEL FOR ILLITE FORMATION IN ARKOSIC SANDSTONES

The experiments have confirmed that the precipitation of illitic clays commonly observed in arkosic sandstones containing feldspars and kaolinite is the thermodynamic consequence of the chemical convergence of the fluid chemistry by the reaction of pore fluid with arkosic sandstone. The absence of illitization in such an arkosic sandstone is attributed to specific constraints. Illitization may not occur in arkosic sandstone at shallow depths due to the high-flow rate of the non-illitization fluids, such as meteoric water, or to some kinetic limitation arising either from a slow convergence of the brine chemistry or from a slow precipitation rate of illite.

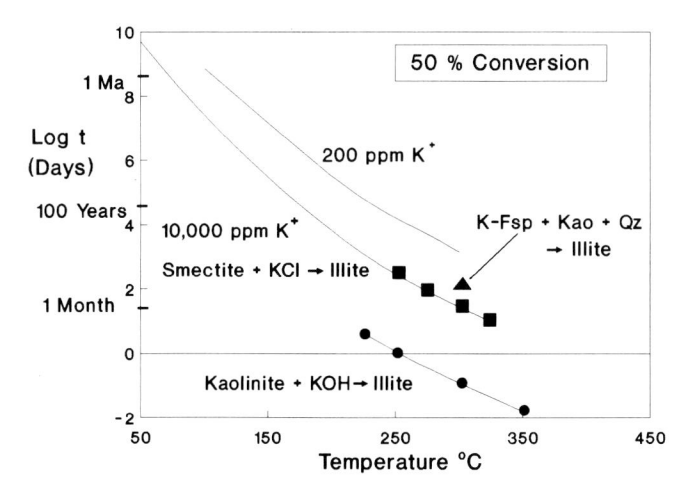

FIG. 10.—Temperature-age diagram showing that illite can be formed over a wide range of reaction rates depending on the fluid chemistry as well as the precursor minerals. Sources of data: illite forms from kaolinite in KOH solution (Huang and Otten, 1985); from K-feldspar + kaolinite + quartz in KCl solution (present study); and from smectite in KCl solution (Huang and others, in prep.).

The presence of bicarbonate ions or a weak acid (e.g., acetic acid) and its conjugate base may provide a buffer for the pH of the pore fluid (Surdam and others, 1984; Kharaka and others, 1986). These external pH buffers may significantly retard or accelerate the rate of the evolution of pore-fluid composition for a_{K+}/a_{H+} depending on the pH of the buffer solutions. In case of the bicarbonate ion, the change of fluid chemistry to the illite field is retarded and the fluid/rock ratio required for illite formation in the presence of the buffer should be significantly lower (Fig. 7).

Although there are still many questions to be resolved about the mechanism of illite formation, particularly the nature of smectite/illite and the kinetics, a qualitative conceptual model for illite formation can now be formulated based on the major factors that are identified in this study as well as in the published literature. These factors include mineralogy, pore-fluid compositions, hydrologic conditions, temperatures and kinetics.

Controlling Factors for Clay Formation in Sandstones

The experiments show that the optimum conditions for authigenic illitic-clay formation occur in arkosic sandstones containing both K-feldspar and kaolinite, and that Na-feldspar in the sandstone plays the same role as K-feldspar as long as there is enough K$^+$ in pore fluid. The same conclusion may also be true if K-feldspar is replaced by unstable detrital K-mica. Field observations seem to support the importance of the detrital mica on authigenic-illite formation (Huggett, 1984; Bjorlykke and Brendsdal, 1986; Dutta and Suttner, 1986). However, the actual role of detrital mica on the model is still under laboratory investigation. Carbonate cements or carbonate ions in pore fluids may provide an additional effect on buffering the pH of fluid chemistry (Surdam and others, 1984) as shown from our results obtained for KHCO$_3$ solutions.

The pore fluid may be either connate or of a composition resulting from influx into the formation during diagenesis. In this model (Fig. 11), the initial subsurface fluid compositions have been categorized into three types (acid, alkaline and neutral) according to the activity ratios (a_{K+}/a_{H+} and a_{K+}/a_{Na+}) of each fluid; other species of Al and Si may not be important in determining the type of clays formed because the amount of solute, particularly the Al and Si, that can be kept in pore fluids at any given time is negligible. In this model, it is assumed that the Si(OH)$_4$ of the fluid is near saturation with quartz and that the aluminum species is near saturation with the coexisting aluminosilicate minerals. The "acidic" fluids or "fresh" water are defined as having an activity ratio (a_{K+}/a_{H+}) low enough to stabilize kaolinite. "Alkaline" fluids, on the other hand, are defined as having a high a_{K+}/a_{H+} ratio, which favors the formation of K-feldspar. "Neutral" fluids have a moderate a_{K+}/a_{H+} ratio, which favors the formation of illite. The boundaries that delimit the three types of fluids depend on temperature and pressure, and therefore on the depth of burial or the geothermal gradient. The concentration of other cations (Ca^{2+}, Mg^{2+}, Fe^{2+}, etc.) determines the types of neoformed clays, such as smectite, S/I, illite or chlorite.

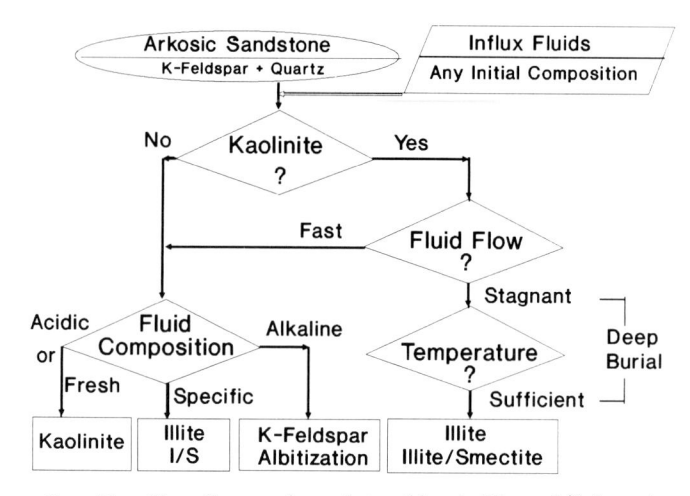

FIG. 11.—Flow diagram shows that authigenic illite or I/S formation in arkosic sandstones is determined primarily by the presence or absence of kaolinite and the dynamics of fluid flow, and secondly by the initial fluid compositions and temperatures. The initial fluid composition becomes crucial only when the flow rate, relative to fluid/rock reaction rate, is high. See text for detailed explanation.

Although the initial fluid composition may determine the incipient authigenesis, the evolution of the fluid chemistry through fluid/rock reaction during diagenesis may or may not significantly modify the authigenic-fluid composition, depending on the flow rate. This relation seems especially important for the formation of illite because the fluid tends to shift to the illite stability field in arkoses, which is most pronounced when the flow rate of the fluid is either stationary or much slower than the mineral reaction rates (i.e., rock-dominant system; designed as "stagnant"). For a fluid with a relatively high-flow rate or a large total flux, the fluid chemistry (e.g., a_{K+}/a_{H+} ratio) remains nearly constant because of the continuous replenishment of the fluid (fluid-dominant system). This condition is designated as "fast" flow in the diagrams. The boundary between the "stagnant" and "fast" flows depends on mineral reaction rates and temperatures. An accurate prediction of authigenic minerals in a reservoir, therefore, requires reliable hydrologic information about the reservoir throughout its diagenetic history. Conversely, the paragenesis and distribution of authigenic minerals in a reservoir, if known, can be valuable in determining the hydrologic history of the reservoir (Whitney and Roy, 1984).

The influence of temperature on illitization may involve two aspects: thermodynamics (equilibrium) and kinetics. Thermodynamically, temperature can shift the illite stability field on a phase diagram and, in this respect, the composition of illitization fluids depends on temperature. However, the chance for an influxing fluid to have a composition in the illite stability field is so small that initial fluid composition cannot explain the widespread illitization at deep burial. Our model suggests that the thermodynamic effect of temperature on illite formation is not important because the chemistry of stagnant pore fluids in arkosic sandstone containing kaolinite should converge to the stability field of illite regardless of temperature. The kinetic effect of temperature, however, is significant. The analysis of formation water from a late Triassic-Jurassic sandstone reservoir, offshore Norway, suggests that the formation water is in general saturated with illite and chlorite at all depths but illitization is not observed until the temperature reaches 100 to 120°C (Aagaard and Egeberg, 1987). Similar observations have been made from petrographic studies of North Sea sandstones, which show kaolinite and K-feldspar are metastable down to about 3.7 km and, below this depth, there is a significant increase in the amount of authigenic illite (Bjørlykke and others, 1986). This suggests that kinetics should be considered in predicting illitization in sandstones. Experimental results show that the illitic clays can be grown over a wide range of reaction rates. Illite precipitates from feldspar in acidic solutions or kaolinite in initially alkaline solutions more quickly than from feldspar and kaolinite in near-neutral solutions (Fig. 10). In addition, illite may precipitate directly from a continuously replenished "neutral" illitization fluid at a faster rate than from reaction with feldspar or kaolinite, although the influx of this illitization fluid may not be common in nature.

Unfortunately, little quantitative information is available on the kinetics of the formation of illite from smectite (Eberl and Hower, 1976; Huang and others, in prep.) and from kaolinite (Huang and Otten, 1985; Chermak and Rimstidt, 1990). The predictive capability of the model will be significantly improved as more complete information on kinetics formation becomes available.

Model for Authigenic-Clay Formation in Arkosic Sandstones

The conceptual illite model is presented as a self-explanatory flow diagram (Fig. 11). The model illustrates that the paths and the final products of diagenesis depend on the initial fluid compositions, the presence or absence of pre-existing kaolinite, the dynamics of fluid flow, and the kinetics of illite formation.

Note from Figure 11 (the pathway shown by the right branch of flow diagram) that diagenesis of arkosic sandstone containing feldspar, kaolinite and "stagnant" pore fluid will precipitate I/S or illite regardless of initial fluid composition. This diagenetic pathway, which results from the control of the feldspar + kaolinite + quartz assemblage, is responsible for the frequent occurrence of authigenic I/S or illite and perhaps that of chlorite in arkosic sandstone. This pathway particularly prevails during deep burial diagenesis because of the decrease of flow rate and increase of temperature. The rise of temperature can increase both the convergence rate of pore fluids into the illite field and the precipitation rate of illitic clays.

The model also predicts the possible variations of clay types formed during diagenesis. Precipitation of discrete illite requires a fluid composition with high K^+/H^+ and K^+/Na^+ activity ratios. The proportion of smectite (expandable layers) in I/S increases with the decrease of activity ratios of K^+/H^+ and K^+/Na^+ or with the increase of concentrations of Na^+, Ca^{2+}, and Mg^{2+}. Illite with a more fibrous habit will form if the pore fluid is oversaturated with respect to quartz. In addition, the longer the sandstones have been subjected to illitization, the more fibrous illite will form. The model also suggests that, in the presence of significant amounts of Fe^{2+} and Mg^{2+} either in pore fluid or in the volcanic lithic fragments of a sandstone reservoir, chlorite may be formed in association with illite at conditions of illitization. The close association of chlorite and illite in deeply buried sandstone (Glennie and others, 1978; Burley, 1984; Whitney and Roy, 1984) implies that the two clay types may precipitate under similar conditions.

In contrast, illitic clays will less likely form in arkosic sandstones lacking kaolinite and/or if there is a high fluid-flow rate. In either of these circumstances, the final diagenetic minerals would depend greatly on the initial composition of the pore fluid. The influx of acidic or fresh fluids into arkosic sandstone at high-flow rate leads to the formation of kaolinite. However, if the flow rate later decreases, a significant amount of illite may start to form at the expense of feldspar and kaolinite until one of these two minerals disappears. If feldspar is completely consumed before illitization, kaolinite may survive at great depths (Bjørlykke, 1989).

The influx of neutral (illitization) fluids into arkosic sandstones with or without kaolinite at high-flow rates (left branch of Fig. 11) can precipitate much illite. However, if the flow rate of this fluid is not fast enough and the arkosic sandstone contains no kaolinite, illitization may cease unless there are some other sources of aluminum such as unstable plagioclase or volcanic fragments; then illite and K-feldspar may precipitate together.

The influx of alkaline fluids into arkosic sandstones containing kaolinite at high-flow rate can lead to K-feldspar overgrowth or albitization, depending on the activity ratio (a_{K+}/a_{Na+}) (Fig. 11). In an arkosic sandstone without kaolinite, but with other sources of aluminum (e.g., unstable plagioclase or volcanic lithic fragments), K-feldspar overgrowth or albitization can occur in an alkaline fluid with high-flow rate, or precipitation of both feldspar and illite can occur at a low-flow rate.

APPLICATION OF THE MODEL TO CASE STUDIES

The Middle Jurassic sandstones (Brent in the Viking Graben and its equivalent in the Haltenbanken area, offshore Norway) and the Permian Rotliegende Sandstone in the northern and southern North Sea, respectively, are two major authigenic illite-bearing reservoirs (Hancock and Taylor, 1978; Seemann, 1979; McHardy and others, 1982; Lee and others, 1985; 1989; Bjørlykke and others, 1986). In spite of the diversity in depositional environments, lithology, age, and diagenetic history, the two sandstone formations have been found to have a common authigenic mineral paragenesis. The major diagenetic stages include kaolinitization of feldspar and detrital K-mica during shallow burial, and illitization of kaolinite and/or feldspar at greater depths. Similar parageneses have been found in other arkosic-sandstone formations (Wilson, 1982; Dutta and Suttner, 1986). Although the characteristics of the diagenetic process may vary from one place to another, the onset of illitization at greater depths suggests that illite growth is controlled mainly by temperature, pressure and other depth-related factors, such as the flow rate and fluid composition. The observed unique paragenesis implies that a single dominant process universally controls the illitization of arkosic sandstone.

The illite parageneses and depth distributions have been attributed to the influx of potassium-bearing pore fluid (Hurst and Irwin, 1982; Rossel, 1982) from great depths. A sudden change in the composition of pore fluids from "acidic" or "fresh" to "illitization"-type fluid provides a condition thermodynamically favorable for illite formation (Kantorowicz, 1984). However, for illite to form, the a_{K+}/a_{H+} ratio in the pore fluid must fall within a well-defined, small window, which depends greatly on temperature and pressure. The influx of a specific illitization fluid seems too coincidental to explain the universal occurrence of illitization at depth. The paragenesis also has been attributed to the increase of temperature (Hancock and Taylor, 1978), which shifts the illite stability field to trigger illitization, while the fluid composition remains constant. This also requires a specific pore-fluid composition that causes illitization only in a small depth interval.

CONCLUSIONS

The present model suggests that kaolinitization at shallow depths results from the alteration of feldspars by a continuous replenishment of acidic or fresh meteoric water. As depth increases, the flow rate decreases because of the reduced permeability of the rock. The decrease of the flow rate enhances the driving force of the arkosic-mineral assemblage and causes the fluid chemistry change to thermodynamically favor the formation of illite. In other words, the rock controls the fluid composition. The increase of temperatures with depth may increase the rates at which pore-fluid composition changes to illite stability field, as well as accelerate the precipitation rate of illitic clays. This interpretation is consistent with the field observation made by Bjørlykke and others (1986) for the Jurassic sandstone from offshore mid-Norway. The illitization can also be triggered by the influx of hot compaction fluid (Glasmann and others, 1989). However, the initial composition of this influx fluid is not necessarily an illitization type since the fluid-arkose reaction will change fluid composition and precipitate illite if flow rate is not too high (Fig. 11). Higher temperature of the compaction fluid also enhances the kinetics of illite formation.

ACKNOWLEDGMENTS

The approval from Exxon Production Research company to release this paper is greatly appreciated. I thank my colleagues: A. M. Bishop and R. W. Brown, G. A. Otten for their assistance in the laboratories, J. M. Longo for valuable discussions in developing the model, W. J. Harrison for using the results from EQ3/6 modeling and A. Cochran and L. L. Summa for evaluating the model in diagenetic studies. My thanks extend to professor N. Güven of Texas Tech University for TEM work. I would like to thank Drs. P. Aagaard and K. Bjørlykke, M. T. Heald, I. Hutcheon and J. M. Longo and D. R. Pevear for reviewing the manuscript.

REFERENCES

AAGAARD, P., AND EGEBERG, P. K., 1987, Formation water chemistry in the late Triassic-Jurassic reservoirs offshore Norway (abs.): American Association of Petroleum Geologists Research Conference, Park City, Utah, p. 42.

BJØRLYKKE, K., 1983, Diagenetic reaction in sandstones, in Parker, A., and Sellwood, B. W., eds., Sediment Diagenesis: Reidel Publishing Company, Dordrecht, Netherlands, p. 169–213.

BJØRLYKKE, K., 1989, Closed versus open system in silicate diagenesis – a review, v. 1: Program and Abstracts, 28th International Geological Congress, Washington, D. C., p. 157–158.

BJØRLYKKE, K., AAGAARD, P., DYPVIK, H., HASTINGS, D. S., AND HARPER, A. S., 1986, Diagenesis and reservoir properties of Jurassic sandstones from the Haltenbanken area, offshore mid-Norway, in Spencer, A. M., ed., Habitat of Hydrocarbons on the Norwegian Continental Shelf: Norwegian Petroleum Society Symposium, p. 275–286.

BJØRLYKKE, K., AND BRENDSDAL, A., 1986, Diagenesis of the Brent Sandstone in the Statfjord field, North Sea, in Gautier, D. L., ed., Roles of Organic Matter in Sediment Diagenesis: Society of Economic Paleontologists and Mineralogists Special Publication 38, p. 157–167.

BJØRKUM, P. A., AND GJELSVIK, N., 1988, An isochemical model for formation of authigenic kaolinite, K-feldspar and illite in sediments: Journal of Sedimentary Petrology, v. 58, p. 506–511.

BURLEY, S. D., 1984, Distribution and origin of authigenic minerals in the Triassic Sherwood Sandstone Group, UK: Clay Minerals, v. 19, p. 403–440.

CHERMAK, J. A., AND RIMSTIDT, J. D., 1990, Hydrothermal transformation rate of kaolinite to muscovite/illite: Geochimica et Cosmochimica Acta, v. 54, p. 2979–2990.

DUTTA, P. K., AND SUTTNER, L. J., 1986, Alluvial sandstone composition and paleoclimate. II. Authigenic mineralogy: Journal of Sedimentary Petrology, v. 56, p. 346–358.

EBERL, D. D., AND HOWER, J., 1976, Kinetics of illite formation: Geological Society of America Bulletin, v. 87, p. 1326–1330.

GALLOWAY, W. E., 1979, Diagenetic control of reservoir quality in arc-derived sandstones: implications for petroleum exploration, in Scholle, P.A., and Schluger, P. R., eds., Aspects of Diagenesis: Society of Economic Paleontologists and Mineralogists Special Publication 26, p. 251–262.

GARRELS, R. M., 1984, Montmorillonite/illite stability diagrams: Clays and Clay Minerals, v. 32, p. 161–166.

GLASMANN, J. R., CLARK, R. A., BRIEDIS, N. A., AND LUNDEGARD, P. D., 1989, Diagenesis and hydrocarbon accumulation, Brent Sandstone (Jurassic), Bergen High area, North Sea: American Association of Petroleum Geologists, Bulletin, v. 73, p. 1341–1360.

GLENNIE, K. W., MUDD, G. C., AND NAGTEGAAL, P. J. C., 1978, Depositional environment and diagenesis of Permian Rotliegendes sandstones in Leman Bank and Sole Pit areas of the U.K., southern North Sea: Journal of Geological Society of London, v. 135, p. 25–34.

GÜVEN, N., HOWER, W. F., AND DAVIES, D. K., 1980, Nature of authigenic illite in sandstone reservoirs: Journal of Sedimentary Petrology, v. 5, p. 761–766.

GÜVEN, N., LAFON, G. M., AND LEE, L. J., 1982, Experimental alteration of albite to clays: preliminary results: Proceedings, 7th International Clay Conference, Amsterdam, Elsevier, p. 495–511.

HANCOCK, N.J., AND TAYLOR, A. M., 1978, Clay mineral diagenesis and oil migration in the Middle Jurassic Brent Sand Formation: Journal of Geological Society of London, v. 135, p. 69–72.

HELGESON, H. C., BROWN, T. H., AND LEEPER, R. H., 1969, Handbook of theoretical activity diagram depicting chemical equilibria in geological systems involving an aqueous phase at one atm and 0 to 300°C: San Francisco, Freeman, Cooper and Co., 253 p.

HUANG, W. L., 1986, Buffering capability of mineral assemblage (feldspar + kaolinite + quartz) and its application to predicting illite occurrence in arkosic sandstones (abs.): American Association of Petroleum Geologists Bulletin, v. 75, p. 602.

HUANG, W. L., BISHOP, A. M., AND BROWN, R. W., 1986, Effect of fluid/rock ratio on albite dissolution and illite formation at reservoir conditions: Clay Minerals, v. 21, p. 585–601.

HUANG, W. L., AND OTTEN, G. A., 1985, Kinetics of K-mica and K-natrolite formation as a function of temperature: Program and Abstracts, 2nd International Symposium on Hydrothermal Reactions, p. 34.

HUGGETT, J. M., 1984, Controls on mineral authigenesis in coal measures sandstones of the East Midlands, UK: Clay Minerals, v. 19, p. 343–357.

HURST, A., AND IRWIN, H., 1982, Geological modelling of clay diagenesis in sandstones: Clay Minerals, v. 17, p. 5–22.

INOUE, A., KOHYAMA, N., KITAGAWA, R., AND WATANABE, T., 1987, Chemical and morphological evidence for the conversion of smectite to illite: Clays and Clay Minerals, v. 35, p. 111–120.

KANTOROWICZ, J., 1984, The nature, origin and distributions of authigenic clay minerals from middle Jurassic Ravenscar and Brent group sandstones: Clay Minerals, v. 19, p. 359–375.

KELLER, W. D., REYNOLDS, R. C., AND INOUE, A., 1986, Morphology of clay minerals in the smectite-to-illite conversion series by scanning electron microscopy, Clays and Clay Minerals, v. 34, p. 187–197.

KHARAKA, Y. K., LAW, L. M., CAROTHERS, W. W., AND GOERLITZ, D. F., 1986, Roles of organic species dissolved in formation waters from sedimentary basin in mineral diagenesis, in Gautier, D. L., ed., Roles of Organic Matter in Sediment Diagenesis: Society of Economic Paleontologists and Mineralogists Special Publication 38, p. 111–122.

LEE, M., ARONSON, J. L., AND SAVIN, S. M., 1985, K/Ar dating of time of gas emplacement in Rotliegendes Sandstone, Netherlands: American Association of Petroleum Geologists Bulletin, v. 69, p. 1381–1385.

LEE, M., ARONSON, J. L., AND SAVIN, S. M., 1989, Timing and condition of Permian Rotliegende Sandstone diagenesis, southern North Sea: K/Ar and oxygen isotopic data: American Association of Petroleum Geologists Bulletin, v. 73, p. 195–215.

MCHARDY, W. J., WILSON, M. J., AND TAIT, J. M., 1982, Electron microscope and X-ray diffraction studies of filamentous illitic clay from sandstones of the Magnus field: Clay Minerals, v. 17, p. 23–39.

MONTOYA, J. W., AND HEMLEY, J. J., 1975, Activity relations and stabilities in alkalfeldspar and mica alteration reactions: Economic Geology, v. 70, p. 577–594.

MEUNIER, A., AND VELDE, B., 1989, Solid solution in I/S mixed-layer minerals and illite: American Mineralogist, v. 74, p. 1106–1112.

NADEAU, P. H., WILSON, M. J., MCHARDY, W. J., AND TAIT, J. M., 1984, Interparticle diffractions: a new concept for interstratified clays: Clay Minerals, v. 19, p. 757–769.

POLLASTRO, R. M., 1985, Mineralogical and morphological evidence for the formation of illite at the expense of illite/smectite: Clays and Clay Minerals, v. 33, p. 265–274.

REX, R. W., 1965, Authigenic kaolinite and mica as evidence for phase equilibria at low temperatures: Clays and Clay Minerals, v. 13, p. 95–104.

REYNOLDS, R. C., 1980, Interstratified clay minerals, in Brindley, G. W., and Brown, G., eds., Crystal Structures of Clay Minerals and their X-ray Identification: Mineralogical Society of London, p. 249–303.

ROBERSON, H. E., AND LAHANN, R. W., 1981, Smectite to illite conversion rates: Effects of solution chemistry: Clays and Clay Minerals, v. 29, p. 129–135.

ROSSEL, N. C., 1982, Clay mineral diagenesis in Rotliegend aeolian sandstones of the southern North Sea: Clay Minerals, v. 17, p. 69–77.

SASS, B. M., ROSENBERG, P. E., AND KITTRICK, J. A., 1987, The stability of illite/smectite during diagenesis: an experimental study: Geochimica et Cosmochimica Acta, v. 51, p. 2103–2115.

SCHMIDT, V., AND MCDONALD, D. A., 1979, The role of secondary porosity in sandstones, in Scholle, P. A., and Schluter, P. R., eds., Aspects of Diagenesis: Society of Economic Paleontologists and Mineralogists Special Publication 26, p. 209–225.

SEEMANN, U., 1979, Diagenetically formed interstitial clay minerals as a factor in Rotliegend Sandstone reservoir quality in the North Sea: Journal of Petroleum Geology, v. 1, p. 55–62.

SEEMANN, U., 1982, Depositional facies, diagenetic clay minerals and reservoir quality of Rotliegend sediments in the southern Permian Basin (North Sea): a review: Clay Minerals, v. 17, p. 35–67.

SOMMER, F., 1978, Diagenesis of Jurassic sandstones in the Viking Graben: Journal of Geological Society of London, v. 135, p. 63–67.

STALDER, P. J., 1973, Influence of crystallographic habit and aggregate structure of authigenic clay minerals on sandstone permeability: Geologie en Mijnbouw, v. 52, p. 217–219.

SURDAM, R. C., BOESE, S. W., AND CROSSEY, L. J., 1984, The chemistry of secondary porosity: American Association of Petroleum Geologists Memoir 37, p. 127–149.

THOMAS, M., 1986, Diagenetic sequences and K/Ar dating in Jurassic sandstones, central Viking graben: effects on reservoir properties: Clay Minerals, v. 21, p. 695–710.

VELDE, B., AND WEIR, A. H., 1979, Synthetic illite in the chemical system $K_2O-Al_2O_3-SiO_2$ at 300°C and 2 kb, in Mortland, M. M., and Farmer, V. C., eds., Developments in Sedimentology 27: Amsterdam, Elsevier, p. 395–404.

WAUGH, G., 1978, Authigenic K-feldspar in British Permo-Triassic sandstones: Journal of Geological Society of London, v. 135, p. 51–56.

WHITNEY, G., AND ROY, N. H., 1984, Reconstructing and ancient fluid-flow regime using the mineralogy and stable geochemistry of diagenetic clays in sandstone: Geological Society of America Annual Meeting, Program and Abstracts, v. 15, p. 692.

WILSON, M. D., 1982, Origin of clays controlling permeability in tight gas sands: Society of Petroleum Engineers Paper 9843, 8 p.

WILSON, M. D., AND PITTMAN, E. D., 1977, Authigenic clays in sandstones: recognition and influence on reservoir properties and paleoenvironmental analysis: Journal of Sedimentary Petrology, v. 47, p. 3–31.

WOLERY, T. J., 1979, Calculation of chemical equilibrium between aqueous solution and minerals: the EQ3/6 software package: Lawrence Livermore Laboratory Publication, UCRL-52658.

CLAY MINERALS IN NORTH SEA SANDSTONES

KNUT BJØRLYKKE AND PER AAGAARD
Department of Geology, Box 1047, University of Oslo, 0316, Oslo 3, Norway

ABSTRACT: The most abundant clay-mineral cements in North Sea reservoir sandstones are kaolinite, illite and mixed-layer minerals. Authigenic chlorite commonly is present but rarely is abundant. Permian and Triassic sandstones were deposited in an arid to semiarid climate and contain mostly illite/smectite with subordinate kaolinite. Isotopic and textual evidence suggests that authigenic kaolinite in these sandstones formed after tectonic uplift and meteoric-water flushing. In fluvial and shallow marine Lower and Middle Jurassic reservoirs, kaolinite is the dominant clay mineral and feldspar dissolution forms abundant secondary porosity even in the shallowest reservoirs. Stable-isotope analyses of authigenic kaolinite suggest crystallization at relatively low temperature and from pore water of meteoric origin. Upper Jurassic sandstones representing shallow marine facies also contain secondary porosity, due to feldspar leaching, and authigenic kaolinite. In turbiditic sandstones and sandstones interbedded with the main source rock (Kimmeridge Clay Formation), however, diagenetic kaolinite is rare. The sandstones also show little evidence of feldspar leaching, probably because this distal facies was not effectively leached by meteoric water. Organic acids or carbon dioxide that may have been released from source rocks seem to have had little effect on adjacent sandstones in terms of feldspar leaching and precipitation of kaolinite. Dissolved feldspar and authigenic kaolinite are also relatively uncommon in the Cretaceous and Tertiary turbiditic sandstones. The main control on feldspar leaching and distribution of authigenic kaolinite appears to have been the degree of meteoric-water flushing, which depends on climate, depositional environment and continuity of sandstone beds. Later tectonic uplifts may also have caused meteoric-water recharge from exposed areas into the basin.

The geochemistry of formation water from reservoir rocks suggests that pore waters in North Sea reservoirs are mostly in the stability field of illite during burial diagenesis. In Jurassic reservoirs, illite can be observed to replace kaolinite in the deeper wells if K-feldspar is available as a source of potassium. A strong increase in the degree of illitization can be observed below 3.7 to 3.8 km burial depth in several oil fields. K/Ar dating of illites from reservoirs in the North Sea and Haltenbanken areas give a wide range of estimated ages, many between 30 and 50 Ma. Many of these sandstones would only have been buried to about 2 km during those early Tertiary times. If these dates for illite precipitation are correct, they might indicate high geothermal gradients at that time. The present depth distribution of illite suggests, however, that illitization of kaolinite takes place at greater depth (3.5–4 km). Illite may also form from a smectite precursor and illite of this origin is particularly abundant in Triassic and Permian reservoirs.

Chlorite may replace kaolinite starting at about 90 to 100°C, but the amount is limited by the supply of iron and magnesium from dissolving mafic minerals and rock fragments. Co-existing illites and chlorites appear to follow regular compositional trends with increasing burial temperatures. These trends are in covariance with the stability of the endmember components of the clay minerals. Crystal-size distributions may be explained in terms of Ostwald-ripening mechanisms. Chlorite crystals show compositional variations from core to rim.

Burial diagenesis in the North Sea basin is interpreted to be relatively isochemical and our diagenetic models do not require large-scale transport of solids in solution. Theoretical models of compactional porewater flow and observed compositional stratifications of the pore water in this basin also constrain large-scale mass transfer by advection.

INTRODUCTION

The North Sea Basin is a highly productive oil and gas province and is now at a relatively mature stage in terms of exploration. It is a good laboratory for studying diagenetic processes as a function of increasing burial. Both exploration and production wells are extensively cored. A summary of the North Sea geology has been published by Glennie (1990). Permian and Triassic sandstones of the southern part of the North Sea contain large volumes of gas generated from underlying Carboniferous coals. Permian eolian- and fluvial-desert sediments are overlain by the Zechstein evaporites. Triassic deposits are continental sediments, mostly sandstones and mudstones, deposited in an arid to semiarid climate. Jurassic sandstones, particularly the Middle Jurassic Brent Group, are the most important oil reservoirs in the North Sea, producing at the Statfjord, Brent, Oseberg and Gullfaks fields (Fig. 1).

Large reservoirs are also present in the Cretaceous chalks (e.g., Ekofisk field) and in Lower Tertiary sandstones. Late Jurassic rifting produced tectonic uplift of rotated fault blocks bounded by listric faults. Most of the traps in the central and northern part of the North Sea formed during this phase. The main source rock in the North Sea, the Kimmeridge Clay Formation, was deposited under stagnant bottom-water conditions during this rifting phase. During the Cretaceous, deeper water mudstones that serve as important cap rocks were deposited. Lower Tertiary turbiditic sandstones derived from tectonically uplifted areas around Scotland and Shetland are important reservoir sandstones in Forties field and several other fields in that area. Frigg and Heimdal are important gas fields.

There has been continued subsidence in most areas since Cretaceous time and the burial history, and to a certain extent the temperature history, can be reconstructed relatively reliably. The Pliocene/Pleistocene sequence is about 1 km thick in the central parts of the basin and represents accelerated subsidence in the late Cenozoic (Lovell, 1990). Around the edges of the North Sea Basin, in the southern North Sea Basin, and onshore in Holland, Germany, and Britain, there have been several periods of basin inversion including late Tertiary uplifts, which make the diagenetic history more difficult to interpret.

The precipitation of authigenic clay minerals is an important part of diagenesis, which strongly influences reservoir properties. The growth of relatively small volumes of authigenic-clay minerals in the primary pore space may drastically change the properties of reservoir sandstones. Clay minerals interact more strongly with drilling fluid or water injected into the reservoir during production than most

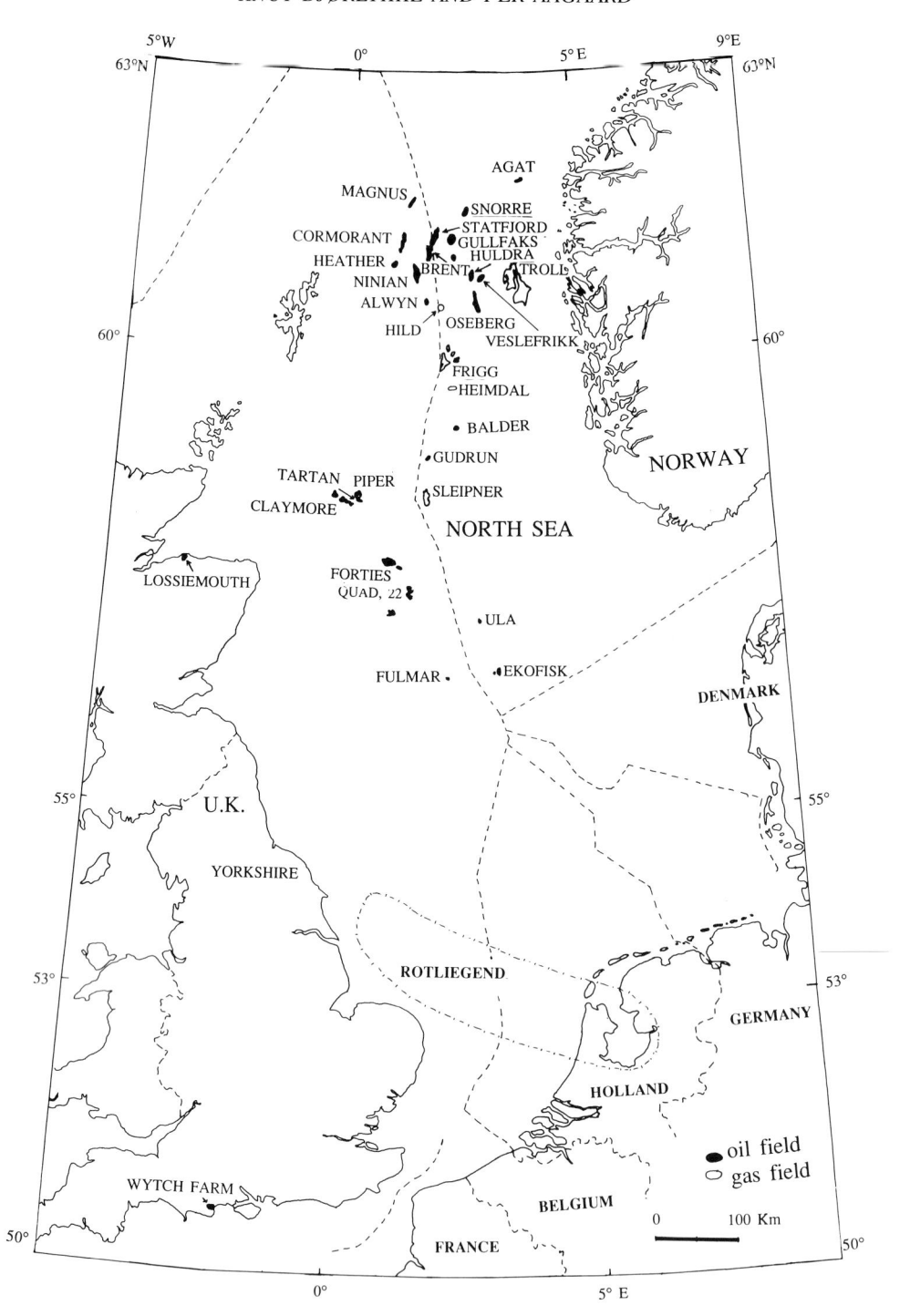

FIG. 1.—Map of North Sea oil fields referred to in this paper.

other types of mineral cements. It is therefore important that we are able to establish predictive models for the distribution of such minerals.

The study of clay minerals in reservoir rocks requires careful petrographic, textural and mineralogical analyses. Recently, isotopic analyses have helped to determine the temperature and isotopic composition of the pore water from which they precipitated. In addition, theoretical modeling of porewater flow and transport of solids in solution should be used to constrain diagenetic models and to improve their predictive power. The clay-mineral reactions, however, cannot be studied in isolation from other diagenetic reactions. It is important to integrate diagenetic studies with other aspects of basin analyses, such as the subsidence history and tectonics.

The aim of this paper is to summarize the clay mineralogy of North Sea reservoir sandstones and try to recognize general trends in the distribution of authigenic clay minerals. Predictions of clay-mineral distributions require, however, an understanding of the processes that cause clay-mineral precipitation. Discussions of the importance of climate, depositional environment and flow of pore fluids are included in this paper.

The reservoir sandstones will be treated in stratigraphic order. We will demonstrate, based on recent literature from the North Sea Basin and new data from this study (Table 1), that depositional environments and climate at the time of deposition played important roles in early diagenesis and indirectly influenced diagenetic processes at greater burial.

DISTRIBUTION OF CLAY MINERALS IN NORTH SEA SANDSTONES OF DIFFERENT AGES

Clay Mineralogy of Permian and Triassic Reservoirs

Permian and Triassic reservoir sandstones were deposited in a continental setting under arid to semiarid conditions and represent environments that were distinctly different from those of Jurassic reservoir rocks. This has important implications for the early diagenetic history.

The Lower Permian Rotliegend sandstones in the southern North Sea were deposited in sabkha, eolian and wadi types of environments (Glennie and others, 1978). Illite and mixed-layer minerals make up most of the authigenic clay. Authigenic kaolinite replacing leached plagioclases may locally be an important component, but is commonly absent. In eolian dune sandstones from the southern North Sea, illite is the only important clay mineral in reservoirs buried below 3.5 km (Pye and Krinsley, 1986). Chlorite (Fe-rich) is much less common and is mostly found in association with detrital mica and lithic grains. In a study of eolian sandstones from the Sole Pit/Leman Bank area, the Indefatigable area, and the Broad Fourteens Basins area, kaolinite is relatively abundant in shallower reservoirs, whereas reservoirs that have been buried to more that 3.5 km (paleo-burial depth) contain mostly illite (Rossel, 1982). Growth of diagenetic illite at the expense of kaolinite and feldspar caused reduction in permeability from a few hundred millidarcies to close to 1 md. Chlorite, however, may occur as grain coats (Rossel, 1982). Seemann (1982) noted that thin clay rims are common on the grains of Permian eolian sandstones and that they played an important role in burial diagenesis. The clay is interpreted to result from mechanical infiltration into the dunes.

In the Rough Gas field, the Rotliegend sandstones contain dominantly illite and smaller amounts of diagenetic kaolinite, which is interpreted to have formed by late leaching by acids released from Carboniferous coaly source rocks (Goodchild and Whitaker, 1986). Small amounts of Fe-rich chlorite are also found. A study of the clay minerals in the Permo-Triassic redbed sequences of the Western Approaches and the Devon coast showed that a detailed clay-mineral stratigraphy, mainly based on the occurrence of chlorites, smectites, and corrensites, could be correlated over long distances (500 km) (Fisher and Jeans, 1982). This suggests that the clay mineralogy is controlled by changes in detrital input, which in this case is laterally extensive.

Rotliegend sandstones from wells in the Netherlands contain abundant authigenic illite and smaller amounts of kaolinite (Lee and others, 1989). The estimated formation temperatures of illite are consistently about 80 to 85°C in the Leman and Indefatigable gas fields, corresponding to burial depths of 2 to 2.5 km. Formation temperatures are close to 100°C in the central Netherlands Basin and are more variable in other basins, reaching maximum values of as much as 140°C at 4-km depth in the Broad Fourteens Basin. Illite probably is formed by illitization of smectite and mixed-layer minerals. Coarse-grained kaolinite is estimated to have formed shortly after deposition at temperatures of 15 to 38°C in meteoric water with $\delta^{18}O$ values between -7 and $-2‰$. A second generation of finer grained kaolinite is interpreted to have formed later, during meteoric-water recharge as the Permian sediments were tectonically uplifted in Late Jurassic times. At that time, the pore water was less depleted with respect to $\delta^{18}O$ (-3 to $+3$) and the estimated temperature was higher (48°C). Lee and others (1989) interpreted illite and quartz to have been precipitated from pore waters with more positive $\delta^{18}O$ values ($+2$) and at greater depth, mostly at temperatures between 80°C and 140°C (Table 2). K/Ar dates of illite and I/S range between 100 and 175 Ma and are interpreted to have formed in two major phases, the first during Late Jurassic/Early Cretaceous (Cimmerian) orogenic movements and the second during Cretaceous-early Tertiary inversion. Lee and others (1989) also found that the isotopic composition of illite was influenced by meteoric-water recharge as far as 50 to 60 km down dip from the exposed Texel-Ijsselmeer High.

Upper Triassic sandstones (Skagerrak Formation) from the Central Graben have chlorite cements, which are Fe-rich clinochlores (Humphreys and others, 1989). These chlorites are interpreted to have been formed by direct precipitation from pore water following dissolution of detrital silicates and dolomites. Kaolinite was not identified in these samples. Farther north in the Claymore oil field in British territory, Spark and Trewin (1986) found equal concentrations of kaolinite and smectite in fluvial deposits of the upper part of the Skagerrak Formation. This may reflect a more humid climate in the Late Triassic (Frostic, 1990), causing more meteoric-water flushing.

Upper Triassic rocks of the Lunde Formation make up parts of the reservoir in the Snorre field in the Norwegian sector, where fluvial sandstones have well-developed soil profiles that may reflect a more humid climate toward the end of the Triassic. Evidence of feldspar leaching and early diagenetic kaolinite precipitation is common in the Lunde Formation, but these features are much more pronounced in the overlying Lower Jurassic Statfjord Formation (Knarud and Bergan, 1990).

Triassic sandstones from onshore Britain (Sherwood Sandstone) have been examined from outcrops and boreholes. Because the sandstones are an important aquifer, recent meteoric-water diagenesis can be studied. In the East Midland aquifer, fresh water flows at a rate of 3m/day in a well at 500 m depth and 20 km down dip from the nearest outcrop (Bath and others, 1987a). Authigenic kaolinite in these sandstones is interpreted to postdate illite and to have precipitated during meteoric water-flushing resulting from the recent tectonic uplift and erosion of the Triassic sequence. In a borehole farther down dip, more saline waters

TABLE 1.—RELATIVE ABUNDANCE OF AUTHIGENIC-CLAY MINERALS IN NORTH SEA RESERVOIR ROCKS BASED ON PUBLISHED DATA AND OUR OWN UNPUBLISHED DATA. MOST OF THE STUDIES DO NOT PRESENT GOOD QUANTITATIVE DATA ON THE CLAY-MINERAL DISTRIBUTION, BUT HAVE GIVEN A SEMI-QUANTITATIVE ESTIMATE OF THE RELATIVE CONCENTRATION OF AUTHIGENIC-CLAY MINERALS

Age/ Strat. Unit	Field/ Locality	Illite/ Smectite	Kaolinite	Chlorite	Source
Tertiary:					
Eocene	16/1-1			**	This work
Paleocene	Balder	***		***	Malm and others, 1984
Paleocene	Heimdal	**		**	Hurst and Buller, 1984
Paleocene	Sleipner		*	***	This work
Cretaceous:					
Agat	Agat (35/5-9)	*	**	*	Saigal and Bjørlykke, unpublished
Upper Jurassic:					
Magnus Fm.	Magnus	***	*		McHardy and others, 1982
Fulmar	Fulmar	**	-	-	Stewart, 1986
Piper	Tartan	***	**	-	Burley, 1986
Piper	Tartan	**	**	-	Burley and others, 1989
Claymore	Claymore	*	**	-	Spark and Trewin, 1986
Fulmar	Fulmar	*	-	-	Johnsen and others, 1986; Saigal and Bjørlykke, unpublished
Draupne	Gudrun	***	*		This work
Middle Jurassic:					
Brent	Heather	*	***		Glasmann and others, 1989a
Brent	Alwyn	***	*	-	Jourdan and others, 1987
Brent	Statfjord	*	***	-	Bjørlykke and Brendsdal, 1986
Brent	Ninian	*	***	*	Kantorowicz, 1984
Brent	Cormornat	***	**	-	Kantorowicz, 1990
Brent	Gullfaks	*	***	-	Bjørkum and others, 1990
Brent	Hild	*	***	-	Lønøy and others, 1986; This work
Brent	Heimdal 3200 m	*	***	-	Thomas, 1986
Brent	Heimdal 3800 m	***	**	-	Thomas, 1986
M. Jurassic Yorkshire and Brent unspecified		***	*		Kantorowicz, 1984
Hugin	Gudrun	***	-		This work
Lower Jurassic:					
Bridgeport Sand	Whytch Farm	***	***	*	Morris and Shepperd, 1982
L. Jurassic	N.E. Scotland	-	***	-	Hurst, 1985
Triassic:					
Lunde	Snorre	**	**	-	Morad and others, 1990
Skagerrak	Claymore	***	***	-	Spark and Trewin, 1986
Skagerrak	Quad. 22	**	**	-	Humphreys and others, 1989
Sherwood	(Onshore Britain)	***	*	*	Burley, 1984
Sherwood		**	**	-	Bath and others, 1987b
Permian:					
Rotliegend	Southern North Sea	***	*		Lee and others, 1985, 1989
Rotliegend	Southern North Sea	***	**		Pye and Krinsley, 1986
Rotliegend	Southern North Sea	***	*	-	Glennie and others, 1978
Rotliegend	Southern North Sea	***	*	-	Goodchild and Whitaker, 1986
Rotliegend	Southern North Sea	***	**	-	Rossel, 1982

*** High concentrations, ** Intermediate concentrations,
* Low concentrations, - trace concentrations.

TABLE 2.—PUBLISHED ISOTOPIC DATA ON DIAGENETIC KAOLINITE FROM NORTH SEA RESERVOIR ROCKS

Field	Strat./ Age	$\delta^{18}O$ SMOW	Temp.°C (inferred)	Author
Heather	Brent	14	<45-60	Glasmann and others, 1989c
Huldra	Brent	12		Glasmann and others, 1989a
Veslefrikk	Brent	11-15		Glasmann and others, 1989a
Southern North Sea	Permian	19	15-38	Lee and others, 1989

Kaolinite is interpreted by these authors to have precipitated from meteoric water with a $\delta^{18}O$ value of -2 to -7.

were encountered. In the East Yorkshire-Lincolnshire Basin, recent meteoric-water recharge has caused dissolution of dolomite and precipitation of calcite in isotopic equilibrium with the present ground water (Bath and others, 1987b). Burley (1984), in a study of the Sherwood Sandstone in the Wessex Basin, Britain, also found evidence of meteoric-water recharge causing dissolution of carbonates and feldspar and precipitation of kaolin (kandite) minerals. This authigenic kaolin is not present in the more deeply buried parts of the Wessex Basin, which supports the interpretation that the kandites were not formed prior to the recent inversion of the basin and the attendant meteoric-water flushing. These are modern examples of unconformity-related leaching.

The dominance of smectite and illite over kaolinite in the Permian and Triassic sandstones in the North Sea can, to a large extent, be explained by climatic conditions. In desert environments, the flow of meteoric water into the subsurface is relatively limited, and dissolution of evaporitic minerals and amorphous silica in the sediments rapidly increases the concentration of silica and cations such as Na^+, K^+, Ca^{++} and Mg^{++} in the pore water. The composition would then fall in the stability field of smectite rather than kaolinite (Figs. 2A and B). Smectite is interpreted to have formed during early diagenesis in these sandstones, and these clay minerals are also the dominant weathering products in modern desert environments (Chamley, 1989). The fact that detrital kaolinite is rare in the Permian and Triassic sandstones (Jeans, 1989; Humpreys and others, 1989) is an indication of the weathering condition in the southern North Sea Basin.

Diagenetic kaolinite may have formed shortly after deposition under the influence of meteoric water where there was a high recharge of ground water and the content of evaporitic minerals was low. It also may have formed later, subsequent to uplift and basin inversion in the southern North Sea during late Jurassic and early Cretaceous times (Lee and others, 1989) or during late Cenozoic uplift of Britain (Bath and others, 1987b).

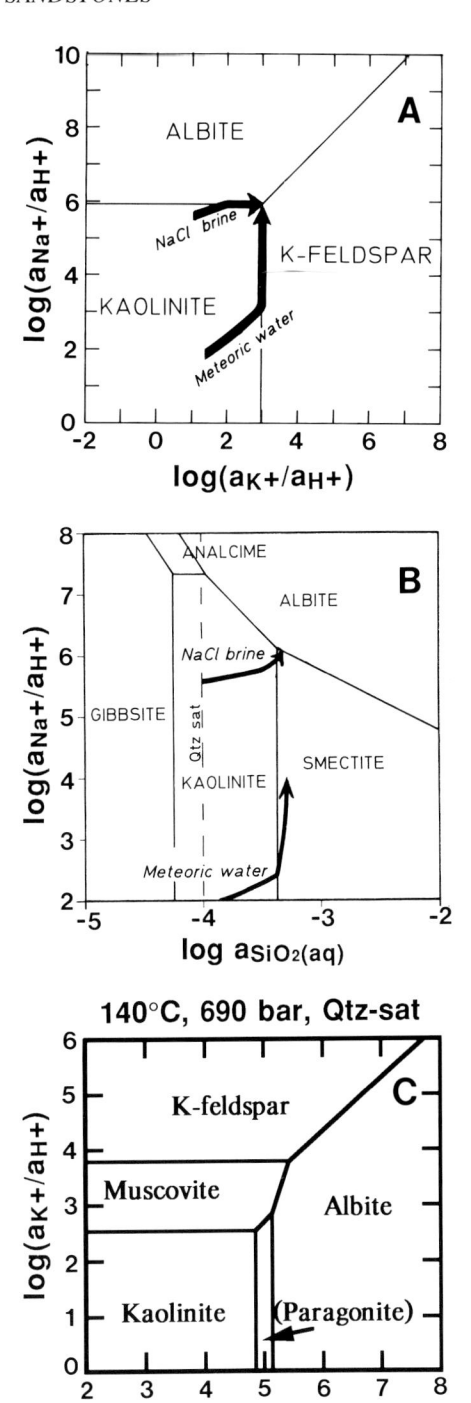

FIG. 2.—Stability diagrams depicting phase relations and reactions in feldspathic sandstones. Thermodynamic data from Helgeson and others (1978). In (A) and (B), reaction paths of aqueous solutions reacting with a feldspathic sandstone at 25°C and 1 bar are indicated by arrows. An equal amount of albite and K-feldspar is assumed released into solution during hydrolysis. The reacting waters, "meteoric water" and "NaCl brine" (1M), were originally charged with CO_2 at a partial pressure of $10^{-1.5}$ bar. In both cases, kaolinite will be the first to precipitate. The "NaCl brine" will soon reach equilibrium with albite, whereas the meteoric water will approach equilibrium with K-feldspar first. Thus, meteoric water will tend to dissolve more albite than K-feldspar and NaCl brines will stabilize albite and dissolve K-Feldspar. (C) Note that kaolinite and K-feldspar cannot co-exist and illite (muscovite) will precipitate whereas albite and kaolinite may be stable in the absence of K-feldspar.

Lower Jurassic Sandstones

The Statfjord Formation is a fluvial and shallow marine sandstone. It exhibits evidence of well-developed feldspar leaching and authigenic kaolinite (Dypvik and others, pers. commun.,1990). Lower Jurassic sandstones cored in an onshore well at Lossiemouth, northeast Scotland, show increasing kaolinite content in a shallowing-upward-marine sequence. Kaolinite has been interpreted to have precipitated during meteoric-water flushing and to have formed prior to precipitation of illite (Hurst, 1985).

In the onshore Wytch Farm in Dorset, Britain, the Lower Jurassic Bridport sand contains abundant authigenic kaolinite (Morris and Shepard, 1982). This kaolinite causes significant permeability reduction, partly due to blocking of pore throats. In addition, mixed-layer illite-smectite clays are important in these sandstones. The friable reservoir sandstones contain 10 to 15% carbonate cement, whereas the carbonate content may exceed 50% in the less porous calcareous sandstones. This example demonstrates that the formation of kaolinite does not require acid pore waters. The abundance of carbonate in this sequence must have caused the pore water at all times to be in equilibrium with carbonate. Thus, the authigenic kaolinite must have precipitated from pore waters that were not highly acid. The increase in the amount of dissolution of feldspar and precipitation of diagenetic kaolinite from Triassic to Jurassic sandstones probably reflects a change toward a more humid climate.

Middle Jurassic Sandstones

A review of the diagenesis of the Brent Group was recently compiled (Bjørlykke and others, pers. commun., 1991). The reservoir sandstones of the Brent Group are characterized by relatively high concentrations of authigenic kaolinite and secondary porosity due to dissolution of feldspar (Fig. 3A). Mica is commonly partly replaced by kaolinite. In this process, the primary sheet of mica is ex-

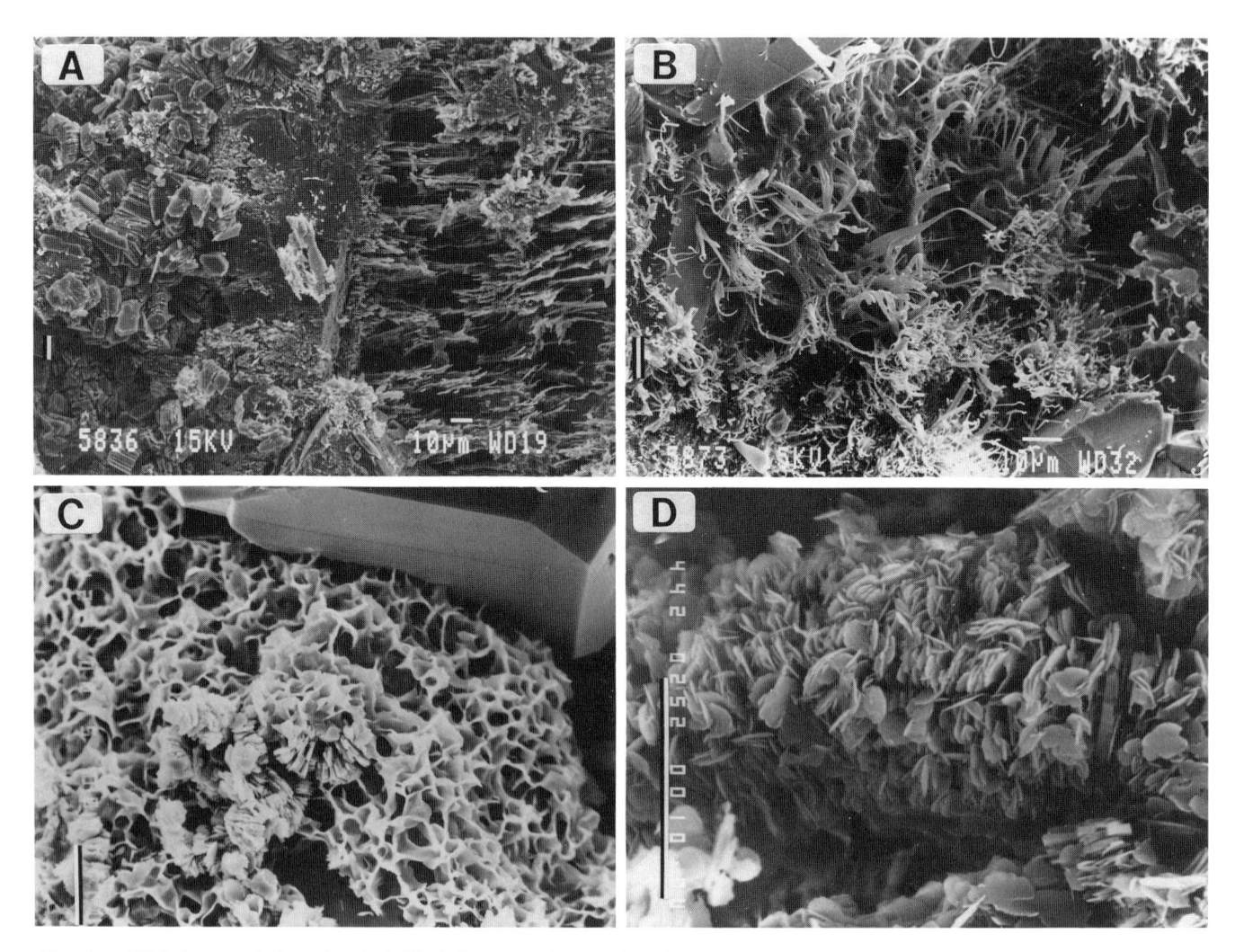

FIG. 3.—SEM pictures of clay minerals in North Sea reservoirs. (A) Dissolved K-feldspar (to the right) with surrounding authigenic kaolinite (Brent Group, Oseberg field, 2,940 m, RBK). (B) Illite formed from kaolinite precursors (Gullfaks, 4,261 m, RBK). (C) Corrensite formed from a smectite precursor (Snorre Field, 2,951 m, RBK) (Sørlie and Mellem, 1990). (D) Chlorite grown on dissolving vermicular kaolinite (Snorre Field 2,631 m, RBK) (Sørlie and Mellem, 1990).

panded into available pore space. Where a grain of mica is squeezed between quartz grains, the mica tends to be less altered to kaolinite and not expanded (Bjørlykke and Brendsdal, 1986).

Smectite is rare and illite occurs only in minor amounts at burial depths shallower than 3.5 km. Chlorite is commonly present, but in relatively low concentrations. Progressive changes in the composition of chlorite can be observed with increasing burial (Jahren and Aagaard, 1989).

Because the Brent Group was deposited in shallow marine, deltaic and fluvial environments under humid-climate conditions, we can assume that it was subjected to relatively high flux of meteoric water after deposition. The Rannoch-Etive sandstone unit was deposited as a sheet of high-permeability sand, which even at moderate to low hydraulic head could be expected to transmit high fluxes of meteoric pore water.

In the Ness Formation, the flux of meteoric water was probably controlled by the permeability and lateral continuity of the channel sandstone facies. Based on point-count data from the Huldra field, Nedkvitne and Bjørlykke (pers. commun., 1991) showed that channel sandstones from the Ness Formation had less feldspar leaching than those of the Rannoch-Etive Formation with the same permeability. This difference is probably attributable to the lower connectivity of the Ness channel facies compared to that of the delta-front facies of the Rannoch-Etive Formations. During Cimmerian (Late Jurassic/Early Cretaceous) uplift and erosion, the sandstones may have been subjected to a second period of meteoric-water flushing (Bjørlykke, 1983). However, the amount of kaolinite does not always show an increase upward toward the unconformity and it is difficult the assess the importance of the leaching phase associated with uplift and exposure. Bjørkum and others (1990) observed a decrease in the concentration of kaolinite in the Rannoch Formation toward the unconformity in the Gullfaks field and interpreted much of the kaolinite to be detrital. They concluded that the Cimmerian unconformity played a minor part in the distribution of kaolinite. However, the Rannoch Formation is characterized by high-mica content and would be expected to have lost much of its permeability by compaction at depths of 0.5 to 1 km prior to uplift and exposure below the unconformity, and therefore may not have been a good aquifer. The well-sorted sandstones of the Etive Formation would probably have been the main aquifer below the unconformity, but firm evidence of the extent of meteoric-water leaching after uplift is lacking. The rotated fault blocks would have been exposed as islands and the extent of feldspar leaching below this unconformity would depend, in part, on the subaerial relief of those islands. Because these sediments must have been poorly indurated, erosion may have kept pace with uplift, thus producing rather low subaerial relief, which limited the extent of meteoric-water recharge.

The pervasive leaching of feldspars and the abundance of diagenetic kaolinite, which is so typical of sandstones from the Brent Group, are not present to the same extent in other sandstones of the North Sea except in the Statfjord Formation, which also represents fluvial and shallow marine deposits.

The Ness Formation commonly contains coal beds several meters thick, which represent a humic source rock interbedded in the reservoir. The sandstones directly overlying these coal beds do not, however, show any higher level of feldspar leaching than the rest of the sequence, suggesting that any CO_2 or organic acids released had little effect on silicate diagenesis (Bjørlykke and Brendsdal, 1986).

A strong increase in the amount of diagenetic illite is observed below 3.7 to 3.8 km burial depth (Thomas, 1986; Scotchman and others, 1989). In the Gullfaks field, the Brent reservoir occurs at depths ranging from 1,800 to more than 4,000 m. Bjørlykke and others (in prep.) have shown that kaolinite is abundant in the shallowest parts, and below 4,000 m there is evidence of replacement of kaolinite by illite (Fig. 3B). A similar trend has been found in Middle Jurassic reservoir sandstones from Haltenbanken (Bjørlykke and others, 1986; Ehrenberg and Nadeau, 1989). K/Ar dating of illite in the Brent Group (Table 3) yields a rather wide range of apparent ages from 100 to 30 Ma (Thomas, 1986; Hamilton and others, 1989; Glasmann, 1989a and c). K/Ar dating of illite in Mesozoic and Tertiary shales from Veslefrikk and Huldra fields yields Early Tertiary ages (Glasmann, 1989b). Jourdan and others (1987) obtained K/Ar dates on illite from the Alwyn field also indicating Early Tertiary ages. In some cases, these dates may be too old due to contamination by clastic feldspars or mica. The K-Ar dates indicate that the illites formed relatively early, mostly in the Early Tertiary at rather shallow burial depth (2 km).

In a study of the Brent Group sandstones from the Cormorant Field, Kantorowicz (1990) observed an extensive increase in diagenetic illite replacing kaolinite at about 3.8 to 4.0 km. The reduction in permeability resulting from the presence of illite was found to be significantly higher when cores were analyzed after critical-point drying rather than air drying.

The rapid increase in the concentration of illite below 3.5 to 4.0 km is related to illitization of kaolinite, which seems to require temperatures of about 130 to 140°C (Bjørlykke and others, 1986; Ehrenberg and Nadeau, 1989). This relation suggests that illite has been forming continuously at that depth to today, but this interpretation is in conflict with most of the K/Ar dates. Ehrenberg and Nadeau (1989) found K/Ar ages ranging from 55 to 31 Ma in the Garn Formation from Haltenbanken. However, they found that the older ages probably were due to contamination and that illitization could have been as recent as 3 Ma. This implies that illitization occurred in a closed system in water-wet pores after oil emplacement.

The reaction:

$$Al_2Si_2O_5(OH)_4 + KAlSi_3O_8 =$$
Kaolinite   K-feldspar

$$KAlSi_3O_{10}(OH)_2 + 2SiO_2 + H_2O$$
Illite   Quartz   Water

requires no supply or removal of ions. It is essentially a kinetically controlled reaction (Bjørlykke, 1983). Kaolinite and K-feldspar are not co-stable at quartz saturation at any diagenetic conditions (compare Figs. 2B and C). At temperatures lower than 130 to 140°C, they will persist as

TABLE 3.—K/Ar DATES FOR JURASSIC RESERVOIR ROCKS (BRENT GROUP) FROM THE NORTH SEA BASIN

Field	Depth	K-AR Age	Number of Anal.	Authors
Heather	3,350-3,650	30-57 Ma	20	Glasmann and others, 1989c
Heather	2,950	40-45	2	Glasmann and others, 1989c
Heather	2,900	90-94	3	Glasmann and others, 1989c
Unknown		56-59	5	Hamilton and others, 1987
25/4-1	3,200	40-44	3	Thomas, 1986
25/4-5	3,700	38-40	2	Thomas, 1986
NW Hutton	3,650	39-49		Scotchman and others, 1989
Alwyn		75-35	12	Jourdan and others, 1987

metastable minerals and at higher temperatures the reaction will proceed and cause precipitation of illite. This explains the rapid increase in the concentration of diagenetic illite in these reservoirs at depths corresponding to this temperature threshold. Where no K-feldspar is present, as in the Hild field, kaolinite occurs and appears to be stable at much higher temperatures, as high as 150 to 160°C (Lønøy and others, 1986).

From the current distribution of illite in the Jurassic sandstones, it seems unlikely that significant quantities of illite should have formed at shallow depth (<3 km). High-temperature events due to inflow of compactional water are, in general, not important unless the porewater flow is focused (Bethke, 1985; Hermanrud, 1986). This means that large volumes of rock must be drained through small cross sections in the overlying sequence. Therefore, focused compactional-porewater flow cannot explain diagenetic features that are commonly observed in larger volumes of rocks. Faults may represent possible conduits for porewater flow, but there is very little fault activity in the Cenozoic in the North Sea Basin (Lovell, 1990). There is no reason why compaction-driven waterflow should have been particularly active in Eocene times when sedimentation rates were low compared to those of the Late Cenozoic. In the North Sea Basin, the present geothermal gradients are remarkably constant despite active Quaternary subsidence (Hermanrud, 1986). In the Veslefrikk field, the present pore water has a low salinity and negative oxygen isotope composition. This precludes the introduction of water from the deeper parts of the basin, where the pore water would be more saline and have a more positive oxygen isotope composition (Glasmann, 1989a). Several other shallow reservoirs from the North Sea are also characterized by low salinities and $\delta^{18}O$ values of about -4 (Bjørlykke and others, 1988; Egeberg and Aagaard, 1989).

Many of the K/Ar dates are of Eocene age and coincide with the volcanism that led to deposition of Eocene tuffs (Malm and others, 1984). A hydrothermal event associated with this volcanism would explain the observed ages, but there is no other evidence of Eocene igneous intrusions in the Viking Graben or East Shetland Basin. If illites were formed by a short episode of heating, however, K/Ar dates would not display relation to the timing of oil emplacement in the reservoirs.

Upper Jurassic Sandstones

Upper Jurassic sandstones form important reservoirs in several parts of the North Sea. They represent depositional environments that were different from those prevailing in Middle Jurassic times, particularly in the northern parts of the North Sea Basin where the Brent Group was deposited. Upper Jurassic sandstones are typically marine shelf or slope deposits, but more proximal facies also exist. These sandstones are most closely associated with the main source rock, the Kimmeridge Clay Formation (Draupne Formation).

The Piper Sandstone Formation from the Tartan and Piper fields has been studied by Burley (1986), who interpreted diagenetic kaolinite to have been formed by flushing of acid pore water generated from the Kimmeridge Shale Formation during the early stages of maturation. Burley estimated the main phase of acid generation to have occurred at 2,000 m depth. The acid solution is thought to have migrated about 1 km vertically. Burley and others (1989) claimed that the precipitation of kaolinite was not related to detrital-feldspar alteration, but that aluminium was transported in solution from deeper parts of the basin. In the argillaceous sandstones and mud rocks interbedded with the Piper sandstones, kaolinite that may be partly detrital is relatively abundant in the shallower part of the reservoir, but is almost absent below 4,000 m due to illitization.

An interesting study, which shows the dependence of claymineral distribution on depositional environment in Upper Jurassic sandstones, was carried out on the Claymore field (Spark and Trewin, 1986). In the Piper Formation, which is a paralic deposit, authigenic kaolinite is the dominant clay mineral and is interpreted to have formed by dissolution of feldspar during meteoric-water flushing. The Claymore Sandstone Member, which represents a more distal turbidite environment, directly overlying the Kimmeridge Clay, contains very little diagenetic kaolinite. The Ten Foot Sandstone is a thin turbiditic sandstone occurring within the Kimmeridge Clay at 2,500 m depth. Spark and Trewin (1986) noted that kaolinite replacing feldspar grains is rare in the Ten Foot Sandstone. This unit only contains a small amount of authigenic clay, mostly illite-smectite. Spark and Trewin (1986) pointed out that the distribution of clay minerals can be best explained in terms of variation in depositional facies.

Similarly, in the Gudrun field, sandstones interfingering with the Draupne (Kimmeridge Clay Formation) at 4,263 to 4,305 m depth contain illite as the only important clay mineral (Table 1). Illite occurs partly as a clay coating and SEM and XRD analyses did not detect any kaolinite or evidence of illitized kaolinite (Ramm, pers. commun., 1990). At this burial depth, the source rock would have reached high maturity and organic acids should have been generated (Surdam and others, 1984; Burley, 1986). To the extent that organic acids or CO_2 are expelled from the shales, these acids should have been expected to leach adjacent sandstones more strongly than in reservoirs farther way from the source rock. The fact that these Upper Jurassic sandstones, which are interbedded with the source rocks, show very little evidence of feldspar leaching and precipitation of diagenetic kaolinite suggests that the organic acids contribute little to feldspar dissolution and precipitation of authigenic kaolinite.

The Upper Jurassic sandstones in the Fulmar field are interpreted to have been deposited in a partly turbiditic shelf environment (Johnsen and others, 1986). The most notable features of these sandstones are the absence of diagenetic kaolinite and the lack of secondary porosity formed by feldspar dissolution (Johnsen and others, 1986; Saigal, Bjørlykke and Larter, pers. commun., 1990). Steward (1986), on the other hand, found textures suggestive of secondary porosity formed by dissolution of feldspar and bivalve shells. He interpreted this dissolution to be caused by early leaching and noted the conspicuous absence of kaolinite. The Fulmar sandstone is surrounded by shales and was poorly connected to the meteoric-water lens. This may explain the low degree of feldspar leaching and the absence of authigenic kaolinite in this unit.

At 3,400 m, the reservoir is above the critical depth for illitization and little illite has formed. However, because the Fulmar sandstone does not contain kaolinite or mixed-layer clays, we predict that significant amounts of illite will not develop at greater depth. It seems reasonable to relate the absence of diagenetic kaolinite in the Fulmar field to depositional environment, as in the case of the Claymore and the Ten Foot Sandstone. Sandstones deposited as intermediate to distal turbidite facies surrounded by shales probably received relatively small amounts of meteoric-water flushing, and this may explain why there is little evidence of leaching.

The Ula sandstone in the Ula field contains some diagenetic kaolinite, but only moderate amounts in the uppermost part (Nedkvitne, pers. commun, 1990). The sandstones of the Troll field, which represent a more proximal facies, have more abundant diagenetic kaolinite. This suggests that these sandstones, which may represent a more proximal shelf facies, were subjected to fluxes of meteoric water that were more voluminous than the turbiditic Upper Jurassic sandstones, but less voluminous than most Middle Jurassic sandstones.

Tertiary Sandstones

Tertiary sandstones are buried to depths as great as 3 km in the central parts of the North Sea. Paleocene/Eocene sandstones are reservoirs in several oil and gas fields in the British and Norwegian sectors. These sandstones were deposited mostly as turbidites on submarine slopes dipping to the southeast. Tertiary sandstones commonly contain significant amounts of authigenic kaolinite, but generally less than the Middle Jurassic sandstones. In part of the Norwegian sector, Hurst and Buller (1984) found that clays in dish structures consist dominantly of diagenetic chlorite with minor illite. Kaolinite was not present in these samples. Interbedded mudstones contain mostly illite-smectite (Malm and others, 1984). Chlorite coatings on sand grains in the dish structures plug pore throats and reduce vertical permeability. Our unpublished analyses of Lower Tertiary sandstones from the Sleipner field and Heimdal field also show little diagenetic kaolinite (Table 1). Lower Tertiary tuffaceous sediments (Balder Formation) are composed mostly of altered volcanic ash and siliceous microfossils. Authigenic clays in tuff beds and siliceous shales include smectite, mixed-layer clays, and chlorite. This clay-mineral assemblage is probably related to the high content of volcanic debris and diatomaceous ooze. The amorphous silica produced high-silica concentrations in the pore water and the volcanic debris contained unstable Fe- and Mg-rich minerals, which may have served as precursors for smectite and chlorite. Volcanic material is otherwise a rather minor component in the North Sea sedimentary sequence. Smectite is also common in Mesozoic and Tertiary shales and is transformed into mixed-layer clays and illite at 2 to 3 km burial depth (Dypvik, 1983).

CLAY-MINERAL REACTIONS

Stability of Clay Minerals

The formation of authigenic-clay minerals in sandstones involves nucleation and growth from a supersaturated aqueous solution and, consequently, occurrences have thermodynamic and kinetic constraints. Minerals will not form from an undersaturated solution, but kinetic constraints may preclude precipitation from a supersaturated solution. The saturation state of a mineral, at a specified pressure and temperature, is determined by the chemical composition of the co-existing pore water. The nucleation and growth rates also depend on the degree of supersaturation as well as temperature (Nielsen, 1964; McLean, 1965; Ridley and Thompson, 1986). In addition, the porewater composition will have an effect on the mineral-water interfacial properties, which are critical for nucleation and growth.

The solubility of aluminium-bearing silicate minerals is low under diagenetic conditions (Berner, 1981). Thus, only insignificant quantities of minerals can precipitate from pore water without simultaneous dissolution of other minerals or amorphous phases. Even if the porewater composition controls the formation of authigenic minerals, the pore water itself is merely a transport medium between the dissolving and the precipitating phases.

The thermodynamic stability of clay minerals in weathering and diagenetic environments is a controversial issue (Hower and others, 1976; Lippmann, 1981 and 1982; Aagaard and Helgeson, 1983; Garrels, 1984; Jiang and others, 1990). Low-temperature experiments have not convincingly demonstrated reversals of equilibrium (May and others, 1986). Due to their small grain size, crystal imperfections, and common compositional heterogeneity, clay minerals may be metastable relative to their coarser grained, rock-forming mineral analogs.

Egeberg and Aagaard (1989) and Aagaard and others (1989) discussed the stability of detrital and authigenic minerals relative to formation-water chemistry of Jurassic clastic reservoirs, offshore Norway. They showed that in the temperature range of 55 to 160°C, most diagenetic minerals were close to equilibrium with formation waters. This applies to kaolinite, albite, K-feldspar, illite (muscovite), and calcite. Silica content in the formation water was close to quartz saturation at temperatures exceeding 60 to 80°C. Dolomite appeared to shift from being saturated with respect to pore waters at lower temperatures to being unstable above 120 to 140°C. Dioctahedral smectites were unstable relative to formation water at all temperatures when silica concentration was at quartz saturation.

The silica concentration in the pore water is a critical parameter in determining the diagenetic-mineral assemblage. At low temperatures, the growth rate of quartz is too slow to keep aqueous-silica concentrations at the level of quartz saturation. Volcanic glass, biogenic silica (diatoms, radiolaria) and unstable silicate minerals dissolve, thereby increasing silica concentrations above quartz saturation and causing precipitation of smectites and zeolites. In clean diatom ooze, opal-CT will normally precipitate instead. Hence, smectites, zeolites and opal-CT are the minerals typical of low diagenetic grade (i.e., below 70°C). As the rate of quartz precipitation increases at higher temperatures, these will become unstable and the pore waters approach equilibrium with quartz.

Smectite-to-Illite Transformation

Smectites, which formed at lower temperatures, either in dry continental environments or in sandstones with a relatively high content of volcanoclastic and/or biogenic silica, will tend to transform to illite with deeper burial. The lower Tertiary Balder Formation is a good example of such a lithology with a primary high amorphous-silica content. Because of the widely different chemical compositions of smectite and illite, the illitization process will normally involve other mineral reactants/products, and can only take place by a complete dissolution of the original smectite precursor (Boles and Franks, 1979).

At temperatures exceeding 60 to 80°C, quartz forms more readily and maintains the concentration of aqueous silica close to the level of quartz saturation. Smectites are then no longer stable relative to the pore water (Aagaard and Helgeson, 1983; Sass and others, 1987) and there is a considerable chemical affinity (10–25 Kcal/mole) to dissolve these minerals. The metastable persistence of smectite is kinetically controlled and depends on the dissolution rate of smectite, nucleation and growth rate of illite (muscovite), and diffusive/advective transport in the pore water. The extent of illite formation may be limited by the potassium supply (Hower and others, 1976), i.e., the content of potassium minerals such as K-feldspar and mica. Albitization of K-feldspar (Aagaard and others, 1990) is a possible source of potassium for illitization. Further illitization of smectite-illite layers appears to be a grain-coarsening process, where the coarser crystals grow at the expense of smaller crystals (Inoue and others, 1988; Eberl and others, 1990).

Formation of Chlorite

Illite is not the only mineral that may form from smectite precursors. Transformation of trioctahedral smectite to chloritic mixed-layer clays and chlorites is a prominent feature in the Triassic Lunde and Skagerrak Formations. Chlorite is also abundant in the volcaniclastic units of the Tertiary Balder Formation (Malm and others, 1984; Hurst and Buller, 1984; Spark and Trewin, 1986; Humphreys and others, 1989). The formation of chlorite appears to follow different pathways depending on the local chemical environment. The fine-grained sandstones of the Upper Lunde Formation from the Snorre Field contain both corrensite (Fig. 3C) and chlorite (Fig. 3D) (Sørlie and Mellem, 1990). The chlorite (chamosite) forms from an iron-rich smectite, whereas the corrensite replaces magnesium-rich smectites.

In Jurassic clastic reservoirs, most of the chlorites appear to form from a kaolinite precursor. This reaction requires sources of iron and magnesium, and the occurrence of chlorite is commonly associated with biotite and other ferromagnesium minerals. Jurassic feldspathic arenites contain these minerals in low abundance and this explains the generally low concentrations of chlorite.

Detailed TEM studies on these chlorites provide insight into the dynamic nature of chlorite growth during burial diagenesis. The chemical composition of authigenic chlorites exhibits a definite temperature dependence (Jahren and Aagaard, 1989). With increasing burial temperature, there is an increase in tetrahedral Al as well as a decrease in octahedral vacancy. With further burial, earlier formed chlorites become unstable and adjust to the new conditions by dissolution/precipitation. The chemical driving force for this process is rather limited, but because the chlorite crystals are undergoing grain coarsening by Ostwald ripening (Jahren, pers. commun., 1991), smaller crystals dissolve and reprecipitate as rims on larger crystals. In this way, chemical adjustments to increasing burial temperature and changing porewater composition are recorded from core to rim in large chlorite crystals (Jahren, pers. commun., 1991).

TEM studies of illites occurring with chlorite in Jurassic sandstones (Jahren and Aagaard, 1989) also have demonstrated compositional-temperature dependence. An increasing muscovite substitution is observed with higher temperatures. Compositional data on co-existing illite/chlorite pairs and thermodynamic data on coupled Tchermak substitution indicate that these illites and chlorites formed close to equilibrium.

DISCUSSION

The Dependence of Early Clay-Mineral Diagenesis on Climate and Depositional Environment

Early diagenesis in continental sediments is a direct continuation of surface weathering and is therefore dependent on climate. In a humid climate with a rainfall of 100 cm/yr and 10% subsurface infiltration into the ground, the shallow subsurface would receive a meteoric-porewater flux of 10 cm^3/cm^2 a year. This amounts to 10^7 cm^3/cm^2 in one million years. The percentage of this flux that will invade units below the sea floor depends on the hydraulic head and on sandbody geometry and connectivity. It would also depend on the thickness of the aquifer compared to the area of meteoric-water infiltration. The purpose of these crude calculations is to indicate the approximate magnitude of the porewater flux that can be expected to occur in fluvial and shallow marine sediments shortly after deposition. The total flux of meteoric water though each volume of sediments depends on the amount of time it remains in the zone flushed by meteoric water. Keeping other factors constant, the meteoric-water flux will therefore be inversely related to sedimentation rate. Similar fluxes can be received after tectonic uplifts, but the permeability of the sediments then may be reduced due to compaction depending on the depth of burial prior to uplift. A sequence with sandstones and shales dipping into the basin will produce confined aquifers, which may extend the meteoric-water flow deep into the basin and far offshore. The meteoric-water flow along permeable sandstones will gradually dissipate into poorly compacted mudstones and flow to the surface. The development of even slight overpressure due to compaction, particularly in rapidly deposited muddy sediments, can prevent infiltration of meteoric water. Fluvial and shallow marine sediments will therefore normally undergo much higher meteoric water fluxes than more offshore shelf sandstones and turbidites. As meteoric-water flows through a sedimentary sequence, the pore waters become more reducing as organic components or minerals are oxidized. The pore waters will gradually approach equilibrium, first with any carbonate minerals that might be present and then, much later, with the more slowly reacting silicate minerals (Lasaga, 1984). Marine sandstones will first undergo early marine diagenesis on the sea floor and in the sulfate-reducing zone, and may then be flushed with meteoric water if the sandstones are connected to a meteoric-water lens from a land area.

This examination of North Sea reservoir sandstones (Table 1) has shown that clay mineralogy is a function of climate and depositional environment (Table 4). The initial composition of the clastic material is certainly important, particularly the content of feldspar and mica and the amount of unstable rock fragments. In the North Sea, the volcanic sediments are of relatively minor importance with the exception of the Lower Tertiary Balder Formation.

Permo-Triassic sandstones were deposited in a dry continental environment. In this environment, the weathering products are dominantly smectite (Chamley, 1989). Since much of the surface water may have been close to equilibrium with smectite and illite due to reactions with amorphous silica and evaporites, these minerals also would form during early diagenesis. Some kaolinite may have formed during meteoric-water flushing in sandstones. Meteoric-water leaching may be particularly efficient in alluvial fans and wadi-fill deposits, where there would be a focused flow of meteoric water. It is difficult, based on the available data,

TABLE 4.—STRATIGRAPHIC AND ENVIRONMENTAL CONTROL ON CLAY-MINERAL COMPOSITION OF NORTH SEA SANDSTONES

Age	Climate	Facies	Clay minerals
Pal/Eocene Volcanics	Humid	Deep	Smectite/illite Chlorite
L.Tertiary Sandstones	Humid	Deep	Chlorite, Illite
U.Jurassic to Tertiary marine sst	Humid	Shallow Marine and Deep Water	Kaolinite mostly in shallow marine facies. Little diagenetic kaolinite in turbidite facies.
L.-M.Jurassic	Humid	Deltaic and Shallow Marine	Kaolinite Minor Illite and Chlorite (<4,000 m)
Permian and Triassic	Dry	Fluvial Eolian	Mixed-layer min. Smectite and Chlorite.

to assess how much early diagenetic kaolinite could have formed in Permian and Triassic sandstones, but the studies cited later suggest that most of the kaolinite precipitated after later tectonic uplift and meteoric-water flushing.

Oxygen-isotopic data and textural evidence indicate that kaolinite in the Permian sandstones formed at low temperatures from meteoric water introduced during late tectonic uplift, whereas illite formed at higher temperatures from isotopically heavier water (Table 2). Triassic sediments, more recently uplifted onshore in Britain, are subjected to meteoric-water flushing and precipitation of kaolinite (Burley, 1984; Bath and others, 1987b).

Kaolinite is abundant in most Lower and Middle Jurassic sandstones in the North Sea, particularly in the Statfjord Formation and the Brent Group, which represent fluvial and shallow marine environments. Upper Jurassic sandstones have variable but generally lower kaolinite content. In sandstones representing shallow-turbidite environments, such as the Fulmar and Claymore fields, diagenetic kaolinite may be absent, probably as a result of insufficient meteoric-water flushing. It is worth noting that these sandstones show no evidence of late diagenetic kaolinite associated with maturation of the adjacent Kimmeridge Clay Formation. Even the Ten Foot Sandstone, which is interbedded in Kimmeridge Clay Formation in the Claymore field, contains no diagenetic kaolinite and displays little evidence of secondary-porosity development. Generally, Jurassic sandstones representing deeper water environments contain less diagenetic kaolinite and less extensive secondary porosity than the sandstones deposited in fluvial and shallow marine environments. This has been particularly well demonstrated from the Claymore field by Spark and Trewin (1986).

Meteoric-Water Diagenesis

Meteoric water is driven into sedimentary basins by the groundwater head and may cause extensive leaching of feldspar and development of secondary porosity (Bjørlykke, 1984). Modeling of meteoric-porewater flow into sedimentary basins suggests that it many extend 2 to 4 km below sea level and that the flow is sensitive to sea-level changes (Bethke, 1989). The most effective leaching, however, probably occurs at burial depths less than a few hundred meters. The pattern of feldspar leaching in North Sea reservoirs shows that the degree of leaching is also facies dependent.

Meteoric water is somewhat acidic initially, and it may become more acidic due to addition of CO_2 in the soil horizon. However, this acidity is rather rapidly neutralized by small amounts of carbonates, which are almost always present, particularly in marine sediments. Hydrolysis of silicate minerals like feldspar and mica does not require acidic waters, and diagenetic kaolinite replacing feldspar is found in highly calcareous rocks. A good example of this is the Lower Jurassic Bridport Sandstone in Britain. Most of the reservoir sandstones from the North Sea examined in this paper contain significant amounts of carbonate.

The pH of pore water in equilibrium with carbonate is highly dependent on the pCO_2. The most important characteristics of meteoric water are its low ionic strength and low chlorinity. The concentrations of alkali (K^+, Na^+) and earth alkali ions (Ca^{++}, Mg^{++}) are, in most meteoric waters, limited and balanced by the bicarbonate (HCO_3^-) counter anion. At high pH, CO_3^{--} and OH^- also contribute. Meteoric water will increase its content of alkali ions and pH along the flow and reaction path (compare Figs. 2A and B) and may, after some reaction, reach saturation with carbonate and silicate minerals. At low temperatures, the pH is likely to be carbonate buffered due to higher reaction rates for carbonate minerals than silicate minerals (Lasaga, 1984). The pH necessary to saturate albite is, however, considerably higher than is required for K-feldspar saturation. The critical K^+/H^+ ratio required to approach equilibrium with K-feldspar is likely to build up in the pore water before the Na^+/H^+ value corresponding to albite saturation is reached. This will cause selective leaching of albite, which is often observed (Bjørlykke and others, 1986). Dissolution of feldspar and mica, and precipitation of kaolinite, can take place at relatively high pH, but requires a supply of protons. Bjørkum and others (1990) argued that the protons available for leaching will be exhausted a few meters below the soil horizon. Because meteoric-water leaching occurs at relatively low temperature when reaction rates are low, the porewater may remain in the stability field of kaolinite for a long time. This allows leaching to take place over long distances along the pathway of porewater flow. The rate of dissolution of mica and feldspar is surface controlled and, therefore, not directly proportional to the flux (Aagaard and Helgeson, 1983), but a minimum flux is required to remove potassium so that the pore water remains in the stability field of kaolinite. In addition to the acids generated in soil horizons, CO_2 may also be generated from organic matter in sediments at shallow depth, thus increasing the potential for leaching of feldspar and mica.

The Effects of Carbon Dioxide and Organic Acids on Clay-Mineral Diagenesis

The hypothesis, that carbon dioxide and organic acids released from source rocks have a major influence on clay-mineral diagenesis, has been put forward by several authors, notably Schmidt and McDonald (1979) and Surdam and others (1984). In the North Sea, Burley (1986) has interpreted grain dissolution and precipitation of kaolinite in the Piper and Tartan fields to be due to generation and upward migration of organic acids at about 2,000 m depth, considerably in advance of maturation and expulsion of hydrocarbons. Burley (1986) also assumed that the acidic pore waters would migrate up dip along an extended pathway over a vertical distance of about 1 km before reaching the reservoir. We expect acids generated in the source rocks, however, to react with carbonate and feldspar minerals in sandstones immediately adjacent to the source rocks and along the flow path, and thus become neutralized before reaching the reservoir rock. Experimental work indicates that complexing of aluminum is only significant at very low pH conditions (Surdam and others, 1984). Data from the North Sea show no relation between distance to the source rock, degree of leaching, and amount of diagenetic kaolinite. Upper Jurassic sandstones closely associated with the Kimmeridge Clay Formation, such as the Fulmar Sandstone and the Claymore Formation, show little leaching and commonly contain no diagenetic kaolinite. If dissolution of feldspar and precipitation of diagenetic kaolinite are related to generation of acids in source rocks, it is difficult to explain why sandstones so closely associated with mature source rocks are not affected by this process. Most North Sea sandstones contain some carbonate cement, mainly calcite. Even where this is a minor component, the pore water must have been close to equilibrium with respect to calcite. The Kimmeridge Clay Formation usually contains carbonates (Irwin, 1981), which can be expected to neutralize much of the acid generated. Silicate minerals in the source rock, and along the pathways of porewater flow, should also be expected to react with the acid generated before it reaches the reservoir units. The source rock (Kimmeridge Clay Formation) generally is rather sapropelic and the amount of acids generated would be even lower than in the Gulf Coast, where the generated acids were shown to be insufficient to explain the observed secondary porosity (Bjørlykke, 1984; Lundegard and others, 1984).

Oil migrates as a separate phase driven by buoyancy and is normally not associated with porewater flow (England and others, 1987). Also, porewater flow generated by compaction has an average upward component that is lower than the subsidence rate (Caritat, 1989; Bjørlykke and others, 1989). In typical North Sea reservoirs, where permeable reservoir sandstones are truncated by low-permeability cap rocks, localized focused porewater flow is not likely to be channeled through reservoir rocks. Once hydrocarbons start to accumulate, the top of the trap becomes even more impermeable with respect to upward water flow. Modeling of porewater flow suggests that convection currents are not important in most sedimentary basins. Vertical changes in porewater geochemistry indicate that pore waters are commonly stratified in a way that precludes vertical mixing (Bjørlykke and others, 1988). This suggests that there normally is no flow of pore water from the deeper parts of basins, which could transport acids into the shallow reservoirs.

It is clear that the degree of feldspar leaching can be related to facies and climate. Secondary porosity is more common in sandstones representing humid fluvial and shallow marine environments, such as Lower and Middle Jurassic sandstones, than in Permian and Triassic sandstones. Sandstones representing deep-water and offshore-shelf environments display less leaching. This pattern of selective leaching is consistent with expected differences in meteoric-water flushing and cannot be explained by burial diagenetic reactions.

Recent studies suggest that the pH of pore waters in sedimentary basins is buffered by clay minerals, and that organic acids and carbonate minerals have a much lower buffering capacity (Hutcheon, 1989; Smith and Ehrenberg, 1989). New experimental data suggest that complexing of aluminum may only be significant at pH values lower than those likely to be encountered in naturally occurring pore water (Stoessell and Pittman, 1990).

Burial Diagenesis and Prediction of Illite and Kaolinite Distribution in Reservoir Rocks

In Permian and Triassic sandstones, illite formed diagenetically from smectite and mixed-layer clays. Jurassic sandstones, where buried to less than 3.5 km, contain little illite or smectite. The present distribution of illite may suggest that illitization of kaolinite is depth and temperature controlled, and that it occurred during the rapid subsidence in late Cenozoic times. K/Ar dates on authigenic illite, however, generally indicate Early Tertiary ages (Hogg and others, 1987; Jourdan and others, 1987; Hamilton and others, 1989; Glasmann and others, 1989a). Ehrenberg and Nadeau (1989) suggested that these ages may be too old due to contamination. If the precipitation of illite from dissolving kaolinite occurs at 3.5 to 4 km, most of the fluids released by compaction have already been expelled. The reaction would therefore take place at low water ratios where illite would be the stable mineral (Huang and others, 1986).

The increase in illite in these sandstones from 3.5 to 4 km burial depth causes a significant reduction in permeability (Bjørlykke and others, in prep.). In sandstones where kaolinite or K-feldspar are absent, high concentrations of illite are not likely to form where buried to the critical depth (3.5–4 km). Thus, the primary clastic input of K-feldspar and the amount of kaolinite generated by feldspar leaching will determine the reservoir quality at greater depth. Given the low solubility of aluminum, it is difficult to form illite by supplying Al^{3+} from outside sources. Illite, therefore, requires a precursor mineral. Illite will not form in the presence of smectite, mixed-layer clay, or kaolinite, even if feldspar is present. Dissolution of K-feldspar and precipitation of illite would require removal of K^+ from the rocks. Basic plagioclases may be more suitable as precursor minerals for illite, but plagioclases in North Sea reservoir rocks, particularly those of Jurassic age, are generally sodic in composition (< An 20). In Upper Jurassic sandstones, (e.g., Fulmar and Claymore sandstones) that have little early diagenetic kaolinite or smectite, there is no evidence of extensive illite growth, probably because a suitable illite precursor is not present. The stability field for kaolinite (Fig. 2) becomes restricted to lower K^+/H^+ with increasing temperature. In the presence of K-feldspar, kaolinite will dissolve to form illite at about 120 to 130°C (Bjørlykke and others, 1986), but albite and kaolinite can co-exist at higher temperatures (e.g., in the Hild Field) in the absence of a potassium source (Lønøy and others, 1986). The dependence on a local source of K+ for illitization of kaolinite is evidence of limited mobility of potassium by porewater flow over longer distances.

In the absence of thermal convection, the flow of pore water during burial diagenesis depends on dewatering of the underlying sequence. If the basement rocks or underlying evaporites are at about 6 km depth, which would be typical for the North Sea, the porewater flux generated from the underlying sequence (1.5–2.5 km) by compaction would be very limited. The accumulated porewater expulsion from this sequence would be less than 10^4 cm^3/cm^2 for 1 to 2 km of subsidence. During burial diagenesis, there is (on average) no net upward porewater flow relative to the sea floor when the porosity-depth function remains constant (Caritat, 1989). Only during periods of non-sedimentation (hiatus), when compaction continues without sedimentation, will a net upward flow across the water/sediment interface occur. Because the pore water is not flowing upward relative to the sea floor, it is not subject to cooling and will not precipitate silicate minerals if the geothermal gradient remains constant. Focused flow of compactional water may occur along permeable sand beds or faults. However, fluids released by compaction of large volumes of rocks would have to be focused through small cross sections if significant amounts of cements are to be supplied by compactional pore waters. Therefore, this mechanism cannot be invoked except to explain local phenomena. Because reservoir sandstones are overlain by low-permeability cap rocks, they should be less likely to be flushed by focused compactional water than sandstones beyond the limits of the reservoir structure. The low salinity and the negative isotopic composition of oxygen in the formation waters of several shallow North Sea oil fields show that they have not been flushed by more saline compactional water from the deeper parts of the basin (Bjørlykke and others, 1988). Aluminium removal from, or supply to, sandstone reservoirs requires high porewater fluxes given the low solubility of aluminium (<1 ppm). One of the highest Al concentrations reported in pore water is 5 ppm (Fisher and Boles, 1990). This would require a vertical flow of 2.10^7 cm^3/cm^2 of pore water through a 100-m-thick sandstone to remove or supply 1% Al, assuming 100% efficiency in precipitation or dissolution. A flux of this magnitude would have to drain pore water from a large volume of underlying rock through a small volume of rocks. The volume of sedimentary rocks drained of all its pore water by compaction must be 10^3 to 10^4 times the volume cemented.

In the North Sea Basin, most faulting took place in Late Jurassic to Early Cretaceous time (Glennie, 1990). There was mostly passive subsidence without faulting in the Cenozoic when illitization and quartz cementation occurred. It is therefore unlikely that these faults should be conduits for compactional waters at that time. Along the margins of the basin, particularly in the Netherlands and Germany, Cenozoic faulting has taken place due to more recent uplifts and rifting (Glennie, 1990). In large parts of the North Sea, the Jurassic and Triassic section is underlain by Permian evaporites that would serve as a floor for compactional flow, and all pore water derived from these deeper portions of the basin would have high salinities and would probably fall outside the stability field of kaolinite. From this discussion, we argue that burial diagenesis below about 2 km in most of the North Sea Basin would have had to be relatively isochemical due to the limited flux generated by compactional-porewater flow. Locally, particularly around evaporites, there is greater potential for transport of solids in solution by advection and diffusion. The potential for moving solids in solution is higher in the shallow parts of the basins (<1 km), where meteoric-water flow could produce high porewater fluxes and where slow kinetic-reaction rates would allow higher degrees of supersaturation and undersaturation with respect to the mineral phases.

CONCLUSIONS

1. Diagenetic-clay minerals in North Sea reservoir sandstones are dominantly kaolinite and illite. Smectite and mixed-layer clays are restricted to shallow Triassic and Permian sandstones and Tertiary volcanic-ash beds. Kaolinite is most abundant in fluvial and shallow marine sandstones deposited in a humid climate, particularly those of Early and Middle Jurassic age. Diagenetic kaolinite is also common in Upper Jurassic sandstones representing shallow marine facies, but sandstones of the same age deposited in deeper and more distal marine environments contain less kaolinite. Cretaceous and Tertiary turbiditic sandstones show little evidence of feldspar leaching and have low contents of diagenetic kaolinite. Illite and mixed-layer clays, probably to a large extent replacing smectites, are most common in Permian and Triassic sandstones that were deposited in arid to semiarid climates. Chlorite is commonly present, but mostly in low concentrations. Authigenic chlorite may replace smectite or kaolinite. Its abundance probably reflects the primary content of mafic minerals and rock fragments. The distribution of early diagenetic-clay minerals can, to a large extent, be explained in terms of climate and depositional environment.

2. Petrographic and isotopic evidence suggests that diagenetic kaolinite formed at rather shallow depth and at low temperatures from meteoric water. Illite forms at higher temperatures from pore water with a more positive $\delta^{18}O$.

3. Acids released from source rocks seem to have little effect on the reservoir quality in terms of feldspar leaching and precipitation of kaolinite. Sandstones interbedded with source rocks (Kimmeridge Clay Formation) approaching maturity show little evidence of feldspar leaching and diagenetic kaolinite may be absent.

4. Diagenetic growth of illite at the expense of kaolinite and feldspar may cause significant reduction in permeability at 3.5 to 4 km current burial depth, particularly in Jurassic reservoir rocks. The dissolution of K-feldspar related to this process causes formation of secondary porosity, but does not add significantly to the overall porosity.

5. Burial diagenesis, particularly at temperatures exceeding 80 to 100°C, can largely be explained in terms of isochemical mineral reactions, where the precipitating phase is strongly controlled by the dissolving phases. Porewater compositions at these temperatures will tend to occur in the stability field of illite.

ACKNOWLEDGMENTS

This research has been supported by the Norwegian Research Council (NAVF) and by VISTA, a research cooperation between the Norwegian Academy of Science and Letters and Statoil. Thanks are due to Albert Matter, Jim Boles, David Pierce, Mark Feldman and Girish Saigal for their useful comments on the manuscript. The suggestions and editorial corrections made by the reviewers Richard Lahann and Michael D. Wilson are also gratefully acknowledged.

REFERENCES

AAGAARD, P., EGEBERG, P. K., SAIGAL, G. C., MORAD, S., AND BJØRLYKKE, K., 1990, Diagenetic albitization of detrital K feldspar in Jurassic, Lower Cretaceous and Tertiary clastic reservoir rocks from offshore Noway, II. Formation water chemistry and kinetic considerations: Journal of Sedimentary Petrology, v. 60, p. 575–582.

AAGAARD, P., AND HELGESON, H. C., 1983, Activity/composition relations among silicates and aqueous solutions: II. Chemical and thermodynamic consequences of ideal mixing of atoms of homological sites in montmorillonites, illites, and mixed-layer clays: Clays and Clay Minerals, v. 31, p. 207–217.

AAGAARD, P., JAHREN, J. S., AND EGEBERG, P. K., 1989, Thermodynamic stability of clay minerals with relevance to the diagenetic regime (abs.): Mineralogical Society of London, Conference on Stability of Minerals, London, England, p. 24.

BATH, A. H., MILODOWSKI, A. E., AND SPIRO, A. E., 1987b, Diagenesis of carbonate cements in Permo-Triassic sandstones in the Wessex and East Yorkshire-Lincolnshire Basins, UK: a stable isotope study, in Marshall, J. D., ed., Diagenesis of Sedimentary Sequences: Geological Society of London, Special Publication 36, p. 173–190.

BATH, A. H., MILODOWSKI, A. E., AND STRONG, G. E., 1987a, Fluid flow and diagenesis in the East Midland Triassic sandstone aquifer, in Goff, J. C., and Williams, B. P. J., eds., Fluid Flow in Sedimentary Basins and Aquifers: Geological Society of London Special Publication 34, p. 127–140.

BERNER, R. A., 1981, Kinetics of weathering in diagenesis, in Lasaga, A. C., and Lirkpatric, R. J., eds., Kinetics of Geochemical Processes: Reviews in Mineralogy, v. 8, p. 111–134.

BETHKE, C. M., 1985, A numerical model of compaction driven ground water flow and heat transfer and its application to the paleohydrology of intercratonic sedimentary basins: Journal of Geophysical Research, v. 90., p. 6817–6828.

BETHKE, C. M., 1989, Modeling subsurface flow in sedimentary basins: Geologische Rundschau, v. 78, p. 129–154.

BJØRKUM, P. A., MJØS, R., WALDERHAUG, O., AND HURST, A., 1990, The role of the late Cimmerian unconformity for the distribution of kaolinite in the Gullfaks Field, northern North Sea: Sedimentology, v. 37, p. 396–406.

BJØRLYKKE, K., 1983, Diagenetic reactions in sandstones, in Parker, A., and Sellwood, B. W., eds., Sediment Diagenesis: NATO ASI Series, Reidel Publishing Company, p. 169–213.

BJØRLYKKE, K., 1984, Formation of secondary porosity: How important is it?, in McDonald, D. A., and Surdam, R. C., eds., Clastic Diagenesis: American Association of Petroleum Geologists Memoir 37, p. 285–292.

BJØRLYKKE, K., AAGAARD, P., DYPVIK, H., HASTINGS, D. S., AND HARPER, A. S., 1986, Diagenesis and reservoir properties of Jurassic sandstones from the Haltenbanken area, offshore mid-Norway, in Spencer, A. M., ed., Habitat of Hydrocarbons on the Norwegian Continental Shelf: Norwegian Petroleum Society, London, Graham and Trotman, p. 275–386.

BJØRLYKKE, K., AND BRENDSDAL, A., 1986, Diagenesis of the Brent Sandstone in the Stalfjord field, North Sea, in Gautier, D. L., ed., Roles of Organic Matter in Sediment Diagenesis: Society of Economic Paleontologists and Mineralogists Special Publication 38, p. 157–166.

BJØRLYKKE, K., MO, A., AND PALM, E., 1988, Modeling of thermal convection in sedimentary basins and its relevance to diagenetic reactions: Marine and Petroleum Geology, v. 5, p. 338–351.

BJØRLYKKE, K., RAMM, M., AND SAIGAL, G. C., 1989, Sandstone diagenesis and porosity modification during basin evolution: Geologische Rundschau, v. 78, p. 243–268.

BOLES, J. R., AND FRANKS, S. G., 1979, Clay diagenesis in Wilcox sandstones of southwest Texas: implications of smectite diagenesis on sandstone cementation: Journal of Sedimentary Petrology, v. 49, p. 55–70.

BURLEY, S. D., 1984, Patterns of diagenesis in the Sherwood Sandstone Group (Triassic), United Kingdom: Clay Minerals, v. 19, p. 403–440.

BURLEY, S. D., 1986, The development and destruction of porosity within Upper Jurassic reservoir sandstones of the Piper and Tartan fields, Outer Morey Firth, North Sea: Clay Minerals, v. 21, p. 649–694.

BURLEY, S. D., MULLIS, J., AND MATTER, A., 1989, Timing diagenesis in the Tartan Reservoir (UK, North Sea): Constraints from combined cathodoluminescence microscopy and fluid inclusion studies: Marine and Petroleum Geology, v. 6, p. 97–119.

CARITAT P. D., 1989, Note on the maximum upward migration of pore water in response to sediment compaction: Sedimentary Geology, v. 65, p. 371–377.

CHAMLEY, H., 1989, Clay Sedimentology: New York, Springer Verlag, 623 p.

DYPVIK, H., 1983, Clay mineral transformations in Tertiary and Mesozoic Sediments from North Sea: American Association of Petroleum Geologists Bulletin, v. 67, p. 160–165.

EBERL, D. D., SRODON, J., KRALIK, M., TAYLOR, B. E., AND PETERMAN, Z. E., 1990, Ostwald ripening of clays and metamorphic minerals: Science, v. 248, p. 474–477.

EGEBERG, P. K., AND AAGAARD, P., 1989, Origin and evolution of formation waters from oil fields on the Norwegian shelf: Applied Geochemistry, v. 4, p. 131–142.

EHRENBERG, S. N., AND NADEAU, P. H., 1989, Formation of diagenetic illite in sandstones of the Garn Formation, Haltenbanken Area, mid-Norwegian Continental Shelf: Clay Minerals, v. 24, p. 233–253.

ENGLAND, W. A., MACKENZIE, A. S., MANN, D. M., AND QUIGLEY, T. M., 1987, The movement and entrapment of petroleum fluids in the subsurface: Journal of the Geological Society of London, v. 144, p. 327–347.

FISHER, J. B., AND BOLES, J. R., 1990, Water-rock interaction in Tertiary sandstones, San Joaquin Basin, California, USA: Diagenetic controls on water composition: Chemical Geology, v. 82, p. 83–101.

FISHER, M. J., AND JEANS, C. V., 1982, Clay Mineral Stratigraphy in the Permo-Triassic red bed sequences of BNOC 72/10–1A, Western Approaches and the south Devon Coast: Clay Minerals, v. 17, p. 79–89.

FROSTIC, L., 1990, Tectonic and climatic controls of Triassic Rift Basin sediments in the northern North Sea (abs.): 13th International Sedimentological Congress, Nottingham, England, p. 176.

GARRELS, R. M., 1984, Montmorillonite/illite stability diagrams: Clays and Clay Minerals, v. 32, p. 33–48.

GLASMANN, J.R., CLARK, R. A., LARTER, S., BRIEDIS, N.A., AND LUNDEGARD, P. D., 1989a, Diagenesis and Hydrocarbon accumulation, Brent Sandstone, (Jurassic), Bergen High area, North Sea: American Association of Petroleum Geologists Bulletin, v. 73, p. 1341–1360.

GLASMANN, J. R., LARTER, S., BRIEDIS, N. A., AND LUNDEGARD, P. D., 1989b, Shale diagenesis in the Bergen High area, North Sea: Clays and Clay Minerals, v. 37, p. 97–112.

GLASMANN, J. R., LUNDEGARD, P. D., CLARK, R. A., PENNY, B. K., AND COLLINS, I. D., 1989c, Geochemical evidence for the history of diagenesis and fluid migration: Brent Sandstone, Heather field, North Sea: Clay Minerals, v. 24, p. 255–284.

GLENNIE, K. W. 1990, Introduction to the Petroleum Geology of the North Sea: Oxford, Blackwell, 402 p.

GLENNIE, K. W., MUDD, G. C., AND NAGTEGAL, P. J. C., 1978, Depositional environment and diagenesis of Permian Rotliegendes sandstones in Leman Bank and Sole Pit areas of the UK, southern North Sea: Journal of Geological Society of London, v. 135, p. 25–34.

GOODCHILD, M. W., AND WHITAKER, J. H. McD., 1986, A petrographic study of the Rotliegendes Sandstone reservoir (Lower Permian) in the Rough Gas field: Clay Minerals, v. 21, p. 439–477.

HAMILTON, P. J., FALLICK, A. E., MACINTYRE, R. M., AND ELLIOT, S., 1987, Isotopic tracing of the provenance and diagenesis of Lower Brent Group sands, North Sea, in Brooks, J., and Glennie, K., eds., Petroleum Geology of North West Europe: London, Graham and Trotman, p. 939–949.

HAMILTON, P. J., KELLY, S., AND FALLICK, A. E., 1989, K/Ar dating of illite in hydrocarbon reservoirs: Clay Minerals, v. 24, p. 215–231.

HELGESON, H. C., DELANY, J. M., NESBITT, J. W., AND BIRD, D. K., 1978, Summary and critique of the thermodynamic properties of rock-forming minerals: American Journal of Science, v. 278-A, 229 p.

HERMANRUD, C., 1986, On the importance to the petroleum generation of heating effects from compaction driven water: an example from the northern North Sea, in Burrus, J., ed., Thermal Modeling of Sedimentary Basins: Paris, Editions Technip, p. 247–269.

HOGG, A. J. C., PEIRSON, M. J., FALLICK, A. E., HAMILTON, P. J., AND MACINTYRE, R. M., 1987, Clay mineral and isotopic evidence for control on reservoir properties of Brent Group sandstones, British North Sea (abs.): Terra Cognita, v. 7, p. 342.

HOWER, J., ESLINGER, E. V., AND PERRY, E. A., 1976, Mechanism of burial metamorphism of argillaceous sediments: I. Mineralogical and Geochemical evidence: Geological Society of America Bulletin, v. 87, p. 725–737.

HUANG, W. L., BISHOP, A., AND BROWN, R. W., 1986, Effects of fluid/rock ratio on albite dissolution and illite formation at reservoir conditions: Clay Minerals, v. 21, p. 585–601.

HUMPHREYS, B., SMITH, S. A., AND STRONG, G. E., 1989, Authigenic chlorite in late Triassic sandstones from the Central Graben, North Sea: Clay Minerals, v. 24, p. 427–444.

HURST, A., 1985, Mineralogy and diagenesis of Lower Jurassic sediments of the Lossiemouth borehole, north-east Scotland: Proceedings, Yorkshire Geological Society, v. 45, p. 189–197.

HURST, A., AND BULLER, A. T., 1984, Dish structures in some Paleocene deep-sea sandstones (Norwegian sector, North Sea): Origin of the dish-forming clays and their effect on reservoir quality: Journal of Sedimentary Petrology, v. 54, p. 1206–1211.

HUTCHEON, I. E., 1989, Application of chemical and isotopic analyses of fluid to problems in sandstone diagenesis, in Hutcheon, I. E., ed., Short Course in Burial Diagenesis: Montreal, Mineral Association of Canada, p. 279–310.

INOUE, A., VELDE, B., MEUNIER, A., AND TOUCHARD, G., 1988, Mechanism of illite formation during smectite-to-illite conversion in a hydrothermal system: American Mineralogist, v. 73, p. 1325–1334.

IRWIN, H., 1981, On calcic dolomite-ankerite from the Kimmeridge Clay: Mineralogical Magazine, v. 44, p. 105–107.

JAHREN, J. S., AND AAGAARD, P., 1989, Compositional variations in diagenetic chlorites and illites, and relationships with formation water chemistry: Clay Minerals, v. 24, p. 157–170.

JEANS, C. V., 1989, Clay diagenesis in sandstones and shales: an introduction: Clay Minerals, v. 24, p. 127–136.

JIANG, W. T., ESSENE, E. J., AND PEACOR, D. R., 1990, Transmission electron microscopic study of coexisting pyrophyllite and muscovite: direct evidence for the metastability of illite: Clays and Clay Minerals, v. 38, p. 225–240.

JOHNSEN, H. D., MACAY, T. A., AND STEWART, D. J., 1986, The Fulmar Oil Field (central North Sea): geological aspects of its discovery, appraisal and development: Marine and Petroleum Geology, v. 3, p. 99–125.

JOURDAN, A., THOMAS, M., BREVART, O., ROBSON, P., SOMMER, F., AND SULLIVAN, M., 1987, Diagenesis as the control of the Brent Sandstone reservoir properties in the Greater Alwyn area (East Shetland Basin), in Brooks, J., and Glennie, K., eds., Petroleum Geology of North West Europe: Oxford, Blackwell, p. 951–961.

KANTOROWICZ, J. D., 1984, Nature, origin and distribution of authigenic clay minerals from Middle Jurassic Ravenscar and Brent Group sandstones: Clay Minerals, v. 19, p. 359–375.

KANTOROWICZ, J. D., 1990, The influence of variations in illite morphology on the permeability of Middle Jurassic Brent Group sandstones, Cormorant field, UK North Sea: Marine and Petroleum Geology, v. 7, p. 66–74.

KNARUD, R., AND BERGAN, M., 1990, Diagenetic history of Upper Triassic-Lower Jurassic alluvial sandstones and mudstones of the Lunde and Statfjord Formations, Snorre Field, Norwegian North Sea (abs.): 13th International Sedimentological Congress, Nottingham, England, p. 276–277.

LASAGA, A. C., 1984, Chemical kinetics of water-rock interactions: Journal of Geophysical Research, v. 89, p. 4409–4025.

LEE, M., ARONSON, J. L., AND SAVIN, S. M., 1985, K/Ar dating of time of gas emplacement in Rotliegendes Sandstone, Netherlands: American Association of Petroleum Geologists Bulletin, v. 69, p. 1381–1385.

LEE, M., ARONSON, J. L., AND SAVIN, S. M., 1989, Timing and conditions of Permian Rotliegend Sandstone diagenesis, southern North Sea: K/Ar and oxygen isotopic data: American Association of Petroleum Geologists Bulletin, v. 73, p. 195–215.

LIPPMANN, F., 1981, Stability diagrams involving clay minerals, in Konta, J., ed., 8th Conference on Clay Mineralogy and Petrology, Teplice, Czechoslavakia, 1979, p. 153–171.

LIPPMANN, F., 1982, The thermodynamic status of clay minerals, in van Olphen, H., and Veniale, F., eds., Proceedings, International Clay Conference, Bologna and Pavia, Italy, 1981, p. 475–485.

LØNØY, A., AKSELSEN, J., AND RØNNING, K., 1986, Diagenesis of a deeply buried sandstone reservoir: Hild Field, northern North Sea: Clay Minerals, v. 21, p. 497–511.

LOVELL, J. P. B., 1990, Cenozoic, in Glennie, K. W., ed., Introduction to the Petroleum Geology of the North Sea: Oxford, Blackwell, p. 273–293.

LUNDEGARD, P. D., LAND, L. S., AND GALLOWAY, W. E., 1984, Problem of secondary porosity: Frio Formation (Oligocene), Texas Gulf Coast: Geology, v. 12, p. 399–402.

MALM, O. A., CHRISTENSEN, O. B., FURNES, H., LØVLIE, R., RUESLÅTTEN, H., AND ØSTBY, K. L., 1984, The Lower Tertiary Balder Formation: an organogenic and tuffaceous deposit in the North Sea region, in Spencer, A. M., Johnsen, S. O., Moerk, A., Nysaether, E., Songstad, P., and Spinnangr, A., eds., Petroleum Geology of the North European Margin: London, Graham and Trotman, p. 149–170.

MAY, H. M., KINNIBURGH, D. G., HELMKE, P. A., AND JACKSON, M. L., 1986, Aqueous solutions, solubilities of thermodynamic stabilities of common aluminosilicate minerals: kaolinite and smectite: Geochimica et Cosmochimica Acta, v. 50, p 1667–1677.

MCHARDY, W. J., WILSON, M. J., AND TAIT, J. M., 1982, Electron microscope and X-ray diffraction studies of filamentous illitic clay from sandstones of the Magnus Field: Clay Minerals, v. 17, p. 23–40.

MCLEAN, D. D., 1965, The science of metamorphism in metals, in Pitcher, W. S., and Flinn, G. W., eds., Controls of Metamorphism: Edinburgh, Oliver and Boyd, p. 103–108.

MORAD, S., BERGAN, M., KNARUD, R., AND NYSTUEN, J.P., 1990, Albitization of detrital plagioclase in Triassic reservoir sandstones from the Snorre Field, Norwegian North Sea: Journal of Sedimentary Petrology, v. 60, p. 411–425.

MORRIS, K. A., AND SHEPPARD, C. M., 1982, The role of clay minerals in influencing porosity and permeability characteristics in the Bridport sands of Wytch Farm, Dorset: Clay Minerals, v. 17, p. 41–54.

NIELSEN, A. E., 1964, Kinetics of Precipitation: New York, MacMillian, 151 p.

PYE, K., AND KRINSLEY, H., 1986, Diagenetic carbonate and evaporite minerals in Rotliegend aeolian sandstones of the southern North Sea: their nature and relationship to secondary porosity development: Clay Minerals, v. 21, p. 443–457.

RIDLEY, J., AND THOMPSON, A. B., 1986, The role of mineral kinetics in the development of metamorphic microtextures, in Thompson, A. B., and Rubie, D.C., eds., Metamorphic Reactions: Kinetics Textures and Deformation: Advances in Physical Chemistry, v. 4, p. 154–193.

ROSSEL, N. C., 1982, Clay mineral diagenesis in Rotliegend aeolian sandstones of the southern North Sea: Clay Minerals, v. 17, p. 69–77.

SASS, B. M., ROSENBERG, P. E., AND KITTRICK, A., 1987, The stability of illite/smectite during diagenesis: an experimental study: Geochimica et Cosmochimica Acta, v. 51, p. 2103–2115.

SCHMIDT, V., AND MCDONALD, D. A., 1979, The role of secondary porosity in the course of sandstone diagenesis, in Scholle, P. A., and Schluger, P. R., eds., Aspects of Diagenesis: Society of Economic Paleontologists and Mineralogists Special Publication 26, p. 209–225.

SCOTCHMAN, I. C., JONES, L. H., AND MILLER, R. S., 1989, Clay diagenesis and oil migration in Brent sandstones of NW Hutton field, UK, North Sea: Clay Minerals, v. 24, p. 339–374.

SEEMANN, U., 1982, Depositional facies, diagenetic clay minerals and reservoir quality of Rotliegend sediments in the southern Permian Basin (North Sea): a review: Clay Minerals, v. 17, p. 55–67.

SMITH, J. T., AND EHRENBERG, S. N., 1989, Correlation of carbon dioxide abundance with temperature in clastic hydrocarbon reservoirs: relationship to inorganic chemical equilibrium: Marine and Petroleum Geology, v. 6, p. 129–135.

SØRLIE, R., AND MELLEM, T. R., 1990, Leirmineralogiske studier av øvre triassiske/nedre jurassiske slam og sandsteiner fra Lunde og Statfjordformasjonen, Snorrefeltet (abs. 19): Nordiske Geologiske Vintermøtet, Stavanger, Norway, p. 38.

SPARK, I. S. C., AND TREWIN, N. H., 1986, Facies related diagenesis in the main Claymore oilfield sandstones: Clay Minerals, v. 21, p. 479–496.

STEWART, D. J., 1986, Diagenesis of the shallow marine Fulmar Formation in the central North Sea: Clay Minerals, v. 21, p. 537–564.

STOESSELL, R. K., AND PITTMAN, E. D, 1990, Secondary porosity revisited: the chemistry of feldspar dissolution by carboxylic acid and anions: American Association of Petroleum Geologists Bulletin, v. 74, p. 1795–1805.

SURDAM, R. C., BOESE, S. W., AND CROSSEY, L. J., 1984, The chemistry of secondary porosity, in McDonald, D. A., and Surdam, R. C., eds., Clastic Diagenesis: American Association of Petroleum Geologists Memoir 37, p. 127–150.

THOMAS, M., 1986, Diagenetic sequences and K/Ar dating in Jurassic sandstones, Central Viking Graben: effects on reservoir properties: Clay Minerals, v. 21, p. 695–710.

AUTHIGENIC CLAYS, DIAGENETIC SEQUENCES AND CONCEPTUAL DIAGENETIC MODELS IN CONTRASTING BASIN-MARGIN AND BASIN-CENTER NORTH SEA JURASSIC SANDSTONES AND MUDSTONES

STUART D. BURLEY AND JOE H. S. MACQUAKER
Department of Geology, Manchester University, Oxford Road, Manchester M13 9PL, United Kingdom

ABSTRACT: Jurassic plays in the North Sea are sandstones deposited in fluvial, shallow marine or submarine-fan environments located in the footwalls or hanging walls of rotated fault blocks. After two decades of exploration, a large data base is available on the diagenesis of Middle Jurassic Brent and Upper Jurassic Piper-Claymore-Brae-Fulmar plays. There is still no consensus, however, on whether the diagenesis of these sandstones is dominated by meteoric-water flushing or by the influence of thermobaric waters released during burial.

Diagenetic assemblages in mudstones and sandstones of the basin-margin sequences currently exposed on the mainland UK can be used as modern analogs for the inferred Mesozoic subaerial exposure of rotated fault-block crests in the North Sea. In this respect, mudstones are particularly sensitive indicators of subaerial exposure and rapidly develop soil profiles. Back-scatter electron microscopic (BSEM) observation of mudstones from outcrops indicates that kaolinite is largely authigenic. Basin-margin sandstones commonly preserve the early diagenetic clay-mineral assemblage (kaolinite, smectite and chlorites) but are generally dominated by vermiform authigenic kaolinite. Analytical transmission electron microscopy (ATEM) analyses of these clay minerals indicate large variations in their chemistries.

Each Jurassic reservoir in the North Sea is characterized by a particular diagenetic-mineral assemblage, of which authigenic clays constitute an important component. Early clays in sandstones may coexist with later authigenic clays, although the earlier formed clays are often replaced. In general, authigenic smectites are present only in the shallower reservoirs; kaolinites typically occur at the crest of structures above 4 km depth and close to faults; illite is most abundant in the deeper reservoirs exceeding 4 km depth; authigenic chlorite is rare. Detailed petrographic observation, supported with SEM and TEM investigation, indicates that clay-mineral authigenesis was typically multiphase. K-Ar dating of illites supports this interpretation with youngest ages in the Tertiary. Oxygen isotope analyses of illites and kaolinites remain problematical. ATEM analysis of authigenic clays in mudstones and sandstones documents considerable uniformity of chemical composition at depth. The authigenic mineralogy evolves toward a widespread and uniform phengitic illite-quartz-albite-ankerite assemblage at depths below 4 km.

Authigenic clay-mineral assemblages support the argument for Mesozoic flushing with meteoric waters in some reservoirs but not in others. No single 'diagenetic model' can thus be applied to Jurassic North Sea fault-block plays; there is probably a spectrum of scenarios from reservoirs that have been extensively flushed with fresh water to those that have not experienced meteoric-water ingress. The potential for freshwater flushing depends upon a combination of depositional environment (fluvial and barrier-beach sandstones having greater potential than shallow and deep marine sandstones) and structural location (footwall crests have a greater potential for flushing than the hanging walls). Meteoric-water flushing is most likely where footwall erosion is greatest at the site of maximum displacement along a fault. Not all footwall crests, however, need have experienced subaerial exposure; the width of an individual fault block controls height of the scarp, whereas rate of erosion dictates fault-scarp persistence.

INTRODUCTION

Despite an enormous wealth of literature regarding the diagenesis of North Sea Jurassic sediments, there is still no consensus regarding typical diagenetic assemblages, representative diagenetic sequences and the genetic interpretation of their origins. Theories to explain the prevailing diagenetic assemblage range from the influence of meteoric water following either Kimmerian or early Cretaceous uplift to the influence of thermobaric waters released during burial (Bjørlykke and others, 1989; c.f. Burley and others, 1989).

Perhaps the diversity that is immediately apparent from only a cursory examination of the literature is not surprising; should we expect a single, all encompassing diagenetic model to account for the diagenetic variation that results from burial in a large sedimentary basin, with a complex and variable sediment fill, and considerable variation in burial and thermal histories? Moreover, to further complicate the potential for diversity, marked local variation of structural setting, particularly in terms of fault juxtaposition of lithologies, development of overpressures and pressure gradients for fluid migration, is a characteristic feature of Jurassic fault-block plays in the North Sea. Nevertheless, if the system is viewed from afar, common factors should be apparent.

Studies that consider the diagenetic modifications of only specific reservoirs are unlikely to be able to address the overall diagenetic evolution in the context of basin development. If a regional, basin-scale perspective is required, it is essential to be able to trace the diagenetic evolution of a given stratigraphic unit from near-surface conditions, to determine the initial starting components, through to deep burial, to determine the sequence and timing of reactions. We advocate comparing basin-margin sequences at outcrop with on-structure, crestal-reservoir sandstones and deep, graben-center sandstones as a means of examining shallow, intermediate and deeply buried diagenetic reactions. Our data set is based partly on literature pertaining to authigenic clay-mineral assemblages, their chemistry, isotopic composition and time of formation in the context of the comparison between basin-margin and basin-center settings. We also include data on other cements and fabrics because authigenic clays alone do not provide a complete diagenetic history. In essence, we argue the case that neither the meteoric nor the thermobaric diagenetic models can be applied unilaterally to all the North Sea Jurassic plays. Rather, the present diagenetic assemblages reflect the interplay between these two extremes. On the one hand, reservoirs that were subaerially exposed during either the Kimmerian uplift or during the early Cretaceous may have been flushed with meteoric waters. At the other extreme, reservoirs that have experienced a prolonged, deep burial diagenetic evolution may have been subjected to episodic flushing with hot compactional waters during the Tertiary. Under conditions of deep burial, we also show that the diagenetic assemblage evolves toward a low-grade metamorphic assemblage composed of phengitic illite, quartz, albitic feldspar and ankeritic carbonate.

Diagenetic studies in North Sea Jurassic sandstones and mudstones are riven with controversy, not the least because many authors do not state what they mean in terms of concepts (basin hydrodynamics, mineralogic components, diagenetic processes, etc.), rarely give details of analytical methods (particularly in terms of error margins), rarely tabulate results, and give little or no information regarding inherent assumptions in either technique or logic. What, for example, do $\delta^{18}O$ and δD values of present and ancient waters (as determined from studies of authigenic minerals) tell us in terms of the origin of waters? Simply because light $\delta^{18}O$ values are obtained, it is often assumed that such a result indicates the presence of a Jurassic or Cretaceous water of meteoric origin. Whether this means that the sequence under consideration was actively flushed by meteoric-water recharge driven by topographic head or whether formation waters, with meteoric characteristics, inherited from when the sediment was deposited and subsequently buried, have been expelled under a compactional drive is not considered. The distinction may be rather subtle and extremely difficult to document; but its implications are far reaching in terms of diagenetic processes. The contrast is essentially between massive, meteoric-water flushing during early burial and slow evolution of a system initially dominated by meteoric water that is gradually, although not necessarily continuously, modified by ingress of deep compactional and dehydration waters.

FAULTING AND NORTH SEA JURASSIC PLAYS

The North Sea forms the southern end of a failed graben system that experienced extensional tectonics from the Devonian to the end of the Jurassic (Ziegler, 1982; Glennie, 1984). Its structural development is best considered in terms of flexural isostatic models of sedimentary-basin formation (Kusznir and others, 1991; Roberts and Yielding, 1991). Most of the major Jurassic oil fields occur in syn-rift sediments deposited on the foot- and hanging walls of rotated fault-blocks. In detail there are many variations to the structural style of the fault-block play (see for examples, Parsley, 1984; Gabrielsen and others, 1986; Ziegler and others, 1986). Subsequent to extension, the fault blocks were buried by Cretaceous and Tertiary sediments during a period of thermal subsidence, which persists to the present (Badley and others, 1988). The apparent onset of thermal subsidence differs between the various individual fault blocks, and this detail has an important bearing on the diagenetic evolution of the sediments under consideration.

A characteristic of all current tectonic-fault models (flexural-cantilever, Kusznir and others, 1991; rigid-domino, Barr, 1987; elastic-dislocation, Walsh and Watterson, 1991) as opposed to gravity-driven listric-fault models (see Gibbs, 1984), is elevation of the footwall crest above the pre-faulting base level. In the flexural-cantilever model, deformation in the upper crust takes place along major planar faults by simple shear, which is balanced by plastic deformation in the lower crust. Footwall uplift is produced by isostatic rebound resulting from mass deficiency in the hanging-wall basin. In the rigid-domino model, extension is achieved by the rotation of a series of rigid blocks along normal planar faults, all of similar dip direction, which rotate together. Footwall uplift is the geometric consequence of rotating a rigid block and the elevation is a function of the distance between the major faults. In reality, however, blocks of sediment do not behave in a rigid manner but accommodate fault movement through internal deformation and are better considered as 'soft' dominoes. The elastic-dislocation model thus achieves footwall uplift by distributing displacement between footwall uplift and hanging-wall subsidence. Uplift is contemporaneous with each slip increment and is a maximum at the point of maximum displacement along the fault trace. The extent to which the different characteristics of each of these models needs to be taken into account depends upon the spatial and temporal scale under consideration. On the scale of the reservoir or individual block, the rigid-domino and elastic-dislocation models are most appropriate. Whereas 'soft' dominoes do not 'rotate' *sensu stricto*, we use the term 'rotated fault block' in its loosest sense to describe the geometric appearance of the North Sea Jurassic sandstone play.

Two aspects of rotated fault-block development are of prime importance to the subsequent diagenetic evolution of Jurassic plays. Firstly, the degree of footwall uplift and, therefore, the potential for subaerial exposure, is a function of the width of the fault block (Yielding, 1990) as well as the fault dip (Walsh and Watterson, 1991). Secondly, faults may continue to influence the movement of fluids within, and adjacent to, their displacement volumes throughout subsidence burial (Watterson, 1986; Barnett and others, 1987; Burley and others, 1989). We shall return to both these points to discuss the regional aspects of diagenesis.

TYPICAL BURIAL PROFILES OF JURASSIC FAULT-BLOCK PLAYS

The nature of the rotated fault block results in characteristic burial histories for the basin margin and basin centers. Figure 1 illustrates idealized burial profiles for Upper Jurassic sandstones in onshore basin-margin, crestal on-structure basin-margin and a deep basin-center sequence. The contrast between the basin-margin and basin-center wells is striking. Throughout extension, the basin-margin and onstructure crestal sequences remain within a few hundred meters of the sediment/atmosphere/water interface for as much as 60 Ma. In terms of diagenesis, these sequences are thus exposed to bacterial-sulphate reduction and fermentation for long periods of time in marine sequences, whereas continental sequences (and uplifted marine sequences) may potentially be repeatedly flushed by meteoric waters. The onshore basin margin is effectively an extreme form of the crestal on-structure sequence that remains close to the surface throughout its burial history. Several onshore basin-margin sequences have probably not exceeded 2.5 km of burial (Hebridean Basins, Morton, 1987; Yorkshire Basin, Hemingway and Riddler, 1982; Dorset Basin, Scotchman, 1991a and b), whereas the Helmsdale fault inlier at the western edge of the Inner Moray Firth Basin has probably only experienced approximately 1 km of post-Jurassic burial (Lam and Porter, 1977). By contrast, the basin-cen-

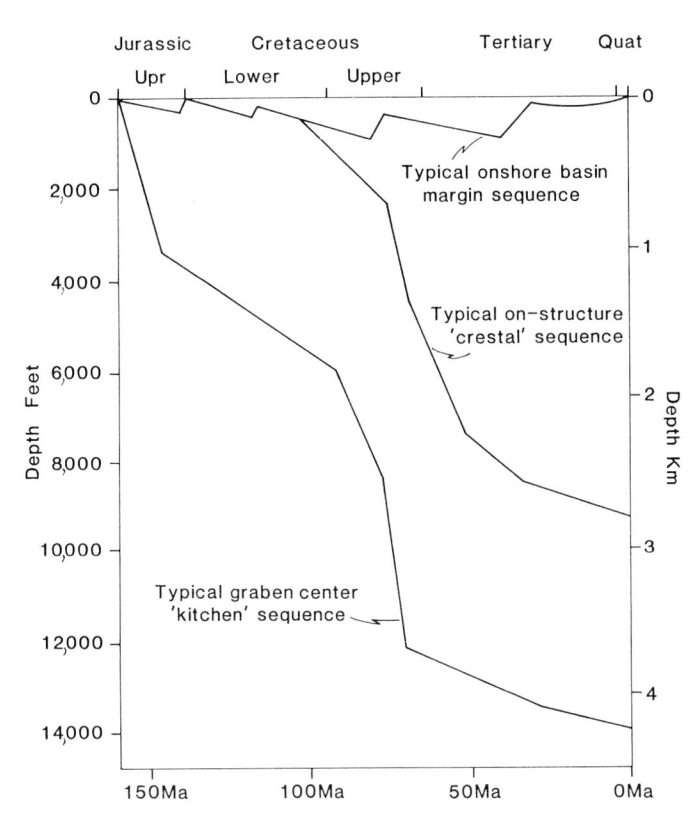

FIG. 1.—Idealized burial curves constructed for basin-margin sequences, on-structure crestal sequences and basin-center sequences in a typical Jurassic rotated fault-block play. By studying the same stratigraphic horizon in different structural settings, the effect of burial history on diagenesis can be assessed.

ter sequences undergo extremely rapid burial, typically reaching depths of 2 km by the mid-Cretaceous. Organic matter-rich sediments will be rapidly transported through the initial zones of bacterial alteration (the ideal scenario for a good source rock) and the potential for meteoric-water ingress is minimal. Overpressures may locally be developed and, if compactional pore waters cannot escape from the rapidly subsiding system, may be available for later burial diagenetic reactions.

The timing of structural development has important implications for the diagenetic evolution and hydrocarbon charging of the Jurassic fault-block plays. Structures typical of Jurassic plays were in existence by the mid-Cretaceous, long before the peak of hydrocarbon generation in the Tertiary. At that time, the potential reservoir may have been at a burial depth of <1 km, whereas the kitchen source rock may have reached burial depths exceeding 3 km. A characteristic feature of the Jurassic fault-block play must, therefore, be the vertical migration of hydrocarbons and pore fluids. If such migration can be effected rapidly, there is considerable potential for transporting hot fluids from the deeper parts of the basin onto the shallow structural highs. Theoretically, providing suitable migration pathways are available, oil accumulation in shallow Jurassic reservoirs could take place from deep kitchen source rocks by the end of the Cretaceous. However, if reservoirs still contain formation waters of meteoric origin, or were charged with hydrocarbons early, then they must have remained closed systems since the onset of rapid Tertiary burial and should now be overpressured.

With the onset of thermal-relaxation subsidence, both on-structure and basin-center sequences subside together. Around 2 km of Tertiary burial are typical for both reservoir and source rock. Most important is the extremely rapid burial characteristic of the early Tertiary—some 2 km in 10 Ma—which often takes the source-rock sequence from just above the oil window to depths where peak oil generation occurs.

These burial profiles also illustrate why it is important to consider the onshore basins in a study of burial diagenesis. Despite the potential for lateral facies changes between basin margin and basin center and the potential for meteoric-water incursion, the onshore basins are the only sequences that are likely to preserve the initial mineralogic and organic components that burial diagenesis has modified in the offshore sequences.

STRATIGRAPHIC SEQUENCE AND SEDIMENTOLOGIC SETTING

There are essentially two stratigraphic horizons in the North Sea that constitute the target for the Jurassic reservoir, which we consider here; the Middle Jurassic Brent Sandstone and the Upper Jurassic Piper-Claymore-Fulmar-Brae sandstone formations (Parsley, 1984). Figure 2 provides an outline structure map of the North Sea that illustrates the basin-margin and main basin-center locations described in the text. There is a wealth of literature that details the stratigraphy, sedimentology and structure of these hydrocarbon plays. Figure 3 summarizes the stratigraphy of the main Jurassic sandstone plays across selected North Sea locations in the context of the regional stratigraphy. Brent Group sandstones, comprising the Broom, Rannoch, Etive and Ness Formations of Bajocian age and the Tarbert Formation of Bathonian age, are the deposits of a northerly prograding delta. By contrast, Upper Jurassic sandstone plays include shallow marine, submarine-fan and turbidite deposits. The Middle Jurassic Ravenscar Group of the north Yorkshire coast provides a good analog of the Brent Sandstone Formation at outcrop (Hancock and Fisher, 1981), whereas the Helmsdale Fault Inlier of northeastern Scotland provides a useful analog for many of the Upper Jurassic sandstone plays (Pickering, 1984). Other onshore Jurassic sequences in England (Dorset coast, Selley and Stoneley, 1985) and Scotland (Hebridean Basins, Hudson, 1983) have attributes that are of some relevance as analogs to the North Sea Jurassic plays.

Unless closed-system diagenetic reaction is invoked (difficult to advocate in a petroleum province), all other sedimentary formations within the North Sea basin can potentially contribute solute and pore fluids to the Jurassic play in consideration. This point was made by Curtis (1978) and further highlighted by Burley and others (1985; see their Fig. 8). The diagenesis of enclosing, interbedded, adjacent, juxtaposed or deeper sediments must be considered in relation to the Jurassic sandstone play. A précis of the regional North Sea sedimentary infill is thus an essential pre-requisite to any discussion of the burial diagenetic re-

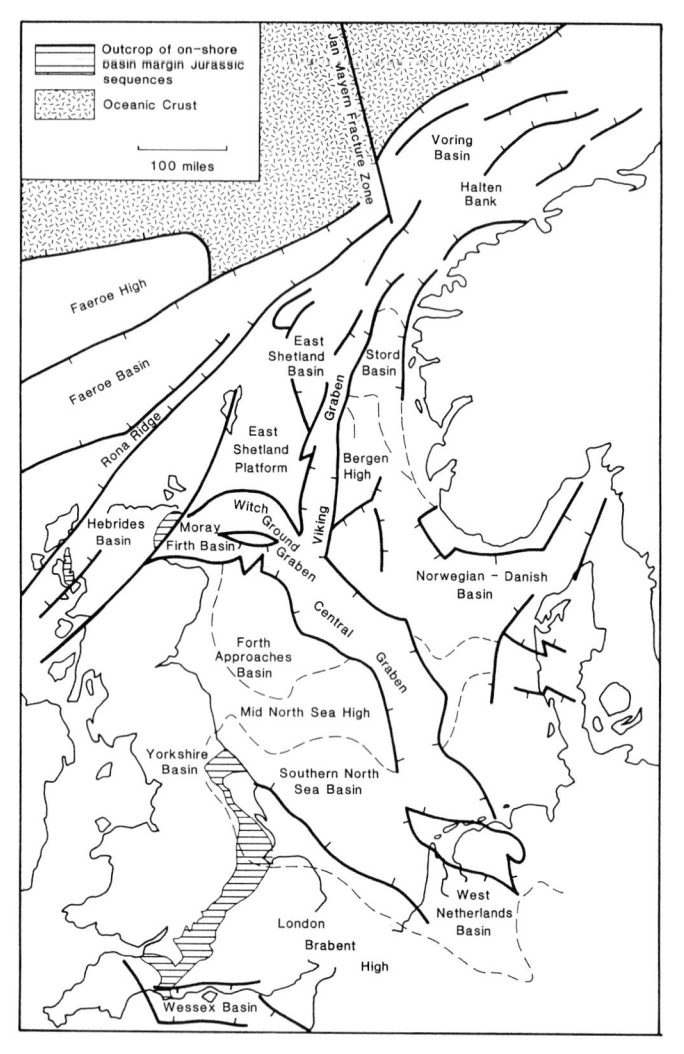

FIG. 2.—Simplified structural location map of the North Sea showing the basin margins and main depositional basins referred to in the text.

actions. Excellent summaries are published in Glennie (1984) and details of many sequences are available in the Institute of Petroleum volumes on the Petroleum Geology of the Continental Shelf of Europe (Woodland, 1975; Illing and Hobson, 1981; Brooks and Glennie, 1987) and the equivalent Norwegian Petroleum Society volumes (Thomas, 1985; Spencer and others, 1986; Kleppe and others, 1987; Spencer, 1987). Generalized stratigraphic sequences are presented in Figure 3 and we draw attention here only to the major features of interest and highlight some of the more fundamental differences between the main basins with Jurassic sandstone plays.

A Permo-Triassic basin underlies much of the Moray Firth, the Central Graben and the Viking Graben. These deposits include a Rotliegend-equivalent aeolian-sandstone sequence, Zechstein dolomites, anhydrites and Na-K salts capped by a thick Triassic evaporitic mudstone-siltstone sequence that interdigitates with coarse-grained fluvial deposits. This Permo-Triassic sequence was deposited under evaporitic, tropical-desert conditions, and provides a source of sulphate and K-Na-Cl for basinal brines. The mudstone units of the Permo-Triassic are also dominated by smectitic clays, which are a potential source of heavy $\delta^{18}O$ water following burial-dehydration reactions. An important point to note here is that, whereas most of the central North Sea and Moray Firth are underlain by this evaporitic sequence, in the northern North Sea it is only the deep graben that contains significant Permo-Triassic deposits. Sulphate, K-Na-Cl and heavy-water sources are absent from most of the sub-basins that flank the Viking Graben. Onset of fully marine conditions in the North Sea occurred during latest Triassic and earliest Jurassic times. The Lower Jurassic, however, is largely absent over the eastern Outer Moray Firth, South Viking Graben and Central Graben, where regional doming of the North Sea was centered. Elsewhere, transgressive, essentially marine mudstones were deposited across the North Sea, which, in the graben, attain up to 1 km of total thickness. The Middle Jurassic Brent Group is renowned for its development of arenaceous, nonmarine and transgressive, paralic deposits and also includes interbedded mudstones and coals. These deltaic coastal environments persisted throughout the Middle Jurassic of the northern North Sea, with up to 300 m of sandstones and interbedded mudstones being deposited in the North Viking Graben and some 150 m in the Moray Firth. Marine conditions were maintained in the extreme north and south of the North Sea, and gradually encroached and flooded the Middle Jurassic deltas. Fully marine conditions returned in the Upper Jurassic, with a variety of mudstone lithologies being deposited across much of the northern North Sea. Active faulting raised the sea floor at fault-block margins and produced a series of intra-basinal footwall highs. High-energy shallow marine sands of the Piper and Fulmar Formations accumulated as beach-bar deposits along the crests of these fault blocks. Fan deltas and submarine fans of the Brae Formation were deposited on the hanging walls and pass laterally into organic matter-rich Kimmeridgian mudstones. Turbidite fans of the Magnus Sandstone, Claymore Formation and parts of the Brae Formation were shed from the footwall marine shelves down into the hanging-wall Kimmeridgian basins.

In many places, the organic-rich facies of the Kimmeridge Clay Formation can be followed into the overlying Ryazanian with no apparent stratigraphic break. Cretaceous sedimentation spans the change from the tectonically controlled deposition associated with the extensional fault block to the quieter conditions of relaxation subsidence centered over the graben systems that characterize the Upper Cretaceous and Tertiary. Some of the fault-block structures were still active throughout the Lower Cretaceous, but by the Campanian, even the larger, more active fault blocks had been transgressed and sealed from the influence of depositional pore waters. It is the fact that many structures were not sealed until the Campanian (e.g., the cross sections of the Piper Field in Maher, 1980 and the Oseberg Field in Nipen, 1987), which has led some authors (e.g., see Bjørlykke and others, 1989) to suggest that such structures were exposed as islands in the Early Cretaceous seas.

The North Sea has been a major sediment sink throughout the Upper Cretaceous and Tertiary with an excess of

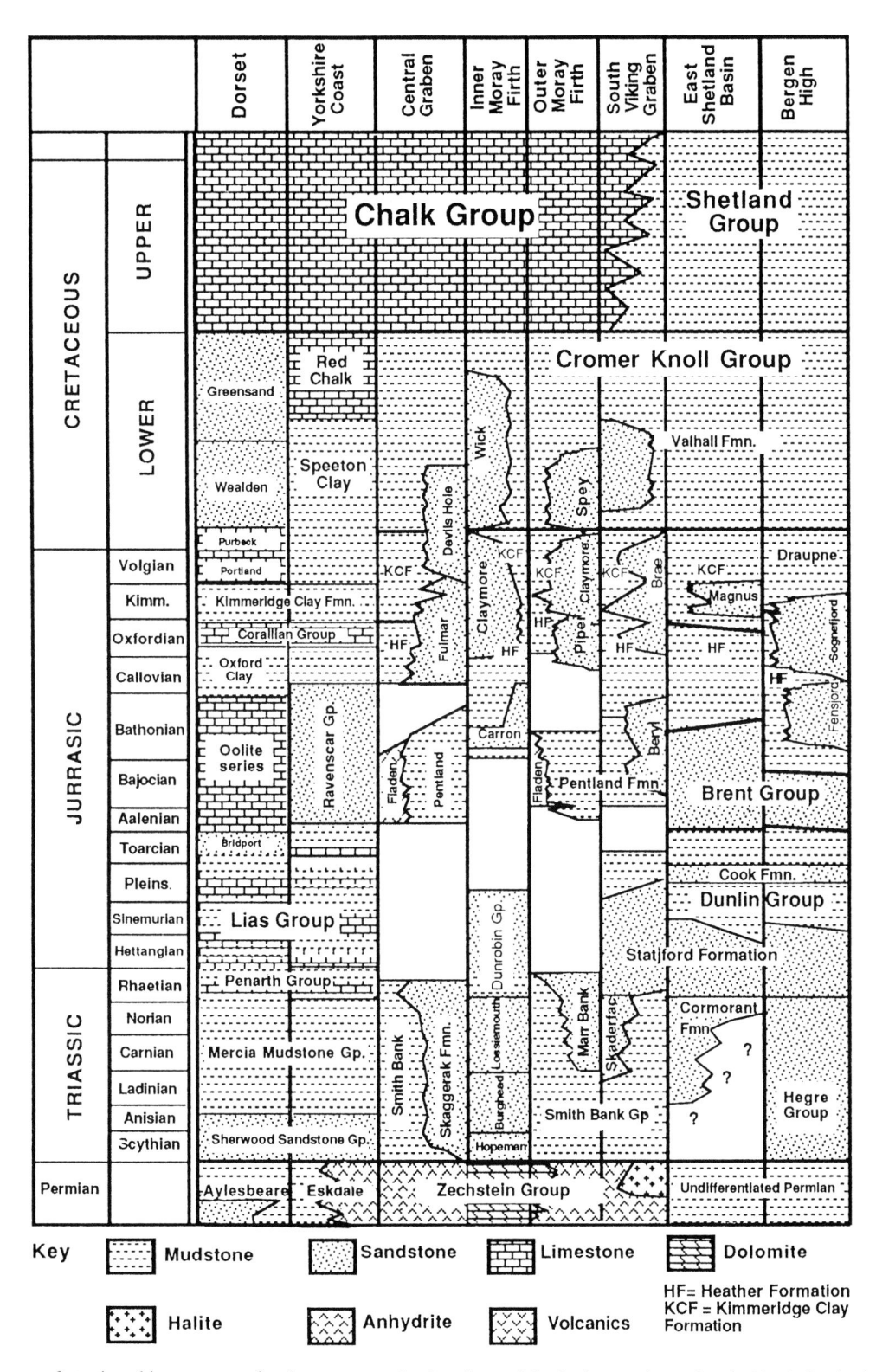

FIG. 3.—Summary of stratigraphic sequences for the representative locations of the basin margins and main North Sea basins. Compiled from various sources.

3 km of sediment accumulation in the depocenters that mirror the north-south axis of the Mesozoic rift systems. As a result, not only has the Chalk-Tertiary sequence provided a regional seal for hydrocarbons, but it has also inhibited the upward migration of basinal pore fluids. Consequently, across large areas of the main graben, the Jurassic and basal Cretaceous sediments are presently overpressured (Chiarelli and Duffaud, 1980; Harris and Fowler, 1987; Buhrig, 1989). In terms of hydrodynamic regime, therefore, the Cenozoic is effectively separated from the Mesozoic and has thus been so since the early Tertiary. The only places where these two regimes can interact are above salt domes or gas chimneys and where major faults penetrate the base Tertiary (Lønøy and others, 1986; Jensensius and Munksgaard, 1989).

The Kimmeridge Clay Formation is generally considered to be the source for most of the hydrocarbons now accumulated in the Jurassic reservoirs, and has, consequently, also been considered as the main Jurassic potential source of deep subsurface pore fluids, CO_2 and organic acids. However, in many fields, the on-structure Kimmeridge has only just attained maturity. Deeper kitchens in the adjacent grabens are usually inferred as the source. It is also worth recalling that considerable thicknesses of gas- and oil-prone Liassic, Bajocian-Bathonian and Oxfordian mudstones are additionally present beneath or interbedded with the Jurassic sandstone plays (Cornford, 1984).

DIAGENETIC-MINERAL ASSEMBLAGES AND THEIR DISTRIBUTION

Diagenetic assemblages that are typically developed in the contrasting basin-margin and basin-center settings are described in the following account with particular reference to the authigenic-clay mineralogy. Figure 4 summarizes the authigenesis and persistence of the main clay-mineral species present in North Sea Jurassic sediments.

Onshore Basin Margins

Despite their obvious relevance as analogs for offshore sequences, the onshore Jurassic basins have attracted relatively little attention from a diagenetic perspective. Kantorowicz (1985, 1987, 1990b) described diagenetic modification of the Ravenscar Group from Yorkshire, and Hurst (1982, 1985a and b) has published on aspects of the Helmsdale inlier. Bryant and others (1988) discussed cementation in the Lower Jurassic Bridport Sandstones from Dorset, Hudson (1970) and Hudson and Andrews (1987) documented early diagenetic minerals from Jurassic sediments in the Hebrides, whereas Searle (1989) described similar diagenetic modification to the Jurassic of Applecross. Irwin and others (1977) and Scotchman (1987, 1989, 1991a and b) provided extensive documentation of early diagenetic reactions in the onshore Kimmeridge Clay Formation.

The onshore mudstones are generally dominated by the presence of illite/smectite (I/S) clays, according to XRD data, although Hurst (1982, 1985a) showed that a significant proportion (commonly >30%) of the early clay-mineral assemblage is kaolinite in the onshore Kimmeridge Clay at Helmsdale. Detailed BSEM studies support Hurst's contention that kaolinite is a common component of the on-

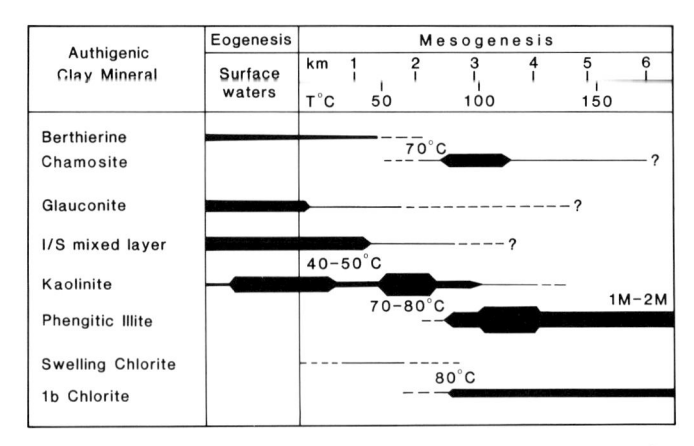

FIG. 4.—Summary bar diagram showing the appearance and persistence of the clay-mineral species reported from the North Sea. Width of the persistence arrows approximates the abundance of each clay-mineral species over a particular burial interval. Compiled from various sources cited in the text.

shore Kimmeridge Clay assemblage but also document a remarkable diversity of clay-mineral fabrics in the mudstones (Fig. 5). ATEM studies show the clay assemblages to be complex mixtures that are composed of kaolinite, I/S and illitic clays (Fig. 6). Kaolinite varies in appearance from anhedral to euhedral and may locally dominate the assemblage. The euhedral kaolinite occurs as a cement within microfossil tests (Fig. 5A) and, in the siltier Kimmeridge Clay facies, as a pore-filling component between detrital-silt grains (Figs. 5B and 5D). The I/S clays are extremely difficult to characterize, and rarely display any recognizable, regular morphology. The euhedral nature and pore-filling habit of a significant proportion of the kaolinite suggest that it is at least partly authigenic in origin and precipitated *in situ* within the mudstones during early diagenesis. However, the major proportion of the mudstone-clay fraction is detrital in origin, and includes both illitic and smectitic material as well as some kaolinite that was transported into the Kimmeridgian sea from adjacent hinterlands. Hurst (1985a) argued that the detrital-clay material was derived directly from erosion of Carboniferous and older rocks of the Scottish land mass without experiencing any intervening Jurassic weathering. This interpretation is not supported by K-Ar age determinations on shallow Kimmeridge Clay mudstone samples that give ages of 140 to 120 Ma (Burley and Flisch, 1989), indicating that the smectitic clays were weathered in the Jurassic and have not inherited a Carboniferous nor Caledonian K-Ar signature. During the Jurassic, the North Sea lay broadly within subtropical latitudes and experienced a distinct seasonality of climate (Hudson, 1980; Hallam, 1984). Although total annual rainfall was probably too meager for widespread regolith development, surface weathering was sufficiently intense to initiate local kaolinitic seat earths to accumulate (Hurst, 1985a; Kantorowicz, 1990b). Kaolinitization of detrital feldspars and micas during unconformity weathering is also inferred by Bjorkum and others (1990) at the crests of rotated North Sea fault blocks as a source of detrital kaolinite in the Rannoch Formation. Short transport paths between weathered hinter-

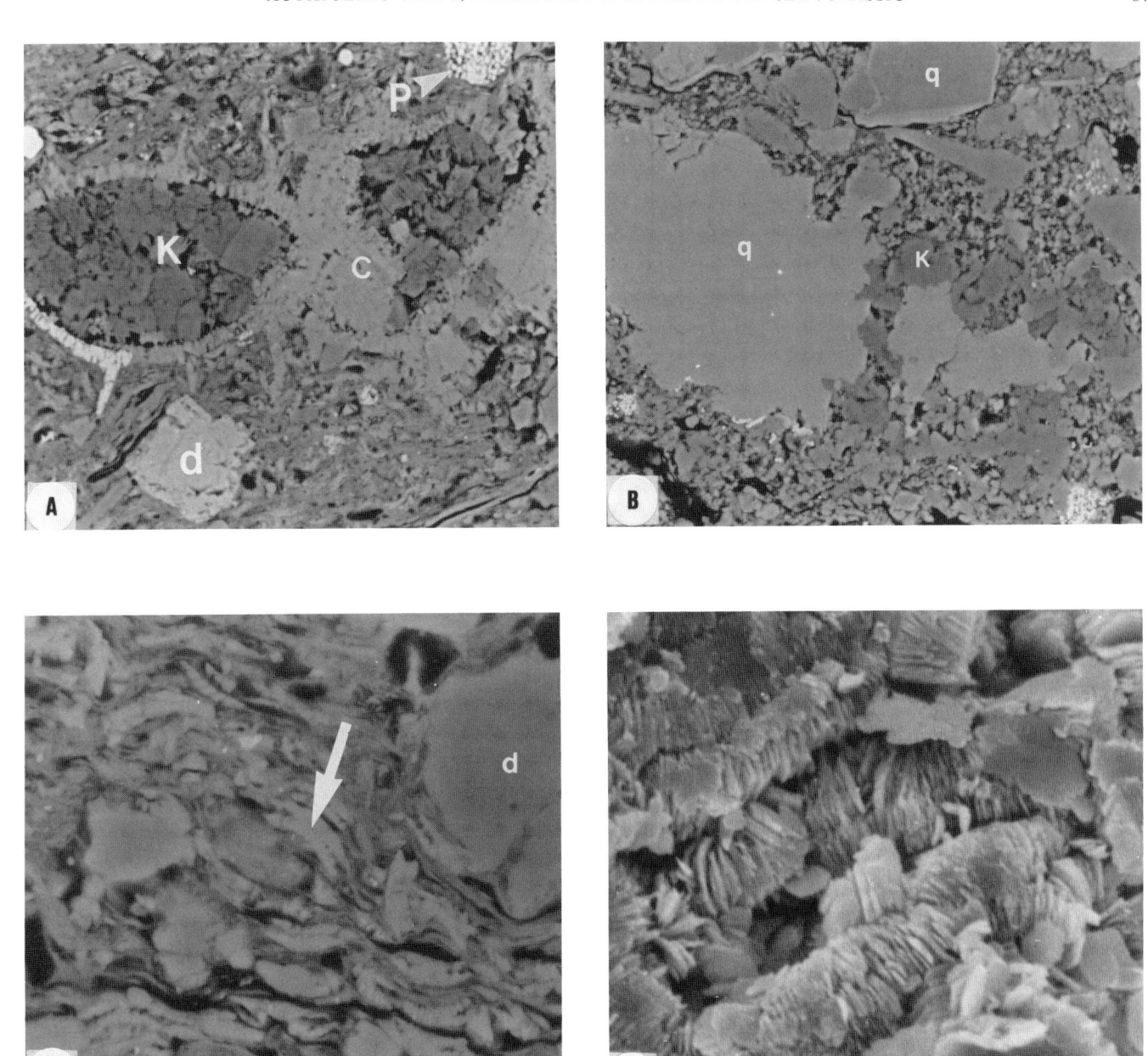

FIG. 5.—Onshore, basin-margin mudstones. (A) Chambered foraminifera in a fine-grained detrital clay and coccolith-rich matrix. Shelter intragranular porosity within the foraminiferal test is filled with calcite cement (c) and aggregates of authigenic kaolinite (K). Framboidal pyrite (arrowed P) and rhombic, ferroan dolomite (d) are also common, early diagenetic precipitates. Back-scattered electron micrograph, field of view 100 μm, onshore Kimmeridge Bay. (B) Angular, silt-size quartz grains (q) in a fine-grained matrix comprised of illitic clays, organic matter and coccolith debris. Authigenic kaolinite forms distinct aggregates (k), which are enclosed in ferroan dolomite (d). Back-scattered electron micrograph, field of view 80 μm, onshore Kimmeridge Bay. (C) Detail of "matrix" showing detrital coccolith plates (arrowed) associated with illitic clays and organic matter. Note the authigenic ferroan dolomite (d). Back-scattered electron micrograph, field of view 26 μm, onshore Kimmeridge Bay. (D) Detail of authigenic-kaolinite aggregates infilling microporosity. Secondary electron micrograph, field of view 30 μm, onshore Inner Moray Firth.

lands and the depositional site may, however, render the petrographic distinction between detrital and authigenic kaolinite extremely difficult.

In addition to the detrital-clay minerals, marine mudstones contain a significant detrital-carbonate component, comprised of disarticulated coccolith plates (Fig. 5C) and reworked molluscan shells. All these detrital carbonates are composed of non-ferroan calcite. A variety of authigenic carbonates has precipitated in the mudstones, primarily as a consequence of anaerobic bacterial activity. The depth-related sequence of sulphate reduction and methanogenic-carbonate cementation was first described from the onshore Kimmeridge Clay Formation of southern England (Irwin and others, 1977). Carbon stable isotopes show that the meta-

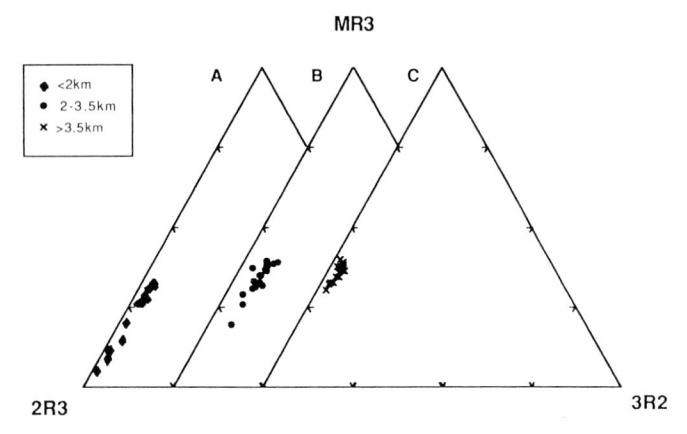

FIG. 6.—Chemical composition of Kimmeridge Clay Formation mudstone-matrix material over the depth interval (A) <2.5 km; (B) 2 km to 3.5 km and (C) >3.5 km determined from ATEM plotted on "Velde" ternary diagrams, where MR3 = K + Na + 2Ca, 2R3 = (Al^{3+} + Fe^{3+} − MR3)/2 and 3R2 = (Mg^{2+} + Fe^{2+} + Mn)/3.

bolic activity of sulphate-reducing bacteria result in the precipitation of non-ferroan calcite, whereas methanogens cause the iron-rich carbonates, ferroan calcite, ferroan dolomite and ankerite to precipitate (Figs. 5A, 5B) (Irwin and others, 1977; Astin and Scotchman, 1988; Scotchman, 1989, 1991a). In the nonmarine mudstones, a somewhat different carbonate assemblage appears to be characteristic of early diagenesis (Hurst, 1985a; Kantorowicz, 1990b) with siderite and ferroan calcite being the dominant cements. The siderite may be related to soil development. Dolomite, sulphate and halite cements are also recorded from the Middle and Upper Jurassic and formed under extreme evaporitic conditions in coastal lagoons (Hudson, 1970; West, 1975; Andrews, 1985; Hudson and Andrews, 1987).

Jurassic sandstones, exposed on the basin margins, display a comparatively simple authigenic mineralogy, dominated by kaolinite, pyrite and carbonate cements, together with minor I/S clays, berthierine, 14Å chlorite and quartz overgrowths. Subtle variation in the assemblage can be closely correlated with variation in the sedimentologic environment (Hurst, 1985b; Hudson and Andrews, 1987; Kantorowicz, 1990b). Kaolinite is, by far, the most abundant clay-mineral cement as determined from thin-section modal analysis and by XRD of clay-fraction separates. Thin-section and SEM studies document that the kaolinite is authigenic. Texturally, it may be related to oversize pores, partially corroded feldspars or altered detrital micas. Despite the abundance of this kaolinite, there are no definitive studies on its origin; does it represent the products of eogenesis in Jurassic soils, paleo-unconformity weathering during the Cretaceous (or Tertiary), or the ingress of Recent groundwaters into the aquifer? This omission is remarkable given that the presence of widespread authigenic-kaolinite cement in Jurassic sandstones is frequently quoted as evidence for subaerial exposure (Bjørlykke, 1983). The obvious place to test such an hypothesis is in a Jurassic basin-margin sequence, which has undergone subaerial exposure.

Importantly, none of the shallow basin-margin sequences definitively document the presence of early authigenic illite. Although illite is reported by some authors as an early diagenetic mineral, such assertions are typically based on SEM observation of authigenic-clay morphology, sometimes supported with semi-quantitative clay-fraction XRD data. There are no cases where illite has been conclusively shown with either careful BSEM or ATEM analyses to be an early diagenetic precipitate in fluvial or shallow marine Jurassic sequences.

Most onshore Jurassic sandstones are variably cemented with carbonate cements. This observation is remarkable in that subaerial exposure is commonly argued to result in the generation of secondary porosity through the dissolution of carbonate cements (Giles and Marshall, 1986). Many of the carbonate cements reported from the onshore Jurassic sandstones are clearly early in origin and have been preserved since the Jurassic (Bryant and others, 1988; Wilkinson, 1991). Subaerial exposure does not ensure carbonate-cement dissolution and thus cannot be generally invoked as a means of generating secondary porosity.

On-Structure, Offshore Graben Margins

We distinguish here shallow reservoirs (~2.0 km to ~3.5 km) from deeper reservoirs (3.5 km to ~4.5 km). The shallow reservoirs are characterized by a clay-mineral assemblage similar to that of the onshore, basin-margin sequences. In the mudstones, kaolinite and I/S clays are present in approximately equal proportions according to clay-fraction XRD studies, and are associated with detrital micas, pyrite and variable amounts of carbonate detritus. Diagenetic modification includes recrystallization of early carbonates as well as a decline in the amount of kaolinite, expandable proportion of the clay fraction and in the total amount of calcite with depth (Fig. 7; Figs. 8A-B). In contrast to the basin-margin samples, there is less fine-grained matrix kaolinite and the I/S clays contain more potassium (Fig. 6).

In the sandstones at shallow depths, both kaolinite and I/S clays are also present, although kaolinite is usually predominant (e.g., Larese and others, 1984; Morton and Humphreys, 1983; Bjørlykke and others, 1986; Burley, 1986; Scotchman and others, 1989; Harris, 1992). Petrographic studies suggest that the kaolinite occurs in different associations, which in turn are generally interpreted to have formed by different processes (Plate 2; Fig. 9). In the micaceous Rannoch Formation Sandstones of the Brent Group and in similar fine-grained, micaceous sandstones that typically occur in the lower part of the Piper Sandstone Formation, most authigenic kaolinite is associated with the alteration of detrital muscovite (Fig. 9A; Bjørlykke and Brendsal, 1986). Some authors have even argued that much of this kaolinite might be detrital in origin (Bjorkum and others, 1990), although the thin-section textures they use to illustrate this interpretation (see their Fig. 4) are very similar to the appearance of kaolinite replacement of muscovite. Kaolinite is rarely directly associated with the dissolution of detrital feldspar; no published photomicrographs document a genetic textural association with partially dissolved feldspar relics and authigenic kaolinite in the North Sea Jurassic. This contrasts with the interpretation offered

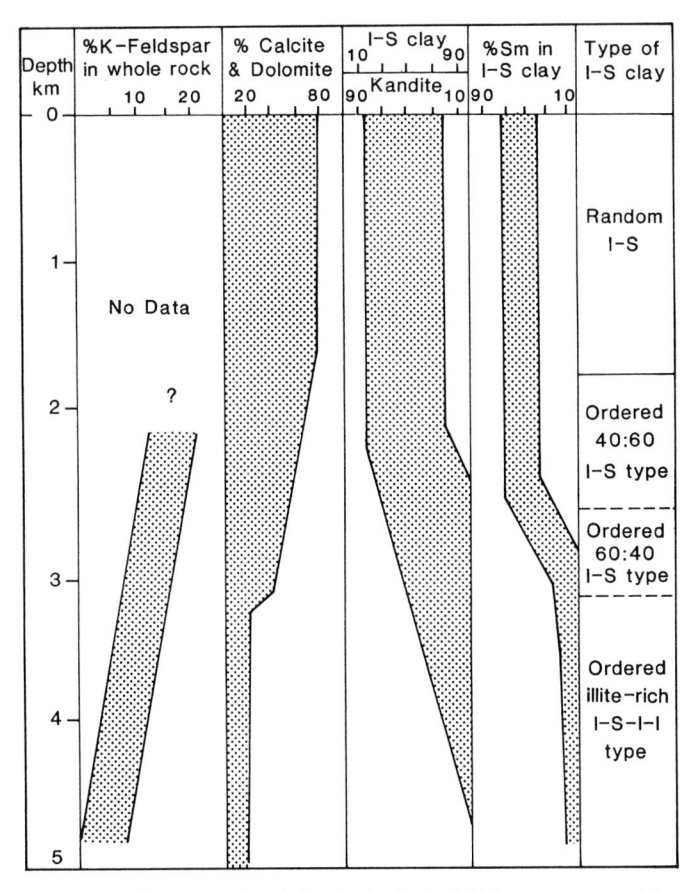

FIG. 7.—Summary of variation in the detrital feldspar content, calcite and dolomite content and clay mineralogy of separates from the Kimmeridge Clay Formation (compiled from Scotchman, 1987, 1991b; Burley and Flisch, 1989; Glasmann and others, 1989b; and Macquaker, unpublished data).

by several authors (see, for examples, Hancock and Taylor, 1978; Bjørlykke and others, 1979) that much of the kaolinite in Jurassic North Sea reservoir sandstones results from the alteration of detrital feldspar according to the reaction:

$$2KAlSi_3O_8 + 2H^+ + 9H_2O$$
$$\rightarrow Al_2Si_2O_5(OH)_4 + 2K^+ + 4H_4SiO_4.$$

In contrast, petrographic evidence suggests, at least on the scale of the thin section and probably much further (although this is difficult to actually document), that Al has been mobile in subsurface pore waters. Feldspar dissolution is widespread in many Jurassic reservoirs and the reaction is initiated during shallow burial diagenesis (Plates 2A and 2B, Fig. 10; e.g., Larese and others, 1984; Lønøy and others, 1986; Harris, 1989). In most cases, secondary pores resulting from feldspar dissolution are devoid of kaolinite (see Plate 2B). More typically, the kaolinite occurs in intergranular pores, some of which are argued to be secondary in origin after the removal of pore-filling carbonates (e.g., Olaussen and others, 1984; Burley, 1986; Lønøy and others, 1986). I/S clays and, at deeper burial, illite appear to be the reaction products of feldspar alteration (Plates 2A and 2G) (Burley and Flisch, 1989; Ehrenberg, 1990, 1991). These I/S clays also appear to be multiphase in origin, and may be enclosed in quartz-overgrowth cements or, in turn, may coat the overgrowths (Fig. 9F). When viewed with the SEM, these I/S clays display a variety of habits and might be mistaken for illite. There are, however, no combined mineralogical and chemical studies of these clays and it is probable that they are not true illites. Berthierine is both chemically and structurally unstable at temperatures approaching 70°C (Jahren and Aagaard, 1989) and appears to recrystallize to chamosite. Similarly, early diagenetic 14Å

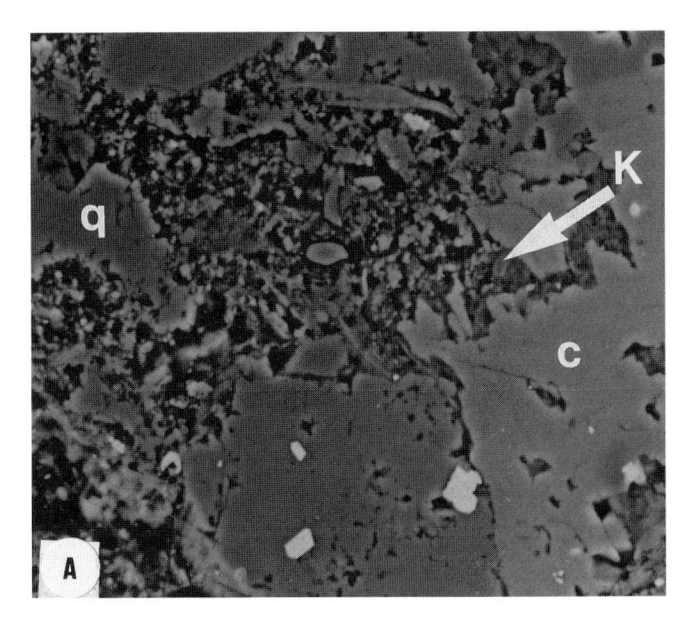

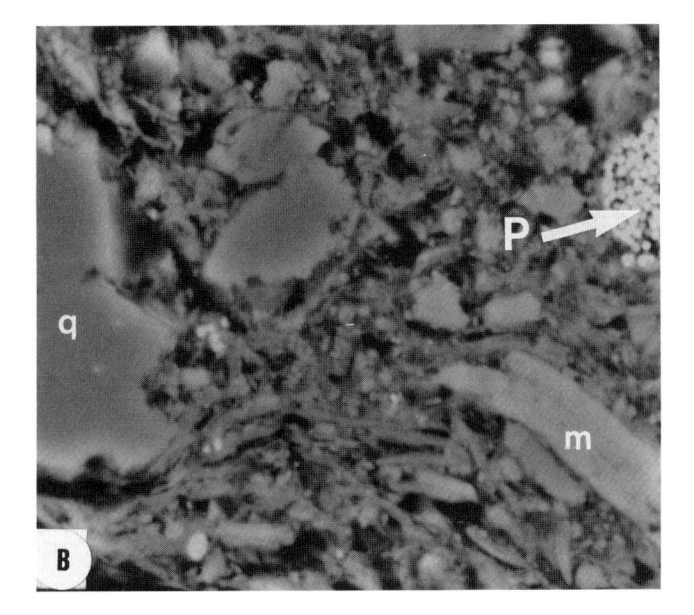

FIG. 8.—On-structure, offshore basin-margin mudstones. (A) Detrital quartz grains (q) in a fine-grained, illitic clay and organic-rich matrix. Both authigenic kaolinite (K, arrowed) and calcite (c) are present but note the corroded appearance of the calcite cement. Back-scattered electron micrograph, field of view 62 μm, onshore Inner Moray Firth. (B) Detail of the illitic-clay and organic-matter matrix, showing its poorly organized appearance. Note the presence of detrital quartz (q) and mica (m), together with a framboidal pyrite (P, arrowed). Back-scattered electron micrograph, field of view 25 μm, onshore Inner Moray Firth.

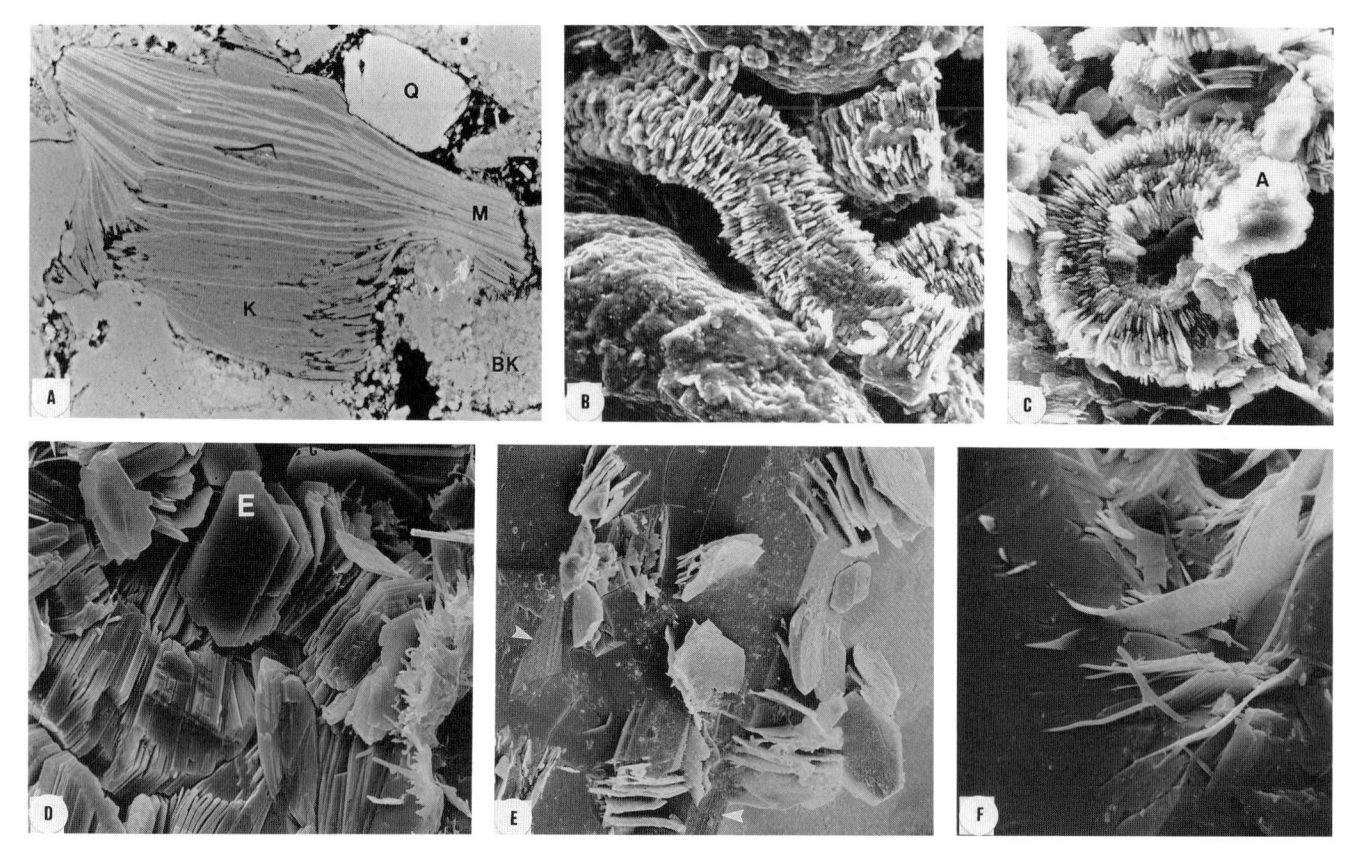

FIG. 9.—Scanning electron micrographs of kaolinite and illite in sandstones. (A) Precipitation of authigenic kaolinte (K, medium grey) within spaces left between expanded detrital muscovite (M). Note the blocky kaolinite (BK) in adjacent intergranular pores. Bright mineral is detrital quartz (Q). Back-scattered electron micrograph, field of view 220 μm, 3.9 km depth, South Viking Graben. (B) Large, vermiform aggregate of stacked kaolinite crystals. Secondary electron micrograph, field of view 60 μm, basin-margin outcrop. (C) Curled "roulette" of stacked, authigenic kaolinite crystals. Note the poor morphological development of individual crystals (example annotated A), giving an anhedral shape. Secondary electron micrograph, field of view 40 μm, basin-margin outcrop. (D) Densely packed aggregates of euhedral kaolinite (example annotated E). Secondary electron micrograph, field of view 150 μm, 3.2 km depth, Brent Province. (E) Euhedral crystals of authigenic kaolinite enclosed within the outer part of a quartz overgrowth. Note the micron-size growth striations visible on the quartz overgrowth faces that are parallel to the overgrowth-face edges (arrowed). Secondary electron micrograph, field of view 160 mm, 3.8 km depth, Witch Ground Graben. (F) Authigenic illite fibers enclosed within quartz overgrowth. Secondary electron micrograph, field of view 20 μm, 3.9 km depth, Witch Ground Graben.

chlorites undergo chemical reaction during burial and evolve toward a tetrahedral Al-rich species with increasing temperature, although there are no structural changes associated with this reaction (Jahren and Aagaard, 1989).

Several other reactions are initiated at these shallow burial depths and involve other detrital and authigenic minerals. Albitization of detrital K-feldspar first becomes apparent at burial depths of around 2 km and temperatures of ~70°C (Fig. 11; Saigal and others, 1988; Aagaard and others, 1990). Variable but generally minor amounts of authigenic-quartz overgrowths are reported by many authors and ferroan dolomite/ankerite carbonate cements may also appear in small amounts.

As a result of these reactions, a diagenetic assemblage dominated by blocky kaolinite, together with fibrous I/S clays, variable but generally minor amounts of quartz overgrowths and feldspar dissolution porosity is typically developed in most on-structure reservoir sandstones between burial depths of ~2.0 to 3.5 km. Associated with this diagenetic assemblage may be relics of early diagenetic assemblages that have survived burial to moderate depths. Endmember 10Å, non-expanding authigenic illite is still not present at these burial depths. In terms of bulk sandstone composition, the assemblage is essentially arkosic, being dominated by quartz-feldspar-kaolinite.

Deep Structures and Basins

Below depths of around 3 km, authigenic illite becomes the dominant clay-mineral species (Figs. 4, 7 and 12; e.g., Hancock and Taylor, 1978; Bjørlykke and others, 1986; Liewig and others, 1987; Ehrenberg, 1990). A number of studies has documented that there is a sudden and dramatic increase in the total amount of illite and proportion of illite in the Jurassic sediments at around this depth in the North Sea (see Fig. 12). The depth itself is probably not critical; more important is the temperature represented by this depth, which corresponds approximately with the 100°C isotherm. The illite is a 10Å, non-expanding 1M polytype species that has a phengitic mica chemistry with a high tetrahedral Si

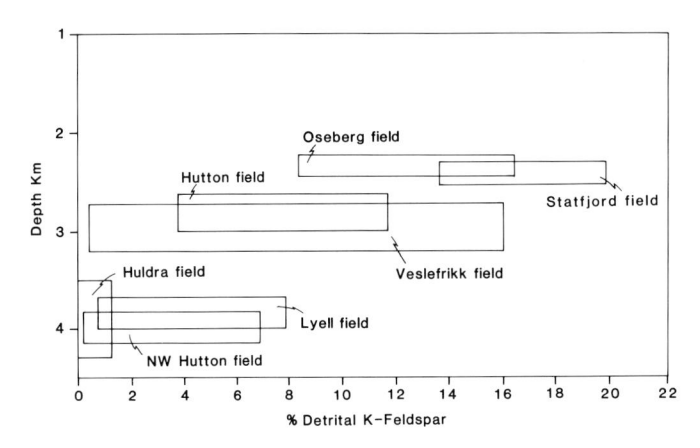

FIG. 10.—Summary diagram of detrital K-feldspar abundance in North Sea Jurassic reservoir sandstones illustrating its gradual loss with present burial depth (compiled from Scotchman and others, 1989; Glasmann and others, 1989c; Harris, 1991).

occupancy (Burley and Flisch, 1989; Ehrenberg and Nadeau, 1989; Jahren and Aagaard, 1989) and shows little variation in its chemical composition. This uniformity of composition is further highlighted by ATEM analyses of deep burial illites in North Sea sandstone reservoirs of Carboniferous, Permian and Triassic age that all provide similar chemical analyses (Warren and Curtis, 1989; Macchi and others, 1990).

The arkosic assemblage is a disequilibrium one and serves as a buffer for a fluid composition that favors the precipitation of illite (Huang and others, 1986; Bjorkum and Gjelsvik, 1988; Huang, 1990). Most likely, this illite does not form from a single reaction. There are probably numerous precursors and reaction mechanisms that give rise to illite (see Plate 2). Illite-generating reactions include alteration of detrital muscovite (Bjørlykke and Brendsall,

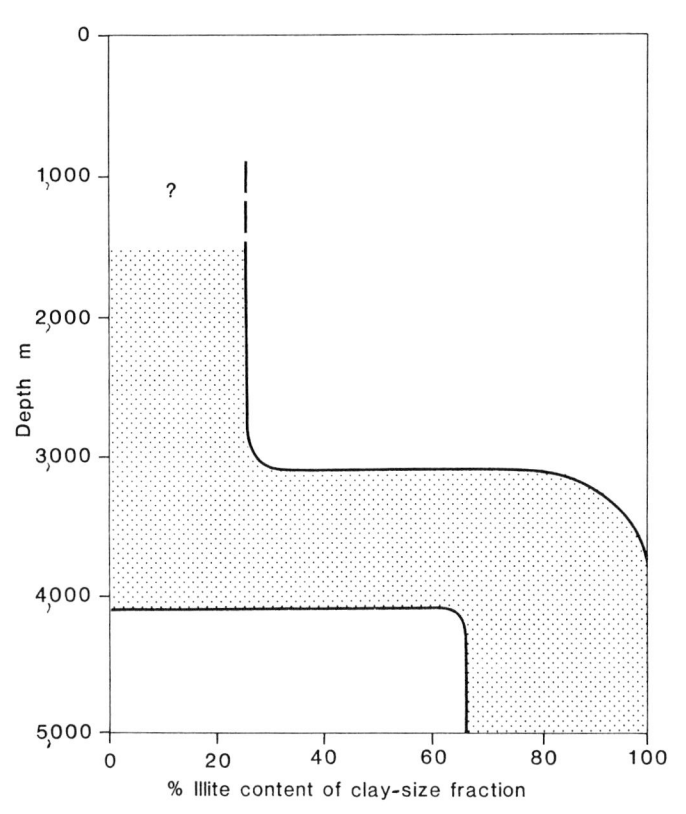

FIG. 12.—Compilation of the proportion of authigenic illite in the clay fraction separated from various Jurassic reservoir sandstones of the UK and Norwegian North Sea (data summarized from Bjørlykke and others, 1986; Burley and Flisch, 1989; Scotchman and others, 1989; Ehrenberg, 1990; Harris, 1991).

1986), dissolution of detrital K-feldspar (Burley, 1986; Ehrenberg and Nadeau, 1989), replacement of earlier formed authigenic I/S clays and the replacement of authigenic kaolinite (Hancock and Taylor, 1978; Ehrenberg and Nadeau, 1989), as well as forming as a direct precipitate from pore fluids (e.g., Harris, 1991). Variation in the morphology of this deep burial illite (fibrous vs. platey) is probably a result of the ambient, pore-fluid chemistry during precipitation (Huang, 1990). Fibrous illite forms slowly in neutral solutions oversaturated with respect to quartz, whereas platey illite precipitates from acid or alkaline solutions.

The same studies that document the increase in illite with depth also document a reciprocal decline in the abundance of kaolinite and I/S clays with depth (see Fig. 7; Jourdan and others, 1987; Scotchman and others, 1989; Glasmann and others, 1989b; Harris, 1991). In most sandstones, kaolinite and I/S clays are minor components of the diagenetic assemblage at depths below 4 km. However, in some sequences, kaolinite is recorded as persisting to depths approaching 5 km at temperatures of 140°C (Lønøy and others, 1986; Jourdan and others, 1987; Ehrenberg, 1991). This is probably due either to potassium deficiency or to the presence of hydrocarbons that inhibit the illitization reaction and preserve the kaolinite to depths below which it would not normally survive.

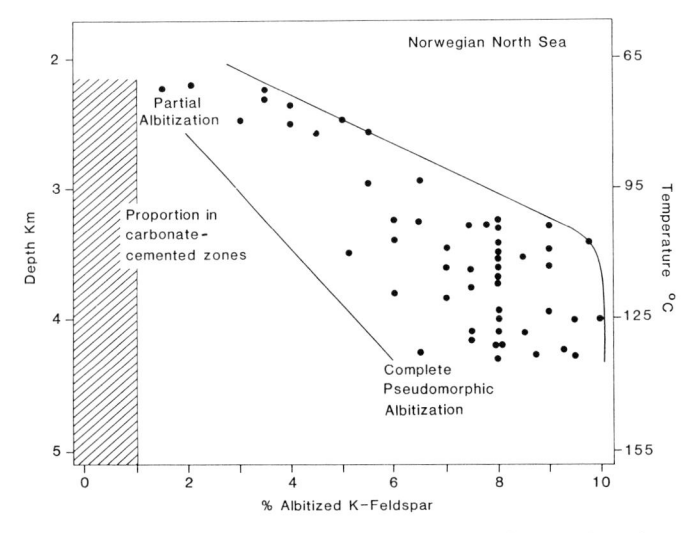

FIG. 11.—The progressive increase in the extent of albitization of detrital K-feldspar in Jurassic sandstones from the Norwegian North Sea (after Saigal and others, 1988). Note how albitization is inhibited in sandstones pervasively cemented with early carbonate.

Chlorites that first appear in the diagenetic assemblage at burial depths of around 2 km continue to be present to depths of 5 km but undergo changes in chemical composition with increasing burial depth and temperature (Jahren and Aagaard, 1989). Tetrahedral Al increases, while the octahedral vacancy decreases correspondingly over the temperature range of 100 to 160°C in chlorite. This change appears to involve continuous recrystallization of the chlorite, an inference supported by the Fe^{2+}/Mg ratio of the chlorites, which appears to be close to equilibrium with that of the present pore waters.

Jurassic reservoir sandstones at burial depths in excess of 3.5 km are also deficient in detrital feldspar (Fig. 10). A significant proportion of the porosity available at depth results from grain dissolution (typically >50% of the total porosity available) and has largely been generated through the dissolution of detrital feldspar (Plates 2A and 2B; Ehrenberg, 1990, 1991; Harris, 1991). Many of the sandstones at these burial depths are now quartz arenites; between 5 and 15% of the bulk sandstone volume may have been dissolved in some cases. Much of the feldspar that does remain is albitic in composition and BSEM studies show that detrital K-feldspar has undergone extensive albitization at depths of >3.5 km (Fig. 11; Saigal and others, 1988). These authors suggest that the reaction is a one-to-one replacement, which can be written as

$$Na^+ + KAlSi_3O_8 \rightarrow NaAlSi_3O_8 + K^+$$

and involves the simultaneous dissolution of K-feldspar and precipitation of authigenic albite.

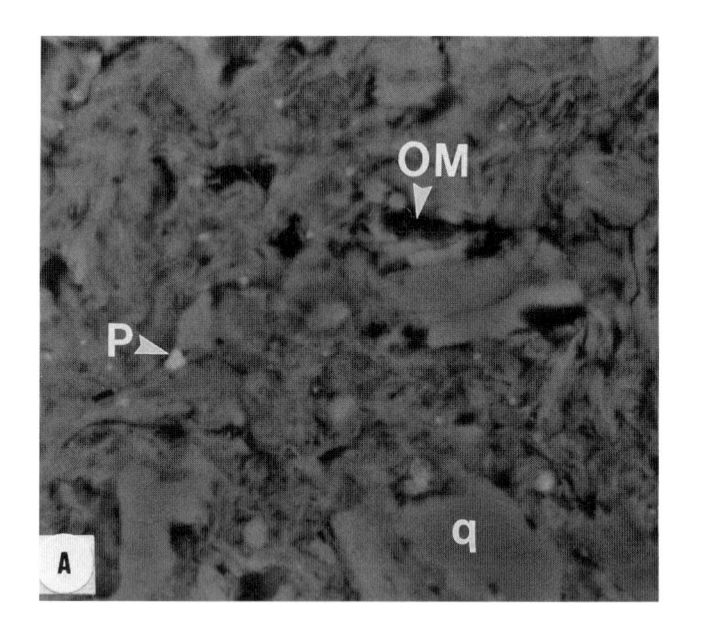

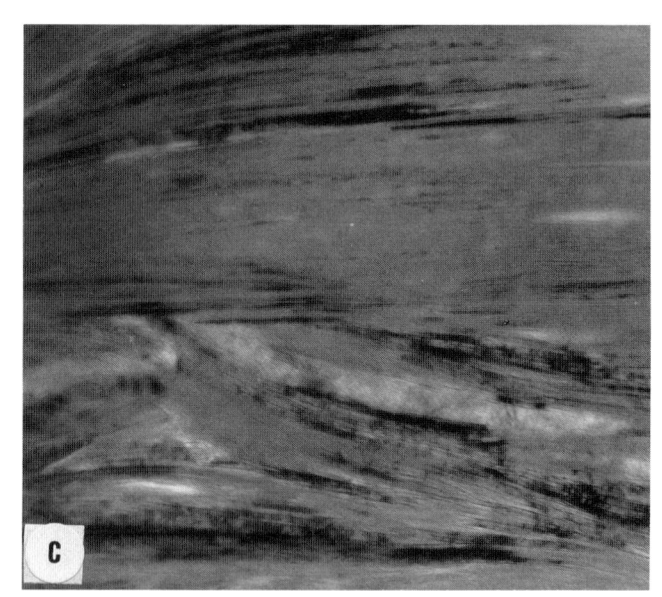

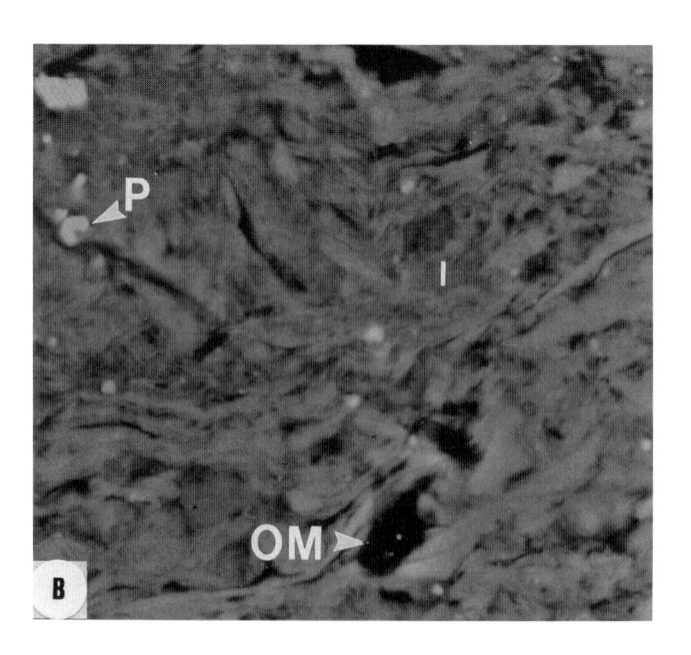

FIG. 13.—Deep structure and basin-center mudstones. (A) Typical fabric of illitic clay-rich matrix containing small amounts of organic matter (arrowed OM) and pyrite (arrowed P). Back-scattered electron micrograph, field of view 26 μm, 4.4 km depth, Outer Moray Firth. (B) Detail of clay-rich matrix showing the development of discrete bundles of illite fibers (I), which are arranged into packages. Note also the presence of pyrite (arrowed p) and organic matter (arrowed om). Back-scattered electron micrograph, field of view 12 μm, 4.4 km depth, Outer Moray Firth. (C) High-resolution image of well-crystalized bundles of clay fibers. Transmission electron micrograph, field of view 0.8 mm, 4.4 km depth, Outer Moray Firth.

In the mudstones, illite dominates the clay-mineral assemblage (Fig. 12; see also Fig. 13; Burley, 1986; Pearson and Small, 1988; Burley and Flisch, 1989; Shaw and Primmer, 1991), although chlorite is locally recorded (Hurst, 1985c). TEM observations show that the chaotic fabrics typical of the shallow, on-structure mudstones are replaced by texturally more ordered fabrics in which discrete oriented packages of 10Å illitic clay dominate (Lindgreen and Hansen, 1991; see Fig. 13C). Chemically, this illite is similar to that present within the sandstones and, according to K-Ar data, has formed during the Tertiary (Burley and Flisch, 1989; Glasmann and others, 1989b) via a distinct dissolution-reprecipitation process.

The overall result of burial diagenetic reactions at depths of around 4.5 to 5 km in North Sea Jurassic reservoirs is to move the bulk-sandstone mineralogy from an arkosic composition toward an authigenic assemblage comprised of authigenic quartz-phengitic illite-albitized feldspar and ferroan carbonate. This assemblage is remarkably similar to a low-grade metamorphic quartz-mica-albite assemblage.

DIAGENETIC SEQUENCES

Basin-margin sandstones are characterized by considerable diversity in the type and mineralogy of diagenetic sequence developed (Fig. 14). In most cases, the early diagenetic sequence conforms to clay—quartz—carbonate. Diversity appears to reflect subtle variation in the depositional environment and can, in some cases, be related to contemporary pedogenic processes and degree of freshwater recharge (Kantorowicz, 1990b; Wilkinson, 1991). As a generalization, nonmarine pore waters typical of stagnant lagoons and marshes between the main river channels are characterized by the sequence chloritic clay—minor quartz overgrowth—siderite. This sequence reflects the rapid attainment of anoxia below the sediment/water interface. By contrast, the main fluvial-channel sandstones often remained oxic because of continued freshwater recharge, and an assemblage of kaolinite, minor quartz overgrowth, and non-ferroan calcite is typically developed. Sequences in marine sandstones begin with a variety of early authigenic clays (illite/smectites, berthierine, glauconite) that are followed by minor quartz overgrowth and pyrite. Ferroan calcite is typically a widespread late carbonate cement that is related to bacterial fermentation or the initiation of burial diagenetic reactions.

The onshore basins at outcrop are exposed to the effects of modern meteoric-water ingress and, in effect, are being weathered. Some of the basin-margin sequences may have been subjected to such weathering for a considerable period of time, depending upon the timing of structural inversion (see Fig. 1). Those sequences that have experienced intense weathering show the development of carbonate-dissolution porosity and kaolinite precipitation. Dissolution of iron oxides leaves an iron-oxide residue, whereas chlorites are oxidized to vermiculites and pyrite to sulphates (Kantorowicz, 1985).

Many of the Jurassic mudstones show similar diagenetic sequences to interbedded and associated sandstones, al-

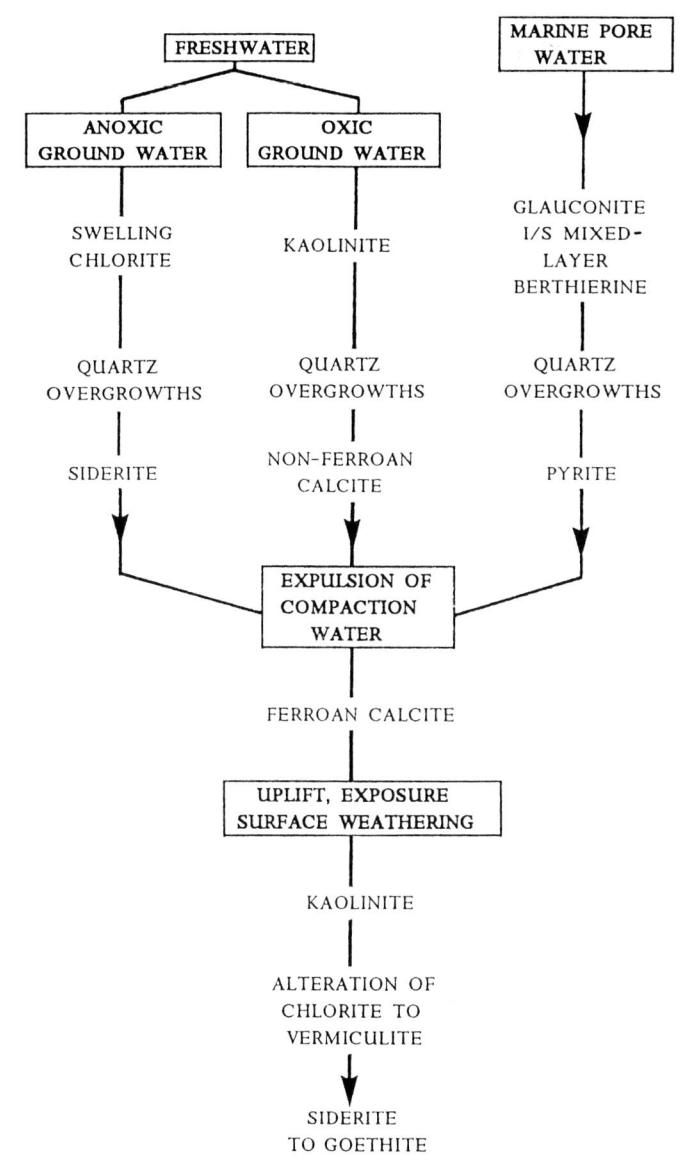

FIG. 14.—Simplified diagenetic sequence for basin-margin Jurassic sandstones (based in part on Burley and others, 1985, Bryant and others, 1988), illustrating the influence of depositional environment on resulting early diagenetic assemblage.

though this is not always recognized. There is widespread evidence for early authigenic pyrite, kaolinite, quartz and carbonate precipitation in most marine mudstones. This similarity reflects the high initial porosity of the mudstones. There is no reason why mudstones should not be as reactive as sandstones during early diagenesis; indeed, because of their fine grain size they should be more reactive.

Sequences reported in deeply buried sandstones are more complex. Virtually every paper published on North Sea diagenesis includes a summary sequence of diagenetic events and processes (e.g., Bjørlykke and Brendsal, 1986; Burley, 1986; Lonoy and others, 1986; Stewart, 1986; Jourdan and in Figure 15. The degree and type of complexity of diagenetic sequences, however, varies drastically depending upon structural position. Reservoir sandstones at the crest of structures and sandstones close to major faults tend to have most complex sequences. Down-flank, off-structure and graben-center wells show the least complex sequences, even though comparable mineral assemblages may be present (see Burley, 1986; Riches and others, 1986; Jourdan and others, 1987). The most complex sequences developed at the crest of reservoirs or adjacent to major fault zones can be considered as only an extension of the simple burial sequence. The complexity can be reduced to a cyclicity (Fig. 16), which involves dissolution, cementation and compaction. Multiple generations of kaolinite, quartz overgrowths, illite or carbonate cements may be present and the exact order (where this can be established) may vary. Such "diagenetic cycles" may be repeated several times, although each cycle need not show the full development of all diagenetic minerals or events. A further degree of complexity is developed where reservoir sandstones are juxtaposed with either Permian or Triassic evaporite sequences as a result of faulting. In such cases, extensive baryte and sulphide "mineralization" occurs in and adjacent to the fault zone and at the crest of the major structural highs (Burley, 1988; Baines and others, 1991). In effect, the "mineralization" is part of the diagenetic sequence, representing an extreme geochemical environment where an evaporitic sulphate brine has mixed and reacted with an evolved brine, of unknown composition or type, expelled from the Kimmeridge Clay Formation mudstones.

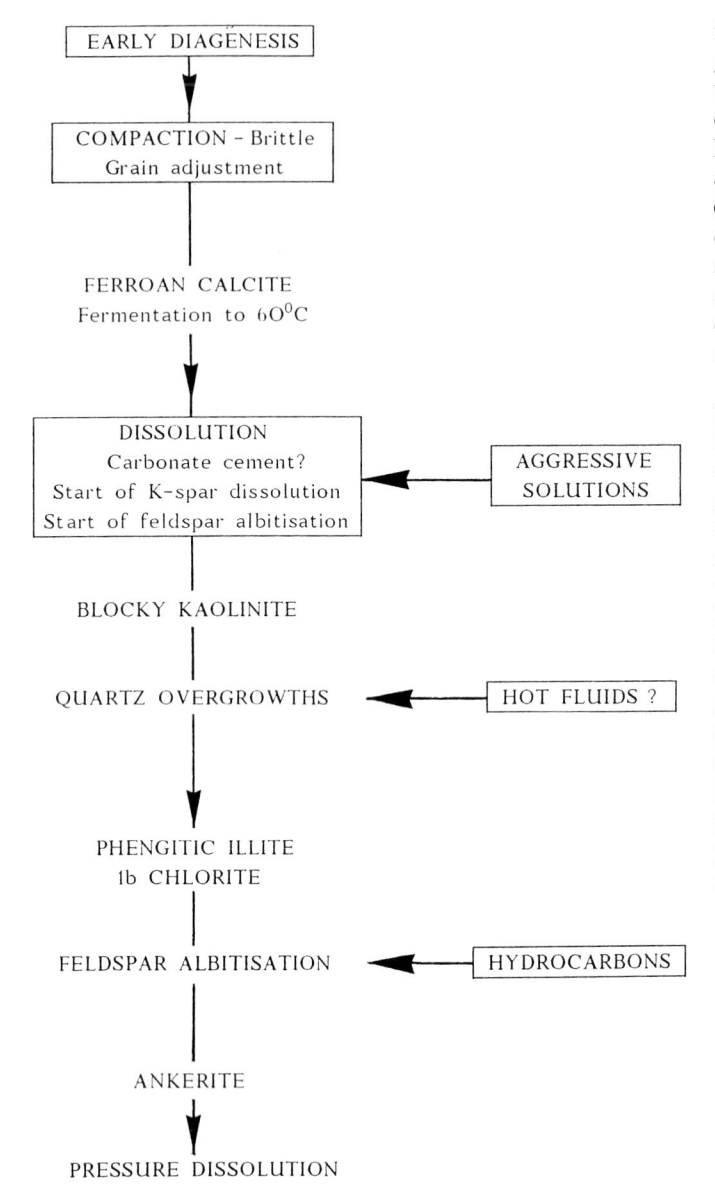

FIG. 15.—The typical diagenetic sequence for deeply buried, North Sea Jurassic reservoir sandstones after removal of local and small-scale variability. This sequence can be broadly recognized in most deeply buried North Sea Jurassic reservoirs.

others, 1987; Scotchman and others, 1989; Glasmann and others, 1989a and c; Ehrenberg, 1990). For each reservoir described in the literature, there is a different diagenetic sequence established. Such apparent diversity and complexity are bewildering, and in some cases, the sequences reported for seemingly similar sandstones in comparable settings are simply inconsistent. The situation is even worse than this synopsis would indicate because in the case of some reservoirs that have been studied by different workers, different sequences have been erected. Despite the confusion, if the minor cements are left aside and some of the complexity reduced, a remarkable degree of uniformity in the diagenetic sequence subsequent to eogenesis can be recognized in virtually all Jurassic reservoirs, as summarized

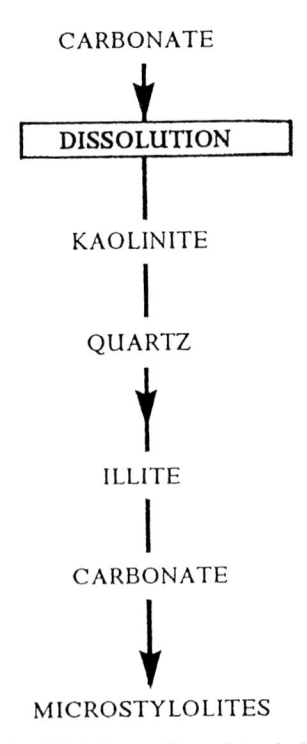

FIG. 16.—The simplified diagenetic cycle typical of on-structure Jurassic reservoirs.

ISOTOPIC STUDIES OF NORTH SEA AUTHIGENIC CLAYS

Oxygen and Hydrogen Stable Isotopes

Despite the abundance of authigenic clays in North Sea Jurassic reservoirs, there are few studies that report the results of their stable isotopic composition. All the available data are compiled in Table 1.

Most data are available for oxygen isotopes in kaolinite (Glasmann and others, 1989a and c; our own unpublished data, courtesy of G. McAulay). Values range from +11‰ SMOW to around +17‰ SMOW, and show a general decrease with increasing present burial depth (Fig. 17). Glasmann and others (1989c) argued that the apparent uniformity of δO isotope compositions across the Heather Field indicates field-wide flushing with meteoric water. These authors then assumed kaolinite precipitation at temperatures of around 45 to 60°C (as constrained from burial-history modeling for an early Cretaceous meteoric-flow regime) to deduce that the kaolinite precipitated from isotopically light $\delta^{18}O$ waters (−6 to −8‰) typical of Mesozoic rain water (Fig. 18). Such circular arguments are extremely dangerous and rely on several assumptions. As with all mineral oxygen isotope systems, either the precipitation temperature or water $\delta^{18}O$ composition must be independently known to solve the isotopic-temperature equation for the unknown parameter. What if kaolinite precipitation took place at elevated temperatures? Kaolinite of 17‰ forming at 80°C, for example, would have precipitated from a water of around 0‰. The trend of kaolinite oxygen isotope data with depth could be interpreted to reflect precipitation at elevated temperatures, and not from meteoric waters. Such an argument is supported by the crossplot of the limited $\delta^{18}O$ vs. δD data available for the North Sea (Fig. 19). Compared with Savin and Epstein's (1970) oxygen isotope data for kaolinites, the North Sea kaolinites are unlike kaolinites formed from meteoric water and show marked similarity to hydrothermal dickites. Moreover, kaolinite in most North Sea Jurassic reservoirs is multiphase in origin. Oxygen isotope analyses have thus been carried out on size separates that are mixtures of kaolinite generations. A component of the kaolinite separates must, therefore, be very heavy.

The argument of Glasmann and others (1989c) that the isotopic composition of kaolinite is remarkably uniform is not valid under close scrutiny either. If the kaolinite is precipitated from meteoric water, then its composition should

TABLE 1.—COMPILATION OF OXYGEN AND HYDROGEN STABLE ISOTOPIC DATA FROM NORTH SEA KAOLINITES AND ILLITES (GLASMANN AND OTHERS, 1989A AND B).

KAOLINITE

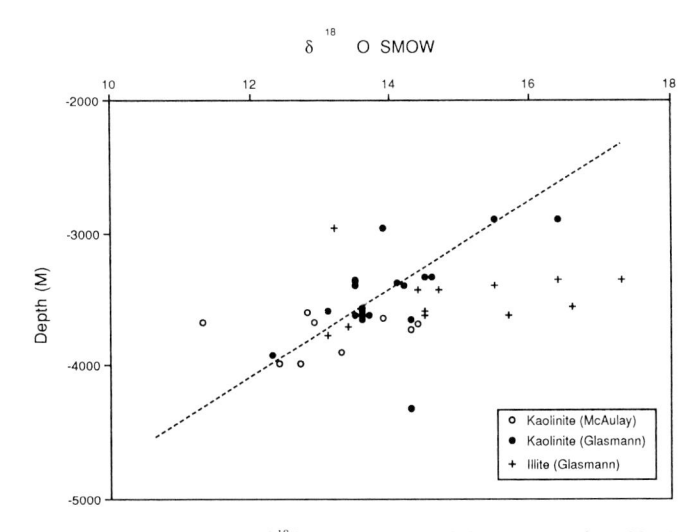

Well	Depth (m)	Kaolinite $\delta^{18}O$	Kaolinite δD
2/5-3	3353	13.5	-15.4
2/5-3	3364	13.5	-52.4
2/5-3	3374	14.1	
2/5-3	3389	14.2	
2/5-3	3390	13.5	
2/5-3	3918	12.3	
2/5/04	2963	13.9	-52.3
2/5/09	3326	14.6	-36.2
2/5/09	3326	14.5	-43.6
2/5-8b	2892	16.4	
2/5-8b	2892	15.5	
2/5-12a	3570	13.6	-42.8
2/5-12a	3582	13.1	
2/5-12a	3590	13.6	-54.3
2/5-H4	3620	13.7	-54.3
2/5-H4	3622	13.5	
2/5-H4	3623	13.6	-53
2/5-H4	3647	13.6	-54.2
2/5-H4	3655	14.3	-55.3
2/5-H34	4325	14.3	-58.4

ILLITE

Well	Depth (m)	Illite $\delta^{18}O$	Illite δD
2/5-3	3389	15.5	
2/5-3	3422	14.4	
2/5-3	3422	14.7	-59.5
2/5/04	2963	13.2	
2/5/09	3345	16.4	-52.5
2/5/09	3345	17.3	-49.8
2/5-12a	3550	16.6	
2/5-12a	3582	14.5	-59.6
2/5-13	3618	15.7	
2/5-H4	3622	14.5	
2/5-H4	3626		-54.4
30/2-1	3705	13.4	
30/2-1	3775	13.1	

FIG. 17.—Variation of $\delta^{18}O$ in kaolinite and illite separates from North Sea Jurassic sandstones with present depth of burial. Dashed line is approximate trend with depth. Data compiled from Glasmann and others (1989a and c), together with unpublished data of G. McAulay (Manchester).

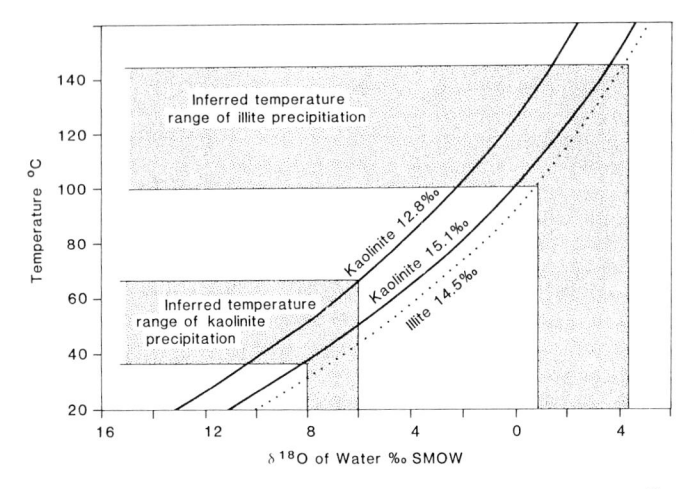

FIG. 18.—The inferred temperature of precipitation and porewater $\delta^{18}O$ composition for Brent Group kaolinite and illites (from Glasmann and others, 1989c). Given a determined oxygen isotope composition for the mineral separate (in this case kaolinite or illite), then either a temperature of precipitation or water oxygen isotope composition can be inferred by assuming a value for the other. Unless temperature or porewater oxygen isotope composition can be independently constrained, there is no unique solution to the conditions of mineral authigenesis.

be uniform across the whole northern North Sea region; Figure 17 shows that at 3.5 km burial depth, the recorded variation in $\delta^{18}O$ composition is 3.5‰. There is even less oxygen isotope data available for authigenic illites. In general, the oxygen isotopes of authigenic illite are heavier than those of kaolinite, but there is considerable variation in $\delta^{18}O$ values, ranging from +13 to +17‰ (Fig. 17; Table 1). Despite this overlap of oxygen isotope composition, illite

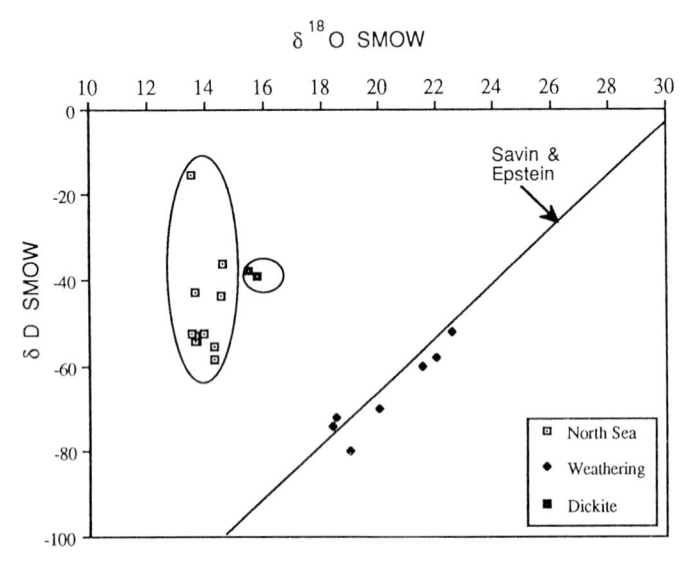

FIG. 19.—The isotopic composition of North Sea kaolinites compared with Savin and Epstein's (1970) isotopic data for weathering kaolinites and hydrothermal dickites. North Sea kaolinite data from Glasmann and others (1989a and c).

is generally inferred to have precipitated at higher temperatures than kaolinite and from heavier, more evolved waters (Fig. 18; Glasmann and others, 1989c). This latter interpretation is supported by the diagenetic sequence recorded from petrography and by K-Ar radiometric-age data.

Oxygen isotope studies of authigenic clays in North Sea Jurassic reservoirs are in their infancy and the data base does not even begin to match that gathered by Longstaffe and his co-workers for the Western Canadian sedimentary basins (see Longstaffe, 1989, for a summary and references therein). Extreme caution is still required when attempting to extrapolate conditions of mineral formation, particularly on a field-wide basis, from the limited North Sea data presently available.

K-Ar Radiogenic Studies of Illite

Illite K-Ar ages have been reported from 19 oil fields and two exploration wells from the North Sea Jurassic (Table 2). Very few of these studies, however, give full details of analytical procedure or results. Only Burley and Flisch (1989), Ehrenberg and Nadeau (1989), and Glasmann and others (1989a and c) provided sufficient detail to enable significant conclusions to be drawn.

Several trends are apparent from the whole data set. Illite ages generally decrease with increasing depth of burial (Fig. 20), although depth of burial alone is almost certainly not the controlling factor on illite authigenesis; temperature and pore-fluid chemistry are more likely. Most illite ages fall in the range of 55 to 30 Ma, suggesting illite authigenesis generally coincides with the onset of rapid burial in the late Cretaceous-early Tertiary. Using burial curves, we infer that most illite authigenesis takes place at depths exceeding 3,600 m; this broadly coincides with petrographic and XRD observations that illite increases in total abundance at this depth and also in the proportion of the clay fraction. Older ages (as reported in Burley and Flisch, 1989, and Glasmann and others, 1989a, c) are due to contamination with either detrital illitic clays or mixtures of younger generations with older authigenic I/S clays or illite *sensu stricto*. In general, younger ages are recorded from water-zone illites than from the corresponding oil-zone illites. However, very recent ages (<10 Ma) are not reported because the youngest illite forming at depth tends to precipitate on earlier formed illite crystals.

Interpretation of these K-Ar illite ages is not, therefore, straightforward, but in most cases K-Ar ages are considered to record the actual age of illite authigenesis as a direct precipitate from pore fluids. Various workers have addressed the problems associated with the systematics of the K-Ar system applied to authigenic illite (Aronson and Douthitt, 1986; Glasmann, 1987; Burley and Flisch, 1989; Ehrenberg and Nadeau, 1989; Hamilton and others, 1989). There is agreement that the $^{40}K/K$ ratio is constant over geological time, that below temperatures of >200°C diffusive loss of Ar will be insignificant and that the finest clay fractions (0.2 μm) will give the youngest ages. It is also clear that contamination with only small amounts of other K-bearing minerals, such as detrital K-feldspar and

TABLE 2.—COMPILATION OF ILLITE K-AR AGES FROM NORTH SEA JURASSIC OIL FIELDS (BURLEY AND FLISCH, 1989; EHRENBERG AND NADEAU, 1989; GLASMANN AND OTHERS, 1989A AND B; SCOTCHMAN AND OTHERS, 1989).

Well	Depth (m)	Age (Ma)	Well	Depth (m)	Age (Ma)
15/17-6	2593	143	2/5-H4	3622	34
15/17-6	2596	118	2/5-H4	3626	39
15/17-6	2602	120	2/5-H4	3647	31
15/17-7	2801	131	2/5-H34	4319	31
15/17-7	2808	122	2/5-12A	3570	31
15/17-7	2832	129	2/5-12A	3591	35
15/17-11	3017	134	2/5-13	3615	32
15/17-11	3018	143	2/5-8b	2892	89
15/17-11	3028	103	2/5-8b	2895	94
15/16-T4	3695	86	2/5-8b	2912	90
15/16-T4	3700	68	211/27	3650	43
15/16-T4	3710	58	211/27	3680	41
15/16-T4	3730	62	30/2-2	4030	69
15/16-10	3700	64	30/2-1	3705	58
15/16-10	3709	57	30/2-1	3775	38
15/16-10	3713	29	30/3-4	2902	34
15/16-15	3705	92	30/3-4	2921	34
15/16-15	3707	74	30/3-4	2936	51
15/16-15	3710	55	6407/1-2	3367	41
15/16-15	3713	63	6407/1-3	3334	46
15/16-15	3716	56	6506/12-3	3566	41
2/5-3	3374	31	6506/12-5	3679	31
2/5-3	3375	34	6406/6-1	3850	33
2/5-3	3417	28	6407/4-1	3664	37
2/5-3	3422	31	6506/12-1	3710	39
2/5-4	2963	40	6506/12-4	3708	37
2/5-9	3326	40	6507/7-1	3953	55
2/5-9	3345	51	6507/7-1	3971	45

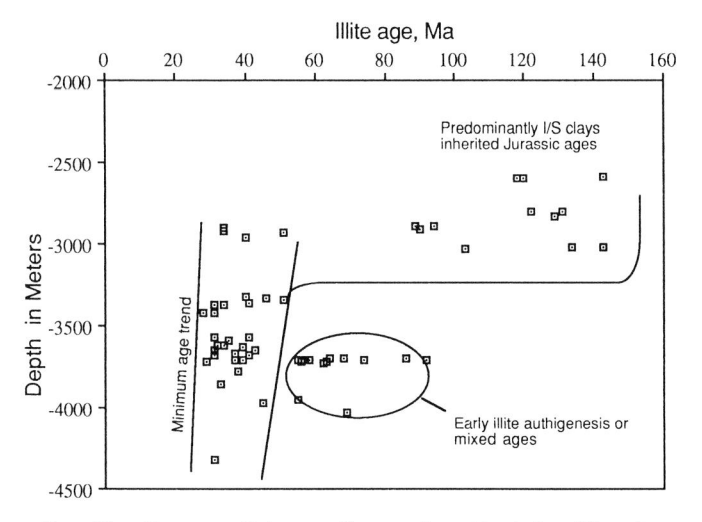

FIG. 20.—Summary K-Ar age diagram from North Sea illites from Jurassic reservoir sandstones. (Data compiled from Thomas, 1986; Jourdan and others,1987, Macaulay and others, 1987; Burley and Flisch, 1989; Ehrenberg and Nadeau, 1989; Kantorowicz and others, 1989; Scotchman and others, 1989; Glasmann and others, 1989a and c.)

detrital micas (with ages in excess of 300 Ma; see Hamilton and others, 1987; and Burley and Flisch, 1989), will significantly alter authigenic-illite ages. Moreover, illite authigenesis is not a single event in many North Sea Jurassic reservoirs. There is evidence, reported in several studies, for multiple generations of authigenic illite (see Burley and Flisch, 1989; Kantorowicz, 1990a). Mixtures of authigenic illite are impossible to separate. Radiometric K-Ar dating of illites inherently suffers from this problem of mixing and many of the published ages must be considered as mixed ages, and are, therefore, only maximum ages of the last generation of illitization.

Differences in maximum age are generally considered to represent real variation in timing of hydrocarbon charge. In the East Shetland Basin and the Bergen High fields, for example, the variation in ages is interpreted to reflect differential reservoir sealing, ingress of compactional fluids and charging with hydrocarbons (Thomas, 1986; Jourdan and others, 1987; Glasmann and others, 1989a). The time span in illite ages recorded across oil/water contacts in several reservoirs is very short (typically in the order of only 5 Ma), suggesting the actual hydrocarbon-charging event is geologically rapid. Within the more comprehensive data sets, there is a trend of younger illite ages with increasing, present depth of burial. Burley and Flisch (1989) showed in the Tartan Field that this is due to the greater proportion of the youngest, newly formed, authigenic-illite generation in the deepest water zone.

EVIDENCE FROM RELATED QUARTZ CEMENTS

In many Jurassic North Sea fields, both authigenic kaolinite and illite are observed to be enclosed in quartz overgrowths and, certainly in the case of the illite, are also observed to postdate the quartz overgrowths. Detailed cathodoluminescence of the quartz cements indicates that the quartz overgrowths are characteristically growth zoned (Burley and others, 1989; Hogg, Sellier, and Jourdan, pers. commun., 1991) and that both the kaolinite and illite postdate earliest quartz cements (Fig. 9E and 9F; Fig. 21). Whereas this is not necessarily evidence that quartz and kaolinite nor quartz and illite were precipitated simultaneously (although Bjorkum and Gjelsvik, 1988, document that this is at least a theoretical possibility), it does mean that the conditions of quartz cementation can be used to further constrain the conditions of kaolinite and illite formation.

The conditions of quartz cementation can be constrained by the combined use of fluid inclusions and oxygen isotope geothermometry. Microthermometric results from fluid inclusions are reported by several workers, although there are very few data presently available on the O-isotopic composition of the quartz overgrowths. Six studies from the UK sector of the North Sea (Beatrice Field, Inner Moray Firth, Haszeldine and others, 1984, a and b; East Shetland Basin, including N.W. Hutton, Heather and the Alwyn Fields specifically named, Jourdan and others, 1987; Scotchman and others, 1989; Glasmann and others, 1989a; in addition to a regional Brent Sandstone study, see Brint and others, 1987; and Haszeldine and others, 1990; Tartan and Piper Fields, Outer Moray Firth, Burley and others, 1989) give paleo-salinities and homogenization temperatures (T_H) for quartz cements, although in most of these studies the data presented are somewhat less than comprehensive. There are only four studies from the Norwegian sector that provide fluid-inclusion data (Huldra Field, Bergen High, Glasmann and others, 1989a; Hammerfest Basin, Barents Sea, Berglund and others, 1986; and unnamed wells from Haltenbanken, Konnerup-Madsen and Dypvik, 1988; Walderhaug, 1990), although Ehrenberg (1990) also quoted unpublished T_H from the Garn Formation.

Despite the paucity of well-described data on the fluid inclusions in quartz cements, a pattern is apparent in the data set. Several authors report more than one generation of inclusions on the basis of paleo-salinities. In the Brent Sandstones of the East Shetland Basin and the Bergen High areas, an earlier, low-salinity, aqueous fluid (0.5 wt % NaCl equivalents in N.W. Hutton, up to ~5 wt % NaCl equivalents in Alwyn, Heather and Huldra) is succeeded by a

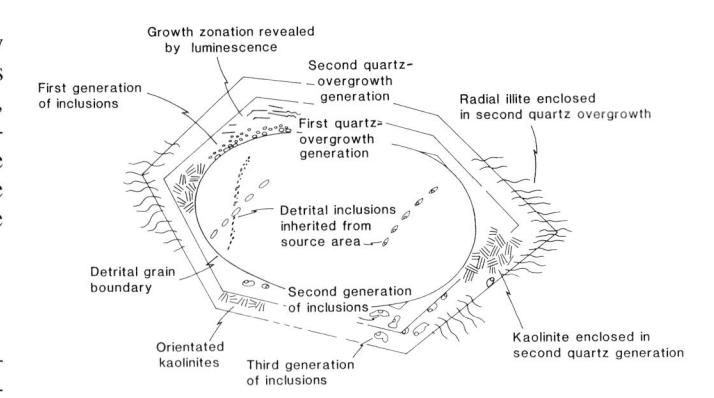

FIG. 21.—Schematic representation of authigenic quartz-overgrowth zones to show their relation to fluid inclusions and authigenic clays in Jurassic reservoirs from the North Sea (based partly on Burley and others, 1989).

more saline brine (up to 11.5 wt % NaCl equivalents reported by Brint and others, 1987 from Brent Sandstones); in Piper Formation sandstones of the Outer Moray Firth, the opposite trend is reported by Burley and others (1989); a high-salinity brine of up to 22 wt % NaCl equivalents for the earlier generation of quartz cement is succeeded by a lower salinity brine (~10–15 wt % NaCl equivalents). In all these cases, however, the later generations of quartz cement have T_H higher than the earlier generation. Uncorrected T_H reported by these authors varies from 68°C to around 140°C and, in some cases, reach temperatures of 170°C. Because the quartz cements precipitated at depth under pressure, a temperature correction normally has to be applied to the laboratory-measured T_H values, unless methane saturation can be documented (see Potter, 1977; and Burley and others, 1989, for a fuller description of the logic behind this concept). In the absence of methane, corrections to the measured T_H values range between 10°C and 20°C; actual trapping temperatures may therefore be as high as 160°C and may approach 200°C in some cases.

On their own, it is difficult to use fluid-inclusion trapping temperatures of quartz to constrain the timing of quartz authigenesis; paleogeothermal gradients vary with time and, in any case, only take account of conductive-heat transfer (McKenzie, 1978). However, if the authigenic illite that has been dated with K-Ar techniques is considered broadly contemporary with quartz precipitation, then the temperature at the time of quartz cementation can be defined on the burial curve. In all the cases where this approach has been applied, the temperatures suggested by the fluid-inclusion T_H data are considerably higher than would be predicted by a thermal model that assumes relaxation-subsidence cooling of the upper crust (see Jourdan and others, 1987; Burley and others, 1989; Scotchman and others, 1989; Glasmann and others, 1989a and c). Moreover, in some cases, the measured T_H data are hotter than the present downhole temperatures (see Haszeldine and others, 1984a and b; Berglund and others, 1986; Glasmann and others, 1989a and c; Burley and others, 1989). On the basis of this common observation, most authors have concluded that hot, migrating fluids were responsible for the precipitation of the widespread, authigenic quartz typical of many North Sea Jurassic reservoirs. This inference is supported by other lines of evidence; fault-related cements in the Tartan field (Burley and others, 1989; Baines and others, 1991) precipitated at higher temperatures (up to 20°C) than the corresponding cements distal from the main faults that delineate the structure.

Anomalous T_H values can be recorded if the fluid or vapor trapped within inclusions undergoes any volume change (stretching) or suffers leakage after formation (Roedder, 1984). In such cases, T_H will not equate to approximate trapping temperature. Fluid inclusions are well known to undergo stretching and leakage, particularly in minerals with strong cleavage (Prezbindowski and Tapp, 1991). Osborne and Haszeldine (1990) have also argued that fluid inclusions in authigenic quartz throughout the North Sea have been reset as a result of the internal overpressures that are generated within inclusions during burial subsequent to trapping. Some inclusions may have reset. Large inclusions along the detrital grain/overgrowth boundary and those trapped along microfractures, for example, may be particularly susceptible to resetting. However, unless quartz cementation took place at very shallow burial depths, stretching and leakage are unlikely to occur on a regional scale in quartz overgrowths because of the high yield strength of quartz (Sterner and Bodnar, 1986; Bodnar and others, 1989). Unfortunately, the available fluid-inclusion data base for North Sea Jurassic sandstones is too small to objectively assess the extent of T_H resetting in authigenic quartz.

Oxygen isotope data are only reported from from two studies that give fluid-inclusion T_H data (Brint and others, 1987; Haszeldine and others, 1990). In both these cases, the quartz overgrowths are reported to have $\delta^{18}O$ values of +15 to +19‰ SMOW. Using pressure-corrected, fluid-inclusion temperatures determined in the overgrowths to constrain the temperature of quartz precipitation, the quartz-H_2O fractionation equation gives a porewater isotopic composition of between 0 and +2‰ SMOW.

CHEMICAL AND ISOTOPIC COMPOSITION OF THE PRESENT FORMATION WATERS

Only Egeberg and Aagaard (1989) provided a comprehensive data set of chemical and O/D isotopic analyses of a regional suite of water samples, although Bjørlykke and others (1986), Burley and others (1989), Glasmann and others (1989b), and Coleman and others (1990) gave chemical and isotopic analyses of present formation waters directly associated with sandstone reservoirs.

Chemically, all the available water compositions are of the Na-Ca-Cl type, with the concentration of chloride varying between 11 and 185 g/l. These extreme chloride concentrations effectively define two porewater groups; a first group with total dissolved solid (TDS) concentrations close to or less than that of sea water, and a second group, with salinities 4 or 5 times greater than that of sea water, that are often close to halite saturation. Remarkably, these present pore-fluid compositional extremes are similar to those recorded from fluid inclusions in quartz-overgrowth cements in the associated sandstones. TDS generally increases with depth, Na-Cl cross-plots define a linear distribution, whereas Cl-Br ratios show a positive correlation (Egeberg and Aagaard, 1989). We have also compiled all available $\delta^{18}O$ and δD data for North Sea pore fluids (Table 3). The oxygen isotope composition varies between ~ −4 and +10‰ SMOW, displaying a progressive shift toward heavier compositions with depth (Fig. 22; c.f. Bjørlykke and others, 1989) and defining several possible mixing lines. The cross-plot of $\delta^{18}O$ against δD (Fig. 23) also shows a good linear relation between a pore fluid similar to meteoric water and a highly evolved, heavy, evaporitic pore fluid

TABLE 3.—COMPILATION OF STABLE ISOTOPIC OXYGEN AND HYDROGEN DATA FOR NORTH SEA PORE FLUIDS (BJØRLYKKE AND OTHERS, 1989; BURLEY AND OTHERS, 1989).

Well	Depth (m)	Temp. °C	$\delta^{18}O$	δD
SF1	4225	154	3.9	-30.6
SH2	4294	152	2.2	-46
SH2	4258	151	2.6	-32
SI1	2267	93	-1	-31.3
SK2	2640	101	-2.1	-28.5
SN1	1579	55	-2.5	-25.8
SN2	1608	68	-1.5	-23.7
SO3	3600	107	-0.2	-28.2
SO5	3750	103	-0.2	-24.2
SP1	2795	91	-0.3	-24.8
SQ2	2622	96	-1.1	-31.8
SQ3	2588	100	-1.3	-30
SR1	2508	90	-5	-38.4
SR2	2683	93	-4.1	-29.2
SR3	2730	82	-0.5	-25
SU1	4365	160	1.3	-18.8
CA1	3468	136	6.7	-30.6
CA2	4018	157	6.9	-32.4
CA3	3269	128	3.6	-31.6
CA4	4132	161	9.3	-21.9
CA5	3632	142	7.6	-26.5
CA6	3820	149	6.3	-30.3
CA7	3221	126	7.1	-29.3
CB1	4150	162	6.7	-31.9
CC1	4403	171	8.1	-34.4
CD1	3481	134	6.3	-30
CD2	3431	134	4.4	-29.3
6507/7-1	4300	145	1.3	-18.8
30/3/04	2904	145	-1.5	-31
30/3/04	2940	145	-2.2	-33
--	2453		1.1	-18.6
--	2483		3.8	-17.6
2/5-H44	3100	100	0	-21
15/17-P2	2600	80	-0.6	-26
15/16-11	3100	100	0.1	-4.3
15/16-T9	3800	112	0.1	-5

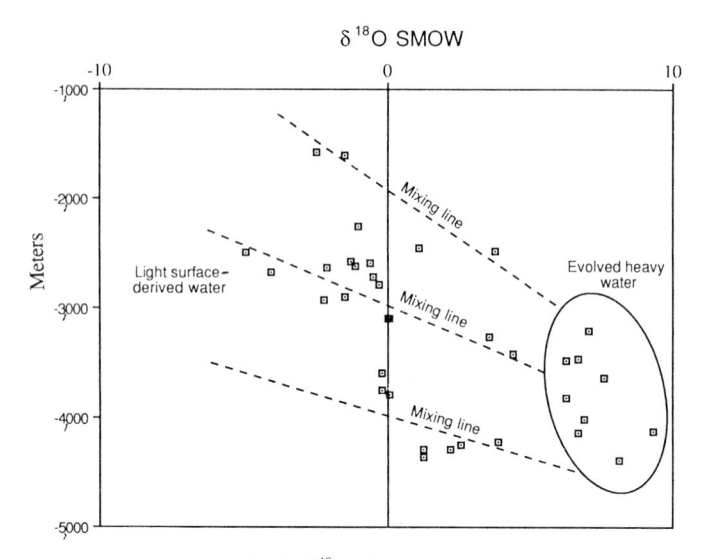

FIG. 22.—The variation in $\delta^{18}O$ of North Sea formation waters with present burial depth. (Data compiled from Bjørlykke and others, 1986; Burley and others, 1989; Egeberg and Aagaard, 1989; Glasmann and others, 1989b.) Dashed lines show possible mixing trends.

that has been through the gypsum-to-anhydrite dehydration reaction. These trends in pore-fluid chemistry and isotope composition suggest that the two endmember pore fluids have different sources and hydrological histories; the shallow, relatively dilute pore fluid probably originated from meteoric recharge, whereas the generally deeper, more concentrated brine is the result of expulsion of highly evaporitic fluids from the underlying or fault-juxtaposed Permian sediments, and may be related to the gypsum-anhydrite dehydration reaction. Given these two fluid sources, trends in TDS and $\delta^{18}O$ with depth and $\delta^{18}O$ against δD probably reflect mixing of the two extreme fluid compositions as a result of subsurface fluid migration.

A REGIONAL PERSPECTIVE OF DIAGENESIS IN NORTH SEA JURASSIC PLAYS

The Influence of Depositional Environment

Variation in the depositional environment is often reflected in the resulting diagenetic assemblage (Burley and others, 1985 and references therein). This variation has been well documented in many instances worldwide, and is clearly seen in the diagenetic assemblages and textures recorded from onshore Jurassic sandstones and their corresponding reservoir sandstone analogs. However, the depositional environment also determines the potential for early, effectively eogenic, cementation and dissolution from meteoric waters. Jurassic soils, fluvial channels, delta-plain sand bodies and beach-bar deposits tend to be deposited on the footwall block and are all likely to have been cemented with early authigenic minerals that should have meteoric isotopic signatures and have trapped fluid inclusions that contain fresh water. By contrast, submarine-fan and turbidite sandstones tend to be deposited on the hanging-wall block and are unlikely to experience contemporary subaerial exposure. Early authigenic minerals in these sandstones should retain characteristics of marine cements.

Whether sandstones that are subaerially exposed contemporaneously with deposition can be said to be flushed with meteoric water is arguably a point of semantics. Effectively, such waters are the depositional pore fluids. If the sediments are exposed at the surface for sufficient periods of time for true soils to form, then there may be a case for arguing meteoric-water flushing, although not in the sense of telogenic modification as defined by Schmidt and MacDonald (1979).

Footwall Uplift of the Rotated Fault Block

Current tectonic models of fault development predict uplift of the footwall above the pre-faulting base level, which is commonly taken to be sea level. Crests of extensional fault blocks thus have the potential for subaerial exposure. The implications of this subaerial exposure related to footwall uplift along the crest of fault blocks are profound. Here is a mechanism for allowing weathering and soil development, together with the associated dissolution of detrital feldspars (as well as carbonate cements?) and precipitation of kaolinite, across the crest of potential reservoir sandstones, as a result of pervasive flushing of the subcrop Jurassic aquifer via outcrop recharge. Such meteoric-water penetration could take place at any time, while extension, and therefore, faulting, were proceeding. Figure 24 is a conceptual model of the diagenetic processes that might be

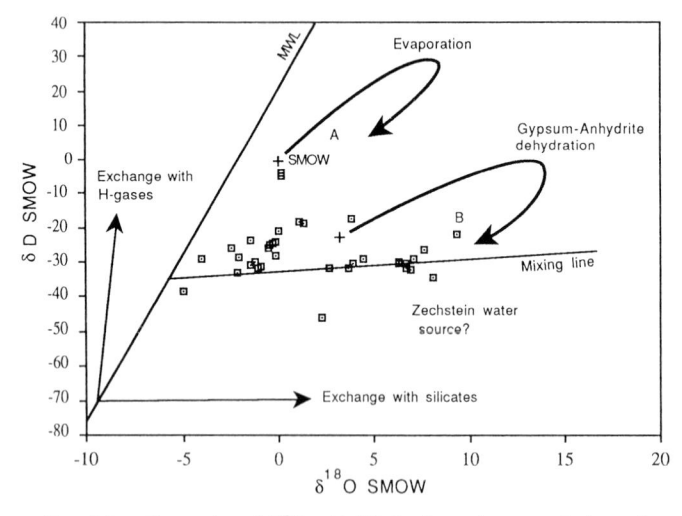

FIG. 23.—Cross-plot of $\delta^{18}O$ with δD for Jurassic reservoir formation waters illustrating possible processes for producing evolved, heavy water. (Data compiled from Burley and others, 1989 and Egeberg and Aagaard, 1989. Processes based on Knauth and Beeunas, 1986.) Curve A is an estimate of isotopic evolution after evaporation 10 times the salinity of sea water (at point A) and curve B represents the isotopic evolution of hydration water in gypsum crystallized along curve A.

expected to take place if prolonged, Jurassic or early Cretaceous subaerial exposure were to be maintained. This is true telogenesis in the sense of Schmidt and MacDonald (1979).

The extent and penetration of such meteoric-water ingress is a function of local topographic head, the geometry and hydrologic properties of the aquifer being flushed and the length of time the aquifer is exposed to such ingress (Hurst and Irwin, 1982). On a geologic time scale, meteoric-water flushing can be very rapid and deep groundwater penetration to depths of kilometers can be achieved in tens to hundreds of thousands of years. Compared to the millions of years that Jurassic extensional fault blocks may have been either subaerially exposed or maintained close to the depositional surface (Fig. 1), a few tens of thousands of years can be considered geologically instantaneous.

Theoretically, meteoric-water ingress could penetrate the width of individual fault blocks and, provided that hydrologic connectivity was established, migrate into adjacent blocks. Meteoric-water penetration to depths of ~1–2 km could result in warming of the circulating pore fluids to temperatures of 45 to 70°C, thus approaching the conditions envisaged, for example, by Glasmann and others (1989c) for kaolinite precipitation in the Heather Field (Fig. 18) without having to invoke porewater migration down active faults (see their Fig. 14). This is the endmember mechanism of meteoric flushing presumably envisaged by Bjørlykke in many Norwegian North Sea reservoirs, but particularly for the Oseberg Field (see Bjørlykke, 1989, his Fig. 4).

Tectonic-fault models not only predict whether meteoric flushing is probable or not, but also relate the potential for flushing to specific geometric characteristics of the fault blocks and faulting. The rigid-domino model predicts that the degree of fault-scarp uplift is directly related to the width across strike of the fault block undergoing rotation (Fig. 25). Larger fault blocks, with widths in excess of 20 km, such as Magnus, Snorre and Gullfaks, can be expected to experience greatest uplift and therefore have the greatest potential for meteoric-water ingress (Fig. 26). The soft-domino model can also be used to predict potential for meteoric-water flushing. Footwall uplift in this model is contemporary with each slip event and is a function of the fault displacement, with maximum uplift coinciding with the point of maximum displacement along the fault trace. Areas of maximum fault displacement, in footwall reservoirs showing evidence of footwall erosion, are therefore the most probable sites of meteoric-water ingress.

These predictions appear valid for the Magnus fault block (Macaulay and others, 1987; Emery and others, 1990). Turbidite sandstones in this footwall reservoir were subaerially exposed as a result of footwall uplift during the early Cretaceous, according to restored, reverse-modeled

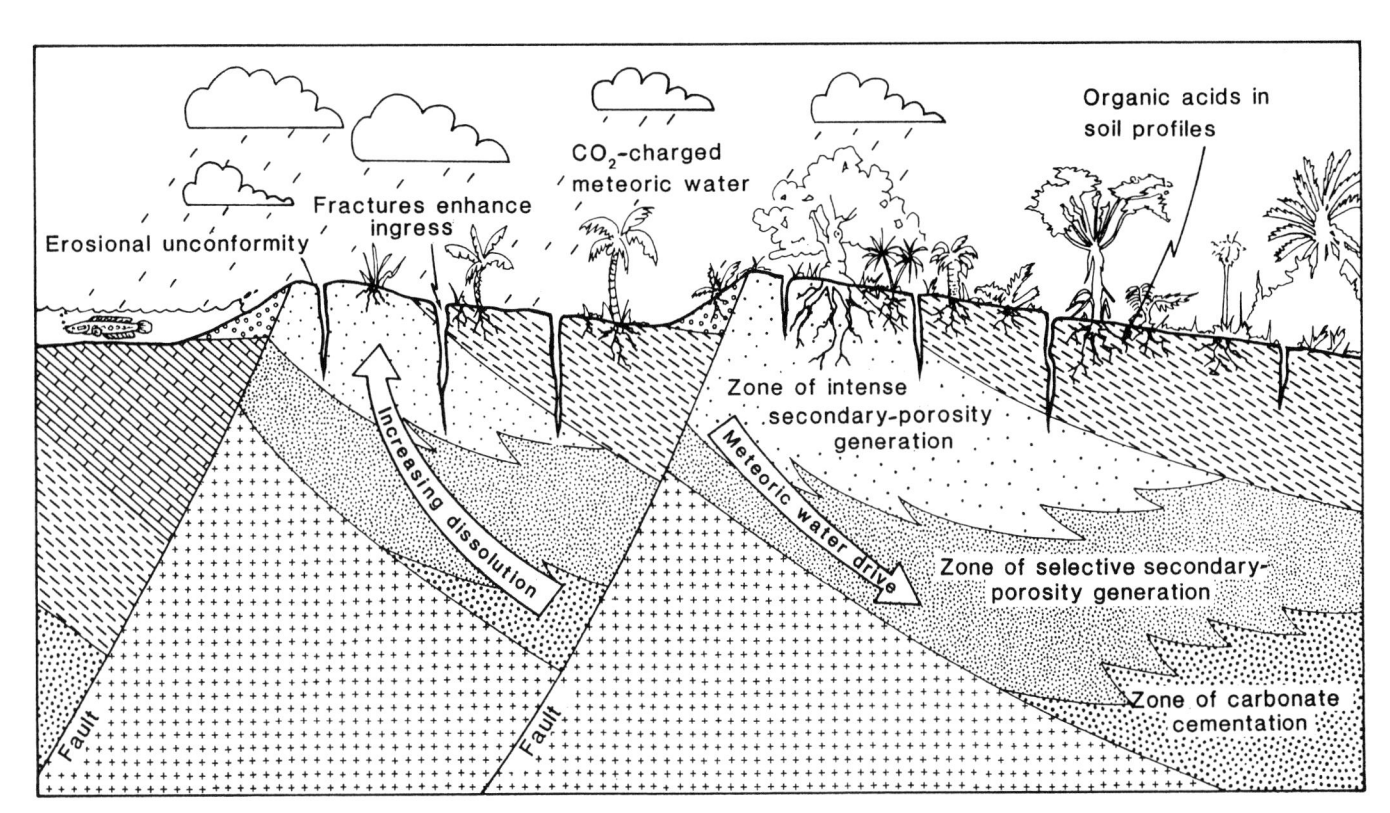

FIG. 24.—Conceptual model of diagenetic modification in a rotated Jurassic fault block as a result of prolonged meteoric-water flushing (depth zonation based on the soil profile development in Bjorkum and others, 1990). Note how the zones of diagenetic porosity modification (and kaolinite development) are related to the unconformity and each structural block. Contemporary, active footwall uplift will probably remove the zone of intense alteration as a result of erosion.

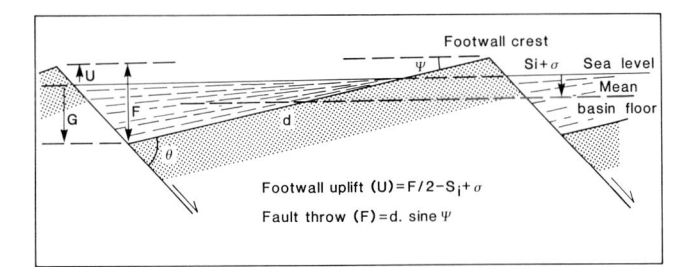

FIG. 25.—The relation between the amount of footwall uplift and the width of the fault block (across strike), according to the rigid-domino model of fault-block rotation (after Yielding and Roberts, 1991). Faults have an initial spacing d and dip i. The average level of the basin floor undergoes subsidence S_i that is a function of the stretching factor β, and there is further subsidence σ due to sedimentary loading. After fault-block tilt ψ, fault throw F comprises footwall uplift U and hanging-wall subsidence G.

seismic sections. Formation-water analysis indicates that a thin lens of fresh water is trapped beneath the mid-Cretaceous caprock (De'ath and Schuyleman, 1981) and the diagenetic mineralogy is consistent with reaction of the original arkose with meteoric water. However, the predictions apparently are not valid for the Gullfaks block, where there is no evidence for extensive meteoric-water penetration preserved beneath the Kimmerian unconformity (Bjorkum and others, 1990). This may be because any aquifer flushed with meteoric water was removed by contemporary or subsequent uplift and erosion; such an interpretation is supported by the flat crests to some of the fault-block footwalls, suggesting that contemporary erosion was broadly equal to uplift. The smaller fault blocks, such as Heather, Tartan and Brent, where the amount of footwall uplift is much less, show evidence for the erosion of unconsolidated crestal sediment, which was redeposited in the hanging-wall graben. In these cases, the poorly consolidated strata of fault-block footwall crests were rapidly submerged at the end of the Jurassic, underwent marine erosion and were never subaerially exposed.

Problems with the "Meteoric-Water Flushing Model"

The concept of using meteoric water as a means of generating the early diagenetic-mineral assemblages and fabrics suffers from the problem that not all meteoric water has the ability to leach feldspars and carbonates, nor to precipitate kaolinite. The composition and abundance of meteoric water are largely a function of climate, with the most acidic and prolific rainfall being associated with tropical climates (Birkeland, 1974; Drever, 1982). Paleogeographic and paleo-oceanographic evidence suggests that the late Jurassic and early Cretaceous climates in the North Sea were subtropical and had only limited potential for kaolinitic-soil development (Hudson, 1980; Hallam, 1984; Hurst, 1985a; Miller, 1990).

Recognition of the diagenetic products of meteoric-water invasion of uplifted footwalls and their distinction from eogenic modification is difficult. The isotopic signature of authigenic kaolinites developed under telogenic conditions may be impossible to distinguish from eogenic kaolinite. However, the distribution of telogenic kaolinite and secondary porosity may be diagnostic and mappable at outcrop or from oilfield core coverage. They should be related to the unconformity surface and, therefore, cross-cut depositional facies, although to our knowledge weathering profiles related to the late Upper Jurassic unconformities have not yet been described from in or around the margins of the North Sea.

Petrographic evidence suggests that kaolinite often postdates an earlier generation of quartz-overgrowth cement that, according to fluid-inclusion homogenization temperatures, precipitated at around 70°C from low-salinity fluids. It could be argued that this is warmed, deep circulating, meteoric water. However, the potential for deep meteoric recharge in fault blocks is restricted, being essentially limited to the depth the Jurassic sandstones dip into the subsurface on a given fault block. In most cases, this will not exceed 1 km. A good analog is provided by the basin-margin, East Midlands Triassic sandstone aquifer that is presently recharged along the western margin of the southern North Sea Basin. Here, meteoric waters penetrate for some 20 km laterally and to depths of around 1 km before mixing with much older brines, presumably being expelled from the southern North Sea (Bath and others, 1987).

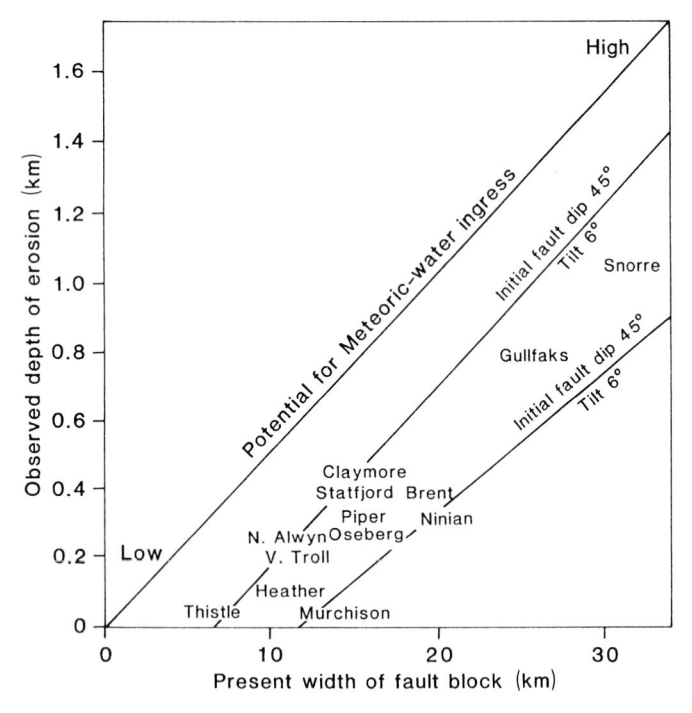

FIG. 26.—Relation between footwall uplift (expressed as the observed depth of erosion) and fault-block width for selected North Sea Jurassic reservoirs showing the potential for meteoric-water ingress (based on Yielding and Roberts, 1991). The sloping lines indicate predicted footwall uplift according to the rigid-domino model for ranges of initial fault dip and block tilt.

Perhaps the strongest objection to meteoric-water flushing as a means of generating secondary porosity and kaolinite cements close to the surface is the need to maintain the enhanced porosity from near-surface to moderate or deep burial conditions. This objection can be overcome if the potential-reservoir sandstone is sealed shortly after meteoric-water flushing, preventing compactional-fluid escape and allowing overpressure development during rapid Tertiary burial. This appears to have occurred in the Fulmar Sandstone Formation of the Central Graben (Stewart, 1986) and has also been argued to be the case for the Piper Formation sandstones of the Outer Moray Firth (Maher, 1980). Many equivalent, Upper Jurassic reservoir sandstones that are presently deeply buried in the Viking, Witch Ground and Central grabens are all highly overpressured and may have been so for long periods of time (Harris and Fowler, 1987; Buhrig, 1989). The nature of the rotational fault block would also allow relatively early hydrocarbon migration into shallow structures under a buoyancy drive, because of the differential burial history between structural highs and source-rock kitchens. Recall from Figure 1 that while the typical, crestal, reservoir sandstone sequence is at burial depths of <1 km, the corresponding source-rock kitchen may be at depths of 3 km and generating hydrocarbons.

We are left, therefore, with half a mechanistic solution. Some fault blocks (that is the larger ones) can, and probably did, experience meteoric flushing during the late Jurassic-early Cretaceous, which may have resulted in modification of early diagenetic assemblages prior to burial. In these larger fault blocks meteoric recharge may have penetrated for considerable distances and resulted in the widespread dissolution of feldspars and carbonates. Where these potential reservoir sandstones were sealed by a marine transgressive-mudstone sequence, the expulsion of compactive fluids was prevented. Migration of buoyant hydrocarbons into such structures subsequently took place, preserved the telogenic mineral assemblages and resulted in the development of overpressure. Other fault blocks (that is, some of the larger ones and many of the smaller ones) cannot have experienced meteoric-water flushing because they were never subaerially exposed and were not connected to aquifers undergoing recharge. An alternative mechanistic model is therefore needed.

Diagenesis During Late-Cretaceous and Early-Tertiary Rapid Burial

There is considerable evidence, documented in the diagenetic studies of many Jurassic sandstone reservoirs, that suggests the present reservoir characteristics are related to deep burial diagenesis and not shallow meteoric-water flushing. Indeed, it may well be that the behavior of normal faults and the character of the rotational block persist in influencing the extent and type of diagenetic modification throughout the subsurface. The very nature of the extensional fault-block structure provides an alternative mechanistic model for the generation of diagenetic fluids and conditions that are capable of providing solutes for cementation and mineral dissolution.

The differential subsidence characteristic of the extensional fault block establishes a vertical separation between the crestal structure and source rock very early on in the burial history of the fault block (see Fig. 1). This necessitates a major vertical component for hydrocarbon migration from mature source rocks in the kitchen to the structural highs. In fact, in many fault-block reservoirs, the Kimmeridge Clay overlying the reservoir and forming the cap rock seal is immature at present burial depths; only off-structure in the adjacent deep graben is the Kimmeridge Clay of oil-generating maturity (Cornford, 1984). The vertical-migration distance involved may exceed 2 km (see Fig. 1). To facilitate such vertical-migration distances, faults must be part of the fluid transport system.

Such fluid migration need not be restricted to passive flow through open fractures. In subsiding basins or in rock sequences experiencing a regional tectonic strain, faults have the potential for active fluid transport either across or along the plane of the fault. Displacement on faults in the slip direction gives rise to volume changes in the rock surrounding the fault (Barnett and others, 1987). Compressed and dilated volumes are distributed according to position relative to the fault (Fig. 27). Intermittent seismic slip results in rapid pressure changes and high hydraulic gradients, capable of leading to fluid movement or persistent pore-pressure differentials if fluid migration is inhibited. During periods of fault displacement or development of fluid overpressure, faults can thus actively pump or valve fluid movement (Sibson, 1981; Burley and others, 1991). Through these mechanisms, large volumes of fluids can be moved rapidly from one formation to another laterally across a fault or vertically through thousands of meters along the plane of the fault (Halfman and others, 1984; Meyer and McGee, 1985; Wood, 1985; Anderson and others, 1991).

In the context of a subsiding, compacting sedimentary basin, seismic valving, rather than seismic pumping, is the probable fluid-transport mechanism (Burley and others, 1989). Regional or local overpressure results in fault dilation and fluid movement (Fig. 28). Fault relaxation and fluid movement need not be associated with displacement, but are commonly intimately linked with extreme pressure and temperature variation. Rapid variation in pressure and temperature, together with the potential for mixing of subsurface fluids adjacent to fault planes, means that faults are probable foci of zones of intense diagenetic modification. During movement along the fault plane, fluids have little opportunity to interact with the host rocks so that both temperature and chemical disequilibrium can be readily introduced to high structural levels. Dissolution of detrital and authigenic components may result because of chemical aggressivity or, in the case of minerals with retrograde solubility (e.g., carbonates), as a result of fluid cooling (Giles and de Boer, 1989). Conversely, pressure decrease and fluid cooling may result in precipitation of minerals with prograde solubilities (e.g., clay minerals, authigenic quartz).

The evidence for fault-related fluid flow and diagenetic modification is compelling in many North Sea Jurassic reservoir sandstones. The concentration of cementation adjacent to faults and at the crest of structures, the presence of

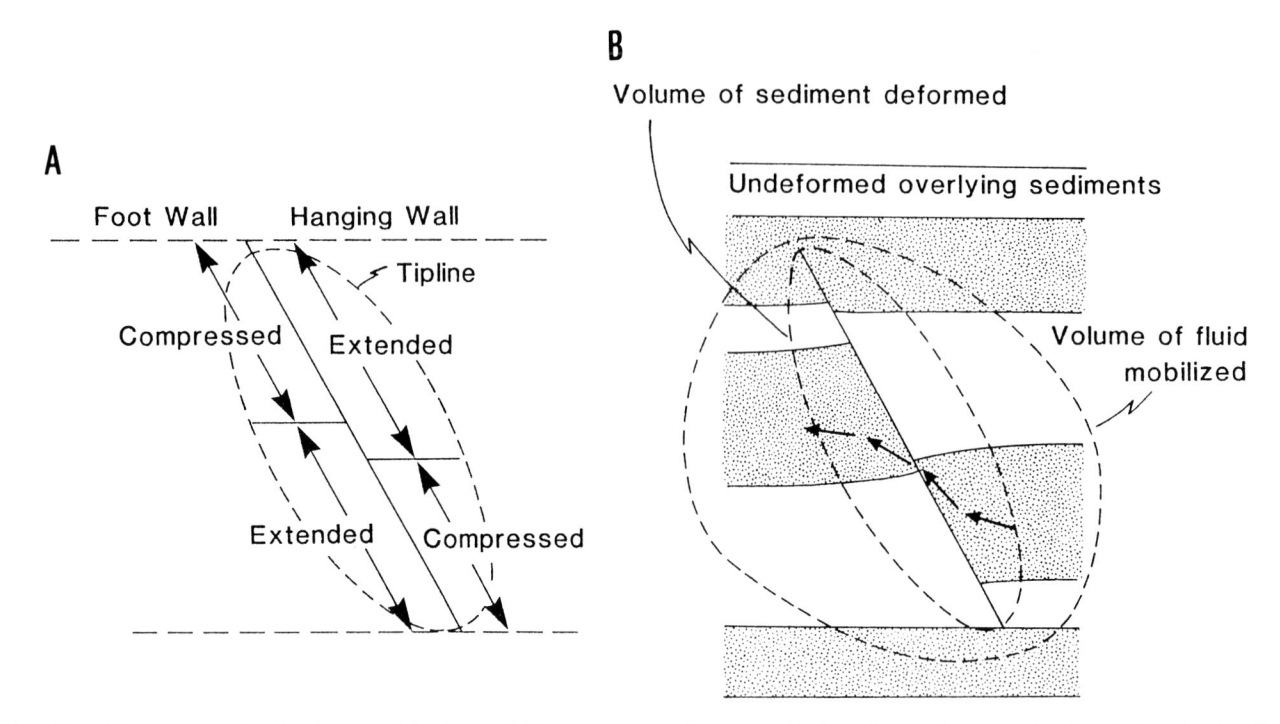

FIG. 27.—The concept of active fault participation in fluid movement as shown by (A) the volume of compressional and dilatational sediment deformation associated with normal faulting (based on Barnett and others, 1987) and (B) the potential for vertical and lateral fluid movement in such deformed sediments around faults as a result of the establishment of hydraulic gradients (Burley and others, 1991).

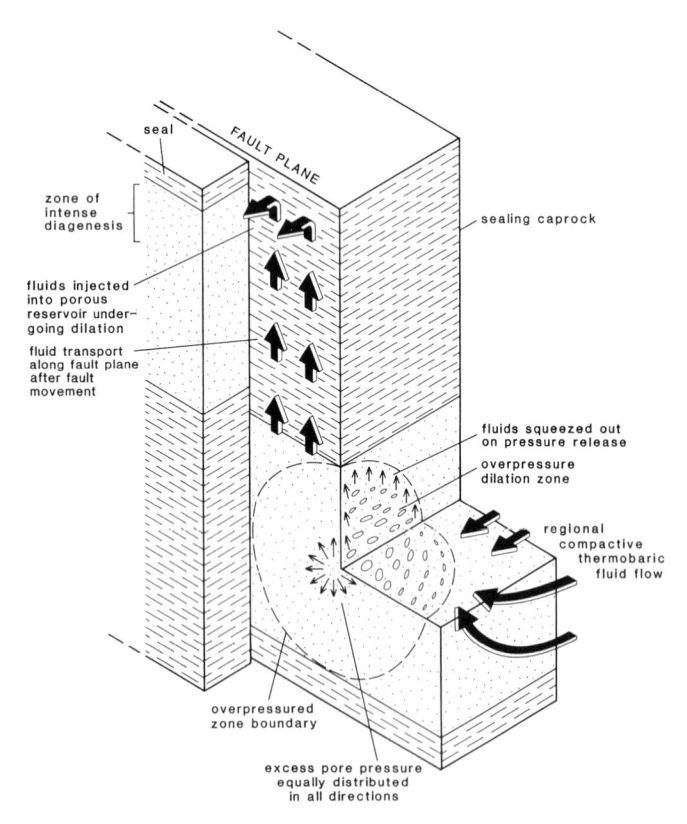

FIG. 28.—Conceptual model of fault-related fluid movement driven by compactional fluid flow (from Burley and others, 1989). Note how the zone of most intense diagenetic modification is located in the structurally high, porous-sandstone lithology beneath a mudstone seal.

temperature anomalies from fluid inclusions and the cyclical nature of cementation all suggest that faults exert a major influence on deep subsurface diagenesis by controlling fluid flow, fluid mixing and localization of cements and porosity enhancement (e.g., Blackbourne, 1984; Jourdan and others, 1987; Samways and Marshall, 1988; Burley and others, 1989). Moreover, variation in local pore-fluid pressure adjacent to the fault plane may control development of cementation cycles and development of compaction fabrics (Fig. 29). When the fault seals, compartmentalizing the structural block, pore pressures increase and grain-inflation fabrics may result (Gretener, 1976; Hunt, 1990). Chemical diagenetic reactions during this time will be diffusion controlled. At some point the fault relaxes, the pore pressure drops rapidly, dilated fractures and pores close, forcing fluid movement upward along the plane of the fault (Sibson and others, 1975). Temperature and chemical disequilibria induced in the structurally higher, porous lithologies on the plane of the fault, or in the footwall where the fluids can enter lithologies still experiencing dilation (see Figs. 27 and 28), result in the development of a dissolution-cementation cycle (see Fig. 16). Release of overpressure causes the pore pressure to fall toward the hydrostatic-pressure gradient (although it is unlikely to reach it in a burial setting) and compaction fabrics (brittle fracturing, grain shattering, shearing or pressure dissolution) will develop (Phillips, 1972). The cycle is repeated as the fault reseals and overpressure gradually builds again.

This model implies that the diagenetic system is open and mass transfer of solute takes place from deep sources to

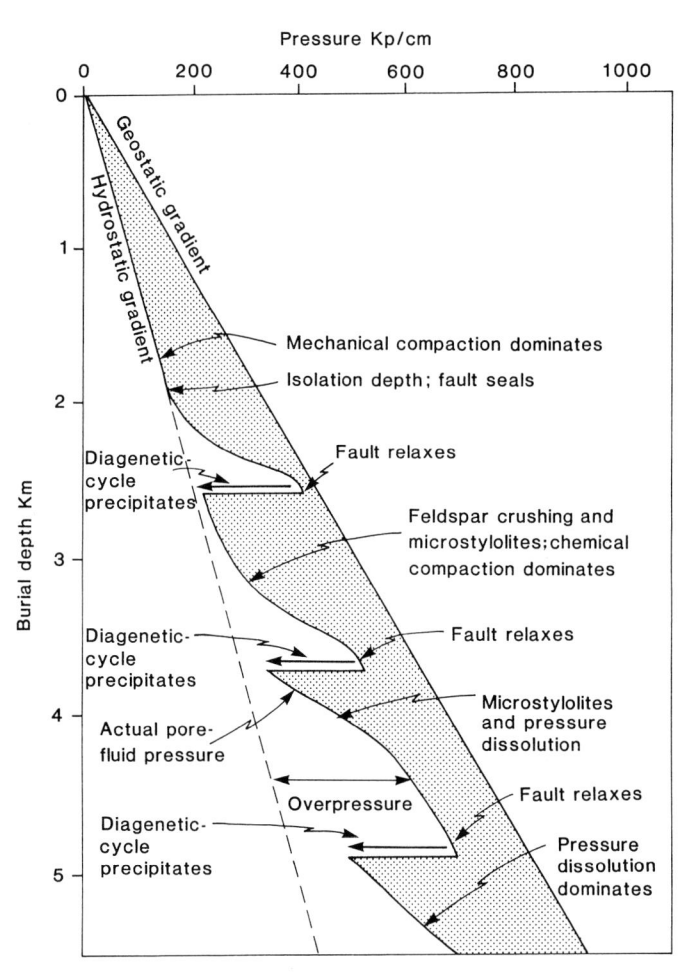

FIG. 29.—Interpretative genetic relation of pore pressure to diagenetic processes. Fault sealing begins at the isolation point and overpressuring continues until the rock strength is exceeded. At this point, the sealing fault relaxes and pore fluids can escape along the fault. Chemical and temperature disequilibrium during fluid movement result in precipitation of a 'diagenetic cycle' and compaction processes operate as the pore pressure drops toward hydrostatic pressure. The fault re-seals and the process is repeated.

structural highs. The type of fluids transported will depend on the lithologies juxtaposed along the fault, hence the importance of a knowledge of the stratigraphic infill of the basin, and the diagenetic grade or maturity of the juxtaposed lithologies. In this latter aspect, the timing of fluid transport is all important; it is the relative timing of secondary-porosity generation, mesogenic cementation and hydrocarbon migration that determines the reservoir quality at depth. We have summarized data from the North Sea that document that the Kimmeridge Clay Formation mudstones have undergone a series of gradual diagenetic reactions that proceed with increasing depth (although we do not wish to imply here that depth is the controlling parameter on these reactions). These could result in liberation of aqueous pore fluids, bicarbonate, Mg-Ca-Na-K cations, silica, alumina and base metals, in addition to hydrocarbons. Most of these solute- and fluid-releasing reactions are initiated in the kitchen source rocks during the early Tertiary (Fig. 1) and are broadly coincident with the timing of reactions in the reservoirs at structural highs. The temporal and spatial relations of kitchen and structural high diagenetic reactions are too closely linked for them not to be genetically associated. We accept, however, that there is as yet no definitive evidence to document which solutes actually do escape from the Kimmeridge Clay Formation or what the composition of the expelled pore fluids might be.

The thick Permo-Triassic sequences that underlie the Jurassic reservoir sandstones include continental, redbed mudstones and sandstones, carbonates and evaporites. Triassic mudstones are known from shallow basin-margin sequences (see Fig. 3; Jeans, 1978; Burley, 1984) and modern analogs (Walker and others, 1978) to be dominated by smectitic and highly expanded I/S clays. The Zechstein evaporites in the Central Graben and Witch Ground Graben include thick anhydrite and dolomite sequences. Here is a potential source for the deep, saline and isotopically heavy, evolved brines that are recorded from some of the deep reservoirs and in fluid inclusions from late cements. This argument is supported by the structural juxtaposition of Triassic mudstones or Zechstein evaporites against the major structure-delineating faults wherever intense sulphide mineralization is present in Jurassic reservoir sandstones. The fault-induced model of diagenesis therefore predicts that zones of most intense diagenetic modification will occur wherever there is the greatest potential for the introduction into reservoir lithologies of temperature or chemical disequilibrium as a result of fluid propagation along faults.

This does not imply all mesogenic cementation is fault related. In the Ness Formation of the Brent Group, for example, many mouth-bar and delta-top sandstone bodies are completely enclosed in marine or nonmarine mudstones. Such depositional systems are more likely to have been dominated by the effects of porewater expulsion from the enclosing mudstones. Similarly, sandstones of the Brae Formation submarine-fan facies interdigitate distally with marine-source rocks. Lateral and updip migration are the most probable fluid-migration pathways in such settings (Mackenzie and others, 1987), even though the Brae Formation is banked against bounding faults of the South Viking graben. There are obviously some depositional systems where there is no need to invoke fault-related fluid transport to account for the observed diagenetic modifications. Furthermore, even in the highly diagenetically modified reservoirs that are bounded by major faults, much of the observed hydrofracturing, brecciation and sulphide-vein cementation clearly postdates most of the mesogenic quartz, clay and carbonate cementation, implying that other burial-related diagenetic processes have been operative. Much of the observed mesogenesis may thus reflect slow, gradual burial evolution of pore fluids that appears episodic because each diagenetic reaction is limited by a kinetic threshold.

Equally, we readily accept that this view of burial diagenesis is not without its problems, many of which are well known and long standing. The limited volume of compactional water available for solute transport, whether solute is expelled from inferred mudstone sources, and the problems of silica mobility, all remain unresolved and awaiting attention.

SYNTHESIS: WHERE TO NEXT

Most of the ideas and hypotheses presented in this paper are controversial and untested; they need to be carefully assessed in the light of detailed regional and field-based diagenetic studies. Whereas we argue two endmember mechanistic models of diagenesis for the Jurassic fault-block play in the North Sea, meteoric-water flushing vs. fault-related compactional fluid flow, it is also probable that in many reservoirs both processes may have been operative to varying degrees. In some reservoirs, the evidence suggests that one model may have been dominant over the other in shaping the present reservoir characteristics. There is even the scope for other variation; early burial fabrics and authigenic-mineral assemblages may have been preserved through the development of overpressures at shallow depths and shallow-oil accumulation from deeper source rocks; diagenetic fluids, solutes and hydrocarbons may have as their source interbedded, distal or enclosing mudstones and have only undergone local migration. We conclude that no single 'diagenetic model' can be applied to the Jurassic North Sea fault-block play. However, all of the potential variation is inherent in the very nature of the rotated fault block and should be predictable given a working knowledge of the depositional environment, structural location, local stratigraphic-basin infill and burial history. There is clearly a need for detailed diagenetic studies of many other North Sea Jurassic reservoirs that not only carefully document the diagenetic assemblages, textures and sequences, but also try to place quantitative constraints on the absolute timing of the reactions and processes.

ACKNOWLEDGMENTS

This paper draws heavily on the data of many other authors and our co-workers, particularly the team of research students and research fellows in the Clastic Diagenesis Research Group in Manchester; Shelagh Baines, Gavin McAulay, Ian Hawthorne, Myles "better without" Jordan, Darren McAulay, Simon Guscott, Patrick McKeever and Colin Hughes, all led by Charles Curtis. Grateful thanks are extended to all these individuals, many off whom we have also bounced our ideas. Similarly, thoughtful and provocative reviews were kindly provided by John Kantorowicz, Juan Watterson and Graham Yielding, all of whom gave the benefit of their expertise and experience to improve the manuscript. SEPM referees, Lori Suma and Dennis Prezbindowski, painstakingly struggled through an overlong first version of the paper; their careful reviews clarified our expression and, we hope, the paper. Our apologies to those people whose data and ideas we have used and abused; misinterpretations of previously published data are solely our responsibility. Stuart Burley is the British Gas Geology Research Fellow at Manchester and expresses his gratitude to British Gas for generously supporting diagenetic research, while Joe MacQuaker is funded by the NERC Petroleum Directorate Controlled Program in Petroleum Science. Publication of a color plate was made possible through funds provided by British Gas.

REFERENCES

AAGAARD, P., EGEBERG, P. K., SAIGAL, G. C., MORAD, S., AND BJØRLYKKE, K., 1990, Diagenetic albitization of detrital K-feldspars in Jurassic, Lower Cretaceous and Tertiary clastic reservoir rocks from offshore Norway. II. Formation Water Chemistry and Kinetic Considerations: Journal of Sedimentary Petrology, v. 60, p. 575–581.

ANDERSON, R. N., HE, W., HOBART, M. A., WILKINSON, C. R., AND NELSON, H. R., 1991, Active fluid flow in the Eugene Island area, offshore Louisiana: Geophysics: The Leading Edge of Exploration, v. 1991, p. 12–17.

ANDREWS, J. E., 1985, The sedimentary facies of a late Bathonian regressive episode; the Kilmaluag and Skidiburgh Formations of the Great Estuarine Group, Inner Hebrides, Scotland: Journal of the Geological Society of London, v. 142, p. 1119–1137.

ARONSON, J. L., AND DOUTHITT, C. B., 1986, K/Ar systematics of acid-treated illite/smectite: implications for evaluating age and crystal structure: Clays and Clay Minerals, v. 34, p. 473–482.

ASTIN, T. R., AND SCOTCHMAN, I. C., 1988, The diagenetic history of some septarian concretions from the Kimmeridge Clay, England: Sedimentology, v. 35, p. 349–368.

BADLEY, M. E., PRICE, J. D., RAMBECH, D. C., AND AGDESTEIN, T., 1988, The structural evolution of the northern Viking Graben and its bearing upon extensional modes of basin formation: Journal of the Geological Society of London, v. 145, p. 455–472.

BAINES, S. J., BURLEY, S. D., AND GIZE, A. P., 1991, Sulphide mineralization and hydrocarbon migration in North Sea oilfields, in Pagel, M., and Leroy, J., eds., Source, Transport and Deposition of Metals, Proceedings, 25th Anniversary Mineral Deposits Meeting, Nancy, France, p. 87–91.

BARNETT, J. A. M., MORTIMER, J., RIPPON, J. H., WALSH, J. J., AND WATTERSON, J., 1987, Displacement geometry in the volume containing a single normal fault: American Association of Petroleum Geologists Bulletin, v. 71, p. 925–937.

BARR, D., 1987, Structural/stratigraphic models for extensional basins of half-graben type: Journal of Structural Geology, v. 9, p. 491–500.

BATH, A. H., MILODOWSKI, A. E., AND STRONG, G., 1987, Fluid flow and diagenesis in the East Midlands Triassic sandstone aquifer, in Goff, J. C., and Williams, B. P. J., eds., Fluid Flow in Sedimentary Basins and Aquifers: Geological Society of London Special Publication, v. 34, p. 127–140.

BERGLUND, L. T., AUGUSTSON, J., FAERSETH, R., GJELBERG, J., AND RAMBERG, M. H., 1986, The evolution of the Hammerfest Basin, in Spencer, A. M., ed., Habitat of Hydrocarbons on the Norwegian Continental Shelf: Stavanger, Norwegian Petroleum Society, Statoil, p. 319–338.

BIRKELAND, P. W., 1974, Pedology, Weathering and Geomorphological Research: London, Oxford University Press, 285 p.

BJORKUM, P. A., AND GJELSVIK, N., 1988, An isochemical model for formation of authigenic kaolinite, K-feldspar and illite in sediments: Journal of Sedimentary Petrology, v. 58, p. 506–511.

BJORKUM, P. A., RUNE, M., WALDERHAUG, O., AND HURST, A., 1990, The role of late-Cimmerian unconformity for the distribution of kaolinite in the Gullfaks Field, northern North Sea: Sedimentology, v. 37, p. 395–406.

BJØRLYKKE, K. O., 1983, Diagenetic reactions in sandstones, in Parker, A., and Sellwood, B. W., eds., Sediment Diagenesis, NATO ASI: Dordrecht, Reidel Publishing Company, p. 169–205.

BJØRLYKKE, K. O., 1989, Sandstone diagenesis and porosity during basin evolution: Geologische Rundschau, v. 78, p. 243–268.

BJØRLYKKE, K., AAGAARD, P., DYPVIK, H., HASTINGS, D. S., AND HARPER, A. S., 1986, Diagenesis and reservoir properties of Jurassic sandstones from the Haltenbanken area, offshore mid-Norway, in Spencer, A. M., Holter, E., Campbell, C. J., Hanslien, S. J., Nelson, P. H. H., and Ormaasen, E. G., eds., Habitat of Hydrocarbon on the Norwegian Continental Shelf: Norwegian Petroleum Society, London, Graham and Trotman Ltd., p. 275–286.

BJØRLYKKE, K.O., AND BRENDSAL, A., 1986, Diagenesis of the Brent Sandstone in the Statford Field, North Sea, in Gautier, D. L., ed., Roles of Organic Matter in Sediment Diagenesis: Society of Economic Paleontologists and Mineralogists Special Publication 38, p. 157–167.

BJØRLYKKE, K. O., ELVERHOI, A., AND MALM, A. O., 1979, Diagenesis in Mesozoic sandstones from Spitsbergen and the North Sea—a comparison: Geologische Rundschau, v. 68, p. 1152–1171.

BJØRLYKKE, K. O., RAMM, M., AND SAIGAL, G. C., 1989, Sandstone diagenesis and porosity modification during basin evolution: Geologische Rundschau, v. 78, p. 243–268.

BLACKBOURNE, G., 1984, Diagenetic history and reservoir quality of a Brent sand sequence: Clay Minerals, v. 19, p. 377–390.

BODNAR, R. J., BINNS, P. R., AND HALL, D. L., 1989, Synthetic fluid inclusions VII: Re-equilibration of fluid inclusions in quartz during laboratory-simulated metamorphic burial and uplift: Journal of Metamorphic Geology, v. 7, p. 243–260.

BRINT, J. F., HAMILTON, P. J., HASZELDINE, R. S., FALLICK, A. E., AND BROWN, S., 1987, Quartz overgrowth authigenesis: an isotopic and fluid inclusion study, Brent sands, northern North Sea: Abstract Volume, British Sedimentological Research Group, Aberdeen, p. 17.

BROOKS, J., AND GLENNIE, K., 1987, Petroleum Geology of North-West Europe: London, Graham and Trotman, 1211 p.

BRYANT, I. D., KANTOROWICZ, J. D., AND LOVE, C. F., 1988, Depositional and diagenetic models to explain origin of laterally continuous carbonate-cemented layers, Upper Lias sands, southern England (abs.): American Association of Petroleum Geologists Bulletin, v. 72, p. 166.

BUHRIG, C., 1989, Geopressured Jurassic reservoirs in the Viking Graben; modelling and geological significance: Marine and Petroleum Geology, v. 6, p. 31–48.

BURLEY, S. D., 1984, Patterns of diagenesis in the Sherwood Sandstone Group (Triassic), United Kingdom: Clay Minerals, v. 21, p. 649–694.

BURLEY, S. D., 1986, The development and destruction of porosity within Upper Jurassic reservoir sandstones of the Piper and Tartan oilfields, Outer Moray Firth, North Sea: Clay Minerals, v. 21, p. 649–694.

BURLEY, S. D., 1988, Sulphide Mineralization in North Sea reservoirs; a consequence of burial diagenesis: Abstract Volume, Mineral Deposits Studies Group Annual Meeting, Oil and Ore Symposium, Royal Holloway and Bedford New College, London, United Kingdom, 1988, p. 9.

BURLEY, S. D., AND FLISCH, M., 1989, K/Ar geochronology and the timing of detrital I/S clay illitization and authigenic illite precipitation in the Piper and Tartan fields, outer Moray Firth, UK North Sea: Clay Minerals, v. 24, p. 285–315.

BURLEY, S. D., KANTOROWICZ, J. D., AND WAUGH, B., 1985, Clastic Diagenesis, in Brenchley, P. J., and Williams, B. P. J., eds., Sedimentology: Recent and Applied Aspects: Geological Society of London Special Publication 18, p. 189–226.

BURLEY, S. D., MULLIS, J., AND MATTER, A., 1989, Timing diagenesis in the Tartan reservoir (UK, North Sea); constraints from combined cathodoluminescence microscopy and fluid inclusion studies: Marine and Petroleum Geology, v. 6, p. 98–120.

BURLEY, S. D., WALSH, J. J., AND WATTERSON, J., 1991, Faults, fluids and diagenesis: active fault participation in reservoir porosity modification, in Hardman, R. F. P., ed., Exploration Britain–Into the Next Decade: Petroleum Group Conference Abstract Volume, Bath, England, April, p. 7.

CHIARELLI, A., AND DUFFAUD, F., 1980, Pressure origin and distribution in Jurassic of Viking Basin (United Kingdom, Norway): American Association of Petroleum Geologists Bulletin, v. 64, p. 1245–1250.

COLEMAN, M. L., JONES, M. R. O., AND COX, M. A., 1990, Analysis of formation water sampled from core, in Kleppe, J., Berg, E. W., Buller, A. T., Hjelmeland, O., and Torsafeter, O., eds., North Sea Oil and Gas Reservoirs: Norwegian Institute of Technology, London, Graham and Trotman, p. 165–171.

CORNFORD, C., 1984, Source rocks and hydrocarbons of the North Sea, in Glennie, K. W., ed., Introduction to the Petroleum Geology of the North Sea: Oxford, Blackwell, p. 171–204.

CURTIS, C. D., 1978, Possible links between sandstone diagenesis and depth-related geochemical reactions occurring in enclosing mudrocks: Journal of the Geological Society of London, v. 135, p. 107–117.

DE'ATH, N. G., AND SCHULYMAN, S. F., 1981, The geology of the Magnus Oilfield, in Illing, L. V., and Hobson, G. D., eds., Petroleum Geology of the Continental Shelf of North-West Europe: London, Heydon and Sons Ltd., p. 342–251.

DREVER, J. I., 1982, The Geochemistry of Natural Waters: Englewood Cliffs, N.J., Prentice-Hall Inc., 388 p.

EGEBERG, P. K., AND AAGAARD, P., 1989, Origin and evolution of formation waters from oil fields on the Norwegian shelf: Applied Geochemistry, v. 4, p. 131–142.

EHRENBERG, S. N., 1990, Relationship between diagenesis and reservoir quality in sandstones of the Garn Formation, Haltenbanken, mid-Norwegian Continental Shelf: American Association of Petroleum Geologists Bulletin, v. 74, p. 1538–1558.

EHRENBERG, S. N., 1991, Kaolinized, potassium-leached zones at the contacts of the Garn Formation, Haltenbanken, mid-Norwegian continental shelf: Marine and Petroleum Geology, v. 8, p. 250–269.

EHRENBERG, S. N., AND NADEAU, P. H., 1989, Formation of diagenetic illite in sandstones of the Garn Formation, Haltenbanken area, mid-Norwegian continental shelf: Clay Minerals, v. 24, p. 233–253.

EMERY, D., MYERS, K. J., AND YOUNG, R., 1990, Ancient subaerial exposure and freshwater leaching in sandstones: Geology, v. 18, p. 1178–1181.

GABRIELSEN, R. H., EKERN, O. F., AND EDVARDSEN, A., 1986, Structural development of hydrocarbon traps, Block 2/2, Norway: in Spencer, A. M., ed., Habitat of Hydrocarbons on the Norwegian Continental Shelf: London, Graham and Trotman, p. 129–141.

GIBBS, A. D., 1984, Structural evolution of extensional basin margins: Journal of the Geological Society of London, v. 141, p. 609–620.

GILES, M. R., AND DE BOER, R. B., 1989, Secondary porosity: creation of enhanced porosities in the subsurface from the dissolution of carbonate cements as a result of cooling formation waters: Marine and Petroleum Geology, v. 6, p. 261–269.

GILES, M. R., AND MARSHALL, J. D., 1986, Constraints on the development of secondary porosity in the subsurface; re-evaluation of processes: Marine and Petroleum Geology, v. 4, p. 261–269.

GLASMANN, J. R., 1987, Comments on "The evolution of illite to muscovite; mineralogical and isotopic data from the Glarus Alps, Switzerland": Contributions to Mineralogy and Petrology, v. 96, p. 72–74.

GLASMANN, J. R., CLARK, R. A., LARTER, S., BRIEDIS, N. A., AND LUNDEGARD, P. D., 1989a, Diagenesis and hydrocarbon accumulation, Brent Sandstone, (Jurassic), Bergen High area, North Sea: American Association of Petroleum Geologists Bulletin, v. 73, p. 1341–1360.

GLASMANN, J. R., LARTER, S., BRIEDIS, N. A., AND LUNDEGARD, P. D., 1989b, Shale diagenesis in the Bergen High area, North Sea: Clays and Clay Minerals, v. 37, p. 97–112.

GLASMANN, J. R., LUNDEGARD, P. D., CLARK, R. A., PENNY, B. K., AND COLLINS, I. D., 1989c, Geochemical evidence for the history of diagenesis and fluid migration: Brent Sandstone, Heather field, North Sea: Clay Minerals, v. 24, p. 255–284.

GLENNIE, K. W., 1984, The structural framework and the pre-Permian history of the North Sea area, in Glennie, K. W., ed., Introduction to the Petroleum Geology of the North Sea: Oxford, Blackwell, p. 17–39.

GRETENER, P. E., 1976, Pore pressure: fundamentals, general ramifications and implications for structural geology: American Association of Petroleum Geology, Continuing Education Course Note Series, v. 4, 87 p.

HALFMAN, S. E., LIPPMANN, M. J., ZELWER, R., AND HOWARD, J. H., 1984, Geologic interpretation of geothermal fluid movement in Cerro Prieto Field, Baja California, Mexico: American Association of Petroleum Geologists, v. 68, p. 18–30.

HALLAM, A., 1984, Continental humid and arid zones during the Jurassic and Cretaceous: Palaeogeography, Palaeoclimatology, Palaeoecology, v. 47, p. 195–223.

HAMILTON, P. J., FALLICK, A. E., MACINTYRE, R. M., AND ELLIOT, S., 1987, Isotopic tracing of the provenance and diagenesis of Lower Brent Group Sands, North Sea, in Brooks, J., and Glennie, K. W., eds., Petroleum Geology of North-West Europe: London, Graham and Trotman, p. 939–949.

HAMILTON, P. J., KELLY, S., AND FALLICK, A. E., 1989, K-Ar dating of illite in hydrocarbon reservoirs: Clay Minerals, v. 24, p. 215–231.

HANCOCK, N. J., AND FISHER, M. J., 1981, Middle Jurassic North Sea Deltas with particular reference to Yorkshire, in Illing, L. V., and Hobson, G. D., eds., Petroleum Geology of the Continental Shelf of North-West Europe: Proceedings, 2nd Conference, Institute of Petroleum, London, United Kingdom, p. 186–195.

HANCOCK, N. J., AND TAYLOR, A. M., 1978, Clay mineral diagenesis and oil migration in the Middle Jurassic Brent Sand Formation: Journal of the Geological Society of London, v. 135, p. 69–72.

HARRIS, J., AND FOWLER, R., 1987, Enhanced prospectivity of the Mid-Late Jurassic sediments of the South Viking Graben, northern North Sea, in Brooks, J., and Glennie, K., eds., Petroleum Geology of North-West Europe, v. 2: London, Graham and Trotman, p. 879–889.

HARRIS, N. B., 1989, Diagenetic quartzarenite and destruction of secondary porosity; an example from the Middle Jurassic Brent Sandstone of Northwest Europe: Geology, v. 17, p. 361–364.

HARRIS, N. B., 1992, Burial diagenesis of Brent sandstones: a study of Stattfjord, Hutton and Lyell fields, in Haszeldine, S., and Morton, A. C., eds., Geology of the Brent Group: Geological Society of London Special Publication, v. 61, p. 109–134.

HASZELDINE, R. S., BRINT, J. F., FALLICK, A. S., HAMILTON, P. J., AND BROWN, S., 1990, Diagenesis, porosity and fluid evolution in the Brent Sandstone, United Kingdom, North Sea (abs.): American Association of Petroleum Geologists Bulletin, v. 74, p. 671.

HASZELDINE, R. S, SAMSON, I. M., AND CORNFORD, C., 1984a, Dating diagenesis in a petroleum basin, a new fluid inclusion method: Nature, v. 307, p. 354–357.

HASZELDINE, R. S., SAMSON, I. M., AND CORNFORD, C., 1984b, Quartz diagenesis and convective fluid movement; Beatrice Oilfield, UK, North Sea: Clay Minerals, v. 19, p. 391–402.

HEMINGWAY, J. E., AND RIDDLER, J., 1982, Basin inversion in North Yorkshire: Transactions, Institute of Mining and Metallurgy (B), v. 91, p. 175–186.

HUANG, W. L., 1990, Illitic clay formation during experimental diagenesis of arkoses: Program and Abstracts, Clay Minerals Society 27th Annual Meeting, Columbia, Missouri, p. 62.

HUANG, W. L., BISHOP, A. M., AND BROWN, R. W., 1986, The effect of fluid/rock ratio on feldspar dissolution and illite formation under reservoir conditions: Clay Minerals, v. 21, p. 585–602.

HUDSON, J. D., 1970, Algal limestones with pseudomorphs after gypsum from the Middle Jurassic of Scotland: Lethia, v. 3, p. 11–40.

HUDSON, J. D., 1980, Aspects of brackish water facies and faunas from the Jurassic of northwest Scotland: Proceedings, Geological Association of London, v. 91, p. 99–105.

HUDSON, J. D., 1983, Mesozoic sedimentation and sedimentary rocks in the Inner Hebrides: Proceedings, Royal Society of Edinborough, v.. 83B, p. 47–63.

HUDSON, J. D., AND ANDREWS, J. E., 1987, The diagenesis of the Great Estuarine Group, Middle Jurassic, Inner Hebrides, Scotland, in Marshall, J. D., ed., Diagenesis of Sedimentary Sequences: Geological Society of London Special Publication 36, p. 259–276.

HUNT, J. M., 1990, Generation and migration of petroleum from abnormally pressured fluid compartments: American Association of Petroleum Geologists, v. 74, p. 1–12.

HURST, A., 1982, The clay mineralogy of Jurassic shales from Brora, N.E. Scotland: Proceedings, 7th International Clay Conference, Doorwerth, Netherlands, v. 35, p. 677–684.

HURST, A., 1985a, The implications of clay mineralogy to palaeoclimate and provenance during the Jurassic in NE Scotland: Scottish Journal of Geology, v. 21, p. 143–160.

HURST, A., 1985b, Mineralogy and diagenesis of Lower Jurassic sediments of the Lossiemouth borehole, northeast Scotland: Proceedings, Yorkshire Geological Society, v. 45, p. 189–197.

HURST, A., 1985c, Diagenetic chlorite formation in some Mesozoic shales from the Sleipner area of North Sea: Clay Minerals, v. 20, p. 69–80.

HURST, A., AND IRWIN, H., 1982, Geologic modelling of clay diagenesis in sandstones: Clay Minerals, v. 17, p. 5–22.

ILLING, L. V., AND HOBSON, G. D., 1981, Petroleum geology of the continental shelf of North-West Europe: Proceedings, 2nd Conference, Institute of Petroleum, London, United Kingdom, 521 p.

IRWIN, H., CURTIS, C. D., AND COLEMAN, M. L., 1977, Isotopic evidence for source of diagenetic carbonates formed during burial of organic-rich sediments: Nature, v. 269, p. 209–213.

JAHREN, J. S., AND AAGAARD, P., 1989, Compositional variations in diagenetic chlorites and illites, and relationships with formation-water chemistry: Clay Minerals, v. 24, p. 157–170.

JEANS, C. V., 1978, The origin of the Triassic clay assemblages of Europe with special reference to the Keuper Marl and Rhaetic of parts of England: Philosophical Transactions of the Royal Society, v. 289B, p. 549–639.

JENSENSIUS, J., AND MUNKSGAARD, N. C., 1989, Large-scale hot water migration systems around salt diapirs in the Danish central trough and their impact on diagenesis of chalk reservoirs: Geochimica et Cosmochimica Acta, v. 53, p. 79–88.

JOURDAN, A., THOMAS, M., BREVART, O., ROBSON, P., SOMMER, F., AND SULLIVAN, M., 1987, Diagenesis as the control of Brent sandstone reservoir properties in the Greater Alwyn area, East Shetland Basin, in Brooks, J., and Glennie, K. W., eds., Petroleum Geology of North-West Europe, v. 2: London, Graham and Trotman, p. 951–961.

KANTOROWICZ, J. D., 1985, The petrology and diagenesis of Middle Jurassic clastic sediments, Ravenscar Group, Yorkshire: Sedimentology, v. 32, p. 833–853.

KANTOROWICZ, J. D., 1987, The significance of variations in illite morphology in North Sea reservoir sandstones: Abstracts, Society of Economic Paleontologists and Mineralogists, Annual Midyear Meeting, v. 4, p. 40.

KANTOROWICZ, J. D., 1990a, The influence of variations in illite morphology on the permeability of Middle Jurassic Brent Group sandstones, Cormorant Field, UK North Sea: Marine and Petroleum Geology, v. 7, p. 66–74.

KANTOROWICZ, J. D., 1990b, Lateral and vertical variations in pedogenesis and other early diagenetic phenomena, Middle Jurassic Ravenscar Group, Yorkshire: Proceedings, Yorkshire Geological Society, v. 48, p. 61–74.

KANTOROWICZ, J. D., LIVERA, S. E., AND HAMILTON, P. J., 1989, Illite age dates in the Pelican Field, UK, North Sea: constraints on hydrocarbon migration and implications for reservoir quality prediction: Abstract Volume, British Sedimentological Research Group Annual Meeting, Leeds, England, p. 81.

KLEPPE, J., BERG, E. W., BULLER, A. T., HJELMELAND, O., AND TORSAETER, O., 1987, Proceedings, North Sea Oil and Gas Reservoirs Seminar: Norwegian Institute of Technology, Trondheim, London, Graham and Trotman, 356 p.

KNAUTH, L. P., AND BEEUNAS, M. A., 1986, Isotope geochemistry of fluid inclusions in Permian halite with implications for the isotope history of ocean water and the origin of saline formation waters: Geochimica et Cosmochimica Acta, v. 50, p. 419–433.

KONNERUP-MADSEN, J., AND DYPVIK, H., 1988, Fluid inclusions and quartz cementation in Jurassic sandstones from Haltenbanken, offshore mid-Norway: Bulletin Mineralogique, v. 111, p. 401–411.

KUSZNIR, N. J., MARSDEN, G., AND EGAN, S. S., 1991, A flexural-cantilever simple-shear/pure-shear model of continental lithospheric extension: applications to the Jeanne d'Arc Basin, Grand Banks and Viking Graben, North Sea, in Roberts, A. M., Yielding, G., and Freeman, B., eds.,The Geometry of Normal Faults: Geological Society of London Special Publication 56, p. 41–60.

LAM, K., AND PORTER, R., 1977, The distribution of palynomorphs in the Jurassic rocks of the Brora Outlier, NE Scotland: Journal of the Geological Society of London, v. 134, p. 45–55.

LARESE, R. E., HASKELL, N. L., PREZBINDOWSKI, D. R., AND BEJU, D., 1984, Porosity development in selected Jurassic sandstones from the Norwegian and North Seas, Norway—an overview, in Holter, E., Spencer, A. M., Johnsen, S. O., Moerk, A., Nysaether, E., Songstad, P., and Spinnangr, A., eds., Petroleum Geology of the North European Margin: Norwegian Petroleum Society, London, Graham and Trotman, p. 81–95.

LIEWIG, N., CLAUER, N., AND SOMMER, F., 1987, Rb-Sr and K-Ar dating of clay diagenesis in Jurassic sandstone oil reservoir, North Sea: American Association of Petroleum Geologists Bulletin, v. 71, p. 1467–1474.

LINDGREEN, H., AND HANSEN, P. L., 1991, Ordering of illite smectite in Upper Jurassic claystones from the North Sea: Clay Minerals, v. 26, p. 105–125.

LONGSTAFFE, F.J., 1989, Stable isotopes as tracers in clastic diagenesis, in Hutcheon, I. C., ed., Burial Diagenesis: Mineralogical Association of Canada, Short Course Handbook, v. 15, p. 201–277.

LØNØY, A., AKSELSEN, J., AND RONNING, K., 1986, Diagenesis of a deeply buried sandstone reservoir: Hild Field, northern North Sea: Clay Minerals, v. 21, p. 497–511.

MACAULAY, C. I., HASZELDINE, R. S., HAMILTON, P. J., FALLICK, A. E., AND RAINEY, S., 1987, Isotopic constraints on the diagenetic history of the Magnus oilfield reservoir: Abstract Volume, British Sedimentological Research Group, Aberdeen, Scotland, p. 89.

MACCHI, L., CURTIS, C. D., LEVISON, A., WOODWARD, K., AND HUGHES, C. R., 1990, Chemistry, morphology, and distribution of illites from

Morecambe gas field, Irish Sea, offshore United Kingdom: American Association of Petroleum Geologists Bulletin, v. 74, p. 296–308.

MacKenzie, A. S., Price, I., Leytheuser, D., Muller, P., Radke, M., and Schafer, R. G., 1987, The expulsion of petroleum from Kimmeridge clay source-rocks in the area of the Brae oilfield, UK continental shelf, in Brooks, J., and Glennie, K., eds., Petroleum Geology of North-West Europe: London, Graham and Trotman Ltd., p. 856–877.

Maher, C. E., 1980, The Piper Oilfield, in Halbouty, M. T., ed., Giant Oil and Gas Fields of the Decade 1968–1978: American Association of Petroleum Geologists Memoir 30, p. 131–172.

McKenzie, D. P., 1978, Some remarks on the development of sedimentary basins: Earth and Planetary Science Letters, v. 40, p. 25–32.

Meyer, H. J., and McGee, H. W., 1985, Oil and gas fields accompanied by geothermal anomalies in Rocky Mountain region: American Association of Petroleum Geologists Bulletin, v. 69, p. 933–945.

Miller, R. G., 1990, A paleoceanographic approach to the Kimmeridge Clay Formation, in Huc, A. Y., ed., Deposition of Organic Facies: American Association of Petroleum Geologists, Studies in Geology, v. 30, p. 13–26.

Morton, A. C., and Humphreys, B., 1983, The petrology of the Middle Jurassic sandstones from the Murchison field, North Sea: Journal of Petroleum Geology, v. 5, p. 245–260.

Morton, N., 1987, Jurassic subsidence history in the Hebrides, N.W. Scotland: Marine and Petroleum Geology, v. 4, p. 226–242.

Nipen, O., 1987, Oseberg, in Spencer, A. M., Holter, E., Campbell, C. J., Hanslien, S. J., Nelson, P. H. H., and Ormaasen, E. G., eds., Geology of the Norwegian Oil and Gas Fields: Norwegian Petroleum Society, London, Graham and Trotman Ltd., p. 379–387.

Olaussen, S., Dalland, A. Gloppen, T. G., and Johannessen, E., 1984, Depositional environment and diagenesis of Jurassic reservoir sandstones in the eastern part of Troms I area, in Spencer, A. M., Johnsen, S. O., Moerk, A., Nysaether, E., Songstad, P., and Spinnangr, A., eds., Petroleum Geology of the North European Margin: Norwegian Petroleum Society, London, Graham and Trotman, p. 61–79.

Osborne, M., and Haszeldine, R. S., 1990, Fluid inclusions in diagenetic quartz record oilfield burial temperatures, not precipitation temperatures: British Sedimentological Research Group, Reading, England, Abstract Volume, p. 65.

Parsley A. J., 1984, North Sea hydrocarbon plays, in Glennie, K. W., ed., Introduction to the Petroleum Geology of the North Sea: Oxford, Blackwell, p. 205–230.

Pearson, M. J., and Small, J. S., 1988, Illite/smectite diagenesis and palaeotemperatures in northern North Sea Quaternary to Mesozoic shale sequences: Clay Minerals, v. 23, p. 109–132.

Phillips, W. J., 1972, Hydraulic fracturing and mineralization: Journal of the Geological Society of London, v. 128, p. 337–359.

Pickering, K. T., 1984, The Upper Jurassic "boulder beds" and related deposits: a fault-controlled submarine slope, NE Scotland: Journal of the Geological Society of London, v. 141, p. 357–374.

Potter, R. W., 1977, Pressure corrections for fluid-inclusion homogenization temperatures based on the volumetric properties of the system Na-Cl-H$_2$O: Journal of Research of the U. S. Geological Survey, v. 5, p. 603–607.

Prezbindowski, D., and Tapp, J. B., 1991, Dynamics of fluid inclusion alteration in sedimentary rocks: a review and discussion: Organic Geochemistry, v. 17, p. 131–142.

Riches, P., Traub-Sobott, I., Zimmerlie, W., and Zinkernagel, U., 1986, Diagenetic peculiarities of potential Lower Jurassic reservoir sandstones, Troms I area, off northern Norway, and their tectonic significance: Clay Minerals, v. 21, p. 565–584.

Roberts, A. M., and Yielding, G., 1991, Deformation around basin margin faults in the North Sea/Norwegian Rift, in Roberts, A. M., Yielding, G., and Freeman, B., eds., The Geometry of Normal Faults: Geological Society of London Special Publication 56, p. 79–89.

Roedder, E., 1984, Fluid Inclusions: Reviews in Mineralogy, v. 12: Washington, D.C., Mineralogical Society of America, 644 p.

Saigal, G. C., Morad, S., Bjørlykke, K., Egeberg, P. K., and Aagaard, P., 1988, Diagenetic albitization of detrital K-feldspars in Jurassic, Lower Cretaceous and Tertiary clastic reservoir rocks from offshore Norway. I. Textures and origin: Journal of Sedimentary Petrology, v. 58, p. 1003–1013.

Samways, G., and Marshall, J., 1988, A seismic pumping model to account for complex diagenesis in faulted sediments: Society of Economic Paleontologists and Mineralogists, Midyear Annual Meeting, Columbus, Ohio, Abstract Volume, p. 47.

Savin, S. M., and Epstein, S., 1970, The oxygen and hydrogen isotope geochemistry of clay minerals: Geochimica et Cosmochimica Acta, v. 34, p. 25–42.

Schmidt, V., and MacDonald, D. A., 1979, The role of secondary porosity in the course of sandstone diagenesis, in Scholle, P. A., and Schluger, P. R., eds., Aspects of Diagenesis: Society of Economic Paleontologists and Mineralogists Special Bulletin, v. 26, p. 175–208.

Scotchman, I. C., 1987, Relationship between clay diagenesis and organic maturation in the Kimmeridge Clay Formation, onshore U.K., in Brooks, J., and Glennie, K., eds., Petroleum Geology of North-West Europe: London, Graham and Trotman Ltd., p. 251–261.

Scotchman, I. C., 1989, Diagenesis of the Kimmeridge Clay Formation, onshore UK: Journal of the Geological Society of London, v. 146, p. 285–303.

Scotchman, I. C., 1991a, The geochemistry of concretions from the Kimmeridge Clay Formation of southern and eastern England: Sedimentology, v. 38, p. 79–106.

Scotchman, I. C., 1991b, Kerogen facies and maturity of the Kimmeridge Clay Formation in southern and eastern England: Marine and Petroleum Geology, v. 8, p. 278–295.

Scotchman, I., Johnes, L. H., and Miller, R. S., 1989, Clay diagenesis and oil migration in Brent Group sandstones of NW Hutton field, UK North Sea: Clay Minerals, v. 24, p. 339–374.

Searle, A., 1989, Sedimentology and early diagenesis of the Lower Jurassic, Applecross, Wester Ross: Scottish Journal of Geology, v. 25, p. 45–62.

Selley, R. C., and Stoneley, R., 1985, Petroleum Habitat in South Dorset, in Brooks, J., and Glennie, K. W., eds., Petroleum Geology of North-West Europe, v. 1: London, Graham and Trotman, p. 139–148.

Shaw, H. F., and Primmer, T. J., 1991, Diagenesis of mudrocks from the Kimmeridge Clay Formation of the Brae Area, UK, North Sea: Marine and Petroleum Geology, v. 8, p. 270–277.

Sibson, R. H., 1981, Fluid flow accompanying faulting: field evidence and models, in Simpson, D. W., and Richards, P. G., eds., Earthquake Prediction: An International Review: American Geophysical Union, Washington, D.C., Maurice Ewing Series, v. 4, p. 593–603.

Sibson, R. H., Moore, J. McM., and Rankin, A. H., 1975, Seismic pumping–a hydrothermal fluid transport mechanism: Journal of the Geological Society of London, v. 131, p. 653–659.

Spencer, A. M., 1987, Habitat of Hydrocarbons on the Norwegian Continental Shelf: Norwegian Petroleum Society, Stavanger, Graham and Trotman Ltd., 402 p.

Spencer, A. M., Holter, E., Campbell, C. J., Hanslien, S. H., Nelson, P. H. H., Nysaether, E., and Ormaasen, E. G., 1986, Geology of the Norwegian oil and gas fields: Norwegian Petroleum Society, Stavanger, Graham and Trotman Ltd., 464 p.

Sterner, S. M, and Bodnar, R. J., 1986, Re-equilibration of fluid inclusions in quartz at elevated temperature and pressure: the role of H_2O diffusion (abs.): EOS, American Geophysical Union, Washington, D.C., v. 67, p. 407.

Stewart, D. J., 1986, Diagenesis of the shallow marine Fulmar Formation in the central North Sea: Clay Minerals, v. 21, p. 537–564.

Thomas, B. M., 1985, Petroleum Geochemistry in Exploration of the Norwegian Shelf: Norwegian Petroleum Society, London, Graham and Trotman Ltd., 350 p.

Thomas, M., 1986, Diagenetic sequences and K/Ar dating in Jurassic sandstones, central Viking graben: effects on reservoir properties: Clay Minerals, v. 21, p. 695–710.

Walderhaug, O., 1990, A fluid inclusion study of quartz-cemented sandstones from offshore mid-Norway; possible evidence for continued quartz cementation during oil emplacement: Journal of Sedimentary Petrology, v. 60, p. 203–210.

Walker, T. R., Waugh, B., and Crone, A. J., 1978, Diagenesis in first cycle desert alluvium of Cenozoic age, southwestern United States and northwestern Mexico: Geological Society of America, v. 89, p. 19–32.

Walsh, J., and Watterson, J., 1991, Geometric and kinematic coherence and scale effects in normal fault systems, in Roberts, A. M., Yielding, G., and Freeman, B., eds., The Geometry of Normal Faults: Geological Society of London Special Publication 56, p. 193–203.

WARREN, E. A., AND CURTIS, C. D., 1989, The chemical composition of authigenic illite within two sandstone reservoirs as analyzed by ATEM: Clay Minerals, v. 24, p. 137–155.

WATTERSON, J., 1986, Fault dimensions, displacements and growth: Pure and Applied Geophysics, v. 124, p. 365–373.

WEST, I. M., 1975, Evaporite diagenesis in the lower Purbeck beds of Dorset: Proceedings, Yorkshire Geological Society, v. 35, p. 47–58.

WILKINSON, M., 1991, The concretions of the Bearreraig Sandstone Formation: geometry and geochemistry: Sedimentology, v. 38, p. 408–412.

WOOD, S. H., 1985, Regional increase in groundwater discharge after the 1983 Idaho earthquake: coseismic strain release, tectonic and natural hydraulic fracturing: U. S. Geological Survey Open-File Report 85–290, p. 573–592.

WOODLAND, A. W., 1975, Petroleum and the Continental Shelf of North-West Europe, v. 1: Institute of Petroleum, London, Applied Science Publishers, Halsted Press, 501 p.

YIELDING, G., 1990, Footwall uplift associated with Late Jurassic normal faulting in the northern North Sea: Journal of the Geological Society of London, v. 147, p. 219–222.

YIELDING, G., AND ROBERTS, A., 1991, Footwall uplift during normal faulting–implications for structural geometries in the North Sea, in Larson, R. M., ed., Structural and Tectonic Modelling and its Application to Petroleum Geology: Stavanger, Norwegian Petroleum Society, p. 85–98.

ZIEGLER, P. A., 1982, Geological Atlas of Western and Central Europe: Shell Reprographics, Amsterdam, Elsevier, 130 p.

ZIEGLER, W. H., DOERY, R., AND SCOTT, J., 1986, Tectonic habitat of Norwegian oil and gas, in Spencer, A. M., ed., Habitat of Hydrocarbons on the Norwegian Continental Shelf: Stavanger, Graham and Trotman Ltd., p. 3–19.

VOLUMETRIC RELATIONS BETWEEN DISSOLVED PLAGIOCLASE AND KAOLINITE IN SANDSTONES: IMPLICATIONS FOR ALUMINUM MASS TRANSFER IN THE SAN JOAQUIN BASIN, CALIFORNIA

MICHAEL J. HAYES* AND JAMES R. BOLES
Department of Geological Sciences, University of California, Santa Barbara, California 93106

ABSTRACT: Mass transfer of aluminum is investigated on a thin-section scale by comparing volumes of dissolved plagioclase and authigenic kaolinite in quartzofeldspathic sandstones from the San Joaquin Basin. Samples include Oligocene marine-shelf sandstones, which have been infiltrated by meteoric water, and Late Miocene turbidite sandstones, which contain diluted sea water. Other aluminum sources and sinks are volumetrically minor in these sandstones.

Dissolved plagioclase and kaolinite presently appear from 600 m to depth of sample control (30–70°C present temperature) in the meteoric zone and from 2,100 m to depth of sample control (75–130°C present temperature) in the marine zone. Leached plagioclase and kaolinite are rare in the matrix-rich or carbonate-cemented sandstones, but appear in more than 80% of the uncemented turbidite sandstones and in up to 60% of the uncemented sandstones in some meteoric-zone reservoirs.

Point-counted volumes of plagioclase porosity and kaolinite in all sandstones are compared with relative volumes calculated from a mass-balance reaction in which aluminum is conserved between An_{30} plagioclase and kaolinite (25 to 50% microporosity). Aluminum is conserved on a centimeter scale in shale-encased turbidite sandstones exposed to limited fluxes of marine pore water, despite enrichment in organic-acid anions, which potentially may mobilize aluminum in soluble complexes. The average marine-zone sandstone has a volume of kaolinite approximately equal to that calculated for plagioclase porosity, based on relative volumes of the mass-balance reaction.

In contrast, the average sandstone with meteoric pore water has a kaolinite shortfall of 0.4 ± 0.3 volume percent of total rock relative to plagioclase porosity. This average aluminum loss is 0.2 ± 0.1 gm/100 cm³ rock volume. Complementary zones of aluminum import are not found in the meteoric zone. A small amount of aluminum is mobilized beyond a centimeter scale in shelf sandstones flushed by low-temperature, dilute waters.

Plagioclase dissolution and kaolinite precipitation in sandstones from both pore-water settings result in compositional shifts of less than 0.4 weight percent Al_2O_3, too small to discriminate from the natural bulk chemical variation of 1 to 2 weight percent Al_2O_3. Data from this study do not support models proposed for transfer of large masses of aluminum over significant distances.

INTRODUCTION

Burial diagenesis reflects complex interrelations between detrital mineralogy, pore-water chemistry, hydrology, depositional environment, temperature, and pressure. Diagenesis involves mass transfer of chemical components and can have a significant impact upon reservoir quality. Many geochemical reactions are written with aluminum conserved in solid reactant and product phases. Is this valid on a thin-section scale? Models proposed for large-scale aluminum mass transfer (Fischer and Surdam, 1988; Surdam and others, 1989) are based largely on results of solubility experiments, in which aluminosilicate minerals have been reacted with organic acids and anions (Surdam and others, 1984; Hansley, 1987; MacGowan and Surdam, 1988). In the experiments of Surdam and associates, feldspar dissolution in organic-acid anions has produced the high-aluminum solubilities necessary for increased aluminum mobility, although different experiments have not produced uniform results (Stoessell and Pittman, 1990). Aluminum solubility may increase upon complexation with organic-acid anions. However, aluminum *mobility* in sedimentary basins is a product of concentration *and* aluminum flux.

The purposes of this study are to investigate the potential for mass transfer of aluminum in meteoric and marine porewater settings by comparing volumes of leached plagioclase and authigenic kaolinite in sandstones, and to assess the sources of error in such mass-balance studies. Other aluminum sources (e.g., dissolved potassium feldspar and rock fragments) and sinks (e.g., smectite, chlorite, laumontite, and feldspar overgrowths) are volumetrically minor in the sandstones examined and have minimal impact on the aluminum mass balance. The relatively uniform sandstone detrital composition and simple burial history of the southeastern San Joaquin Basin reduce the diagenetic complexity that is encountered when attempting to isolate the roles of water chemistry, hydrology, and temperature in mass transfer. In addition, petrographic data are related to formation-water analyses and to results of solubility experiments, many of which have involved San Joaquin Basin formation waters and core material, in order to underscore the distinction between aluminum solubility and aluminum mobility. Finally, petrographic data are compared with published data from other sedimentary basins in a search for consistent patterns of aluminum mass transfer in similar chemical and hydrologic settings.

GEOLOGIC SETTING

The southern San Joaquin Valley of central California (Fig. 1) contains 7 km of Cenozoic strata (Zieglar and Spotts, 1984). Sediments in the east-central basin have a simple subsidence history and presently are at maximum burial depth and temperature (Fig. 2). The southern and western basin margins are folded, faulted, and uplifted (Harding, 1976; Namson and Davis, 1988a). Paleocene through Miocene strata are primarily shallow marine and deep marine in origin (Bandy and Arnal, 1969; Graham, 1987; Nilsen, 1987). Marginal, alluvial and fluvial sediments grade basinward to shelf sandstones, and ultimately to basinal shales and submarine-fan sandstones (e.g., Bartow and McDougall, 1984; Nilsen, 1984; Olson, 1988). Provenances include the Sierra Nevada to the east, the Tehachapi and San Emigdio mountains to the south, and the Salinian block to the southwest

---

*Present Address: Exxon Production Research Co., P.O. Box 2189, Houston, Texas 77252-2189

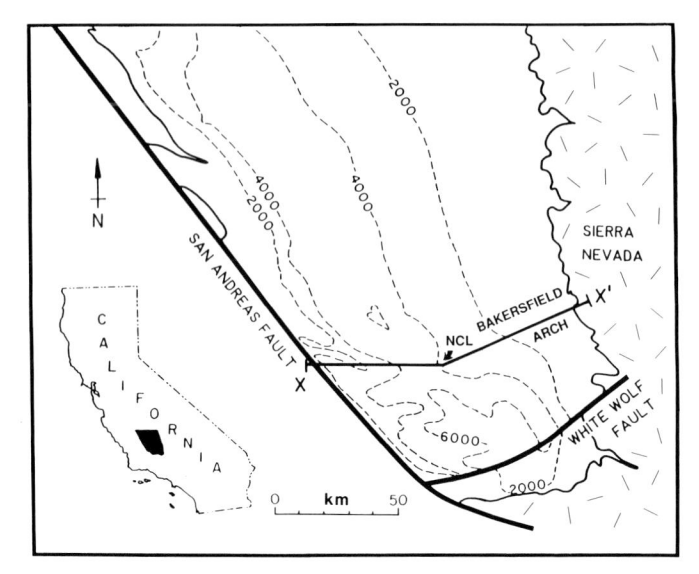

FIG. 1.—Isopach map of study area, location of North Coles Levee (NCL), and location of transect X-X' on map modified from Boles (1987). Thickness of sedimentary rocks (in meters) after Callaway (1971).

(Bent, 1988). Late Miocene Stevens turbidite sandstones are encased in organic-rich, pelagic and hemipelagic shales of the Monterey Formation (Graham and Williams, 1985). Rapid marginal uplift and sedimentation since the Pliocene have resulted in basinal shoaling and infilling by shallow marine and nonmarine sediments (Callaway, 1971; Crowell, 1987).

STUDY AREA

This study focuses on Oligocene Vedder sandstones (600–1,500 m depth) and Late Miocene Stevens sandstones (2,100–3,300 m depth) along a transect (Fig. 2) that follows the Bakersfield arch from North Coles Levee oil field to the eastern basin margin. Vedder deposits thicken basinward from 100 m at Mount Poso to 300 m at Kern River oil field (Fig. 2). The Vedder section at Mount Poso consists of more than 80% marine-shelf sandstone. The predominantly fine- to medium-grained, moderately well sorted sandstones are up to 20 m thick and are interbedded with silty and sandy mudstones as thick as 10 m. Vedder sandstones in this region typically are correlated easily between wells (Olson, 1988). Inner neritic and outer neritic foraminifera appear in Vedder strata at Mount Poso and Kern River oil fields, respectively (Olson, 1988). Meteoric water

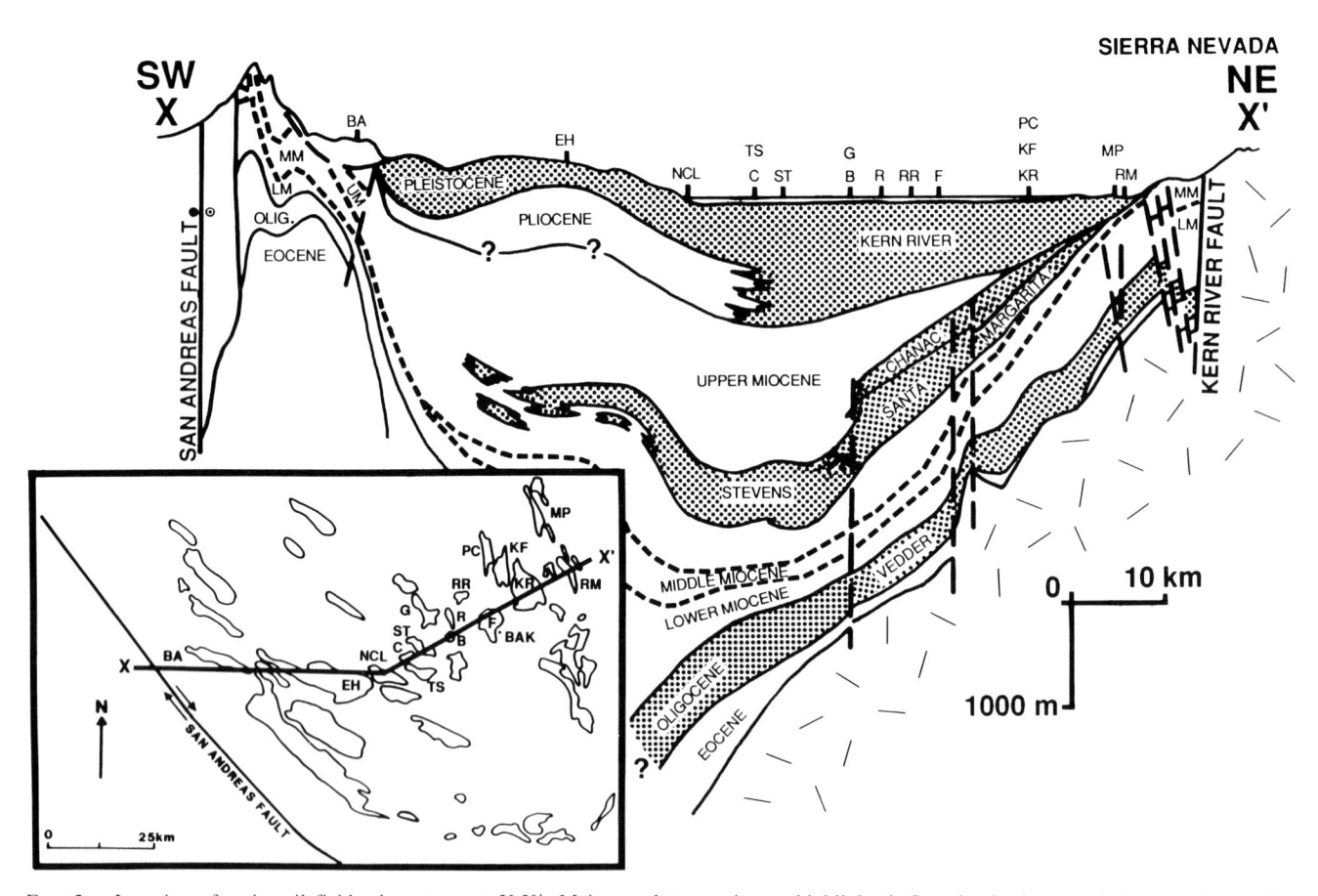

FIG. 2.—Location of major oil fields along transect X-X'. Major sandstone units are highlighted. Samples in the mass-balance study are from Mount Poso (MP), Kern River (KR), Fruitvale (F), Rosedale Ranch (RR), Greeley (G), Bellevue (B), Strand (ST), Canal (C), Ten Section (TS), and North Coles Levee (NCL). Cross section modified from Boles and Ramseyer (1987).

has infiltrated these marine sandstones to depths of 1,500 m in the eastern basin (Fisher and Boles, 1990).

Stevens turbidite reservoirs are predominantly medium- to coarse-grained, moderately to moderately well sorted, massive sandstones up to 150 m thick. MacPherson (1978) and Webb (1981) interpreted these submarine-fan sandstones as braided-channel deposits diagnostic of upper midfan environments. The Stevens is more than 1,200 m thick locally, typically contains 75 to 80% sandstone (Webb, 1981), and is encased in hundreds of meters of marine shale (Fig. 2). Amalgamated-channel sandstones have lateral continuity more than 600 m across the structure at the Greeley oil field (Webb, 1981). Stevens deposits interfinger with Santa Margarita shelf deposits updip (Fig. 2); however, it is uncertain whether individual sand beds are continuous between formations.

SANDSTONE PETROLOGY

Provenance

Approximately 200 thin sections of Eocene to Pleistocene sandstones from depths of 280 to 4,350 m were point counted in an initial, regional survey of detrital composition and diagenesis. San Joaquin Basin sandstones were uniformly quartzofeldspathic (Fig. 3) and had plutonic provenances in the Sierra Nevada and Salinian block (MacPherson, 1978; Webb, 1981; Bent, 1988). Plio-Pleistocene sandstones in the eastern basin had mean compositions that resembled those of Mount Whitney suite granodiorites reported by Moore (1987) and Hirt (1989). Sandstones downdip and downsection were slightly quartz rich and plagioclase poor relative to Sierran granodiorites, in part due to plagioclase dissolution. Relative volumes of feldspar and rock fragments in Figure 3 were restored to original values by including volumes of leached grains.

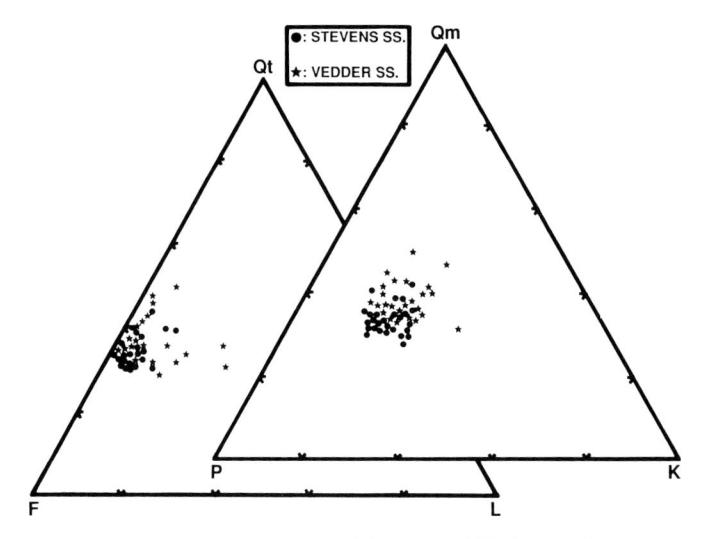

FIG. 3.—Detrital compositions of Stevens and Vedder sandstones are plotted on ternary diagrams after Dickinson and Suczek (1979). Total quartz (Qt), feldspar (F), and rock fragments (L) are plotted on diagram at left. Monocrystalline quartz (Qm), plagioclase (P), and potassium feldspar (K) are plotted on diagram at right. Data represent 65 sandstones from the east-central basin. Three hundred points were counted per sample.

Diagenesis

The relatively homogeneous detrital composition of sandstones resulted in fairly uniform diagenetic trends, although very few samples contained all of the features listed in the regional paragenetic sequence. This composite paragenetic sequence included: early carbonate cementation; subsequent dissolution of detrital grains and precipitation of authigenic clays; late cementation and replacement of detrital grains by carbonate and laumontite; and albitization of plagioclase (Boles, 1984, 1987; Boles and Ramseyer, 1987, 1988; Schultz and others, 1989).

Leached plagioclase and authigenic kaolinite were absent from early, high-volume carbonate-cement zones, but appeared in interbedded uncemented sandstones. Some carbonate infilled skeletal plagioclase after the onset of plagioclase dissolution; however, late cements lacked hydrocarbon inclusions and apparently preceded hydrocarbon emplacement (Boles, 1987; Boles and Ramseyer, 1987). Plagioclase dissolution and kaolinite precipitation largely followed carbonate cementation, preceded or accompanied hydrocarbon emplacement, and probably began about 2.5 Ma at temperatures of approximately 80°C in Stevens sandstones at North Coles Levee (Boles, 1987; Boles and Ramseyer, 1987; Schultz and others, 1989).

The mass-transfer study focused on dissolved plagioclase and authigenic kaolinite in 33 sandstone samples (Table 1), which lacked other significant aluminum sources and sinks. Sandstones from the albitization zone were excluded from this study. Trace amounts (0.3 volume percent) of potassium-feldspar porosity appeared in only two sandstone samples. In these two samples, plagioclase porosity was two to six times greater than potassium-feldspar porosity. There was no evidence for significant leaching of potassium feldspar. The ratio of plagioclase to total feldspar (restored for feldspar porosity) was 0.66 to 0.69 in granodiorites and 0.62 to 0.69 in cemented and uncemented Stevens and Vedder sandstones. Potassium-feldspar overgrowths, a minor aluminum sink, had an average volume of 0.3% in only three samples.

Volcanic-rich sandstones were omitted from this study because plagioclase and volcanic detritus altered to smectite, chlorite, and laumontite. Most sandstones contained 5 to 10 volume percent of relatively stable granite and gneiss fragments, hypabyssal-rock fragments, chert, quartz-mica schist, and other meta-sedimentary lithics. Volcanic detritus almost always constituted less than 2% of total rock volume. In samples where rock fragments were partially leached, skeletal-plagioclase grains were five to ten times more abundant than skeletal-rock fragments.

Although heavy minerals represented potential aluminum sources, hornblende dissolved well before the onset of plagioclase dissolution and biotite was crushed but not leached in Stevens sandstones (Boles, 1984). Clays other than kaolinite represented potential aluminum sinks; however, most sandstones had less than 1 volume percent detrital or authigenic-clay matrix or clay coats. A few samples contained up to 3% clay matrix.

Intergranular porosity was reduced by cementation and compaction from 40 to 45% at deposition to 10 to 15% in

TABLE 1.—SAMPLES EXAMINED IN ALUMINUM MASS-TRANSFER STUDY

WELL	FIELD	UNIT	SEC-T-R	INTERVAL(m)	SAMPLES
Vedder Rall 431	Mount Poso	Vedder	09-27S-28E	597.3-626.5	6
Apollo 51-X	Kern River	Vedder	04-29S-28E	1508.2-1517.1	2
Shell. Sheldon 1	Fruitvale	Vedder	30-29S-27E	3027.4	1
Cont. KCLB-1	Rosedale Ranch	Vedder	12-29S-26E	2899.4	1
Cont. KCLB-1	Rosedale Ranch	Santa Marg.	12-29S-26E	2250.0	1
Std. Wegis Comm. 1	Greeley	Stevens	22-29S-25E	2388.1	1
Superior KCL-9	Bellevue	Stevens	03-30S-26E	2125.0-2308.5	3
A-21-13	Canal	Stevens	14-30S-25E	2646.7	1
KCLA-21	Canal	Stevens	13-30S-25E	2538.4	1
KCLE-23	Canal	Stevens	14-30S-25E	2466.8-2470.6	2
MacFarland 18R-3	Canal-Pioneer	Stevens	03-30S-25E	2732.4-3211.0	3
Feykert 2	Strand	Stevens	?	2358.1	1
KCLA 78-30	Ten Section	Stevens	30-30S-26E	2493.6	1
Ohio KCLG-1	Wildcat Well	Vedder	36-29S-26E	3197.9	1
NCL 487-29	N. Coles Levee	Stevens	29-30S-25E	2713.7-2730.5	2
NCL 488-29	N. Coles Levee	Stevens	29-30S-25E	2745.6	1
NCL 28-29	N. Coles Levee	Stevens	29-30S-25E	2741.3	1
68-29 RD-1	N. Coles Levee	Stevens	29-30S-25E	3335.7-3336.3	2
Richfield A-52	N. Coles Levee	Stevens	34-30S-25E	2609.8	1
CLA 63-32	N. Coles Levee	Stevens	32-30S-25E	2654.9	1

sandstones at 2,750 to 3,000 m depth. During plagioclase dissolution and kaolinite precipitation, relatively high-permeability intergranular pores were converted to relatively low-permeability intragranular pores and micropores in leached grains and pore-filling clay. Microporosity registered as porosity in geophysical logs but was isolated and discontinuous. Distinction of pore types and recognition of the impact of diagenesis on reservoir quality were important in Stevens sandstones, where microporosity represented up to one-third of total porosity. Ehrenberg (1990) documented a similar situation in North Sea sandstones. Permeability was reduced after diagenesis because the aluminum released during feldspar dissolution reprecipitated locally as kaolinite or illite (Ehrenberg, 1990).

ALUMINUM MASS-TRANSFER STUDY

Methods

Mass transfer of aluminum was investigated by measuring volumes of kaolinite and intragranular porosity within plagioclase during high-density point counts (1,000 counts per analysis) of 33 matrix-free, uncemented sandstones (Table 1). Potential sources of error included: detrital-plagioclase composition; kaolinite microporosity; interpretation of oversized pores; and measurement of trace amounts of plagioclase porosity and kaolinite by standard petrographic methods.

Average composition of detrital plagioclase in the eastern San Joaquin Basin was assumed to be An_{30} based on a regional albitization study (Boles and Ramseyer, 1988), although plagioclase ranged from approximately An_{20} to An_{40}. The mean composition of detrital plagioclase in individual samples ranged from An_{25} to An_{31}, with standard deviations of ±4 to ±14 mole percent anorthite (Boles and Ramseyer, 1988). For every 1 cm^3 (2.65 gm) An_{20} dissolved, 0.6 cm^3 (1.57 gm) of kaolinite could have precipitated. For every 1 cm^3 (2.69 gm) An_{40} dissolved, 0.7 cm^3 (1.83 gm) of kaolinite could have precipitated. The most calcium-rich plagioclase grains perhaps dissolved first, as Milliken and others (1989) observed, at least on a large scale, in Gulf Coast sandstones.

Point-counted kaolinite volume was corrected for microporosity estimated by inspection with the petrographic microscope (Neofluor 100X oil-immersion objective lens) and scanning electron microscope (JEOL JSM-2 model, 500 to 3,000X magnification). Estimated microporosity was 50% and 25% for kaolinite found in meteoric-zone and marine-zone sandstones, respectively. These estimates were consistent with measurements of microporosity by image analysis of back-scattered electron micrographs (Nadeau and Hurst, 1991). Vermicular kaolinite contained 40 to 60% microporosity and blocky kaolinite had 15 to 30% microporosity (Nadeau and Hurst, 1991).

Leached plagioclase and kaolinite were assumed to be *in situ* diagenetic by-products. There was no evidence for detrital kaolinite or early kaolinite cement in these sandstones, based on petrographic inspection of high-volume carbonate-cement zones. Skeletal plagioclase was absent from cemented and uncemented sandstones updip and upsection of Vedder sandstones at the Mount Poso oil field. There was no evidence that feldspars were transported and deposited with relict skeletal textures in the eastern San Joaquin Basin, as suggested by Passaretti and Eslinger (1987) for Gulf Coast sediments. San Joaquin Basin sediments probably experienced minimal weathering at the source area and were eroded, transported, and deposited rapidly relative to material examined by Passaretti and Eslinger (1987).

Measured volumes of intragranular porosity within plagioclase (plagioclase porosity) and kaolinite (corrected for microporosity) were compared with relative volumes in a reaction where aluminum released from plagioclase reprecipitated as kaolinite (Boles, 1984):

$$Na_{0.7}Ca_{0.3}Al_{1.3}Si_{2.7}O_8 + H^+ + 3.45H_2O + 0.3HCO_3^- =$$
(plagioclase)
$$0.7Na^+ + 1.4H_4SiO_4 + 0.65Al_2Si_2O_5(OH)_4 + 0.3CaCO_3$$
(kaolinite)

For every 1 cm^3 (2.62 gm) of An$_{30}$ plagioclase dissolved, 0.65 cm^3 (1.7 gm) of kaolinite could have precipitated. Dissolution of 1 volume percent plagioclase also would have produced 0.1 volume percent calcite. Measured volumes of late carbonate cement and grain replacement were less than 5% in Stevens sandstones. Late calcite apparently was supplied by calcium sources in addition to leached plagioclase within a thin section. Finally, the excess silica released from plagioclase in the reaction did not precipitate as quartz overgrowth or microcrystalline-silica cement on a centimeter scale.

Distribution

Leached plagioclase and authigenic kaolinite almost always appeared together in Stevens and Vedder sandstones (Fig. 4). Skeletal plagioclase was found with other aluminum sinks such as smectite or laumontite; however, kaolinite was found only with associated skeletal plagioclase in sandstones from the study area.

Most kaolinite infilled intergranular pores near skeletal-plagioclase grains, not intragranular pores within plagioclase, indicating that aluminum was mobilized on a millimeter scale. The distribution of kaolinite could imply that kaolinite precipitated before plagioclase dissolved. However, some kaolinite did crystallize in feldspar porosity, and kaolinite always appeared with leached plagioclase. Kaolinite was not found in early carbonate-cement zones, but

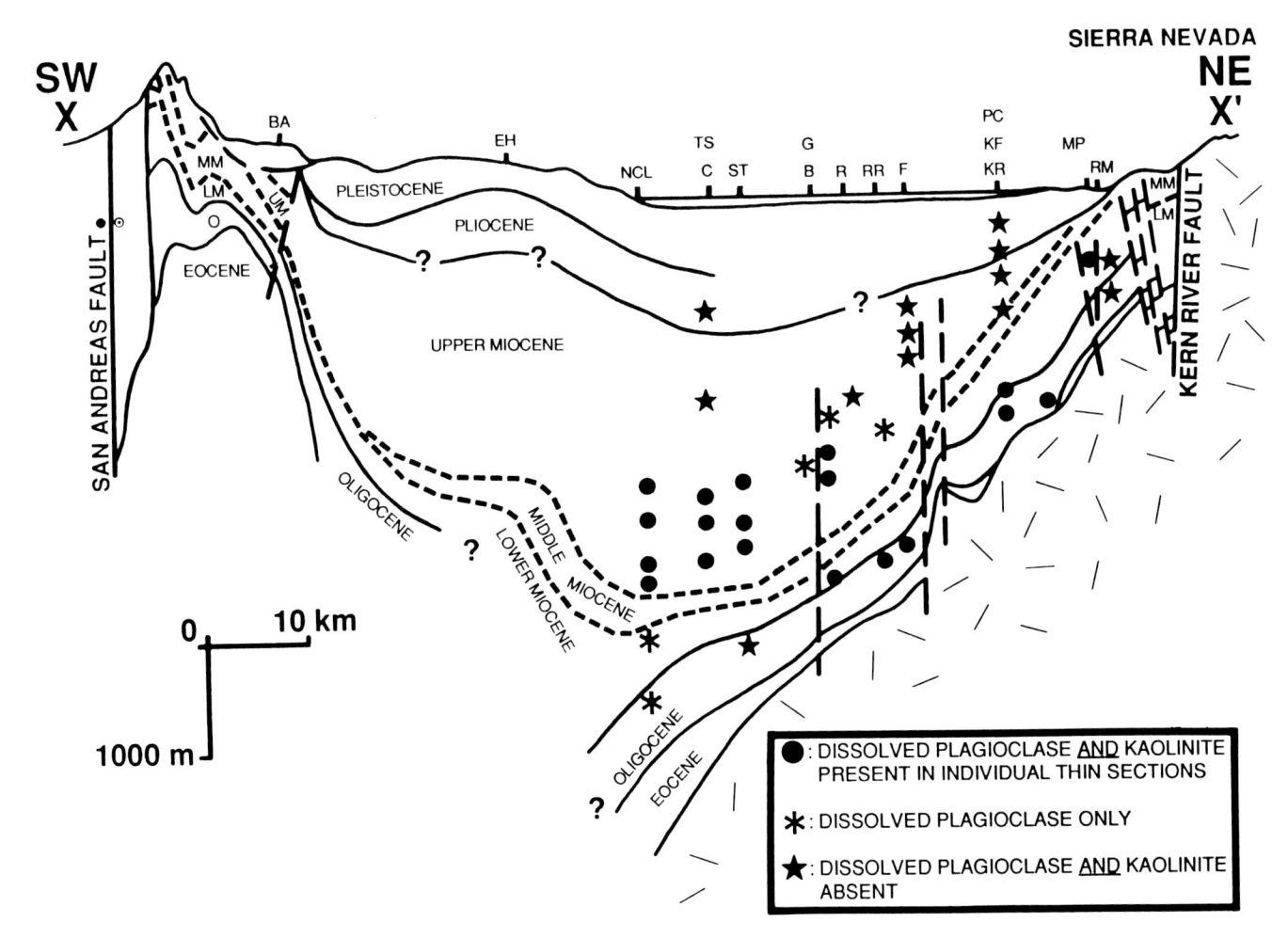

FIG. 4.—Distribution of leached plagioclase and authigenic kaolinite in sandstones along transect X-X'. Each point represents observations from one or more thin sections. Samples are from Mount Poso (MP), Round Mountain (RM), Poso Creek (PC), Kern Front (KF), Kern River (KR), Fruitvale (F), Rosedale Ranch (RR), Rosedale (R), Greeley (G), Bellevue (B), Strand (ST), Ten Section (TS), Canal (C), and North Coles Levee (NCL).

kaolinite and dissolved plagioclase appeared in uncemented sandstones immediately across sharp cement-zone contacts (Boles, 1984). These observations all support a source-sink relation between plagioclase and kaolinite. Boles and Johnson (1983) suggested that variable chemical environments on a microscale perhaps led to selective precipitation of carbonate rhombs between cleavage flakes of crushed biotite. Perhaps the distribution of kaolinite also reflected variable porewater chemistry within and outside of dissolving feldspar grains.

Plagioclase porosity and kaolinite formed relatively recently, near maximum burial temperatures, based on petrographic evidence presented earlier. Plagioclase porosity and kaolinite appeared in meteoric-zone Vedder sandstones at depths of 597 to 1,521 m (depth of sample control) and at present temperatures of 30 to 70°C, although plagioclase porosity was most abundant from 30 to 50°C (Fig. 5). The decrease in plagioclase porosity with increasing depth perhaps reflected reduced leaching power of meteoric water downdip, or shorter reaction time between plagioclase and meteoric water at depth. This decrease probably did not reflect the collapse of intragranular pores. Skeletal feldspars with delicate textures would have been crushed if significant compaction had taken place after grain dissolution.

Skeletal plagioclase and kaolinite appeared in Stevens sandstones at depths of 2,125 to 3,336 m (depth of sample control) and at present temperatures of 75 to 130°C, although these features were concentrated from 85 to 105°C (Fig. 5). Plagioclase dissolution was incipient at depths less than 2,125 m and temperatures less than 75°C. Plagioclase dissolution was less extensive in relatively poorly sorted, matrix-rich lower Stevens sandstones than in upper Stevens sandstones.

Morphology

Kaolinite.—

Kaolinite formed booklets composed of stacked pseudohexagonal platelets. Kaolinite found in meteoric-zone sandstones was relatively coarse and had approximately 50% microporosity. Platelets of 10 to 30 μm diameter formed booklets 30 to 75 μm thick with aspect ratios of two to three. Kaolinite of marine origin was more finely crystalline, had approximately 25% microporosity (one-half that of meteoric kaolinite), and was more densely packed than meteoric kaolinite. Platelets of 8 to 10 μm diameter formed booklets 8 to 12 μm thick with an aspect ratio of one. Hurst and Irwin (1982) observed a similar morphologic contrast in kaolinite in North Sea sandstones from zones of meteoric recharge and marine pore waters.

Dissolved plagioclase.—

Leached plagioclase had gridlike, blocky, or fluted textures, which commonly were defined by fracture-filling albite. Dissolution often began at fractures and cleavages, which provided fresh surfaces for reaction. Gridlike-porosity networks characterized slightly dissolved grains and were distinct from fracture porosity. Blocky textures often developed at intermediate stages of leaching, and continued porosity enhancement along fractures and cleavages resulted in jagged, sawtooth-porosity patterns. Clay coats and skeletal-albite fracture fills outlined some extensively leached grains. Although clay coats and fracture-filling albite represented potential aluminum sinks, textural evidence indicated that these features formed prior to dissolution of plagioclase grains.

Results

Plagioclase porosity and kaolinite had volumes less than 3% in Stevens and Vedder sandstones (Fig. 6). Stevens sandstones with marine pore waters plotted near aluminum mass balance. Many samples, in fact, appeared to have excess aluminum. Plagioclase porosity and kaolinite had average volumes of 0.9 ± 0.6% and 0.7 ± 0.5%, respectively, in Stevens sandstones (Table 2). In contrast, all Vedder sandstones from the meteoric zone were aluminum deficient. Only about 40% of the aluminum released from plagioclase was accounted for as kaolinite in the average meteoric-zone sandstone. Average volumes of plagioclase porosity and kaolinite were 1.1 ± 0.7% and 0.3 ± 0.4%, respectively, in meteoric-zone sandstones (Table 2). The uncertainty in these point-counted volumes approached ±1% at 1,000 counts per thin section, equivalent to 33 to 50% of the measured volumes. Six Stevens sandstones were recounted (Fig. 7) to illustrate the error involved in measuring volumetrically minor features. Individual samples plotted as aluminum enriched or aluminum deficient simply by shifting the point-count grid, without altering the grid density or area of coverage. The errors involved in the mass-balance study are evaluated later.

Volume percentages of plagioclase porosity and kaolinite measured in thin sections with volumes of 0.018 cm^3 (2 cm × 3 cm × 30 μm) were converted to masses of aluminum dissolved and precipitated in a 100 cm^3 rock volume (Table 2). Plagioclase (An$_{30}$) had a formula weight of 267 gm and a molar volume of 100 cm^3. Kaolinite had a formula weight of 258 gm and a molar volume of 98.6 cm^3. Aluminum imbalances in individual Stevens and Vedder samples were

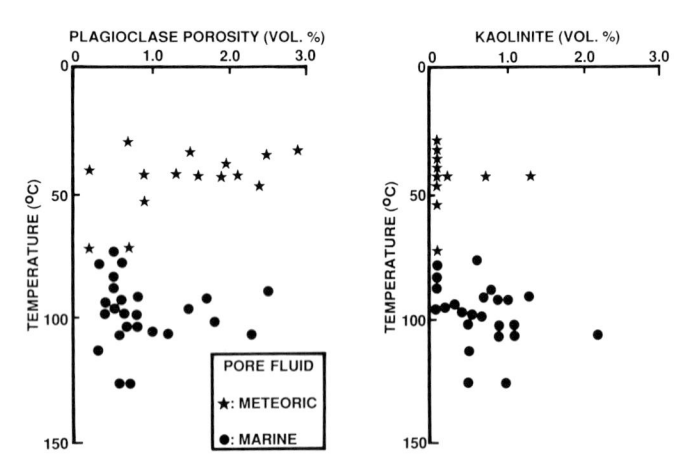

FIG. 5.—Distribution of intragranular porosity in plagioclase (plagioclase porosity) and kaolinite (corrected for microporosity) with present burial temperature in sandstones along transect X-X'. Sandstones with meteoric and marine pore waters are identified. Present temperature is determined from bottom-hole temperatures in some wells, and is estimated from burial depth using local geothermal gradients in other wells.

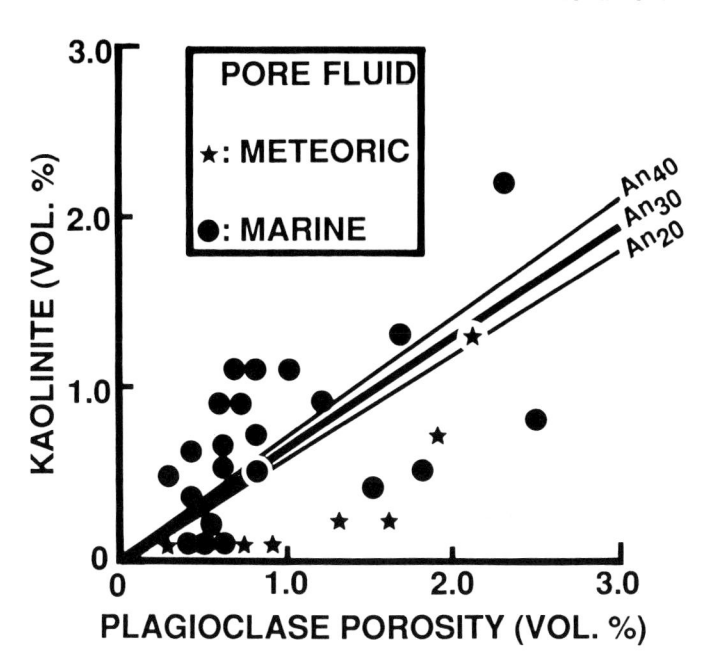

FIG. 6.—State of aluminum mass balance in 33 sandstones. Point-counted volumes of intragranular porosity in plagioclase (plagioclase porosity) and kaolinite (corrected for microporosity) are cross plotted for sandstones with meteoric and marine pore waters. Volumetric relations in mass-balance reactions between kaolinite and An_{20}, An_{30}, and An_{40} plagioclase are depicted as aluminum conservation lines. The correlation coefficient for the data set is 0.56 along the line:

(vol. % kaolinite) = 0.46 (vol. % plagioclase ø) + 0.13.

less than or equal to 0.5 gm/100 cm^3, equivalent to kaolinite imbalances of less than or equal to 0.8 volume percent of total rock relative to observed volumes of plagioclase porosity (Table 2). The mass of aluminum in kaolinite was 0 to 150% of that released from plagioclase in individual samples. About 40% of the aluminum released from plagioclase reprecipitated in kaolinite in the average meteoric-zone sandstone. The average aluminum shortfall was 0.2 ± 0.1 gm/100 cm^3 rock volume, or 0.4 ± 0.3 volume percent kaolinite. Although the error was proportionally high, it was notable that *no* meteoric-zone sandstone had excess kaolinite. In contrast, about 120% of the aluminum released from plagioclase reprecipitated in kaolinite in the average sandstone with marine pore waters. The average aluminum gain of 0.05 ± 0.2 gm/100 cm^3 rock volume, or 0.1 ± 0.4 volume percent kaolinite, was insignificant within the precision of point counts. Aluminum imbalances were significant proportionally in San Joaquin Basin sandstones, but were small in absolute terms.

DISCUSSION

Error Analysis

Potential sources of error in the mass-balance study include: detrital-plagioclase composition, kaolinite microporosity, and interpretation of oversized pores. The magnitudes of these errors are compared to the precision of point counts below.

Given the range of detrital plagioclase composition reported by Boles and Ramseyer (1988), mass balance cal-

TABLE 2.—ALUMINUM MASS-BALANCE DATA

DEPTH (m)	VOLUME % PLAG Φ	VOLUME % KAOLINITE	VOLUME % IMBALANCE[1]	gm ALUMINUM DISSOLVED[2]	gm ALUMINUM PRECIPITATED[3]	gm ALUMINUM IMBALANCE[4]
625.3 [5]	1.6	0.2	-0.8	0.6	0.1	-0.5
2358.1 [5]	2.5	0.8	-0.6	0.9	0.4	-0.5
2745.6 [5]	2.3	2.2	+0.7	0.8	1.2	+0.4
Average Meteoric[6]	1.1 ± 0.7	0.3 ± 0.4	-0.4 ± 0.3	0.4 ± 0.2	0.2 ± 0.2	-0.2 ± 0.1
Average Marine[6]	0.9 ± 0.6	0.7 ± 0.5	+0.1 ± 0.4	0.3 ± 0.2	0.4 ± 0.3	+0.1 ± 0.2

[1] Aluminum imbalance expressed as the volumetric excess or deficiency of kaolinite, relative to the volume of kaolinite expected from observed plagioclase porosity.
[2] Mass of aluminum released from plagioclase in a 100 cm^3 rock volume.
[3] Mass of aluminum incorporated in kaolinite in a 100 cm^3 rock volume.
[4] Aluminum imbalance expressed as the excess or deficiency of aluminum in kaolinite, relative to the mass of aluminum released from plagioclase.
[5] Samples which plot furthest from mass balance on crossplot.
[6] Mean values and standard deviation of volumetric and mass imbalances in sandstones with meteoric (n=8) and marine (n=25) pore-waters.

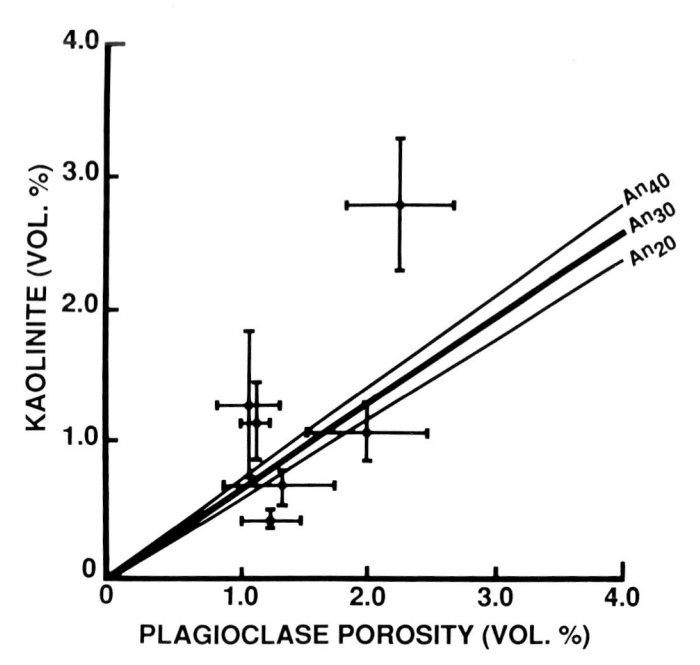

FIG. 7.—Six Stevens sandstones were point counted three times each (1,000 counts per analysis) to illustrate some of the error in measured volumes of plagioclase porosity and kaolinite. The mean and standard deviation are plotted for each sample. Aluminum conservation lines appear as in Figure 6.

culations (based on point-counted volumes) are completed for reactions involving An_{20} and An_{40} plagioclase. The average mass of aluminum released from plagioclase ranges from 0.36 to 0.44 gm/100 cm^3 rock volume in meteoric-zone sandstones, and from 0.29 to 0.33 gm/100 cm^3 rock volume in marine-zone sandstones. Corresponding aluminum imbalances are 0.19 to 0.25 gm/100 cm^3 in the average meteoric-zone sandstone, and 0.03 to 0.07 gm/100 cm^3 in the average marine-zone sandstone. The error involved in point counting is up to 10 times greater than the error in mass-balance calculations due to variable plagioclase composition.

Nadeau and Hurst (1991) found kaolinite microporosity to be 43 ± 10%. We assume that our estimates of microporosity have a similar uncertainty (i.e., ±10% of point-counted kaolinite volume). This error is as much as 40% of the corrected kaolinite volume in individual samples, but falls within the precision of point counts.

The interpretation of oversized pores is potentially a major source of error in mass-balance studies. In our conservative approach, only pores within recognizable feldspar remnants are counted as feldspar porosity. In our view, pores that have diameters greater than grain diameters are not necessarily a by-product of feldspar dissolution. The volume of oversized pores is 1 to 1.5% in six compacted Stevens sandstones and up to 8% in two relatively uncompacted Vedder sandstones. These volumes are equivalent to 50 to 400% of the definite feldspar porosity in individual samples. The error involved when oversized pores are counted as feldspar porosity is up to an order of magnitude greater than the error involved in point counting. Many samples, which plot near aluminum mass balance in Figure 6, can be interpreted as grossly aluminum deficient if oversized pores are counted as plagioclase porosity.

Replacement carbonate represents another potential source of error. Carbonate, which appears to infill leached plagioclase, is not included in feldspar porosity, because the origin of the replaced material cannot always be confirmed. The volume of replacement carbonate usually is less than 20% and at most 50% of definite plagioclase porosity in eight Stevens sandstones. This error falls within the precision of point counts. If the volumes of replacement carbonate and plagioclase porosity are added, then some Stevens sandstones shift closer to aluminum mass balance, but others shift further from aluminum conservation.

Bulk Chemistry

Aluminum mass transfer also can be investigated by bulk chemical analysis. The Al_2O_3 content varies from 13.1 to 14.6 weight percent in cemented and uncemented Stevens sandstones (Boles, pers. commun., 1990) and from 14.4 to 15.8 weight percent in Mount Whitney suite granodiorites (Moore, 1987; Hirt, 1989). If 3 volume percent An_{30} plagioclase dissolves and no kaolinite precipitates, then less than 7% of the aluminum in the rock is lost. Stevens and Vedder sandstones have kaolinite imbalances of less than, or equal to, 0.8 volume percent and shifts in Al_2O_3 are less than, or equal to, 0.4 weight percent. Changes in Al_2O_3 due to plagioclase dissolution and kaolinite precipitation cannot be isolated in sandstones derived from granodiorites in which Al_2O_3 varies by 1 to 2 weight percent.

Relation of Petrographic Data to Pore Waters

Meteoric water has infiltrated shallow marine Vedder sandstones in the eastern basin to depths of 1,500 m (Fisher and Boles, 1990). Stevens formation waters are slightly diluted sea water. Carboxylic acids and anions have concentrations up to 5,100 mg/l in San Joaquin Basin formation waters (Carothers and Kharaka, 1978; MacGowan and Surdam, 1988). Concentrations of aluminum and total organic species (Fig. 8) are below detection limits in Vedder pore waters (Fisher and Boles, 1990) at Mount Poso (509–524 m) and Kern River (1,488–1,501 m) oil fields. Stevens formation waters locally are enriched in aluminum or organic-acid anions, mainly acetate. Aluminum concentration reaches 840 μg/l and total organic-acid anion concentration reaches 3,432 mg/l (Fisher and Boles, 1990); however, only North Coles Levee water samples are enriched in both components (Fig. 8).

Plagioclase is dissolving at North Coles Levee, where Stevens formation waters (2,621–2,779 m) are acetate rich (Fig. 8); however, plagioclase is not leached and formation waters (1,621–1,796 m) are organic-acid poor at Rosedale oil field (Fig. 8). In summary, plagioclase dissolves in the presence of organic-acid anions in Stevens reservoirs, but plagioclase also is leached from Vedder sandstones by meteoric waters, which lack measurable amounts of organic species. Aluminum concentrations vary by an order of magnitude in these formation waters and are not clearly related to volumes of dissolved plagioclase or to concentrations of organic-acid anions. The measured aluminum con-

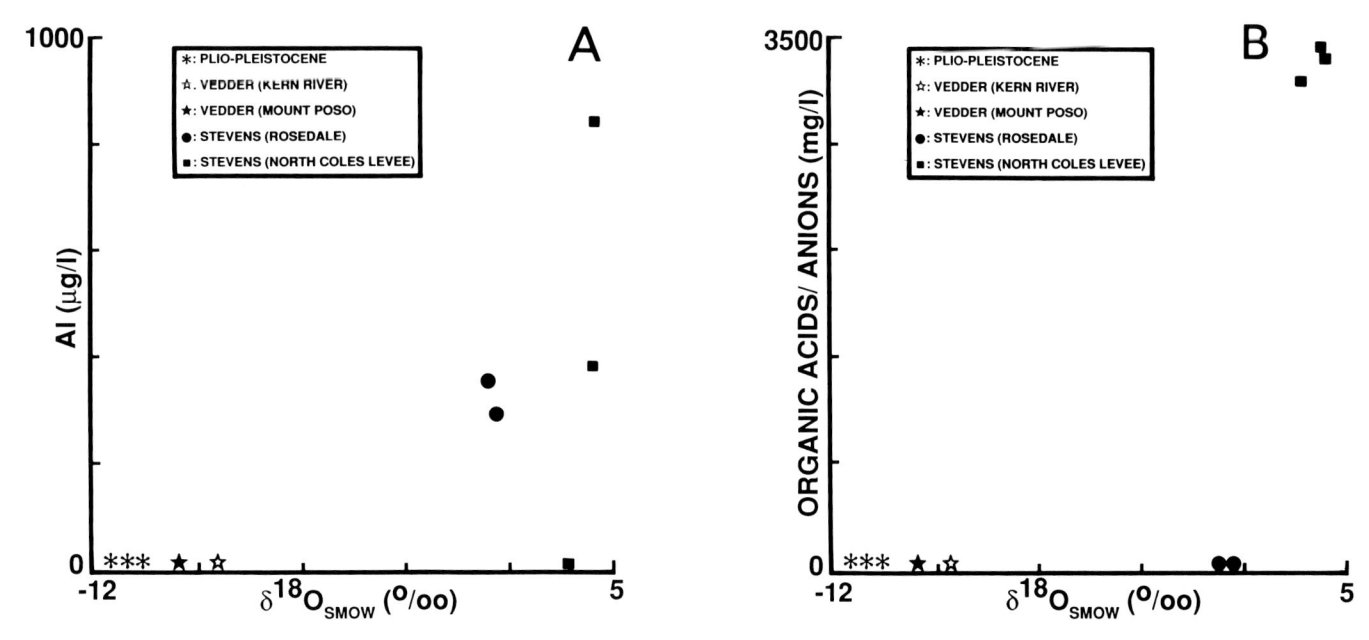

FIG. 8.—(A) Aluminum concentration (μg/l) is plotted against oxygen-isotopic composition for formation waters from individual Stevens and Vedder reservoirs. Data from Fisher and Boles (1990). (B) Total concentration of organic acids and anions (mg/l) is plotted against oxygen-isotopic composition for formation waters from individual Stevens and Vedder reservoirs. Data from Fisher and Boles (1990).

centrations also are 10^2 to 10^3 less than those recorded in solubility experiments, where feldspars are reacted with organic acids and anions of comparable strength and composition (Surdam and others, 1984; MacGowan and Surdam, 1988).

Aluminum Flux and Diffusion Calculations

Porewater, petrographic, and geohistory data are integrated in a flux calculation involving the main Stevens reservoir at North Coles Levee. This sandstone is assumed to be 150 m thick across the 15 km^2 oil field and to have 2.2 × 10^{14} cm^3 (10 volume percent) porosity. The maximum aluminum concentration is 840 μg/l in Stevens formation waters at North Coles Levee (Fisher and Boles, 1990). An estimated 10^4 to 10^5 pore volumes are required to leach 2.5 volume percent An$_{30}$ plagioclase from each thin-section-size volume. Although fluid flow may be focused through specific intervals in the reservoir, approximately 2.4 × 10^{19} cm^3 of water is required to leach 2.5 volume percent plagioclase from the entire reservoir. If plagioclase dissolves over 2.5 × 10^6 years (Boles, 1987), then 0.4 cm^3/year of fluid must flow through each 100 cm^3 of rock. In a similar calculation, Boles and Ramseyer (1987) assume an aluminum solubility of 10 mg/l and dissolution time of 3.5 × 10^6 years, then calculate that a flux of 0.02 cm^3/year per 100 cm^3 of rock is required to leach 2 volume percent plagioclase at North Coles Levee. In a final example, a maximum aluminum concentration of 75 μg/l (detection limit of Fisher and Boles, 1990) and porosity of 20% are assumed for a meteoric-zone sandstone with dissolved plagioclase at Mount Poso. A flux of 4.6 cm^3/year per 100 cm^3 of rock is required to leach 2.5 volume percent plagioclase from this interval (7.1 × 10^{13} cm^3 total volume) over 2.5 × 10^6 years. Order of magnitude differences in the three calculated flux rates mainly reflect order of magnitude differences in assumed aluminum solubility. The aluminum imbalances found in San Joaquin Basin sandstones can be produced by limited fluid fluxes acting over 10^6 years.

Aluminum also may diffuse over time. The diffusion coefficient (D$_{fluid}$) for Al^{+3} in dilute water is 5.59 × 10^{-6} cm^2/s at 25°C (Li and Gregory, 1974), approximately 4 × 10^{-5} cm^2/s at 100°C, and may be an order of magnitude larger in organic solvents (Cussler, 1976). Sandstones with 10 to 20% porosity have diffusion coefficients equal to one-fourth to one-eighth of D$_{fluid}$, based on calculations outlined in Berner (1980) and Manheim (1970). An aluminum concentration gradient of 1 mg/l loss over 3 cm is assumed in diffusion calculations because: (1) 1 mg/l approximates the aluminum concentration in Stevens pore waters at North Coles Levee, and (2) aluminum apparently reprecipitates as kaolinite within a thin-section-size volume in Stevens sandstones.

"Instantaneous" dissolution of 2.5 volume percent plagioclase (An$_{30}$) in a sandstone with 10% porosity results in an aluminum concentration of 3.3 × 10^{-3} M. Based on calculations using Fick's second law of diffusion (e.g., summary of Berner, 1980), aluminum can diffuse beyond a centimeter scale in 5 × 10^3 to 4 × 10^4 years, and up to 14 m in 2.5 × 10^6 years at temperatures of 25 to 100°C. Thus, the aluminum imbalances observed in thin sections of Stevens and Vedder sandstones can be accounted for by diffusion alone. Moving fluids are not required to export small masses of aluminum beyond a centimeter scale if diffusion operates at diagenetic temperatures over geologic time.

Factors in Aluminum Mass Transfer

Mass transfer of aluminum is dependent on porewater chemistry, hydrology, and temperature. Depositional environment determines initial porewater chemistry, sandstone texture and geometry, and sandstone-shale relations, and, together with adjacent facies relations, influences fluid-migration pathways, flux rates, and degree of compartmentalization of sedimentary basins. Aluminum is conserved on a thin-section scale in shale-encased Stevens sandstones that are subject to relatively low fluid fluxes. Aluminum may be mobilized in soluble organic complexes within a reservoir, but the isolated nature of Stevens sandstones prevents significant aluminum export from the reservoir. Aluminum is transferred only on a millimeter scale, assuming that the kaolinite found in intergranular pores incorporates aluminum released from nearby leached grains. The clustering of Stevens porewater data also implies compartmentalization.

In contrast, small amounts of aluminum are exported from meteoric-zone Vedder sandstones. Relatively high fluxes of dilute water and slow reaction rates at low temperatures may result in undersaturated waters and open-system diagenesis. Vedder shelf sandstones are more continuous laterally than Stevens turbidite sandstones and have greater potential for large-scale mass transport. Although compartmentalization of Vedder sandstones is not ruled out, compositional differences between shallow and deep Vedder formation waters probably reflect variable mixing with meteoric water.

Regional Trends

Volumes of dissolved plagioclase grains and kaolinite are documented in a separate, regional survey of 133 sandstone samples from 44 cores at 23 oil fields; 300 points were counted per sample. Dissolved plagioclase and kaolinite appear in more than 80% of the matrix-free, uncemented Stevens sandstones inspected from the central basin. Volumes of skeletal-plagioclase grains (not plagioclase porosity) are 0 to 5% from 2,125 to 3,336 m, with maximum volumes of 6 to 8% appearing in Stevens sandstones from 2,609 to 2,899 m. The volume of skeletal-plagioclase grains is 1 to 5% at 1,864 to 2,122 m and 3,269 to 4,348 m in Eocene to Late Miocene sandstones in the southern basin, but rises to 7 to 13% from 2,634 to 2,827 m. Volumes of leached plagioclase grains are 0 to 8% in Oligocene sandstones from 1,384 to 3,059 m in the second-cycle western basin. Leached plagioclase and kaolinite are absent at depths less than 1,864 m in sandstones with marine pore waters.

Leached plagioclase and kaolinite are absent in some meteoric-zone reservoirs at basin margins, but appear in up to 60% of matrix-free, uncemented samples in other meteoric-zone reservoirs. Kaolinite precipitates in east-flank Vedder sandstones, but smectite precipitates in volcanic-rich Santa Margarita and Chanac sandstones in the southern basin. Volumes of skeletal plagioclase grains are 0 to 5% in Vedder sandstones from 597 to 1,521 m and 2 to 7% in Santa Margarita and Chanac sandstones from 436 to 1,024 m. Plagioclase dissolution at depths less than 1,500 m is a basin-margin feature associated with meteoric recharge. Relatively low volumes (1–3%) of skeletal grains appear in less than 30% of uncemented samples inspected from depths of 1,864 to 2,122 m in the southern basin. These sandstones are too deep to have undergone meteoric flushing, yet too shallow to have experienced substantial non-meteoric plagioclase dissolution.

In summary, the San Joaquin Basin contains two environments of plagioclase dissolution and kaolinite precipitation: (1) deep sandstone reservoirs (deeper than 2 km) with modified marine pore waters, and (2) shallow sandstone reservoirs at basin margins subjected to meteoric recharge. Plagioclase dissolution in the marine zone apparently is temperature or kinetic controlled, and perhaps is affected by organic-acid anions, which form during hydrocarbon maturation. Plagioclase dissolution in meteoric waters apparently is controlled by inorganic-water chemistry, flux rates, and low-temperature reaction kinetics.

Relation to Other Mass-Balance Calculations and Models

Fischer and Surdam (1988) propose a model that implies transfer of potentially large masses of aluminum in the San Joaquin Basin. In this model, plagioclase dissolves and aluminum is exported from Stevens sandstones at 80 to 120°C and from shallow, meteoric-zone Santa Margarita and Chanac sandstones. Soluble alumino-organic complexes migrate updip from the Stevens and downdip from the meteoric zone. These complexes break down and aluminum precipitates as kaolinite in the mixing zone between meteoric and marine pore waters (Fischer and Surdam, 1988).

Petrographic data of this study are inconsistent with the scenario of Fischer and Surdam (1988), even though this study involves many of the same rocks in a nearby study transect. Aluminum is not exported from Stevens reservoirs. Fischer and Surdam (1988) assumed that turbidite sandstones and shelf sandstones are laterally continuous but did not establish the continuity of individual sandstone beds across formational contacts. Fischer and Surdam (1988) suggested that aluminum migrates downdip through the meteoric zone and precipitates as kaolinite in a mixed-water zone from 2,100 to 2,400 m. In this study, only small amounts of aluminum are exported from meteoric-zone sandstones beyond a centimeter scale and kaolinite-rich, plagioclase porosity-poor zones are not found down dip.

Mass-balance data from Stevens sandstones are consistent with other calculations, which do not support transfer of large masses of aluminum by organic acids and anions. For example, the volume of organic acids generated from Gulf Coast source rocks could produce only 1 to 4% total secondary porosity regionally, much less than the 10 to 13% observed (Lundegard and others, 1984; Lundegard and Land, 1986; Kharaka and others, 1986).

Relation to Other Basins

Meteoric water has infiltrated Gulf Coast sandstones to depths of 2 km (Galloway, 1984). Boles (1982), Franks and Forester (1984), Galloway (1984), Land (1984), and Loucks and others (1984) find dissolved feldspar and kaolinite in Cenozoic sandstones at depths of 1.2 to 3.6 km and present temperatures of 75 to 140°C. The average volume of leached feldspar (plagioclase + potassium feldspar) is 5% in Cenozoic Gulf Coast sandstones and feldspar porosity is comparable in meteoric and marine porewater zones (Lundegard and others, 1984; Lundegard and Land, 1986, 1989; Milliken, 1988; Milliken and others, 1989). Authigenic clays have volumes less than 1% (Milliken 1988; Milliken and others, 1989). These findings imply aluminum export, not aluminum conservation, as is found in Stevens sandstones. Kaolinite volumes are comparable in the San Joaquin and Gulf Coast basins, but feldspar porosity is more extensive in the Gulf Coast if reported volumes are accurate on a regional scale. Potassium feldspar dissolves and feldspar overgrowths precipitate in trace amounts in Gulf Coast sandstones. Calculating the aluminum mass balance is more difficult in the Gulf Coast than in the San Joaquin Basin due to these additional aluminum sources and sinks.

Meteoric water has infiltrated marine sandstones to depths of hundreds of meters in the North Sea (Bjørlykke, 1984). Feldspar dissolves and kaolinite precipitates in Jurassic sandstones upon uplift and meteoric recharge at subaerially exposed unconformities (Blanche and Whitaker, 1978; Hancock and Taylor, 1978; Sommer, 1978). Potassium feldspar dissolves, potassium-feldspar overgrowths precipitate, micas alter to kaolinite, and kaolinite alters to illite in North Sea sandstones (Bjørlykke, 1984; Bjørlykke and Brendsdal, 1986). Calculating the aluminum mass balance is more difficult in the North Sea than in the San Joaquin Basin because multiple aluminum sources and sinks are present. Despite these differences, volumes of leached feldspar are comparable in both basins. Emery and others (1990) reported aluminum conservation in marine sandstones flushed by meteoric water, not aluminum loss as is encountered in a similar setting in the San Joaquin Basin. Aluminum also is conserved as kaolinite and illite in deep North Sea reservoirs (Ehrenberg, 1990), as in Stevens sandstones.

In summary, the state of aluminum mass balance in San Joaquin Basin sandstones is consistent with that in some North Sea and Gulf Coast sandstones, but inconsistent with others, based on our data from the San Joaquin Basin and on our interpretation of published data for North Sea and Gulf Coast sandstones. Aluminum is conserved in deep reservoirs with marine pore waters in the San Joaquin Basin, but apparently not in the Gulf Coast. Aluminum is exported from meteoric zones in the San Joaquin and Gulf Coast basins, but not in the North Sea. Although processes and environments of plagioclase dissolution and kaolinite precipitation may be similar in basins worldwide, many variables are involved and any broad generalizations or predictions of aluminum mass transfer and associated reservoir enhancement are suspect. Careful volumetric measurements are essential in mass-balance studies involving volumetrically minor features.

CONCLUSIONS

Mass transfer of aluminum in the San Joaquin Basin is dependent on porewater chemistry, hydrology, depositional environment, and temperature. Aluminum is conserved on a thin-section scale in Stevens sandstones with marine pore waters, despite enrichment in organic-acid anions capable of forming soluble complexes with aluminum. Large-scale aluminum export from these shale-encased turbidite sandstones is difficult to envision. In contrast, relatively small amounts of aluminum are exported beyond a thin-section scale from meteoric-zone Vedder sandstones, although complementary intervals of aluminum import are not found downdip. Relatively high flux rates, slow reaction rates, and more sheetlike sandstone geometry all favor mass transfer during meteoric recharge into shelf sandstones.

Results of this detailed petrographic study do not support models that predict large-scale aluminum mass transfer. Shifts in aluminum content due to plagioclase dissolution and kaolinite precipitation are too small to discriminate from the natural bulk chemical variability of the sandstones examined. Although solubility experiments demonstrate that aluminum released from feldspar can be mobilized in organic complexes, petrographic and porewater data are not clearly related in San Joaquin Basin reservoirs. Plagioclase dissolves in acetate-rich Stevens formation waters, but aluminum is not mobilized beyond a centimeter scale in Stevens sandstones. In contrast, some aluminum is exported from Vedder sandstones by meteoric waters, which lack detectable amounts of organic-acid anions. Aluminum mobility in sedimentary basins is a function of aluminum solubility *and* aluminum flux.

ACKNOWLEDGMENTS

This study was funded by NSF grant EAR-17013, by PRF grant 18438-AC2, and by a grant from Chevron Oil Field Research Company. The authors thank employees of ARCO, Shell, and Texaco in Bakersfield for providing access to cores sampled for this study. Thoughtful reviews by D. Houseknecht, E. Pittman, T. Dunn, and K. Milliken improved the quality of the manuscript. Discussions with K. Bjørlykke, K. Ramseyer, J. Hickey, J. Wood, J. Schultz, and M. Feldman were enlightening. Dave Pierce provided instruction on the SEM.

REFERENCES

BANDY, O. L., AND ARNAL, R. E., 1969, Middle Tertiary basin development, San Joaquin Valley, California: Geological Society of America Bulletin, v. 80, p. 783–820.

BARTOW, A. J., AND MCDOUGALL, K., 1984, Tertiary stratigraphy of the southeastern San Joaquin Valley, California: U. S. Geological Survey Bulletin 1529-J, p. J1–J41.

BENT, J. V., 1988, Paleotectonics and provenance of Tertiary sandstones of the San Joaquin Basin, California, *in* Graham, S. A., and Olson, H. C., eds., Studies in the Geology of the San Joaquin Basin: Pacific Section, Society of Economic Paleontologists and Mineralogists Book 60, p. 109–127.

BERNER, R. A., 1980, Early Diagenesis–A Theoretical Approach: Princeton Series in Geochemistry: Princeton, New Jersey, Princeton University Press, 241 p.

BJØRLYKKE, K., 1984, Formation of secondary porosity: how important is it?, *in* McDonald, D. A., and Surdam, R. C., eds., Clastic Diagenesis: American Association of Petroleum Geologists Memoir 37, p. 277–286.

BJØRLYKKE, K., AND BRENDSDAL, A., 1986, Diagenesis of the Brent Sandstone in the Statfjord field, North Sea, *in* Gautier, D. L., ed., Roles of Organic Matter in Sedimentary Diagenesis: Society of Economic Paleontologists and Mineralogists Special Publication 38, p. 157–167.

BLANCHE, J. B., AND WHITAKER, J. H., 1978, Diagenesis of part of the Brent Sand Formation (Middle-Jurassic) of the northern North Sea basin: Journal of the Geological Society of London, v. 135, p. 73–82.

BOLES, J. R., 1982, Active albitization of plagioclase, Gulf Coast Tertiary: American Journal of Science, v. 282, p. 165–180.

BOLES, J. R., 1984, Secondary porosity reactions in the Stevens sandstone, San Joaquin Valley, California, in McDonald, D. A., and Surdam, R. C., eds., Clastic Diagenesis: American Association of Petroleum Geologists Memoir 37, p. 217–224.

BOLES, J. R., 1987, Six million year diagenetic history, North Coles Levee, San Joaquin Basin, California, in Marshall, J. D., ed., Diagenesis of Sedimentary Sequences: Geological Society of London Special Publication 36, p. 191–200.

BOLES, J. R., AND JOHNSON, K. S., 1983, Influence of mica surfaces on pore-water pH: Chemical Geology, v. 43, p. 303–317.

BOLES, J. R., AND RAMSEYER, K., 1987, Diagenetic carbonate in Miocene sandstone reservoir, San Joaquin Basin, California: American Association of Petroleum Geologists Bulletin, v. 71, p. 1475–1487.

BOLES, J. R., AND RAMSEYER, K., 1988, Albitization of plagioclase and vitrinite reflectance as paleothermal indicators, San Joaquin Basin, in Graham, S. A., and Olson, H. C., eds., Studies in the Geology of the San Joaquin Basin: Pacific Section, Society of Economic Paleontologists and Mineralogists Book 60, p. 129–139.

CALLAWAY, D. C., 1971, Petroleum potential of San Joaquin Basin, California, in Cram, I. H., ed., Future Petroleum Provinces of the United States–Their Geology and Potential: American Association of Petroleum Geologists Memoir 15, p. 239–253.

CAROTHERS, W. W., AND KHARAKA, Y. K., 1978, Aliphatic acid anions in oil-field waters–implications for origin of natural gas: American Association of Petroleum Geologists Bulletin, v. 62, p. 2441–2453.

CROWELL, J. C., 1987, Late Cenozoic basins of onshore southern California: complexity is the hallmark of their tectonic history, in Ingersoll, R. V., and Ernst, W. G., eds., Cenozoic Basin Development of Coastal California (Rubey Volume VI): Englewood Cliffs, Prentice-Hall, p. 207–241.

CUSSLER, E. L., 1976, Multicomponent diffusion: Chemical Engineering Monographs, v. 3: New York, Elsevier, 176 p.

DICKINSON, W. R., AND SUCZEK, C. A., 1979, Plate tectonics and sandstone compositions: American Association of Petroleum Geologists Bulletin, v. 63, p. 2164–2182.

EHRENBERG, S. N., 1990, Relationship between diagenesis and reservoir quality in sandstones of the Garn Formation, Haltenbanken, Mid-Norwegian continental shelf: American Association of Petroleum Geologists Bulletin, v. 74, p. 1538–1558.

EMERY, D., MYERS, K. J., AND YOUNG, R., 1990, Ancient subaerial exposure and freshwater leaching in sandstones: Geology, v. 18, p. 1178–1181.

FISCHER, K. J., AND SURDAM, R. C., 1988, Contrasting diagenetic styles in a shelf/turbidite sandstone sequence: the Santa Margarita and Stevens sandstones, southern San Joaquin Basin, California, in Graham, S. A., and Olson, H. C., eds., Studies in the Geology of the San Joaquin Basin: Pacific Section, Society of Economic Paleontologists and Mineralogists Book 60, p. 233–247.

FISHER, J. B., AND BOLES, J. R., 1990, Water-rock interaction in Tertiary sandstones, San Joaquin Basin, California, U.S.A.: diagenetic controls on water composition: Chemical Geology, v. 82, p. 83–101.

FRANKS, S. G., AND FORESTER, R. W., 1984, Relationships among secondary porosity, pore-fluid geochemistry and carbon dioxide, Texas Gulf Coast, in McDonald, D. A., and Surdam, R. C., eds., Clastic Diagenesis: American Association of Petroleum Geologists Memoir 37, p. 63–80.

GALLOWAY, W. E., 1984, Hydrogeologic regimes of sandstone diagenesis, in McDonald, D. A., and Surdam, R. C., eds., Clastic Diagenesis: American Association of Petroleum Geologists Memoir 37, p. 3–13.

GRAHAM, S. A., 1987, Tectonic controls on petroleum occurrence in central California, in Ingersoll, R. V., and Ernst, W. G., eds., Cenozoic Basin Development of Coastal California (Rubey Volume VI): Englewood Cliffs, Prentice-Hall, p. 47–63.

GRAHAM, S. A., AND WILLIAMS, L. A., 1985, Tectonic, depositional, and diagenetic history of Monterey Formation (Miocene), central San Joaquin Basin, California: American Association of Petroleum Geologists Bulletin, v. 69, p. 385–411.

HANCOCK, N. J., AND TAYLOR, A. M., 1978, Clay mineral diagenesis and oil migration in the Middle Jurassic Brent Formation: Journal of the Geological Society of London, v. 135, p. 69–72.

HANSLEY, P. S., 1987, Petrologic and experimental evidence for the etching of garnets by organic acids in the Upper Jurassic Morrison Formation, northwestern New Mexico: Journal of Sedimentary Petrology, v. 57, p. 666–681.

HARDING, T. P., 1976, Tectonic significance and hydrocarbon trapping consequences of sequential folding synchronous with San Andreas faulting, San Joaquin Valley, California: American Association of Petroleum Geologists Bulletin, v. 60, p. 356–378.

HIRT, W. H., 1989, The petrological and mineralogical zonation of the Mount Whitney intrusive suite, eastern Sierra Nevada, California: Unpublished Ph.D. Dissertation, University of California, Santa Barbara, 278 p.

HURST, A., AND IRWIN, H., 1982, Geological modelling of clay diagenesis in sandstones: Clay Minerals, v. 17, p. 5–22.

KHARAKA, Y. K., LAW, L. M., CAROTHERS, W. W., AND GOERLITZ, D. F., 1986, Role of organic species dissolved in formation waters from sedimentary basins in mineral diagenesis, in Gautier, D. L., ed., Roles of Organic Matter in Sediment Diagenesis: Society of Economic Paleontologists and Mineralogists Special Publication 38, p. 111–122.

LAND, L. S., 1984, Frio sandstone diagenesis, Texas Gulf Coast: a regional isotopic study, in McDonald, D. A., and Surdam, R. C., eds., Clastic Diagenesis: American Association of Petroleum Geologists Memoir 37, p. 47–62.

LI, Y-H., AND GREGORY, S., 1974, Diffusion of ions in seawater and in deep-sea sediments: Geochimica et Cosmochimica Acta, v. 38, p. 703–714.

LOUCKS, R. G., DODGE, M. M., AND GALLOWAY, W. E., 1984, Regional controls on diagenesis and reservoir quality in Lower Tertiary sandstones along the Texas Gulf Coast, in McDonald, D. A., and Surdam, R. C., eds., Clastic Diagenesis: American Association of Petroleum Geologists Memoir 37, p. 15–45.

LUNDEGARD, P. D., AND LAND, L. S., 1986, Carbon dioxide and organic acids: their role in porosity enhancement and cementation, Paleogene of the Texas Gulf Coast, in Gautier, D. L., ed., Roles of Organic Matter in Sediment Diagenesis: Society of Economic Paleontologists and Mineralogists Special Publication 38, p. 129–146.

LUNDEGARD, P. D., AND LAND, L. S., 1989, Carbonate equilibria and pH buffering by organic acids–response to changes in p_{CO_2}: Chemical Geology, v. 74, p. 277–287.

LUNDEGARD, P. D., LAND, L. S., AND GALLOWAY, W. E., 1984, Problem of secondary porosity: Frio Formation (Oligocene), Texas Gulf Coast: Geology, v. 12, p. 399–402.

MACGOWAN, D. B., AND SURDAM, R. C., 1988, Difunctional carboxylic acid anions in oilfield waters: Organic Geochemistry, v. 12, p. 245–259.

MACPHERSON, B. A., 1978, Sedimentation and trapping mechanism in Upper Miocene Stevens and older turbidite fans of southeastern San Joaquin Valley, California: American Association of Petroleum Geologists Bulletin, v. 62, p. 2243–2274.

MANHEIM, F. T., 1970, The diffusion of ions in unconsolidated sediments: Earth and Planetary Science Letters, v. 9, p. 307–309.

MILLIKEN, K. L., 1988, Loss of provenance information through subsurface diagenesis in Plio-Pleistocene sandstones, northern Gulf of Mexico: Journal of Sedimentary Petrology, v. 58, p. 992–1002.

MILLIKEN, K. L., MCBRIDE, E. F., AND LAND, L. S., 1989, Numerical assessment of dissolution versus replacement in the subsurface destruction of detrital feldspars, Oligocene Frio Formation, south Texas: Journal of Sedimentary Petrology, v. 59, p. 740–757.

MOORE, J. G., 1987, Mount Whitney Quadrangle, Inyo and Tulare Counties, California–Analytic Data: U. S. Geological Survey Bulletin 1760, 10 p.

NAMSON, J. S., AND DAVIS, T. L., 1988a, Seismically active fold and thrust belt in the San Joaquin Valley, central California: Geological Society of America Bulletin, v. 100, p. 257–273.

NAMSON, J. S., AND DAVIS, T. L., 1988b, Structural transect of the western Transverse Ranges, California: implications for lithospheric kinematics and seismic risk evaluation: Geology, v. 16, p. 675–679.

NADEAU, P. H., AND HURST, A., 1991, Application of back-scattered electron microscopy to the quantification of clay mineral microporosity in sandstones: Journal of Sedimentary Petrology, v. 61, p. 921–925.

NILSEN, T. H., 1984, Oligocene tectonics and sedimentation, California: Sedimentary Geology, v. 38, p. 305–336.

NILSEN, T. H., 1987, Paleogene tectonics and sedimentation of coastal

California, *in* Ingersoll, R. V., and Ernst, W. G., eds., Cenozoic Basin Development of Coastal California (Rubey Volume VI): Englewood Cliffs, Prentice-Hall, p. 81–123.

OLSON, H. C., 1988, Oligocene-Middle Miocene depositional systems north of Bakersfield, California: eastern basin equivalents of the Temblor Formation, *in* Graham, S. A., and Olson, H. C., eds., Studies in the Geology of the San Joaquin Basin: Pacific Section, Society of Economic Paleontologists and Mineralogists Book 60, p. 189–205.

PASSARETTI, M. L., AND ESLINGER, E. V., 1987, Dissolution and relic textures in framework grains of Holocene sediments from the Brazos River and Gulf Coast of Texas: Journal of Sedimentary Petrology, v. 57, p. 94–97.

SCHULTZ, J. L., BOLES, J. R., AND TILTON, G. R., 1989, Tracking calcium in the San Joaquin Basin, California: a strontium isotopic study of carbonate cements at North Coles Levee: Geochimica et Cosmochimica Acta, v. 53, p. 1991–1999.

SOMMER, F., 1978, Diagenesis of Jurassic sandstones in the Viking graben: Journal of the Geological Society of London, v. 135, p. 63–67.

STOESSELL, R. K., AND PITTMAN, E. D., 1990, Secondary porosity revisited: the chemistry of feldspar dissolution by carboxylic acids and anions: American Association of Petroleum Geologists Bulletin, v. 74, p. 1795–1805.

SURDAM, R. C., BOESE, S. W., AND CROSSEY, L. J., 1984, The chemistry of secondary porosity, *in* McDonald, D. A., and Surdam, R. C., eds., Clastic Diagenesis: American Association of Petroleum Geologists Memoir 37, p. 127–149.

SURDAM, R. C., CROSSEY, L. J., HAGEN, E. S., AND HEASLER, H. P., 1989, Organic-inorganic interactions and sandstone diagenesis: American Association of Petroleum Geologists Bulletin, v. 73, p. 1–23.

WEBB, G. W., 1981, Stevens and earlier Miocene turbidite sandstones, southern San Joaquin Valley, California: American Association of Petroleum Geologists Bulletin, v. 65, p. 438–465.

ZIEGLAR, D. L., AND SPOTTS, J. H., 1984, Reservoir and source bed history in the Great Valley of California, *in* Demaison, G., and Murris, R. J., eds., Petroleum Geochemistry and Basin Evaluation: American Association of Petroleum Geologists Memoir 35, p. 193–204.

AUTHIGENIC MINERALOGY OF SANDSTONES INTERCALATED WITH ORGANIC-RICH MUDSTONES: INTEGRATING DIAGENESIS AND BURIAL HISTORY OF THE MESAVERDE GROUP, PICEANCE BASIN, NW COLORADO

LAURA J. CROSSEY AND DANIEL LARSEN
Department of Geology, University of New Mexico, Albuquerque, New Mexico 87131

ABSTRACT: Twenty-nine sets of sandstone and stratigraphically adjacent, fine-grained, organic-rich samples representing diverse depositional environments (fluvial, coastal, paludal, and marine) have been obtained from cores of the Mesaverde Group within the Piceance Basin, Colorado. These units have been buried to a sufficient depth that organic maturation has progressed to an advanced state (vitrinite-reflectance values range from 1.0 to 2.2 in the study area). The authigenic-mineral paragenesis and organic data are integrated with burial and thermal-history modeling to place diagenetic events and hydrocarbon generation into a temporal framework. Three phases of diagenesis (early, late, and post-hydrocarbon) are characterized on the basis of petrography, XRD, and geochemistry of authigenic phases.

Clay-mineral distributions of sandstone/mudstone pairs indicate that mixed-layer illite/smectite dominates mudstone mineralogy, although subordinate amounts of chlorite are observed in the marine interval. Sandstone mineralogy also includes mixed-layer clays and chlorite, with the addition of kaolinite. Additional aspects of sandstone diagenesis examined include feldspar albitization and dissolution, and carbonate mineralogy.

Organic analyses indicate the presence of a type III kerogen component in all samples. The paludal interval contains the most oxygen-rich organic material and exhibits the highest sandstone porosities. Water-soluble organic compounds released from organic material as burial progresses have been invoked as agents affecting the course of mineral diagenesis in clastic sediments and sedimentary rocks. Our results indicate the potential for the field evaluation of the effects of thermal maturation of organic matter on the diagenesis of closely adjacent sandstones.

INTRODUCTION

One of the primary goals of diagenetic studies is to understand the processes operating on sedimentary materials as they undergo progressive burial. Once the processes, or pathways, of diagenesis are known, the results can be integrated with burial and thermal modeling to evaluate the timing of events of regional significance (e.g., the development of porosity and the production and migration of hydrocarbons). A process-oriented approach is essential if models are to be applied to a variety of sedimentary basins. This approach has been used with great success in the evaluation of liquid-hydrocarbon production from organic-rich source rocks (see reviews in Demaison and Murris, 1984; Waples, 1985). A similar approach applied to potential reservoir rocks should enable prediction of spatial and temporal porosity variations within a basin.

A key aspect of a process-oriented approach to clastic diagenesis is the incorporation of organic-maturation reactions (including the generation and release of water-soluble organic compounds) into diagenetic models. Numerous workers have shown experimentally and theoretically that water-soluble organic compounds can have a profound effect on stability of aluminosilicates and carbonates through pH control and complexation (Huang and Keller, 1970; Surdam and others, 1984; Crossey, 1985; Kharaka and others, 1986; Surdam and Crossey, 1987), although others question the overall importance of these effects in diagenetic processes (e.g., Lundegard and Land, 1986; Lundegard and Kharaka, 1990; Stoessell and Pittman, 1990). Both the type of organic material initially present and the extent of thermal exposure during burial are critical to the evaluation of the impact of organic-maturation reactions on 'inorganic' diagenesis (Surdam and others, 1984; Kawamura and others, 1985, 1986; Kawamura and Kaplan, 1987; Lundegard and Senftle, 1987; Surdam and MacGowan, 1987). Water-soluble organic compounds contain oxygen-bearing functional groups; thus, a terrigenous (type III, oxygen-rich) kerogen would have a greater capacity for the production of these reactive compounds than a marine (type II, more hydrogen-rich) kerogen (Surdam and others, 1984; Crossey and others, 1986a).

Authigenic clay minerals and clay-mineral transformations are important to organic-inorganic models for diagenesis. These minerals are sensitive to variations in pore-fluid composition, and also can serve as thermal-maturity indicators (e.g., Hower and others, 1976).

APPROACH

Whereas much remains to be learned about the chemistry of organic-inorganic interactions through controlled experiments, detailed examinations of natural systems are essential to guide such studies. Several studies have described the effects of organic-inorganic interactions during burial diagenesis in actual field settings (e.g., Fischer and Surdam, 1988; Dixon and others, 1989; Moraes, 1989; Surdam and others, 1989a; Hayes and Boles, 1990), but have focused on sandstone reservoirs at a field or regional scale, where fluid movements are complex and reservoir rocks are stratigraphically removed from source rocks. Sandstones intimately associated with organic material are the ideal testing ground for the hypothesis that water-soluble organic compounds, released from kerogen as burial depth increases, significantly influence subsequent mineral diagenesis. Ideally, the authigenic mineralogy and geochemistry (including mass-balance constraints), paragenetic sequence, and structure of the remaining organic matter must be examined in order to elucidate these interactions. In addition, detailed burial- and thermal-history models are required to provide a temporal framework. This study attempts to examine the paragenetic sequence in genetically related depositional units where some initial variation in type of buried organic matter is present and reasonable thermal-history constraints are available.

The initial type and quantity of organic matter present

Origin, Diagenesis, and Petrophysics of Clay Minerals in Sandstones, SEPM Special Publication No. 47
Copyright © 1992, SEPM (Society for Sedimentary Geology), ISBN 0-918985-95-1

are anticipated to exert a degree of control over the diagenesis of adjacent sandstones. The Department of Energy's multiwell experiment (MWX) site in the Piceance Basin (near Rifle, Colorado) may provide an excellent opportunity for testing of this proposed hypothesis. In brief, extensive core coverage (from three closely spaced drill holes) of a series of sandstone reservoirs within the Upper Cretaceous Mesaverde Group (MVG) is available, as well as a wealth of subsurface data generated by the Department of Energy's continued investigation of the site.

The MVG is interpreted to record a series of minor marine transgressive and regressive events, followed by marginal marine and coastal-plain deposition associated with a major regression and the subsequent infilling of an epeiric seaway. Depositional environments of the MVG at the MWX site have been described in detail (summarized in Lorenz, 1989). The diverse environments represented within the MVG are ideal for the examination of sandstones intimately associated with organic material of various types (predominantly humic [type III] and marine [type II]). Investigation of the authigenic mineralogy and paragenetic sequence of sandstones, coupled with organic characterization of adjacent organic-rich mudstones, may allow correlation of organic and inorganic diagenesis. Despite the inherent complexity of the overall diagenetic picture, several factors indicate that the diagenetic effects of the maturation of organic matter may be isolated to some degree within sandstones of the MVG: (1) the source area throughout the time of deposition is interpreted to have been relatively consistently from the west (Sevier Orogenic Belt), minimizing inherited differences in lithology; and (2) the MVG has been interpreted to have been relatively impermeable since the onset of gas accumulation in Tertiary time (Johnson, 1989), preserving burial diagenetic effects and minimizing overprinting during the last 10 million years as formation temperatures decreased (associated with regional downcutting of the Colorado River system).

REGIONAL SETTING

The depositional and structural setting of the MVG are closely intertwined, and will be discussed together. The setting presented here is abstracted from the overview provided by Johnson (1989). Figures 1 and 2A depict the location of critical elements of the geologic setting of the Piceance Basin.

The Piceance Basin is one of several subsidiary sedimentary basins within the Rocky Mountain foreland basin formed by Laramide tectonism from latest Cretaceous through Paleocene time (Fig. 2A). The MVG represents the deposits accumulated during cycles of marine transgression and regression, as well as the subsequent infilling of the basin with marginal marine and coastal-plain sediments. Regional uplift at the end of Cretaceous time produced an unconformity separating the MVG from overlying strata (Fig. 2B). Renewed subsidence during Paleocene and Eocene time resulted in the deposition of a thick nonmarine sequence, although regional relief is interpreted to have been subdued (as evidenced by extensive sediment-starved lacustrine deposits of the Eocene Green River Formation). Between Late

FIG. 1.—Location of the Cretaceous epeiric seaway during Cretaceous time. From Johnson (1989).

Eocene and Late Miocene time, little sedimentation occurred. Two periods of rapid downcutting associated with the Colorado River system, which occurred between 10 and 8 Ma and from 1.5 Ma to the present, are interpreted to have resulted from renewed pulses of regional uplift (Larsen and others, 1975). This depositional scenario is used as the basis for a burial-history reconstruction (discussed and illustrated later).

Sedimentology and Diagenesis of the Mesaverde Group

A comprehensive summary of sedimentology of the MVG in the Piceance Basin has been provided by Lorenz (1989). The petrography and mineralogy of the MVG at the MWX site has been summarized by Pitman and others (1989) and Lorenz and others (1989).

Four major depositional settings are recognized within the MVG at the MWX site and are referred to according to the 'interval' designations listed below.

(1) *Marine (shoreline-coastal) interval (2,272–2,545 m).* This interval (at the base of the MVG) is composed of fine-grained, cross-bedded, laterally continuous sandstones interbedded with shales and coals. The sandstones are interpreted as shallow marine and shoreline deposits associated with a wave-dominated deltaic system. As a group, these sandstones are the most quartz rich of the MVG (generally >75%) and fall in the subarkose and sublitharenite categories of Folk (1968) (Fig. 3E). Subequal amounts of feldspar grains and lithic fragments are present. The dominant feldspar is plagioclase, ranging from relatively unaltered to albitized

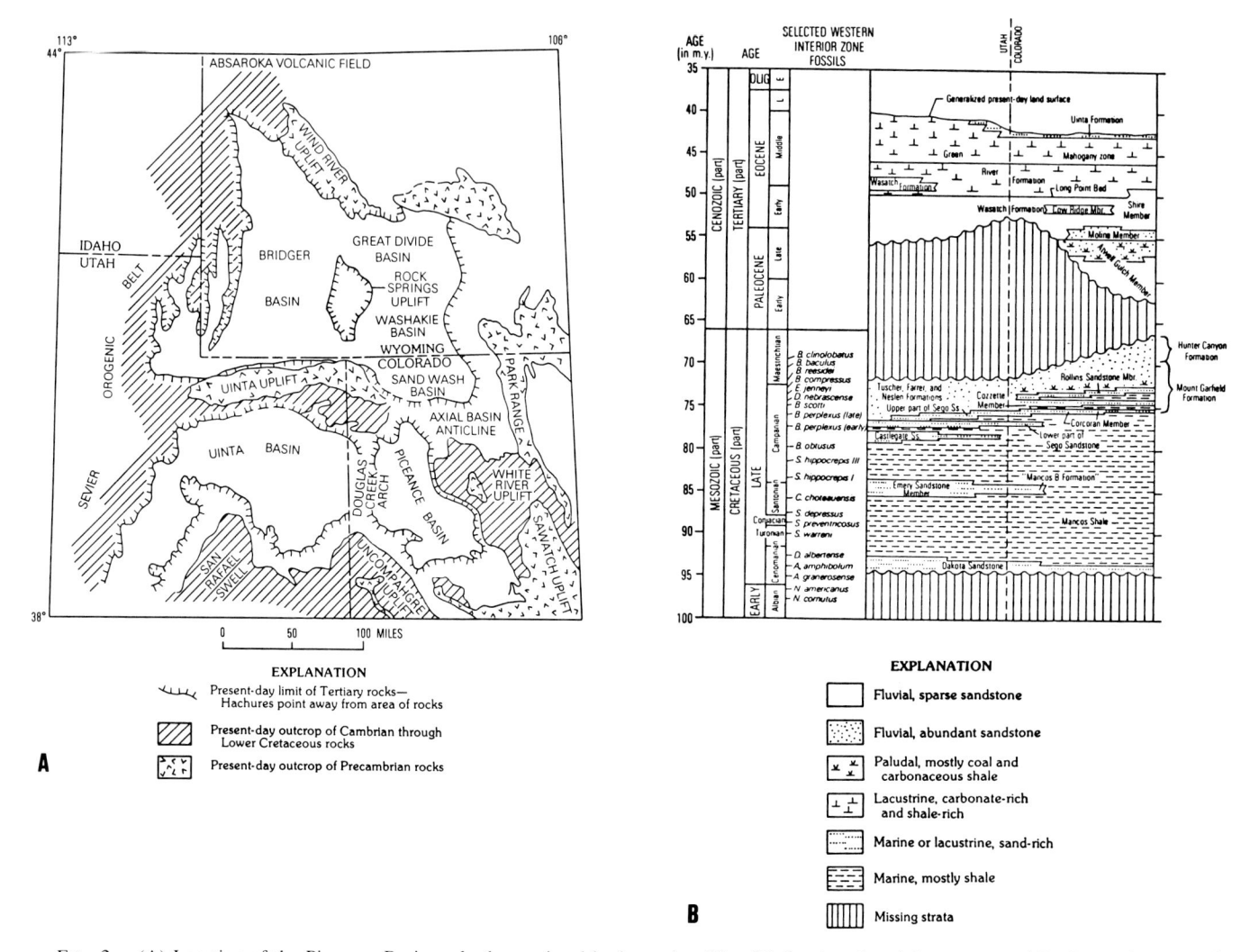

FIG. 2.—(A) Location of the Piceance Basin and other regional basins and uplifts. (B) Stratigraphy of Cretaceous and Tertiary units across the Uinta and Piceance Basins. The Hunter Canyon and Mount Garfield Formations comprise the Mesaverde Group. Both from Johnson (1989).

and/or replaced by clays and carbonate. Chert is a common sedimentary lithic constituent, as are intraclasts of mudstone and siltstone exhibiting compaction deformation. Detrital dolomite also is present. Calcite- and dolomite-cement contents range up to 20%. Overall porosities (range 5.6–7.3%, average 6.4%) and permeabilities (range 0.5–1.5 μd, average 1.0 μd) are relatively low (Table 1).

(2) *Paludal (or lower delta plain) interval (2,012–2,272 m)*. This interval contains lenticular, distributary-channel and overbank-splay sandstones interbedded with coals, carbonaceous mudstones, mudstones, and siltstones. Framework-grain mineralogy of this interval is more variable than that of the underlying marine interval; as a result, sandstone classifications range from sublitharenite to feldspathic litharenite to litharenite (Fig. 3D). Again, the dominant feldspar is plagioclase, with minor potassium feldspar. Deformed mudstone and siltstone clasts are the major lithic component; detrital-carbonate, coal, and chert fragments also are present. Abundant carbonate cements (including ankerite and dolomite) and authigenic-clay minerals are present. Highest average porosities are observed in the paludal interval (9.4%; range 7.7–11.4%). Permeabilities are low, ranging from 0.6–8.3 μd (average 2.8 μd; see Table 1).

(3) *Coastal (or upper delta plain) interval (1,829–2,012 m)*. This interval is lithologically similar to the underlying lower delta-plain deposits, but coal beds are much less abundant. It is interpreted to have been deposited in an upper delta-plain environment. The detrital- and authigenic-mineral assemblages of these sandstones also are similar to those of the underlying paludal deposits, although a slightly higher lithic component is present (Fig. 3C). Porosity ranges from 4.4 to 7.7% (average 5.9%) and permeabilities are consistently the lowest of the depositional environments at the study site (range 0.1–2.0 μd, average 1.0 μd; see Table 1).

(4) *Fluvial interval (1,340–1,830 m)*. This interval is dominated by stacked point-bar sequences, with rare in-

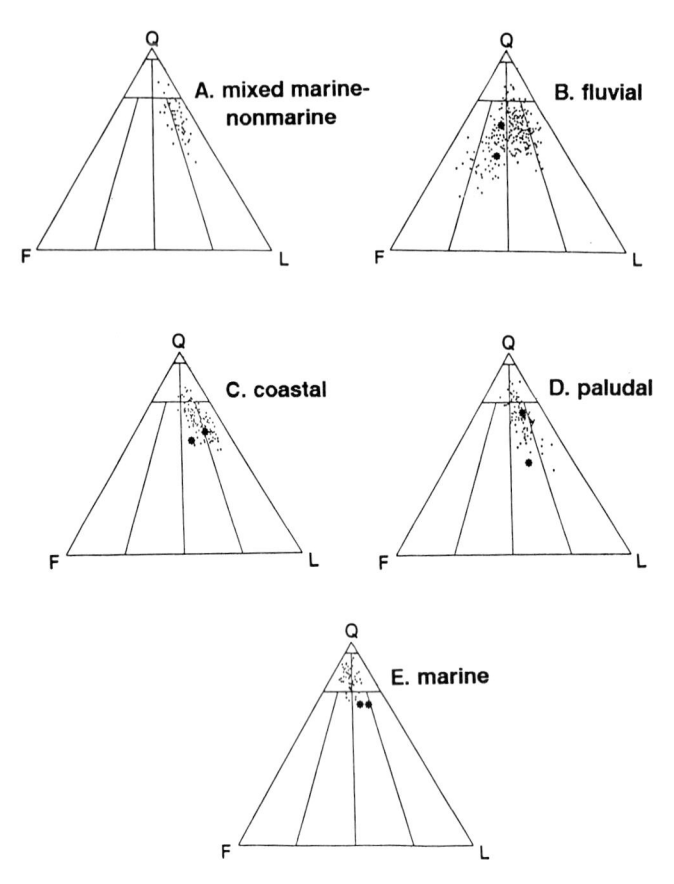

FIG. 3.— QFL diagrams for five depositional intervals of the MVG (sandstone classification boundaries as described by Folk, 1968). Points represent results presented in Pitman and others (1989); asterisks represent samples from this study. Environmental designations interpreted by Lorenz (1989).

terbeds of carbonaceous shale, mudstone, and siltstone. It is interpreted to represent deposits of a meandering fluvial system. The detrital mineralogy of the fluvial interval is the most diverse within the MVG at the MWX site. Most sandstones are characterized as lithic arkoses or feldspathic litharenites, but sublitharenites and litharenites are also noted (a few samples would be classified as subarkoses and arkoses as well; Fig. 3B). Quartz is the dominant framework constituent, but both sodium and potassium feldspars are present. Detrital carbonate (dolomite) and volcanic and sedimentary lithic clasts occur in minor amounts. Mudstone and siltstone fragments are extensively deformed. Authigenic quartz, carbonate, chlorite, and kaolinite are noted. Feldspar dissolution is commonly observed in this interval. Porosity ranges from 4.6 to 10.8%, with an average of 7.4%; permeabilities range from 0.1 to 4.9 µd, averaging 2.6 µd (Table 1).

In addition to these environments, the uppermost portion of the MVG at the MWX site represents mixed marine and nonmarine settings, and is interpreted to represent deposition during a marine transgression associated with the Upper Cretaceous Lewis Shale. The interval was not sampled for the current study. A summary of framework-grain mineralogy is shown in Figure 3A; note lithologic similarity to the delta-plain deposits of the paludal and coastal intervals.

METHODS

Sampling

Twenty-nine sets of sandstone/organic-rich pairs were obtained from slabbed cores. The sampling goal was to obtain closely adjacent sandstone-mudstone pairs. Relative thicknesses of sandstone and mudstone units typically vary with depositional environment, but generally sandstones are meters to tens of meters in thickness, with interbedded mudstones and coals generally as thick as 1 meter. Sample sets are given a 'type' designation according to the following association:

Type A: adjacent pair–sandstone and associated fine-grained unit are in direct contact (sampled interval ranges up to 10 cm).
Type B: interbedded pair–sandstone and mudstone lithologies are finely interbedded (at approximately a 1:1 ratio) to laminated (sampled interval represents one hand specimen; scale of centimeters).
Type C: transitional–usually a set of three samples was obtained from a fining-upward sequence (sampled interval ranges up to 1 m).

Figure 4 depicts location of sample sets within the environmental designations of Lorenz (1989). Table 2 presents actual depths and sample-set 'types'.

Analysis

Organic Material.—

Samples were ground in a mortar and pestle; organic determinations were performed on aliquots of the ground material. Total carbon (TC) was measured on a Carlo Erba 1106 Elemental Analyzer. Inorganic carbon (IC) values were obtained from coulometric titration of CO_2 evolved from acidification of a sample aliquot. Total organic carbon (TOC) was determined by difference. Programmed pyrolysis also was performed on at least one sample per sample set. Volatile reduced-carbon species evolved during sample pyrolysis (300–650°C over 15 min.) were detected as carbon by flame-ionization detection. Calibrated peak areas for S_1 (volatile hydrocarbons already present in the sample) and S_2 (hydrocarbons generated during pyrolysis) were determined. CO_2 (S_3) evolved during pyrolysis was not analyzed. Two parameters were calculated from the pyrolysis data. Hydrogen Index (HI) is defined as the quantity of hydrocarbon compounds evolved during pyrolysis (S_2), normalized to the TOC content of the sample. The Production Index (PI) refers to the ratio of hydrocarbons measured as S_1 to the total quantity of hydrocarbons released by the sample ($S_1 + S_2$).

Petrography.—

Standard thin sections were prepared for at least one sample per sample set and detailed point-count analyses (500

TABLE 1.—ORGANIC- AND CLAY-MINERAL RESULTS

Sample Set #	Well (MWX)	Sample Type[a]	Depth (m)	Lith.[b]	TOC (%C)	HI[c]	PI[d]	ML[e] S/I	Ch[f]	Kaol[g]	I[h]	ML[i] Ch/S	%[j] ML	R[k]
Fluvial														
1	1	B	1430.5	ss	0.45	197.8	0.28	2	2	0	0	0	82	1
2	1	C	1479.2	ss	.[l]	-	-	2	2	0	0	0	86	1
"	1	C	1479.8	sl	0.31	96.8	0.25	1	2	1	0	0	86	1
"	1	C	1480.6	md	0.45	40.0	0.14	2	1	0	1	0	84	1
3	1	B	1493.8	ss	0.56	64.3	0.25	2	1	1	0	0	86	1,3
4	2	B	1497.8	ss	13.63	105.0	0.20	2	2	0	2	0	89	1,3
5	2	A	1507.8	ss	-	-	-	2	3	2	0	0	66	-
"	2	A	1507.8	sl	0.34	185.3	0.42	3	2	0	0	0	86	1
6	1	B	1537.3	ss	5.89	69.1	0.42	1	2	0	0	0	90	-
7	1	A	1554.3	ss	-	-	-	1	2	1	0	0	-	-
"	1	A	1554.3	md	1.08	55.6	0.26	2	1	0	1	0	89	1,3
8	1	A	1633.8	ss	-	-	-	2	3	0	0	0	90	1,3
"	1	A	1633.8	md	1.24	87.1	0.41	3	2	0	2	0	88	1
9	2	B	1674.1	ss	4.61	85.2	0.20	2	1	1	0	0	90	3
10	2	A	1765.1	ss	-	-	-	2	2	2	0	0	86	3
"	2	A	1765.2	md	1.63	62.6	0.38	3	2	0	0	0	90	3
11	2	B	1773.7	ss	0.78	34.6	0.70	2	3	1	1	0	90	1
Coastal														
12	1	A	1841.6	sl	6.19	106.8	0.24	1	0	0	0	0	-	-
"	1	A	1841.7	ss	-	-	-	1	2	1	0	0	90	1,3
13	1	A	1871.9	md	6.38	83.2	0.19	1	0	1	0	0	-	-
"	1	A	1871.9	ss	-	-	-	2	0	2	2	0	84	1
14	1	A,C	1903.6	md	11.77	83.7	0.17	2	0	0	0	0	-	-
"	1	A,C	1903.9	ss	1.13	35.4	0.44	2	0	0	1	1	90	1
"	1	C	1904.4	ss	-	-	-	2	1	3	0	1	89	1
15	1	B	1939.1	ss	4.92	63.4	0.61	2	0	0	1	0	86	1,3
Paludal														
16	1	A	2024.3	ss	-	-	-	1	0	2	0	0	90	-
"	1	A	2024.5	sl	1.72	57.0	0.36	2	0	2	0	0	88	1,3
17	3	A	2096.7	ss	-	-	-	1	0	1	0	0	90	-
"	3	A	2097.0	md	7.62	33.2	0.31	1	1	0	1	0	-	-
18	3	B	2161.0	ss	0.44	43.2	0.51	2	0	0	1	0	86	3
19	2	B	2168.9	ss	4.04	28.7	0.51	2	0	2	0	0	91	3,1
20	3	B	2172.2	ss	2.82	25.9	0.44	1	0	0	0	0	88	3
21	2	A	2178.1	ss	-	-	-	2	0	1	0	0	89	3
"	2	A	2178.3	co	43.78	33.7	0.08	-	-	-	-	-	-	-
22	2	B	2218.6	ss	1.96	10.7	0.80	2	0	0	0	0	89	3,1
23	2	B	2234.3	ss	6.05	23.0	0.61	2	0	0	0	0	90	3
Marine														
24	3	(B)	2305.5	ss	-	-	-	3	0	0	0	0	84	3
25	1	B	2407.0	ss	2.25	17.8	0.49	2	1	0	2	0	90	3
26	1	A	2414.9	ss	-	-	-	2	1	0	0	0	90	-
"	1	A	2414.9	md	2.71	22.1	0.69	2	1	0	0	0	90	-
27	1	A	2415.8	ss	-	-	-	2	1	0	0	0	90	-
"	1	A	2415.9	md	2.08	11.5	0.74	1	1	0	0	0	88	3
28	2	B	2472.4	ss	1.67	35.9	0.77	2	1	1	0	0	90	3
29	2	B	2474.9	ss	0.90	12.2	0.92	3	1	0	1	0	89	3

[a]sample type designations defined in text.
[b]ss = sandstone, sl = siltstone, md = mudstone, co = coal.
[c]hydrogen index = mg hydrocarbon generated per gm organic carbon in sample.
[d]production index = mg hydrocarbon generated at 300°C/total hydrocarbons generated during pyrolysis.
[e]proportion of mixed-layer smectite/illite in <1 μm clay fraction (1 = major, 2 = minor, 3 = trace).
[f]proportion of chlorite in <1 μm clay fraction (1 = major, 2 = minor, 3 = trace).
[g]porportion of kaolinite in <1 μm clay fraction (1 = major, 2 = minor, 3 = trace).
[h]proportion of discrete illite in <1 μm clay fraction (1 = major, 2 = minor, 3 = trace).
[i]proportion of mixed-layer chlorite/smectite in <1 μm clay fraction (1 = major, 2 = minor, 3 = trace).
[j]percent non-expandable layers in mixed-layer smectite/illite as determined by XRD.
[k]Reichweite ordering as determined by XRD (see Moore and Reynolds, 1989).
[l]not determined.

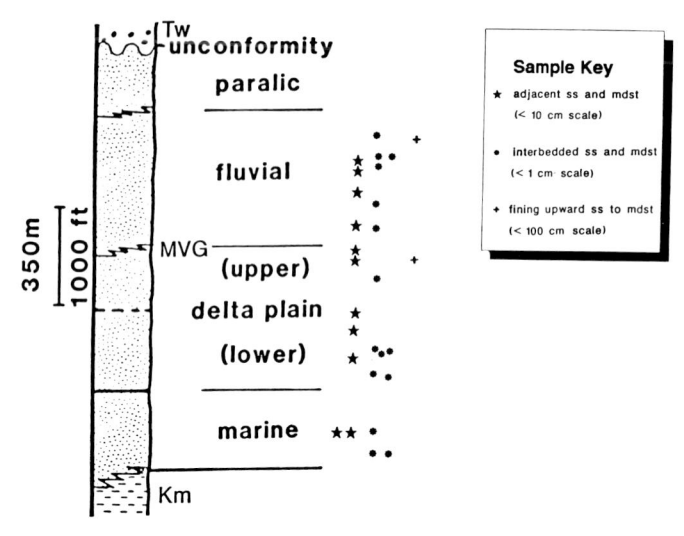

FIG. 4.—Stratigraphy and sampling locations of the MVG at the MWX site. Sample types are as discussed in the text. The paralic unit is equivalent to the mixed marine-nonmarine interval; upper and lower delta-plain units correspond to coastal and paludal intervals, respectively. Km is Mancos Shale, and Tw is Wasatch Formation.

counts per section) were performed on two sandstones from each environment to compare with previous studies. Quartz, feldspar, and lithic components were categorized according to Folk (1968) for classification purposes, although numerous diagenetic and detrital components also were evaluated quantitatively. Paragenetic-sequence and textural data were obtained from petrographic examination of 30 sandstone samples. Cathodoluminescence was used in the qualitative evaluation of carbonate paragenesis, the extent of overgrowths on quartz and feldspar, and to assist in selecting areas for microprobe analysis.

Clay Mineralogy.—

Samples were gently disaggregated in a mortar and pestle, then treated according to Jackson (1979) for carbonate removal. Two size separates were obtained by centrifugation (<1 μm and <2 μm). Following Mg-saturation, oriented mounts for XRD analysis were prepared according to the method of Drever (1973). Samples were subsequently solvated with ethylene glycol. Analyses were performed on a Scintag diffraction unit with Cu-K radiation operated at 30 mA and 40 kV. Clay-mineral identification was per-

TABLE 2.—POINT-COUNT RESULTS

Dep.[a] Int.	Depth (m)	Qm[b]	Qp[c]	Qc[d]	F[e]	Fcl[f]	Fcc[g]	Fd[h]	Mica[i]	Sed[j]	Meta[k]	Volc[l]	Ls[m]
F	1430.5	29.8	7.2	8.2	6.0	9.2	0.8	4.0	3.6	3.6	nd[n]	nd	nd
F	1493.8	34.8	3.4	5.8	2.2	7.4	nd	nd	0.8	6.0	nd	nd	nd
C	1841.7	31.0	8.2	15.0	2.4	nd	7.6	nd	0.4	4.2	1.0	1.0	nd
C	1939.1	39.0	9.0	18.8	2.4	2.6	2.2	2.0	0.4	8.0	0.2	0.2	nd
P	2024.2	17.2	2.8	1.2	0.6	0.4	6.6	0.2	0.2	6.2	0.4	0.4	1.2
P	2161.0	31.8	1.8	19.0	1.8	1.4	1.4	nd	0.2	5.8	nd	nd	0.8
M	2414.9	29.8	3.6	27.6	2.6	1.6	0.4	nd	0.8	4.0	0.2	0.2	nd
M	2472.4	28.2	2.2	12.6	1.2	0.6	0.8	1.0	0.2	5.2	2.4	2.4	nd

Dep.[a] Int.	Depth (m)	Kaol[o]	Chl[p]	S/I[q]	Cmm[r]	Cc[s]	Ank[r]	Dol[u] det	Dol[v] auth	Sid[w]	OM[x]	Por[y]
F	1430.5	3.2	1.8	4.2	8.4	nd	nd	nd	nd	1.0	3.8	4.0
F	1493.8	1.2	1.0	11.8	23.4	nd	nd	nd	nd	nd	1.4	0.2
C	1841.7	0.4	nd	2.6	9.4	9.2	nd	nd	8.8	1.2	nd	1.0
C	1939.1	1.2	nd	3.2	5.6	0.8	1.0	0.2	2.4	0.2	nd	5.6
P	2024.2	nd	nd	nd	nd	0.8	55.4	0.6	2.2	nd	0.6	1.6
P	2161.0	nd	nd	4.0	12.6	2.2	3.6	1.2	4.0	1.2	6.6	1.6
M	2414.9	nd	3.8	1.2	10.4	nd	2.8	0.6	4.8	nd	nd	1.4
M	2472.4	0.8	1.4	5.8	12.0	0.2	1.0	0.4	2.2	0.4	1.4	6.8

[a] depositional intervals: F = fluvial, C = coastal, P = paludal, M = marine.
[b] monocrystalline quartz.
[c] polycrystalline quartz.
[d] quartz cement.
[e] framework feldspar.
[f] clay-replaced feldspar.
[g] carbonate-replaced feldspar.
[h] dissolved feldspar.
[i] detrital biotite + muscovite.
[j] sedimentary-rock fragments (-limestone clasts).
[k] metamorphic-rock fragments.
[l] volcanic-rock fragments.
[m] limestone-rock fragments.
[n] not detected.
[o] kaolinite (including associated microporosity).
[p] chlorite.
[q] mixed-layer smectite/illite.
[r] finely intergrown clay mineral-quartz-carbonate cement.
[s] calcite cement.
[t] ankerite rims + cement.
[u] detrital dolomite.
[v] authigenic dolomite.
[w] siderite.
[x] organic matter.
[y] porosity.

formed on XRD results obtained from the <1-μm size fractions. Mixed-layer compositions were estimated as described by Hower (1981) with no correction for extent of R3 ordering. Relative abundance determinations (qualitatively designated as 'major', 'minor', 'trace', or 'absent') are based on relative peak heights.

Geochemistry.—

Selected samples were analyzed using a JEOL 2000FX electron microprobe operated at 15 kV and a reduced current of 9 nA, using a 3-μm spot and 20-sec. analysis time. Carbonate and feldspar compositions were determined using natural feldspar and pyroxene standards. Average total of the 21 carbonate analyses was 99% with a standard deviation of 2.9%. Average total for 105 feldspar analyses was 99% with a standard deviation of 1.6%. Backscatter-electron (BSE) imaging also was performed to obtain textural information associated with carbonate phases and albitization of feldspars.

RESULTS

Organic Material

The results of anhydrous pyrolysis of mudstones and interbedded sandstones and mudstones are presented in Table 2. TOC and HI values for the sample sets are shown graphically in Figures 5 and 6, along with data for the MWX site from other sources. TOC values range between 0.3 and 3% for most of the mudstone samples, except for those from the coastal and paludal intervals, which have values up to 11.8%. The one coal sampled for this study (from the paludal interval) has a TOC content of 43.8% and was omitted from Figure 5. Numerous coals are present in the paludal interval and have TOC values up to 68.1% (Barker, 1989b). TOC values for interbedded mudstone and sandstone samples of the fluvial interval range from 0.44 to 6.05%, with one outlier of 13.6%.

Fluvial-interval mudstones have low (<300), but variable, HI values (Fig. 6). HI values progressively decrease with depth from the fluvial to the marine interval. The overall low values are interpreted to result primarily from two factors. The first is the mature to overmature nature of the organic material at the MWX site (Barker, 1989b), as an increase in maturation generally causes a decrease in HI relative to that for immature kerogen, regardless of kerogen type (Hunt, 1979; Waples, 1985; Crossey and others, 1986a). Secondly, the abundance of coals in the paludal interval indicates a large terrestrial component (type III kerogen), which has characteristically low HI values regardless of level of maturation (Hunt, 1979; Waples, 1985; Crossey and others, 1986a).

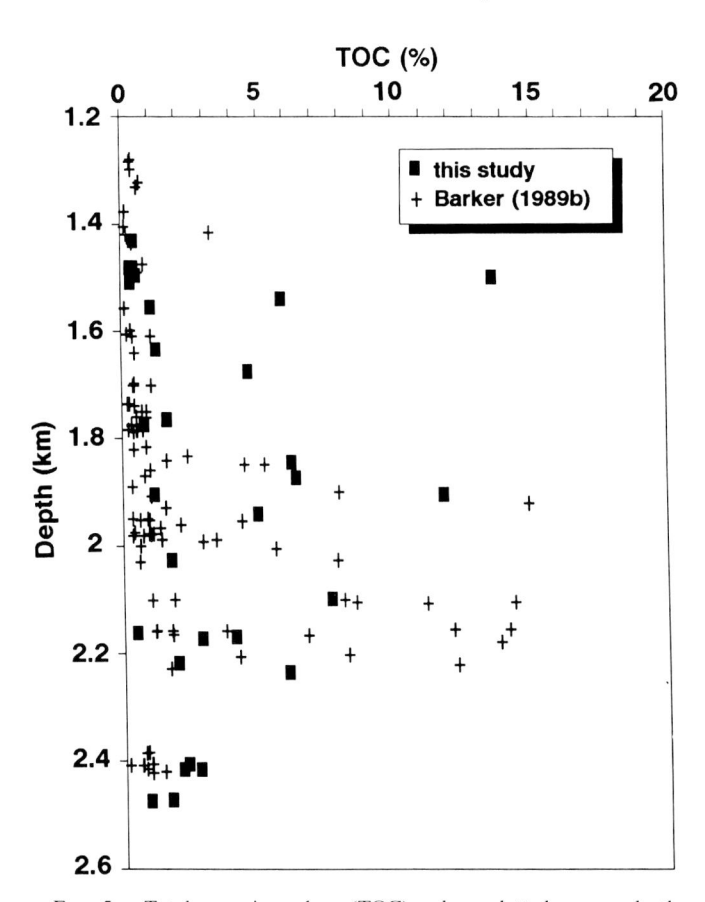

FIG. 5.—Total organic carbon (TOC) values plotted versus depth. Crosses designate data from coals, sandstones, and mudstones presented by Barker (1989b). Squares designate analyses of mudstone and interbedded sandstone and mudstone (type B) samples from this study. One coal sample (TOC = 44%) was excluded from the plot.

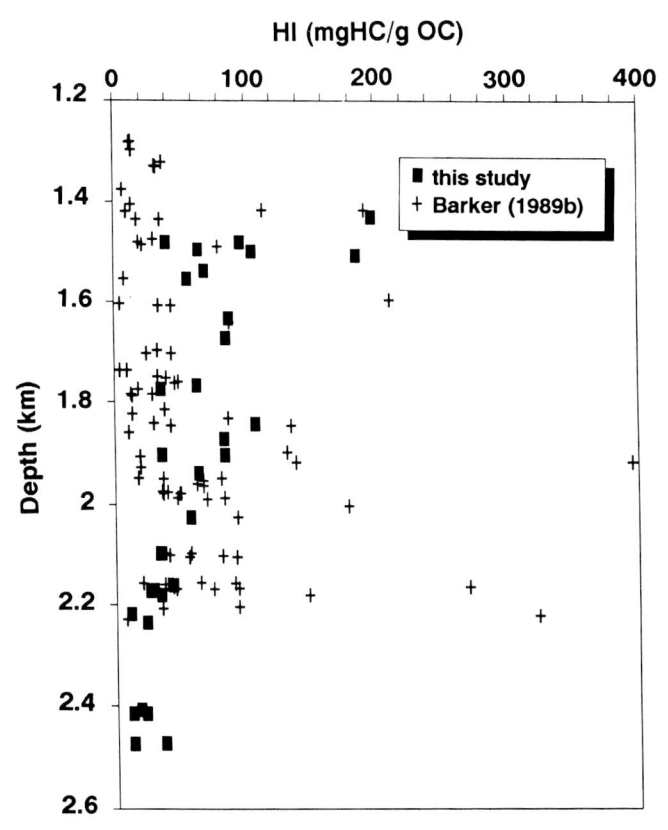

FIG. 6.—Hydrogen index (HI) values plotted versus depth. Crosses designate data from coals, sandstones, and mudstones presented by Barker (1989b). Squares designate analyses of mudstone and interbedded sandstone and mudstone (type B) samples from this study.

The general trend of decreasing HI with increasing depth observed in this study is not apparent in results of previous studies (Fig. 6). In fact, the HI values from the pre-existing data base appear to remain constant or increase slightly with depth. This may be a result of matrix effects induced by analysis of whole-rock samples with variable TOC and lithology (Fig. 5; see Crossey and others, 1986a), as the data presented by Barker (1989b) include sandstones, mudstones, and coals with almost one-third of the samples containing <0.5% TOC. Crossey and others (1986a) have demonstrated that type III kerogens are most susceptible to mineral-matrix effects during pyrolysis. Only mudstones and one coal were analyzed for this study. However, the goal of sampling for this study was not to obtain representative samples of organic material within the MVG; rather to obtain sandstones and adjacent organic-rich mudstones. Thus, any conclusions regarding general downhole trends in organic-matter characteristics are not justified from data obtained from this study. An additional possibility for the variable HI values reported by Barker (1989b) is the mixing of types II and III organic materials in the coastal and paludal intervals; but again, all HI values observed in both studies are fairly low (classified as type III kerogen; Tissot and Welte, 1978; Crossey and others, 1986a).

X-ray Diffraction

The results of X-ray diffraction analysis of the <1-µm size fraction for all samples are given in Table 2. The clays are interpreted to represent a mixture of detrital and authigenic minerals. The clay mineralogy is dominated by four clay species: R1- and R3-ordered smectite/illite mixed-layer clay (S/I ML); Fe-rich chlorite; kaolinite; and illite. An R1-ordered chlorite/smectite mixed-layer clay (corrensite) was noted in two adjacent sandstones of the coastal interval. Quartz and feldspar were occasionally detected in the <1-µm size fraction. Clay-mineral distributions are plotted by lithology and depositional environment in Figure 7.

S/I ML clay is ubiquitous throughout the sampled intervals, although chlorite is more abundant in many of the fluvial-sandstone samples (Fig. 8A). In general, a mixture (either physical or intergrowth) of R1- and R3-ordered S/I ML clays is present in the fluvial through paludal intervals, although a progressive decrease in R1 ordering occurs with increasing depth (Table 2; Fig. 8). Only R3 ordering is exhibited by S/I ML clays in the marine interval. Interstratified illite content ranges from 82 to 90%, with a decrease in the occurrence of values less than 88% with increasing depth.

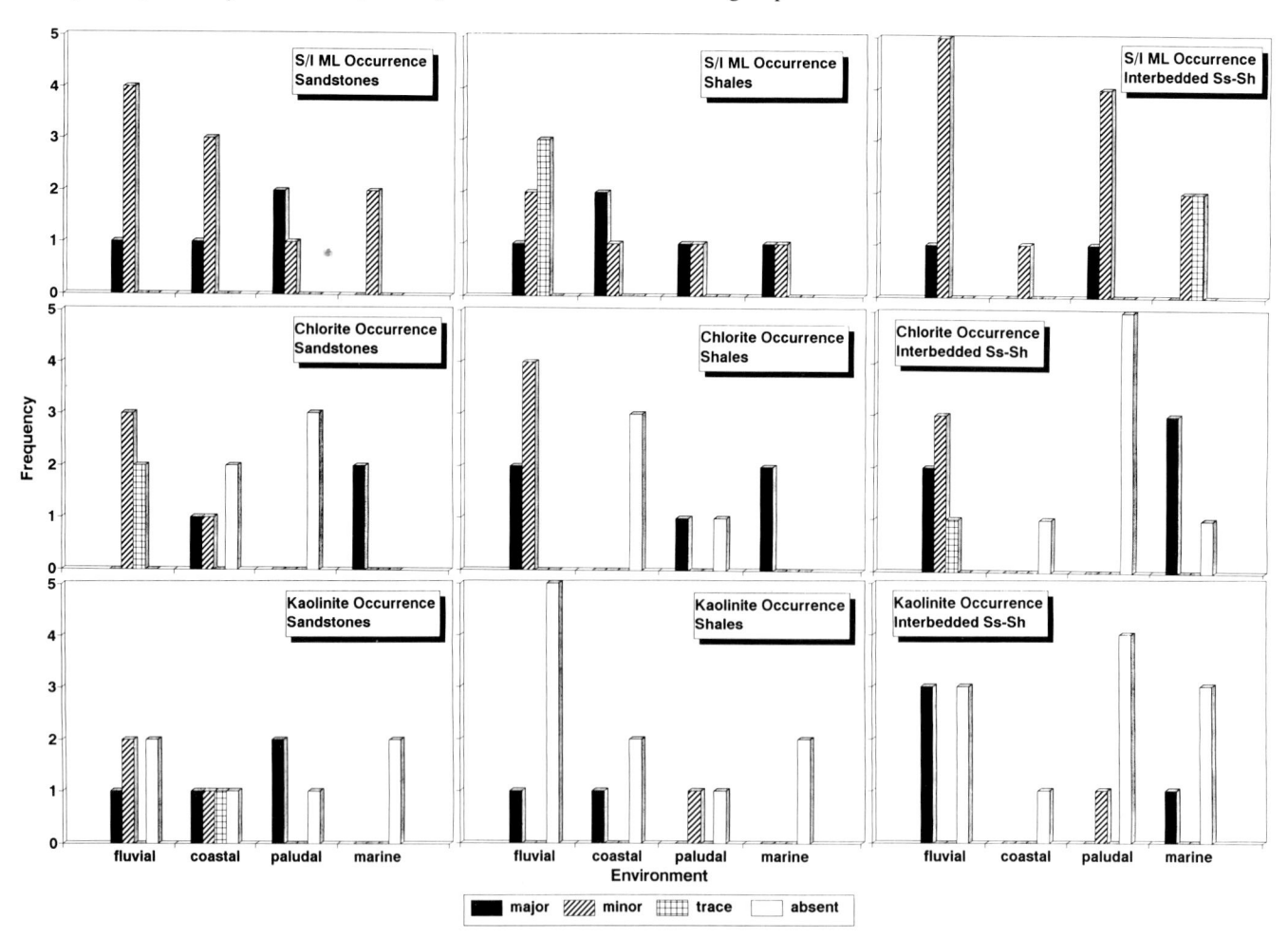

FIG. 7.—Histograms of clay results grouped by clay mineral. Frequency refers to number of samples characterized by clay abundance indicated. Environmental designations interpreted by Lorenz (1989).

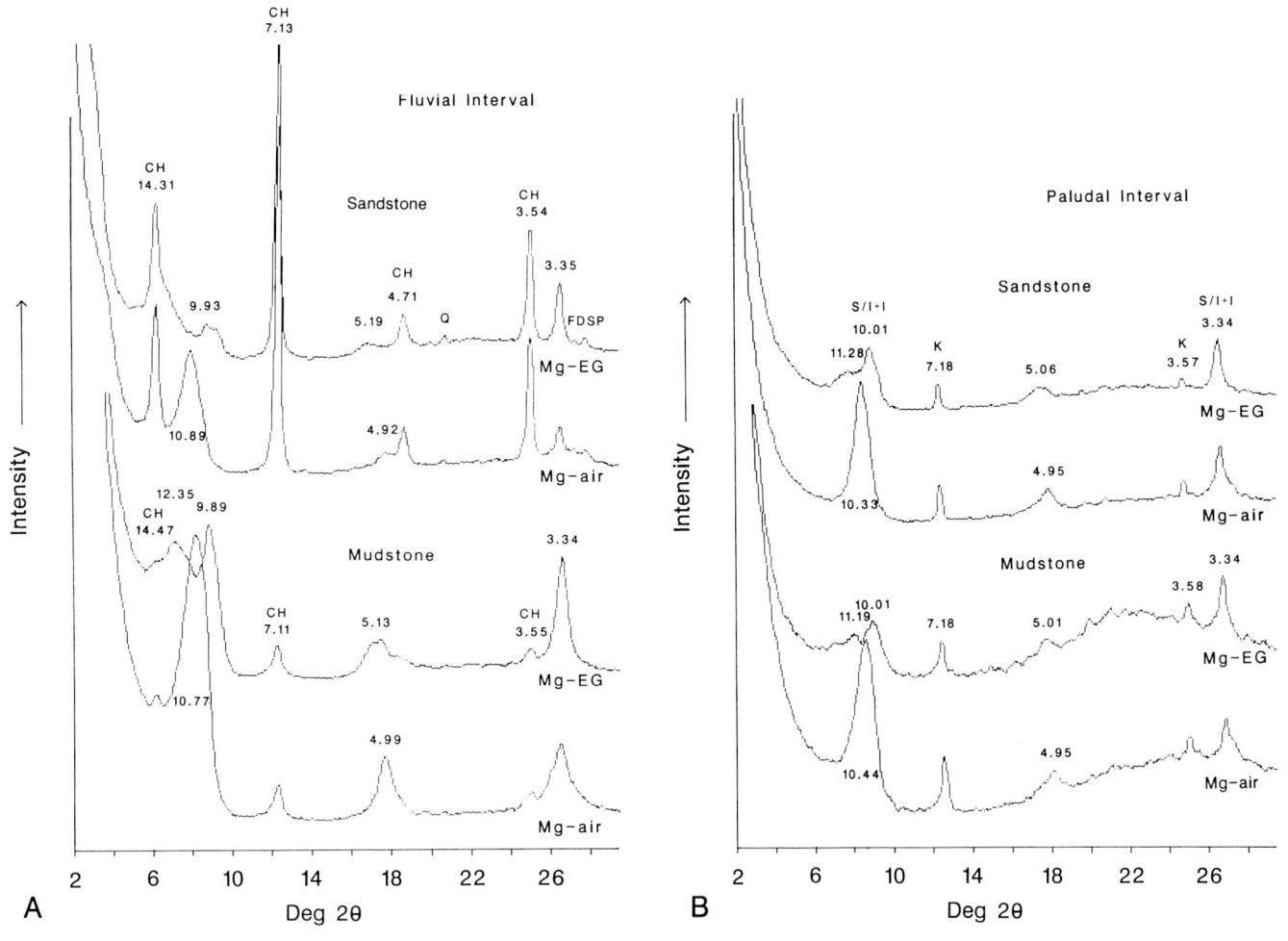

FIG. 8.—Representative XRD patterns of the <1-μm size fractions of closely adjacent sandstone-mudstone (type A) sample sets from the fluvial (A) and paludal (B) intervals. All d spacings are given in angstroms. CH = chlorite, K = kaolinite, S/I + I = composite peak for ML S/I and illite, Q = quartz, FDSP = feldspar. Peaks labeled with d spacing alone are ML S/I. Mg-air indicates samples saturated with magnesium chloride and air dried; Mg-EG indicates samples saturated with magnesium chloride and solvated with ethylene glycol.

Chlorite is abundant in both sandstones and shales of the fluvial interval, common in the marine interval, and generally absent in the coastal and paludal intervals. The reduced intensity of odd-ordered (00l) peaks relative to even-ordered peaks and the approximately 2:1 ratio of the (001) and (003) peaks observed in Figure 8A suggest that the chlorite has a high-iron content concentrated mostly in the octahedral sheet (Moore and Reynolds, 1989). The XRD characteristics of MWX chlorites appear to be similar to other sedimentary grain-coating chlorites (Curtis and others, 1985). Kaolinite occurs sporadically in the fluvial through paludal intervals (Fig. 7), and is the most abundant clay mineral in several sandstones from the fluvial, coastal, and paludal intervals. It is most commonly absent in shales. Illite occurs in a few samples in all of the depositional intervals. It is identified on the basis of a sharp 10Å (001) peak resolvable within the broad S/I ML 10Å band, and by a sharp peak at 3.3Å (003; Fig. 8B).

Several observations are noted regarding the distribution of clay minerals in the <1-μm size fraction in sandstones, mudstones, and interbedded samples as indicated by X-ray diffraction data: (1) coastal and paludal mudstones are dominated by S/I ML clays, although chlorite is abundant in marine mudstones; (2) S/I ML clays also are present in sandstones from all environments, but other clay-mineral species also are important. The mixed-layer clay consists of detrital material (based on petrographic textures) and probably an authigenic component as well. Kaolinite is an additional component observed in all but the marine sandstones (authigenic, based on petrographic textures), whereas chlorite is present mainly in fluvial and marine sandstones, and interbedded type B samples (authigenic, based on petrographic textures); (3) kaolinite is consistently absent in shales from all depositional environments. The interbedded samples (type C) are generally similar in their clay mineral distribution to the shale patterns, with the exception of notable kaolinite in several samples; and (4) variations in clay mineralogy do not appear to correlate with variations in the organic parameters.

Sandstone Petrography

Framework-grain compositions were determined for two samples within each depositional environment. These results are similar to those of other workers (Pitman and others, 1989). Sandstone compositions cover a broad range and vary with depositional environment (Fig. 3). Sandstones of the marine interval are relatively quartz rich, those of the coastal and paludal intervals are predominantly litharenites or feldspathic litharenites, and those of the fluvial interval cover the broadest range and contain the most feldspar. Point-count data are presented in Tables 3A and 3B.

All 30 sandstone samples were examined for characterization of paragenetic sequence and replacement textures. The earliest diagenetic textures observed are associated with the tight packing of grains resulting from early compaction. Some labile lithic and mineral grains show deformation and pseudomatrix pore-filling characteristics as a result of pressure from neighboring grains.

Carbonate-cementation events are mineralogically and texturally distinguishable. The earliest type is a pore-filling calcite and the later carbonates are pervasive ankerites and dolomites. These are discussed in more detail later. Grain-rimming chlorite and S/I ML, and quartz overgrowths and cements are abundant in most samples (Fig. 9A). Albite overgrowths on feldspars are rare in fluvial and marine sandstones, and absent in coastal and paludal sandstones. Chlorite rim growth is generally followed by quartz cementation in fluvial sandstones, but multiple generations of chlorite and quartz may be present locally. In contrast, chlorite commonly follows quartz cementation in the marine interval. In coastal and paludal sandstones grain-rimming chlorite is rare, but partial rims of S/I ML clay are present locally. The calcite cement occurs in many sandstones from the fluvial interval and commonly fills pore space following chlorite and quartz cementation (Fig. 9B). Quartz and calcite cements between grains appear to have retarded compaction.

Where present, intergranular pore-filling clays of the coastal, paludal, and marine intervals include S/I ML clays with intergrown authigenic quartz (Fig. 9C). These mixed-layer clays are interpreted to represent a mixture of altered/transformed detrital clay (as evidenced by fine-grained laminations in many samples; no clay coats on grains or meniscus-clay cements were noted). In the fluvial interval, chlorite and kaolinite are present (textures indicative of authigenic origin). The S/I ML clays are difficult to distinguish from pseudomatrix resulting from deformation of sedimentary lithic fragments. C/S ML clay (corrensite; identified on the basis of XRD analysis) was noted only in two adjacent sandstones of the coastal interval, and was not positively identified petrographically.

Altered, dissolved, and replaced feldspar grains are ubiquitous (Fig. 9D) and are usually rimmed by clay-mineral and quartz cements or authigenic albite. Feldspars in all sandstones are commonly altered to illite or S/I ML clay, especially those in fine-grained or organic-rich sandstones. Feldspar dissolution and its replacement with calcite (fluvial and coastal intervals) or dolomite and ankerite (in the deeply buried intervals) are more common in medium-grained sandstones. Both the dissolution of plagioclase and alkali feldspars and their replacement with albite are evident by

TABLE 3.—BURIAL-MODEL RESULTS

Layer[a]	Present Depth (m)	Max.[b] Depth (m)	Max.[c] T(C)	Final[d] TTI	EVR[e]	TR[f]	Ro[g] (depth)
Top of Wasatch	2	1639	78.9	3	0.50	-[h]	0.6 (surf.)
124 m above Wasatch base	1058	2695	117.0	23	0.72	-	1.06 (1056)
Base of Wasatch	1182	2819	120.9	29	0.75	-	1.06 (1158)
Base of Mixed Marine-Nonmarine Interval	1333	2970	127.6	42	0.86	-	0.88 (1341)
Base of Fluvial Interval	1818	3455	150.6	164	1.30	0.68	1.39 (1828)
Base of Coastal Interval	2000	3637	159.2	277	1.45	0.72	1.43 (2000)
Base of Paludal Interval	2259	3896	173.8	687	1.85	0.75	1.80 (2250)

[a]layer used for thermal-maturation modeling.
[b]maximum depth as determined from burial history described in text (attained between 36 and 24 Ma for all layers).
[c]maximum temperature determined from burial and thermal model as described in text (Model C; at 12 Ma for all layers).
[d]final TTI as determined from maturation model described in text (Model C).
[e]estimated vitrinite reflectance obtained by conversion from TTI (Waples, 1985).
[f]transformation ratio as determined from kinetic-maturation model as described in text (Model C).
[g]measured vitrinite reflectance at closest interval to model layer (depth of measured value in m).
[h]not calculated.

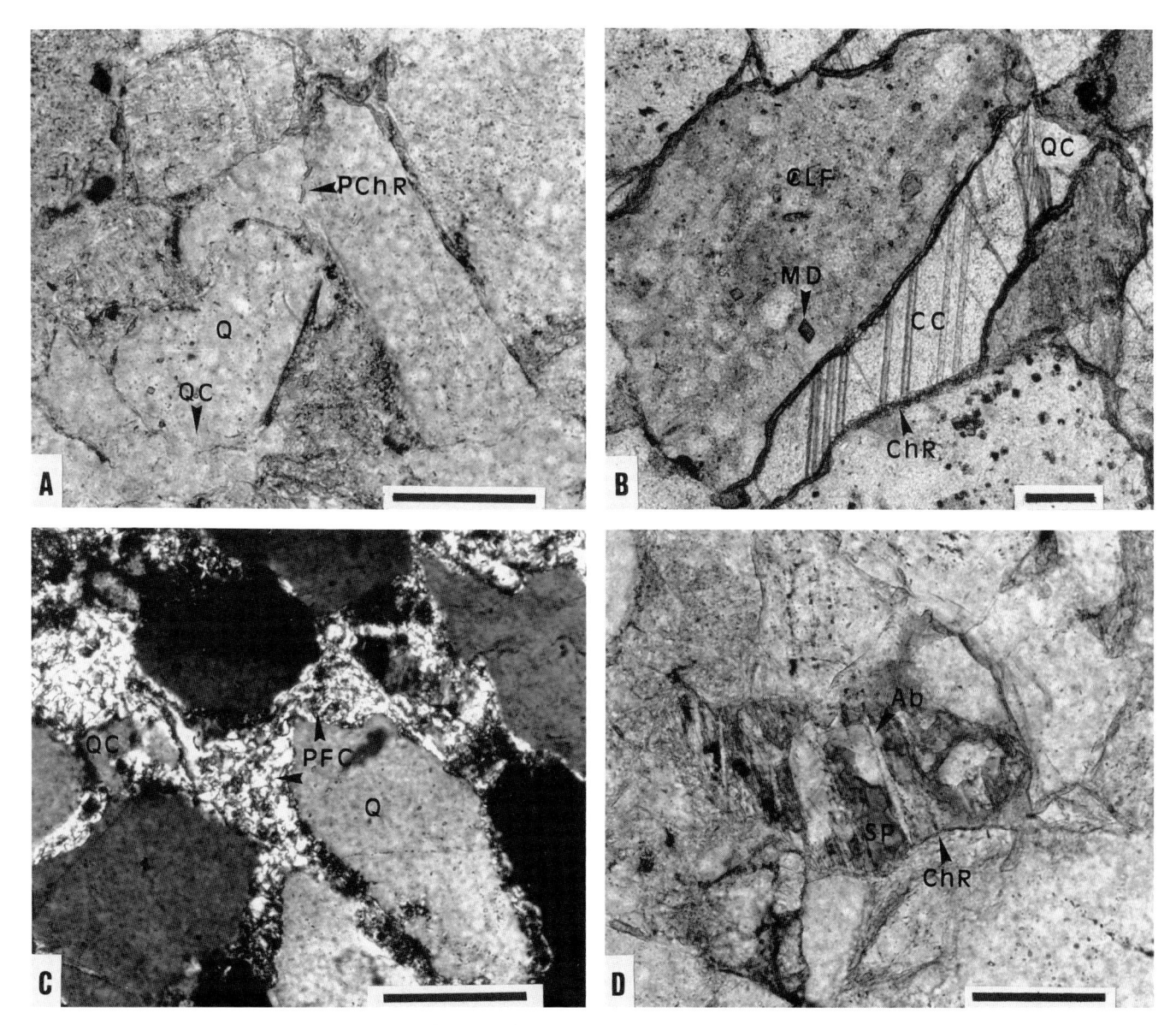

FIG. 9.—Fluvial-sandstone photomicrographs (scale bars = 0.1 mm). (A) A fine- to medium-grained sandstone. Partial chlorite rims (PChR) are present on most quartz (Q) grains, with quartz cement (QC) filling intergranular regions. (B) A medium-grained sandstone. Complete chlorite rims (ChR) on cherty lithic fragments (CLF). Note that pore in center of photo is completely rimmed by chlorite, and filled with calcite cement (CC) and minor quartz cement (QC). Microdolomite inclusions (MD) occur in most of the cherty lithic fragments. (C) A fine-grained sandstone. Pore-filling clay minerals (PFC) (probably a mixture of ML S/I, chlorite, and kaolinite) and quartz cement (QC) occlude all porosity between quartz grains (Q). (D) A fine- to medium-grained sandstone. Partially dissolved plagioclase grain with albitized cleavage remnants (Ab) and secondary porosity (dark − SP) rimmed by chlorite (ChR).

extensive crystallographically controlled inclusions and microporosity. Petrographic observations are verified by microprobe analyses (discussed later) and backscattered electron images (BSE) of andesine and oligoclase replaced by albite; chlorite and S/I ML clay intragranular pore fill; and microporosity (Fig. 10A). Dissolution of plagioclase and partial replacement by clays, carbonates, or albite to produce secondary intragranular porosity occurs in many sandstones, especially in high-porosity sandstones of the more deeply buried intervals.

Dolomite rhombs and anhedral cements are common in sandstones from the coastal through marine intervals (Fig. 11A) and are locally present in sandstones at the base of the fluvial interval. Detrital-dolomite rhombs also are present and are distinguished by subrounded and sorted grains, abundant inclusions, and extensive twinning and recrystallization due to compaction. Partial dolomitization of early calcite cements is observed in the lower part of the fluvial and coastal intervals.

Dolomite precipitation was followed by extensive carbonate and aluminosilicate dissolution to produce oversized pores; ragged, dolomite-grain and cement margins; and oversized pores partially filled with clay minerals and feldspar remnants. These textures are best observed in medium-

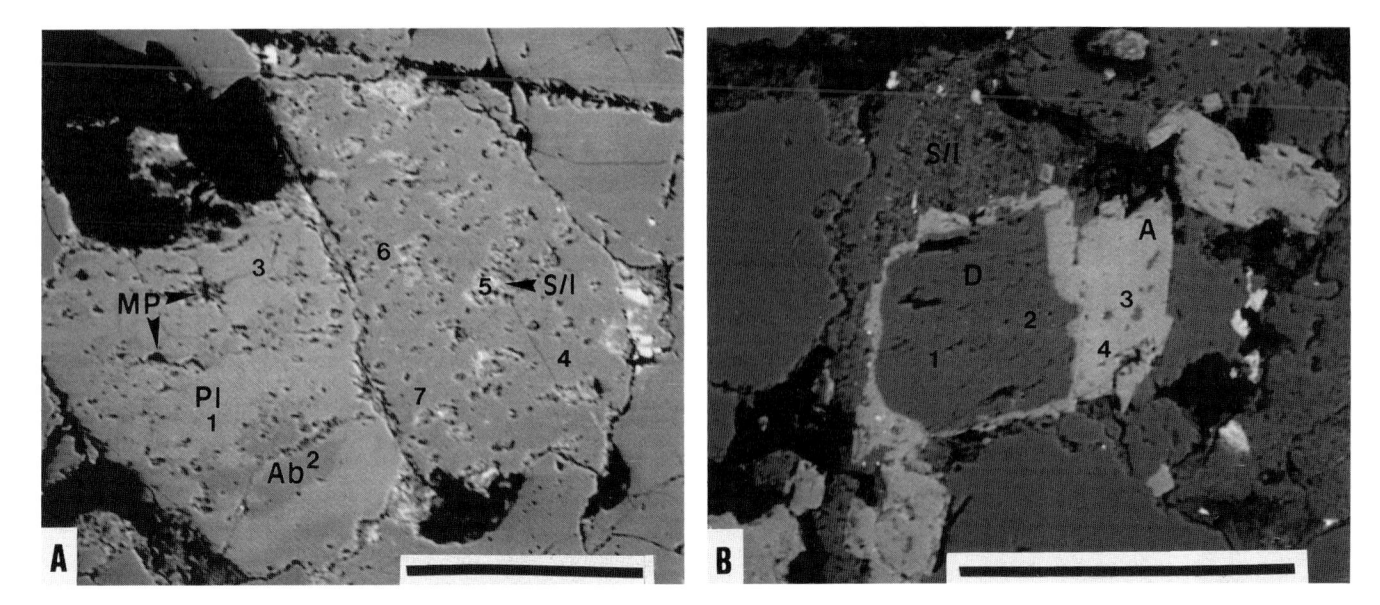

Fig. 10.—Backscattered electron images. (A) Partially and completely albitized feldspar grains in a fine- to medium-grained sandstone from the fluvial interval (scale bar = 0.1 mm). Light gray areas are plagioclase (Pl), slightly darker gray areas are albite (Ab), bright areas are ML S/I (S/I), and black areas within grains are microporosity (MP). Point chemical analyses are as follows: 1. $(Ca_{0.14}Na_{0.76})Al_{1.16}Si_{2.87}O_8$ — andesine; 2. $Na_{0.96}Al_{1.01}Si_{3.00}O_8$ — albite; 3. $(Ca_{0.14}Na_{0.82}K_{0.01})Al_{1.15}Si_{2.85}O_8$ — andesine; 4. $(Ca_{0.02}Na_{0.81}K_{0.02})Al_{1.10}Si_{2.96}O_8$ — albite; 5. $(K_{1.53}Na_{0.07})(Ca_{0.01}Fe_{0.08}Mg_{0.09}Al_{3.32})(Si_{6.06}Al_{1.93})O_{20}(OH)_2$ — S/I ML; 6. $(Ca_{0.02}Na_{0.83}K_{0.01})Al_{1.06}Si_{2.98}O_8$ — albite; 7. $(Ca_{0.03}Na_{0.86}K_{0.01})Al_{1.05}Si_{2.98}O_8$ — albite. (B) Diagenetic dolomite with ankerite rim and cement in a fine-grained sandstone from the paludal interval (scale bar = 0.1 mm). Dark gray dolomite (D) is surrounded by a thin, light gray ankerite rim and patchy ankerite cement (A). ML S/I (S/I) fills pore space adjacent to carbonate phases. Point chemical analyses are as follows: 1. $(Ca_{1.12}Mg_{0.88})(CO_3)_2$ — dolomite; 2. $(Ca_{1.11}Mg_{0.88}Fe_{0.01})(CO_3)_2$ — dolomite; 3. $(Ca_{1.17}Mg_{0.37}Fe_{0.46})(CO_3)_2$ — ankerite; 4. $(Ca_{1.17}Mg_{0.37}Fe_{0.46})(CO_3)_2$ — ankerite.

grained sandstones and sandstones lacking abundant late carbonate cement. Secondary intergranular and feldspar intragranular porosity in coastal and paludal sandstones is commonly occluded by coarsely crystalline (up to 5 μm), vermicular kaolinite (Fig. 11B).

Ankerite rims on dolomite and ankerite cement occur in some sandstones in the coastal through marine intervals and fill porosity almost completely in sandstones at the top of the paludal interval (Fig. 11A). Ankerite cement also occurs in a sandstone lens between two organic-rich laminae in an interbedded-sandstone sample from the fluvial interval. Ankerite rims and cement, as observed with back-

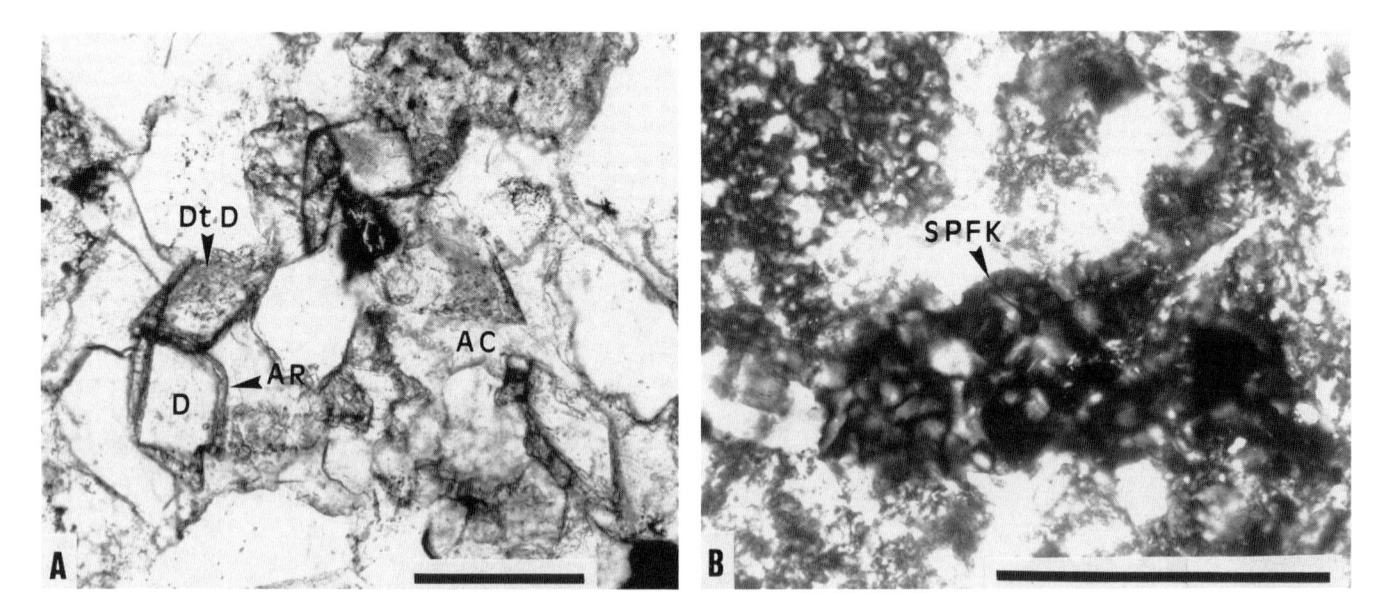

Fig. 11.—Photomicrographs of sandstones from the paludal interval (scale bars = 0.1 mm). (A) A fine-grained sandstone: diagenetic dolomite (D) and inclusion-rich detrital dolomite (Dt D) are rimmed by gray ankerite (AR). Light gray ankerite cement (AC) fills in the intergranular spaces. (B) A fine- to medium-grained sandstone; secondary pore-filling kaolinite (SPFK) occludes porosity within ML S/I pore fill.

scattered-electron images (Fig. 10B), appear to replace the margins of, and surround, dolomitic precursors

As is evident from sandstones of the fluvial and paludal intervals, bitumen migrated into porosity remaining after partial pore filling by dolomite and ankerite, and/or generated by renewed dissolution of carbonates (Fig. 12). Illite partially fills secondary porosity in all of the intervals. Intergranular and secondary pore-filling pyrite and siderite occur sparsely in sandstones throughout the MVG, with pyrite commonly associated with organic matter.

Siltstone, claystone, chert, and silicic volcanic lithic fragments locally are partially dissolved, replaced by S/I ML and chlorite clays, and, in sandstones from the coastal through marine intervals, partially replaced by dolomite rhombs (Fig. 9B). Detrital biotite, muscovite, and chlorite are generally deformed and locally are partially replaced by S/I ML and chlorite clays. Granitic lithics show sericitization, and carbonate and albite replacement of feldspars.

Some general trends in cementation and porosity, based on petrographic examination of 30 samples, are noted within each interval. In the fluvial interval, the abundance of carbonate replacement and cement correlates well with proximal organic material and extensive feldspar alteration and replacement. Fine-grained sandstones of the fluvial interval tend to exhibit low porosity due to abundant clay-mineral matrix (Fig. 9C), whereas medium-grained sandstones have higher porosities and are dominated by authigenic chlorite, quartz, and calcite (Fig. 9B). However, some medium-grained fluvial sandstones have low porosity due to partial pore filling by illite.

In the paludal and coastal intervals, sandstone porosity is low in ankerite- and dolomite-cemented sandstones near the top of each interval, but increases with depth in each interval concurrent with decreases in the degree of carbonate cementation. Qualitatively, the abundance of organic-rich material in mud rocks is correlated with dolomite and ankerite cement in sandstones for sample sets of the coastal and paludal intervals. Factors such as degree of bioturbation also influence the abundance of late carbonate cements.

In sandstones of the marine interval, porosity increases with increasing grain size and decreases with the degree of secondary pore filling by illite. Late carbonate phases are dominated by dolomite with ankerite rims. The extent of cementation shows no apparent relation with the quantity of organic material in adjacent mudstones.

Geochemistry

Compositional data obtained from carbonates and feldspars are presented graphically in Figure 13. Carbonates are classified as dolomite and ankerite according to compositional results (Fig. 13A). Representative point chemical analyses (Fig. 10B) indicate Ca-rich compositions for both ankerite and dolomite, and a low-iron content in the dolomites. Feldspar-grain compositions cluster toward the albite end member (a large number of analyses are >90% Ab), with no plagioclase compositions below 75% Ab observed (Fig. 13B). Representative point chemical analyses (Fig. 10A) indicate excess aluminum in most of the albitized feldspars. Similar depletion of Na^+, K^+, and Ca^{2+} relative to Al^{3+} in microporous diagenetic albite has been reported by Milliken and others (1989). The S/I ML analysis indicates significant Mg and Fe content.

DISCUSSION

Burial and Uplift Model

Prior to the presentation of an overall model for diagenesis of the MVG, a thermal and burial history of the MWX site will be presented. The burial scenario described by Nuccio and Johnson (1989) and Johnson and Nuccio (1986), with stratigraphic modification based on the analyses of Lorenz (1985) and Barker (1989a), is used as the basis for interpreting the diagenetic sequence of the MVG (Fig. 14). Numerous heat-flow conditions, thermal-conductivity variations, and overburden reconstructions have been examined. Models were constrained by attainment of reasonable fits to the present geothermal profile (Blackwell and Steele, 1988) and to measured vitrinite profiles (Barker, 1989b). The significant stratigraphic and thermal configurations used in the model will be described first, followed by a discussion of the organic-maturation model results.

Regional deposition of the MVG is assumed to have occurred at a relatively uniform rate from 72 to 67 Ma. The post-Cretaceous unconformity is modeled essentially as a hiatus from 67 to 60 Ma. From 60 to 51 Ma, the present stratigraphic thickness of Tertiary Wasatch Formation accumulated in the Piceance Basin, followed by deposition of the Green River and Uinta Formations from 51 to 36 Ma. The youngest rocks in the Piceance Basin today are basalts deposited approximately 10 Ma (Marvin and others, 1966), associated with regional volcanism occurring from 12 to 10 Ma (Larsen and others, 1975). The time of maximum burial occurred between 36 and 24 Ma. Uplift of the sequence is interpreted to have occurred in several stages.

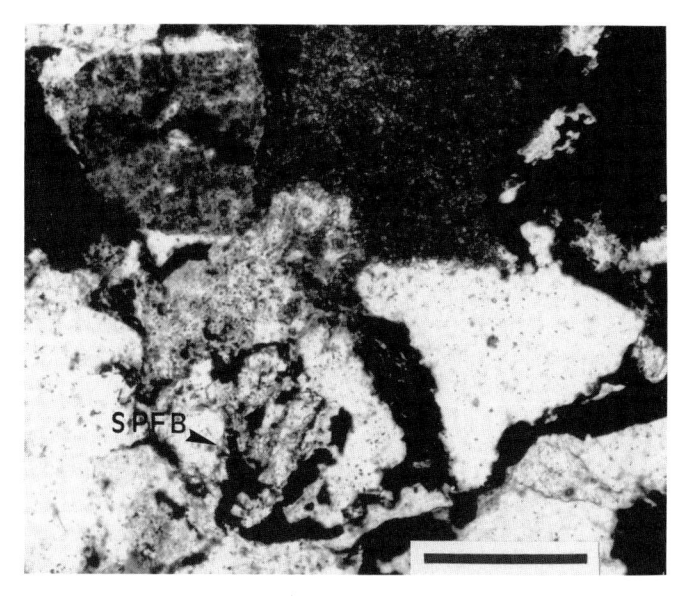

FIG. 12.—Photomicrograph of a type B sample (cm-scale interbedded, fine-grained sandstone and organic-rich mudstone) from the marine interval (scale bar = 0.1 mm). Secondary pore-filling bitumen (SPFB) occluding porosity in partially dissolved feldspar.

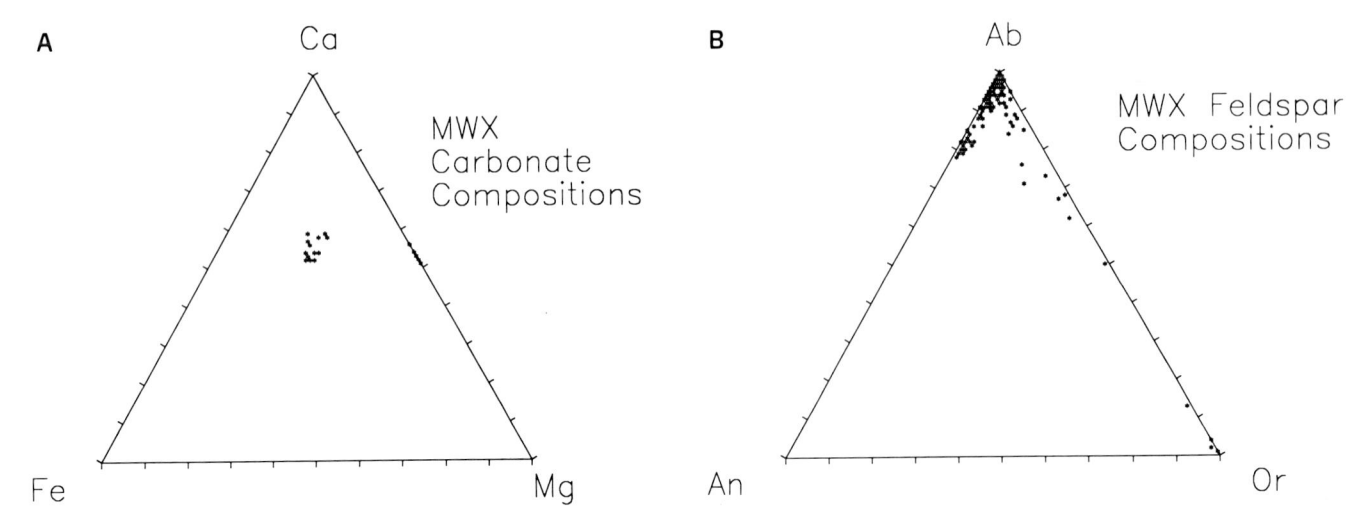

FIG. 13.—(A) Ternary plot of microprobe analyses of carbonates from the coastal interval. Analyses of ankerite rims and cements cluster in the center of the carbonate cation triangle, whereas the dolomite compositions lie along the Ca-Mg line. (B) Ternary plot of microprobe analyses of feldspars (fluvial and coastal intervals). Plagioclase compositions range from An_{22} to An_0, with albite compositions being by far the most common. Alkali feldspar compositions range from nearly pure orthoclase to nearly pure albite, with a higher proportion of analyses toward albite compositions.

Slow rates of uplift occurred from 24 to 10 Ma, followed by a rapid pulse at 10 to 8 Ma, a hiatus from 8 to 2 Ma, and a final small uplift from 2 Ma to the present (Larsen and others, 1975). The pulsed nature of Cenozoic uplift used in this study differs from the model of Nuccio and Johnson (1989) in that the time spent at maximum burial depth is shortened. Because units stratigraphically above the Wasatch Formation have been removed by erosion at the MWX site, regional-isopach reconstructions used by Johnson and Nuccio (1986) were applied to estimate formation thicknesses at the MWX site (including an additional 550 m of Wasatch Formation).

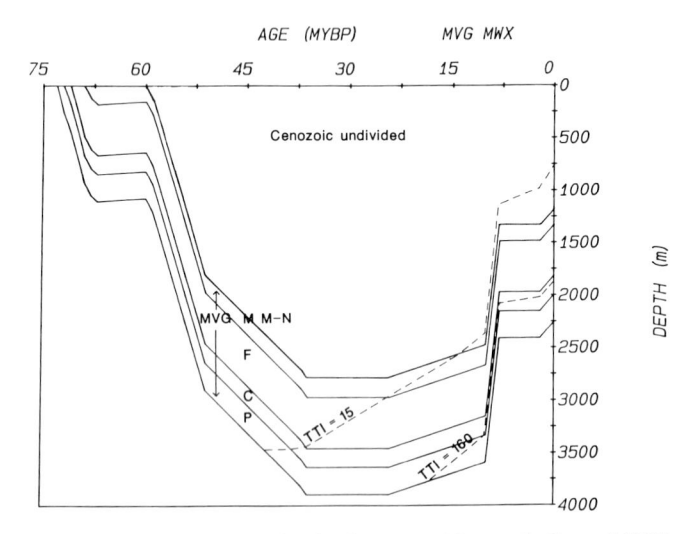

FIG. 14.— Burial history for the Cretaceous Mesaverde Group (MVG) (excluding the basal marine interval) and Tertiary strata at the MWX site. Model C (heat pulse initiated 13 Ma) thermal parameters were used to generate the organic-maturation parameter (TTI). The region between the TTI contours is interpreted to represent the liquid-hydrocarbon window (Waples, 1985). The abbreviations for depositional intervals are as follows: P, paludal; C, coastal; F, fluvial; M M-N, mixed marine and nonmarine.

Heat Flow and Thermal Conductivity

The computer program used for generation of time-temperature profiles and organic-maturation estimates (TTINDEX; written by the senior author) assumes constant stratigraphic thickness of units and allows for variable heat flow and thermal conductivity of units throughout the burial history. This method differs significantly from the downhole-adjusted thermal-gradient approach of Nuccio and Johnson (1989); it provides a means of evaluating complex burial and thermal histories using geologic input (thermal conductivity of lithologic units and heat-flow values) and allows reasonable extrapolations of temperature for lithologic units throughout the stratigraphic section. A modern heat-flow value of 90 mWm^{-2} is obtained using thermal profiles for the MWX site and average thermal conductivities based on lithology (both from Blackwell and Steele (1988); thermal conductivities assumed to be constant throughout the history of the unit). The heat-flow value is high compared to the available published values for the closest site (Rifle, CO; 58 and 62 mWm^{-2}; Decker and others, 1988), but has been confirmed for the MWX site based on thermal well logs and thermal-conductivity measurements (Hagedorn, 1985; Blackwell, pers. commun., 1990). Local heat-flow variations on this order have been observed in other Laramide basins (San Juan Basin, for example; Decker and others, 1988). The 90-mWm^{-2} value is applied to the site from 12 Ma to the present (Model A), 14 Ma to the present (Model B), 13 Ma to the present (Model C), and 25 Ma to the present (Model D). The timing of elevated heat flow at the MWX site is interpreted to be associated with regional basaltic volcanism (known events range from 34 Ma to the present, local basalts are approximately 10 Ma; summarized in Johnson, 1989). High degrees of thermal maturation of organic matter would be noted if such an elevated heat flow had been active throughout the burial history. Thus, the interval of relatively high heat flow (extending to the present) is preceded by average crustal heat flow (60 mWm^{-2}) for the entire history of the MVG. Thermal histories for selected intervals of the MVG are shown in Figure 15.

Maturation Parameters

As noted by Nuccio and Johnson (1989), the actual effect of the assumption of a linear geothermal gradient rather than allowing the gradient to vary with depth (due to lithologic heterogeneity of the section) is, in this instance, negligible in terms of the prediction of the timing of hydrocarbon generation. The time spent at maximum burial depth and the elevated heat-flow application are critical to the degree and timing of kerogen maturation. TTI values were calculated according to Waples' (1980) modification of Lopatin's Time-Temperature Index. Additional calculations based on six parallel reaction rates describing the transformation of kerogen to petroleum (Tissot and Welte, 1978) were also performed. The Transformation Ratio (TR) calculated in this phase of the modeling is the ratio of amount of hydrocarbons generated to the generation potential of the kerogen, and has an advantage over the TTI in that different kinetic parameters may be applied for different kerogen types. Figure 16 demonstrates the model fit to the modern thermal profile, and Figure 17 correlates calculated TTI values (converted to an equivalent vitrinite-reflectance value [EVR] using the correlation presented in Waples, 1985) to measured vitrinite values presented by Law and others (1989). Sweeney and Burnham (1990) have recently proposed a kinetic method for the estimation of vitrinite-reflectance values. Their interpretation indicates that, for typical heating rates associated with burial diagenesis, Waples' (1985) correlation would underestimate vitrinite estimates obtained by their methods. An additional criterion used in evaluating the model results is agreement with the maximum temperature determinations from fluid inclusions in quartz and calcite fracture fillings (Fig. 15C; Pitman and Dickinson, 1989; Barker, 1989a). The exact timing of the fracture fillings are unknown, but are presumed to be post-HC generation on the basis of methane inclusions and were interpreted by Barker (1989a) to have formed between 36 and 9 Ma. The fluid-inclusion temperatures obtained for two stratigraphic levels (1,700 and 2,390 m; within the fluvial interval and 50 m beneath the base of the paludal interval) are indicated on the time-temperature plot for the MVG, demonstrating the differences in thermal-history models A, B, C, and D (Fig. 15B). As discussed previously, had these elevated temperatures been maintained for an extended time at maximum burial maturation, indices for modeled MVG layers would far exceed the measured values. This is shown in the results of Model D, where an EVR value of 3.0 is calculated for the base of the paludal interval, far exceeding the measured value of 1.8. The timing of the heat-flow pulse immediately prior to late Tertiary uplift (rather than at maximum burial, interpreted to occur between 36 and 24 Ma) results in reasonable fits for both vitrinite and fluid-inclusion data. Figure 14 depicts the timing of liquid-hydrocarbon generation, with TTI values of 15 and 160 contoured (interpreted as representing the liquid hydrocarbon 'window'; Waples, 1985). The timing of the onset of liquid-

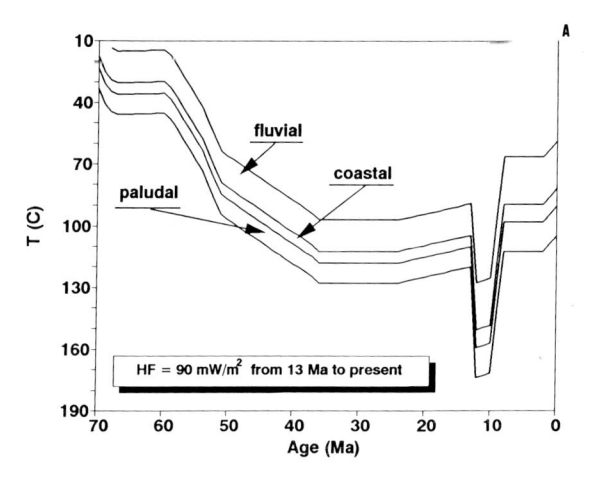

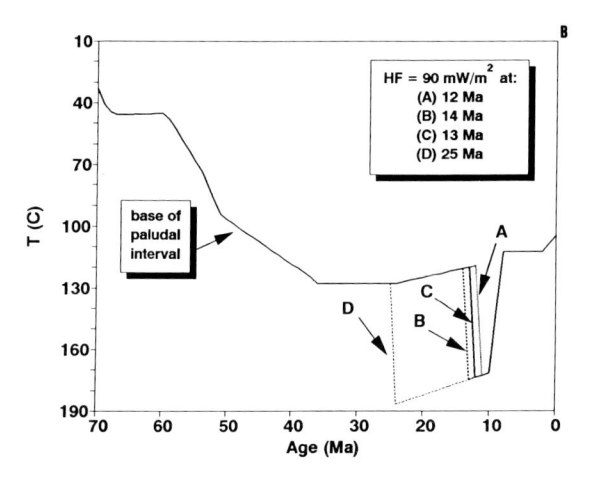

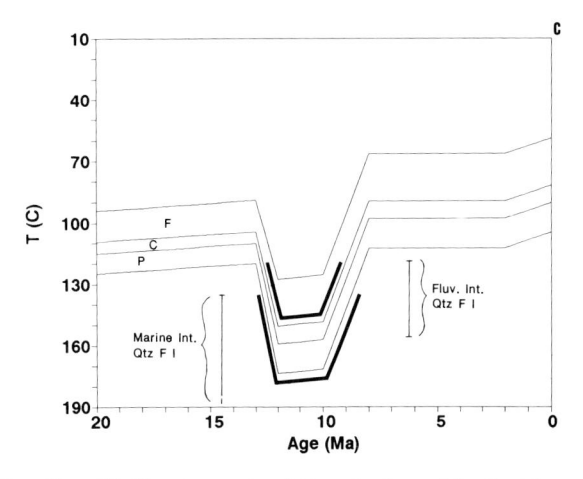

FIG. 15.—(A) Time-temperature history for base of the fluvial, coastal, and paludal intervals of the MVG; model C is shown. (B) Thermal history for the base of the paludal interval, with models A, B, C, and D representing a variation of the timing of the heat pulse (increase in heat flow from crustal average value of 60 mWm^{-2} to 90 mWm^{-2}). (C) Expanded view of model C thermal histories for selected intervals of the MVG (F = fluvial, C = coastal, P = paludal). The vertical labeled lines (F I = fluid inclusion) represent the range of temperatures obtained for quartz-fluid inclusions at the MWX site from the fluvial and marine intervals (present depths of 1,700 and 2,390 m, respectively; Barker, 1989a). The heavy black lines illustrate the time period over which these stratigraphic positions were in the measured thermal range as predicted by the model.

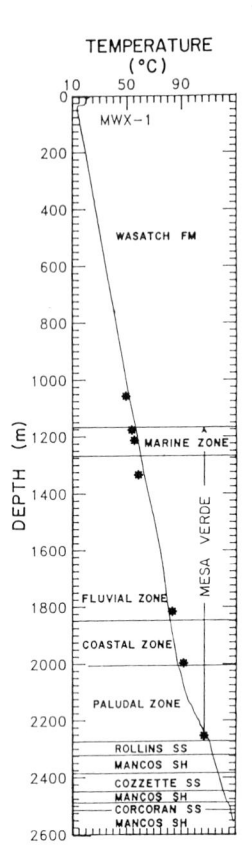

FIG. 16.—Model temperature results for selected intervals (asterisks) compared with measured modern geothermal profile (modified from Blackwell and Steele, 1988).

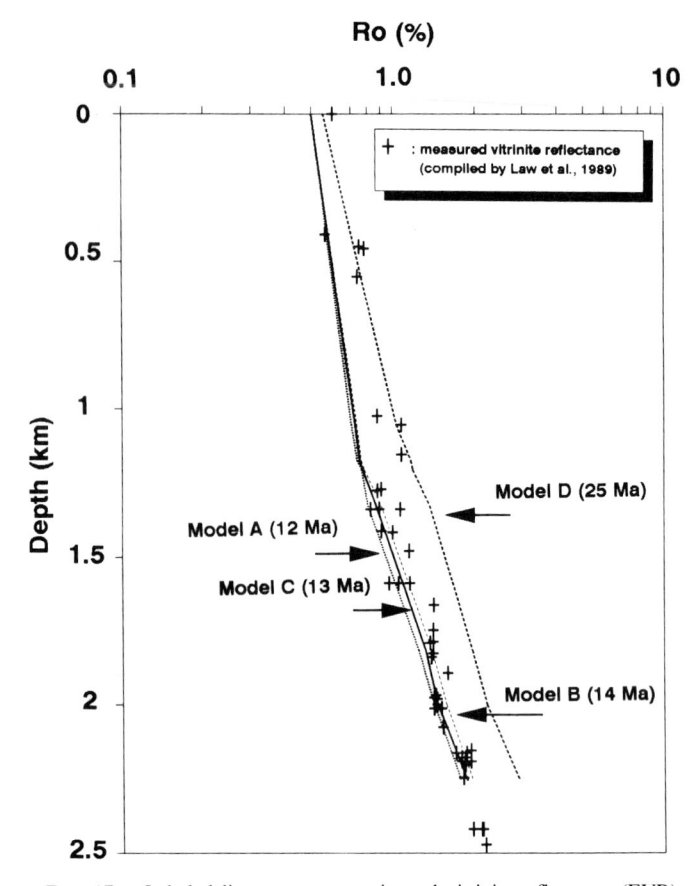

FIG. 17.—Labeled lines represent estimated vitrinite-reflectance (EVR) values calculated from the thermal and burial history for the MWX site: (EVR determined from TTI-Ro correlation of Waples, 1985). Model A, B, and C results compare favorably with measured vitrinite-reflectance values from coals for the MVG (crosses; from compilation by Law and others, 1989). Midpoints of sampling intervals are plotted for depths < 1,340 m.

hydrocarbon production determined by this procedure is used to constrain the timing of diagenetic events in the following section. Model outputs of present depth; estimated maximum burial depths and temperatures; final TTI, EVR, and TR; and measured vitrinite-reflectance values are presented for several intervals in Table 3.

Diagenetic Sequence

A diagenetic sequence can be constructed from the petrographic, SEM, and X-ray diffraction results (Fig. 18). The sequence is divided into three parts (early, late, and post-HC; see Fig. 19 for integration with burial history) based on interpretation of diagenetic events that are thought to be closely related. Note that the actual timing of these divisions differ for each unit, and are related to the degree of thermal exposure. Also, some diagenetic features (e.g., feldspar reactions, smectite-to-illite transformation) are not discrete events but occurred throughout much of the documented diagenetic history.

Early diagenetic events are associated with lithification, and involve cementation and recrystallization or transformation reactions involving unstable detrital components. These events are interpreted to have occurred from time of burial through early Cenozoic burial (timing varies, depending upon position in the MVG section; see Fig. 19). During this period, clay-mineral, quartz, and calcite cements precipitated from mildly alkaline solutions at tem-

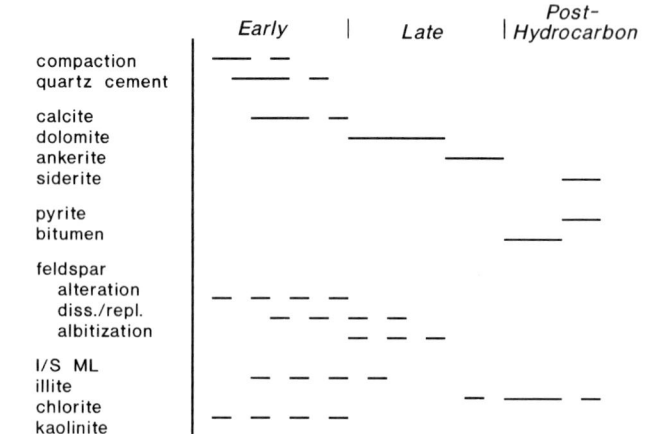

FIG. 18.—Generalized diagenetic sequence for the MVG at the MWX site as interpreted from petrographic analysis. Early, late, and post-hydrocarbon refer to stages of diagenesis discussed in the text. Dissolution of phases other than feldspar are not shown, but is discussed in the text, as is the smectite-to-illite transformation and hydrocarbon generation.

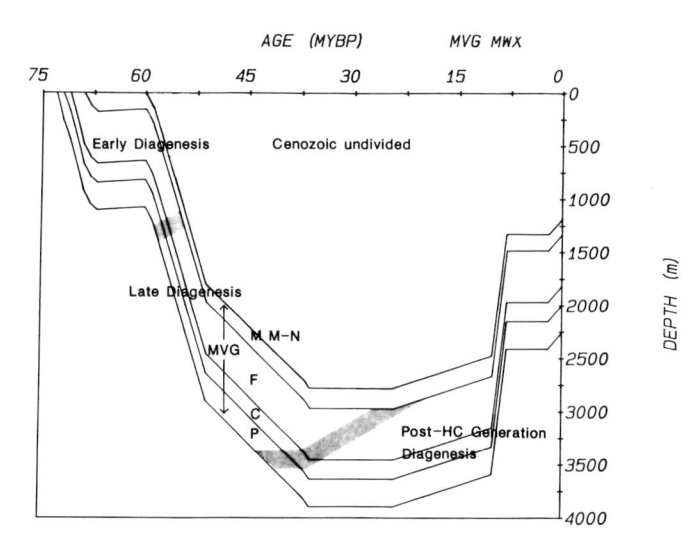

FIG. 19.—Burial history for the Cretaceous Mesaverde Group (MVG) (excluding the basal marine interval) and Tertiary strata at the MWX site. Parameters used to construct the diagram and abbreviations used are the same as those for Figure 14. Timing of the boundary between early and late diagenetic stages is interpreted to represent changes in pore-fluid composition and temperature occurring during rapid burial. Timing of boundary between late and post-hydrocarbon-generation diagenetic stages is interpreted to coincide with entrance into the liquid-hydrocarbon window.

peratures less than 50°C (see Figs. 15A and 19). Other reactions include dissolution and recrystallization of fine oxide rims, dissolution and alteration of feldspar grains, and initial stages of transformation of detrital S/I ML clays (Whitney, 1990).

Late diagenetic events are associated with rapid burial of the section to maximum depth during Laramide basinal subsidence (see Figs. 15A and 19 for timing of these events for various intervals of the MVG). The late diagenetic effects are interpreted to have resulted from changes in temperature and pore-fluid chemistry with respect to conditions developed during early diagenesis. The resulting diagenetic reactions include transformation and dissolution of earlier precipitated diagenetic phases, precipitation of cements, and, to a lesser degree, transformation of unstable detrital components. Events common to all intervals during the late phase of diagenesis include albitization of feldspar, formation of secondary porosity, and much of the S/I ML clay transformation. Dolomitization and precipitation of dolomite, kaolinite, and ankerite cements also occurred, and will be discussed later.

Post-HC diagenetic events are associated with hydrocarbon maturation and thermal decomposition of organic components (see Figs. 15A and 19 for timing). Events occurring during this phase of diagenesis include migration of hydrocarbons into secondary porosity and precipitation of illite, pyrite, and siderite in remaining secondary pores.

The interpreted diagenetic sequence is generally similar to those of Pitman and others (1989) and Lorenz and others (1989); however, several significant differences are worthy of note. Kaolinite is observed to be more abundant in our XRD analyses compared to those of Pitman and others (1989) and Pollastro (1984). This may be due to the focus on sampling of organic-rich mudstone-sandstone associations for this study. The relative timing of some diagenetic phases and events does not correspond to those of Pitman and others (1989) or Lorenz and others (1989). These workers, however, are not always in agreement in terms of relative timing of diagenetic events, a problem commonly encountered when comparing petrographic observations. In addition, some diagenetic aspects and phases incorporated into the sequence determined from this study were not recognized or reported by previous workers (e.g., C/S ML clay, albitization, pyrite, and siderite).

Association of Diagenetic Events with Depositional Facies and Environments

Within a given depositional interval of the cores studied, differences in the proportion and distribution of diagenetic phases are observed. Two consistent relations between depositional facies and distribution of diagenetic minerals are noted. (1) Diagenetic phases in fine-grained sandstones are usually dominated by quartz cements and clay minerals, whereas medium-grained sandstones are more likely to contain a mixture of early clay-mineral, quartz, and calcite cements, with later dolomite and ankerite. This relation is interpreted to result from the greater amount of detrital clay and fine-grained lithic and soil materials deposited with fine-grained sandstones (especially in the fluvial through paludal intervals), coupled with the attendant effect on porosity (Curtis, 1985; Wilson and Pittman, 1977), contrasted with the more permeable nature of the medium-grained sandstones. (2) Proximity to coal or organic-rich mudstones generally results in either abundant secondary porosity or porosity occluded by late diagenetic carbonates (especially ankerite). This relation is interpreted in terms of inorganic-organic reactions resulting from progressive burial (Surdam and others, 1984; Surdam and others, 1989b), and will be discussed later in this section.

Differences in the diagenetic mineralogy and sequence are closely associated with changes in depositional environment and depth of burial. The three major differences in diagenetic relations discussed in this paper are: (1) grain-rimming chlorite is abundant in the fluvial and marine intervals, but rare or absent in the coastal and paludal intervals; (2) kaolinite occurs as early diagenetic pore fill in the fluvial interval, whereas it occurs as late diagenetic, secondary pore fill in the coastal and paludal intervals, and is absent in the marine interval; and (3) the distribution and abundance of authigenic carbonates change from sparse early diagenetic calcite through much of the fluvial interval, to abundant late diagenetic dolomite and ankerite in the coastal and paludal intervals, to common dolomite with ankerite rims in the marine interval.

At present, insufficient quantitative chemical data and mass-balance information are available for rigorous evaluation of the mechanisms and conditions that produced the observed variations in diagenetic mineralogy. A preliminary interpretation regarding the distribution of chlorite is that, rather than formation via reaction with kaolinite or smectite (Kaiser, 1984; Boles and Franks, 1979), chlorite formed by recrystallization of iron oxy-hydroxides and amorphous aluminosilicates in Mg-rich pore fluids. Based on work in Pennsylvanian deltaic strata in Texas, Land and Dutton (1978) suggested that iron oxides and hydroxides, and X-ray-amorphous aluminosilicates may combine with magnesium from pore waters during early diagenesis to produce chlorite-grain coatings. This interpretation is attractive for two reasons: (1) sandstones and mudstones in the fluvial interval probably contained abundant hydrated oxides of iron, aluminum, and silica from soil materials as grain coatings or colloids that could combine to produce the observed abundance of early diagenetic chlorite in sandstones of the fluvial interval; and (2) the lower Eh and pH conditions of pore waters in the organic-rich sediments of the coastal and paludal intervals may have dissolved or complexed more of the oxides and oxy-hydroxides, and allowed iron and aluminum to remain in solution or tied up in the organic phase (Stumm and Morgan, 1981; Curtis, 1985).

The formation of late diagenetic dolomite, secondary porosity, secondary pore-filling kaolinite, and ankerite are interpreted in terms of interactions between organic and inorganic reactions with increased depth of burial (Surdam and others, 1989b). Organic reactions are thought to play a major role in this part of diagenetic history because of: (1) the abundance of dolomite, kaolinite, ankerite, and development/preservation of high porosity in sandstones adjacent to high organic-content mudstones and coals (rich in type III organic materials) in the coastal and paludal intervals, and (2) the ability of type III kerogens to produce water-soluble organic compounds upon maturation that are known to affect pore-fluid compositions and carbonate- and aluminosilicate-mineral stability (Crossey and others, 1984; Surdam and others, 1984; Kharaka and others, 1986).

Diagenetic dolomite is interpreted to have resulted largely from dolomitization of early calcite cements, driven by an increase in Mg^{+2} contents of pore fluids and decrease in the hydration energy of magnesium with increasing temperature during progressive burial. CO_2 and HCO_3^- produced through bacterial degradation of kerogen and kerogen-cleaved organic acids in the subsurface (Kharaka and others, 1986) also may have been a driving force by supplying carbonate to pore fluids. Increased magnesium, and presumably iron, content of pore waters may have been related to release of cations during the transformation of smectite to illite (Boles and Franks, 1979). The iron-poor nature of the dolomite (Fig. 13A) suggests, however, that the origin of the Mg^{+2} may be associated with some other process.

The occurrence of secondary pore-filling, late diagenetic kaolinite in the coastal and paludal intervals implies that secondary porosity was generated prior to precipitation, and that pore-fluid pH was low (Franks and Forester, 1984; Kaiser, 1984). Another important consideration is the fact that kaolinite occurs within intergranular pores as well as intragranular feldspar pores. This observation implies aqueous transport of aluminum, assuming feldspar dissolution is the source of aluminum. The development of the secondary porosity and precipitation of late diagenetic kaolinite are interpreted to have been aided by the corrosive nature and complexing ability of water-soluble organic compounds known to be generated during kerogen maturation. An alternative interpretation for late diagenetic kaolinite has been described by Hansley and Johnson (1980) on the basis of outcrop and core studies in the southern and western regions of the Piceance Basin. These workers related kaolinite formation to the deep weathering of the upper portion of the MVG during post-Cretaceous exposure. More detailed morphologic and isotopic studies of the kaolinite occurrences could assist in assessing the degree to which meteoric waters may have played a role during later diagenesis.

The late diagenetic ankerite shows textures that suggest replacement of dolomite and pervasive cementation. The critical driving force for the replacement reaction is the ratio of Fe^{+2} to Mg^{+2} in solution, implying that iron was released into solution by a concomitant reaction. Cementation by ankerite requires both cation and carbonate sources. Difunctional and long-chain monofunctional carboxylic acids are known to be less stable than acetic and formic acids in the diagenetic environment, and degrade to form CO_2, as well as other products (Crossey, 1991). CO_2 generated by such a mechanism is interpreted to have driven the ankerite-cementation reaction under the influence of the external acetic-acid pH-buffer, where sufficient calcium, iron, and magnesium were available. The close spatial association of pervasive ankerite with high organic contents in the paludal interval (despite the lack of early iron-bearing phases) suggests that iron may have been incorporated into the organic matter, not discernible through routine petrographic observation (see discussion in Surdam and others, 1989a).

Type III organic components occur throughout the MVG section, although late diagenetic dolomite, kaolinite, and ankerite phases occur together only in the coastal and paludal intervals. This suggests that the effects of kerogen-cleaved, water-soluble organic compounds may be localized to those strata with high organic contents.

SUMMARY

Based on organic, petrologic, and mineralogic examinations, the diagenesis of the Mesaverde Group at the multiwell experiment site in the Piceance Basin in northwestern Colorado is integrated with burial and thermal data for the basin. Three phases of diagenesis are described relative to the thermal history of the MVG: an early phase (characterized by compaction, feldspar alteration, chlorite-rim growth, and quartz and calcite cementation), a late phase (with dolomite and ankerite precipitation, feldspar dissolution and albitization, and continued mixed-layer clay transformation), and a 'post-hydrocarbon' phase (characterized by the presence of bitumen, pyrite, siderite, and authigenic illite). Differences in clay-mineral assemblage are noted between associated sandstone/mudstone pairs (kaolinite and chlorite generally more abundant in the sandstones), and among the four depositional environments (chlorite commonly dominating the marine and fluvial intervals; kaolinite prevalent in all sandstones except those of the marine interval). Some aspects of diagenesis may be related to the abundance and nature of organic matter (itself undergoing modification during burial) within the sedimentary sequence.

Whereas the diagenesis of sedimentary materials is a response to a complex interaction of factors (including, but not limited to: inherited characteristics of the detrital grains; early burial processes such as bioturbation, microbial activity and possible meteoric influences; later basin-source fluid migration, uplift, and erosion), the sample selection scheme utilized in this study has attempted to minimize these complexities. Investigation of the Mesaverde Group allows for the examination of the diagenetic modification of organic matter, mudstones, and sandstones deposited in marine and terrestrial settings, while maintaining a degree of constancy of source materials, as well as burial and thermal history. The sampling of closely adjacent pairs of interbedded sandstones and organic-rich mudstones maximizes the probability that local diagenetic effects resulting from organic-rich mudstone/sandstone interactions will be reflected, and may help to delineate specific processes operating during diagenesis (thick, more permeable sandstones will most likely undergo a broader spectrum of diagenetic processes as both meteoric influences and basin-source fluids may participate to a greater degree). Selection of the MWX site itself permits a view of the full spectrum of processes, as the site has attained a high degree of thermal maturity during its burial history. Numerous ancillary studies at this site provide a broad base of supporting information.

Results indicate that sandstones associated with terrestrially derived kerogen exhibit characteristic diagenetic associations, which may be linked to the presence of water-soluble organic compounds generated during the course of progressive diagenesis. This concept could be tested through comparison of diagenetic patterns observed within the MVG at this site with those at sites of lower degrees of thermal maturation.

ACKNOWLEDGMENTS

Support for this study was provided by the Sandia National Laboratory/University Collaborative Research Program. John Lorenz provided encouragement for this study, as well as access to the MWX cores. Larry Smith helped with sampling and XRD analyses, Terry Sewards performed electron microprobe analyses, and David Hicks assisted with petrographic analysis. X-ray diffraction analyses were performed at the X-Ray Diffraction Laboratory of the Department of Geology, University of New Mexico. Microprobe analyses were performed at the Microbeam Analysis Facility in the Institute of Meteoritics, University of New Mexico. Organic analyses were provided by Steven Boese at the University of Wyoming, Laramie, Wyoming. The manuscript was improved by the thoughtful and critical reviews of Stephen Franks and an anonymous reviewer, and helpful comments of Ed Pittman.

REFERENCES

BARKER, C. E., 1989a, Fluid inclusion evidence for paleotemperatures within the Mesaverde Group, Multiwell Experiment Site, Piceance Basin, Colorado: U.S. Geological Survey Bulletin 1886, Chapter M, 11 p.

BARKER, C. E., 1989b, Rock-Eval analysis of sediments and ultimate analysis of coal, Mesaverde Group, Multiwell Experiment Site, Piceance Basin, Colorado: U.S. Geological Survey Bulletin 1886, Chapter N, 11 p.

BLACKWELL, D. D., AND STEELE, J. L., 1988, Thermal conductivity of sedimentary rocks, in Naeser, N. D., and McCulloh, T. H., eds, Thermal History of Sedimentary Basins: New York, Springer-Verlag, p. 13–36.

BOLES, J. R., AND FRANKS, S. G., 1979, Clay diagenesis in Wilcox sandstones of southwest Texas: implications of smectite diagenesis on sandstone cementation: Journal of Sedimentary Petrology, v. 49, p. 55–70.

CROSSEY, L. J., 1985, The origin and role of water-soluble organic compounds in the diagenetic environment: Unpublished Ph.D. Dissertation, University of Wyoming, Laramie, 134 p.

CROSSEY, L. J., 1991, Thermal degradation of aqueous oxalate species: Geochimica et Cosmochimica Acta, v. 55, p. 1515–1527.

CROSSEY, L. J., FROST, B. R., AND SURDAM, R. C., 1984, Secondary porosity in laumontite-bearing sandstones, in McDonald, D. A., and Surdam, R. C., eds., Clastic Diagenesis: American Association of Petroleum Geologists Memoir 37, p. 225–237.

CROSSEY, L. J., HAGEN, E. S., SURDAM, R. C., AND LAPOINT, T. W., 1986a, Correlation of organic parameters derived from elemental analysis and programmed pyrolysis of kerogen, in Gautier, D. L., ed., Roles of Organic Matter in Sediment Diagenesis: Society of Economic Paleontologists and Mineralogists Special Publication 38, p. 35–45.

CROSSEY, L. J., SURDAM, R. C., AND LAHANN, R., 1986b, Application of organic/inorganic diagenesis to porosity prediction, in Gautier, D. L., ed., Roles of Organic Matter in Sediment Diagenesis: Society of Economic Paleontologists and Mineralogists Special Publication 38, p. 147–155.

CURTIS, C. D., 1985, Clay mineral precipitation and transformation during burial diagenesis: Philosophical Transactions of the Royal Society of London, Series A, v. 315, p. 91–105.

CURTIS, C. D., HUGHES, C. R., WHITEMAN, J. A., AND WHITTLE, C. K., 1985, Compositional variation within some sedimentary chlorites and some comments on their origin: Mineralogical Magazine, v. 49, p. 375–385.

DECKER, E. R., HEASLER, H. P., BUELOW, K. L., BAKER, K. H., AND HALLIN, J. S., 1988, Significance of past and recent heat-flow and radioactivity studies in the southern Rocky Mountains region: Geological Society of America Bulletin, v. 100, p. 1851–1885.

DEMAISON, G., AND MURRIS, R. J., eds., 1984, Petroleum Geochemistry and Basin Evaluation: American Association of Petroleum Geologists Memoir 35, 426 p.

DIXON, S. A., SUMMERS, D. M., AND SURDAM, R. C., 1989, Diagenesis and preservation of porosity in the Norphlet Formation (Upper Jurassic), southern Alabama: American Association of Petroleum Geologists Bulletin, v. 73, p. 707–728.

DREVER, J. I., 1973, The preparation of oriented clay mineral specimens for X-ray diffraction analysis by a filter-membrane peel technique: American Mineralogist, v. 58, p. 553–554.

FISCHER, K. J., AND SURDAM, R. C., 1988, Contrasting diagenetic styles in a shelf-turbidite sequence, the Santa Margarita and Stevens Sandstone, southern San Joaquin Basin, California, in Graham, S., ed., Geologic Studies of the San Joaquin Basin, CA: Society of Economic Paleontologists and Mineralogists Special Publication 60, p. 233–248.

FOLK, R. L., 1968, The Petrology of Sedimentary Rocks: Austin, Texas, Hemphill's, 170 p.

FRANKS, S. G., AND FORESTER, R. M., 1984, Relationships among secondary porosity, pore-fluid chemistry and carbon dioxide, Texas Gulf Coast, in McDonald, D. A., and Surdam, R. C., eds., Clastic Diagenesis: American Association of Petroleum Geologists Memoir 37, p. 63–79.

HAGEDORN, D. N., 1985, The calculation of synthetic thermal conductivity logs from conventional geophysical well logs: Unpublished M.S. Thesis, Southern Methodist University, Dallas, 110 p.

HANSLEY, P. L., AND JOHNSON, R. C., 1980, Mineralogy and diagenesis of low-permeability sandstones of Late Cretaceous age, Piceance Creek Basin, northwestern Colorado: Mountain Geologist, v. 17, p. 88–129.

HAYES, M. J., AND BOLES, J. R., 1990, Volumetric relations between dissolved plagioclase and kaolinite in San Joaquin Basin sandstones: implications for aluminum mobility (abs.): American Association of Petroleum Geologists Bulletin, v. 74, p. 672.

HOWER, J., 1981, X-ray identification of mixed-layer clay minerals, in Longstaffe, F. J., ed., Clays and the Resource Geologist: Edmunton, Canada, Co-op Press, p. 39–60.

HOWER, J., ESLINGER, E. V., HOWER, M., AND PERRY, E. A., 1976, Mechanism of burial metamorphism of argillaceous sediments: I. Mineralogical and chemical evidence: Geological Society of America Bulletin, v. 87, p. 725–737.

HUANG, W. H., AND KELLER, W. D., 1970, Dissolution of rock-forming minerals in organic acids: simulated first-stage weathering of fresh mineral surfaces: American Mineralogist, v. 55, p. 2076–2094.

HUNT, J. M., 1979, Petroleum Geochemistry and Geology: San Francisco, Freeman, 617 p.

JACKSON, M. L., 1979, Soil Chemical Analysis–Advanced Course: Madison, Wisconsin, published by the author, 895 p.

JOHNSON, R. C., 1989, Geologic history and hydrocarbon potential of Late Cretaceous-age, low-permeability reservoirs, Piceance Basin, western Colorado: U. S. Geological Survey Bulletin 1787, Chapter E, 51 p.

JOHNSON, R. C., AND NUCCIO, V. F., 1986, Structural and thermal history of the Piceance Creek Basin, western Colorado, in relation to hydrocarbon occurrence in the Mesaverde Group: American Association of Petroleum Geologists Studies in Geology No. 22, p. 165–205.

KAISER, W. R., 1984, Predicting reservoir quality and diagenetic history in the Frio Formation (Oligocene) of Texas, in Gautier, D. L., ed., Roles of Organic Matter in Sediment Diagenesis: Society of Economic Paleontologists and Mineralogists Special Publication 38, p. 195–215.

KAWAMURA, K., HULL, R. W., AND CAROTHERS, W. W., 1985, Water-rock interactions in sedimentary basins, in Kharaka, Y. K., Gautier, D. L., and Surdam, R. C., eds., Relationship of Organic Matter and Mineral Diagenesis: Society of Economic Paleontologists and Mineralogists Short Course 17, p. 79–176.

KAWAMURA, K., AND KAPLAN, I. S., 1987, Dicarboxylic acids generated by thermal alteration of kerogen and humic acids: Geochimica et Cosmochimica Acta, v. 81, p. 3201–3207.

KAWAMURA, K., TANNEBAUM, E., HUIZINGA, B. J., AND KAPLAN, I. S., 1986, Volatile organic acids generated from kerogen during laboratory heating: Geochemical Journal, v. 20, p. 51–59.

KHARAKA, Y. K., LAW, L. M., CAROTHERS, W. W., AND GOERLITZ, D. F., 1986, Role of organic species dissolved in formation waters from sedimentary basins in mineral diagenesis: Society of Economic Paleontologists and Mineralogists Special Publication 38, p. 111–122.

LAND, L. S., AND DUTTON, S. P., 1978, Cementation of a Pennsylvanian deltaic sandstone: isotopic data: Journal of Sedimentary Petrology, v. 48, p. 1167–1176.

LARSEN, E. E., OZIMA, M., AND BRADLEY, W. C., 1975, Late Cenozoic basaltic volcanism in northwestern Colorado and its implications concerning tectonism and the origin of the Colorado River system, in Curtis, B. F., ed., Cenozoic History of the Southern Rocky Mountains: Geological Society of America Memoir 144, p. 155–178.

LAW, B. E., NUCCIO, V. F., AND BARKER, C. E., 1989, Kinky vitrinite reflectance well profiles: evidence of paleopore pressure in low-permeability, gas-bearing sequences in Rocky Mountain foreland basins: American Association of Petroleum Geologists Bulletin, v. 73, p. 999–1010.

LORENZ, J. C., 1985, Tectonic and stress histories of the Piceance Creek Basin and the MWX site, from 75 million years ago to the present: Sandia National Laboratories Report SAND84–2603, 48 p.

LORENZ, J. C., 1989, Reservoir sedimentology of rocks of the Mesaverde Group, Multiwell Experiment Site and east-central Piceance Basin, northwest Colorado: U. S. Geological Survey Bulletin 1886, Chapter K, 24 p.

LORENZ, J. C., SATTLER, A. R., AND STEIN, C. L., 1989, Differences in reservoir characteristics of marine and nonmarine sandstones of the Mesaverde Group, northwestern Colorado: Sandia National Laboratories Report SAND88–1963, 62 p.

LUNDEGARD, P. D., AND KHARAKA, Y. K., 1990, Geochemistry of organic acids in subsurface waters–field data, experimental data, and models, in Melchior, D. S., and Bassett, R. C., eds., Chemical Modeling of Aqueous Systems II: American Chemical Society Symposium Series 416, p. 169–189.

LUNDEGARD, P. D., AND LAND, L. S., 1986, Carbon dioxide and organic acids: their role in porosity enhancement and cementation, Paleogene of the Texas Gulf Coast, in Gautier, D. L., ed., Roles of Organic Matter in Sediment Diagenesis: Society of Economic Paleontologists and Mineralogists Special Publication 38, p. 129–146.

LUNDEGARD, P. D., AND SENFTLE, J. T., 1987, Hydrous pyrolysis—a tool for the study of organic acid synthesis: Applied Geochemistry, v. 2, p. 605–612.

MARVIN, R. F., MEHNERT, H. H., AND MOUNTJOY, W. M., 1966, Age of basalt cap on Grand Mesa: U. S. Geological Survey Professional Paper 550-A, 81 p.

MILLIKEN, K. L., MCBRIDE, E. F., AND LAND, L. S., 1989, Numerical assessment of dissolution versus replacement in the subsurface destruction of detrital feldspars, Oligocene Frio Formation, south Texas: Journal of Sedimentary Petrology, v. 59, p. 740–757.

MOORE, D. M., AND REYNOLDS, Jr., R. C., 1989, X-Ray Diffraction and the Identification and Analysis of Clay Minerals: New York, Oxford University Press, 332 p.

MORAES, M. A. S., 1989, Diagenetic evolution of Cretaceous-Tertiary turbidite reservoirs, Campos Basin, Brazil: American Association of Petroleum Geologists Bulletin, v. 73, p. 598–612.

NUCCIO, V. F., AND JOHNSON, R. C., 1989, Thermal history of selected coal beds in the Upper Cretaceous Mesaverde Group and Tertiary Wasatch Formation, Multiwell Experiment Site, Colorado, in relation to hydrocarbon generation: U. S. Geological Survey Bulletin, Chapter L, 8 p.

PITMAN, J. K., AND DICKINSON, W. W., 1989, Petrology and isotope geochemistry of mineralized fractures in Cretaceous rocks; evidence for cementation in a closed hydrologic system, in Law, B. E., and Spencer, C. W., eds., Geology of Tight Gas Reservoirs in the Pinedale Anticline Area, Wyoming, and at the Multiwell Experiment Site, Colorado: U. S. Geological Survey Bulletin 1787, Chapter J, 15 p.

PITMAN, J. K., SPENCER, C. W., AND POLLASTRO, R. M., 1989, Petrography, mineralogy, and reservoir characteristics of the Upper Cretaceous Mesaverde Group in the east-central Piceance Basin, Colorado: U. S. Geological Survey Bulletin 1787, Chapter G, 31 p.

POLLASTRO, R. M., 1984, Mineralogy of selected sandstone/shale pairs and sandstone from the multiwell experiment; interpretations from X-ray diffraction and scanning electron microscopy analyses, in Spencer, C. W., and Keighin, C. W., eds., Geologic Studies in Support of the U.S. Department of Energy Multiwell Experiment, Garfield County, Colorado: U. S. Geological Survey Open-File Report 84–757, p. 67–74.

STOESSELL, R. K., AND PITTMAN, E. D., 1990, Secondary porosity revisited: the chemistry of feldspar dissolution by organic acids and anions: American Association of Petroleum Geologists Bulletin, v. 74, p. 1795–1805.

STUMM, W., AND MORGAN, J. P., 1981, Aquatic Chemistry: New York, John Wiley and Sons, 780 p.

SURDAM, R. C., BOESE, S. W., AND CROSSEY, L. J., 1984, The chemistry of secondary porosity, in McDonald, D. A., and Surdam, R. C., eds., Clastic Diagenesis: American Association of Petroleum Geologists Memoir 37, p. 127–149.

SURDAM, R. C., AND CROSSEY, L. J., 1987, Integrated diagenetic modelling: a process-oriented approach for clastic systems: Annual Review of Earth and Planetary Sciences, v. 15, p. 141–170.

SURDAM, R. C., CROSSEY, L. J., HAGEN, E. S., AND HEASLER, H. P., 1989a, Organic-inorganic interactions and sandstone diagenesis: American Association of Petroleum Geologists, v. 73, p. 1–23.

SURDAM, R. C., AND MACGOWAN D. B., 1987, Oilfield waters and sandstone diagenesis: Applied Geochemistry, v. 2, p. 613–619.

SURDAM, R. C., MACGOWAN, D. B., AND DUNN, T. L., 1989b, Diagenetic pathways of sandstone and shale sequences: University of Wyoming, Contributions to Geology, v. 27, p. 21–31.

SWEENEY, J. J., AND BURNHAM, A. K., 1990, Evaluation of a simple model of vitrinite reflectance based on chemical kinetics: American Association of Petroleum Geologists Bulletin, v. 74, p. 1559–1570.

TISSOT, B., AND WELTE, D. H., 1978, Petroleum Formation and Occurrence: Berlin, Springer-Verlag, 538 p.

WAPLES, D. W., 1980, Time and temperature in petroleum formation: application of Lopatin's method to petroleum exploration: American Association of Petroleum Geologists Bulletin, v. 64, p. 916–926.

WAPLES, D. W., 1985, Geochemistry in Petroleum Exploration: Boston, International Human Resources Development Corporation, 232 p.

WHITNEY, G., 1990, Role of water in the smectite-to-illite reaction: Clays and Clay Minerals, v. 38, p. 343–350.

WILSON, M. D., AND PITTMAN, E. D., 1977, Authigenic clays in sandstones: recognition and influence on reservoir properties and paleoenvironmental analysis: Journal of Sedimentary Petrology, v. 47, p. 3–31.

CLAY MINERALOGY OF AN INTERBEDDED SANDSTONE, DOLOMITE, AND ANHYDRITE: THE PERMIAN YATES FORMATION, WINKLER COUNTY, TEXAS

J. S. JANKS
Texaco Inc., Exploration and Producing Technology Department, 5901 S. Rice Avenue, Bellaire, Texas 77401, USA
M. R. YUSAS
Maersk Olie OG Gas A/S, 50 Esplanaden, DK-1263, Copenhagen K., Denmark
AND
C. M. HALL*
Department of Physics, University of Toronto, Toronto, Ontario, Canada, M5S 1A7

ABSTRACT: Two cores from the Permian (Guadalupian) Yates Formation in Winkler County, Texas, were analyzed using thin-section petrography, scanning electron microscopy/energy dispersive X-ray, stable-isotope geochemistry, and $^{40}Ar/^{39}Ar$ laser step heating. The Yates was deposited in a coastal-sabkha environment. The sandstone facies is the hydrocarbon reservoir; the dolostone and anhydrite facies are impermeable. Sandstones are very fine- to fine-grained arkoses, subarkoses and lithic arkoses. The major authigenic phases in the sandstones are corrensite and dolomite. Mg-rich sabkha-based pore fluids were responsible for their formation. Other modifications to the sandstones include K-feldspar and quartz overgrowths, and unstable-grain dissolution.

The clay-mineral suite in the Yates Formation consists of corrensite, illite and chlorite. Corrensite (R1 ordered, trioctahedral chlorite/smectite, with approximately 50% smectite layers) is found in the three major facies; however, it is most prominent in the sandstones and is absent from the interbedded black shales. Chlorite-to-smectite ratios do not vary with changes in stratigraphic position or lithology. Clay-mineral suites in dolostone and anhydrite layers, where present, are similar to those in the sandstones. Illite is more prevalent in dolostones than in sandstones, however. Elemental analysis indicates that Mg is a major component in the corrensite, although Fe is also present. The persistence of corrensite in the sandstones is interpreted as the result of relatively uniform porewater salinity and Mg levels.

Stable-isotope values were determined on dolostones and dolomite cement in sandstones. In well #269, dolostone values range from $\delta^{13}C$ +4.97 to +5.94‰ and $\delta^{18}O$ from −1.74 to +1.94‰ PDB. In well #270, $\delta^{13}C$ ranges from +4.46 to +5.97‰ and $\delta^{18}O$ from +0.15 to +2.39‰. Dolomite cements in sandstones from well #269 range from $\delta^{13}C$ +0.67 to +4.92‰ and $\delta^{18}O$ from −1.68 to +0.91‰. In well #270, $\delta^{13}C$ ranges from +0.97 to +3.26‰ and $\delta^{18}O$ from −6.09 to +1.23‰.

The $^{40}Ar/^{39}Ar$ laser step-heating method was used to determine absolute ages on three clay-size mineral separates. In these samples, the clay-mineral suite consists of mixtures of corrensite, illite, and chlorite. Feldspars and mica are also present. The analyses revealed a low-retentivity phase with an apparent age of 275 to 250 Ma, which may represent the age of corrensite formation.

INTRODUCTION

The Yates Formation (Guadalupian) offers a unique opportunity to examine clay mineralogy in an interbedded sandstone-dolostone-anhydrite sequence. This study presents results of our investigation of the sabkha-deposited Yates Formation in Winkler County, Texas. The S. M. Halley field is located on the western edge of the Central Basin Platform (Fig. 1). Sandstones were analyzed using thin-section petrography, scanning electron microscopy/energy dispersive X-ray (SEM/EDS), X-ray diffraction, stable-isotope analysis of carbonates and $^{40}Ar/^{39}Ar$ laser step heating. The purpose of the study is to gain a better understanding of authigenic-corrensite formation in an interbedded sandstone, dolomite, and anhydrite sequence. Chlorite/smectite has been well studied in evaporite, dolostone, and shale, but sparsely studied in sandstone facies interlayered with anhydrite and dolostone.

Yates Formation samples from the S. M. Halley field consist predominantly of sandstone, dolomicrite (dolostone), anhydrite and siltstone. In the S.M. Halley field, the Yates sandstone facies predominates, although the vertical sequence is interlayered and complex (Fig. 2). Typically, the unit is capped by anhydrite. The carbonate layers are medium gray, dolomitized mudstones (Dunham, 1962). Most dolostone layers are essentially nonporous, with only minor occurrences of intercrystalline or vuggy porosity.

Massive, light gray anhydrite, interbedded with anhydritic dolostone and sandstone, is generally found near the top of the cored intervals. Most of the anhydrite has a felted-lath texture, but crystals are tabular. Gypsum is present in trace amounts as well-formed laths and as corroded precursors of anhydrite. Local stringers of dolomitized mudstone are found in the anhydrite, perhaps suggesting replacement of original mudstone by anhydrite.

CHLORITE/SMECTITE IN SEDIMENTARY ROCKS

Mixed-layer chlorite/smectite (C/S) is found in rocks from many depositional and diagenetic environments. C/S is commonly associated with hypersaline and evaporitic deposits (Kopp and Fallis, 1974; Bodine, 1978; 1983; Padan, 1984; Bodine and Madsen, 1987; Palmer, 1987; Fisher, 1988) and bedded dolostone (Hauff and McKee, 1982; Weaver and others, 1990). Chlorite-to-smectite ratios in mixed-layer C/S can be correlated with lithology in some hypersaline (Fisher and Jeans, 1982) and evaporitic settings (Bodine and Madsen, 1987).

Originally, the term corrensite was applied only to a regular 1:1 interstratification of trioctahedral Mg-chlorite and swelling chlorite, but is now applied to regular 1:1 chlorite/smectite (usually saponite) and chlorite/vermiculite (Weaver, 1989). The mineral is not actively forming in any environment today, and is therefore considered to be diagenetic.

In evaporite sequences, the proposed process for C/S formation involves the transformation of detrital clays and micas to Mg-rich species, such as saponite, by interaction with Mg-rich pore fluids. In the first reaction between clay and pore fluid, Mg is preferentially exchanged onto the interlayer surfaces of detrital clays. This reaction is followed

---

*Present Address: Department of Geological Sciences, University of Michigan, 1006 C. C. Little Building, Ann Arbor, Michigan 48109-1063.

146   J. S. JANKS, M. R. YUSAS, AND C. M. HALL

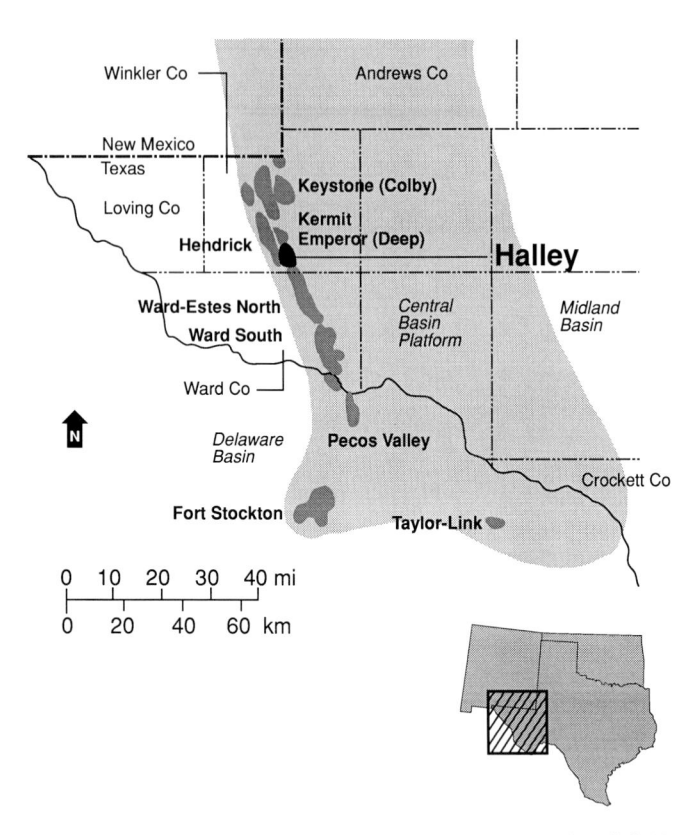

FIG. 1.—Location map showing S. M. Halley and other major oil fields along the western margin of the Central Basin Platform (darkened area).

by Mg-for-Al substitution in octahedral sites, and then by precipitation of brucite interlayers (April, 1980; Palmer, 1987). The parent material for C/S may have been detrital clay, or partially degraded mica or chlorite (Bodine and Madsen, 1987; Palmer, 1987; Fisher, 1988). Alternatively, chlorite may have partially degraded to smectite (Ataman and Gokçen, 1975).

In sandstones, C/S is commonly found in volcanogenic sequences and associated sediments (Wilson, 1971; Almon and others, 1976; Klimentidis and others, 1990), although dioctahedral aluminous species are found in some red beds (Morrison and Parry, 1986). In many sandstones, C/S occurs as a pore-lining and pore-filling clay. Chlorite/smectite can be a basalt alteration product (Kristmannsdottir, 1978). It may also be associated with dike emplacement in shales (Blatter and others, 1973) and hydrothermal activity (April, 1980). Chlorite-to-smectite ratios in shales can increase as the depth of burial increases, similar to the well-known smectite-to-illite transformation (Chang and others, 1986).

METHODS

Samples of whole core from the S.M. Halley wells #269 and #270 (Fig. 2) were prepared for thin-section, SEM/EDS, X-ray diffraction, and stable-isotope analyses. Thin-section samples were impregnated with blue-dyed epoxy and ground to 30 μm. SEM/EDS samples were extracted with solvents to remove hydrocarbons, dried, and gold-coated. X-ray diffraction samples were powdered to 325 mesh and

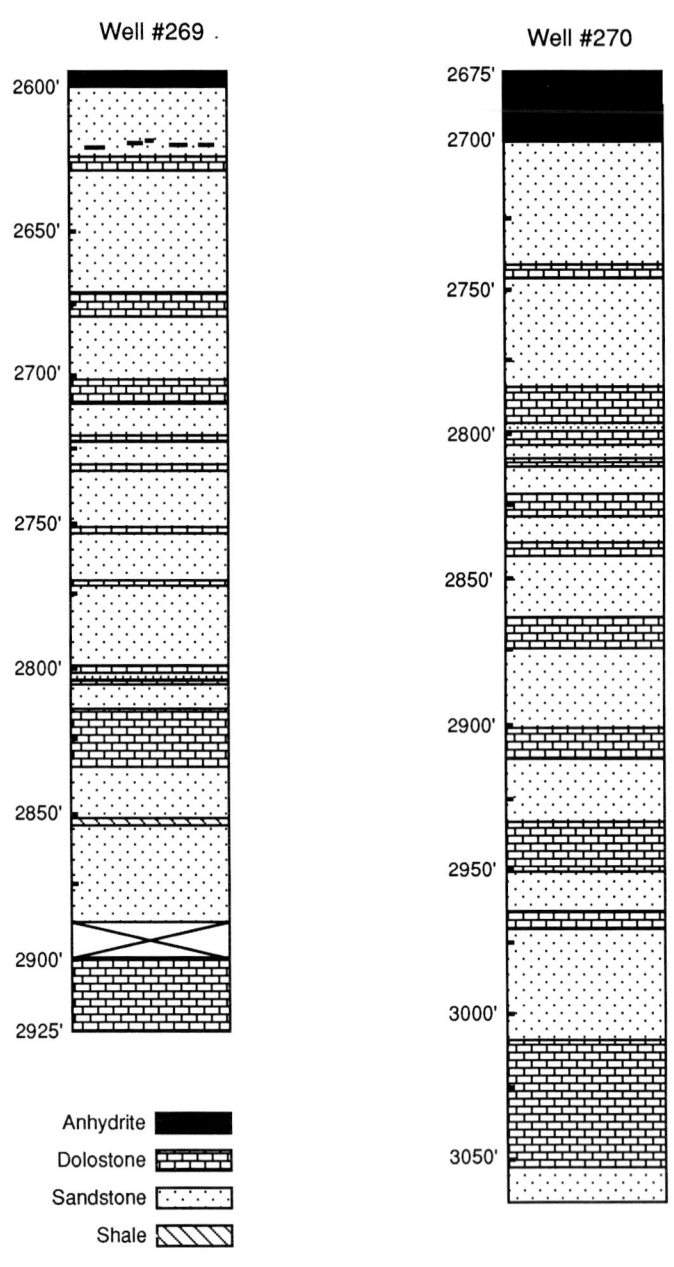

FIG. 2.—Generalized stratigraphic sections of study wells #269 and #270. These sections show only major facies changes.

random mounts were prepared for whole-rock ("bulk") mineralogy. Samples were scanned from 2° to 70° 2θ on a Philips APD 3600 diffractometer at a scanning rate of 2°/min. using CuKa radiation. The method of Schultz (1964) was used to estimate the abundance of minerals present, and intensity factors unique to our laboratory were used to refine the quantification. Comparisons were made between X-ray diffraction data and modal analysis of thin sections as a check on the results.

Clay mounts were prepared by dispersing samples of the rock in deionized water, separating the <2.0-μm fraction, and mounting the separates onto glass slides (Drever, 1973). Both air-dried (50% relative humidity) and ethylene-glycolated (1 hr at 60°C) slides were scanned from 2° to 48°

2θ. Quantitative analyses were performed using the Mineral Intensity Factor method and coefficients of Reynolds (1985).

Three clay samples (<1.0 μm) were separated and dried for $^{40}Ar/^{39}Ar$ analysis. The clays were irradiated for 96 hrs in the McMaster Nuclear Reactor and, after a "cool-down" period of two weeks, loaded into the automated laser-fusion system at the University of Toronto. Step heating was performed using a defocused Ar-ion laser beam and temperatures were monitored with a Barnes Engineering RM2A infrared microscope. Fusion-system blanks were frequently monitored (typically every 3 or 4 sample fractions) and subtracted from the sample data. Typical blanks were 4×10^{-14} cm^3 STP (Standard Temperature and Pressure) at mass 36, 8×10^{-12} cm^3 STP at mass 40. Procedural details are outlined in Layer and others (1987) and Hall and others (1988). The samples were not re-dispersed prior to analysis. This helped reduce net ^{39}Ar loss due to recoil. With the expected average recoil distance of ^{39}Ar being about 0.08 μm (Turner and Cadogan, 1974), recoil might be significant for fine-grained clay minerals. We would expect the recoil effects would remain small as long as clay K concentrations in the samples were fairly uniform over the space of a few tens of microns.

RESULTS

Petrography

Petrographic and SEM analyses were performed to determine porosity types, rock composition, clay fabric, and habit. Clastics in the Yates Formation consist of olive gray and grayish red, very fine- and fine-grained sandstones and red, gray, and black shales. Black-to-dark gray shale layers are interbedded with the sandstones. Sandstones are subarkoses, sublitharenites and lithic wackes (classification of Folk, 1974). Sedimentary structures were difficult to determine, but parallel horizontal and wavy discontinuous laminations were present. Some load structures between sandstone lenses and mudstone laminae were observed. Sandstones were texturally immature, commonly containing clay-size matrix and angular clasts. Framework grains in the sandstones consisted primarily of quartz with lesser amounts of feldspars and rock fragments. Quartz was typically angular to subrounded and very fine-grained, although well-rounded, coarse grains occurred locally. Some grains were very well rounded and frosted. Feldspars (plagioclase and K-feldspar) were very fine-grained and angular to subrounded. Franco (1973) indicated the presence of microcline and orthoclase species, with plagioclase ranging from An^{18} to An^{41}. Plagioclase and K-feldspars displayed various stages of dissolution in some samples (Figs. 3A and B), whereas feldspar overgrowths were present in others (Fig. 3C). K-feldspar also was present as a euhedral, intergranular cement. Low- to moderate-rank meta-sedimentary clasts, chert, metachert, and shale rip-up clasts were the principal lithic fragments. Rock fragments also were observed in various stages of dissolution (Fig. 3C). Muscovite was the primary mica, but minor amounts of detrital chlorite were also present. Opaques included pyrite in the gray sandstones, and hematite and ilmenite in the red sandstones. Garnet, zircon, tourmaline, and apatite formed the heavy-mineral suite.

Dolomite and clay were the principal components of the sandstone matrix, although anhydrite and halite were present in some intervals (Tables 1 and 2). Carbonate cements included calcite and dolomite. Calcite occured sporadically throughout the sandstones as poikilotopic cement, whereas dolomite cement was ubiquitous. Dolomite in sandstones is a dispersed, very finely crystalline, intergranular cement (Fig. 3A). Multiple dolomite cementation phases are possible, as the cement occured in both primary and secondary pores (Figs. 3A and C). Dolomite cement in primary pores was commonly more abundant and massive than in secondary pores. In secondary pores, it was found as clay- and silt-size crystals, commonly on the surface of multiple K-feldspar overgrowths. Typically, the primary dolomite did not have well-defined crystal faces.

The clay fraction of the matrix was composed of detrital and authigenic clays. Detrital clays were deposited in association with coarser clastics but appeared not to have been the result of vadose diagenesis (Walker, 1975). The authigenic clay was mostly corrensite, which occured ubiquitously in the sandstones (Figs. 3D and E). It was undoubtedly authigenic, occurring as grain coatings (Welton, 1984) and associated with authigenic K-feldspar (Fig. 3F). Corrensite occured on the surfaces of partially dissolved grains, suggesting that it formed early in the burial history. Corrensite did not occur along most grain contacts, nor was there any indication that detrital clays or micas are altered to corrensite. No corrensite was observed on K-feldspar overgrowths, indicating that the clay formed before the K-feldspar (Fig. 3F). Figure 4 shows a typical energy dispersive spectrum of corrensite in a Yates sandstone sample. The role of Fe in the corrensite is uncertain because finely divided hematite is common in the Yates and may interfere with the analysis.

Porosity in the sandstones is a combination of primary intergranular and secondary voids. Leaching of rock fragments, feldspars (particularly K-feldspars), dolomite cement, and matrix enhanced porosity in the gray sandstones. However, there was less secondary porosity in the red units. Dolomite cement displayed evidence of partial dissolution in some primary pores, but not in secondary pores. Other intergranular cements included pyrite and hematite (responsible for the red color in the red sandstones).

X-ray diffraction data from all samples (whole-rock and clay-size fraction) are presented in Tables 1 and 2. In sandstone samples, the total clay content ranged from 2 to 38%. Quartz was ubiquitous, ranging from 32 to 74%. K-feldspar and plagioclase occurred in amounts subordinate to quartz (K-feldspar, 3 to 22%; plagioclase, 5 to 16%). Anhydrite occurred sporadically throughout the cores (below limit of detection to 32%), but was more prevalent in well #270. Bedded anhydrite was present at the top of the study interval in both wells, and occurred as thin beds between 841 m and 846 m (2,759 and 2,775 ft) in well #270. Minor amounts of anhydrite cement occurred in sandstone samples from 823 m and 831 m (2,699 to 2,723 ft) in well #270. Dolostone layers commonly contained low percentages of quartz and clay.

148   J. S. JANKS, M. R. YUSAS, AND C. M. HALL

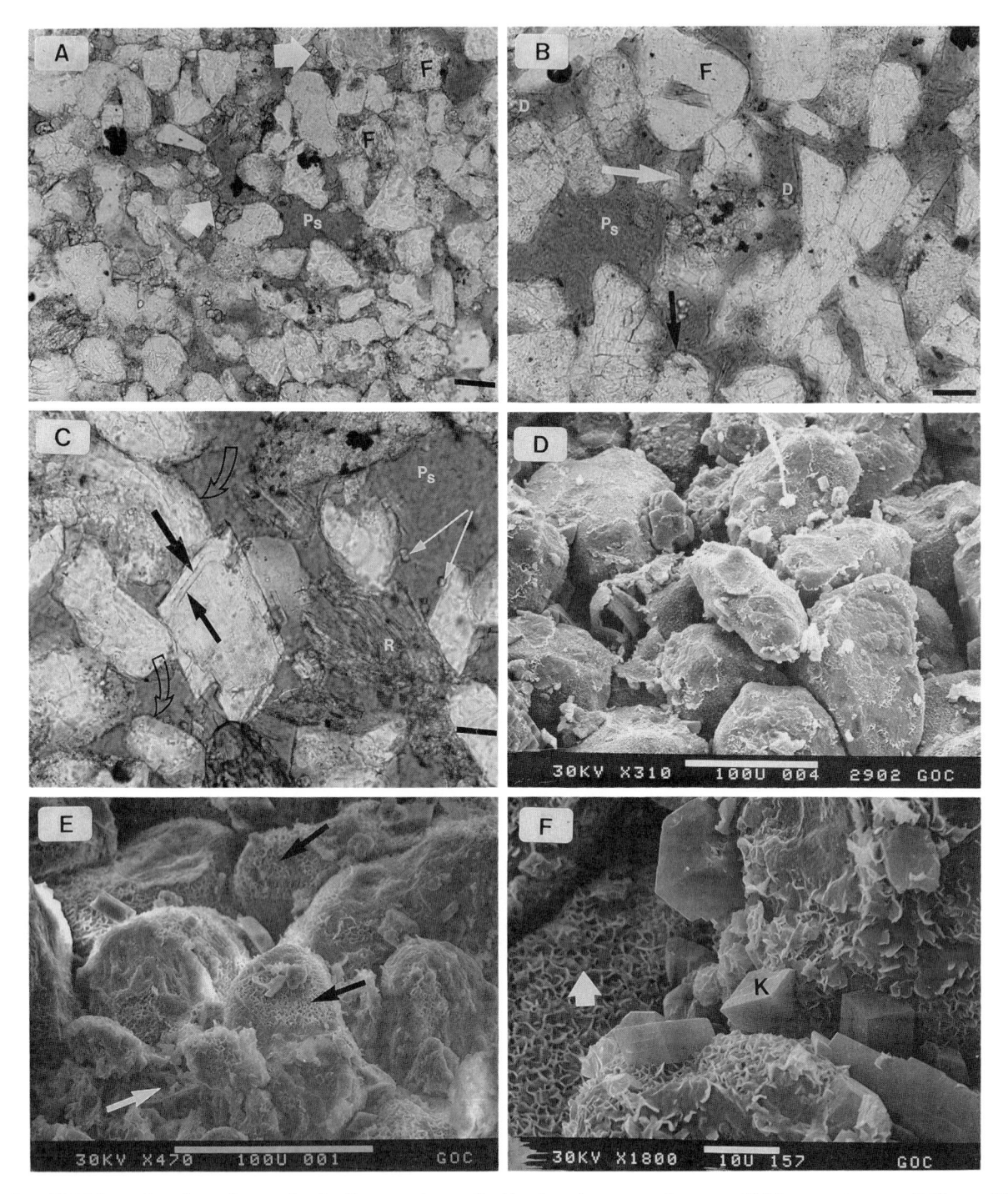

FIG. 3.—(A) Thin-section photomicrograph of a Yates sandstone. In this view, large secondary pore (Ps) is seen at center. Feldspars (F) are in varying stages of dissolution. Authigenic dolomite cement occurs throughout the pore system, but in this view cementation in primary pores is common (arrow). Scale bar 0.1 mm. Plane polarized light. (B) Thin-section photomicrograph showing dissolution porosity (Ps), authigenic-dolomite cement (D), and pore-filling authigenic K-feldspar (white arrow). In this view most of the dolomite is found in primary pores. Corrensite rims line most pores (black arrow). Feldspar grain (F) has partially dissolved. Abundant primary porosity remains. Scale bar 0.05 mm. Plane polarized light. (C) Thin-section photomicrograph showing large secondary pore (Ps) with authigenic dolomite (white arrows), pore-lining corrensite (hollow arrows), a partially dissolved rock fragment (R) and multiple generations of authigenic overgrowths on K-feldspar (solid black arrows). Dolomite cement in secondary pores is commonly smaller and less abundant than in primary pores (compare to A). Scale bar 0.05 mm. Plane polarized light. (D) SEM photomicrograph showing grain and pore relations in a Yates sandstone. Corrensite grain coatings are pervasive. Partially dissolved rock fragment is seen at left center. (E) SEM photomicrograph showing partially dissolved feldspar (white arrow) and abundant grain-coating corrensite (black arrows). Bare patches on some surfaces are grain-to-grain contact points. (F) High-magnification view showing authigenic K-feldspar (K) and corrensite (white arrow) in the pore system.

TABLE 1.—X-RAY DIFFRACTION, STABLE-ISOTOPE, AND LITHOLOGIC DATA FOR WELL #269, S. M. HALLEY FIELD, YATES FORMATION

DEPTH (FT)	TC	QZ	KS	PL	CA	DO	PY	ANH	GYP	HA	SM	IL	CH	C/S	%Exp C/S	δ13C	δ18O	LITHOLOGY
SANDSTONE																		
2634	10	52	13	16	–	9	–	–	–	–	–	7	1	92	50	ND	ND	Gray-green sandstone
2638	17	61	5	6	–	11	–	–	–	–	ND	ND	ND	ND	ND	+4.55	+0.66	Gray-green sandstone
2650	18	60	4	8	–	10	–	–	–	–	ND	ND	ND	ND	ND	+4.18	+0.91	Gray-green sandstone
2693	3	74	9	15	–	–	–	–	–	–	–	3	1	96	50	ND	ND	Gray-green sandstone
2728	17	46	8	16	–	13	–	–	–	–	–	10	3	87	50	+3.64	-1.00	Gray-green shaley sandstone
2737	2	50	14	9	–	23	1	–	–	–	–	7	2	91	50	ND	ND	Gray-green dolomitic sandstone
2755	38	39	7	9	–	7	–	–	–	–	–	14	3	83	50	ND	ND	Gray-green shaley sandstone
2761	16	45	3	13	2	20	–	–	–	–	–	4	1	95	60	ND	ND	Gray-green sandstone
2776	17	58	6	8	–	10	tr	–	–	–	–	16	2	82	50	ND	ND	Gray-green sandstone
2840	9	55	8	11	–	18	–	–	–	–	–	3	2	95	50	+1.01	+0.45	Green-gray, laminated sandstone
2850	8	48	6	13	–	25	–	–	–	–	–	22	3	75	50	ND	ND	Brown sandstone
2866	14	57	6	15	–	8	–	–	–	–	–	10	4	86	50	ND	ND	Brown sandstone
2875	7	54	8	15	–	17	–	–	–	–	–	6	4	90	50	+0.67	-1.29	Dark gray sandstone
2896	6	52	9	13	–	20	–	–	–	–	–	10	2	88	50	ND	ND	Gray-green sandstone
DOLOSTONE																		
2626	4	5	2	2	–	86	–	–	–	–	–	7	1	92	50	+5.91	-1.74	Gray dolostone
2672	3	6	1	1	–	89	–	–	–	–	–	15	1	84	50	+4.97	+1.30	Light gray dolostone
2771	2	3	tr	tr	–	95	–	–	–	–	–	20	1	79	50	+4.98	+1.73	Gray dolostone
2708	8	4	–	1	–	86	–	–	–	–	–	8	2	90	50	+5.50	+1.28	Light gray dolostone
2822	15	19	8	8	–	50	–	–	–	–	–	10	3	89	50	+5.65	+1.13	Light gray dolostone
2926	7	6	1	12	–	74	–	–	–	–	–	56	7	37	50	+5.42	+1.93	Gray dolostone
2934	2	1	–	–	–	97	–	–	–	–	–	99	1	–	–	+5.94	+1.16	Gray dolostone
2962	17	31	8	8	–	37	–	–	–	–	–	46	8	46	50	ND	ND	Red, shaley, sandy dolostone
2973	30	19	9	11	–	31	–	–	–	–	–	19	6	75	50	+4.50	+0.80	Red, sandy, shaley dolostone
ANHYDRITE																		
2596	9	19	2	8	–	9	–	53	tr	–	–	15	1	84	50	ND	ND	White, sandy anhydrite

TC	=	Total Clay	DO	=	Dolomite	
QZ	=	Quartz	PY	=	Pyrite	
KS	=	K-feldspar	ANH	=	Anhydrite	
PL	=	Plagioclase	GYP	=	Gypsum	
CA	=	Calcite	HA	=	Halite	

SM = Smectite
IL = Illite
CH = Chlorite
C/S = Chlorite/Smectite
%EXP C/S = %Smectite in C/S

ND = Analysis not determined
– = Below detection
tr = Trace amounts

X-ray diffraction data are expressed in %; clay mineral data from <2μm fraction. Stable isotopes are relative to PDB. Results are listed by depth and facies.

TABLE 2.—X-RAY DIFFRACTION, STABLE-ISOTOPE AND LITHOLOGIC DATA FOR WELL #270, S. M. HALLEY FIELD, YATES FORMATION

DEPTH (FT)	TC	QZ	KS	PL	CA	DO	PY	ANH	GYP	HA	SM	IL	CH	C/S	%Exp C/S	δ13C	δ18O	LITHOLOGY
SANDSTONE
2699	12	51	6	8	4	11	–	8	–	–	ND	ND	ND	ND	ND	+2.65	-2.76	Gray-green sandstone
2704	8	58	7	8	–	13	–	7	–	–	–	15	3	82	50	+3.17	-0.17	Gray-green sandstone
2715	9	54	4	10	–	18	1	3	–	–	–	18	2	80	50	+1.44	-0.03	Gray-green sandstone
2723	8	67	6	8	–	7	–	3	–	1	ND	ND	ND	ND	ND	+1.80	-4.92	Gray-green sandstone
2730	12	62	9	13	–	4	–	–	–	–	ND	ND	ND	ND	ND	+0.97	-3.55	Gray-green sandstone
2742	14	72	4	8	–	2	–	–	–	–	ND	ND	ND	ND	ND	+2.53	-6.09	Gray-green sandstone
2754	25	43	8	15	–	8	1	–	–	–	–	13	6	81	50	ND	ND	Gray-green sandstone
2759	14	32	4	5	–	12	–	32	–	–	–	40	8	52	50	+3.45	+1.21	Gray, anhydritic sandstone
2794	10	48	5	8	–	29	–	–	–	–	ND	ND	ND	ND	ND	+2.48	+0.64	Gray-green sandstone
2812	14	65	7	7	–	7	–	–	–	–	ND	ND	ND	ND	ND	+2.24	-2.37	Red-to-green sandstone
2852	7	57	22	12	–	2	–	–	–	–	ND	ND	ND	ND	ND	+3.26	-0.83	Red sandstone
2868	10	44	12	10	–	23	tr	–	–	–	–	9	3	88	50	ND	ND	Red sandstone
2879	11	56	10	9	–	15	–	–	–	–	–	10	2	88	50	+3.07	-0.57	Red sandstone
2887	15	64	8	6	–	7	–	–	–	–	–	7	2	91	50	+2.02	-1.53	Red sandstone
2913	15	60	4	8	–	14	–	–	–	–	–	6	1	93	50	+4.32	-3.38	Red sandstone
2950	12	57	6	11	–	14	–	–	–	–	–	9	3	88	50	+3.11	-1.51	Gray-green sandstone
2959	9	66	6	12	–	7	–	–	–	–	–	10	2	88	50	+1.62	-1.32	Gray-green sandstone
2977	4	67	13	13	–	2	1	–	–	–	–	10	6	84	50	ND	ND	Gray-green sandstone
DOLOSTONE
2697	4	5	2	2	–	86	–	–	–	–	–	7	1	92	50	+5.91	-1.74	Gray dolostone
2717	17	9	–	2	–	71	1	–	–	–	ND	ND	ND	ND	ND	+4.46	+1.37	White, shaley dolostone
2751	tr	10	1	1	–	87	–	–	–	–	ND	ND	ND	ND	ND	+4.85	+1.23	Gray dolostone
2792	7	5	2	2	–	84	–	–	–	–	–	14	6	80	50	+5.26	+0.15	Light gray dolostone
2809	–	5	–	tr	–	95	–	–	–	–	ND	ND	ND	ND	ND	+5.40	+1.17	Light gray dolostone
2889	33	17	5	8	–	37	–	–	–	–	91	6	2	–	–	ND	ND	Red, shaley dolostone
2904	18	20	3	6	–	54	–	–	–	–	–	52	9	39	50	+5.62	+0.32	Red and white sandy dolostone
2932	29	24	–	11	–	36	–	–	–	–	–	10	4	86	50	+4.70	+0.59	Brown, shaley, sandy dolostone
2937	2	1	–	–	–	97	–	–	–	–	–	6	2	92	50	+5.60	+0.48	Brown dolostone
2966	2	1	–	–	–	97	–	–	tr	–	–	25	29	46	50	+5.31	+0.92	Vuggy, green dolostone
3029	44	6	–	4	–	45	–	–	–	–	–	18	4	78	50	+4.78	+1.26	Gray, shaley dolostone
3032	8	3	–	1	–	88	–	–	–	–	–	86	14	–	–	+5.97	+2.21	White dolostone
3041	14	6	5	4	–	69	2	–	–	–	–	89	11	–	–	+5.17	+1.43	White dolostone
ANHYDRITE
2678	2	–	–	–	–	26	–	72	–	–	–	20	5	75	50	+5.69	+1.61	Gray anhydrite
2843	3	2	tr	–	–	–	–	95	–	–	–	37	10	53	50	ND	ND	Red and white anhydrite
2767	1	12	4	5	–	–	–	79	–	–	–	3	–	97	50	ND	ND	Gray, sandy anhydrite
2775	3	13	–	3	–	3	tr	78	–	–	–	11	4	85	50	+5.60	+0.48	Gray, sandy anhydrite

TC = Total Clay	DO = Dolomite	SM = Smectite	ND = Analysis not determined
QZ = Quartz	PY = Pyrite	IL = Illite	– = Below detection
KS = K-feldspar	ANH = Anhydrite	CH = Chlorite	tr = Trace amounts
PL = Plagioclase	GYP = Gypsum	C/S = Chlorite/Smectite	
CA = Calcite	HA = Halite	%EXP C/S = %Smectite in C/S	

X-ray diffraction data are expressed in %; clay mineral data from <2μm fraction stable isotopes are relative to PDB. Results are listed by depth and facies.

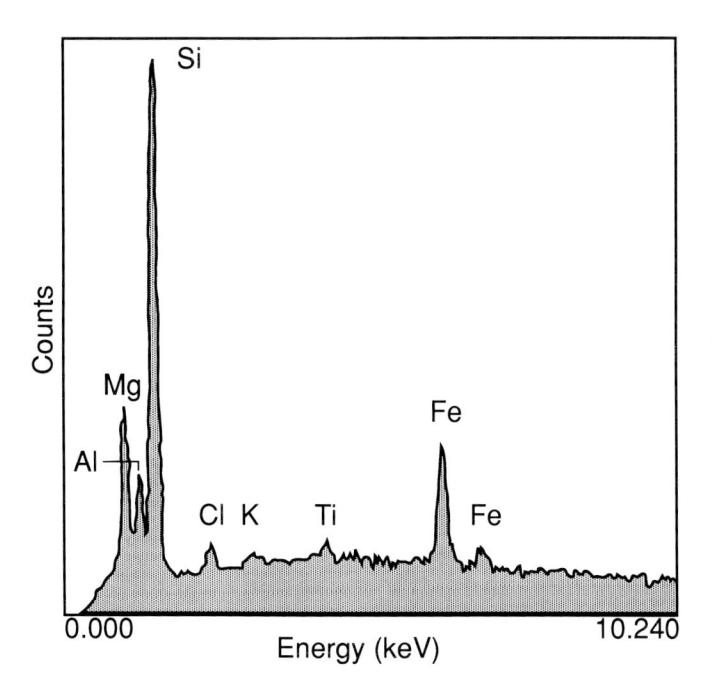

FIG. 4.—Representative energy dispersive spectrum of corrensite (see Fig. 3F) from the Yates sandstone.

Clay Mineralogy

The clay-mineral suite in the Yates Formation was relatively uniform (Tables 1 and 2). In sandstones, corrensite, illite and chlorite were the predominant clay minerals. None of the samples examined was monomineralic. Where present, clay-mineral suites in dolostone and anhydrite layers were similar to those in sandstones. Corrensite was the most abundant clay species, typically ranging from about 50 to 95% of the <2-μm fraction. It was present in all major facies, but was absent from interbedded black shales (Fig. 5). It was more abundant in sandstone facies than in dolostone or anhydrite. In the clay-size fraction in sandstones, corrensite averaged 88.0% in well #269 and 70.4% in well #270. Illite was more abundant in dolostone layers than in sandstones. In dolostones, illite averaged 31.1% in well #269 and 31.3% in well #270. In sandstones, illite was a minor component of the clay fraction, averaging 9.3% in well #269 and 13.3% in well #270. In several dolostone layers (895 m; 2,934 ft in well #269, and 925 m; 3,032 ft and 925 m; 3,034 ft in well #270), illite was the most abundant clay mineral and corrensite was absent. Smectite was the predominant clay mineral in one shaley dolostone sample (881 m; 2,889 ft in well #270).

Corrensite was the only C/S species found in the study wells. We did not observe changes in the ratio of chlorite to smectite in the C/S, regardless of lithology, depth, or cementation in sandstones. In the Yates, the ratio is not related to the amount of dolomite or anhydrite cement present in the sandstone. This uniformity contrasts with the lithology-dependent chlorite-to-smectite ratios found in some evaporitic rocks (Bodine and Madsen, 1987).

Corrensite was recognized by major peak shifts from about 6° and 11.8° 2θ in air-dried samples, to approximately 5.7° and 11.3° upon glycolation. Both the air-dried and ethylene-glycolated samples produced a superstructure d(001) formed from the combination of chlorite (14.2Å) and glycolated smectite (16.9Å) for the ethylene glycol-solvated samples, and chlorite (14.2Å) and hydrated smectite (15Å) for the air-dried samples (Fig. 6). This superstructure is characteristic of R1-ordered chlorite/smectite having a 50/50 composition (Moore and Reynolds, 1989). The presence of the 004 peak at about 11.3° 2θ is considered diagnostic for corrensite (Moore and Reynolds, 1989). The 25.8° glycolated peak was used for quantification of the amount of corrensite present in the <2-μm fraction (Reynolds, 1985).

The 060 reflection was used to differentiate between dioctahedral and trioctahedral types. The interval from 55 to 65° 2θ was slow scanned at a rate of 0.1°/min. Trioctahedral species generally fall in the 59.5 to 60.5° 2θ range and dioctahedral species in the 61 to 62.5° 2θ range (Moore and Reynolds, 1989). Corrensite in the Yates Formation produced a peak at 60.4° 2θ, indicating that it is trioctahedral. A representative EDS spectrum (Fig. 4) showed that the clays contained significant Mg and Fe, but little Al, supporting the trioctahedral interpretation.

Interbedded black shales are almost exclusively mixtures of illite and chlorite, although rare smectitic layers are present—

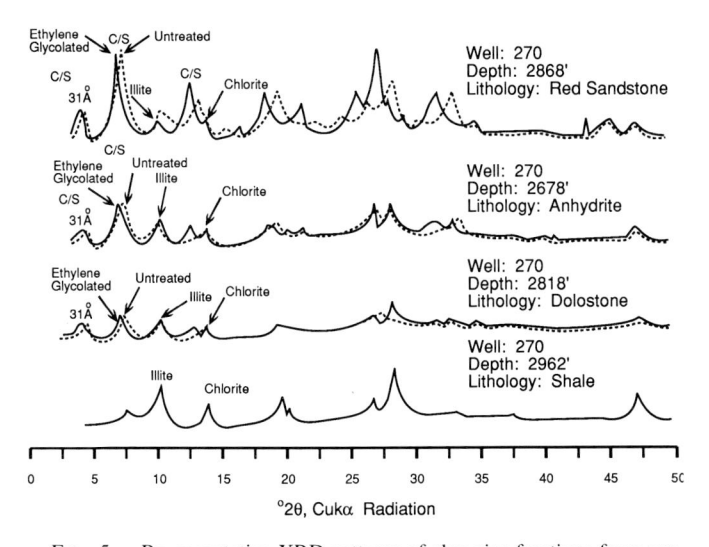

FIG. 5.—Representative XRD patterns of clay-size fractions from various lithologies in the Yates. Corrensite is found in all facies except interbedded shales. Illite and chlorite are usually minor phases. Glycolated and air-dried diffraction patterns for the illite and chlorite mixture in the shale were identical.

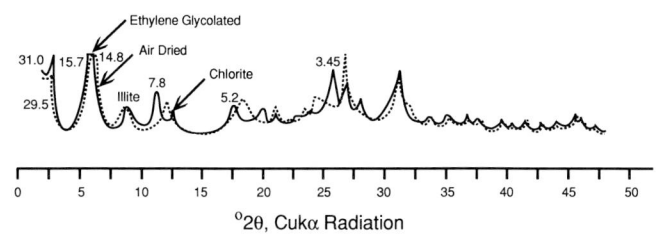

FIG. 6.—Representative XRD pattern of corrensite from a Yates sandstone sample, in air-dried and ethylene glycolated states. The superstructure is present under both conditions (29Å, air dried; 31Å, glycolated). The emergence of the 004 peak upon glycolation (11.3°) is diagnostic of corrensite (Moore and Reynolds, 1989). D spacings are in angstroms.

ent. In shales, the presence of illite is presumed to be an end product of the smectite-to-illite transformation and/or a derivative of a weathered terrain. We found no evidence of illite/smectite in any shale sample. Illite was recognized by the lack of 10Å-peak shift between untreated and glycolated samples. In sandstones, illite is a subordinate clay mineral and generally constitutes from 10 to 50% of the clay-size fraction. In interbedded shales, however, it is the predominant clay mineral. Chlorite was identified by its non-reactivity with ethylene glycol and by the presence of the 7Å and the 3.54Å peaks. It was found in most samples, at levels generally less than 10%.

Stable Isotopes

Stable-isotope values (relative to PDB standard) were determined on dolostone (dolomicrite), anhydrite (where dolomite was present), and dolomite cement from sandstones (Tables 1 and 2). In well #269, layered dolostones were considerably heavier in both $\delta^{13}C$ and $\delta^{18}O$ than dolomite cement in associated sandstones (Fig. 7). In dolostones, $\delta^{13}C$ values ranged from +4.97 to +5.94‰ and $\delta^{18}O$ values from −1.74 to +1.94‰. Dolomite cement in sandstones from well #269 ranged from $\delta^{13}C$ +0.67 to +4.92‰ and $\delta^{18}O$ from −1.68 to +0.91‰. Layered dolostones tended to cluster in the region $\delta^{13}C$ +4.9 to +5.9‰, whereas the sandstones, although exhibiting considerably more scatter, clustered in the region $\delta^{13}C$ +3.5 to +4.9‰ (however, three samples fell in the range +0.5 to +1.00‰).

Stable-isotope values from dolostones in well #270 ranged from $\delta^{13}C$ +4.46 to +5.97‰ and $\delta^{18}O$ from +0.15 to +2.39‰. Dolomite cements in the interlayered sandstones were considerably lighter, ranging from $\delta^{13}C$ +0.97 to +3.26‰ and $\delta^{18}O$ −6.09 to +1.23‰. Dolostone $\delta^{13}C$ values formed a cluster in the range of about +4.5 to +6.0‰, similar to dolostones in well #269; $\delta^{18}O$ values fell in the general range of about 0.0 to +2.4‰. The majority of values from dolomite cement in sandstones fell in the range of about $\delta^{13}C$ +1.0 to +3.5‰, although four samples were between +4.0 and +5.8‰. $\delta^{18}O$ values were significantly lighter in well #270. $\delta^{18}O$ values in the sandstones ranged from about +1.0 to −6.1‰.

$^{40}Ar/^{39}Ar$ Laser Step Heating

Three clay samples of the Yates Formation were analyzed by the $^{40}Ar/^{39}Ar$ step-heating method. The results are shown in the age-spectral diagrams in Figures 8 through 10. In these plots, apparent age, which is deduced from the ratio of radiogenic ^{40}Ar to ^{39}Ar, is plotted against the fraction of ^{39}Ar released for increasing temperature steps. Ideally, for a pure mineral separate that has undergone a simple thermal history with no Ar loss subsequent to crystallization, simple horizontal sequences of ages ("plateaus") should be obtained. By the jagged age-spectra (e.g., Figs. 8 and 9), it is clear that these samples contain a mixture of phases with differing ages and Ar release patterns.

In the three samples tested, there was a distinct, early gas-release fraction with an age of about 275 to 250 Ma. One sample was heated slowly to obtain better resolution of the early degassing phase (Fig. 10). This early fraction degassed at low laboratory temperatures (approximately 300°C), much lower than the expected temperature of gas release for illite (Bray and others, 1987) or mica (Masliwec, 1984; Weaver, 1989; Dallmeyer and Nance, 1990). The higher temperature portions of the age spectra oscillated between Paleozoic and late Precambrian ages, and almost surely represent the degassing of a complex mixture of detrital clays and feldspars.

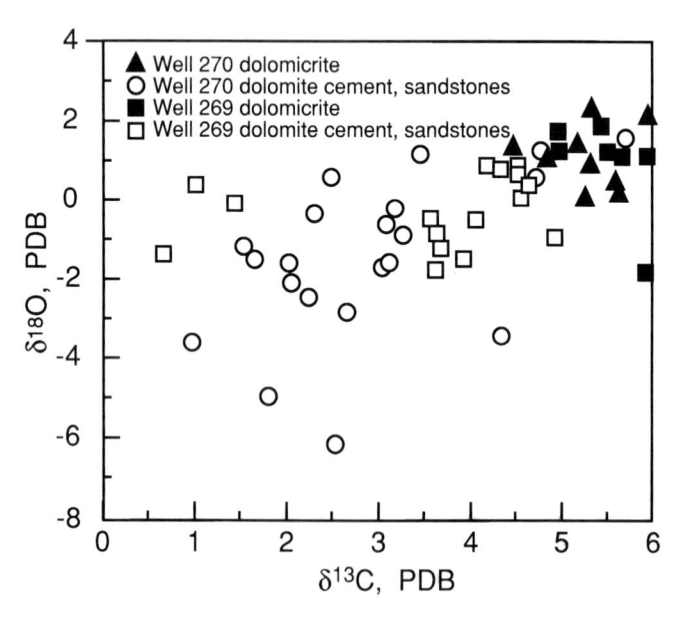

FIG. 7.—Stable-isotope data from wells #269 and #270. Dolostone (dolomicrite, black symbols) data from both wells fall in a narrow range. Dolomite cements in sandstones are generally lighter and more variable than the dolostones. All values are relative to PDB.

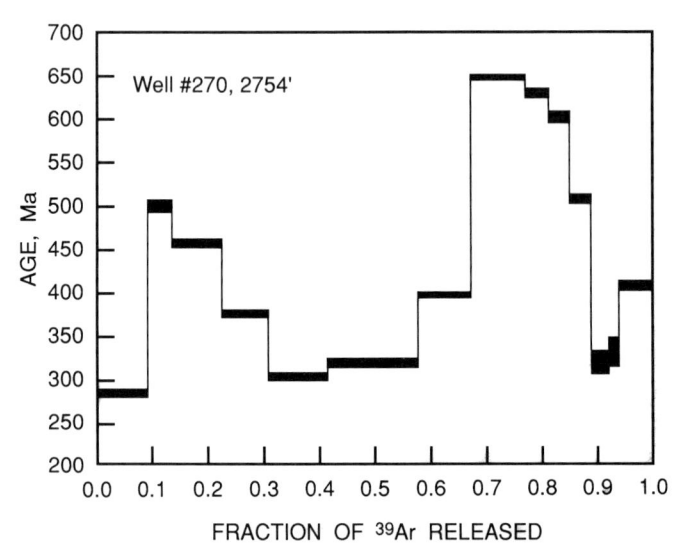

FIG. 8.—$^{40}Ar/^{39}Ar$ step-heating curve for clay fraction from 2,754 ft, well #270. The curve records the absolute age in Ma against the fraction of ^{39}Ar released from the sample until all ^{39}Ar is released (fraction ^{39}Ar = 1.0). One early phase records a Guadalupian-Leonardian age (about 275 Ma). Two events suggest mineral phases about 325 to 310 Ma. The remaining ages are most likely "mixtures" of detrital and possibly authigenic minerals.

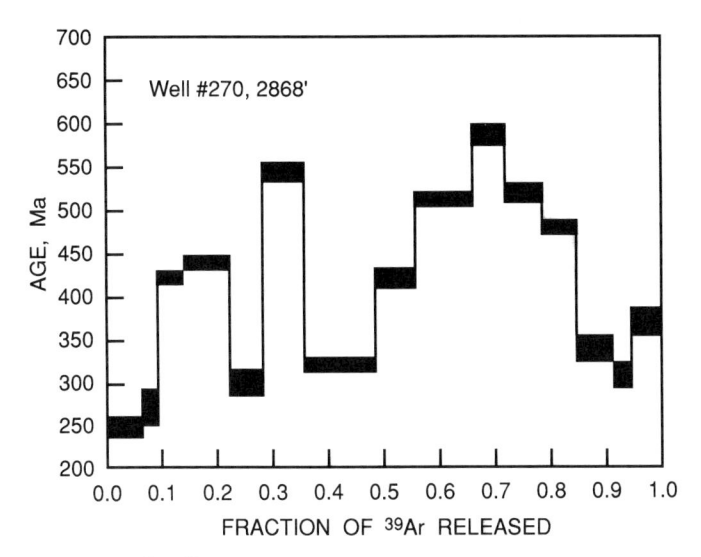

FIG. 9.—^{40}Ar/^{39}Ar step-heating curve from clays at 2,868 ft in well #270. The early phase is shown by the two rectangles about 270 Ma and 250 Ma. These two releases should probably be averaged.

It is interesting to note that several portions of the age spectra in Figures 8 and 9 hint at an event at about 325 to 310 Ma. Also, in both of these spectra, gas fractions yielded apparent ages reaching into the late Precambrian. Unfortunately, because of the likely possibility of mixing occurring among phases of different crystallization ages in this region of the spectra, it is unwise to attach too much significance to these "ages."

DISCUSSION

Diagenesis Of The Yates Sandstones

Sandstones of the Yates Formation have undergone a complex history of cementation, replacement, and leaching during burial diagenesis. The major diagenetic modifications to the sandstones are: (1) mechanical compaction and early precipitation of dolomite and sulfate; (2) cement and unstable-grain dissolution (lithic fragments and feldspars); (3) precipitation of authigenic corrensite; (4) precipitation of authigenic K-feldspar as overgrowths and discrete crystals; (5) minor quartz cementation; and (6) late stage precipitation of dolomite and, locally, halite. Although it is difficult to construct a sequence in which all diagenetic events are placed on a relative time scale, it is possible to identify some early and late events. Earlier events include compaction, dolomite and sulfate cementation, cement and grain dissolution, and the formation of authigenic corrensite. Later events include precipitation of K-feldspar, later stage dolomite, quartz, and halite. Diagenetic events commonly overlap. Dolomite cement precipitated at two distinct times, and distinction between the two phases is sometimes difficult. Dissolution of feldspars and/or rock fragments may be significant in one sandstone layer, but not in another. This generalized paragenesis cannot be identified in all samples.

The predominant authigenic minerals in Yates sandstones are corrensite and dolomite. Both of these minerals formed early in the burial history (although a late-stage dolomite cement also occurs), by the interaction of the sediment with Mg-rich sabkha waters. Later diagenetic events produced authigenic K-feldspar, but this later event is poorly understood.

Corrensite is the only C/S species present in the Yates, despite variability in stratigraphy, lithology and cementation. In sandstones, the chlorite-to-smectite ratio is unaffected by grain size or the amount of anhydrite and dolomite cement (Tables 1 and 2). This indicates that porewater chemistry remained favorable for corrensite stability, as well as occasional anhydrite precipitation, in both sandstone and dolostone facies. In the sandstones, corrensite is clearly authigenic, occurring as a pore lining with individual crystals oriented perpendicular to grain surfaces (Fig. 3F). We did not observe transitions from detrital mica or clay to corrensite, nor corrensite occurring in individual laminae. Such occurrences would suggest that corrensite formed from a pre-existing detrital clay, a mechanism widely proposed for C/S formation in evaporite facies (Padan, 1984; Palmer, 1987; Bodine and Madsen, 1987; Fisher, 1988; Weaver, 1989). Rather, corrensite formed from sediment-pore-fluid interaction, either directly as a precipitate (unlikely) or indirectly as a diagenetic alteration of an earlier precursor Mg-silicate. Direct precipitation of corrensite was unlikely, as its precipitation has not been documented in any modern depositional environment. However, alteration of a precursor Mg-clay by the addition of brucite layers is a possibility. A likely precursor was Mg-smectite, whose precipitation has been reported in modern environments (Gac and others, 1977; Jones and Weir, 1983). The addition of brucite layers to this Mg-smectite (saponite?) would form corrensite. Magnesium levels, which control the formation and stability of brucite layers in clays in contact with pore fluids, generally increase as the salinity of the pore fluid increases (Weaver, 1989). That abundant Mg was available is beyond question, as elsewhere in the Yates even pure magnesite has been reported (Crawford and Dunham, 1983;

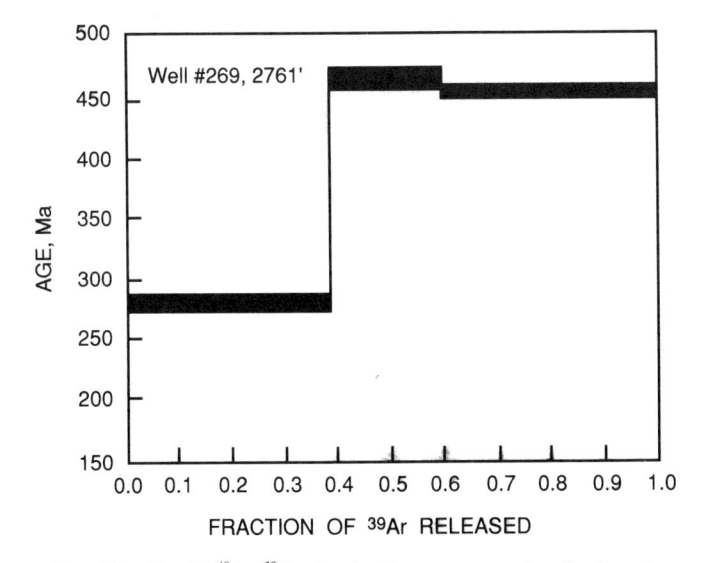

FIG. 10.—Partial ^{40}Ar/^{39}Ar step-heating curve on clay fraction from 2,761 ft, well #269. This analysis was performed at a lower step-heating rate to emphasize the early decomposing phase.

Garber and others, 1990). Pore-fluid salinity (and Mg) levels must have been sufficiently high, and remained relatively constant, over the early diagenetic history of the sandstone for the precipitation of only corrensite. Had conditions been more evaporative, creating more saline, Mg-rich pore fluids, it is likely that other C/S ratios would have developed (Bodine and Madsen, 1987).

Clay-mineral suites and chlorite-to-smectite ratios are known to vary with lithology in evaporite settings (Padan, 1984; Weaver, 1989). In lithologies such as dolomite and anhydrite, corrensite and illite are the predominant clay minerals. Illite, which can precipitate under some mildly evaporitic conditions, is commonly associated with corrensite in dolostones (Weaver, 1989; Weaver and others, 1990). In the Yates, illite is more abundant in dolostone layers than in adjacent sandstone layers; dolostones average about 30% illite, whereas sandstones average about 10% illite. Furthermore, petrographic observations indicate that most, if not all, of the illite in the sandstones is detrital. The most stable phase in the interlayered Yates sandstones is corrensite. Relative changes in corrensite and illite percentages in dolostone and sandstone facies reflect changes in the pore-water chemistry. Small increases in pH and/or Mg in the sandstone pore water would favor corrensite over illite (Weaver, 1989).

In addition to Mg, there must also have been sufficient Si available for corrensite (and its precursor) formation. Although corrensite has yet to be found precipitating directly in any modern environment, some Mg-silicates have. In Lake Chad, Mg-smectite formation has been reported (Gac and others, 1977). Magnesian-clay minerals have also been reported in evaporative and alkaline lakes (Jones and Weir, 1983; Pueyo Mur and Urpinell, 1987). In some of these environments, Si concentrations in pore fluids parallel those of Mg. By precipitation of magnesian clays, the chemistry of the water is partially controlled. We have no ready explanation for the sources of Si necessary to form corrensite in the Yates Formation, although one potential source may be partially dissolved rock fragments and feldspars (Fig. 3F).

The coastal sabkhas of the Mediterranean region are probable modern analogs for deposition of the Yates Formation. In the sabkha environment, sporadic rainfall, sea water, evaporative brine, and subsurface fluids all affect the porewater chemistry. As Ca is precipitated and the Mg/Ca ratio rises, a dense Mg-bearing hypersaline pore fluid forms, sinks, and flows seaward through the sabkha sediments (Warren, 1989). Mg concentrations are highly variable; ground water beneath Trucial Coast sabkhas contains 8,000 to 18,000 ppm Mg (Butler, 1969), whereas other sabkhas can reach as high as 47,000 ppm Mg (Glennie and Evans, 1976). Complicating the picture are periodic influxes of continental water, which mix with waters in the sabkha environment (Kendall and Warren, 1988).

Whereas C/S has been extensively studied in evaporites and certain sandstones, its occurrence in sabkha sandstones has not. Corrensite forms in some redbed and eolian sandstones (Morrison and Parry, 1986), and where volcanogenic debris is present (Dodd and others, 1955; Almon and others, 1976; Klimentidis and others, 1990). Although rock fragments and feldspars are present in the Yates, there is no evidence that volcaniclastic detritus was ever present. Almon and others (1976) noted that, even where volcaniclastic debris was present, authigenic-clay formation was dependent upon depositional facies; deltaic sandstones contained corrensite, whereas shallow marine and beach sandstones contained montmorillonite.

At some point after corrensite formation, pore waters in the sandstones changed so that K-feldspar was the stable phase. K-feldspar occurs as overgrowths and authigenic crystals, which clearly postdate corrensite (Fig. 3F). K-feldspar precipitation represented a change to more acidic conditions, perhaps by an influx of more continental waters. The source of the potassium, while unknown, could have been brines associated with evaporite formation. These become enriched in K (and Na) as Ca and Mg are precipitated. Another source of both Si and Al could have been early dissolution of detrital feldspars, clays and muscovite within the Yates.

Stable Isotopes

The range of stable-isotope values obtained from dolostones differs from the range of values obtained from dolomite cement in sandstones (Fig. 7). Dolostone values fall in a narrow range, whereas the values for dolomite cement from sandstones are more variable. Isotope compositions of neither are correlated with stratigraphy or facies, suggesting isotopic homogenization throughout the sediments during diagenesis. Stable isotopes from dolostones are similar to those of other fine-grained, marine-derived dolomites (Choquette and Steinen, 1980; McKenzie, 1981; Pierre, 1988), but are slightly higher in $\delta^{13}C$. Isotopic signatures of the dolostones are similar to those of Persian Gulf deposits, where dolomites develop by replacive crystal growth after dissolution by Mg-rich fluids and marine-derived brines (Pierre, 1988).

In sandstones, isotope values from dolomite cements are considerably lighter and more variable (Fig. 7). Furthermore, $\delta^{13}C$ values in well #270 are lighter than $\delta^{13}C$ values in well #269. $\delta^{18}O$ values are variable in well #270, but at least 7 samples fall below the lowest value from sandstones in well #269. The relation between the percent dolomite in the sandstone and the $\delta^{18}O$ values suggests that there is either a correlation between the water/rock ratio and the $\delta^{18}O$ values, or that higher dolomite percentages are reflecting an earlier and heavier cement (Fig. 11). Higher dolomite content may indicate a greater abundance of early precipitated, pore-filling dolomite, whereas lower levels may indicate a later stage dolomite. Dolomite cement in secondary pores is commonly smaller and less abundant than in primary pores, supporting the latter interpretation. If the heavier isotopic signature is indicative of early cementation history, the temperatures suggested are close to ambient conditions. The $\delta^{18}O$ values are influenced by complex interaction of atmosphere-water-sediment, temperature, and evaporating solutions. Later dolomite cement may have been affected by porewater mixing, as well as temperature differences. It is unlikely that temperature alone can account for the variability observed in the oxygen isotopes. Rather, the signatures reflect changes in climatic and vegetation

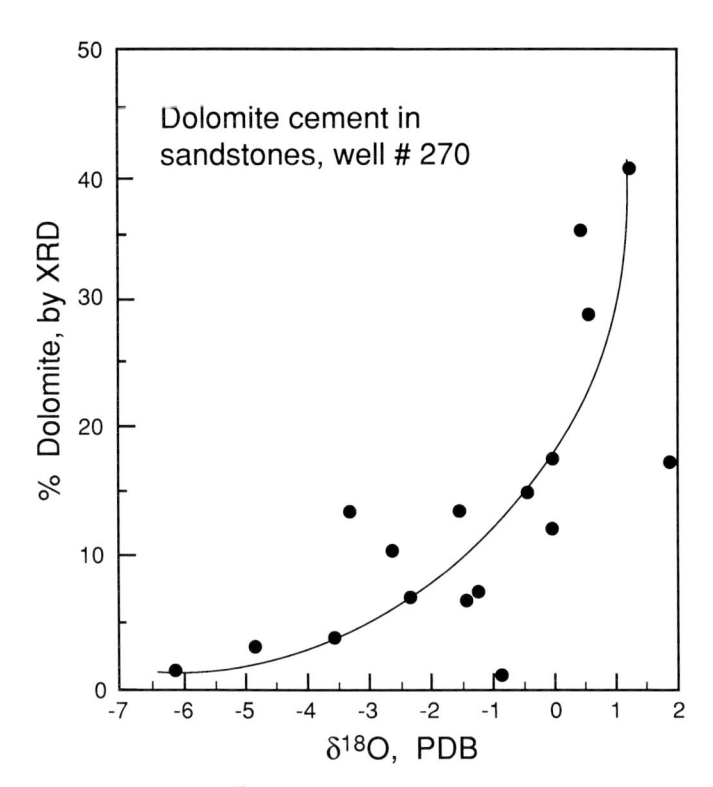

FIG. 11.—Plot of $\delta^{18}O$ vs. percent dolomite (as determined by XRD) in sandstones from well #270. The isotopic signature is a function of the amount of dolomite cement present in the sandstone.

fluctuations or the progressive changes in rock-water interaction (Meyers and Lohmann, 1985). Magaritz (1983) noted that increasing the water/rock ratio decreased the $\delta^{18}O$ value. Evaporation cycles in brines also influence isotopic fractionation (Pierre, 1988).

In sandstones, the lighter carbon-isotopic signature could indicate the incorporation of small amounts of biogenic CO_2 or an influx of continental ground water (Pierre, 1988). Changes in $\delta^{13}C$ in carbonates may also reflect the sediment-water system. If the system is closed for CO_2, the progressive crystallization of carbonates leads to a decrease of $\delta^{13}C$ values in the $\Sigma\ CO_2$ as well as in the precipitated carbonate (Deuser and Degens, 1967).

$^{40}Ar/^{39}Ar$ Age Dating

The timing of clay-mineral formation is important to an understanding of the history of the Yates Formation as a hydrocarbon reservoir. From the measured $^{40}Ar/^{39}Ar$ age spectra (Figs. 8 to 10), it is clear that most of the analyzed clay samples have apparent ages that predate deposition of the Yates, and therefore probably represent detrital phases. There are several degassing stages centering around 325 to 310 Ma, which hint at some event for that period. The remainder of the data probably represent the ages of the detrital fraction. The only portions of the age spectra that might represent a diagenetic phase are the early, low-temperature gas fractions. These analyses consistently yielded apparent ages in the 275 to 250-Ma range, the approximate age of the Yates Formation. We tentatively interpret the 275 to 250-Ma age as the age of corrensite formation, although there is no plateau age supporting this hypothesis. We cannot demonstrate conclusively that corrensite is solely responsible for the low-temperature fractions, but we would expect that Ar would degas in the following approximate order:

smectite, I/S, C/S > illite, mica > feldspar, hornblende.

At the low temperature of the early gas release ($\approx 300°C$), we would not expect a significant contribution from either illite or any fine-grained feldspars. Of course, the presumed diagenetic ages could be an artifact of ^{39}Ar recoil (Turner and Cadogan, 1974). If the low-temperature fraction represents a low-K phase that has received ^{39}Ar from a neighboring high-K illite, its apparent age would be lowered. Thus, the low-K phase might be another detrital component. Alternatively, the low-temperature ages of 275 to 250 Ma might be too high because of net ^{39}Ar loss due to recoil. In either case, special consideration would have to apply for the same measured ages to be obtained. The simpler interpretation, which we prefer, is that the 275 to 250-Ma age is a reasonable first approximation for the time of corrensite formation.

CONCLUSIONS

Samples from two cores from the Yates Formation in the S.M. Halley field, Winkler County, Texas, were analyzed by petrographic, geochemical and $^{40}Ar/^{39}Ar$ methods. The Yates Formation is a series of interlayered sandstone, dolostone, and anhydrite, which was deposited in a coastal sabkha-like environment. Mg-based diagenesis was predominant and responsible for the formation of corrensite and dolomite cement in sandstones. Throughout the Yates, R1-ordered, trioctahedral corrensite is the predominant authigenic clay mineral. The chlorite-to-smectite ratio is invariant with stratigraphic position or lithology. In sandstones, the degree of dolomite or anhydrite cementation does not affect this ratio. The presence of corrensite in the Yates is consistent with clay-mineral suites found in the dolomite-anhydrite facies of evaporitic rocks. Other clay minerals include illite and chlorite. Interbedded black shales consist predominantly of illite and chlorite; corrensite is absent. Stable-isotope data from bedded dolostones and intergranular dolomite cement in sandstones suggest that the dolostone beds formed under evaporitic conditions. Stable-isotope values from dolostones range from $\delta^{13}C$ +4.46 to +5.97 ‰ and $\delta^{18}O$ from −1.74 to +2.39‰. Dolomite cements in sandstones are isotopically lighter and more variable than the dolostones ($\delta^{13}C$: +0.67 to +4.92; $\delta^{18}O$: −6.09 to +1.23‰). Absolute ages of three clay-size fractions, determined by the $^{40}Ar/^{39}Ar$ laser step-heating method, range from 650 to 250 Ma. A more limited 275- to 250-Ma age may represent the absolute age of corrensite formation, whereas the older ages probably represent the age of the source terrain.

ACKNOWLEDGMENTS

The authors express their deep appreciation to Texaco Denmark, Inc., Texaco Inc. and Texaco USA (Western Region) for granting permission to publish this paper. We sincerely thank Drs. James J. Howard and Eric V. Eslinger, who reviewed the manuscript and offered much-needed suggestions and criticisms. The authors also thank their colleagues A. R. Thomas, M.D. Hogg and T. C. O'Hearn for reviewing an earlier version of the manuscript and offering corrections and criticisms. We are grateful to Dr. Derek York for his help and for providing laboratory time and resources for the argon analyses. Financial support for CMH was provided by the Natural Sciences and Engineering Research Council of Canada from an operating grant to Derek York. C. A. Callender and Dan Tolopka (both of Texaco EPTD) assisted in the EDS and X-ray diffraction analyses (respectively). We thank Angela R. Birdow for preparing and proofreading the manuscript.

REFERENCES

ALMON, W. R., FULLERTON, L. B., AND DAVIES, D. K., 1976, Pore space reduction in Cretaceous sandstones through chemical precipitation of clay minerals: Journal of Sedimentary Petrology, v. 46, p. 89–96.

APRIL, R. H., 1980, Regularly interstratified chlorite/vermiculite in contact metamorphosed red beds, Newark Group, Connecticut Valley: Clays and Clay Minerals, v. 28, p. 1–11.

ATAMAN, G., AND GOKÇEN, S. L., 1975, Determination of source and palaeoclimate from the comparison of grain and clay fractions in sandstones: a case study: Sedimentary Geology, v. 13, p. 81–107.

BLATTER, C. L., ROBERSON, H. E., AND THOMPSON, G. R., 1973, Regularly interstratified chlorite-dioctahedral smectite in dike-intruded shales, Montana: Clays and Clay Minerals, v. 21, p. 207–212.

BODINE, M. W., 1978, Clay mineral assemblages from drill core of Ochoan evaporites, Eddy County, New Mexico: New Mexico Bureau of Mines Mineral Research, Circular 159, p. 21–31.

BODINE, M. W., 1983, Trioctahedral clay mineral assemblages in Paleozoic marine evaporite rocks, in Schreiber, B. C., and Harner, H. L., eds., Sixth International Symposium on Salt, v. 1: Salt Institute, Alexandria, VA, p. 267–283.

BODINE, M. W., AND MADSEN, B. M., 1987, Mixed-layer chlorite/smectites from a Pennsylvanian evaporite cycle, Grand County, Utah: Proceedings, International Clay Conference 1985, Denver, CO: Clay Mineral Society, Bloomington, IN, p. 85–96.

BRAY, C. J., SPOONER, E. T. C., HALL, C. M., YORK, D., BILLS, T. M., AND KRUEGER, H. W., 1987, Laser probe $^{40}Ar/^{39}Ar$ and conventional K-Ar dating of illites associated with the McClean unconformity-related uranium deposits, North Saskatchewan, Canada: Canadian Journal of Earth Science, v. 24, p. 10–23.

BUTLER, G. P., 1969, Modern evaporite deposition and geochemistry of coexisting brines, the sabkha, Trucial Coast, Arabian Gulf: Journal of Sedimentary Petrology, v. 39, p. 70–89.

CHANG, H. K., MACKENZIE, F. T., AND SCHOONMAKER, J., 1986, Comparisons between the diagenesis of dioctahedral and trioctahedral smectite, Brazilian offshore basins: Clays and Clay Minerals, v. 34, p. 407–423.

CHOQUETTE, P. W., AND STEINEN, R. P., 1980, Mississippian non-supratidal dolomite, Ste. Genevieve Limestone, Illinois Basin: evidence for mixed-water dolomitization, in Zenger, D. H., Dunham, J. B., and Ethington, R. L., eds., Concepts and Models of Dolomitization: Society of Economic Paleontologists and Mineralogists Special Publication 28, p. 163–196.

CRAWFORD, G. A., AND DUNHAM, J. B., 1983, Sedimentation in the Permian Yates Formation, Dollarhide Field, Central Basin Platform, Andrews County, west Texas: Permian Basin Section, Society of Economic Paleontologists and Mineralogists Core Workshop No. 2, p. 225–271.

DALLMEYER, R. D., AND NANCE, R. D., 1990, $^{40}Ar/^{39}Ar$ ages of detrital muscovite within early Paleozoic overstep sequences, Avalon composite terrane, southern New Brunswick: implications for extent of late Paleozoic tectonothermal overprint: Canadian Journal of Earth Science, v. 27, p. 1209–1214.

DEUSER, W. G., AND DEGENS, E. T., 1967, Carbon isotope fractionation in the system CO_2 (gas)-CO_2 (aqueous)-HCO_3 (aqueous): Nature, v. 215, p. 1033–1035.

DODD, C. G., CONLEY, F. R., AND BARNES, P. M., 1955, Clay minerals in petroleum reservoir sands and water sensitivity effects: National Academy of Science, National Research Council Publication 395, p. 221–238.

DREVER, J. I., 1973, The preparation of oriented clay mineral specimens for X-ray diffraction analysis by a filter-membrane peel technique: American Mineralogist, v. 58, p. 553–554.

DUNHAM, R. J., 1962, Classification of carbonate rocks according to depositional texture: American Association of Petroleum Geologists Memoir 1, p. 108–121.

FISHER, M. J., AND JEANS, C. V., 1982, Clay mineral stratigraphy in the Permo-Triassic red bed sequences of BNOC 72/10-1A, western approaches, and the south Devon coast: Clay Minerals, v. 17, p. 79–89.

FISHER, R. S., 1988, Clay minerals in evaporite host rocks, Palo Duro Basin, Texas Panhandle: Journal of Sedimentary Petrology, v. 58, p. 836–844.

FOLK, R. L., 1974, Petrology of sedimentary rocks: Austin, Texas, Hemphill Publishing Co., 182 p.

FRANCO, L. A., 1973, Deposition and diagenesis of the Yates Formation, Guadalupe Mountains and Central Basin Platform: Unpublished M.S. Thesis, Texas Tech University, Lubbock, TX, 67 p.

GAC, J. Y., DROUBI, A., FRITZ, B., AND TARDY, Y., 1977, Geochemical behavior of silica and magnesium during the evaporation of waters in Lake Chad: Chemical Geology, v. 19, p. 215–222.

GARBER, R. A., HARRIS, P. M., AND BORER, J. M., 1990, Occurrence and significance of magnesite in Upper Permian (Guadalupian) Tansill and Yates Formations, Delaware Basin, New Mexico: American Association of Petroleum Geologists Bulletin, v. 74, p. 119–134.

GLENNIE, K. W., AND EVANS, G., 1976, A reconnaissance of the recent sediments of the Ranns of Kutch, India: Sedimentology, v. 23, p. 625–647.

HALL, C. M., REDMAN, J. D., LAYER, P. W., AND YORK, D., 1988, $^{40}Ar/^{39}Ar$ ages from a fully automated laser microprobe: EOS Transactions, American Geophysical Union, v. 69, p. 520.

HAUFF, P. H., AND MCKEE, E. D., 1982, Clay mineralogy, in McKee, E. D., ed., The Supai Group of the Grand Canyon: U.S. Geological Survey Professional Paper 1173, p. 287–332.

JONES, B. F., AND WEIR, A. H., 1983, Clay minerals of Lake Albert, an alkaline, saline lake: Clays and Clay Minerals, v. 31, p. 161–172.

KENDALL, C. G. ST. C., AND WARREN, J. K., 1988, Peritidal evaporites and their sedimentary assemblages, in Schrieber, B. C., ed., Evaporites and Hydrocarbons: New York, Columbia University Press, p. 66–138.

KLIMENTIDIS, R. E., PEVEAR, D. R., AND ROBINSON, G. A., 1990, Analysis of CMS special clay "CorWa-1," a corrensite pore-filling in an Eocene mafic volcaniclastic sandstone from Washington State: Programs and Abstracts, Clay Minerals Society 27th Annual Meeting, Columbia, Missouri, p. 72.

KOPP, O. C., AND FALLIS, J. M., 1974, Corrensite in the Willington Formation, Lyons, Kansas: American Mineralogist, v. 59, p. 623–624.

KRISTMANNSDOTTIR, K. B., 1978, Alteration of basaltic rocks by hydrothermal activity at 100–300°C: Proceedings, International Clay Conference, Oxford, p. 359–367.

LAYER, P. W., HALL, C. M., AND YORK, D., 1987, Derivation of $^{40}Ar/^{39}Ar$ age spectra–single grains of hornblende and biotite by laser step-heating method: Geophysical Research Letters, v. 14, p. 757–760.

MAGARITZ, M., 1983, Carbon and oxygen isotope composition of Recent and ancient coated grains, in Peryt, T. M., ed., Coated Grains: New York, Springer-Verlag, p. 27–37.

MASLIWEC, A., 1984, Applicability of the $^{40}Ar/^{39}Ar$ method to the dating of ore bodies: Unpublished Ph.D. Dissertation, University of Toronto, Toronto, 187 p.

MCKENZIE, J. A., 1981, Holocene dolomitization of calcium carbonate sediments from the coastal sabkhas of Abu Dhabi, U.A.E.: A stable isotope study: Journal of Geology, v. 89, p. 185–189.

MEYERS, W. J., AND LOHMANN, K. C., 1985, Isotope geochemistry of regionally extensive calcite cement zones and marine components in

Mississippian limestones, New Mexico, *in* Schneidermann, N., and Harris, P. M., eds., Carbonate Cements: Society of Economic Paleontologists and Mineralogists Special Publication 36, p. 223–240.

MOORE, D. M., AND REYNOLDS, JR., R. C., 1989, X-ray diffraction and the identification and analysis of clay minerals: New York, Oxford University Press, 332 p.

MORRISON, S. J., AND PARRY, W. T., 1986, Dioctahedral corrensite from Permian red beds, Lisbon Valley, Utah: Clays and Clay Minerals, v. 34, p. 613–624.

PADAN, A., 1984, Clay mineralogy of the bedded salt deposits in the Paradox Basin, Gibson Well No. 1, Utah: Unpublished Ph.D. Dissertation, Georgia Institute of Technology, Atlanta, Georgia, 272 p.

PALMER, D. P., 1987, A saponite and chlorite-rich clay assemblage in Permian evaporite and red-bed strata, Palo Duro Basin, Texas panhandle: University of Texas Bureau of Economic Geology Circular 87-3, 21 p.

PIERRE, C., 1988, Applications of stable isotope geochemistry to the study of evaporites, *in* Schreiber, B. C., ed., Evaporites and Hydrocarbons: New York, Columbia University Press, p. 300–344.

PUEYO MUR, J. J., AND URPINELL, M. I., 1987, Magnesite formation in recent playa lakes, Los Monegros, Spain, *in* Marshall, J. D., ed., Diagenesis of Sedimentary Sequences: Geological Society of America Special Publication 36, p. 119–122.

REYNOLDS, R. C., Jr., 1985, Principles and techniques of quantitative analysis of clay minerals by X-ray diffraction methods: Preconference Workshop, 1985 International Clay Conference, Denver, Colorado, 148 p.

SCHULTZ, L.G., 1964, Quantitative interpretation of mineralogical composition from X-ray and chemical data for the Pierre Shale: U.S. Geological Survey Professional Paper 391-C, p. C1-C31.

TURNER, G., AND CADOGAN, P. H., 1974, Possible effects of ^{39}Ar recoil in ^{40}Ar/^{39}Ar dating: Proceedings, 5th Lunar Conference, Supplement 5, Geochemica et Cosmochemica Acta, v. 2, p. 1601–1615.

WALKER, T. R., 1975, Diagenetic origin of continental redbeds, *in* The Continental Permian in Central, West and South Europe: Proceedings, NATO Advanced Study Institute, D. Reidel Co., Boston, p. 240–282.

WARREN, J. K., 1989, Evaporite sedimentology: Englewood Cliffs, NJ, Prentice Hall, 285 p.

WEAVER, C. E., 1989, Clays, muds, and shales: Developments in Sedimentology, v. 44: New York, Elsevier, 819 p.

WEAVER, C. E., BECK, K. C., AND CARR, M. K., 1990, Physils also grow in carbonate rocks: Program and Abstracts, Clay Minerals Society 27th Annual Meeting, Columbia, MO, p. 127.

WELTON, J. E., 1984, SEM petrology atlas: American Association of Petroleum Geologists Methods in Exploration Series, 237 p.

WILSON, M. J., 1971, Clay mineralogy of the Old Red Sandstone (Devonian) of Scotland: Journal of Sedimentary Petrology, v. 41, p. 995–1007.

INFILTRATED MATERIALS IN CRETACEOUS VOLCANOGENIC SANDSTONES, SAN JORGE BASIN, ARGENTINA

THOMAS L. DUNN

Department of Geology & Geophysics, University of Wyoming, Laramie, Wyoming 82070

ABSTRACT: Microcrystalline intergranular material of varying compositions occurs as well-developed geopetal structures, cutans, and massive pore fillings within subsurface samples of the fluvial, volcanogenic sandstones of the Comodoro Rivadavia Formation. These intergranular materials occur either as mixtures of chlorite, smectite and iron oxides, or as mixtures thereof that contain abundant quartz. Electron microprobe analyses of the material show that the quartz-rich material contains from 71 to 91% SiO_2.

The textures are interpreted to be the result of infiltration. It is suggested that the microcrystalline quartz-rich material is the alteration product of originally fine-grained volcanic glass, which was infiltrated into the sands from suspended loads of the "Comodoro Rivadavia" rivers during the Cretaceous. This inclusion of fine-grained glassy material is attributed to the erosion of airfall ashes that intermittently blanketed the drainage basin.

The formation of the quartz-rich material requires both a siliceous source and additional silica derived from hydration reactions of adjacent volcanogenic sediment. The erosion of volcanic-ash deposits provides for intermittent, concentrated, suspended loads, resulting in a variable source of solutes during early burial. These variations of suspended-load composition are potentially overlooked features of arid and semiarid floodplain deposits, which are proximal to contemporaneous volcanism (e.g., rift, intermontane, forearc, and intra-arc settings). Infiltrated glassy materials are a potential source of compositional and textural heterogeneity in volcanogenic sandstones that influences the path of later diagenetic alteration and the change of permeability and porosity of those sandstones.

INTRODUCTION

Infiltration within Volcanogenic Regions

Textures in ancient clastic deposits interpreted to be the result of infiltration of clay minerals have been recognized by Crone (1975), Walker and others (1978), Walker (1976), Moraes and De Ros (1990, and this volume), and Matlack and others (1989). Infiltration textures also have been experimentally produced, using a variety of clay types and concentrations of suspended loads, by Crone (1975) and Matlack and others (1989). Within settings where contemporaneous volcanism is prevalent, fluvial systems can be sporadically overwhelmed with volcanic debris (e.g., Davies and others, 1979). Suspended loads derived from volcanogenic deposits can contain a variety of materials. Dethier and others (1981) found that <2-μm size fractions from tephra and flows of the 1980 Mount St. Helens eruption (principally reworked material from the volcanic edifice) contained smectite, mixed-layer smectite-chlorite, chlorite, cristobalite, glass and plagioclase. Volcanoes are sites of widely varying thermal and geochemical regimes, hence, an assortment of clays, glass and other minerals is reasonable. This paper presents observations and discussions of some infiltrated materials whose composition and occurrence suggest that volcanic input was important to their origin. The material for this study was obtained from well cuttings and conventional core taken from fields and wildcat wells within the Amoco Argentina Oil Company (AAOC) Contract Area of the San Jorge Basin, southern Argentina (Figs. 1 and 2). Over 390 samples from 22 wells were examined.

Geologic and Burial Histories of the San Jorge Basin

The Comodoro Rivadavia Formation (CRF) is an upper Cretaceous nonmarine sequence recognized on the northern flank of the San Jorge Basin (Figs. 2 and 3). Fitzgerald and others (1990) also used the name Comodoro Rivadavia to refer to equivalent sequences across the basin. Basin subsidence was initiated during the Early Jurassic as a consequence of the opening of the South Atlantic. The rift cut deep into the interior of the South American continent, placing parts of the basin close to the western convergent margin. The sandstones of the CRF contain abundant volcanogenic and plutonic detritus eroded from arc complexes to the west, along the Pacific convergent margin and the massifs to the north. Discrete volcaniclastic units, such as lahars and airfall tuffs, are recognizable throughout the Cretaceous sequence.

Examination of the extent of clay reactions within the shales, bottom-hole temperatures (BHT), and thermal-history models suggests that the Comodoro Rivadavia Formation has not undergone deep burial or high temperatures within the study area. Bottom-hole temperatures from 20 wells between 1,264 and 2,630 m (4,147 and 8,628 ft) were obtained from electric logs. No information regarding time since circulation was available. The temperature measurements ranged from 52°C (126°F) at 1,264 m (PCD-6) to a maximum of 95°C (203°F) at 2,313 m (PZ-801), with calculated linear temperature gradients ranging from 25.9°C/km (1.4°F/100 ft) to 41.2°C/km (2.3°F/100 ft) and a mean value of 32.3°C/km (1.8°F/100 ft) based on a 10°C surface temperature and assuming the BHTs are representative (Fig. 4).

Shale samples (cuttings) from two wells, PE-1 and PSA-1, were collected for X-ray diffractometry analysis to help ascertain paleotemperature. Shale pieces were hand picked from washed cuttings. Oriented clay mounts of magnesium-saturated and ethylene-glycolated <2-μm size fraction samples were prepared using techniques described by Drever (1973). Each set of samples shows an increase in chlorite peak intensities and a decrease in smectite peak intensities with increasing depth (Fig. 5). The uppermost sample of PE-1 contains kaolinite, which interferes with the chlorite peaks, but the trend remains discernable. Within these samples, the amounts of chlorite are low and ordered smectite-chlorites are not recognized. Chang and others (1986) observed ordered saponite-chlorite in shales from offshore Brazilian rift basins at approximately 70°C. Hoffman and

160   THOMAS L. DUNN

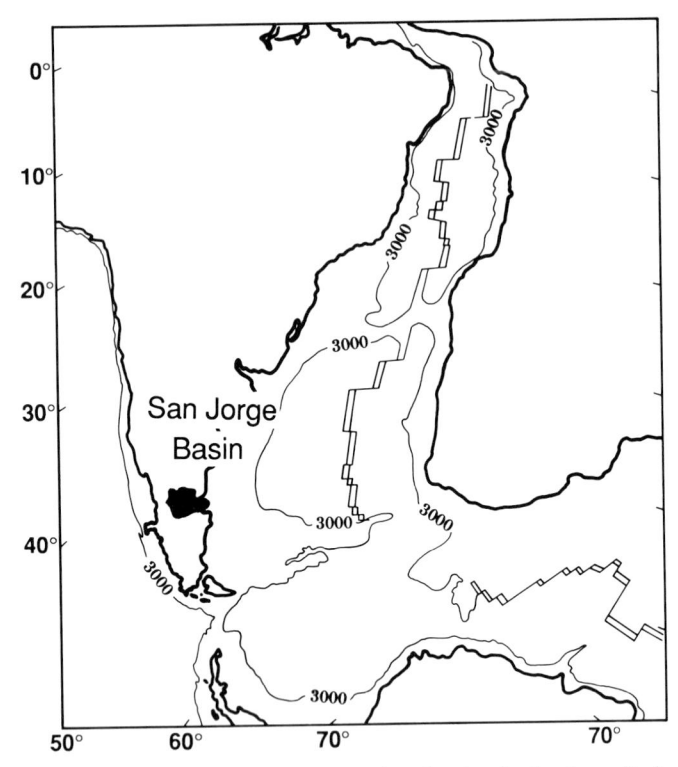

FIG. 1.—Late Cretaceous reconstruction showing the San Jorge Basin with present continental outlines and 3,000-m contours (after Lawver and others, 1984).

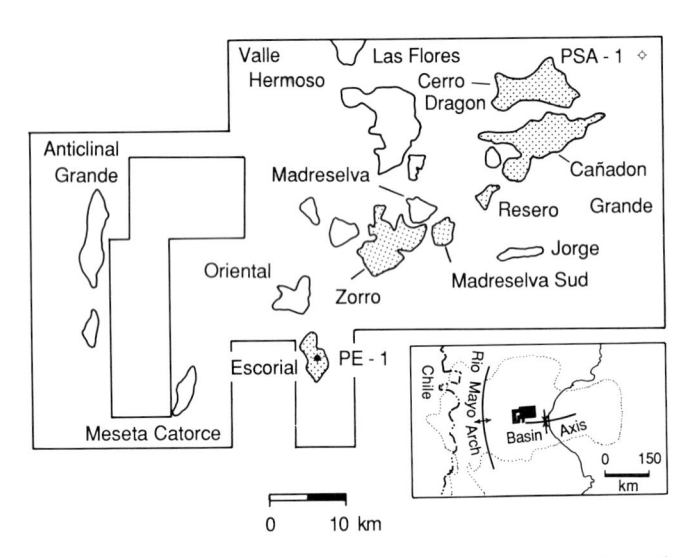

FIG. 2.—Map showing location of fields (samples obtained from stippled fields) and wells PE-1 and PSA-1 within the study area (Amoco Argentina Oil Company Contract Area). Inset at lower right shows location of the study area within the basin and gross structural features.

FIG. 3.—Stratigraphic column of north flank of the San Jorge Basin, courtesy of Amoco Argentina Oil Company.

Hower (1979) suggested similar temperatures for the formation of corrensite within the Disturbed Belt of Montana.

A burial history for the PSA-1 well is shown in Figure 6. The modeled maximum temperature for the base of the Comodoro Rivadavia Formation for the PSA-1 well is 78°C. Thermal-conductivity values were estimated by using values for simple lithologies listed in Sclater and Christie (1980). Relative percentages of the lithologies present were obtained from electric logs. Tops provided by AAOC were used to establish interval thicknesses and ages. Mean interval values for thermal conductivities were calculated using a harmonic average. The mean values were corrected for compaction and porosity loss using an exponential porosity decay described in Sclater and Christie (1980).

Because the rifting event began some 100 million years prior to the deposition of the interval of interest, a constant heat flow was assumed (McKenzie, 1978). Foreland de-

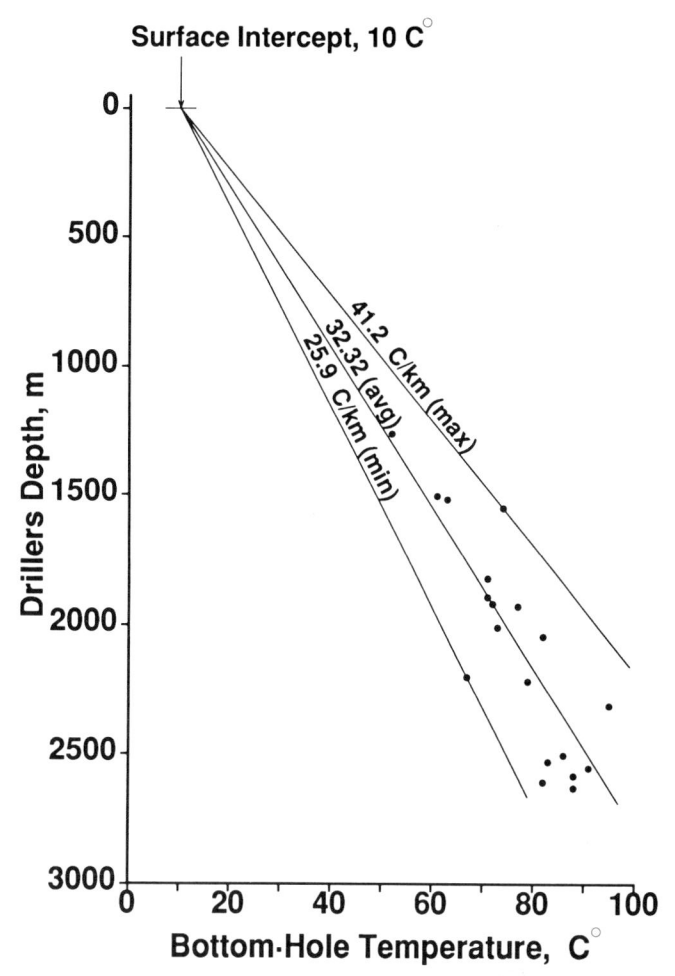

FIG. 4.—Uncorrected bottom-hole temperatures; minimum, maximum and averaged linear geothermal gradients are also shown. Bottom-hole temperatures were obtained from electric-log header data from 20 wells in the study area.

velopment to the west of the basin as a result of convergence along the Pacific margin also would suggest constant heat-flow-with-time curves (Hagen and Surdam, 1989). The impact of the Eocene and Miocene basaltic volcanism on the regional heat flow is unknown and requires more investigation. The Miocene event appears to have extruded the larger volume of magma, although neither appears to have been extensive. A heat-flow value of 50 mW/m^2 (Lachenbruch and Sass, 1977) and a surface intercept of 10°C were chosen. These values, along with the averaged and compaction-corrected interval thermal conductivities, provide a good fit to the observed bottom-hole temperatures. Hence, the model also suggests that, within the study area, the Comodoro Rivadavia Formation has not undergone significant heating.

There are, of course, a few caveats for interpreting this relatively primitive thermal model. The age and interval-thickness data for the Tertiary, critical to the modeling of the Cretaceous, are poor. The burial histories show no uplift or erosion, yet, particularly in the west, uplift is known to have occurred. However, the extent and timing of the uplifts also are poorly known. Thicknesses of lithologies were picked from logs and checked by examination of cuttings. This, however, is troublesome due to the conductivity of the rocks (smectite-rich) and the high resistivity of the formation waters (fresh to brackish; Khatchikian and Lesta, 1973), which significantly reduces the spontaneous potential-log response. Hence, the sandstone-to-shale ratio may be slightly underestimated. For these reasons, the burial history shown in Figure 6 was generated using a single temperature gradient of ~40°C/km (near the maximum observed from the BHTs). Despite these complications, the results of the thermal modeling, the BHTs, as well as the clay mineralogy of shales indicate that temperatures in the CRF within the study area have not exceeded 100°C; nor has the thermal exposure varied greatly across the relatively small study area.

The Petrology of the Comodoro Rivadavia Sandstones

The petrology and diagenesis of these sandstones are the subject of another study and will only be summarized here. Two sandstone petrofacies of the CRF can be distinguished by the amounts of quartz (monocrystalline, polycrystalline and chert) relative to volcanic lithic detritus (Fig. 7). The ratio of feldspar to lithic fragments remains roughly constant at about 1:3 throughout both petrofacies. The San Jorge petrofacies is the more abundant and widespread of the petrofacies. Its mean composition is $Q_{27}F_{21}L_{52}$. The lithic fragments include volcanic and basement-related lithologies (granitics and minor meta-sedimentary detritus). Plagioclase is relatively abundant, whereas potassium feldspar is only locally present. The diagenetic minerals present within the San Jorge petrofacies include smectite, smectite-chlorite, chlorite, kaolinite, albite, analcime, laumontite, heulandite, quartz and carbonates.

The Zorro petrofacies is wholly derived from a volcanic source. The Zorro petrofacies has a mean modal composition of $Q_6F_{25}L_{69}$. Pumiceous fragments are common. The small amounts of quartz detritus are volcanic varieties displaying inclusion-filled embayments and euhedral forms. Sandstones of this petrofacies are relatively rare and are associated with smaller sand bodies (Fig. 8). The Zorro petrofacies is an end-member assemblage in that volcanic debris is present throughout the CRF. Differences in composition and bed thickness noted between sandstones from closely spaced wells illustrate that sands representing different drainage systems were being deposited in close proximity within the basin (Fig. 8). Hence, mixing of detritus likely occurred on the Cretaceous San Jorge flood plain, in addition to the compositional variations attributable to multiple and varied provenance areas. Diagenetic phases present within the Zorro petrofacies are smectite, chlorite, quartz, heulandite, albite and traces of carbonate.

INFILTRATED MATERIAL

Criteria for the Identification of Mechanically Infiltrated Material

Walker (1976) stated that mechanically infiltrated material has a clastic texture, which is "readily recognizable" when viewed using the scanning electron microscope. Many

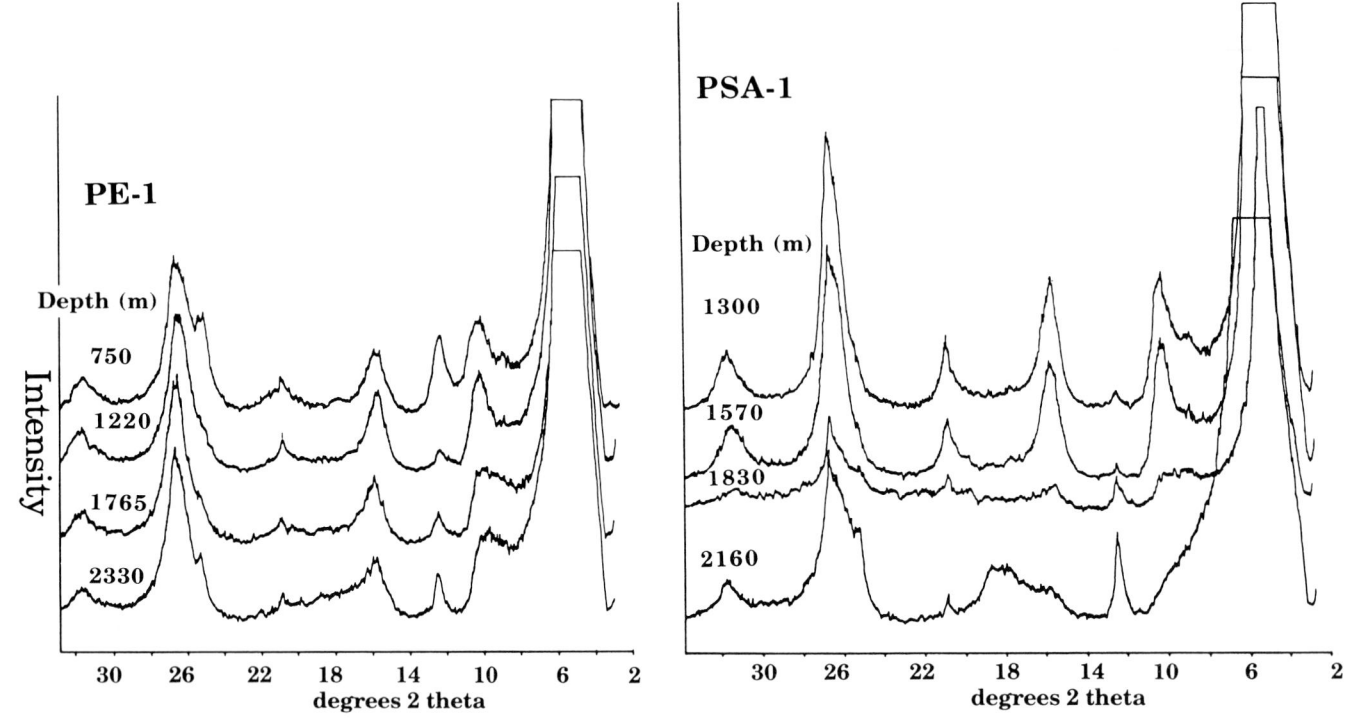

FIG. 5.—X-ray diffractometry for picked shale cuttings for the PE-1 and PSA-1 wells, showing slight increase in chlorite with increasing depth. The doublet for chlorite (004) and kaolinite (002) can be seen in the shallowest sample in the PE-1 well. Samples are <2-µm size fractions, magnesium saturated and glycolated.

of the thicker structures are clearly discernable within thin sections as well. Crone (1975), Walker (1976), Wilson and Pittman (1977), Walker and others (1978), and Matlack and others (1989) describe criteria used for distinguishing infiltrated clays from authigenic clays in sandstones. Most recently, these criteria have been reviewed and expanded by Moraes and de Ros (1990, and this volume). Features indicative of infiltration are ridges and pore bridges, geopetal fabrics, loose aggregates of polymineralic clay within pores, cutans, and massive aggregates. Moraes and de Ros (1990) also list the presence of shrinkage cracks and/or impurities as additional criteria. Similarly described textures are, however, produced by *in situ* growth, such as concentric or pore-lining growths, loose aggregates of kaolinite, or bridging clays. In each case, however, the textures are readily discernable. Cutans are distinguished from authigenic-clay rims by the parallel orientation of the platey

FIG. 6.—Burial-history plot for the PSA-1 well, showing the modeled temperature (dashed line) through time for the base of the Comodoro Rivadavia Formation.

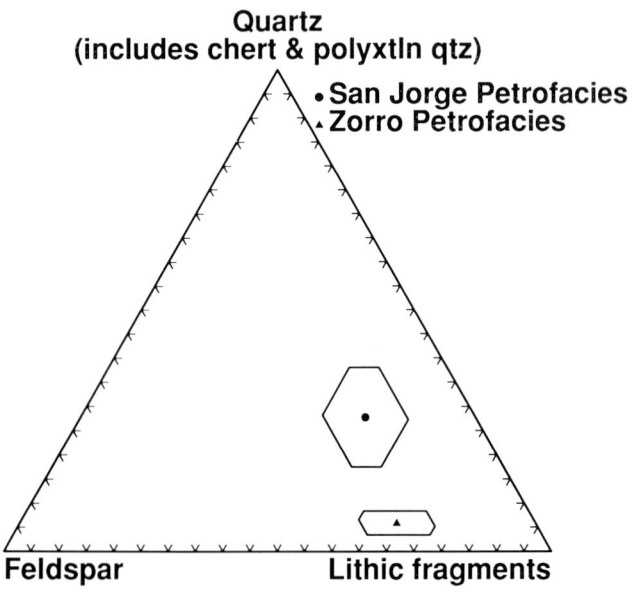

FIG. 7.—Q-F-L ternary diagram showing mean (symbols) and one standard deviation (polygons) modal compositions for the two petrofacies recognized in the Comodoro Rivadavia Formation sandstones. Data are derived from 157 point-counted samples.

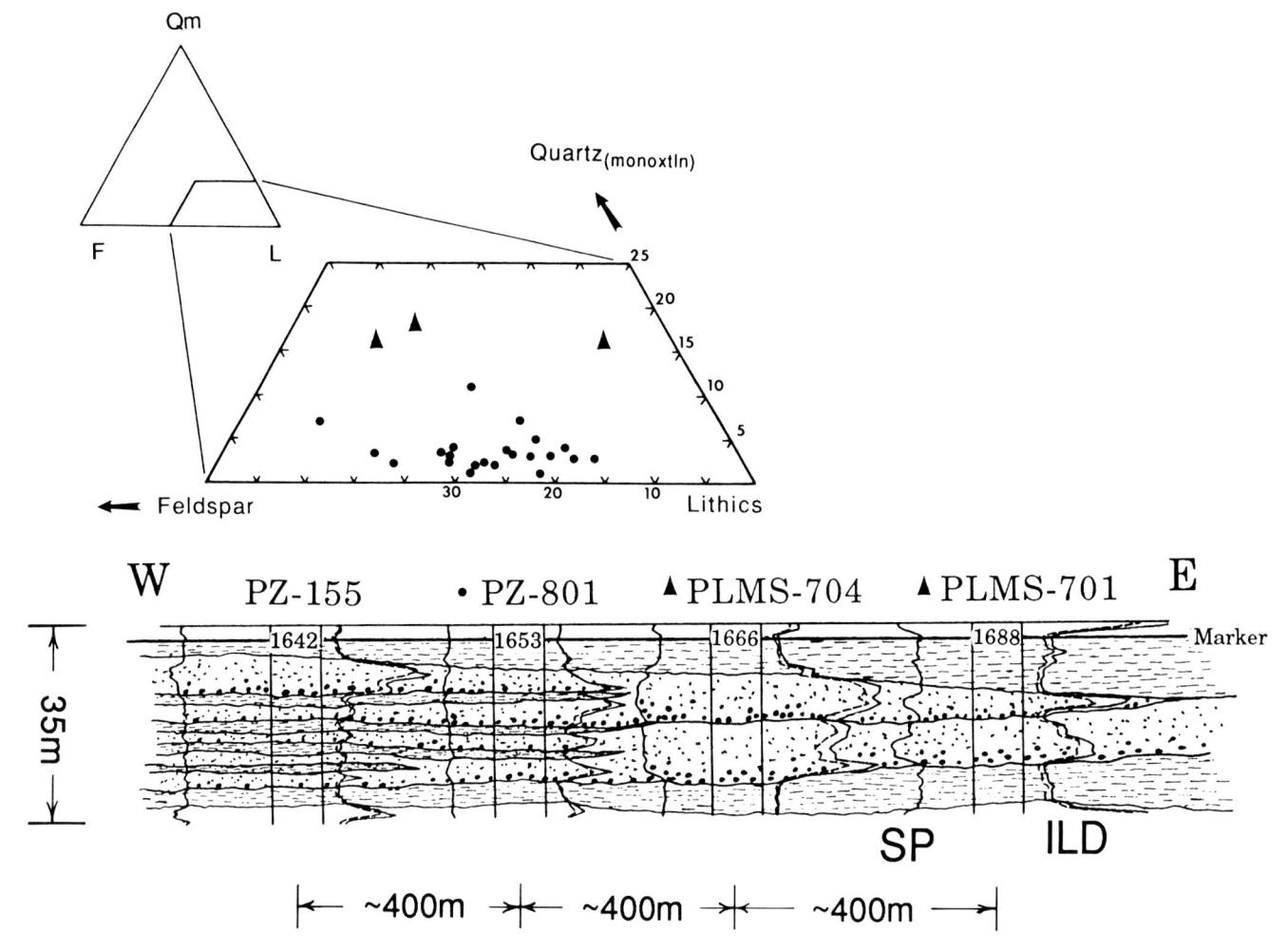

FIG. 8.—Cross section (courtesy of R. Merino, Amoco Argentina Oil Company) and Q-F-L modal compositions of sandstones in two adjoining fields (Zorro and La Madreselva Sud) illustrate the differences in composition and sandstone-body thickness. Wells are spaced approximately 400 m apart. Curves shown on the cross section are electrical-log responses: spontaneous potential (SP) on the left, and an induction tool on the right (ILD).

minerals to the grain surfaces. Authigenic clays are generally oriented perpendicular to grain surfaces. Loose aggregates of monomineralic authigenic euhedra (e.g., kaolinite pore fillings) are readily distinguishable and are not confused with detrital assemblages. Bridging clay structures are well established as occurring as authigenic and infiltrated material. Such criteria as euhedral form and monomineralic composition indicate authigenesis. It is important to point out, however, that infiltrated material undergoes alteration along with the rest of the sediment and that these structures are likely precursors for authigenic phases (e.g., Matlack and others, 1989).

Occurrence and Petrology

The following textures are recognized within the CRF sandstones: ridges and bridges (Figs. 9A and B), geopetal structures (Figs. 9C and D; Plates 3A, B, and D), cutans (Figs. 10A and B; Plate 3C) and massive aggregates (Fig. 11; Plate 3B). Textures and their compositional occurrence are listed in Table 1. Examples of thick, well-developed geopetal structures, cutans and massive pore fillings were recognized in one-fifth of the sandstones examined. This is a conservative estimate of the extent of infiltration in that the ridge and bridge structures are common within many of the CRF sandstones, but are only clearly discernable using scanning electron microscopy, a technique that was not applied to the majority of samples. However, the following discussion concentrates on the thicker features. Half of the samples in which these features were observed contain 5% or greater (bulk volume, point counted) infiltrated material. The maximum amount observed was 32%, with an average of 8.5%. Infiltration material was identified by the presence of internal-fabric orientation (laminae parallel to pore walls), massive pore fillings and well-developed geopetal fabrics.

The infiltrated clay-size material found within Comodoro Rivadavia Formation sandstones is now composed of mixtures of smectite, chlorite, iron oxides and microcrystalline quartz. The infiltrated mixtures fall into two categories: those mixtures that do not contain an abundance of microcrys-

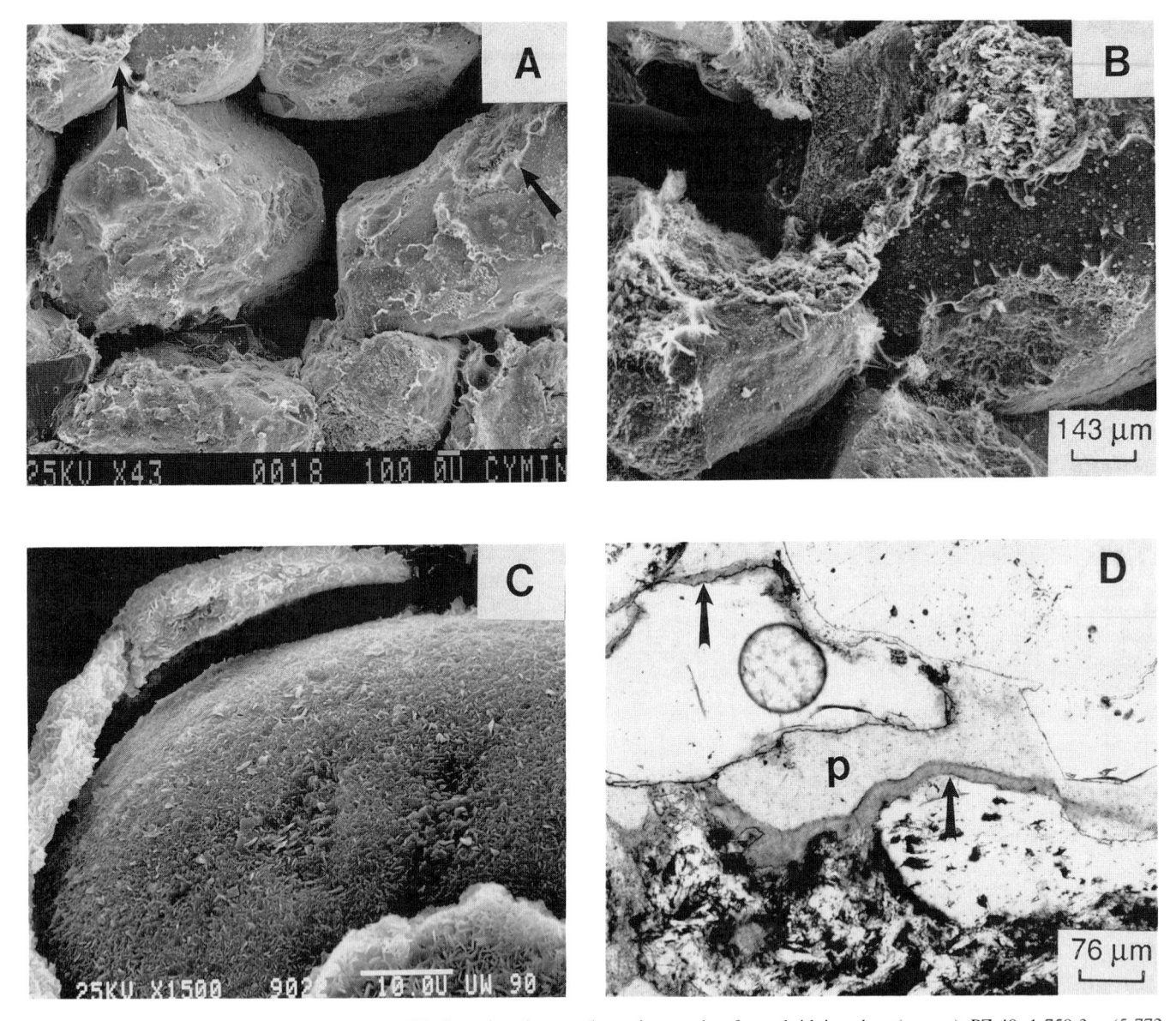

FIG. 9.—Clay-rich infiltration structures: (A and B) Scanning electron photomicrographs of pore-bridging clays (arrows); PZ-40, 1,759.3 m (5,772 ft). (B) Closeup of (A) showing fine textures and sparse distribution of these vadose structures. (C) Scanning electron photomicrograph of geopetal-clay laminae which have pulled away from the grain during sample preparation; PZ-40, 1,760.85 m (5,777 ft). (D) Same sample as in (C), showing geopetal nature of these clays. Note the clay occurs only along the bottoms or "floors" of the pores (arrows).

talline quartz and those mixtures that do contain significant microcrystalline quartz. The variation in relative amounts of quartz, oxides and clay provides for the wide range of optical properties observed, ranging from reddish-brown to brown-opaque features with high relief; to pale green to greenish-blue forms, which have low relief and very low birefringence (first-order grays to pseudoisotropic). The generally low birefringence is attributed, in part, to the abundance of chlorite and quartz as well as the minute-crystal sizes present.

Infiltrated materials become more common within basinward deposits. The basinward sandstones represent channels deposited by meandering or sinuous-stream systems. More of the sandstones found to the north and east within the study area generally are thicker, coarser grained, braided, or low-sinuosity stream deposits, and probably proximal portions of alluvial fans fringing the basin rim. The more common occurrence of infiltrated material within meandering-stream deposits is likely a function of preservation (also recognized in other deposits by Matlack and others, 1989). Braided-stream systems tend to rework the coarse clastics and winnow out fines. Reworking is not as extensive within meander belts. Infiltration material was not observed in the more restricted Zorro (monolithologic volcanic) petrofacies. The lack of infiltration material in the restricted petrofacies may be the result of the difficulty in creating a concentrated suspended load within a localized drainage.

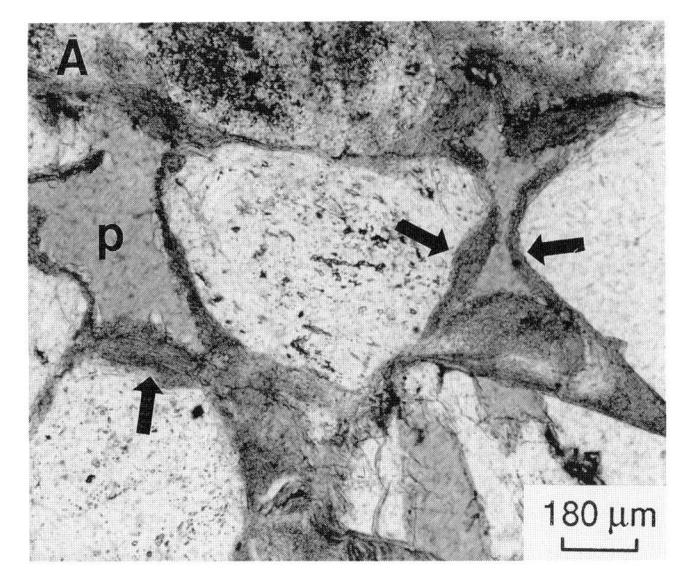

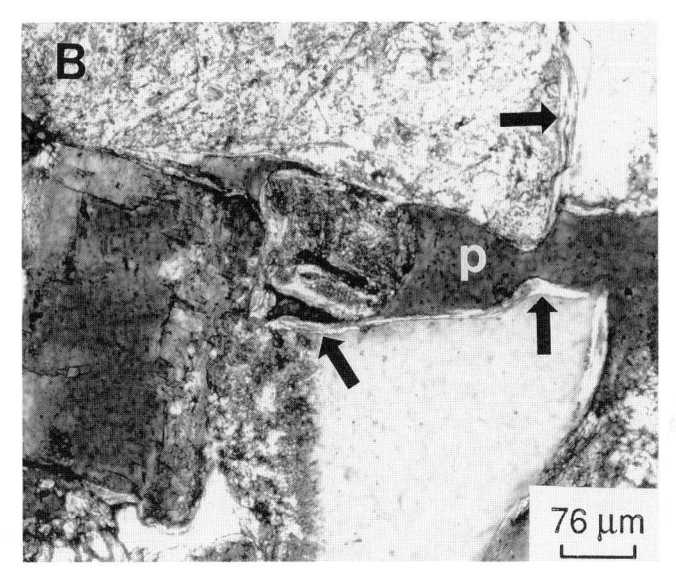

FIG. 10.—Clay-rich cutans: (A) Relatively thick but irregular cutans (arrows) showing faint laminations parallel to the grain surfaces; ICD-6, 1,511.0 m (4,958 ft). (B) Partially crossed polars photomicrograph of thin, birefringent clay cutans (arrows) lining the pore (p) walls. The clay is only thinly developed on the upper surfaces; ICD-6, 1,508.6 m (4,950 ft).

The two types of infiltration material are found within sandstones of the same compositional field; however, they are temporally (vertically) separated from each other. The clay-rich type is the more common of the two. Geopetal structures are the most common fabric, with cutans less common, and massive pore filling the least common. Cutans and massive pore fillings commonly occur with at least some faint geopetal structures present (Fig. 10A; Plates 3B and E). Many of the thicker infiltration structures have been disrupted during compaction (Fig. 12). The quartz-rich infiltration material occurs as bands several centimeters thick whose boundaries do not follow bedding, or as streaks of irregular thickness, varying from a few centimeters to a millimeter in thickness, that follow bedding (Fig. 13).

All of these thicker structures, the geopetal laminae, cutans, and massive pore fillings, are indicative of deposition within the phreatic or saturated zone (Walker, 1976; Moraes and de Ros, 1990). This is supported by the observation that these structures commonly form within the lower portions of point-bar deposits (Fig. 14) within sands that were more likely to remain below or near the water table.

Both the clay-rich and quartz-rich infiltration materials contain features of micro-segregation of opaque minerals radially away from grain boundaries, forming concentrations within the centers of pores and within the central areas of the infiltrated material (Plates 3E and F). In some cases, the texture tends to obscure the faintly visible infiltration (geopetal) structures. The form of the segregation of the oxides suggests a small-scale, diffusion-controlled alteration process dissolving and reprecipitating the iron compounds progressively away from the interfaces with the framework grains.

Mineral and Chemical Composition

The principal phases present within the infiltration structures are quartz, chlorite and smectite. Iron oxides and,

TABLE 1.—INFILTRATED MATERIALS AND STRUCTURE OF THE COMODORO RIVADAVIA FORMATION

Structures	Composition of Material	
	Clay-rich	Quartz-rich
Ridges and bridges	common	not observed
Geopetal	common	rare
Cutans	common	rare
Cutans	common	rare
Massive aggregates	rare	rare

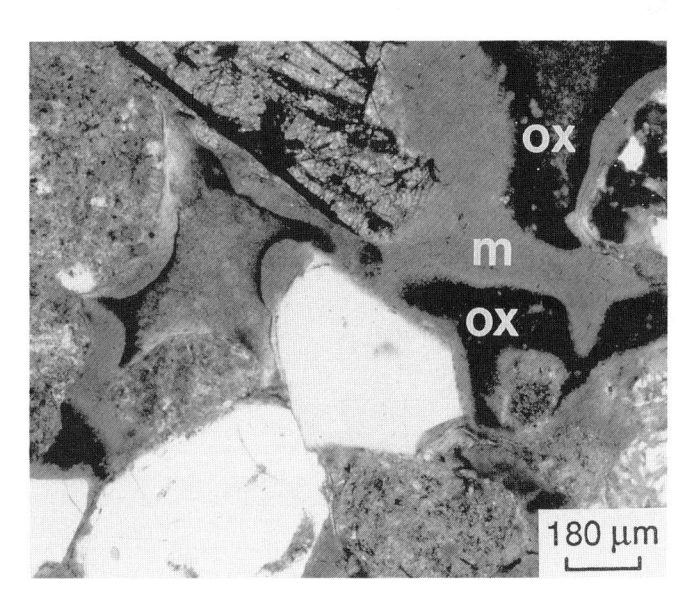

FIG. 11.—Massive quartz-rich pore filling (m) with oxide staining (ox); partially crossed polars, PCD-220, 1,782.96 m (5,850 ft).

166   THOMAS L. DUNN

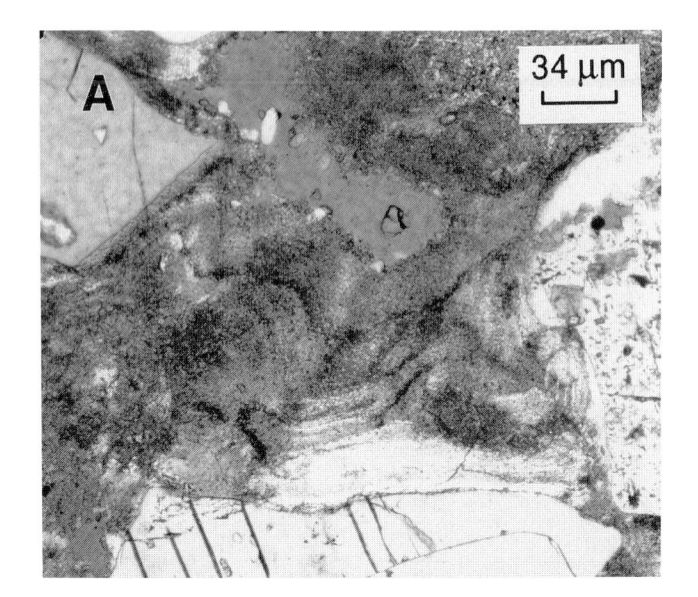

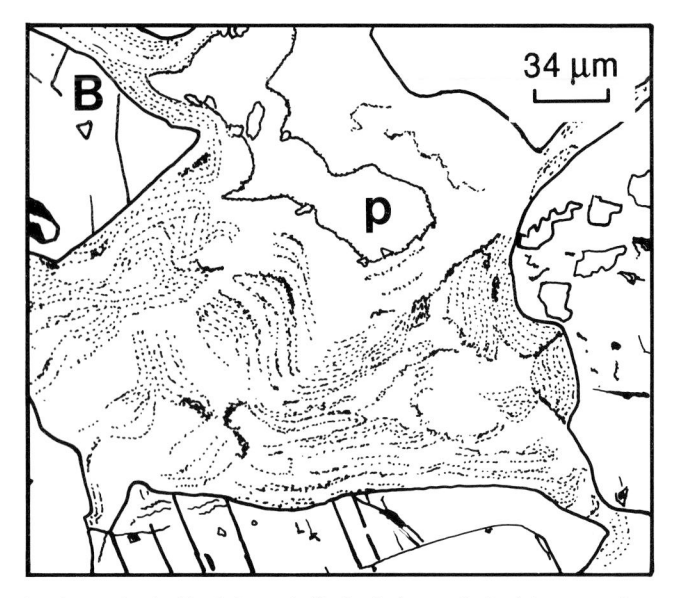

FIG. 12.—(A and B) Companion illustrations of quartz-rich material that has been plastically deformed, likely during early burial compaction. (A) Partially crossed polars photomicrograph. (B) Ink tracing of photomicrograph highlighting (interpreting) the deformation structures within the quartz-rich matrix. Small amounts of intergranular porosity (p) remain. Sample is from PZ-823.

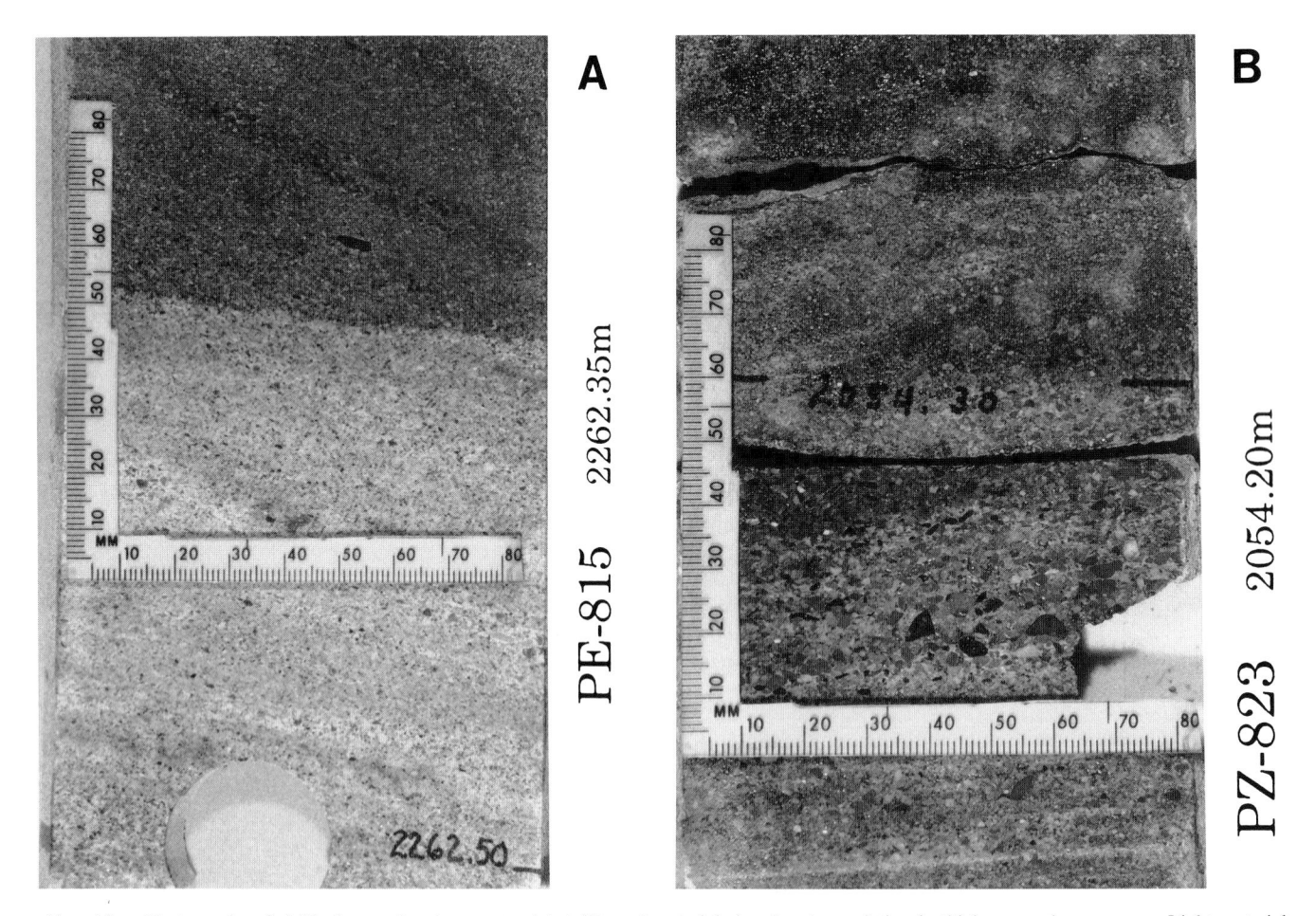

FIG. 13.—Photographs of slabbed cores bearing quartz-rich infiltrated material showing the variation in thickness and occurrence. Light material in (A) is quartz-rich material and is several centimeters thick. Quartz-rich material in (B) is millimeter-thin streaks, which tend to follow bedding. Mottling in uppermost portions of the core are poikilotopic laumontite crystals.

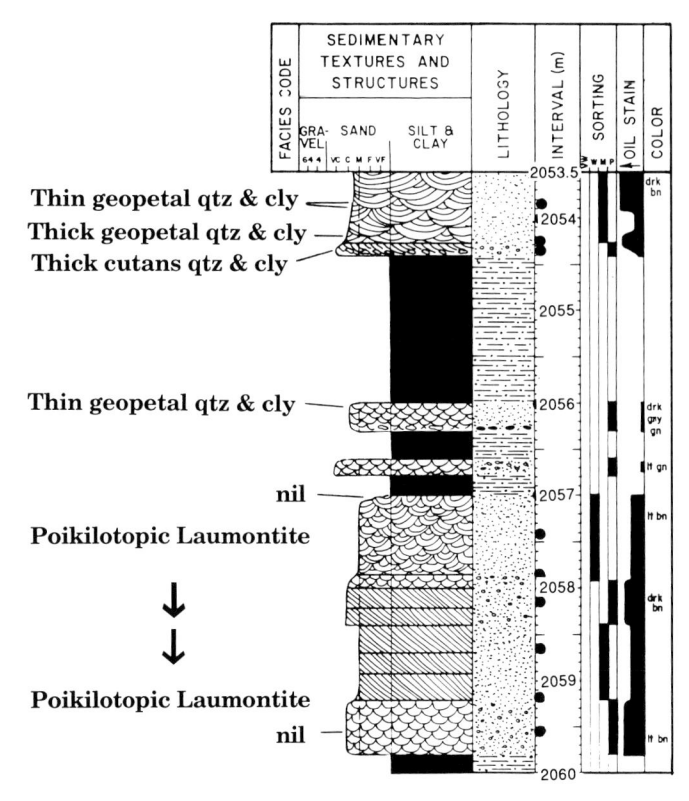

FIG. 14.—Lithologic column of a PZ-823 core illustrating that the quartz and clay infiltration structures within the fluvial sandstones are not consistently found within any particular portions of the point-bar sequences.

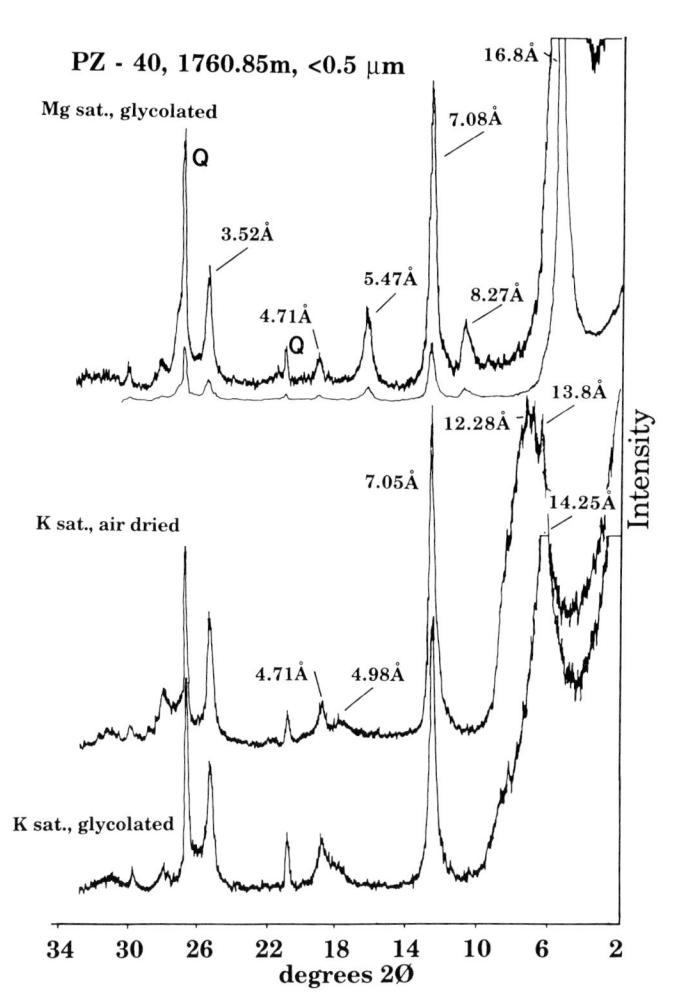

FIG. 15.—X-ray diffractograms (Cu radiation) for a clay-rich infiltrated sandstone, PZ-40. The patterns are from a <0.5-μm size separation, mounted on a porous tile.

rarely, pyrite also are present in small amounts. Diffractograms of the <0.5-μm size fractions of examples of the quartz-rich and clay-rich infiltration types are shown in Figures 15, 16, and 17. Samples were prepared by vacuum suction of clay suspensions onto porous ceramic tiles. Saturations with potassium and magnesium were also performed using vacuum suction of the respective chloride solutions through the samples prepared on the porous plates. The blank, clean, porous plates regrettably result in quartz and feldspar peaks on the diffraction patterns (Fig. 18), hence, the clay separations shown in Figures 15, 16, and 17 are better than they appear to be. Samples were run both air dried and after ethylene glycol solvation (24 hrs at 60°C). Chlorite is present in all samples in widely varying amounts. Kaolinite, although present in the coarser fractions within some sandstones and within a few samples of shales, is not present in appreciable amounts within these finer size separations. Within the coarser fractions (not shown) the (002) peak of kaolinite and the (004) peak of chlorite clearly separate. The (001) 14Å and (003) 4.71Å peaks of chlorite are readily identifiable and, therefore, chlorite accounts for the peak intensities at the 7Å and 3.53Å peaks as well. Kaolinite may be present within the infiltrated material in small amounts; however, it was not detected by scanning electron microscopy. Smectite was recognized operationally by the increase in the (001) peak d spacing to 16.8Å upon glycol solvation. The smectite does not fully re-expand to 16.8Å after glycolation of the potassium-saturated samples, suggesting that the smectite is of an intermediate-charge variety.

Examination of the (060) peaks using random (side pack) mounts shows the presence of both dioctahedral and trioctahedral clays within the clay-rich infiltration material (Fig. 19). The (060) d spacings of 1.55Å indicate chlorites with trioctahedral occupancy. Samples also contain smectite and show (060) d spacings of either 1.498Å or 1.516Å, indicating dioctahedral or trioctahedral occupancy in the smectites, respectively.

Results of chemical analyses of the shales adjacent to the infiltrated sandstones and microprobe analyses of the infiltrated material are shown in Tables 2, 3, and 4 and Figures 20 and 21. Each column of microprobe analyses shown in Tables 2 and 3 is an average of two to nine spot analyses taken as clusters on optically similar material adjacent to one another. The microprobe results reflect the variation in quartz abundance within the infiltrated material. The quartz-rich varieties (Table 3) contain abundant silica relative to other oxides (silica ranges from 91 to 71%); iron (as oxide) ranges widely from 0.3 to 10.2%; alumina ranges from 2.1 to 9.9%. The shale whole-rock analyses appear intermediate in composition between the clay-rich and quartz-rich infiltrated materials (Tables 2, 3, and 4; Fig. 20). The shales contain less

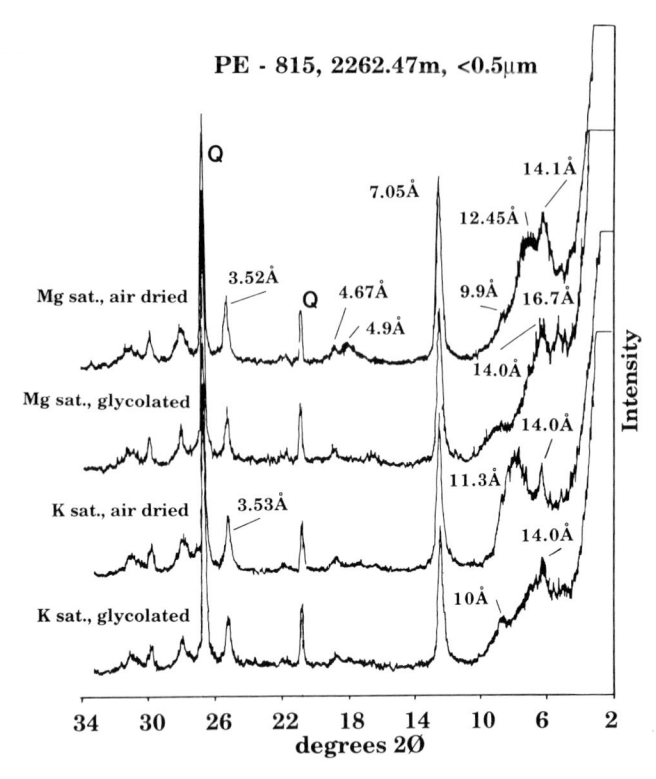

FIG. 16.—X-ray diffractograms (Cu radiation) for a quartz-rich infiltrated sandstone, PE-815. The patterns are from a <0.5-μm size separation, mounted on a porous tile.

silica than the quartz-rich material (59 to 66% silica in the shales) relative to all other elements analyzed. The shales contain less iron and silica than do the clay-rich infiltration materials. The clay-rich infiltration varieties are the most aluminous, reflecting their phyllosilicate mineralogy. Titanium and manganese are present only in trace quantities.

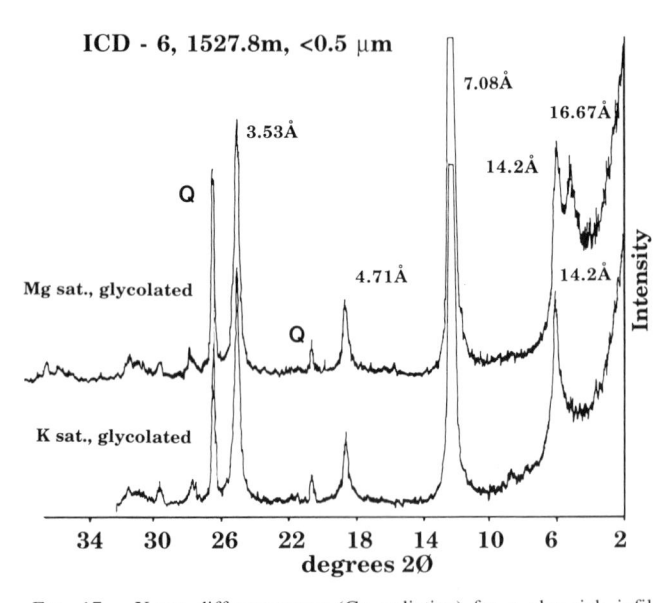

FIG. 17.—X-ray diffractograms (Cu radiation) for a clay-rich infiltrated sandstone, ICD-6. The patterns are from a <0.5-μm size separation, mounted on a porous tile.

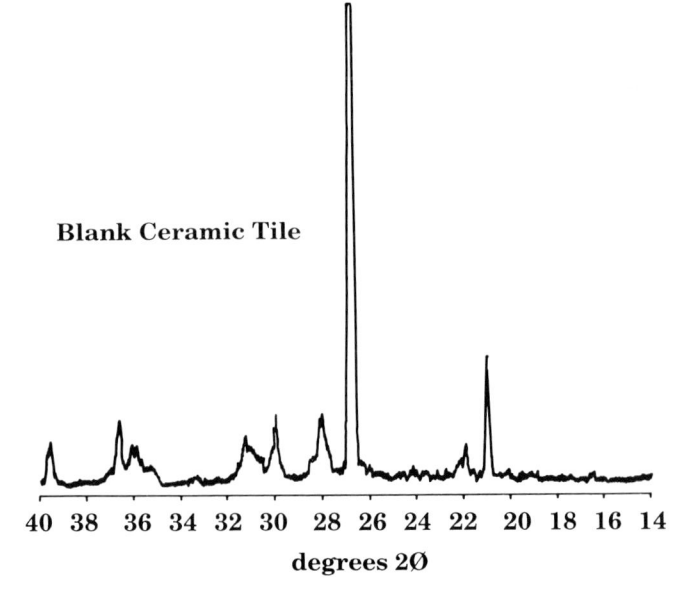

FIG. 18.—X-ray diffractogram (Cu radiation) of a clean, blank, porous tile showing its noisy background, which appears in the patterns shown in Figures 15, 16 and 17.

Composition of some of the infiltrated (geopetal) material from ICD-6 and PZ-40 plot within or near the range of microprobe compositions found for diagenetic chlorite (Figs. 20 and 21). X-ray diffractometry of the ICD-6 material shows chlorite as the dominant clay type (Fig. 17). A series of chlorite structural cell compositions calculated for the ICD-6 infiltration material are included in Table 2. The calculated analyses give Fe-rich compositions for the chlorites, which are compatible with the observed relative intensities of the odd and even (001) reflections (Fig. 17).

Figure 20 also shows the variation of Si, Al and Fe for the adjacent shales, the infiltrated material and a field representing the variation within the calc-alkaline igneous series (rhyolite to basalt; taken from Nockolds, 1954). Much

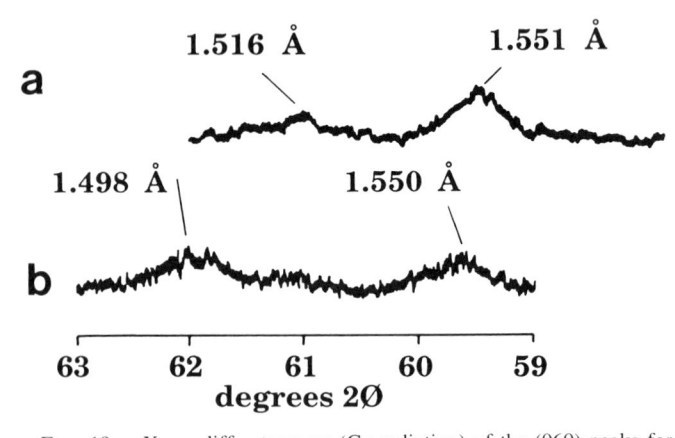

FIG. 19.—X-ray diffractograms (Cu radiation) of the (060) peaks for the <0.5-μm size fractions, (a) ICD-6, 1,527.8 m (5,013 ft) and (b) PZ-40, 1,760.85 m (5,777 ft). The peaks with 1.55Å d spacings are from the chlorite present, showing trioctahedral occupancy. The peaks of 1.498Å and 1.516Å indicate dioctahedral and trioctahedral occupancy for the smectites present in each sample, respectively.

TABLE 2.—MICROPROBE[1] ANALYSES OF CLAY-RICH INFILTRATION MATERIAL

	ICD-6 1577.8m*							PZ-40 1760.85m				
	n=7	n=4	n=3	n=3	n=3	n=3	n=3	n=6	n=8	n=5	n=6	n=5
SiO_2	22.62	24.91	23.01	25.32	20.77	23.83	21.83	37.16	34.75	38.31	37.11	40.62
Al_2O_3	19.49	20.87	20.00	21.03	18.01	20.78	18.54	24.81	24.48	21.13	21.73	21.24
TiO_2	0.03	0.05	0.10	0.06	0.05	0.02	0.04	0.56	0.66	0.68	2.29	1.15
FeO	25.99	28.54	25.66	24.81	24.38	27.27	24.03	11.94	14.16	13.71	12.78	11.20
MgO	4.31	4.65	4.39	4.38	4.07	4.65	4.10	2.75	3.13	3.29	3.02	2.82
CaO	0.57	0.62	0.56	0.52	0.41	0.53	0.51	1.22	1.32	1.22	1.34	1.36
Na_2O	0.26	0.26	0.37	0.31	0.18	0.27	0.40	0.38	0.41	0.36	0.34	0.34
K_2O	0.11	0.10	0.10	0.35	0.08	0.14	0.08	0.57	0.56	0.61	0.71	0.74
MnO	1.14	1.21	0.99	1.04	1.02	1.21	1.03	0.55	0.62	0.61	0.53	0.44
Sum	74.53	81.20	75.17	77.82	68.96	78.69	70.56	79.95	80.08	79.93	79.85	79.90

[1] analyst: T.L. Dunn, CAMECA MBX, sample current: 8 nanoamps, accel. voltage: 15kV, beam diameter: 2 microns.

* structural cell compositions, selected microprobe analyses for chlorites, ICD-6, 1577.8, clay-rich infiltration material:

$$(Fe_{5.54}Mg_{1.64}Ti_{0.01}Al_{3.63}Mn_{0.25})(Si_{5.77}Al_{2.23})O_{20}(OH)_{16}$$

$$(Fe_{5.38}Mg_{1.64}Ti_{0.01}Al_{3.69}Mn_{0.23})(Si_{5.84}Al_{2.16})O_{20}(OH)_{16}$$

$$(Fe_{5.63}Mg_{1.67}Ti_{0.01}Al_{3.59}Mn_{0.24})(Si_{5.73}Al_{2.27})O_{20}(OH)_{16}$$

of the quartz-rich infiltrated material is more siliceous and contains more Fe than the calc-alkaline igneous series, which is likely representative of the composition of much of the detrital lithic-framework material and volcanic glasses, possibly deposited as matrix within the sandstones. Note that the clay-rich infiltration material is more aluminous than the adjacent shales and that some of the quartz-rich varieties of infiltrated material overlap with the shales at their more aluminous values.

Figure 21 is a diagram constructed as described by Velde (1985) and Velde and Meunier (1987) as a means of illustrating variations of clay compositions. The construction of an MR^3–$2R^3$–$3R^2$ ternary projection from the chemical data begins with grouping homologous components, such as the divalent cations Mg^{2+}, Fe^{2+} and Mn^{2+}, (R^2); the trivalent cations Al^{3+} and Fe^{3+}, (R^3); the alkali and alkaline-earth cations, Na^+, K^+, and Ca^{2+}, (M) and silica. The MR^3 coordinate represents the interlayer cation-charge balance ac-

TABLE 3.—MICROPROBE ANALYSES OF QUARTZ-RICH INFILTRATION MATERIAL

	PE-815 2262.3m						PZ-823 2054.32m							
	n=5	n=3	n=4	n=4	n=4	n=6	n=5	n=2	n=5	n=2	n=5	n=6	n=8	n=9	n=8
SiO_2	88.53	84.28	85.71	82.31	90.85	86.25	73.36	73.23	78.99	74.74	71.32	73.58	78.75	79.55	78.54
Al_2O_3	2.12	3.62	2.80	3.61	1.93	3.29	8.14	8.04	6.78	8.86	9.91	9.08	6.54	7.24	8.16
TiO_2	0.14	0.42	0.42	0.76	0.27	0.43	0.17	1.17	0.17	0.22	0.39	0.25	0.22	0.51	1.73
FeO	0.28	0.57	0.31	0.49	0.27	0.50	9.41	7.14	6.38	9.55	10.17	5.24	2.73	3.86	1.92
MgO	1.39	3.13	1.15	2.43	1.43	2.79	1.73	1.56	1.32	1.89	2.08	1.22	0.79	1.01	0.89
CaO	0.08	0.11	0.10	0.15	0.13	0.11	0.32	0.36	0.30	0.36	0.36	0.31	0.33	0.32	0.43
Na_2O	0.12	0.13	0.16	0.20	0.21	0.14	0.19	0.29	0.20	0.17	0.15	0.35	0.36	0.38	0.45
K_2O	0.11	0.16	0.20	0.33	0.14	0.14	0.34	0.60	0.30	0.38	0.47	0.55	0.47	0.49	1.03
MnO	0.02	0.07	0.04	0.07	0.07	0.09	n.a.	n.a.	n.a.	n.a.	n.a.	n.a.	n.a.	n.a.	n.a.
Sum	92.78	92.51	90.90	90.35	95.30	93.72	93.67	92.39	94.44	96.18	94.83	90.57	90.17	93.35	93.14

analyst: T. L. Dunn, CAMECA MBX, sample current: 8 nanoamps, accel. voltage: 15kV, beam diameter: 2 microns.

TABLE 4.—WHOLE-ROCK CHEMICAL ANALYSES OF ADJACENT SHALES

Well no.	ICD-6	ICD-6	PE-815	PE-815	PR-816	PZ-823
Depth(m)	1530.9	1535.8	2269	2263.8	2090.3	2057
SiO_2[1]	59.94	60.43	66.64	60.80	63.20	62.93
TiO_2	0.77	0.81	0.64	0.70	0.87	0.71
Al_2O_3	15.71	15.87	14.10	16.83	15.94	16.07
Fe_2O_3	6.23	6.51	3.53	5.76	4.19	5.40
CaO	1.54	1.96	1.11	1.78	1.56	2.13
MgO	1.63	1.41	1.20	1.37	1.44	1.53
MnO	0.13	0.12	0.07	0.10	0.11	0.13
Na_2O[2]	1.18	1.85	1.27	1.32	1.67	2.28
K_2O	1.63	2.60	2.58	1.84	1.90	2.21
P_2O_5[3]	0.08	0.07	0.02	0.04	0.05	0.04
L.O.I.[4]	9.87	8.63	6.90	7.97	8.76	7.06
Totals	98.70	100.25	98.06	98.49	99.69	100.49

Analyst: S. Boese.
[1] Si, Al, Ca, Mg, Ti, Mn were measured using an ICP.
[2] Na, K were measured using AA.
[3] measured by colorimetry.
[4] measured gravimetrically.

companying R^3 substitution for Si in the tetrahedral layer. The $2R^3$ and $3R^2$ coordinates represent occupancy variation in the octahedral layer (i.e., dioctahedral and trioctahedral, respectively). Framework silicate compositions plot within the MR^3 apex.

A number of points can be illustrated using this ternary space. First, a few samples from ICD-6, PZ-823 and PE-815 have compositions overlapping with those of published diagenetic chlorites. The material from PZ-40 plots be-

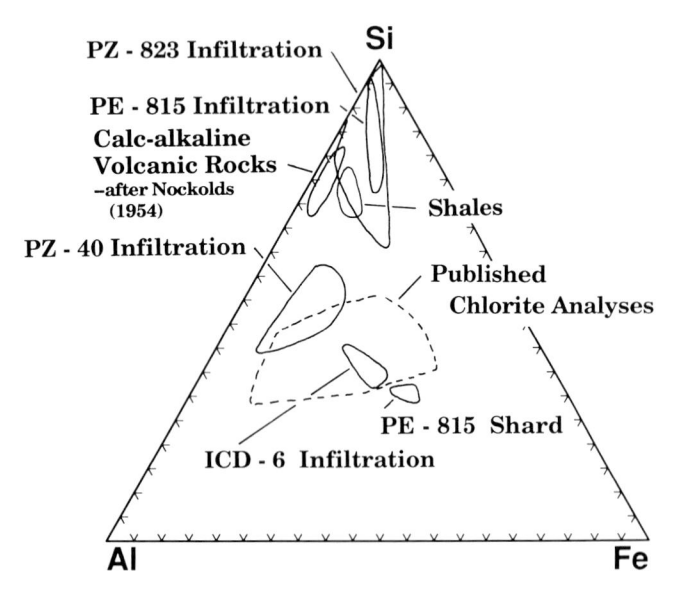

FIG. 20.—Si-Fe-Al ternary diagram showing microprobe analyses of the infiltrated material, in moles, for comparison with calc-alkaline volcanic series (Nockolds, 1954) and whole-rock analyses of the shales adjacent to the infiltrated sandstones. Published chlorite values are from Alford (1983) and those reported in Curtis and others (1984).

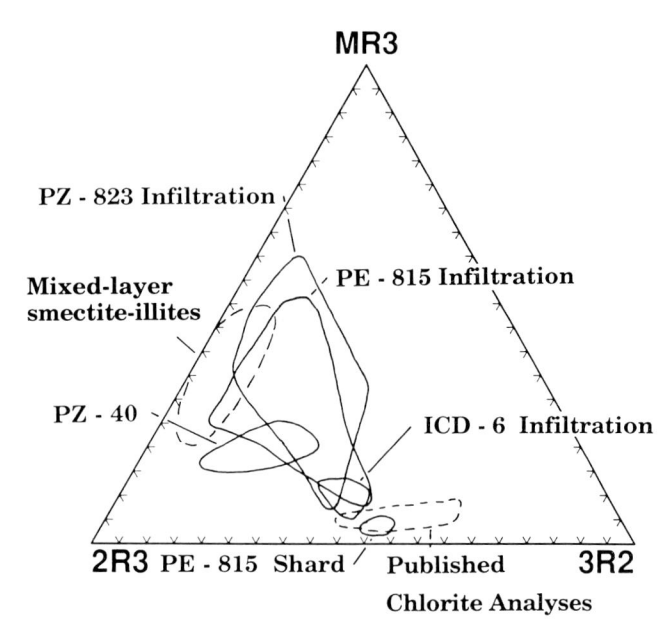

FIG. 21.—Ternary diagram constructed as described by Velde (1985) and Velde and Meunier (1987) showing microprobe analyses of infiltrated material, published chlorite analyses (sources as in Fig. 20) and the field for smectites and mixed-layer smectite-illites generalized from Velde (1985). See text and references cited for discussion of how the diagram is constructed. Briefly, MR^3 represents cation-charge balance for trivalent substitution for Si^{4+} in the tetrahedral layer; $2R^3$ and $3R^2$ represent octahedral-layer compositions and occupancy (dioctahedral and trioctahedral, respectively).

tween montmorillonite (essentially the dioctahedral end $[2R^3]$ of the smectite-illite mixed-layer field) and the chlorite field, a composition that is supported by the X-ray diffraction results showing a mixture of the two clays with smectite as the most abundant phase. The two examples of the quartz-rich material analyzed show great variation from essentially a chlorite composition to one that reaches well into the smectite field. Some of the large variation may be attributed to the dilution by quartz, which makes the plotted component values small and, hence, accentuates any variations.

Alteration of the Infiltrated Material

X-ray analyses reported by Walker (1976) comparing interstitial (infiltrated) clay with surface muds of modern and Holocene alluvium in arid environments showed that those clays had the same mineralogy. Therefore, Walker (1976) inferred that they had been derived from the same provenance. Within the San Jorge Basin, however, the infiltrated materials have a wide range of compositions and differ from the adjacent shales. Also, submicron-size quartz crystals are unlikely major constituents of fluvial suspended loads. Clearly, some alteration has occurred to the original infiltrated material.

The quartz-rich infiltrated material requires a siliceous starting material, as well as an additional source of silica, in order to produce the observed abundance of quartz. The amount of silica present in the infiltrated material is more than that present within either siliceous volcanic glasses or the adjacent shales (Fig. 20). Silica released during the hydration of the surrounding volcanic lithic-framework grains was likely incorporated into the infiltrated material. Many of the remaining volcanic lithic grains are aluminous (i.e., silica depleted), with pumiceous material now having chloritic compositions (e.g., Figs. 20, 21, and 22; Table 5).

Prismatic quartz is prevalent as a minor diagenetic phase throughout much of the Comodoro Rivadavia Formation sandstones, commonly occurring with albite and chlorite within the remaining void space above thick geopetal quartz-rich structures (Plate 3D). However, precipitation of quartz occurs within all of the sandstones regardless of whether or not abundant phreatic infiltration material is present. Clay and adsorbed iron hydroxides were likely present within the suspended load at all times and, as such, were likely present in trace to minor amounts with the infiltrated volcanic glasses. The present mixture of authigenic phases (quartz, chlorite and iron oxides) is postulated to have formed through the following reactions:

infiltrated volcanic glass + iron hydroxides

$$\rightarrow opal \pm smectite$$

With continued burial, the opal reacted to quartz with additional silica:

SiO_2 (from the hydration of volcanic lithic fragments)

$$+ opal \rightarrow quartz$$

Additionally, chlorite formed through conversion from smectite.

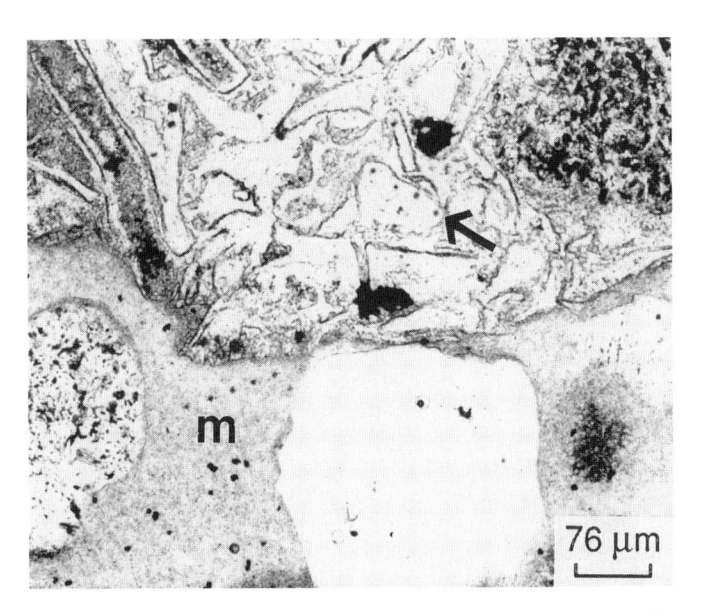

FIG. 22.—Chloritized pumiceous-rock fragment within a sandstone (PE-815, 2,262.3 m; 7,422.1 ft) containing abundant quartz-rich massive pore filling (m). The formation of the quartz-rich intergranular material requires additional sources of silica (e.g., silica derived from the alteration of volcanic-framework detritus). This process ought to produce more aluminous (silica-depleted) framework detritus. Chloritic compositions (and hence, more aluminous) were found within framework volcanic-rock fragments using microprobe analysis (arrow points toward line of four analyses spots on shard relict; see Table 5).

TABLE 5.—MICROPROBE ANALYSES[1] AND STRUCTURAL CELL COMPOSITIONS OF CHLORITES FROM PUMICE-SHARD REPLACEMENTS, PE-815, 2,263.3 m

Spot No.	3	4	5	6	7
SiO_2	22.22	25.30	24.44	24.84	24.56
Al_2O_3	18.52	21.50	20.93	21.83	19.34
TiO_2	0.06	0.04	0.01	0.00	0.04
FeO	3.66	4.81	4.40	5.13	3.87
MgO	36.85	36.74	35.76	36.52	37.66
CaO	0.31	0.23	0.27	0.18	0.23
Na_2O	0.18	0.15	0.07	0.13	0.11
K_2O	0.02	0.02	0.00	0.00	0.02
MnO	0.79	0.92	0.86	1.01	0.96
Total	82.59	89.71	86.76	89.62	86.78

$(Fe_{7.52}Mg_{1.33}Ti_{0.01}Al_{2.75}Mn_{0.16})(Si_{5.42}Al_{2.58})O_{20}(OH)_{16}$

$(Fe_{6.73}Mg_{1.57}Ti_{0.01}Al_{3.09}Mn_{0.17})(Si_{5.54}Al_{2.46})O_{20}(OH)_{16}$

$(Fe_{6.78}Mg_{1.49}Al_{3.13}Mn_{0.17})(Si_{5.54}Al_{2.46})O_{20}(OH)_{16}$

$(Fe_{6.70}Mg_{1.68}Al_{3.10}Mn_{0.19})(Si_{5.45}Al_{2.55})O_{20}(OH)_{16}$

$(Fe_{7.24}Mg_{1.33}Ti_{0.01}Al_{2.89}Mn_{0.19})(Si_{5.65}Al_{2.35})O_{20}(OH)_{1-6}$

[1] analyst: T.L. Dunn, CAMECA MBX, sample current: 8 nanoamps, accel. voltage: 15kV, beam diameter: 2 microns.

The formation of some intermediate forms of silica (opals) likely occurred relatively early, in that samples containing an abundance of the quartz-rich, massive pore filling commonly have open frameworks. The intermediate silica-phase transitions would tend to obscure any relict structures indicative of the pyroclastic origins of the fine-grained glass.

The clay-rich varieties are characterized by mixtures of chlorite and smectite. However, their possible origins are not as clearly discernable as that of the quartz-rich material. The tentative assumption made here is that the initial material was mostly smectite, although Dethier and others (1981) showed that smectite, smectite-chlorite, and chlorite were present in abundance within the suspended-load fractions of the Mount St. Helens 1980 deposits. The extent of burial diagenesis no doubt has influenced the formation of chlorite at the expense of smectite in a manner similar to that invoked by Chang and others (1986). An increase in the amount of chlorite is observed in the shales with increasing burial depth (Fig. 5). However, the chlorites, now representing geopetal infiltrated material from approximately 1,600 m depth in the ICD-6 well, are compositionally similar to those found in the deep Tuscaloosa Formation (most of which now lies at depths near 6 km and temperatures of 200°C) by Alford (1983) and Curtis and others (1984) (Figs. 20 and 21). Yet, as discussed earlier, samples from the PZ-40 well, at similar depths (1,760 m) to the ICD-6 samples, contain abundant smectite. Hence, the relative amount of chlorite to smectite in the sandstones varies widely and does not follow depth of burial. The variation, in part, may be a function of varying suspended-load compositions.

Water analyses reported by Khatchikian and Lesta (1973) and unpublished analyses of Amoco Argentina Oil Company indicate that produced waters from the Cretaceous reservoirs range from 1,000 mg/l to well above 10,000 mg/l total dissolved solids. The origin of the waters is complex in that the formation waters were likely fresh to slightly brackish fluvial waters initially. The Tertiary sequences contain rocks deposited during marine incursions into the basin and presumably contain marine connate waters. Fitzgerald and others (1990) suggested that marine incursions also occurred during the Cretaceous as well. The numerous unconformities within both the Cretaceous and Tertiary suggest that later flushing by meteoric waters also is likely. The presence of extensive residual kaolin deposits within Jurassic volcanic rocks in the Chubut River valley just to the north of the study area is reported by Murray (1988). Hence, pore fluids in contact with the infiltrated material likely varied from fresh to brackish through time and laterally throughout the basin. The large variation in pore-water chemistry may have also contributed to the variation of clay minerals present.

EFFECTS OF THE INFILTRATED MATERIAL ON DIAGENESIS AND RESERVOIR HETEROGENEITY

Two clear effects on later alteration are associated with the presence of infiltrated material. Where the infiltrated material is abundant (regardless of its composition), the rock is relatively impermeable and does not show further extensive dissolution. Where there is only sparse infiltrated material, the composition of the infiltrated material determines the character of later diagenetic alteration. In those sandstones containing quartz-rich infiltrated material, sandstone authigenesis is dominated by the formation of chlorite, quartz and albite within the remaining open-pore space (Plate 3D). Zeolites do not occur within sandstones containing quartz-rich infiltration structures. Where the infiltration material is composed only of clay and oxides, the authigenic assemblages are varied and include quartz, analcime, laumontite, chlorite, smectite and albite.

The presence of horizons of mechanically infiltrated clays within the coarse clastics produces meso-scale vertical-permeability barriers within the reservoirs. Data are insufficient to illustrate adequately the magnitude of the effect within the study area. Additional whole-core (horizontal and vertical) permeability measurements would be necessary to demonstrate the anisotropy properly. Conventional core-plug permeabilities from a core were compared for infiltrated and non-infiltrated samples (Table 6). The extent of the filling of pores by infiltrated material and the extent to which infiltration follows cross-bedding laminae determine the magnitude of the anisotropy. Moraes and de Ros (1990) indicated that for sandstone reservoirs of the Recôncavo Basin, Brazil, infiltration horizons are major features of heterogeneity within those reservoirs. There, the zones of infiltration can reach up to tens of meters in thickness and impose a layering that is exclusive of the fabric of the depositional environment. Those authors also discussed the difficulties encountered in formation evaluation due to low induction-log response in those intervals containing abundant mechanically infiltrated clays. The association of infiltrated material with the more sinuous-stream deposits

TABLE 6.—PERMEABILITY AND POROSITY[1]: INFILTRATION-RICH VERSUS INFILTRATION-POOR SAMPLES

Infiltration-rich		Infiltration-poor	
Permeability (md)			
31.8		17.9	
0.673		49.0	
0.0531		251	
5.45		124	
		111	
Permeability average			
(n=4)		(n=5)	
geometric	arithmetic	geometric	arithmetic
1.58	9.49	78.7	110.6
Porosity average			
(n=4)		(n=5)	
5.2		19.1	

[1]measurements courtesy of Amoco Argentina Oil Company: air permeabilities and Boyle's Law helium porosities.

within the Comodoro Rivadavia Formation suggests that the associated heterogeneities and potential difficulties in evaluation may be found within specific fields and/or reservoirs.

DISCUSSION

Textural evidence indicates that significant amounts of the fine-grained material found within the intergranular pores of the Comodoro Rivadavia Formation sandstones were formed there as infiltration materials, yet the chemical and mineral compositions of these materials varies greatly. There are several possible situations that could explain these observations: (1) the quartz-rich structures are not true infiltration structures at all, but microgeoidal precipitates within the open pores and the variations observed are only an artifact of this silicification process; (2) the quartz-rich structures are siliceous pseudomorphous replacements of pre-existing intergranular material; (3) the composition of the suspended load of Comodoro Rivadavia streams varied and much of the variation observed can be attributed to that initial compositional variation; (4) the composition of the suspended load of Comodoro Rivadavia streams was relatively constant and the observed variations in chemical and mineral composition are the product of soil or shallow burial conditions that varied; and/or (5) some combination of varying suspended-load compositions and soil or shallow burial conditions produced the observed differences.

The microcrystalline quartz-rich structures are unlikely to be merely precipitates since precipitation tends to produce at least some micro-laminations of crystals that are oriented radially (i.e., perpendicular to the planes of laminations). None of the quartz-rich material examined (using both optical and scanning electron microscopy) exhibits radial or botryoidal forms. Also, the composition (mixtures with clays and iron oxides) of the quartz-rich varieties is more compatible with an origin of infiltrated siliceous and clay material, which has been subsequently altered, than with co-precipitation of these various phases.

It is entirely possible that the observed silica is indeed a pseudomorphous replacement. The fluid chemistries are likely similar for either replacement or the alteration of infiltrated volcanic glass. If the textures are replacements, however, then the replacements are highly selective for only those fabrics interpreted as infiltration structures, such as cutans, geopetal structures and massive pore fillings. These masses of submicron-size quartz crystals do not occur as replacements of any of the framework detritus. Nor are there any textures typical of partial replacement (i.e., immediately adjacent material that is not quartzose, but that occupies the same position).

It is also unlikely that the suspended-load composition was constant in light of the widely varying modal-sandstone compositions and abundant evidence for contemporaneous volcanism in the area. The chemistry also is incompatible with a constant, and presumably clay-dominated, suspended load, in that leaching of infiltrated, clay-rich materials to form silica minerals is constrained by the relative insolubility of Al in natural waters. Leaching would likely only make those materials more aluminous, rather than more siliceous. As mentioned earlier, in the discussion of the chemical composition of these materials, incorporation of silica (likely derived from altering detrital grains) into the intergranular material is required to produce the high-Si compositions. Also mentioned earlier, the composition of the infiltrated material is more compatible with temporal and spatial variation in the suspended-load composition. The inclusion of fine-grained volcanic glass derived from the erosion of airfall-ash deposits blanketing the drainage basin would provide a sporadic and highly variable source of siliceous material. However, some subsequent diagenetic alteration of the material is required to produce the observed compositions.

CONCLUSIONS

Infiltrated materials are found within the intergranular pore spaces of volcanogenic fluvial sandstones of the Comodoro Rivadavia Formation. The infiltrated material is most abundant within the high-sinuosity or point-bar deposits and uncommon within the braided-stream deposits. The textures observed include ridges and bridges, geopetal structures, cutans, and massive pore fillings. Two groups of material are recognized: a quartz-rich variety composed of mixtures of microcrystalline quartz, smectite, and/or chlorite and iron oxides; and a clay-rich variety composed of chlorite, smectite and iron oxides. These two types occur within sandstones of the same field, but are separated temporally (vertically). The clay-rich variety is more common.

The clay-rich variety is likely the result of infiltration of the common or normal, clay-dominated suspended load of the drainage basin, and formed from the weathering of the convergent margin-arc terrain to the west and northwest, as well as the Jurassic volcanics and crystalline basement rocks to the north. Occasionally, the drainage basin was blanketed by airfall-ash deposits whose erosion produced suspended loads charged with siliceous volcanic glass. Infiltration of glassy material and its subsequent alteration produced the quartzose varieties of infiltration structures. These features are potentially representative of distal portions of contemporaneous volcanism.

Silica, derived from the hydration of the adjacent volcanic lithic fragments, is required in order to account for the great abundance of quartz within the infiltration structures. As such, these quartz-rich structures are unlikely to form in deposits lacking additional *in situ* sources of silica.

ACKNOWLEDGMENTS

The author thanks Amoco Argentina Oil Company for providing financial support, samples and data for this study. Additional support was provided by grants from the Enhanced Oil Recovery Institute, State of Wyoming Stripper Well Violation Fund, and Professor Ron Surdam, University of Wyoming and Gas Research Institute contract no. 5089-260-1894. Steve Boese performed the whole-rock chemical analyses. Thanks go to Roberto Merino, Jim Hoffman, Marco Moraes and Laurel Babcock for their help and constructive discussions. Earlier versions of the manuscript were read by J. I. Drever, D. B. MacGowan and R. C. Surdam. The manuscript was greatly improved by the constructive reviews of J. R. Boles and A. W. Walton.

REFERENCES

ALFORD, E. V., 1983, Compositional variations of authigenic chlorites in the Tuscaloosa Formation, Upper Cretaceous, of the Gulf Coast Basin: Unpublished M.S. Thesis, University of New Orleans, New Orleans, Louisiana, 66 p.

CHANG, H. K., MACKENZIE, F. T., AND SCHOONMAKER, J., 1986, Comparisons between the diagenesis of dioctahedral and trioctahedral smectite, Brazilian offshore basins: Clays and Clay Minerals, v. 34, p. 407–423.

CRONE, A. J., 1975, Laboratory and field studies of mechanically infiltrated matrix clay in arid fluvial sediments: Unpublished Ph.D. Dissertation, University of Colorado, Boulder, Colorado, 162 p.

CURTIS, C. D., IRELAND, B. J., WHITEMAN, J. A., MULVANEY, R., AND COBLEY, T., 1984, Authigenic chlorites; problems with chemical analysis and structural formula calculations: Clay Minerals, v. 19, p. 471–481.

DAVIES, D. K., ALMON, W. R., BOWIE, S. B., AND HUNTER, B. E., 1979, Deposition and diagenesis of Tertiary-Holocene volcaniclastics, Guatemala, *in* Scholle, P. A., and Schluger, P. R., eds., Aspects of Diagenesis: Society of Economic Paleontologists and Mineralogists Special Publication 26, p. 281–306.

DETHIER, D. P., PEVEAR, D. R., AND FRANK, D., 1981, Alteration of new volcanic deposits, *in* Lipman, P. W., and Mullineaux, D. R., eds., The 1980 Eruptions of Mount St. Helens, Washington: U. S. Geological Survey Professional Paper 1250, p. 649–665.

DREVER, J. I., 1973, The preparation of oriented clay mineral specimens for X-ray diffraction analysis by a filter-membrane peel technique: American Mineralogist, v. 58, p. 553–554.

FITZGERALD, M. G., MITCHUM, JR., R. M., ULIANA, M. A., AND BIDDLE, K. T., 1990, Evolution of the San Jorge Basin, Argentina: American Association of Petroleum Geologists Bulletin, v. 74, p. 879–920.

HAGEN, E. S., AND SURDAM, R. C., 1989, Thermal evolution of Laramide-style basins: constraints from the northern Bighorn Basin, Wyoming and Montana, *in* Naeser, N. D., and McCulloh, T. H., eds., Thermal History of Sedimentary Basins: New York, Springer-Verlag, p. 277–295.

HOFFMAN, J., AND HOWER, J., 1979, Clay mineral assemblages as low grade metamorphic geothermometers: application to the thrust faulted Disturbed Belt of Montana, U. S. A., *in* Scholle, P. A., and Schluger, P. R., eds., Aspects of Diagenesis: Society of Economic Paleontologists and Mineralogists Special Publication 26, p. 55–79.

KHATCHIKIAN, A., AND LESTA, P., 1973, Log evaluation of tuffites and tuffaceous sandstones in southern Argentina: Society of Professional Well Log Analysts, Proceedings, 14th Annual Logging Symposium, Corpus Christi, Texas, p. 1–24.

LACHENBRUCH, A. H., AND SASS, J. H., 1977, Heat flow in the United States and the thermal regime in the crust, *in* The Earth's Crust: American Geophysical Union Memoir 20, p. 626–675.

LAWVER, L. A., MEINKE, L., AND SCLATER, J. G., 1984, Tectonic reconstruction of the South Atlantic: Antarctic Journal of the United States, v. 18, p. 142–145.

MATLACK, K. S., HOUSEKNECHT, D. W., AND APPLIN, K. R., 1989, Emplacement of clay into sand by infiltration: Journal of Sedimentary Petrology, v. 59, p. 77–87.

MCKENZIE, D. P., 1978, Some remarks on the development of sedimentary basins: Earth and Planetary Science Letters, v. 40, p. 25–32.

MORAES, M. A. S., AND DE ROS, L. F., 1990, Infiltrated clays in fluvial Jurassic sandstones of Recôncavo Basin, northeastern Brazil: Journal of Sedimentary Petrology, v. 60, p. 809–819.

MURRAY, H. H., 1988, Kaolin minerals: their genesis and occurrences, *in* Bailey, S. W., ed., Hydrous Phyllosilicates (exclusive of micas), Reviews in Mineralogy: Mineralogical Society of America, v. 19, p. 67–89.

NOCKOLDS, S. R., 1954, Average chemical compositions of some igneous rocks: Geological Society of America Bulletin, v. 65, p. 1007–1032.

SCLATER, J. G., AND CHRISTIE, P. A. F., 1980, Continental stretching: an explanation of the post-mid-Cretaceous subsidence of the central North Sea basin: Journal of Geophysical Research, v. 85, no. B7, p. 3711–3739.

VELDE, B., 1985, Clay minerals, a physico-chemical explanation of their occurrence: Developments in Sedimentology, v. 4: New York, Elsevier, 427 p.

VELDE, B., AND MEUNIER, A., 1987, Petrologic phase equilibria in natural clay systems, *in* Newman, A. C. D., ed., Chemistry of Clays and Clay Minerals: Mineralogical Society Monograph no. 6, p. 423–458.

WALKER, T. R., 1976, Diagenetic origin of continental red beds, *in* Falke, H., ed., The Continental Permian in Central, West, and South Europe: Boston, D. Reidel Publishing Company, p. 240–282.

WALKER, T. R., WAUGH, B., AND CRONE, A. J., 1978, Diagenesis in first-cycle desert alluvium of Cenozoic age, southwestern U.S. and northwest Mexico: Geological Society of America Bulletin, v. 89. p. 19–32.

WILSON, M. D., AND PITTMAN, E. D., 1977, Authigenic clays in sandstones: recognition and influence on reservoir properties and paleoenvironmental analysis: Journal of Sedimentary Petrology, v. 47, p. 3–31.

CLAY MINERALOGY, SPORE COLORATION AND DIAGENESIS IN MIDDLE MIOCENE SEDIMENTS OF THE NIGER DELTA

FRANCISCA E. OBOH
Department of Geology and Geophysics, University of Missouri, Rolla, Missouri 65401

ABSTRACT: The clay-mineral assemblage of Middle Miocene sediments from the E2.0 Reservoir (3,590–3,655 m) in the Kolo Creek field is correlated with the color of palynomorphs in order to understand the level of diagenesis experienced by the sediments. The clay minerals in the mudstones are predominantly detrital kaolinite with significant amounts of illite/smectite, whereas in the sandstones, clays occur in minor amounts as mainly authigenic kaolinite. Quantitative measurements of the color of palynomorphs show that the sediments are immature to marginally mature and this level of maturity can be related to early diagenetic conditions before the emplacement of oil in the reservoir. A plot of palynomorph carbonization (luminance) versus percent illite in illite/smectite shows a fairly good correlation (statistically significant at a 95% level of confidence) between the two diagenetic parameters. The low level of maturity is not obvious from Kubler's illite crystallinity index for the mudrocks, which show a considerable variation over a short depth range.

INTRODUCTION

The Niger delta occupies an area of 75,000 km² in the eastern corner of the Gulf of Guinea in West Africa (Fig. 1) and its evolution has been linked to the opening of the South Atlantic Ocean in Middle Cretaceous time (King, 1950; Stoneley, 1966; Burke and others, 1971). A major regression in the Late Eocene initiated the deposition of a wedge of fluvio-deltaic sediments, which has continued to the present (Short and Stäuble, 1967). The Kolo Creek field is situated in the central part of the onshore delta (Fig. 1) under the concession of the Shell Petroleum Development Company of Nigeria, which, along with the Koninklijke/Shell Exploration and Production Laboratory in The Netherlands, provided core from the Agbada Formation for this study. The 50-m-long cores from the oil-bearing E2.0 Reservoir in Wells 27 and 29 were recovered from a depth interval of 3,590 m to 3,655 m; the interval has been dated as Middle Miocene on the basis of the palynomorph assemblage (Oboh, 1990).

The purpose of this paper is to correlate clay mineralogy and palynomorph (herein called "spore") coloration to diagenesis, as both components are affected by burial (Gutjahr, 1966; Powers, 1967; Grayson, 1975; Velde, 1985). Recent studies relating clay and kerogen evolution under conditions of diagenesis have been undertaken extensively in other regions (Smart and Clayton, 1985; Burtner and Warner, 1986; Scotchman, 1987; Velde and Espitalie, 1989) in comparison to the Niger delta (Lambert-Aikhionbare, 1981). Spore coloration has been chosen as the parameter of maturity for organic matter because it has not been used previously in the basin. It is quantitatively measured here and related to the Commission Internationale de l'Eclairage (CIE) Color System, details of which are described in Chamberlin and Chamberlin (1980), Wyszecki and Stiles (1967), Grum and Bartleson (1980), and MacAdam (1981). The system defines a color spectrum in terms of a red-green-blue color mixture (Fig. 2A) but some colors exist outside this triangle. As a result, the triangle is enlarged to take in three unreal primaries (Fig. 2B) that lie outside the locus of spectral colors in order to accommodate all colors on the diagram. Therefore, all colors can be defined along a two-coordinate system (x and y, the chromaticity coordinates) and the total amount of light represented by absorbance (luminance). These three parameters are calculated from a series of measurements at regular wavelength increments across a transmitted light spectrum.

STRATIGRAPHY

Short and Stäuble (1967) defined three diachronous subsurface lithostratigraphic units (Fig. 3) in the Tertiary of the Niger delta complex. The basal Akata Formation is composed mainly of marine shales with silty and sandy horizons, which were laid down in front of the advancing delta. These shales are overpressured (Weber and Daukoru, 1975; Bruce, 1984) and have an estimated thickness of 6,500 m. The formation varies in age from Eocene in the north to Recent in the offshore parts of the delta.

The Agbada Formation is a coarsening-upward sequence of alternating sandstones, siltstones and shales of delta-front and lower delta-plain origin. The alternations of sandy and argillaceous sediments are the result of differential subsidence, variation in sediment supply and shifts in the depositional lobes of the delta. The formation is up to 4,000 m thick in the central part of the delta, thinning seaward and toward the delta margins, and ranges in age from Eocene to Recent. The sandy parts of the formation constitute the main hydrocarbon reservoirs, which are trapped in rollover anticlines fronting growth faults with the interbedded shales acting as seals for the reservoirs.

The top unit, called the Benin Formation, is a sandstone sequence with a few siltstone and shale intercalations, which increase with depth. These sediments are of continental delta-plain origin. The formation is about 2,000 m thick in the central part of the onshore delta and ranges in age from Oligocene to Recent.

ANALYTICAL METHODS

Clay Minerals

Fifty-three sandstone and mudrock samples from the two wells were analyzed for clay mineralogy. X-ray diffraction (XRD) was carried out on all the samples for bulk-rock composition using randomly oriented mounts of the <63-μm fraction, and on the <2-μm fraction of 25 mudstones. The sandstones were analyzed by optical microscopy and scanning electron microscopy (SEM); some mudstone samples were also examined with the SEM. The <2-μm fraction was obtained by dispersion method and the oriented

Origin, Diagenesis, and Petrophysics of Clay Minerals in Sandstones, SEPM Special Publication No. 47
Copyright © 1992, SEPM (Society for Sedimentary Geology), ISBN 0-918985-95-1

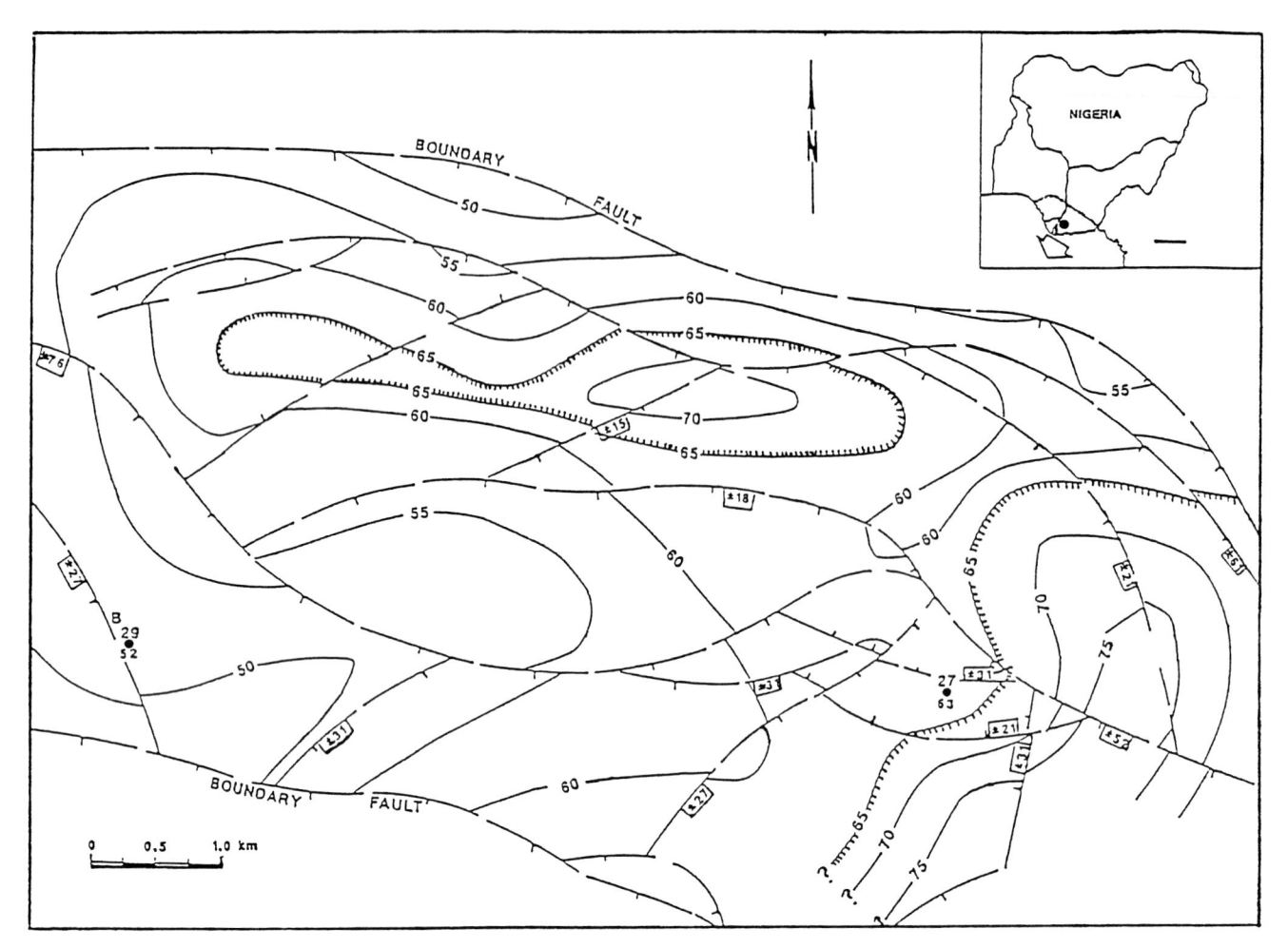

FIG. 1.—Isochore map of the E2.0 Reservoir showing the location of Wells 27 and 29 in the Kolo Creek field; contours in meters. Inset shows the location of the field in the Niger delta, southern Nigeria; scale bar = 200 km.

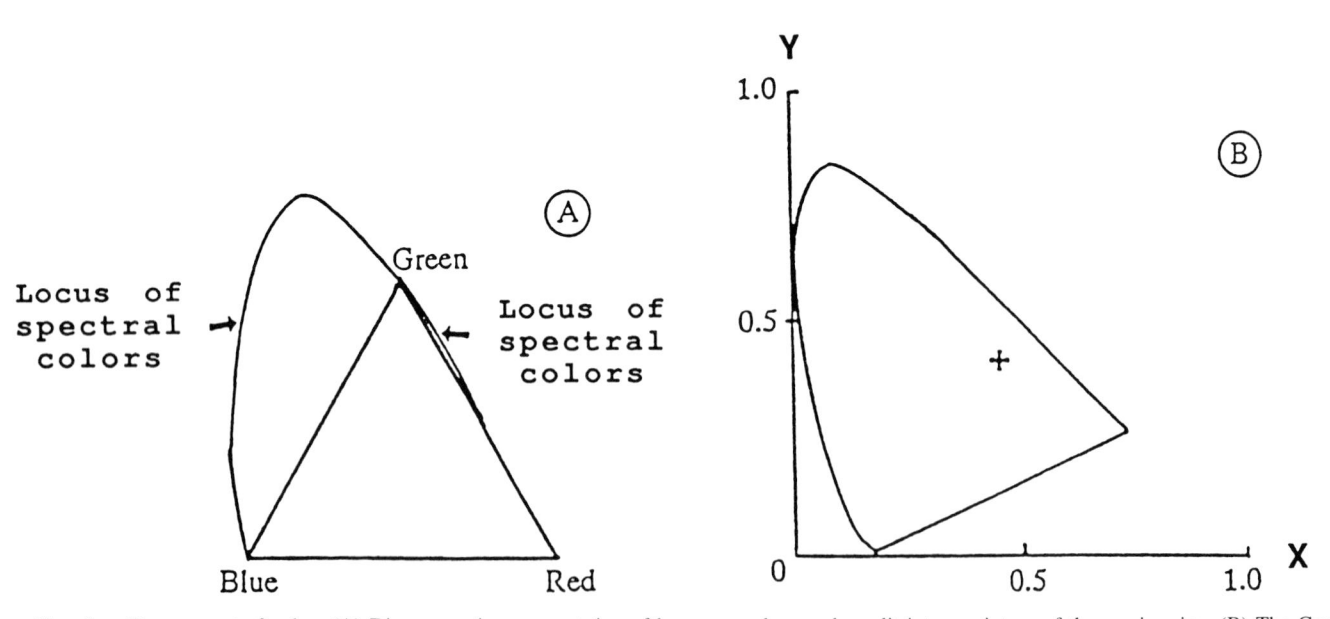

FIG. 2.—The concept of color. (A) Diagrammatic representation of how any color can be split into a mixture of three primaries. (B) The Commission Internationale d'Eclairage 1931 chromaticity diagram showing the locus of spectral colors from (A). The + sign marks the position of illuminant A, which is the position of a transparent object.

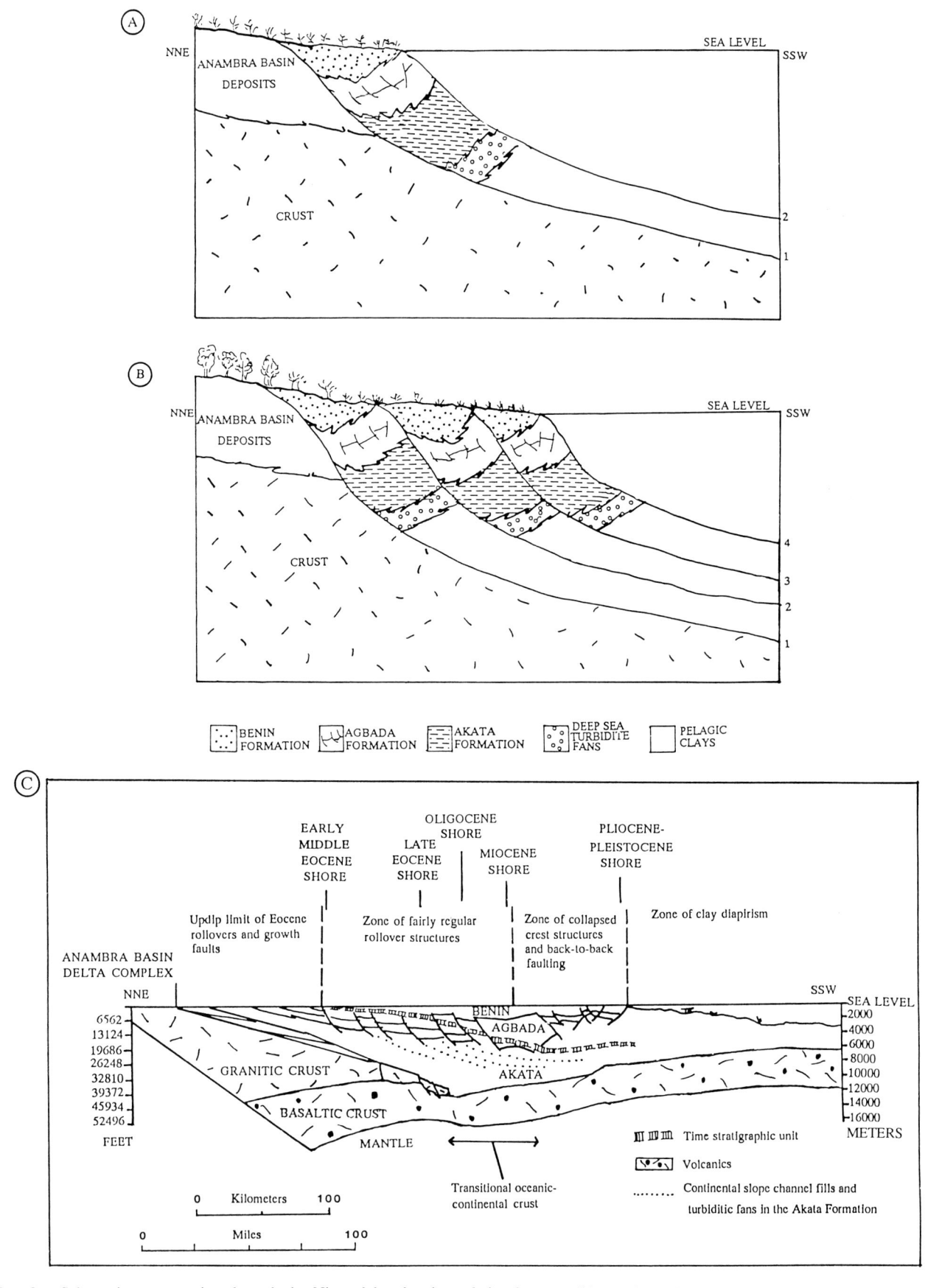

FIG. 3.—Schematic cross section through the Niger delta showing relation between lithostratigraphic units. (A and B) Hypothetical sketches showing progradation of the shoreline and stratigraphic succession; 1–4 are depositional surfaces that are analogous to time lines. (C) incorporates the main tectonic elements with the succession (modified from Whiteman, 1982).

samples were analyzed using CuKα radiation under four conditions: air dried, glycerolated, and heated to 400°C and 550°C (Jeans, 1978). Some samples were subjected to sulphuric-acid treatment to further differentiate between chlorite and kaolinite. The relative proportions of the clay minerals in the <2-μm fraction were estimated as peak-area percentages (Griffin, 1971) and their identification was based on Carroll (1970), Griffin (1971) and Brindley and Brown (1980); the methods described in Reynolds and Hower (1970) and Reynolds (1980) were used to estimate percent illite in illite/smectite. Illite crystallinity data were calculated from diffractograms of glycerolated samples using the technique of Kubler (1968). Kubler's crystallinity index is defined as the width of the 10Å peak at one-half maximum peak height; it normally decreases with increasing illite crystallinity. The Philip's SEM used for studying the sandstones and some mudstones in backscattered and secondary electron-imaging modes was equipped with an energy-dispersive X-ray analyzer (EDX) for the identification of mineral phases; identification followed the methods of Welton (1984).

Palynomorphs

Palynomorphs were extracted by standard palynological techniques of digesting sediments in hydrochloric and hydrofluoric acids. Oxidation with nitric acid was excluded in order to preserve the natural colors of the palynomorphs. The residues were sieved with 10 μm nylon meshes and then strew mounted on glass slides. Color was determined by using microscope equipment at Southampton University, England. The spectra results were produced using a Zeiss UMSP 50 (universal microspectro-photometer) microscope fitted with a 100 W tungsten bulb with a color temperature of 3,400°K. This temperature was modified to the 2,854°K required by standard illuminant A of the CIE system by utilizing a special conversion filter. The microscope was linked to a computer for data acquisition and processing. Light-intensity measurements were recorded at 10 nm steps over the range 400 to 700 nm with a monochromer slit width of 20 nm. The results from spectra, after correction, were converted to CIE coordinates and directly screen plotted on an enlarged section of the chromaticity chart. Measurement was made under a 10x dry objective with a measurement spot of 20 μm and it was always on pollen, 20 of which were measured in each sample (four in Well 27 and six in Well 29). The samples were chosen to represent sandstone and mudstone lithofacies from the top, middle and lower parts of the reservoir.

CLAY MINERALOGY

Mudrocks

The mudrocks of the E2.0 Reservoir are composed predominantly of clay minerals (60–90%) with lesser amounts of quartz, feldspar, siderite, calcite, pyrite and anhydrite. The clay assemblage consists of kaolinite (35–65%), illite (5–15%) and illite/smectite (25–50%). Illite/smectite is regularly interstratified with an 80 to 90% illite component. With the exception of the mixed-layer clays, the values obtained for kaolinite and illite are broadly similar to previously published values of Tertiary and Recent mudrocks from the Agbada Formation (Table 1). This assemblage illustrates that smectite has been interstratified with illite at depths >3,500 m and calls to question the presence of smectite in deeply buried Tertiary sediments analyzed by Lambert-Aikhionbare and Shaw (1982). The collapse of the 7.12Å and 3.56Å basal reflections for kaolinite + chlorite upon heating the clay minerals to 550°C and the persistence of the peaks after treatment with sulphuric acid indicate that there is no chlorite in the sediments. This observation contrasts with the results of Braide and Huff (1986) and Odigi (1987), who obtained chlorite values of up to 13% and 26%, respectively, in the Agbada shales from the eastern part of the delta.

Sandstones

The sandstones are poorly cemented, coarse- to fine-grained quartz arenites and subarkoses with minor amounts (1–7%) of clay minerals. The clay-mineral assemblage is predominantly kaolinite (>95%), which occurs as an authigenic mineral in pore spaces (Fig. 4) and as a thick coating around detrital grains. Minor amounts of illite/smectite were detected by their morphology on the SEM and by their elemental composition from EDX and by XRD. Neither authigenic illite nor smectite was detected using these analytical techniques.

ILLITE CRYSTALLINITY

The values of crystallinity index using the Kubler (1968) technique display a considerable variability within a short depth interval, ranging from 0.4 to 0.71°2θ (7–16 mm) (Fig. 5). There is no clear trend showing increasing crystallinity with depth of burial, which is usually reflective of increasing maturity (Zen and Thompson, 1974; Duba and Williams-Jones, 1983). Such a trend might have been observed if this short interval covered the principal smectite-to-illite conversion temperature. However, the data fall within the limits of diagenesis (0.4–1.0°2θ) because values from 0.25 to 0.4°2θ mark a transitional zone (anchizone) between low-grade metamorphism (epizone) and diagenesis (Frey, 1987).

TABLE 1.—RELATIVE AMOUNTS OF CLAY MINERALS IN THE <2-μm FRACTION OF TERTIARY AND RECENT AGBADA FORMATION SHALES

Clay minerals	E2.0 Reservoir	Tertiary Agbada[1]	Recent Agbada[2]
% kaolinite	35 - 65	40 - 75	35 - 60
% illite	6 - 15	15 - 25	10 - 15
% smectite	-	10 - 35	30 - 50
% illite/smectite	25 - 50*	-	-

[1] Lambert-Aikhionbare and Shaw (1982)
[2] Porrenga (1966)
* Regular interstratification with 80 to 90% illite

FIG. 4.—Backscattered scanning electron micrograph of authigenic kaolinite infilling a pore space in sandstone. Scale bar = 10 μm.

SPORE COLORATION

The variation in the pollen population from single-sample measurements is shown in Figures 6A and B. The + sign marks the position of illuminant A, which is the position of a transparent object, the direction of increasing maturity and spore coloration being to the right from this point. Figure 6A illustrates that the standard deviations of all but one of the samples display a tight to moderate spread of values. The data points in Figure 6B lie in the region of immature to marginally mature sediments on the maturation path defined from the density of the x and y chromaticity coordinates (Marshall, 1991). This maturation path can be compared with the color scale of Fisher and others (1980) to reflect a concentration of the data points around points 1 to 4 on a scale of 10, which are indicative of temperatures of 100° to 180°C. These colors range from yellow to light brown, and the presence of dark brown palynomorphs in some samples has been related to recycling (Oboh, 1990). Quantitative measurements were found to be more representative of the colors of the palynomorphs than visual measurements, although there was a satisfactory correlation between them.

DISCUSSION

The presence of illite/smectite in the mudrocks indicates that there has been diagenetic transformation of smectite, which has been shown to occur from temperatures of 80° to 110°C at depths of 2 km or more in argillaceous sediments of other areas (Burst, 1969; Perry and Hower, 1972; Hoffman and Hower, 1979; Shaw, 1980). The burial temperatures of these Kolo Creek sediments have been reconstructed by using the following equation, derived by Evamy and others (1978):

$$T = a - bS^2 + cD$$

where T = temperature gradient (°F/100 ft); S = sand percent of selected interval (at 100 ft or 30.48 m); D = average depth of selected interval; $a = 1.811 \pm 0.161$; $b = (1.615 \pm 0.146) \times 10^5$; $c = (7.424 \pm 1.749) \times 10^5$; assumed surface temperature = 80°F (26.66°C). The calculated temperatures for the two wells (Fig. 7) give an average geothermal gradient of 2.7°C/100 m for the Agbada Formation (which corresponds largely with that obtained by Evamy and others, 1978) and show that the burial temperatures for the E2.0 Reservoir sediments are around 107°C and 110°C for Wells 27 and 29, respectively. At these temperatures, illite/smectite has regular interstratification in the reservoir, and in similar mudrocks from the Niger delta (Braide and Huff, 1986).

Two factors can be considered from the clay-mineral assemblage in deciphering the level of diagenesis in the sediments. First, the presence of chlorite in any clay-mineral assemblage usually indicates a higher grade of diagenesis, but these sediments lack chlorite, suggesting low-grade diagenesis. Secondly, the similarity in the values of kaolinite and illite between the Tertiary and Recent mudrocks (Table 1) shows that these two minerals have undergone little alteration. The distribution of kaolinite and illite in the Kolo Creek mudrocks appears to be unrelated to the transformation of smectite to illite, and they are considered to be largely the result of primary deposition. Further evidence for early diagenesis can be found in the uncompacted and porous nature of the sandstones at depths in excess of 3,500 m. Concavo-convex and sutured contacts between framework grains in the sandstones are few (4%), in comparison with tangential contacts (52%) and long contacts (44%). Moreover, point-count estimates show that intergranular volume varies from 13% to 33%. There are limited amounts of silica and carbonate cements (see also Lambert-Aikhionbare, 1982), and the relative amounts of the carbonates appear to be related to environment of deposition (Oboh,

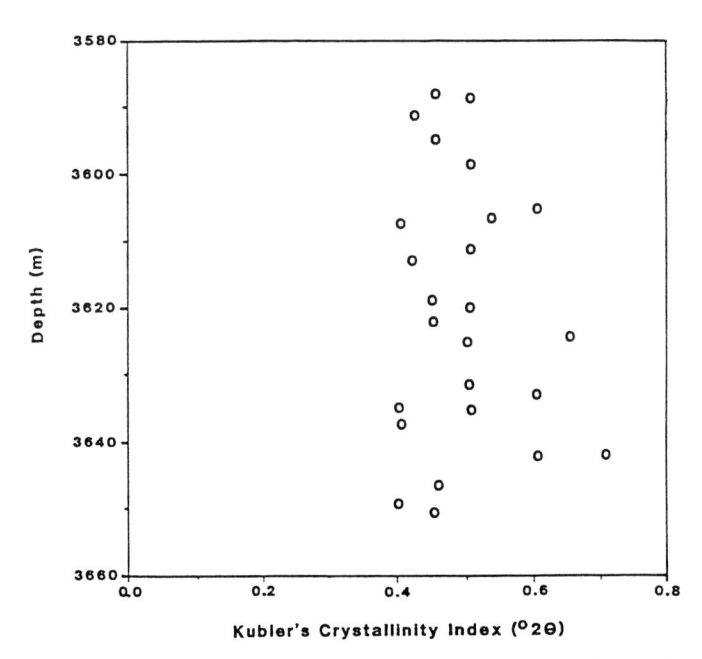

FIG. 5.—Kubler's crystallinity index plotted against burial depth, showing a broad variation within the diagenetic zone >0.4°θ.

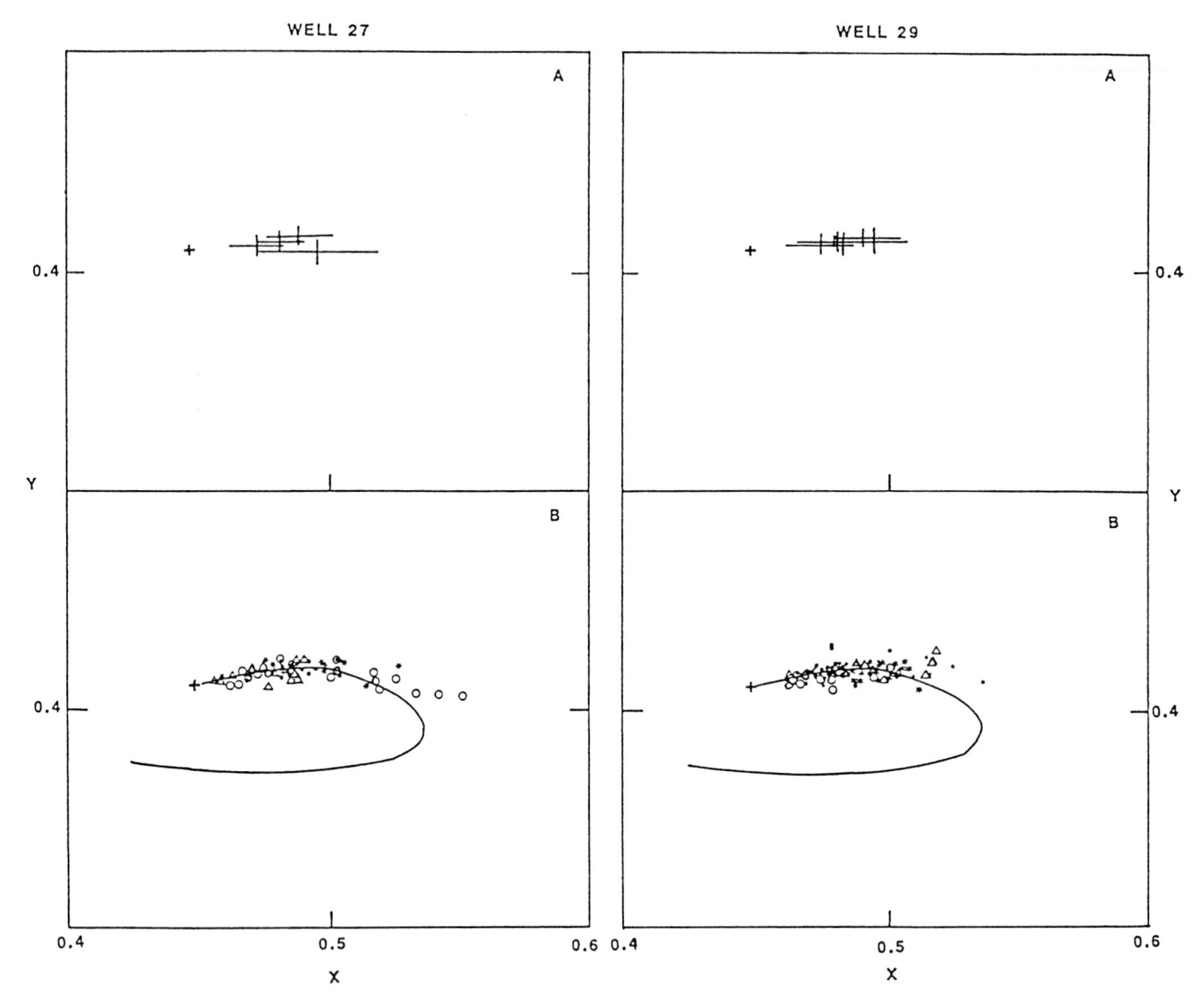

FIG. 6.—Variation in the chromaticity coordinates for a representative group of samples. (A) Standard deviations; x chromaticity coordinates have low standard deviations at low maturities. (B) Variation in individual measurements from single samples. The maturation path (curved line) was defined by Marshall (1991). The + sign marks the position of illuminant A.

1990). Sandstones from the coastal-barrier facies contain more carbonate cements than those from the distributary-channel facies and this observation can be attributed to the availability of carbonate material in the coastal barrier.

Clay minerals are usually considered separately from other cements in sandstones because they do not provide appreciable rigidity to the rocks. Kaolinite is present mainly as an authigenic mineral, characterized by a book-like morphology (Fig. 4). A direct relation exists between kaolinite and feldspars, except in one sample in Well 27 (Fig. 8), indicating that both neoformed and transformed phases are present. The inverse relation indicates the predominance of transformed kaolinite, which was derived from alteration of feldspars, whereas the neoformed phase resulted from the influx of meteoric water into pore spaces. A minor amount of kaolinite is present where extensive carbonate cementation has taken place, and it would appear that its formation was inhibited by calcite cementation. Paragenetically, kaolinite was the first authigenic mineral to be precipitated in the reservoir (Oboh, 1990).

Illite crystallinity measurements (Kubler, 1968) have been used as a measure of thermal maturity by Guthrie and others (1986) and Ungerer and others (1986) on sedimentary rocks that are much older than those of the Niger delta. Crystallinity values for the Kolo Creek mudstones display a variation (Fig. 5) that does not readily give an indication of maturity because of the narrow range of maximum temperatures to which the samples have been exposed. However, low to moderate thermal maturity can be suggested for the sediments from the chromaticity coordinates (Fig. 6) and mean luminance values of 50 to 64 (Table 2; Fig. 9) of palynomorph colors. Because the amount of light ab-

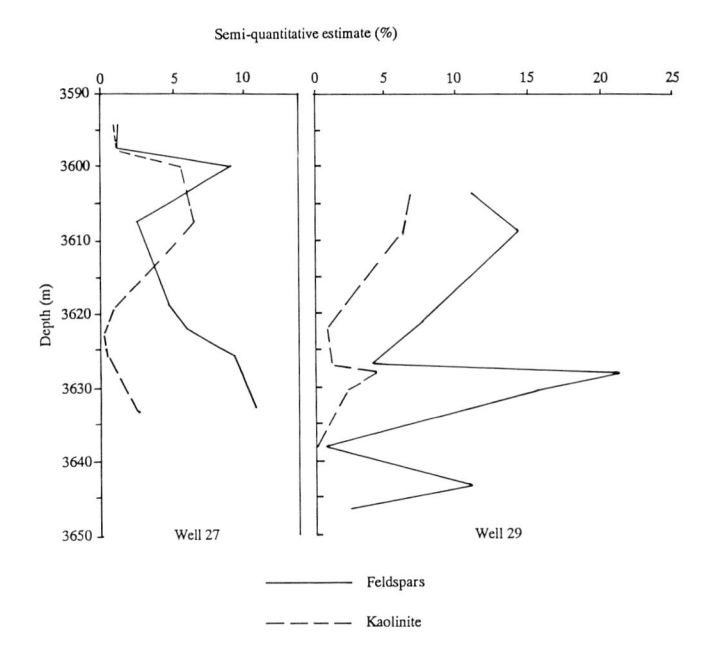

FIG. 8.—Relation between feldspars and kaolinite in sandstones.

sorbed by palynomorphs (luminance) decreases as carbonization increases, this parameter shows a fairly good correlation (statistically significant at a 95% confidence level) with percent illite in illite/smectite, which is also a function of maximum temperature (Fig. 9).

Low thermal maturity of sediments has been used in the past to suggest recent generation of oil in the Niger delta (Weber and Daukoru, 1975; Evamy and others, 1978; Ekweozor and Okoye, 1980). However, the porous and uncompacted nature of sandstones in the E2.0 Reservoir, and the presence of partially altered aragonitic shells (Oboh, 1990) show that the sediments have only experienced minor diagenetic alterations. Lambert-Aikhionbare (1982) suggested early generation and emplacement of oil in the Agbada sandstones on the basis of similar diagenetic inferences, a view recently reiterated by Wilson (1990). Considering the higher amounts of kaolinite and carbonate cements in non-oil-bearing sandstone horizons in the E2.0

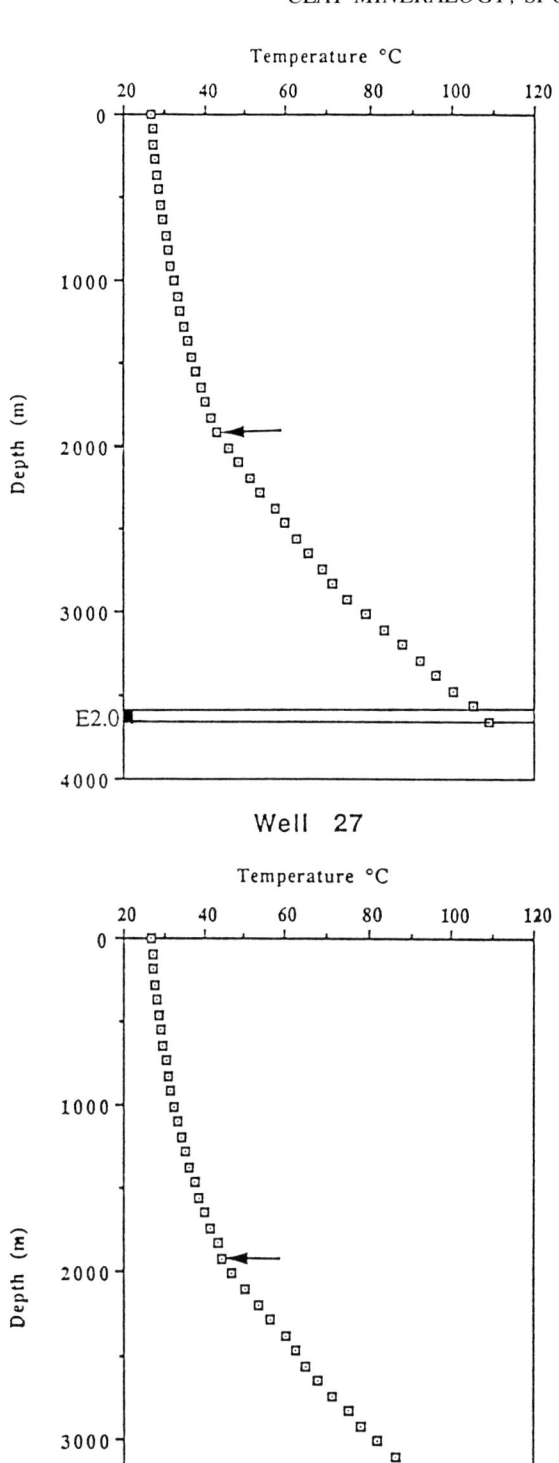

FIG. 7.—Geothermal temperature profiles showing burial temperatures of 107°C and 110°C for the E2.0 Reservoir in Wells 27 and 29, respectively. Each arrow indicates possible boundary between the Benin and Agbada Formations.

TABLE 2.—PERCENT ILLITE IN ILLITE/SMECTITE AND CORRESPONDING SPORE-COLOR VALUES EXPRESSED AS MEAN LUMINANCE

Samples	%I in I/S	Mean Luminance
KC27-5	84.8	55.03
KC27-30	83.6	55.46
KC27-37	88.8	50.09
KC29-21	83.6	58.95
KC29-23	84.8	52.96
KC29-24	88.2	58.55
KC29-26	80.3	62.78
KC29-27	85.2	57.58
KC29-31	81.5	63.30

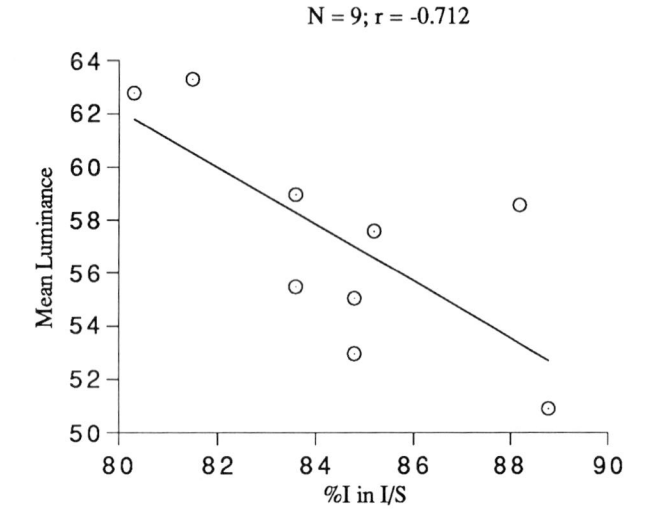

FIG. 9.—Relation between pollen carbonization (mean luminance) and percent illite in illite/smectite of nine mudrock samples for which both data are available.

interval, emplacement of oil might have effectively stopped diagenesis in the reservoir. Because of lack of data on the deeply buried shales of the Akata Formation, the interbedded shales of the Agbada Formation from similar stratigraphic horizons over a wide area in the delta, and on the oil composition in the reservoir, it would be inappropriate here to suggest the source rocks for the oil in the E2.0 Reservoir. It is possible that rapid burial in the delta might have enhanced maturation of organic matter in the area and also played a role in the diagenetic history of the sediments.

CONCLUSIONS

Data from the <2-μm fraction show that the mudrocks of the E2.0 Reservoir (from 3,590–3,655 m) are composed primarily of kaolinite with a substantial amount of mixed-layer illite/smectite; illite occurs in minor amounts. The similarity in the abundance of kaolinite and illite between these Tertiary sediments and recent shales from the Niger delta suggests minimal effects of diagenesis, but smectite has been altered to regularly interstratified illite/smectite. At burial temperatures of 107 to 110°C, the level of maturity is low to moderate, as confirmed by quantitative spore coloration, which has also been correlated with early diagenetic conditions before the emplacement of oil in the reservoir. Low-grade diagenesis is also reflected in the porous and uncompacted nature of sandstones at depths in excess of 3,500 m.

ACKNOWLEDGMENTS

This work was funded by the award of a Commonwealth Scholarship Commission studentship at the University of Cambridge, which is gratefully acknowledged. At all stages, considerable assistance was afforded by Drs. J. L. Chapman and C. V Jeans. My thanks are given to Dr. J. E. A. Marshall for the use of his equipment for spore coloration, for writing the program for computer data acquisition, and for his comments on the initial results. I also thank Dr. J. F. Burst for his comments on an earlier version of the manuscript. This paper is published by permission of the Shell Petroleum Development Company of Nigeria. Cambridge Earth Sciences Series No: 2294.

REFERENCES

BRAIDE, S. P., AND HUFF, W. D., 1986, Clay mineral variation in Tertiary sediments from the eastern flank of the Niger delta: Clay Minerals, v. 21, p. 211–224.

BRINDLEY, G. W., AND BROWN, G., eds., 1980, Crystal Structures of Clay Minerals and their X-ray Identification: London, Mineralogical Society, 495 p.

BRUCE, C. H., 1984, Smectite dehydration: its relation to structural development and hydrocarbon accumulation in northern Gulf of Mexico Basin: American Association of Petroleum Geologists Bulletin, v. 68, p. 673–683.

BURKE, K. C., DESSAUVAGIE, T. F. J., AND WHITEMAN, A. J., 1971, The opening of the Gulf of Guinea and the geological history of the Benue depression and Niger delta: Nature, v. 233, p. 51–55.

BURST, J. F., 1969, Diagenesis of Gulf Coast clayey sediments and its possible relation to petroleum migration: American Association of Petroleum Geologists Bulletin, v. 53, p. 73–93.

BURTNER, R. L., AND WARNER, M. A., 1986, Relationship between illite/smectite diagenesis and hydrocarbon generation in Lower Cretaceous shales of the northern Rocky Mountain area: Clays and Clay Minerals, v. 34, p. 309–402.

CARROLL, D., 1970, X-ray Identification of Clay Minerals: Geological Society of America Special Paper 126, 82 p.

CHAMBERLIN, G. J., AND CHAMBERLIN, D. G., 1980, Colour, Its Measurement, Computation and Application: London, Heyden, 137 p.

DUBA, D., AND WILLIAM-JONES, A. E., 1983, The application of illite crystallinity, organic matter reflectance, and isotopic techniques to mineral exploration: a case study in southwestern Gaspé, Quebec: Economic Geology, v. 78, p. 1350–1363.

EKWEOZOR, C. N., AND OKOYE, N. V., 1980, Petroleum source-bed evaluation of Tertiary Niger delta: American Association of Petroleum Geologists Bulletin, v. 64, p. 1251–1258.

EVAMY, B. D., HAREMBOURE, J., KAMMERLING, P., KNAAP, W. A., MOLLOY, F. A., AND ROWLANDS, P. H., 1978, Hydrocarbon habitat of Tertiary Niger delta: American Association of Petroleum Geologists Bulletin, v. 62, p. 1–39.

FISHER, M. J., BARNARD, P. C., AND COOPER, B. S., 1980, Organic maturation and hydrocarbon generation in the Mesozoic sediments of Sverdrup Basin, Arctic Canada: Lucknow, India, Proceedings, 4th Palynological Conference, v. 2, p. 581–588.

FREY, M., ed., 1987, Low Temperature Metamorphism: London, Blackie, 351 p.

GRAYSON, J. F., 1975, Relationship of palynomorph translucency to carbon and hydrocarbons in clastic sediments, in Alpern, B., ed., Pétrographie de la Matière Organiques des Sédiments, Relations avec la Paléotemperature et le Potential Pétrolier: Paris, Centre National de la Recherche Scientifique, p. 261–273.

GRIFFIN, G. M., 1971, Interpretation of X-ray diffraction data, in Carver, R. E., ed., Procedures in Sedimentary Petrology: New York, Wiley-Interscience, p. 541–569.

GRUM, F., AND BARTLESON, C. J., 1980, eds., Colour Measurement: New York, Academic Press, 372 p.

GUTHRIE, J. M., HOUSEKNECHT, D. W., AND JOHNS, W. D., 1986, Relationships among vitrinite reflectance, illite crystallinity, and organic geochemistry in Carboniferous strata, Ouachita Mountains, Oklahoma and Arkansas: American Association of Petroleum Geologists Bulletin, v. 70, p. 26–33.

GUTJAHR, C. C. M., 1966, Carbonization of pollen grains and spores and their application: Leidse Geologische Mededelingen, v. 38, p. 1–30.

HOFFMAN, H., AND HOWER, J., 1979, Clay mineral assemblages as low grade metamorphic geothermometers–application to the thrust faulted disturbed belt of Montana, U.S.A., in Scholle, P. A., and Schluger, P. R., eds., Aspects of Diagenesis: Society of Economic Paleontologists and Mineralogists Special Publication 26, p. 55–79.

JEANS, C. V., 1978, The origin of Triassic clay assemblages of Europe with special reference to the Keupar marl and Rhaetic of parts of England: Philosophical Transactions of the Royal Society of London, A 289, p. 547–639.

KING, L. C., 1950, Outline and disruption of Gondwanaland: Geological Magazine, v. 87, p. 353–359.

KUBLER, B., 1968, Évaluation quantitative du métamorphisme par la cristallinité; état des progrès réalisés ces dernières annees: Bulletin du Centre de Recherches de Pau, v. 2, p. 385–397.

LAMBERT-AIKHIONBARE, D. O., 1981, Sandstone diagenesis and its relation to petroleum generation and migration in the Niger delta: Unpublished Ph.D. Dissertation, University of London, 280 p.

LAMBERT-AIKHIONBARE, D. O., 1982, Relationship between diagenesis and pore-fluid chemistry in the Niger delta: Journal of Petroleum Geology, v. 4, p. 287–298.

LAMBERT-AIKHIONBARE, D. O., AND SHAW, H. F., 1982, Significance of clays in the petroleum geology of the Niger delta: Clay Minerals, v. 17, p. 91–103.

MACADAM, D. L., 1981, Colour Measurement: New York, Springer-Verlag Series in Optical Sciences, v. 27, 228 p.

MARSHALL, J. E. A., 1991, Quantitative spore colour: Journal of the Geological Society of London, v. 148, p. 223–233.

OBOH, F. E., 1990, Palaeoenvironmental reconstruction of the E2.0 Reservoir in the Kolo Creek field, Niger delta (Nigeria): Unpublished Ph.D. Dissertation, University of Cambridge, 237 p.

ODIGI, M. I., 1987, Mineralogical and geochemical studies of Tertiary sediments from the eastern Niger delta and their relationship to petroleum occurrence: Journal of Petroleum Geology, v. 10, p. 101–114.

PERRY, E. A., AND HOWER, J., 1972, Late stage dehydration in deeply buried pelitic sediments: American Association of Petroleum Geologists Bulletin, v. 56, p. 2013–2021.

PORRENGA, D. H., 1966, Clay minerals in Recent sediments of the Niger delta: Clays and Clay Minerals, v. 14, p. 221–233.

POWERS, M. C., 1967, Fluid-release mechanisms in compacting marine mudrocks and their importance in oil exploration: American Association of Petroleum Geologists Bulletin, v. 51, p. 1240–1254.

REYNOLDS, R. C., 1980, Interstratified clay minerals, in Brindley, G. W., and Brown, G., eds., Crystal Structures of Clay Minerals and their X-Ray Identification: London, Mineralogical Society, p. 249–303.

REYNOLDS, R. C., AND HOWER, J., 1970, The nature of interlayering in mixed-layer illite/montmorillonites: Clays and Clay Minerals, v. 18, p. 25–36.

SCOTCHMAN, I. C., 1987, Clay diagenesis in Kimmeridgian clay formation, onshore UK and its relation to organic maturation: Mineralogical Magazine, v. 51, p. 535–551.

SHAW, H. F., 1980, Clay minerals in sediments and sedimentary rocks, in Hobson, G. D., ed., Developments in Petroleum Geology, v. 2: England, Applied Science Publishers, p. 53–85.

SHORT, K. C., AND STÄUBLE, A. J., 1967, Outline of geology of Niger delta: American Association of Petroleum Geologists Bulletin, v. 51, p. 761–779.

SMART, G., AND CLAYTON, T., 1985, The progressive illitization of interstratified illite-smectite from Carboniferous sediments from northern England and its relationship to organic maturity indicators: Clay Minerals, v. 20, p. 455–466.

STONELEY, R., 1966, The Niger delta region in the light of the history of continental drift: Geological Magazine, v. 103, p. 385–397.

UNGERER, P., ESPITALIÉ, J., MARQUIS, F., AND DURAND, B., 1986, Use of kinetic models for organic matter evolution for the reconstruction of paleotemperatures, in Burrus, J., ed., Thermal Modelling in Sedimentary Basins: Paris, Technip, p. 531–546.

VELDE, B., ed., 1985, Clay Minerals—A Physico-chemical Explanation of their Occurrence: Amsterdam, Elsevier Publishing Company, Developments in Sedimentology, v. 40, 218 p.

VELDE, B., AND ESPITALIÉ, J., 1989, Comparison of kerogen maturation and illite/smectite composition in diagenesis: Journal of Petroleum Geology, v. 12, p. 103–110.

WEBER, K. J., AND DAUKORU, E. M., 1975, Petroleum geology of the Niger delta: Proceedings, 9th World Petroleum Congress, Tokyo, v. 2, p. 209–221.

WELTON, J. E., 1984, SEM Petrology Atlas: Tulsa, Oklahoma, American Association of Petroleum Geologists, 237 p.

WHITEMAN, A. J., 1982, Nigeria: Its Petroleum Geology, Resources and Potentials: London, Graham and Trotman, 394 p.

WILSON, H. H., 1990, The case of early generation and accumulation of oil: Journal of Petroleum Geology, v. 13, p. 127–156.

WYSZECKI, G., AND STILES, W. S., 1982, Color Science: Concepts and Methods, Quantitative Data and Formulae, New York, Wiley, 950 p.

ZEN, E., AND THOMPSON, A. B., 1974, Low grade regional metamorphism—mineral equilibrium relations: Annual Review of Earth and Planetary Sciences, v. 2, p. 179–212.

ORIGIN AND DIAGENESIS OF CLAY-MINERALS IN THE OLIGOCENE SESPE FORMATION, VENTURA BASIN

LORI A. HATHON* AND DAVID W. HOUSEKNECHT
Department of Geological Sciences, University of Missouri, Columbia, Missouri 65211

ABSTRACT: The Oligocene Sespe Formation has undergone differential burial resulting from complex tectonic evolution of the Ventura Basin. Samples collected from more than 5 km of burial depth allow documentation of the origin and depth-dependent diagenesis of clay minerals in texturally variable sandstones.

Clay minerals can be distinguished as depositional matrix, infiltrated clays, or authigenic clays on the basis of petrographic and SEM criteria. "Clean" (<20% ductile components) and "dirty" (>20% ductile components) sandstones reacted differently during burial diagenesis, resulting in contrasting clay-mineral assemblages. In "clean" sandstones, the clay-mineral assemblage changes from smectite plus subordinate kaolinite in shallow samples to chlorite in deep samples, suggesting that the major clay-mineral reaction is the transformation of smectite to chlorite. In "dirty" sandstones, the clay-mineral assemblage changes from smectite plus minor kaolinite at shallow depths, to smectite plus chlorite at intermediate depths, to mixed-layer illite/smectite plus minor chlorite at greatest depths. Formation of illite at the expense of smectite appears to be the predominant clay-mineral reaction. The origin of chlorite and the role that it plays in illite formation are unclear. In both types of sandstones, the presence of kaolinite in shallow samples may reflect meteoric diagenesis following tectonic uplift.

Original texture significantly influenced clay-mineral diagenesis in Sespe sandstones. Low-permeability "dirty" sandstones behaved as closed diagenetic systems, whereas higher permeability "clean" sandstones behaved as relatively open diagenetic systems.

INTRODUCTION

Research on sandstone diagenesis has long sought to establish relations between depth of burial and diagenetic processes. However, limited availability of samples of individual sandstone formations over large ranges of burial depth has restricted the documentation of specific depth-related reactions. Additionally, a large volume of diagenetic literature has concentrated on relatively few basins that contain sandstones of limited compositional variability (e.g., Gulf Coast and North Sea).

The Oligocene Sespe Formation in the Ventura Basin, California, contains texturally and compositionally immature sandstones in which more than 500 million barrels of oil have accumulated (Nagle and Parker, 1971). However, oil recovery has been hampered by tremendous variability in reservoir quality, resulting mostly from the abundance of clay minerals within Sespe sandstones. Recently completed research has documented diagenesis of the Sespe Formation from samples collected over a depth interval of more than 5 km (Hathon, 1991). The objective of this paper is to determine the origin of clay minerals in the Sespe and to relate their occurrence and diagenesis to depth of burial.

GEOLOGIC SETTING

Tectonic History

The Ventura Basin is part of the Transverse Ranges province, a stratigraphically and structurally complex area (Fig. 1) that is the product of convergent and transform tectonic activity. Late Cretaceous through early Miocene sedimentation in the area occurred in a precursor basin ("Santa Ynez Basin"; Nilsen, 1984), which developed in a forearc setting associated with oblique subduction and transform faulting (Crowell, 1987; Nilsen, 1987; Yeats, 1987). The present form of the basin developed during the middle Miocene, apparently as the result of subduction of the Pacific-Farallon spreading ridge and initiation of the Pacific-North American transform margin (Hornafius and others, 1986; Nilsen, 1987). It has been suggested that the entire Transverse Ranges province has undergone tectonic rotation and northward translation since the Miocene as the result of transform motion (Hornafius and others, 1986; Crowell, 1987; Luyendyk and Hornafius, 1987). Compression that began in the Pleistocene continues today, resulting in the present structural configuration of the basin.

Local Structure and Stratigraphy

The Ventura Basin has been divided into several tectonic subprovinces shown in Figure 1 (Nagle and Parker, 1971). Most samples involved in this study are from the Oxnard shelf, as indicated by the line of the sampling traverse (Fig. 1). The northern boundary of the shelf is the Oak Ridge reverse fault, which juxtaposes Oligocene rocks of the shelf against Plio-Pleistocene rocks of the Santa Clara trough (Fig. 2). The southern boundary is a system of high-angle, predominanty normal faults that separates the shelf from the Santa Monica and Simi Uplifts (Figs. 1 and 2). Within the Oxnard shelf subprovince, rocks plunge westward and generally dip northward, although they are folded along axes that are subparallel to the Oak Ridge fault (Figs. 1 and 2).

Strata older than the Sespe comprise Cretaceous through Paleocene marine facies deposited on "basement" rocks of the Jurassic Franciscan Formation. Eocene strata record two episodes of basin filling, the second culminating with deposition of the Sespe Formation. Howard (1988) has informally divided the Sespe into two units, separated by an unconformity. The lower unit (upper Eocene—lower Oligocene) grades upward from delta-plain to braided fluvial facies, reflecting the progradation of shoreline and nonmarine environments. Locally, this unit apparently has been removed by erosion. The upper unit (upper Oligocene—lower Miocene) comprises fluvial deposits that suggest evolution from coarse-grained, bedload-dominated streams to finer grained, mixed-load streams through time (Rarey, 1990).

---

*Present Address: Amoco Production Co., P.O. Box 3092, Houston, Texas 77253.

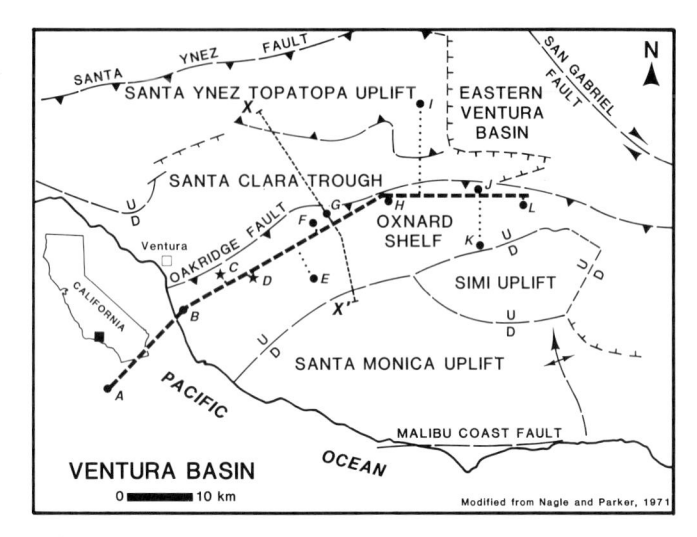

FIG. 1.—Base map of Ventura Basin showing sample locations (indicated by letters) and line of sampling traverse (heavy, dashed line). Dashed line X-X' shows location of cross section shown in Figure 2. Tectonic subprovinces defined by, and map modified from, Nagle and Parker (1971). Key to locations: A = Sockeye field; B = West Montalvo field; C = Saticoy area, Shell Sharp #1; D = Saticoy area, Mobil Butler #1; E = Camarillo Hills area; F = West Mountain field; G = South Mountain field; H = Bardsdale field; I = Sespe field; J = Torrey Canyon field; K = Simi field; and L = South Tapo Canyon field.

Strata younger than the Sespe comprise Miocene through Pleistocene rocks that record syntectonic clastic sedimentation. The Sespe is conformably overlain by Miocene strata that grade upward from shallow to deep marine facies, culminating with the Monterey Formation. Pliocene through Pleistocene strata record filling of the basin, grading upward from submarine-fan through nonmarine facies.

Burial History

This study was designed to document depth-related diagenetic processes within a single formation and, therefore,

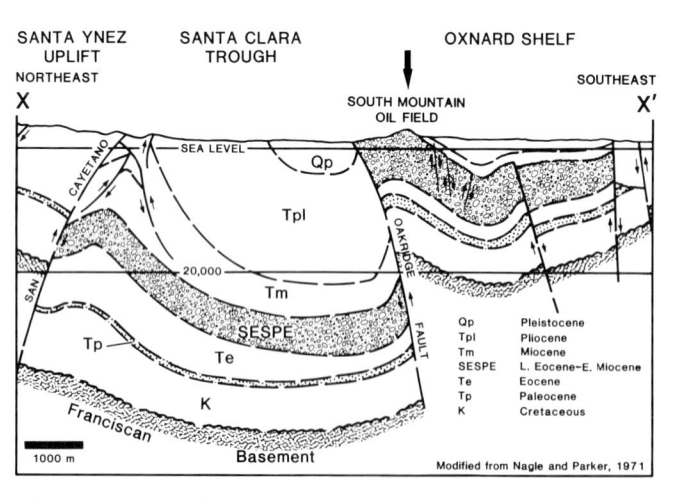

FIG. 2.—Structural cross section of Ventura Basin illustrating major tectonic subprovinces. Most samples were collected from the Oxnard shelf, as indicated by arrow. Location of cross section shown in Figure 1. Modified from Nagle and Parker (1971).

it is important to reconstruct the burial history of the Sespe Formation to determine if modern burial depths are representative of maximum depths to which the formation has been buried. Figure 3 summarizes Sespe burial histories at the western end of the Oxnard shelf and at a central location on the Oxnard shelf, essentially corresponding to the deep and shallow ends of the sampling traverse, respectively.

These burial histories indicate that present sample depths are directly proportional to maximum depths of burial reached by the Sespe. Sample depths represent maximum depths of burial along the western part of the sampling traverse. To the east, sample depths are less than maximum depths of

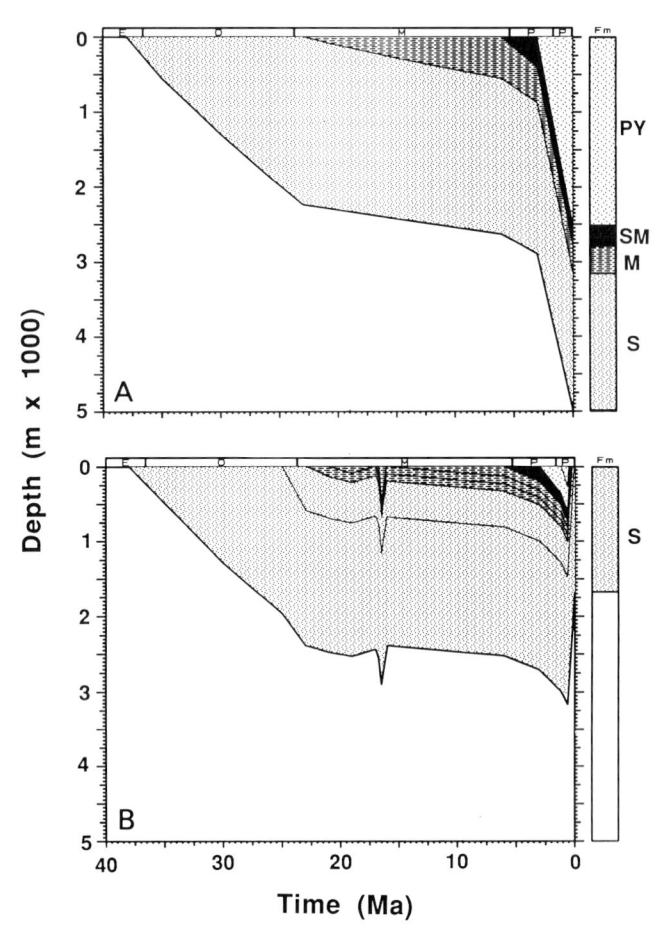

FIG. 3.—Burial-history curves illustrating deep and shallow portions of the sampling traverse. Burial histories constructed using BasinMod™, using the compaction method of Falvey and Middleton (1981). (A) Burial history at west end of Oxnard shelf (location B, Fig. 1) shows that the Sespe underwent slow burial during the Miocene, moderate rates of burial during the latest Miocene and early Pliocene, and rapid burial since the late Pliocene. Current depths of burial equal maximum depths of burial. (B) Burial history at central part of Oxnard shelf (location G, Fig. 1) shows that the Sespe underwent generally slow burial during the Miocene (punctuated by a brief interval of unroofing, rapid burial, and more unroofing during 19–16 Ma), moderate rates of burial during the latest Miocene and early Pliocene, rapid burial during late Pliocene and early Pleistocene, and rapid uplift and erosion since the middle Pleistocene. The Sespe presently extends from outcrop to ~1.7 km depth, and it is estimated that ~450 m of Sespe have been eroded. Key to stratigraphic abbreviations: S = Sespe, M = Monterey (Miocene), SM = Santa Margarita (Lower Pliocene), and PY = Pico and younger strata (Upper Pliocene and younger).

burial because of recent uplift and unroofing. These conclusions are consistent with preliminary results obtained from apatite fission-track analyses (Hathon, 1991).

METHODS

Approximately 300 samples were collected from an east-west traverse along the Oxnard shelf, with sample depths ranging from outcrop (along the eastern half of the Oxnard shelf; Fig. 1) to more than 5,300 m (17,400 ft) of depth at West Montalvo field (location B on Fig. 1), at the western end of the Oxnard shelf. All subsurface samples were collected from pieces of conventional core. Most samples are sandstone; siltstone and shale samples were collected wherever possible, but few were available because only core pieces of sandstone had been archived and most other lithologies had been discarded. Samples were collected from the center of core pieces to minimize the potential for drilling-mud contamination.

Thin sections were prepared from about 200 Sespe samples and, of these, 100 were point counted (minimum of 300 points per thin section). Approximately one-third of the thin sections were stained for feldspar identification. The remaining two-thirds were unstained, polished sections for use in cathodoluminescence microscopy; SEM backscattered electron imaging and EDS analysis; and microprobe analysis. Textural analysis was performed by measuring the long axis of 50 monocrystalline quartz grains per thin section.

X-ray diffraction analysis was performed on a Scintag PadV diffractometer with a germanium-crystal solid-state detector using Cu-K$_\alpha$ radiation and a scan rate of 1° per minute. Oriented aggregates were prepared using the vacuum-filtration method of Drever (1973). XRD analyses were performed on all sandstones that were point counted (100 samples) and on 20 siltstone and claystone samples. For most samples, XRD analyses were performed on both <2.0- and <0.2-μm size fractions to determine if contrasts in the two size fractions would help differentiate between detrital and authigenic clays, and if the finer size fraction would facilitate identification of mixed-layer phases. Only minor differences were observed between the two size fractions, so the <2.0-μm fraction was used for most mineralogical determinations. Air-dried and glycolated aggregates were analyzed for each sample. Selected samples were boiled in 2N HCl and heated to 550°C for 30 min. to determine the relative amounts of chlorite and kaolinite present (Starkey and others, 1984).

Based on petrographic and X-ray results, a suite of 30 samples was selected for SEM/EDS analysis. The nature and distribution of clay minerals, as well as their qualitative chemistry, were analyzed using these techniques. A representative suite of samples was analyzed for bulk chemistry (both major and minor elements) by X-ray fluorescence to help constrain mineralogical trends discussed in this paper.

SANDSTONE PETROLOGY

Detrital Modes

Sespe sandstone samples range from fine- to medium-grained and from poorly to well sorted (Fig. 4). These ranges in mean grain size and sorting are similar over the entire range of sample depths.

Sespe sandstones are mineralogically immature, containing approximately equal volumes of quartz and feldspar, and lesser volumes of lithic fragments (Fig. 5); the mean composition determined from stained thin sections is $Q_{38}F_{37}L_{25}$. Quartz grains are mostly monocrystalline, although polycrystalline quartz is common, particularly in coarser grained samples. Plagioclase and K-feldspar are present in nearly equal volumes. Plagioclase ranges in com-

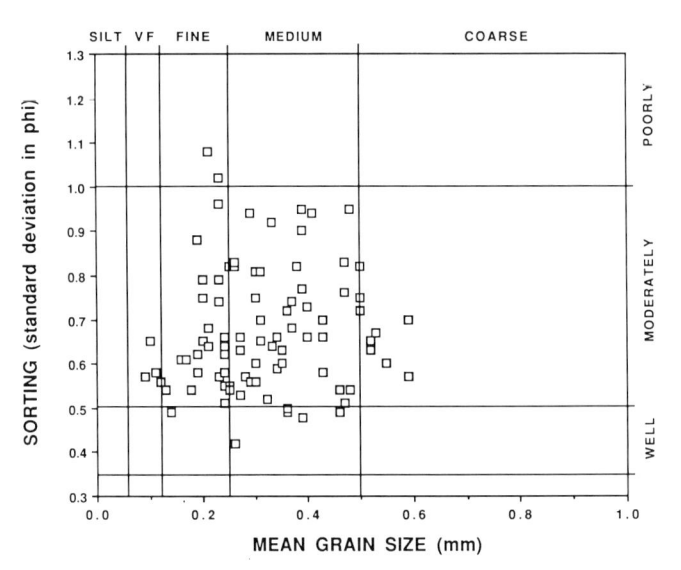

FIG. 4.—Mean grain size vs. sorting of Sespe sandstone samples.

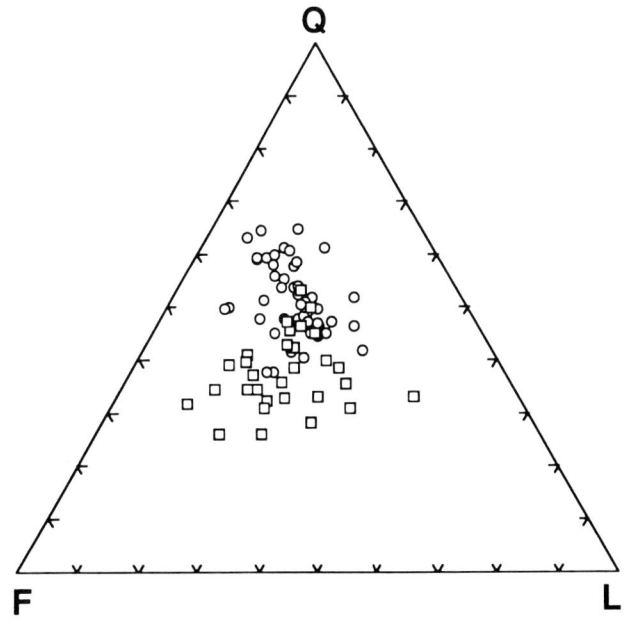

FIG. 5.—Framework-grain composition of Sespe sandstones. Squares represent data collected from stained thin sections (comprehensive feldspar stain) and circles represent data collected from unstained thin sections; the contrast in apparent composition is the result of misidentification of K-feldspar as quartz in unstained thin sections. Q = quartz; F = feldspar; L = lithic fragments. Micas are not included.

position from nearly pure albite (in alkali-granite lithic fragments or in albitized grains) to andesine. K-feldspars include perthite (and locally antiperthite), microcline, sanidine and orthoclase.

Among lithic fragments, plutonic varieties are the most common, making up approximately 50% of the total lithic-fragment component. Volcanic lithic fragments make up approximately 40% of the total. The remaining 10% is composed of chert, siltstone, mudstone, and foliated metamorphic lithic fragments. Accessory minerals, mostly micas, are locally abundant and may account for up to 14% of the whole rock. Micas are mostly biotite, although muscovite and chlorite are present in smaller volumes. Other accessory minerals present in conspicuous amounts include epidote, sphene, zircon and ilmenite. Garnet, apatite, tourmaline and hornblende are locally present.

Diagenesis

Comprehensive discussion of Sespe diagenesis is beyond the scope of this paper. This section summarizes the diagenesis of Sespe sandstones, exclusive of clay minerals, as documented by Hathon (1991).

Sespe sandstones have undergone diagenetic histories that are related to their original textures and maximum depths of burial. Sands with abundant depositional matrix were severely compacted during relatively shallow burial, and subsequent diagenesis was dominated by reactions involving clay minerals.

In contrast, sands that contained relatively little depositional matrix compacted more slowly, and diagenesis involved complex interactions between pore fluids, framework grains, and a variety of cements. Major events documented in these "cleaner" sandstones, listed in order of paragenesis, included local precipitation of calcite cement, mechanical compaction, dissolution of volcanic lithic fragments and ferromagnesian grains (i.e., heavy minerals), and tectonic fracturing associated with faulting (including the Oak Ridge fault; Figs. 1 and 2). As structural plunge was developed on the Oxnard shelf, differences in Sespe burial depths were accentuated and depth-related diagenetic processes proceeded. At the shallow eastern part of the shelf, dissolution of feldspar and ferromagnesian framework grains occurred, as well as pigmentation (reddening). At the deep western part of the shelf, albitization of plagioclase; local precipitation of quartz, carbonate, and laumontite cement; and intergranular pressure solution occurred.

Porosity

Intergranular porosity (inferred to be primary) of Sespe sandstones ranges from 0 to 20%, and two samples display "anomalously" high porosities of 25 and 38%, based on point-count estimates (Hathon, 1991). Most samples also contain microporosity, associated with clay minerals, and dissolution porosity, associated with partly dissolved framework grains or cements. However, these porosity types are volumetrically subordinate to intergranular porosity and do not contribute significantly to reservoir quality.

The amount of intergranular porosity is clearly related to the volume of clay present in Sespe sandstones. Samples that contain less than 5% clay are characterized by low porosity (<5%) because they contain relatively high volumes of cement (mostly calcite); samples that contain 5 to 15% clay are characterized by higher, though variable, porosity (0 to 20%); and samples that contain more than 15% clay are characterized by low, though variable, porosity (0 to 7%). This local variability accounts for the heterogeneity in reservoir quality that hinders oil production from Sespe reservoirs (e.g., Nagle and Parker, 1971).

CLAY MINERALS IN THE SESPE

Distribution and Inferred Origin

Clay minerals in Sespe sandstones are inferred to represent three modes of origin: depositional matrix, infiltrated grain coatings and pore fillings, and authigenic cements.

Depositional matrix.—

Depositional matrix is predominant in relatively fine-grained samples that contain high volumes of clay (>10% by point-count estimate) and display matrix-supported fabrics. In thin section, the matrix appears as a nonporous, heterogeneous mixture of clay minerals, mica, and silt-size grains of quartz and feldspar (Fig. 6A). It is commonly

FIG. 6.—Thin section and scanning electron micrographs illustrating clay minerals in Sespe sandstones. (A) "Dirty" Sespe sandstone with abundant detrital matrix composed of clay minerals, micas, and silt-size grains of quartz and feldspar. Note matrix-supported framework that is typical of most "dirty" samples. Plane polarized light; scale bar 0.1 mm. (B) "Clean" Sespe sandstone with grain coatings composed of infiltrated clays covered with a facade of authigenic clays. Infiltrated clays are oriented parallel to grain surfaces and are draped into grain concavity (solid arrows), whereas authigenic clays are oriented perpendicular to grain surfaces (open arrows); the two clay types are commonly separated by a medial line of Fe-oxide. All clays in these grain coatings are smectite and grain is K-feldspar. Area between grains is porosity (p). Plane polarized light; scale bar 0.05 mm. (C) SEM view of a "dirty" sample showing detrital matrix compacted around a quartz grain. Clay minerals in this relatively shallow sample (870 m) are mostly smectite and kaolinite, based on XRD results. Note ragged, irregular shape of clay particles. Scale bar 10 μm. (D) SEM view of a "dirty" sandstone showing detrital matrix in a deeper sample (5.3 km) illustrating that individual clay particles are difficult to distinguish, and therefore their origin (detrital vs. authigenic) is difficult to interpret. Scale bar 10 μm. (E) SEM view of a "clean" sandstone showing clays inferred to be of infiltration origin, as suggested by their ragged, irregular shapes and their parallel orientation to grain surface. "Bald" spot on grain near center of view is grain-contact scar. Scale bar 100 μm. (F) SEM cross-sectional view of clay coating on a broken grain (upper left) from a "clean" sample from shallow depth (192 m). The clay coating is composed of ragged clay particles oriented parallel to grain surface (inferred to be infiltrated) covered by euhedral clay particles oriented nearly perpendicular to grain surface (inferred to be authigenic). Scale bar 10 μm. (G) SEM view of "clean" sandstone showing chlorite grain coating in a deep sample (5.2 km). The euhedral chlorite crystals occupy the same positions on grains as Fe-smectite grain coatings in shallow samples, suggesting that the chlorite may have formed by high-temperature transformation from smectite. The large, euhedral crystals are authigenic quartz, which clearly postdates chlorite as indicated by chlorite crystals engulfed in the quartz. Scale bar 100 μm. (H) SEM view of "clean" sandstone showing kaolinite in a shallow sample (536 m). The vermicular aggregates of euhedral crystals suggest the kaolinite is authigenic. Scale bar 100 μm.

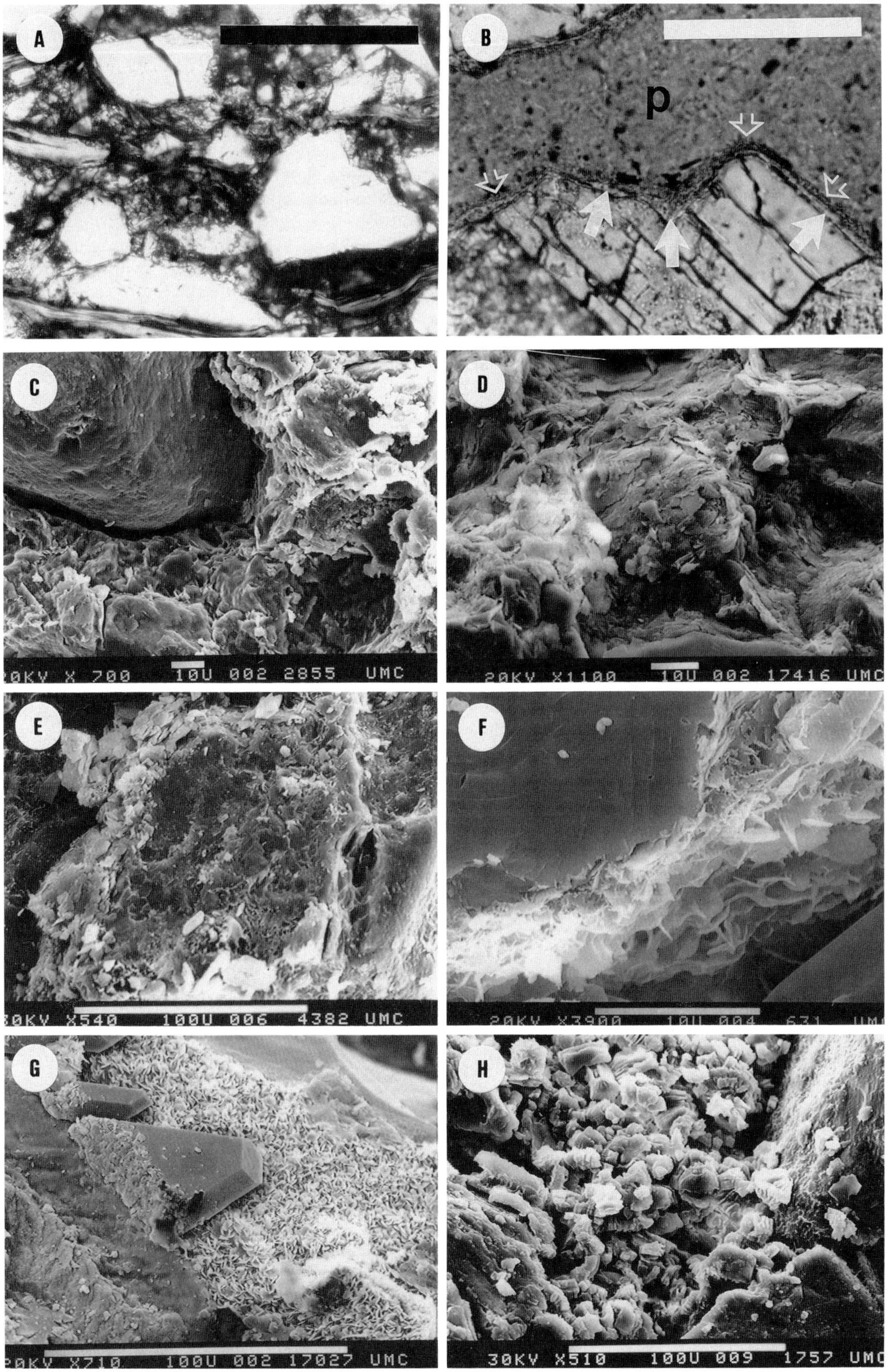
A
B
p
C
X 700 10U 002 2855 UMC
D
20KV X1100 10U 002 17416 UMC
E
X540 100U 006 4382 UMC
F
20KV
G
X710 100U 002 17027 UMC
H
30KV X510 100U 009 1757

stained red or brown and imparts a "dirty" appearance to the rock. In SEM, individual clay particles are irregular in shape and display abraded edges (Figs. 6C, D).

All these characteristics suggest that this matrix is composed of detrital-clay particles, together with mica and silt, that accumulated at the time of deposition. This depositional matrix probably represents deposition from waning ephemeral discharge in braided fluvial environments or from overbank deposition in mixed-load fluvial environments (Rarey, 1990).

Infiltrated clays.—

Infiltrated clays are predominant in relatively coarse-grained samples that contain widely variable volumes of clay and display grain-supported fabrics. In thin section and SEM, these clays occur as grain coatings whose constituent particles are mostly oriented parallel to framework-grain surfaces (Figs. 6B, E, F). Locally, at points of grain contact, clay grain coatings display evidence of having been crushed between framework grains, suggesting their early emplacement on grain surfaces. Where clay coatings are discontinuous, they tend to be concentrated adjacent to points of grain contact. In samples that contain relatively high volumes of clay, yet retain grain-supported fabrics, clay particles also occur as pore-filling material.

Similar clay-particle morphologies and distributions have been attributed to infiltration of clay minerals into sandstones deposited in continental environments (Walker and others, 1978; Molenaar, 1986; Matlack and others, 1989; Moraes and De Ros, 1990, this volume), and we infer a similar origin for this mode of clay occurrence in the Sespe.

Authigenic clays.—

Authigenic clays occur most commonly in samples that contain low to moderate volumes of depositional matrix or infiltrated clays. In thin section, and especially in SEM, these clays are euhedral and appear to be homogeneous in composition (Figs. 6B, F, G, H). They occur as facades on infiltrated-clay grain coats and as pore-filling cements.

The mineralogical purity, euhedral nature, and selective distribution of these clays suggest they are authigenic in origin.

Petrographic subdivisions.—

The volume and type of clay minerals present in Sespe sandstones have exerted important influences on subsequent diagenesis. However, characterization of sandstones on the basis of "matrix" content is confounded by the local abundance of ductile framework grains, which include micas and pelitic lithic fragments. Even though most ductile material can be assigned to one of the three categories described above, Sespe sandstones were divided for further analyses into two groups on the basis of total ductile components (clay minerals plus ductile framework grains). Samples that contain less than 20% (by point count) total ductile components are assigned to a "clean" category and those that contain more are assigned to a "dirty" category. As will be shown in a subsequent section, these two groups behaved differently during burial diagenesis.

Clay Mineralogy

XRD analysis of the clay-size fraction was performed on all samples for which point-count and bulk chemical data were collected. We believe that the clay-mineralogy data reported later are not influenced by drilling-fluid contamination for several reasons. Samples were collected from the centers of core pieces to minimize the possibility of contamination. Where present, mudcakes were sampled from cores and subjected to XRD analysis to provide a comparison for clay-fraction XRD data from the rock. Barite was detected by XRD and SEM-EDS analyses in a few samples, and data from those samples were not included during interpretation of clay mineralogy.

The <2.0-μm fraction commonly contains quartz and feldspar, and locally contains biotite, muscovite, and Mg-chlorite, whereas these minerals generally are absent from the <0.2-μm fraction. We infer that these minerals are present as detrital particles, and they are not emphasized in the following discussion.

"Clean" sandstones.—

The mineralogy of the <2-μm size fraction changes with burial depth. In general, clay minerals in shallow samples are predominantly smectite and subordinate kaolinite (Fig. 7). SEM-EDS analyses indicate that Ca- and Fe-bearing smectite is the most common clay mineral present in shallow samples. Saponitic and nontronitic varieties occur locally, inferred from XRD relative peak intensities and EDS qualitative chemical analyses. D spacing of the 001 peak for glycol-solvated smectites in the Sespe ranges from 16.74Å to 17.01Å. Integral d spacing for higher order peaks is pervasive, except in the most deeply buried samples from the western end of the Oxnard shelf (West Montalvo field), suggesting that mixed-layer phases are not common. This contrasts with previous reports that mixed-layer illite/smectite is predominant in the Sespe (Flemal, 1966; Howard, 1988; Rarey, 1990).

In the deepest samples, the presence of peaks with non-integral d spacings for glycol-solvated samples and the presence of a broad peak between 10 and 12Å for air-dried samples suggests the occurrence of mixed-layer illite/smectite (I/S). The pattern for the glycol-solvated sample from 5,200 m (17,027 ft) shown in Figure 7F exhibits a small 17Å peak, and a shoulder on the 10Å illite peak. Deconvolution of the discrete illite peak and the 001/002 I/S peak yields a peak position of ~8.89Å. The 002/003 I/S peak position (on the high-angle side of the 4.756Å chlorite peak) is ~5.46Å. Using the differential 2θ method (Moore and Reynolds, 1989), the percentage of nonexpandable layers in the I/S is estimated to be between 40 and 50%. The presence of the 17Å peak indicates random mixed layering.

Kaolinite and chlorite are mutually exclusive. Kaolinite is present in small volumes in shallow samples and absent in deep samples. Chlorite is generally absent (except for local occurrence of Mg-chlorite) in shallow samples and Fe-chlorite is present in locally variable volumes in deep samples (Fig. 7). SEM observations reveal that kaolinite is euhedral and occurs in vermicular aggregates (Fig. 6H), suggesting it is authigenic. Moreover, it appears to be non-

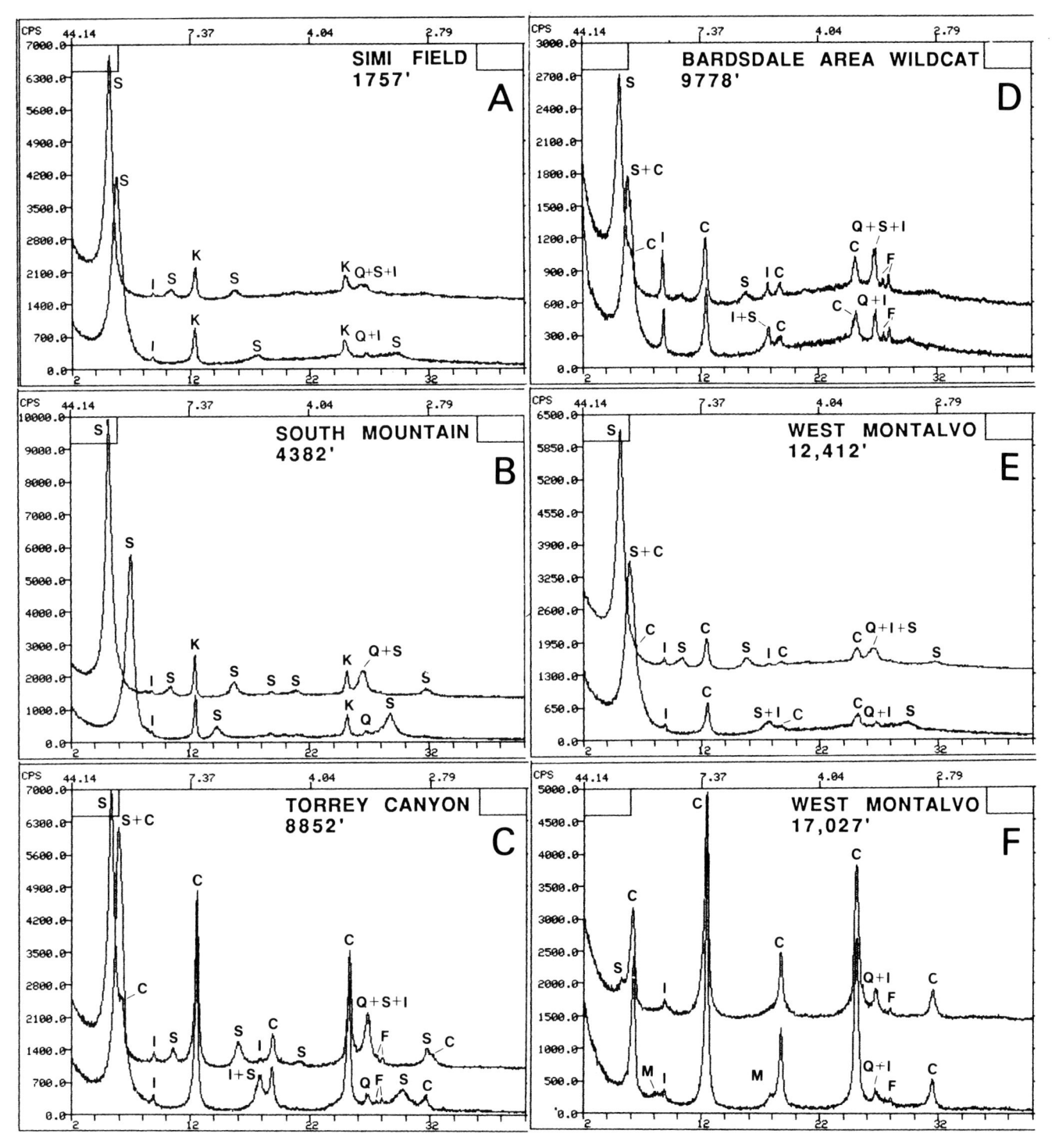

FIG. 7.—XRD patterns for the <2-μm fraction of representative "clean" Sespe samples spanning a range of depths from 536 m to 5.2 km. These patterns illustrate the trends in clay mineralogy relative to depth discussed in the text. Numbers on each pattern refer to sample depth in feet. For each sample, the upper pattern is glycolated and the lower is air dried. C = chlorite; I = illite; K = kaolinite; M = mixed-layer I/S; S = smectite; Q = quartz; F = feldspar; and X = carbonate.

deformed in samples characterized by tectonically shattered framework grains, suggesting a post-uplift origin. Petrographic and SEM observations show that Fe-chlorite is euhedral and occurs mostly as grain coatings (Fig. 6G), suggesting that it, too, is authigenic. This inference is supported by low intensity of the 001 peak at 14.26Å and the 003 peak at 4.756Å, which is typical of Fe-rich authigenic chlorite (Kisch, 1983; Moore and Reynolds, 1989).

Throughout the range of depths sampled, XRD patterns locally show a 10Å peak (Fig. 7), indicating the presence of illite. SEM observations and illite crystallinity index (CI = width of 10Å peak at half maximum height) measurements suggest that some illite is authigenic and some is detrital. The former occurs as delicate, lath-shaped crystals and displays illite crystallinities (mean CI = 0.337°2θ) gradational between the zone of diagenesis and the anchizone (as defined by Kisch, 1983). The latter does not display crystalline morphology and has an illite crystallinity (mean CI = 0.198°2θ) indicative of metamorphic and plutonic micas (Kisch, 1983).

"Dirty" sandstones.—

The mineralogy of the <2-μm size fraction also changes with burial depth in "dirty" sandstones (Fig. 8). Shallow samples contain predominantly smectite, with subordinate and locally variable amounts of kaolinite and illite. Samples from intermediate depths (Camarillo Hills; location E in Fig. 1, and relatively shallow samples from West Montalvo field; location B in Fig. 1) contain approximately equal amounts of smectite and chlorite. Samples from the greatest depths (West Montalvo field; location B in Fig. 1) contain predominantly mixed-layered illite/smectite and subordinate chlorite.

For all "dirty" samples shallower than ~3,985 m (13,000 ft), higher order smectite peaks exhibit integral d spacings, whereas XRD data from deeper samples suggest mixed layering. For example, the sample from 5,035 m (16,510 ft) in Figure 8E exhibits a broad 16.7Å 001 smectite peak, an 001/002 I/S peak at ~8.9Å, and an 002/003 I/S peak at ~5.45Å. These peak positions suggest a random mixed-layer illite/smectite containing 40 to 50% illite layers. In the deepest sample shown in Figure 8F, the mixed-layer phase has an 001/002 peak at ~9.2Å and an 002/003 peak at ~5.33Å, corresponding to an interstratified illite/smectite containing approximately 60% illite layers. Most peak positions described for the I/S phase correspond to an R1-ordered I/S (Moore and Reynolds, 1989). However, the presence of the 001 smectite peak suggests that random mixed layering persists. The increase in the amount of nonexpandable layers between the two deepest samples suggests that the formation of illite at the expense of smectite is an important process in more deeply buried portions of the Oxnard shelf.

The increase in chlorite content from shallow to intermediate depths is consistent with the trend observed in "clean" samples. However, the decrease in chlorite content in the deepest samples is unique. The origin of chlorite is also equivocal. In samples from intermediate depths, relative peak intensities suggest that the chlorite is an Fe-rich diagenetic variety (Kisch, 1983; Moore and Reynolds, 1989), similar to that in "clean" samples. However, attempts to corroborate an authigenic origin by petrographic and SEM examination were frustrated by the abundance of depositional matrix, which precludes visual isolation of individual clay-mineral species (Fig. 6D).

Small volumes of illite are present in "dirty" samples throughout the depth range. As in the "clean" samples, some appears to be of authigenic and some of detrital origin (illite crystallinity values similar to those reported for "clean" samples).

The clay mineralogy of shales from the Sespe Formation appears to be similar to that of "dirty" sandstone samples, although the paucity of samples precludes documentation of shale mineralogy through the entire range of depths.

DISCUSSION

The data summarized earlier demonstrate that depth-dependent changes in clay mineralogy occur in Sespe sandstones, and that these changes are different in "clean" and "dirty" samples. In "clean" sandstones, the predominant clay-mineral assemblage changes from smectite plus subordinate kaolinite in shallow samples to chlorite in deep samples (plus minor mixed-layer I/S in deepest samples) (Fig. 7). In contrast, the predominant clay-mineral assemblage in "dirty" samples changes from smectite plus minor kaolinite at shallow depths, to smectite plus chlorite at intermediate depths, to mixed-layer illite/smectite plus minor chlorite at greatest depths.

Based on clay mineralogy of shallow samples that contain depositional matrix or infiltrated clays, smectite is inferred to have been the predominant detrital-clay mineral in Sespe sandstones. Smectite is also the predominant authigenic-clay mineral in shallow samples, where it occurs mostly as grain coatings. Mixed-layer I/S is the predominant clay mineral and nonexpandable layers increase with depth in "dirty" samples from 3.9 to 5.3 km depth, corresponding to temperatures of 130 to 160°C. However, mixed-layer I/S first appears in "clean" samples at 5.2 km depth and it is present as only a minor constituent. These observations suggest that smectite-to-illite conversion is an ongoing process in the deepest parts of the Oxnard shelf. Moreover, the reaction is a more important process and occurs at lower temperatures (shallower depths) in "dirty" sandstones than in "clean" sandstones. These threshold temperatures are substantially higher than those documented elsewhere (e.g., Boles and Franks, 1979; Kisch, 1983), possibly because the Sespe has only recently reached maximum burial depths and temperatures.

Even though chlorite is abundant in many Sespe samples, its origin is equivocal. In "clean" sandstones, conversion of smectite to chlorite with increasing depth seems to be indicated by XRD data, and this inference is supported by the fact that the two clay minerals are identically distributed, mostly as grain coatings. Conversion of smectite to chlorite, via a mixed-layer corrensite phase, with increases in estimated maximum burial depth has been inferred for sandstones (including the Sespe) in the Santa Ynez-Topa Topa Uplifts by Helmold and Van de Kamp (1984). Although chlorite becomes the predominant clay mineral with increasing depth in "clean" samples from the Oxnard shelf, no mixed-layer chlorite/smectite has been observed in this study (Fig. 7). Interpretation of the origin of chlorite in "dirty" sandstones is confounded by the fact that no systematic relation between smectite and chlorite is suggested by XRD data, and that chlorite is a relatively minor constituent of the clay-mineral fraction in the deepest samples analyzed.

It is not likely that chlorite originated by transformation of kaolinite, a reaction that has been documented in other

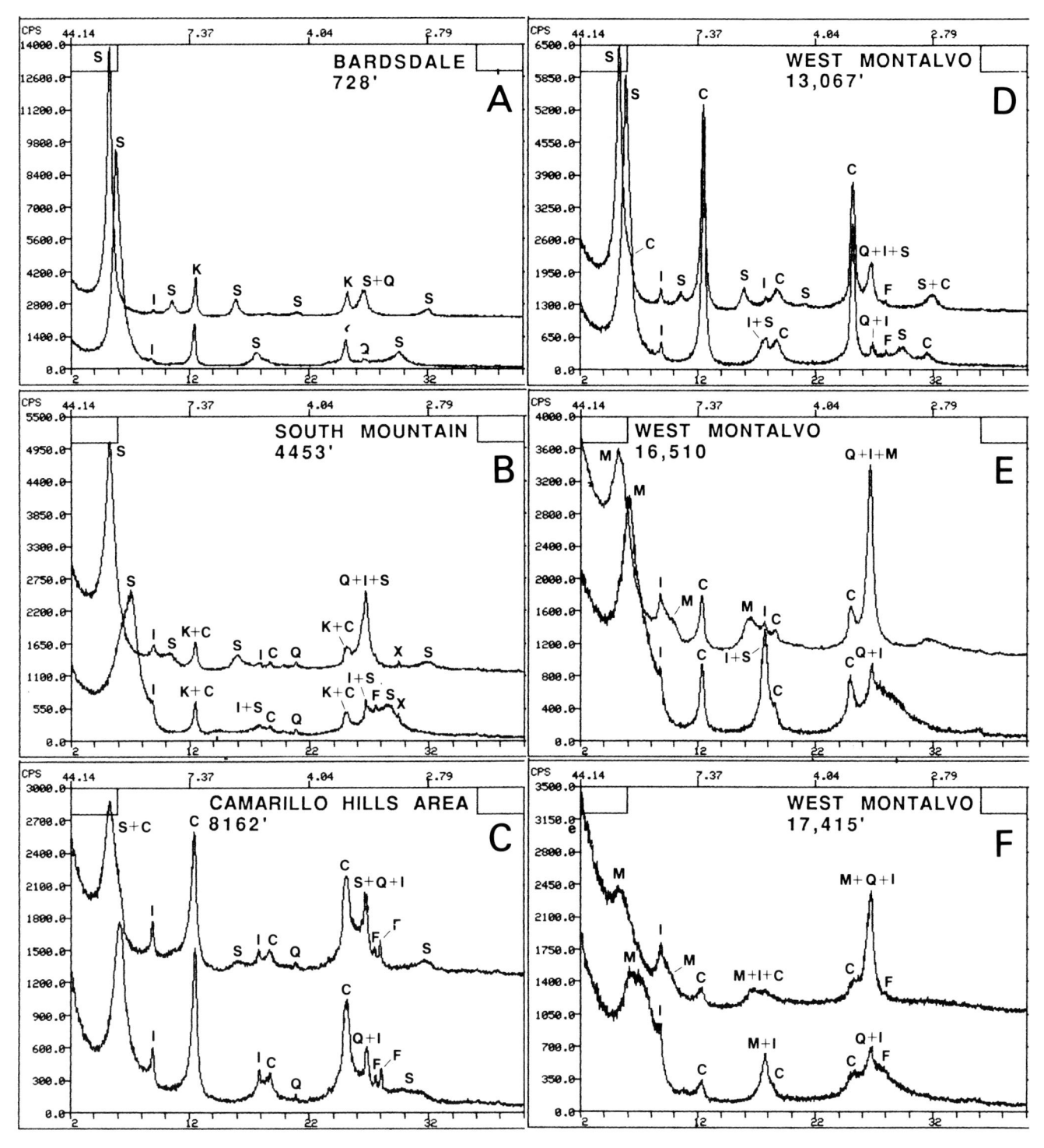

FIG. 8.—XRD patterns for the <2-μm fraction of representative "dirty" Sespe samples spanning a range of depths from 222 m to 5.3 km. These patterns illustrate the trends in clay mineralogy relative to depth discussed in the text. Format and legend same as Figure 7.

basins (e.g., Boles and Franks, 1979). The relative volume of chlorite in many samples is greater than that of kaolinite in any sample (Fig. 7), there is no evidence of gradations from kaolinite to chlorite among the XRD data, and petrographic (including SEM) evidence does not support a kaolinite-to-chlorite transformation because the two clay minerals are not similarly distributed; kaolinite occurs as aggregates of euhedral crystals, whereas chlorite occurs as grain coatings. A more likely scenario is that kaolinite is present only in shallow samples, because it has formed only in the shallow parts of the Oxnard shelf as the result of grain alteration caused by meteoric-water influx following development of the structural plunge described earlier. This inference is supported by the occurrence of nondeformed aggregates of euhedral kaolinite in samples that display pervasive tectonic shattering of brittle framework grains, suggesting that kaolinite precipitation postdated structural deformation.

The contrasts in clay-mineral diagenesis between "dirty" and "clean" sandstones may have been controlled by permeability, which, in turn, was controlled by original texture. "Dirty" sandstones appear to have acted as closed diagenetic systems in which reactions were controlled by local availability of ions. "Dirty" Sespe sandstones contain detrital mica, particularly biotite that commonly displays petrographic evidence of partial dissolution or chloritization, and illitization was likely promoted by local availability of K^+. Locally, precipitation of chlorite or conversion of smectite to chlorite may have occurred where ferromagnesian framework grains were dissolved. In contrast, "clean" sandstones appear to have been characterized by greater ionic mobility. "Clean" Sespe sandstones commonly display evidence of ferromagnesian framework-grain dissolution (especially heavy minerals and volcanic lithics) and contain little mica. Significantly, the abundant K-feldspar in Sespe sandstones displays little evidence of dissolution and, in fact, locally contains K-feldspar overgrowths. Some volcanic lithic fragments contain abundant potassium and locally display evidence of dissolution. These observations suggest that diagenetic reactions involving K^+ mostly involved framework grains and apparently exerted negligible influence on clay-mineral diagenesis. We infer that diagenetic fluids in "clean" sandstones were rich in Fe^{2+} and Mg^{2+} liberated by dissolution of framework grains and were generally deficient in K^+.

The trends in clay-mineral diagenesis discussed in this paper can be used to infer maximum burial depths of specific samples from other locations in the Ventura Basin. For example, a "clean" Sespe sample from shallow depth (192 m; 631 ft) on the Topa Topa Uplift (location I on Fig. 1) contains chlorite and mixed-layer I/S characterized by 50 to 60% nonexpandable layers, similar to some of the deepest samples analyzed from the west end of the Oxnard shelf (Fig. 8). The sample also displays petrographic evidence of deeper burial, including the presence of quartz and laumontite cements. On the basis of clay mineralogy and petrography, it was inferred that the Sespe at this location had been deeply buried and subsequently uplifted during recent tectonic activity. This interpretation was confirmed by apatite fission-track analysis, which suggested the sample had been heated to >110°C (fission tracks completely annealed) and then cooled within the past 3 Ma (Hathon, 1991).

CONCLUSIONS

The Oligocene Sespe Formation, in the tectonically complex Ventura Basin, provides an opportunity to study clay-mineral diagenesis in sandstones that have been exposed to a range of burial depths in excess of 5 km. Sespe sandstones are fine- to medium-grained and poorly to well sorted. Framework-grain compositions are widely scattered around a mean composition of $Q_{38}F_{37}L_{25}$.

Petrographic and SEM observations suggest that clay minerals have three major origins in Sespe sandstones: (1) depositional matrix accumulated contemporaneously with sand-size grains and is responsible for matrix-supported fabrics; (2) infiltrated clays accumulated in intergranular space as the result of infiltration of muddy water into sands with grain-supported fabrics; and (3) authigenic clays precipitated from pore water in sands with grain-supported fabrics. Sespe sandstone samples are segregated into "clean" sandstones that contain <20% ductile material (clay minerals plus ductile lithic fragments) and "dirty" sandstones that contain >20% ductile material. These sandstone types reacted differently to diagenesis during burial and, therefore, developed contrasting diagenetic-mineral assemblages.

In "clean" sandstones, the clay-mineral assemblage changes from smectite plus subordinate kaolinite in shallow samples to chlorite in deep samples, suggesting that the major clay-mineral reaction is the transformation of smectite to chlorite. The presence of kaolinite in shallow samples may reflect meteoric diagenesis.

In "dirty" sandstones, the clay-mineral assemblage changes from smectite plus minor kaolinite at shallow depths, to smectite plus chlorite at intermediate depths, to mixed-layer illite/smectite plus minor chlorite at greatest depths. Illitization of smectite appears to be the predominant clay-mineral reaction. The origin of chlorite and the role that it plays in the illitization reaction are unclear. The presence of kaolinite in shallow samples may reflect meteoric diagenesis.

Original texture is thought to have significantly influenced clay-mineral diagenesis in Sespe sandstones. "Dirty" sandstones, likely characterized by low permeabilities since deposition, behaved as closed diagenetic systems. In contrast, higher permeability "clean" sandstones behaved as relatively open diagenetic systems characterized by greater ionic mobility.

ACKNOWLEDGMENTS

Funding for this work was provided by the National Science Foundation (EAR-8720888). Additional support was provided by Chevron Oil Field Research Company, and we especially thank Alden Carpenter for his advice and assistance. Samples and/or data were provided by the California State Core Repository, Chevron, Texaco, and Unocal. We thank Francisca Oboh and an anonymous reviewer for critiques that strengthened the manuscript.

We acknowledge funding sources for the following facilities within the Department of Geological Sciences: XRD (National Science Foundation, EAR-8708413); SEM-EDS (NSF, EAR-8217931; Amoco Foundation, Chevron U.S.A., Marathon Oil Company, Mobil Oil Corporation, Phillips Petroleum Company, and Texaco Philanthropic Foundation). BasinMod™ software was available through the University of Missouri Arkoma Basin Thermal Maturity Research Program, directed by Houseknecht and funded by Anadarko, Arkla, Conoco, Exxon, Maxus, Mobil, NOMECO, Oxy, Pennzoil, Shell, Texaco, Weyerhaeuser, and Yates petroleum companies.

REFERENCES

BOLES, J. R., AND FRANKS, S. G., 1979, Clay diagenesis in Wilcox Sandstones of southwest Texas: implication of smectite diagenesis on sandstone cementation: Journal of Sedimentary Petrology, v. 49, p. 55–70.

CROWELL, J. C., 1987, Late Cenozoic basins of onshore southern California; complexity is the hallmark of their tectonic history, *in* Ingersoll, R. V., and Ernst, W. G., eds., Cenozoic Basin Development of Coastal California: New York, Prentice-Hall, Inc., p. 207–241.

DREVER, J. I., 1973, The preparation of oriented clay mineral specimens for X-ray diffraction analysis by a filter-membrane technique: American Mineralogist, v. 58, p. 553–554.

FALVEY, D. A., AND MIDDLETON, M. F., 1981, Passive continental margins: evidence for a pre-breakup deep crustal metamorphic subsidence mechanism: Oceanologic Acta, SP, p. 103–114.

FLEMAL, R. C., 1966, Sedimentology of the Sespe Formation, southwestern California: Unpublished Ph.D. Dissertation, Princeton University, Princeton, 230 p.

HATHON, L. A., 1991, Burial diagenesis of the Sespe Formation, Ventura Basin, California: Unpublished Ph.D. Dissertation, University of Missouri, Columbia, 250 p.

HELMOLD, K. P., AND VAN DE KAMP, P. C., 1984, Diagenetic mineralogy and controls on albitization and laumontite formation in Paleogene arkoses, Santa Ynez Mountains, California, *in* MacDonald, D. A., and Surdam, R. C., eds., Clastic Diagenesis: American Association of Petroleum Geologists Memoir 37, p. 239–276.

HOWARD, J. L., 1988, Paleoenvironments, provenance and tectonic implications of the Sespe Formation, southern California: Unpublished Ph.D. Dissertation, University of California, Santa Barbara, 306 p.

HORNAFIUS, P. R., LUYENDYK, J. S., TERRES, B. P., AND KAMERLING, M. J., 1986, Timing and extent of Neogene tectonic rotation in the western Transverse Ranges, California: Geological Society of America Bulletin, v. 97, p. 1476–1487.

KISCH, H. J., 1983, Mineralogy and petrology of burial diagenesis (burial metamorphism) and incipient metamorphism in clastic rocks, *in* Larsen, G., and Chilingar, G. V., eds., Diagenesis in Sedimentary Rocks, v. 2: New York, Elsevier, p. 289–494.

LUYENDYK, B. P., AND HORNAFIUS, J. S., 1987, Neogene crustal rotations, fault-slip and basin development in southern California, *in* Ingersoll, R. V., and Ernst, W. G., eds., Cenozoic Basin Development of Coastal California: New York, Prentice-Hall, Inc., p. 207–241.

MATLACK, K. S., HOUSEKNECHT, D. W., AND APPLIN, K. R., 1989, Emplacement of clay into sand by infiltration: Journal of Sedimentary Petrology, v. 59, p. 77–87.

MOLENAAR, N., 1986, The interrelation between clay infiltration, quartz cementation, and compaction in Lower Givetian terrestrial sandstones, northern Ardennes, Belgium: Journal of Sedimentary Petrology, v. 56, p. 359–369.

MOORE, D. M., AND REYNOLDS, R. C., JR., 1989, X-ray diffraction and the identification and anlysis of clay minerals: New York, Oxford University Press, 332 p.

MORAES, M. A. S., AND DE ROS, L. F., 1990, Infiltrated clays in fluvial Jurassic sandstones of Recôncavo basin, northeastern Brazil: Journal of Sedimentary Petrology, v. 60, p. 809–819.

NAGLE, H. E., AND PARKER, E. S., 1971, Future oil and gas potential of onshore Ventura Basin, California, *in* Cram, I. H., ed., Future Petroleum Provinces of the United States–Their Geology and Potential: American Association of Petroleum Geologists Memoir 15, p. 254–297.

NILSEN, T. H., 1984, Oligocene tectonics and sedimentation, California: Sedimentary Geology, v. 38, p. 305–336.

NILSEN, T. H., 1987, Paleogene tectonics and sedimentation of coastal California, *in* Ingersoll, R. V., and Ernst, W. G., eds., Cenozoic Basin Development of Coastal California: New York, Prentice-Hall, Inc., p. 207–241.

RAREY, P. J., 1990, Sedimentological and reservoir characteristics of the upper Sespe Formation at Sockeye field, offshore California, *in* Keller, M. A., and McGowen, M. K., eds., Miocene and Oligocene Petroleum Reservoirs of the Santa Maria and Santa Barbara-Ventura Basins, California: Society of Economic Paleontologists and Mineralogists, Core Workshop No. 14, p. 12–38.

STARKEY, H. C., BLACKMON, P. D., AND HAUFF, P. L., 1984, The routine mineralogical analysis of clay-bearing samples: U.S. Geological Survey Bulletin 1563, 32 p.

WALKER, T. R., WAUGH, B., AND CRONE, A. J., 1978, Diagenesis in first-cycle desert alluvium of Cenozoic age, southwestern United States and northwestern Mexico: Geological Society of America Bulletin, v. 89, p. 19–32.

YEATS, R. S., 1987, Changing tectonic styles in Cenozoic basins of southern California, *in* Ingersoll, R. V., and Ernst, W. G., eds., Cenozoic Basin Development of Coastal California: New York, Prentice-Hall, Inc., p. 207–241.

DEPOSITIONAL, INFILTRATED AND AUTHIGENIC CLAYS IN FLUVIAL SANDSTONES OF THE JURASSIC SERGI FORMATION, RECÔNCAVO BASIN, NORTHEASTERN BRAZIL

MARCO A. S. MORAES AND LUIZ F. DE ROS*

Petrobrás/Cenpes/Diger, Ilha do Fundão Cid. Universitária Qd. 7, Rio de Janeiro RJ 21910, Brazil

ABSTRACT: The Sergi Formation, a Jurassic pre-rift sequence composed mostly of fluvial sandstones, is one of the major hydrocarbon reservoirs of the Recôncavo Basin in northeastern Brazil. Interstitial clays are important components of sandstones and exert significant control on reservoir properties, including permeability, irreducible water saturation and residual-oil saturation. These clays can be grouped into three types: (1) depositional clays; (2) mechanically infiltrated (MI) clays; and (3) authigenic (neoformed) clays. Each type shows a characteristic petrographic aspect that permits recognition and quantification using thin sections.

Depositional clays were incorporated into the rocks as mud intraclasts resulting from reworking of overbank fines by fluvial processes. Early mechanical compaction crushed the mud clasts among more rigid grains, forming a compaction matrix. Mechanically infiltrated (MI) clays occur chiefly as coatings of tangentially accreted particles (cutans) or, locally, as complete pore fills. MI clays appear to be concentrated within the upper part of the formation. These clays can modify the pore geometry of sandstones. Shrinkage porosity, developed by diagenetic transformation of clays, is the dominant porosity type in the upper part of the Sergi Formation.

Authigenic clays are kaolinite and chlorite. Kaolinite occurs as pore fills in large secondary pores and, where present in large amounts, may generate high microporosity in the reservoirs. Chlorite occurs as pore linings and, locally, as pore fills. In the reservoirs, chlorite causes permeability reduction and is related to the presence of low resistivity in water-free, oil-producing zones. These authigenic clays show a distinct distribution within the basin. Kaolinite dominates in the western portion, where Sergi reservoirs are found at shallow depths (above 1,000 m), whereas chlorite is dominant in the eastern portion, where Sergi reservoirs are found at greater depths. The distribution of these clay minerals is the result of differences in burial/temperature histories, which are still reflected by present depths.

INTRODUCTION

Fluvial-sandstone reservoirs of the Sergi Formation contain the largest oil-in-place volume (approximately 1.5 billion bbl) among the hydrocarbon reservoirs of the Recôncavo Basin, northeastern Brazil (Fig. 1). Interstitial clays are important elements of these rocks and are recognized as exerting significant influence on reservoir properties (Bruhn and De Ros, 1987; Moraes and De Ros, 1990). Distinguishing the types of clays in the Sergi Formation sandstones is necessary, not only for a better understanding of the origin and diagenetic evolution of the rocks, but also for a more accurate assessment of reservoir performance. The purposes of this paper are to present petrographic criteria used to distinguish the different types of interstitial clays found in Sergi Formation sandstones and to discuss the influence of each clay type on reservoir quality. The study included the description and quantification of 350 thin sections, most of them corresponding to the same points where standard laboratory petrophysical analyses (porosity and permeability) had been performed. In addition, 120 selected samples were analyzed by scanning electron microscopy and X-ray diffractometry (XRD). Special petrophysical analyses (e.g., relative permeability and capillary-pressure tests), wire-line logs and production data were also available. Sample depths range from 0 to 4,200 m and cover most of the major oil fields of the basin. Some outcrop samples were also studied.

GEOLOGIC BACKGROUND

The Recôncavo Basin (Fig. 1) is an early Cretaceous continental rift that developed when the South Atlantic opened (Milani and Davison, 1988). Stratigraphy of the basin (Fig. 2) can be divided into pre-rift, rift-fill and post-rift sections. The pre-rift section (Jurassic and Lower Cretaceous) consists of red beds of the Aliança Formation, fluvial sandstones of the Sergi Formation, fluvial-lacustrine rocks of the Itaparica Formation and lacustrine shales of the Tauá Member of the lower part of the Candeias Formation (Fig. 3). The rift-fill section is Lower Cretaceous and includes lacustrine shales and sandstone turbidites of the middle and upper parts of the Candeias Formation, the coarse clastic wedge of the Salvador Formation, deltaic sediments of the Ilhas Group and fluvial sandstones of the São Sebastião Formation (Fig. 3). The post-rift geologic record is poorly preserved and consists of thin Tertiary and Quaternary fluvial deposits.

The Sergi Formation consists of blanket sandstones (Fig. 3). These units commonly show a sheet-like geometry and were deposited by a prograding braided fluvial system on a flat cratonic sag prior to the rifting event. The Sergi section displays a coarsening-upward trend, indicating increasing fluvial energy. Shale intercalations occur mainly in the lower part of the formation. The common occurrence of eolian reworking and caliche layers suggests that an arid/semiarid climate prevailed during the time of Sergi deposition. Sergi sandstones produce hydrocarbons, which were generated in the lacustrine Candeias Formation, from rift-related structural traps in the Recôncavo Basin.

DIAGENETIC EVOLUTION OF SERGI SANDSTONES

The Sergi sandstones are subarkoses with a predominance of potassium feldspar (Table 1). Lithic grains are rare, and commonly are chert or phyllite fragments. These rocks have undergone a complex and extensive diagenetic evolution (Bruhn and De Ros, 1987), which is summarized in Figure 4. Despite the complexity, the sequence shows similarities with diagenetic sequences observed in other sandstones of arid/semiarid environments (Kessler, 1978). The most important diagenetic phases (Table 1) include, in the inferred temporal succession: (1) mechanically infiltrated (MI) clays; (2) mechanical compaction; (3) quartz and feld-

---

*Present Address: Universidade Federal do Rio Grande do Sul, Instituto de Geociências, Departamento Mineralogia e Petrologia, Avenida Bento Gonçalves 9500, Porto Alegre RS 9500, Brazil

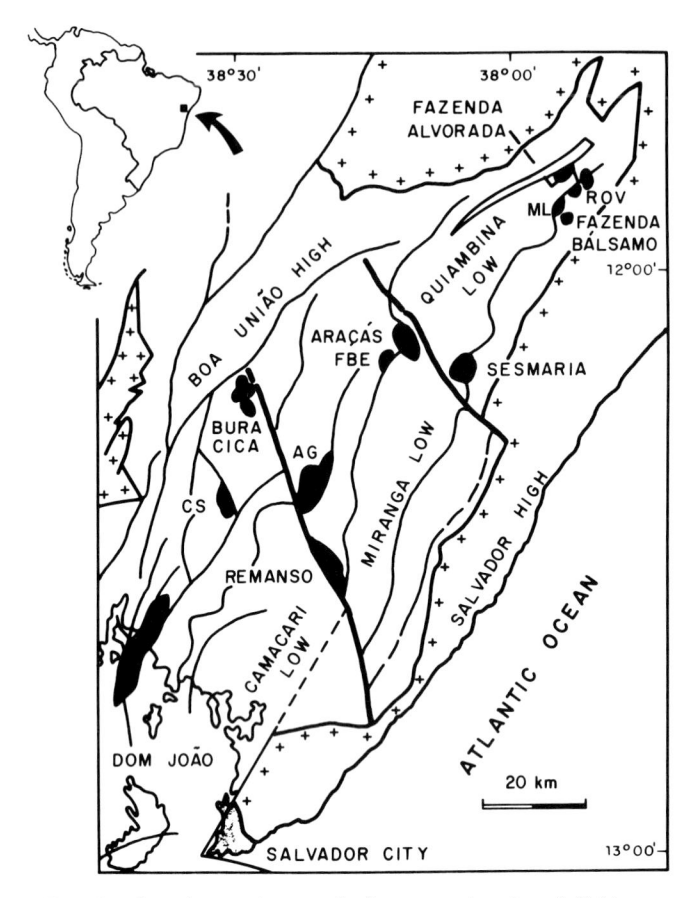

FIG. 1.—Location, main tectonic features and major oil fields producing from the Sergi Formation in the Recôncavo Basin. SW-NE lines represent major normal faults; NW-SE lines represent transfer faults. Crosses are crystalline-rock outcrops. Oil-field codes: AG-Agua Grande; ML-Malombê.

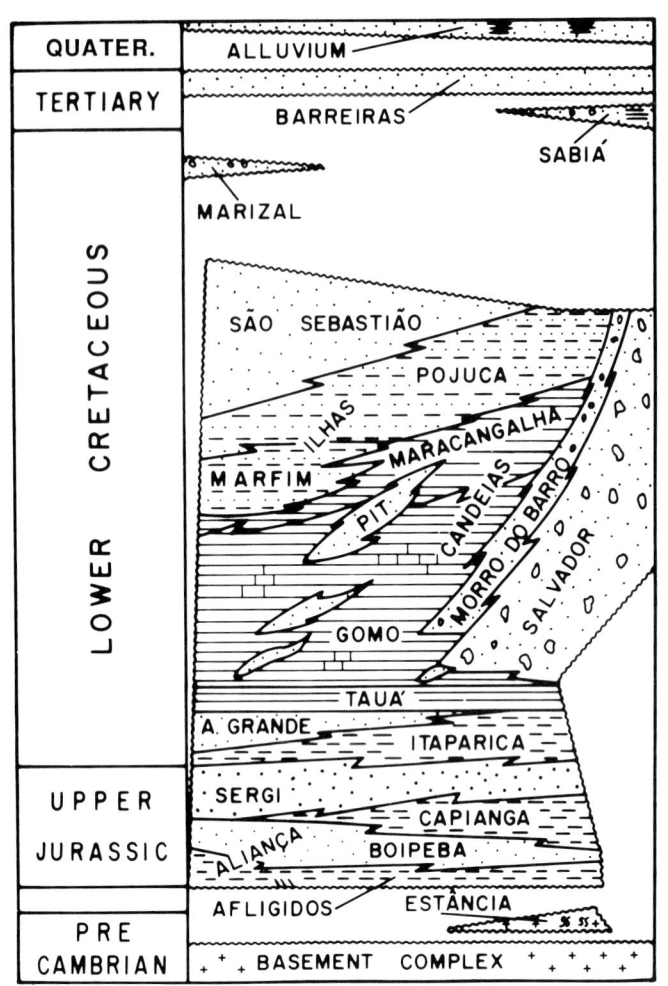

FIG. 2.—General stratigraphic framework of Recôncavo Basin (representative of west to east lithologic relations).

spar overgrowths; (4) calcite; (5) calcite and framework-grain dissolution; (6) kaolinite; (7) dolomite; (8) chlorite; and (9) pyrite and Ti-minerals.

During early diagenesis (referring to phases developed prior to the dissolution phase), the amount of MI clays emplaced into the sandstones defined two distinct diagenetic pathways (Fig. 4). The MI clay-rich rocks are characterized by slight development of quartz and feldspar overgrowths and less calcite cement. Also, in these rocks, shrinkage porosity developed as a result of the diagenetic transformation of the MI clays and is a significant component of the pore system.

The generation of secondary porosity, involving dissolution of grains and cements, was a significant phase in the diagenetic evolution of the sandstones. The process has been interpreted as resulting from the action of fluids released by compacting organic-rich shales (Bruhn and De Ros, 1987). Evidence of calcite and framework-grain dissolution, the latter involving mostly potassium feldspar, commonly are observed in the basin. In general, the amount of secondary pores increases with depth. The secondary porosity was reduced and modified by the introduction of kaolinite, dolomite, and chlorite cements, and late quartz and feldspar overgrowths. Reservoir quality in Sergi sandstones with low MI-clay content is essentially related to the type and distribution of the authigenic clays, although dolomite and quartz and feldspar overgrowths are volumetrically more important.

INTERSTITIAL CLAYS

Average amount of interstitial clays in the Sergi Formation sandstones, as determined by point counting, is 9% of the bulk rock volume (BRV). Considering that average thin-section porosity in the formation is 16% of the BRV (see Table 1) and that minus-cement porosity is approximately 27%, the significance of the clays is clear. Petrographic observations show that these clays display different textures and distribution patterns, related to their origin or diagenetic modifications (Fig. 5). In this section, each major type of interstitial clay in Sergi sandstones will be described in terms of petrographic aspect, distribution within the rocks, and influence on the diagenetic evolution.

Depositional Clays

The hydraulic characteristics of fluvial processes tend to segregate sand and mud. Commonly, sand is deposited within

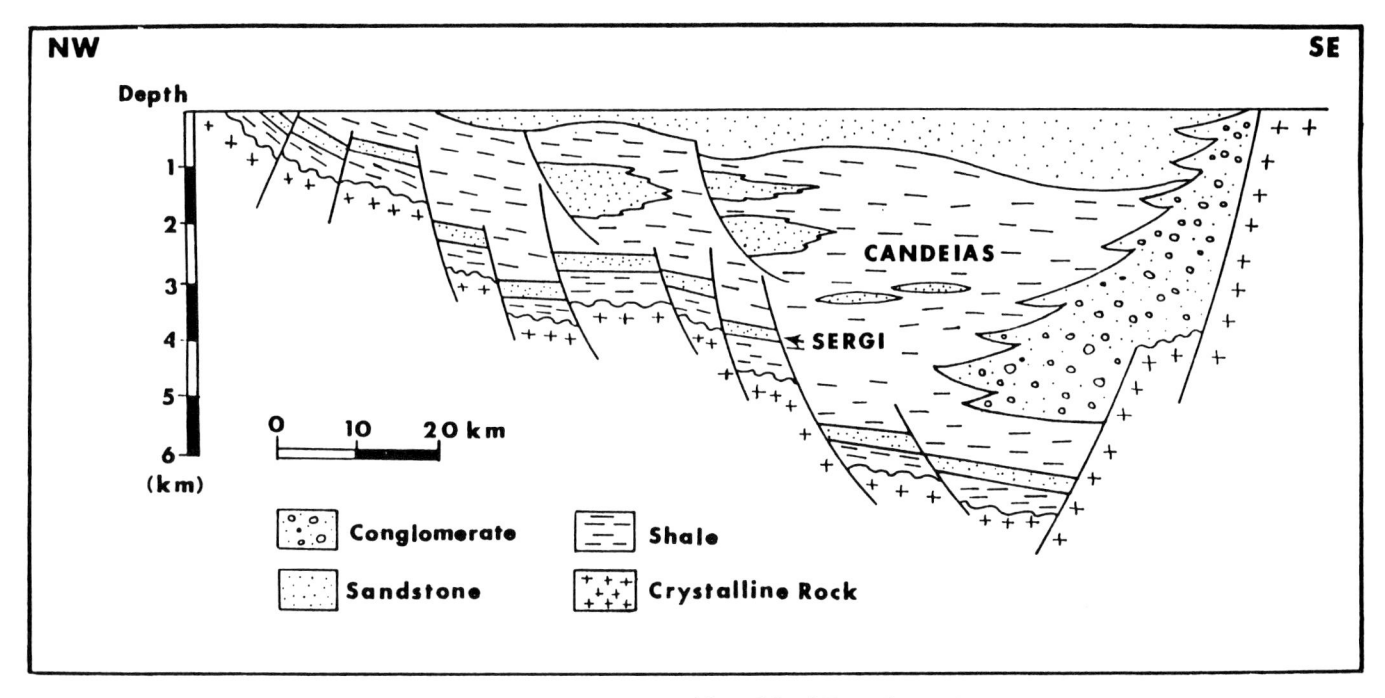

FIG. 3.—Geologic cross section of Recôncavo Basin showing the position of Sergi Formation sandstones and of Candeias Formation lacustrine shales.

channels by lateral accretion and mud is accumulated in overbanks by vertical accretion (Collison, 1986). However, in the Sergi Formation, part of the interstitial clay in the sandstones was incorporated as mud clasts derived from erosion of overbanks. Some of these clasts were already cemented by iron hydroxide or carbonate and therefore were rigid at the time of deposition. Most of them, however, were soft and were crushed among the more rigid grains during mechanical compaction, generating compaction matrix (pseudomatrix). This type of interstitial clay averaged 1% of the BRV (Table 1). Mud clasts commonly are associated with erosional features at the base of fluvial cycles and more rarely concentrated in discrete stratification planes. Compaction matrix can be recognized under the petrographic microscope by the dense structure, the general lack of internal organization, and the detrital aspect of the clay (Fig. 6). Because this matrix is produced by compaction of originally discrete particles, slight contrasts in color and composition among the deformed intraclasts occur. In addition, heterogeneity in the packing of rigid grains (as observed in Fig. 6) is typical of zones containing compaction matrix. In sandstones with large amounts of compaction matrix (>20% of the BRV), porosity was obliterated early, with no significant further development of diagenesis.

Mechanically Infiltrated (MI) Clays

In his study of the Cenozoic alluvium of northwestern Mexico and the southwestern United States, Walker (1976) recognized three mechanisms leading to the accumulation of significant amounts of MI clays: (1) below-surface reworking in vadose zones (Fig. 7A); (2) within a fluctuating water table in phreatic zones (Fig. 7B); and (3) above impermeable barriers in phreatic zones (Fig. 7C). He also suggested that these mechanisms are more effective in coarse-grained alluvium of arid/semiarid settings, where lowered water tables favor the deep penetration of muddy waters (Crone, 1975). Occurrence of this type of clay in the Sergi Formation was described by Moraes and De Ros (1990), applying criteria developed by Crone (1975), Walker (1976) and Matlack and others (1989). In this section, their observations will be summarized, with further discussion on the relations between MI clays and other diagenetic elements, particularly late diagenetic clays.

In the Sergi sandstones, interstitial clays interpreted as the product of mechanical infiltration, averaged 6% of the BRV (Table 1). These clays are characterized by the following textures.

Ridges and bridges.—

Ridges and bridges are aggregates of clay platelets oriented roughly normal to the grain surfaces (Fig. 8A). They

TABLE 1.—AVERAGE BULK COMPOSITION OF SERGI FORMATION SANDSTONES BASED ON THIN-SECTION POINT COUNTING* (N = 350)

	%		%
Quartz	61	Overgrowths	1
Feldspars	8	Calcite	3
Lithics	1	Kaolinite	1
Mud clasts	1	Chlorite	2
MI clays	6	Porosity	16

* 200 points per thin section

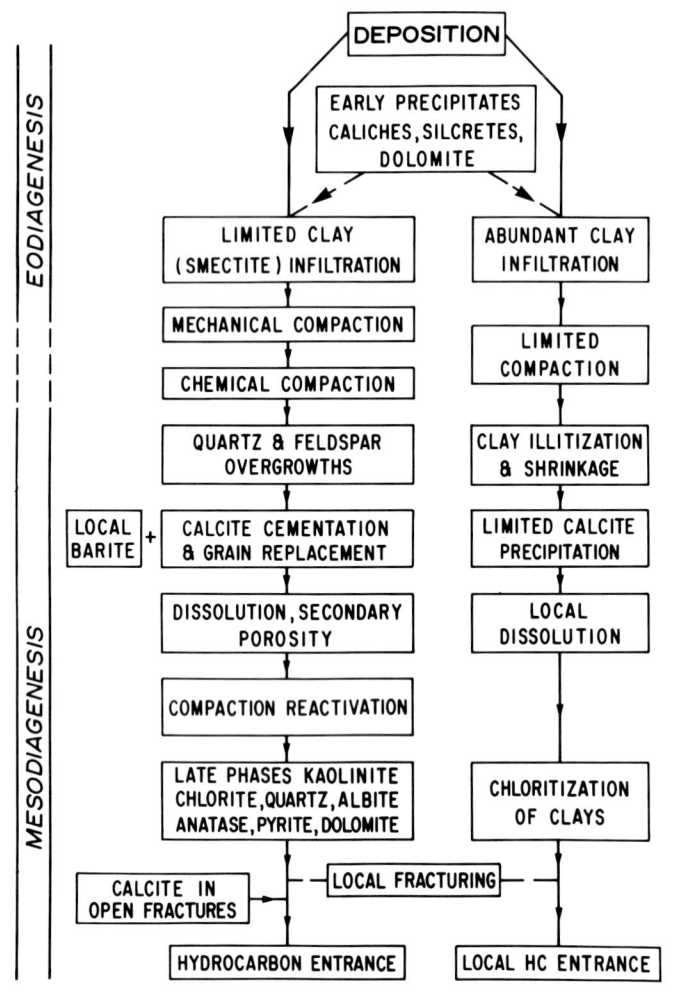

FIG. 4.—Generalized diagenetic sequence for the Sergi sandstones. The two main diagenetic pathways are related to the amount of mechanically infiltrated (MI) clay. (modified from Bruhn and De Ros, 1987).

can consist of small ridges projecting from the grain surface or, more commonly, elongated aggregates connecting two adjacent grains (bridges). Ridges and bridges, which are interpreted as products of infiltration within the vadose zone (Moraes and De Ros, 1990), are not common in the Sergi sandstones.

Geopetal fabric.—

Geopetal fabrics consist of clay accumulations occurring at the bottom of large pores or pendular aggregates (i.e., discontinuous aggregates of detrital clay) attached to the lower surface of the grains. These fabrics, considered typical of the vadose zone, also are rare in Sergi sandstones.

Loose aggregates.—

Loose aggregates include chaotically flocculated aggregates in which clay platelets constitute an open framework without any recognizable orientation. This texture is rare in the Sergi sandstones.

Cutans.—

The term cutan has been used by soil scientists to refer to textural features such as clay platelets completely covering the grains and oriented tangentially to the grain surfaces (Andreis, 1981). Cutans are commonly observed in samples from the Sergi Formation (Figs. 8B and C). In sandstone petrology, clay cutans have the same connotation as clay coatings. However, the term coating has been used for both detrital and authigenic (neoformed) aggregates (Heald and Larese, 1974; Wilson and Pittman, 1977; Eslinger and Pevear, 1988). One important feature to distinguish cutans from authigenic-clay coatings (or, as we prefer, clay rims) is the orientation of MI clay platelets parallel to the grain surfaces, an orientation that seems to reflect their origin as detrital particles decanted from the water and attached by electrostatic forces to the surface of the grains (Matlack and others, 1989). Neoformed clay rims usually consist of crystals oriented normally to the grain surface (Wilson and Pittman, 1977). Cutans are interpreted as produced under phreatic conditions (Moraes and De Ros, 1990), where the complete water saturation of the pores favored extensive covering of the grain surfaces.

Massive aggregates.—

Massive aggregates completely fill intergranular pores (Fig. 8D). The internal structure consists of thick coatings near grain surfaces and a dense, chaotic mass of clay platelets in the center of pores. Because they cover cutans, massive aggregates also are interpreted as being produced in phreatic conditions. The distinction between these aggregates and compaction matrix is possible mainly because of the association of massive aggregates with cutans.

Dehydration and shrinkage.—

The MI clays of Sergi sandstones mostly originated during weathering in a relatively dry environment. Therefore, they originally had a predominant smectite composition, as has been commonly described for such conditions (Keller, 1970; Velde, 1985). The diagenesis of smectites includes an initial phase of regeneration, and then the progressive transformation to more stable clay minerals with increasing depth of burial (e.g., chlorite or illite; Hower, 1981; Eslinger and Pevear, 1988). Through the diagenetic realm, intermediate stages of mixed-layer illite-smectite or chlorite-smectite occur. A change in composition from smectite to illite-smectite or chlorite-smectite, and then to illite or chlorite is registered with increasing depth in Sergi Formation sandstones (Table 2).

Chemically, these reactions cause the release of water and some cations, along with the incorporation of Al, K and/or Mg (Boles and Franks, 1979). Such effects are physically expressed by a bulk-volume decrease. As a consequence, the diagenetic evolution of the MI clays produces shrinkage of the aggregates. Among the features related to shrinkage in the Sergi sandstones are detachment of cutans from the grain surfaces and fragmentation of massive aggregates (Figs. 9A and B). These features can appear as contracted replicas of the original pore fill or as curled sheets.

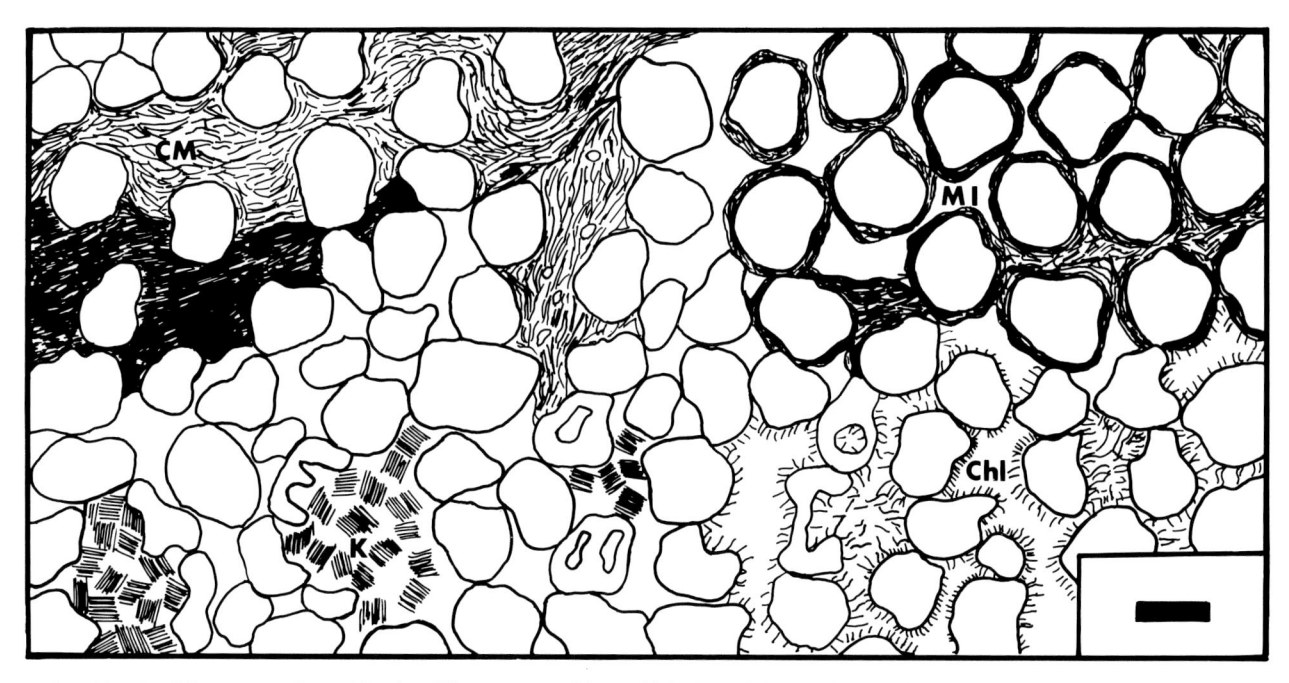

FIG. 5.—Sketch of the textures formed by the different types of interstitial clays of the Sergi Formation sandstones as seen through a petrographic microscope. Upper left: Compaction matrix (CM); upper right: mechanically infiltrated clay (MI); lower left: authigenic kaolinite (K); and lower right: authigenic chlorite (Chl). Dimensions of authigenic-clay crystals are enlarged to facilitate visualization. Scale bar approximately 300 μm.

Shrinkage features can be produced artificially near the surface by drying of samples during storage or sample preparation. In the case of the Sergi Formation, several lines of evidence suggest that shrinkage porosity is significant in the subsurface. These include: (1) presence of contracted aggregates in samples preserved for special petrophysical analyses (e.g., sample of Fig. 9B, in which present depth, not maximum depth, is 256.05 m); (2) presence of contracted aggregates in the interior of calcite cement (Fig. 10); (3) occurrence of shrinkage features in sandstones where MI clay aggregates have a predominant illitic composition, as in the reservoirs of the Araças field (see Table 2); and (4) general agreement among thin-section observations, core petrophysical data and production data (i.e., intervals with dominant shrinkage porosity are effective hydrocarbon producers).

Authigenic (Neoformed) Clays

Late authigenic clays, kaolinite and chlorite, reduce intergranular porosity or replace framework grains. The products of kaolinization, illitization and chloritization of matrix

FIG. 6.—Compaction matrix (arrows) produced by the crushing of mud intraclasts. Photomicrograph, plane polarized light.

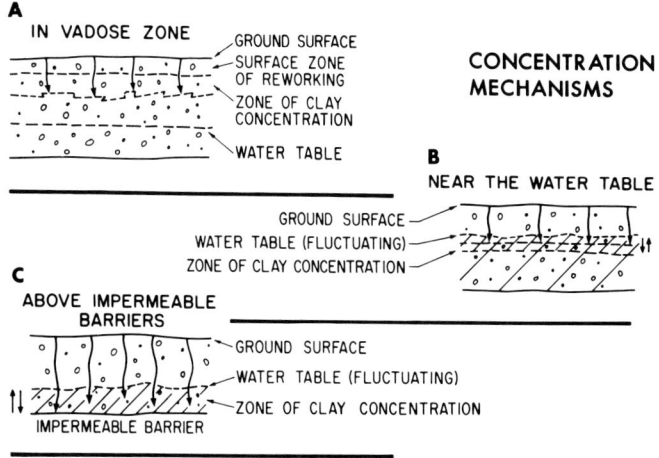

FIG. 7.—Mechanisms of clay infiltration in alluvial sediments (modified from Walker, 1976).

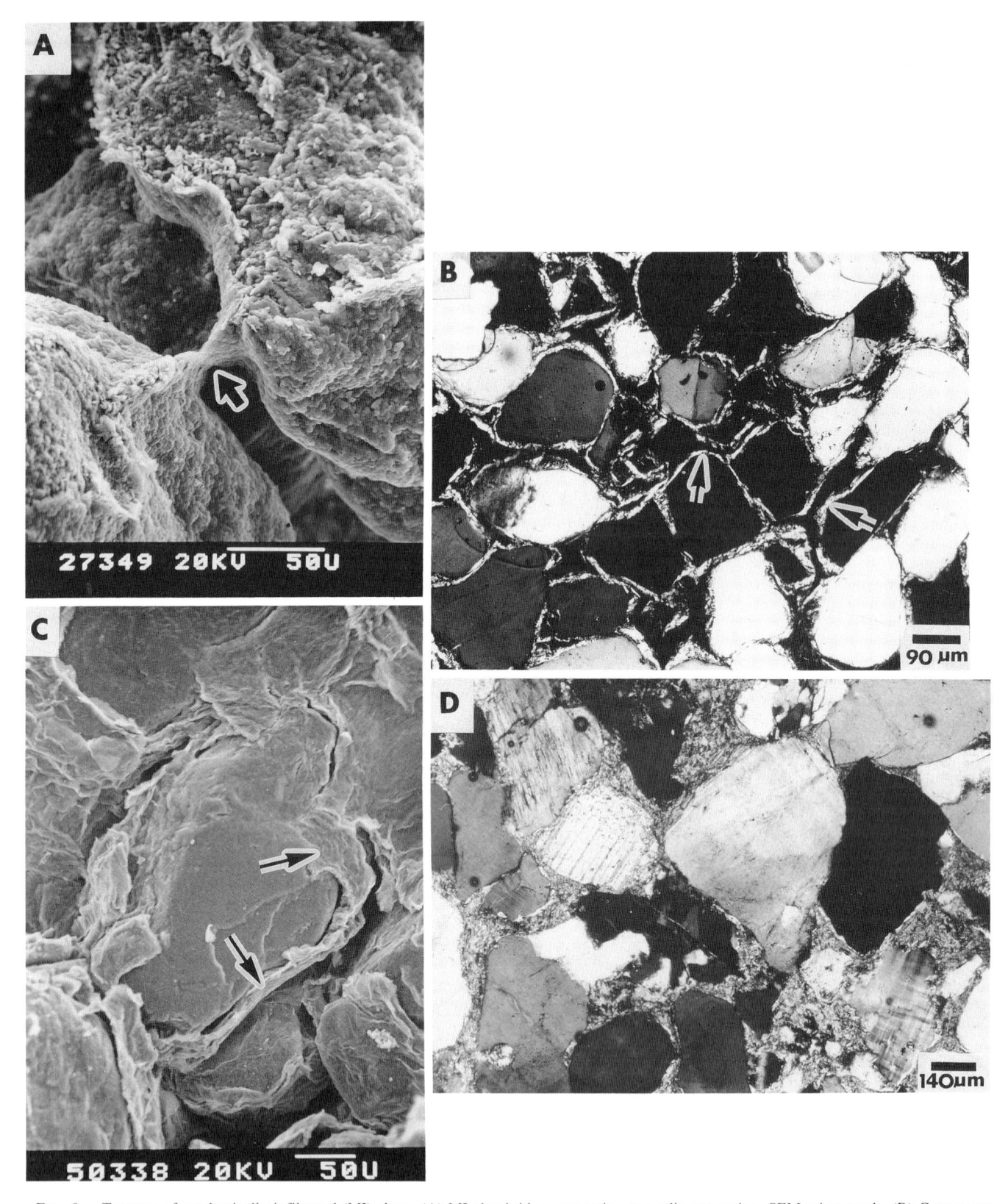

FIG. 8.—Textures of mechanically infiltrated (MI) clays. (A) MI-clay bridge connecting two adjacent grains. SEM micrograph. (B) Cutans consisting of clay platelets oriented parallel to the surface of the grains. Photomicrograph, crossed nicols. (C) Detail of a cutan (arrows) showing the smooth (detrital) aspect of the MI clays. SEM micrograph. (D) Massive aggregates of MI clays with a chaotic internal organization. Photomicrograph, crossed nicols.

TABLE 2.—COMPOSITION OF THE INFILTRATED CLAYS IN
THE SANDSTONE RESERVOIRS OF SERGI FORMATION

Field	Depth Range (m)	Clay Composition
Dom João	200-400	Smectite
Buracica	700-800	I-S (40% illite)
Agua Grande	1,400-1,450	I-S (58% illite)
Araçás	2,700-2,750	I-S (80% illite) and C-S

or infiltrated clays are not included in this category, but are considered to be transformed detrital clays. Both precipitated and grain-replacement clays are quite homogeneous, as indicated by XRD data and EDAX analyses. Also, the presence of well-formed crystals composing the clay aggregates suggests the authigenic nature (Wilson and Pittman, 1977). Authigenic clays averaged 3% of the BRV based on point counts (Table 1), and typically occur in greater abundance in zones with low MI-clay content, where they are included among the more important controls on the reservoir properties of the sandstones.

In the Sergi diagenetic system, alteration of potassium feldspar seems to be a major source of material for late diagenetic clays. A typical reaction would be as follows:

$2KAlSi_3O_8 + 14 H_2O + 5Mg^{2+}$
(K-feldspar)

$\rightarrow Mg_5Al_2Si_3O_{10}(OH)_8 + 2K^+ + 3H_4SiO_4 + 8H^+$
(chlorite)

This reaction indicates that quartz (possibly in the form of late overgrowths) is an expected by-product. Petrographic analyses of Sergi sandstones indicate (Table 3) that the amount of intragranular porosity present in potassium feldspar is always sufficient, at any given depth, to account for the total amount of authigenic silicate phases present in the rocks. Therefore, it is not necessary to invoke external sources to explain late authigenic clays. The only exception is the magnesium present in chlorite, which is inferred to have diffused from surrounding shales.

Kaolinite and chlorite show a distinct distribution in the basin. Kaolinite predominates in shallow sandstones (above 1,000 m), whereas chlorite predominates in deeper sandstones. Possible controls on this distribution are discussed later.

Kaolinite.—

Kaolinite occurs as discontinuous and isolated pore fills occupying large pores produced by dissolution of calcite and framework grains. The typical morphology of authigenic kaolinite is booklets of relatively large crystals forming an open and loose structure (Fig. 11A). Individual crystals, as seen by SEM (Fig. 11B), have a characteristic pseudohexagonal form, well-formed crystalline shapes and no evidence of dissolution. In general, booklet morphology and homogeneous, low birefringence of aggregates are the basic criteria for identification of authigenic kaolinite. The high degree of purity of these aggregates, as indicated by XRD data (Petrobras, unpublished internal reports), is confirmed under the petrographic microscope by the uniform aspect of the clay. Kaolinite averages 0.2% of the BRV in the Sergi sandstones as a whole, and 0.5% of the BRV in sandstones shallower than 1,000 m. Locally, kaolinite accounts for 5% of the BRV.

Late diagenetic kaolinization of feldspar is more difficult to detect. Alteration of feldspars is commonly recognizable by the darkening of the grains. However, because kaolinite apparently was also a product of meteoric alteration prior to deposition or during early diagenesis, it is difficult to quantify the amount of kaolinization produced during late diagenesis.

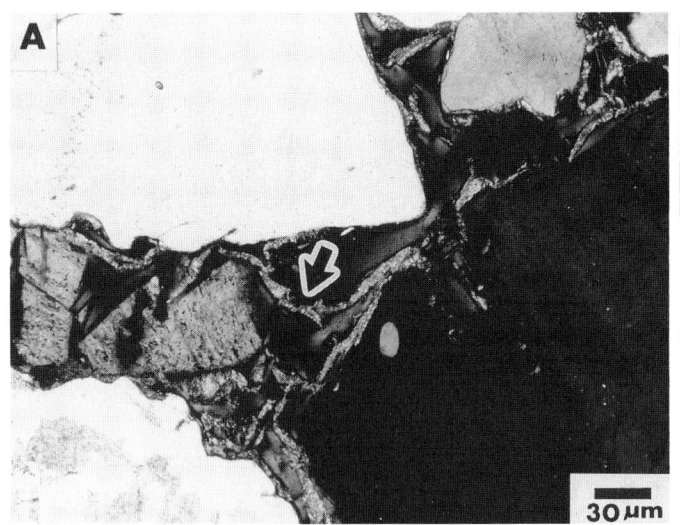

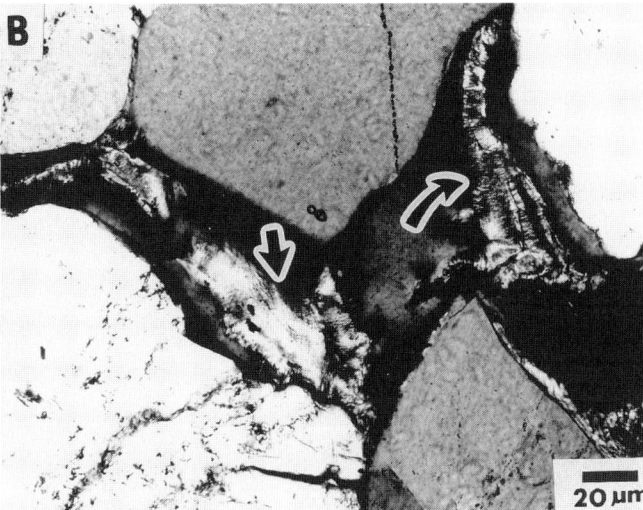

FIG. 9.—Shrinkage features in the sandstones. (A) MI-clay aggregates affected by shrinkage, with clay cutans (arrows) detached from the surface of the grains. Photomicrograph, plane-polarized light. (B) Shrunk MI-clay aggregates (arrows) preserving the original pore outline. A 50% shrinkage can be inferred by comparing pore- and clay-aggregate sizes. Photomicrograph, crossed nicols.

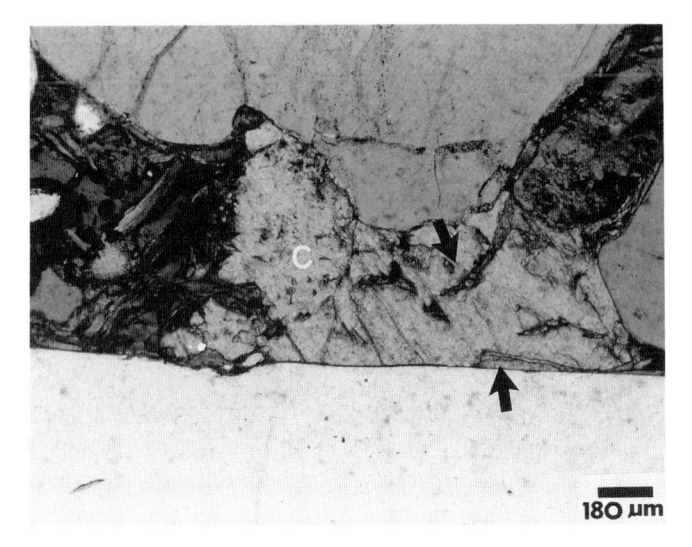

FIG. 10.—Detached cutans engulfed by the calcite cement (arrows); C-calcite; crossed nicols + mica plate.

TABLE 3.—TOTAL AMOUNTS OF DISSOLVED FELDSPARS AND AUTHIGENIC-SILICATE MINERALS IN THE SANDSTONES OF SERGI FORMATION AT DIFFERENT DEPTHS

Field	Depth (m)*	Diss Fld** %	Diag*** %
Dom João	300	4	1.0
Buracica	750	5	1.5
Malombê	1,000	5	1.5
Remanso	1,200	6	3.5
Ag. Grande	1,400	7	2.0
Sesmaria	2,300	6	6.0
Araças	2,700	6	2.5

*Average reservoir depth
** Intragranular pores in feldspar (% BRV)
*** Amount of diagenetic silicates (% BRV)

Chlorite.—

Chlorite is the most common late diagenetic clay in the Sergi sandstones, especially in rocks below 1,000 m. Chlorite appears as pore linings (rims) composed of crystals perpendicularly oriented to grain surfaces (Fig. 12A) and locally evolves into complete pore fillings (Fig. 12B). In thin section, chlorite is recognizable by the homogeneous green to brownish green color. Under the SEM, chlorite commonly appears as rosettes of pseudohexagonal crystals (Fig. 12C). EDAX analysis indicates the presence of both magnesium- and iron-rich types. XRD data reveal that, locally, authigenic mixed-layer chlorite-smectite also is present. Chlorite also occurs as a replacement of MI clays, micas and potassium feldspar (Fig. 12D). Chlorite occurs in most of the Sergi sandstones located below 1,000 m, although in variable amounts. Chlorite averaged 2% of the BRV (Table 1), ranging from 5% of the BRV (average in the Sesmaria field, see Fig. 1) to 0.5% of the BRV (average in the Agua Grande field). There is a trend, although erratic, of increasing chlorite with increasing depth.

INFLUENCE OF THE CLAYS ON THE RESERVOIRS

Sergi sandstones lacked clays at the time of deposition, as expected for braided-stream deposits. The introduction of interstitial clays increased the degree of heterogeneity of the rocks, greatly reducing reservoir quality. Table 4 shows average porosity, permeability and typical diagenetic products of Sergi reservoirs at different depths. In general, the recognition of the type, amount and distribution of interstitial clays is a critical factor for the adequate geologic characterization of the reservoirs.

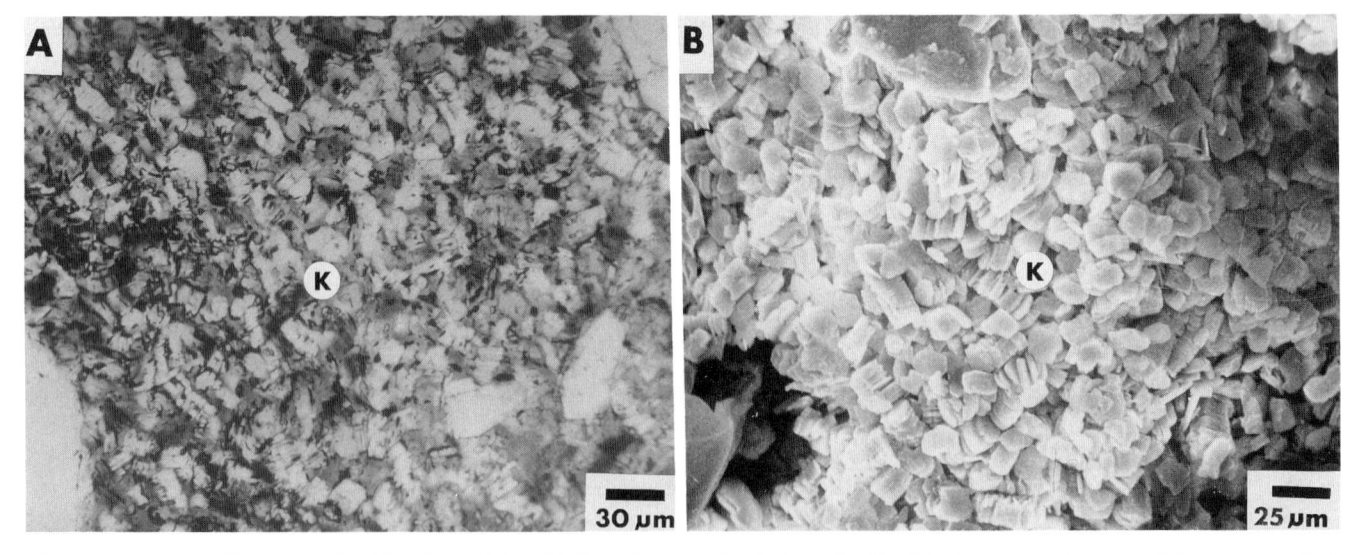

FIG. 11.—Petrographic aspect of authigenic kaolinite. (A) Crystals of authigenic kaolinite (K) fill a large secondary pore. The presence of micropores (dark grey) within the clay aggregate is clearly visible. Photomicrograph, plane polarized light. (B) Detailed view of kaolinite crystals (K) showing pseudohexagonal form. SEM micrograph.

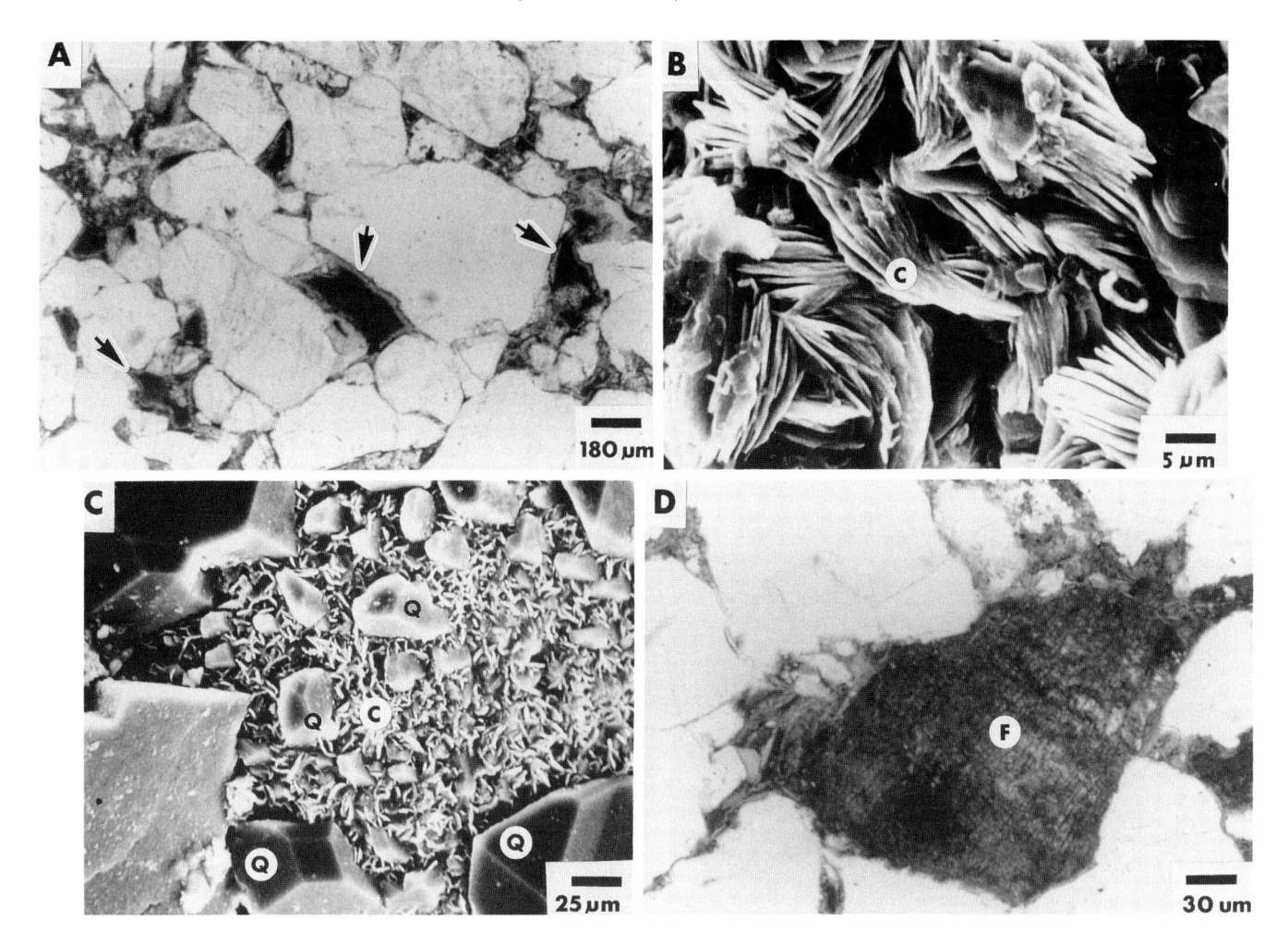

FIG. 12.—Textural expression of authigenic chlorite. (A) Rims (pore-lining) consisting of chlorite crystals (arrows) oriented perpendicularly to the grain surfaces. Photomicrograph, plane polarized light. (B) Detailed view of authigenic-chlorite rosettes composed of pseudohexagonal flakes. SEM micrograph. (C) Chlorite crystals (C) associated with late quartz overgrowths (Q). See text for discussion. SEM micrograph. (D) Chloritization of a potassium-feldspar grain (F). Photomicrograph, plane polarized light.

Compaction matrix is either dispersed in sandstones or concentrated in discrete stratification planes. In the first case, it tends to generate a homogeneous porosity and permeability reduction (Rittenhouse, 1971; Benson, 1981). In the latter case, however, it generates a significant reservoir heterogeneity related to the permeability contrast between adjacent layers with or without compaction matrix. Effective permeability in high-angle cross-bedded zones is commonly determined by the lowest permeability value (Weber, 1982; Kortekaas, 1985). Reduction of the effective permeability by thin, low-permeability layers is not detectable by standard petrophysical evaluation. Compaction matrix, however, is only locally an important factor on the quality of Sergi sandstone reservoirs.

Mechanically infiltrated (MI) clays predominate in the upper part of the Sergi Formation (Fig. 13). In many Sergi reservoirs with thin, but continuous cutans of MI clays, porosity was preserved because of the inhibition of overgrowth development, a process described for clay coatings in general by Heald and Larese (1974). Thick cutans or massive aggregates of MI clays also inhibited calcite cementation. These zones later developed significant porosity because of the shrinkage process. Even considering the possible positive effects, MI clays are, however, generally harmful to reservoir quality. The presence of clay cutans usually causes significant permeability reduction. Even thin

TABLE 4.—AVERAGE POROSITY, PERMEABILITY AND AMOUNTS (% BRV) OF DIAGENETIC ELEMENTS FOR SERGI RESERVOIRS AT DIFFERENT DEPTHS

Field	Depth (m)*	Poros. %	Perm. (md)	MI Clays %	Overg. %	Calcite %	Aut. Clays %
Dom João	300	21	87	8.7	0.3	1.8	0.5
Buracica	750	22	129	4.2	0.6	2.0	0.8
Malombê	1,000	16	40	3.0	1.4	1.8	0.1
Remanso	1,200	15	36	3.5	1.9	2.6	1.7
Agua Grande	1,400	18	68	6.0	1.8	0.8	0.2
Sesmaria	2,300	13	1	2.8	1.4	4.3	4.4
Araças	2,700	12	3	5.2	2.0	3.0	0.2

* Average reservoir depth

7-SI-10-BA

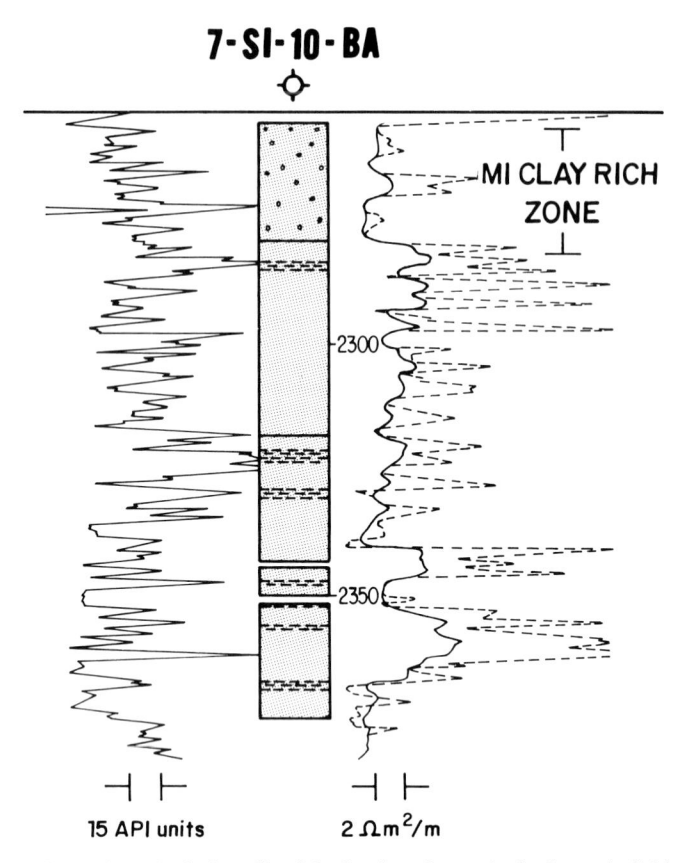

FIG. 13.—Vertical profile of the Sergi sandstones in the Sesmaria field (see Fig. 1 for location). The upper part is characterized by a predominance of MI clays. Within the lower part, clay-free zones are intercalated with chlorite-rich zones. The entire interval is oil prone. See text for additional discussion.

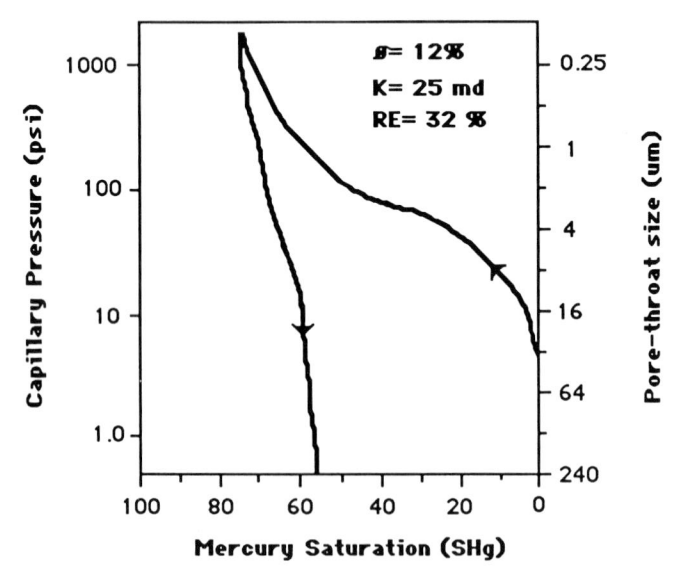

FIG. 14.—Typical capillary-pressure test in Sergi sandstone rich in MI clay. Curves represent injection (right), and withdrawal (left). Non-wetting phase recovery (RE—in this case of mercury) is 32%. Reservoir depth: 2,542 m.

cutans, which may preserve most of the intergranular porosity, severely reduce permeability because of the obstruction of pore throats. Typical permeability values in the upper Sergi reservoirs of the Sesmaria field (average depth of 2,250 m) range from 50 to 100 md for medium- to coarse-grained sandstones with low MI-clay content (less than 5% of the BRV) and are less than 20 md for sandstones rich in MI clay (5 to 20% of the BRV). In addition, in zones of higher clay content (up to 35% of the BRV), MI clays form permeability barriers that are important macroscale heterogeneities in many Sergi reservoirs. In addition, microporosity developed within the MI-clay aggregates contains high volumes of irreducible water. As a consequence, Sergi reservoirs with well-developed clay cutans typically display low induction-log resistivity (commonly below 4 ohms m^2/m), which causes problems for well-log evaluation. Another problem, which occurs in shallow reservoirs where smectite is still predominant, is related to the reactivity of smectite, especially the swelling of the clay aggregates upon contact with low-salinity fluids.

The shrinkage process generated significant amounts of secondary porosity in the MI clay-rich zones. However, such pore systems typically present irregular permeability systems and high pore/pore-throat ratios, factors that combine to produce low-recovery efficiency at the microscopic level (Wardlaw and Cassan, 1979; Wardlaw, 1980). Figure 14 displays mercury capillary-pressure curves including mercury injection and ejection. In this system, mercury is the non-wetting phase, and is thought to adequately represent oil in water-wet reservoir systems (Wardlaw and Cassan, 1979; Jennings, 1987). The recovery of the non-wetting phase after withdrawal (recovery efficiency–RE in Fig. 14) is calculated by subtracting the mercury saturation (SHg), at the end of the withdrawal process, from the maximum saturation obtained at the highest pressure. The result for Figure 14 data is that, despite having moderate permeability, the sandstone is characterized by low recovery of the non-wetting phase because of the geometric characteristics of the pore system.

Authigenic clays have different impacts on reservoirs. Kaolinite, which is irregularly distributed in relatively small amounts, apparently is not a significant cause of permeability reduction. High irreducible water saturation, related to the microporosity in the loosely packed kaolinite aggregates, has been observed in samples where kaolinite averaged 5% or more of the BRV. Such amounts of kaolinite are rare in Sergi reservoirs. Chlorite rims cause significant permeability reduction, even where present in small amounts. Typical permeability values in lower Sergi reservoirs of the Sesmaria field (average depth of 2,350 m) are around 50 md for fine- to medium-grained sandstones with low-chlorite content (less than 1% of the BRV) and less than 10 md, commonly around 1 md, for sandstones with high-chlorite content. Microscopic-recovery efficiency in chlorite-rich reservoirs tends to be higher than in MI clay-rich reservoirs (Fig. 15), because chlorite rims commonly cause no increase of pore-system heterogeneity. Evaluation problems related to low electrical resistivity developed by oil-prone, chlorite-rich zones are common. In the Sesmaria field, zones producing water-free oil have resistivities as low as

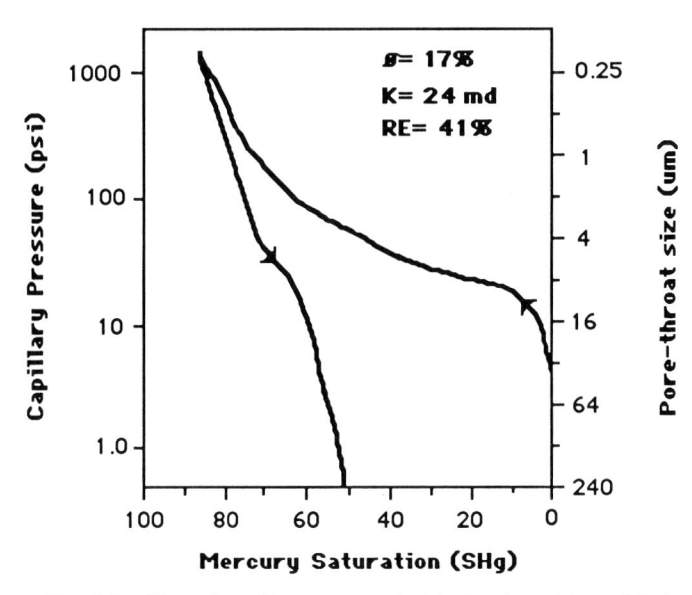

FIG. 15.—Typical capillary-pressure test in Sergi sandstone rich in chlorite. Curves represent injection (right), and withdrawal (left). Non-wetting phase recovery (RE—in this case of mercury) is relatively high (41%). Reservoir depth: 2,359 m.

6 ohms m²/m (see Fig. 13). Chlorite, especially the iron-rich variety, is also a reactive mineral. When contacted by acidic fluids commonly used for well stimulation, chlorite can dissolve and iron hydroxide precipitate, causing permeability reduction.

DISTRIBUTION OF AUTHIGENIC CLAYS

In the Recôncavo Basin, kaolinite predominates in Sergi sandstones above 1,000 m and chlorite predominates below 1,000 m. Because the Sergi Formation in the Recôncavo Basin dips eastward (see Fig. 3), such a configuration results in the predominance of kaolinite in the reservoirs located in the western portion of the basin (Dom João and Buracica fields; see Fig. 1) and in the predominance of chlorite in the reservoirs located in the eastern portion of the basin (Araças, Sesmaria and other smaller fields). In the sandstone reservoirs of the Malombê, Remanso and Agua Grande fields, which are located at depths of 1,000 to 1,500 m, both types of clays occur.

Present distribution of authigenic clays is inferred to reflect burial history and, as a consequence, thermal exposure of the two zones. The burial history of Sergi reservoirs in the Recôncavo Basin is shown in Figure 16. A generalized stratigraphic-lithologic section for the area was used for the burial-history reconstruction. The section was based on a number of wells from or between the Agua Grande and Araças fields. Chronological information was obtained from paleontological studies of several wells in the area. The burial history shows rapid Early Cretaceous syn-rift subsidence followed by post-rift uplift, which is considered to average 500 m based on the extrapolation of porosity-depth curves (obtained from unpublished Petrobras internal reports). Thus, the occurrence of clay type is interpreted as being related to depth or, more specifically, to the temperature reached by the sandstones during late diagenesis.

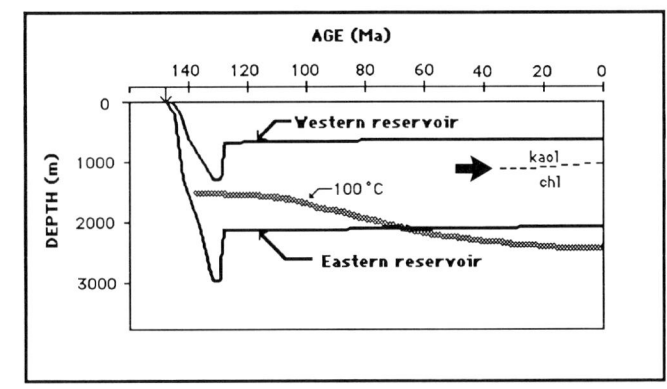

FIG. 16.—Burial history of Sergi reservoirs in western and eastern parts of the Recôncavo Basin. Inferred position of the 100°C isotherm also is indicated. Dashed line on the right (large arrow) indicates the present position of the kaolinite-chlorite (kaol-chl) boundary in the basin.

Based on the distinct burial/temperature histories, we suggest that the diagenetic evolution of the Sergi sandstones was not characterized by the transformation of kaolinite into chlorite with progressive depth, but rather by the formation of distinct late diagenetic clays, depending on the thermal exposure of each portion of the basin. The maximum depths achieved by the rocks is the present depth plus the inferred uplift: a total of 1,500 m. Presently, the temperature in the basin at 1,500 m is 70°C, which is not likely to have been the temperature during the rift phase, when the thermal gradient should be expected to have been higher. Evaluation of thermal-maturation data from Daniel and others (1989), using a modified one-dimensional rift-basin model (McKenzie, 1978), suggests that paleotemperature at 1,500 m during the rift phase was approximately 100°C (Fig. 16). The actual thermal boundary between the two diagenetic phases was between these two values, perhaps around 90°C. The differential distribution of the authigenic clays in the Sergi Formation illustrates how differences in the burial history can have a significant effect on the reservoir quality of otherwise similar sandstones.

CONCLUSIONS

The use of petrographic criteria permitted the distinction and quantification of the different types of interstitial clays found in the Sergi Formation sandstones. Each type of clay, including depositional clay (compaction matrix), mechanically infiltrated (MI) clay, and authigenic kaolinite and chlorite, has a different significance for the reservoir properties of the sandstones. MI clays and chlorite are the most harmful for reservoir quality. MI clays cause permeability reduction and produce pore systems characterized by low microscopic recovery of the non-wetting phase. Chlorite typically causes significant permeability reduction and reservoir evaluation problems related to the low electrical resistivity of chlorite-rich zones.

The distribution of compaction matrix and of MI clays was essentially controlled by depositional processes, whereas distribution of authigenic clays is mainly a function of differences in the burial/temperature histories of the rocks. The correct identification of clay types is essential for precise determination of origin, diagenetic transformations and reservoir characteristics of the sandstones.

ACKNOWLEDGMENTS

We thank Carlos H. L. Bruhn, Jose M. Caixeta, Alberto S. Barroso, Gerson J. S. Terra and Antonio S. T. Netto from Petrobras for discussions concerning the petrology of Sergi Formation sandstones. The manuscript was greatly improved by comments and suggestions by Thomas L. Dunn, Shirley P. Dutton and an anonymous reviewer. We thank Petrobras for permission to publish this paper.

REFERENCES

ANDREIS, R. R., 1981, Identificación e importancia geológica de los paleosuelos: Porto Alegre, Universidade Federal do Rio Grande do Sul, 67 p.

BENSON, D. J., 1981, Porosity reduction through ductile grain deformation: an experimental assessment: Transactions, Gulf Coast Association of Geological Societies, v. 31, p. 235-237.

BOLES, J. R., AND FRANKS, S. G., 1979, Clay diagenesis in Wilcox sandstones of southwest Texas: implications of smectite diagenesis on sandstone cementation: Journal of Sedimentary Petrology, v. 49, p. 55-70.

BRUHN, C. H. L., AND DE ROS, L. F., 1987, Formação Sergi: evolução de conceitos e tendências na geologia de reservatórios: Boletim de Geociencias da Petrobras, v. 1, p. 25-40.

COLLISON, J. D., 1986, Alluvial sediments, in Reading, H. G., ed., Sedimentary Environments and Facies: Blackwell Scientific Publications, p. 20-62.

CRONE, A. J., 1975, Laboratory and field studies of mechanically infiltrated matrix clay in arid fluvial sediments: Unpublished Ph.D. Dissertation, University of Colorado, Boulder, 162 p.

DANIEL, L. M. F., SOUZA, E. M., AND MATO, L. F., 1989, Geoquímica e modelos de migração de hidrocarbonetos no Campo de Rio do Bu-integração com o compartimento nordeste da Bacia do Recôncavo, Bahia: Boletim de Geociencias da Petrobras, v. 3, p. 159-169.

ESLINGER, E., AND PEVEAR, D., 1988, Clay minerals for petroleum geologists and engineers: Society of Economic Paleontologists and Mineralogists Short Course Notes No. 22, 413 p.

HEALD, M. T., AND LARESE, R. E., 1974, Influence of coatings on quartz cementation: Journal of Sedimentary Petrology, v. 44, p. 1269-1274.

HOWER, J., 1981, Shale diagenesis, clays and the resource geologist, in Longstaffe, F. J., ed., Short Course Handbook 7: Mineralogical Association of Canada, p. 60-80.

JENNINGS, J. B., 1987, Capillary pressure techniques: application to exploration and development geology: American Association of Petroleum Geologists Bulletin, v. 71, p. 1196-1209.

KELLER, W. D., 1970, Environmental aspects of clay minerals: Journal of Sedimentary Petrology, v. 40, p. 788-813.

KESSLER II, L. G., 1978, Diagenetic sequence in ancient sandstones deposited under desert climatic conditions: Journal of the Geological Society of London, v. 135, p. 41-49.

KORTEKAAS, T. F. M., 1985, Water/oil displacement characteristics in cross-bedded reservoir zones: Transactions, Society of Petroleum Engineers, v. 279, p. 917-926.

MCKENZIE, D. P., 1978, Some remarks on the development of sedimentary basins: Earth and Planetary Science Letters, v. 40, p. 25-32.

MATLACK, K. S., HOUSEKNECHT, D. W., AND APPLIN, K. R., 1989, Emplacement of clay into sand by infiltration: Journal of Sedimentary Petrology, v. 59, p. 77-87.

MILANI, E. D., AND DAVISON, I., 1988, Basement control and tectonic transfer in the Recôncavo-Tucano-Jatobá Rift, Brazil: Tectonophysics, v. 154, p. 41-70.

MORAES, M. A. S., AND DE ROS, L. F., 1990, Infiltrated clays in fluvial Jurassic sandstones of Recôncavo Basin, northeastern Brazil: Journal of Sedimentary Petrology, v. 60, p. 809-819.

RITTENHOUSE, G., 1971, Mechanical compaction of sands containing different percentages of ductile grains: a theoretical approach: American Association of Petroleum Geologists Bulletin, v. 55, p. 92-96.

VELDE, B., 1985, Clay minerals: a physico-chemical explanation of their occurrence: Developments in Sedimentology, v. 40: Amsterdam, Elsevier, 427 p.

WALKER, T. R., 1976, Diagenetic origin of continental red beds, in Falke, H., ed., The Continental Permian in Central, West and South Europe: Dordrecht, D. Reidel, p. 240-482.

WARDLAW, N. C., 1980, The effects of pore structure on displacement efficiency in reservoir rocks and in glass micromodels: Society of Petroleum Engineers of American Institute of Mining Engineers, Preprint SPE 8843, p. 345-352.

WARDLAW, N. C., AND CASSAN, J. P., 1979, Oil recovery efficiency and the rock-pore properties of some sandstone reservoirs: Bulletin of Canadian Petroleum Geology, v. 27. p. 117-138.

WEBER, K. S., 1982, Influence of common sedimentary structures on fluid flow in reservoir models: Journal of Petroleum Technology, March, p. 665-672.

WILSON, M. D., AND PITTMAN, E. D., 1977, Authigenic clays in sandstones: recognition and influence on reservoir properties and paleoenvironmental analysis: Journal of Sedimentary Petrology, v. 47, p. 3-31.

INHERITED GRAIN-RIMMING CLAYS IN SANDSTONES FROM EOLIAN AND SHELF ENVIRONMENTS: THEIR ORIGIN AND CONTROL ON RESERVOIR PROPERTIES

MICHAEL D. WILSON
1674 Tamarac Dr., Golden, Colorado 80401

ABSTRACT: Inherited grain-rimming clays are common in sandstones from eolian dune and marine-shelf environments. Inherited clay rims are defined as clay coats that form on framework grains prior to their deposition. Development of such clays requires that these clays become attached to framework grains at some other location and that they then be recycled with the host grain to form the present deposit. Such clays have been recognized in a number of major eolian-reservoir units in North America and the North Sea. Inherited clay rims also are extensive in major shelf-sandstone reservoirs. The relatively high levels of reservoir quality of many of these dune and shelf sandstones are at least partially attributable to the presence of inherited clay rims.

The distribution and composition of inherited clay rims allow them to be distinguished from other types of detrital clays and from neoformed clays. Characteristics most useful in identifying inherited clay rims include: (1) presence at points of contact between framework grains; (2) widely varying rim thickness; (3) increased thickness in embayments; (4) absence on the surfaces of diagenetic components; (5) clay mineralogy similar to that of clay interbeds; (6) enhanced development in finer grained sandstones where clay-filled depressions are more abundant; and (7) tendency to be developed and preserved only in selected environments. Recognition can be difficult where significant regeneration has occurred, where rims are extremely thin, or where rims are masked by detrital matrix or by authigenic cements.

In eolian settings, clay rims initially form in sabkha and interdune environments. The coatings are interpreted to form by infiltration of clay-charged waters or by adhesion to wetted sand-grain surfaces in eolian-soil and sabkha environments. Mild reworking allows for preservation of complete clay rims, whereas extended transport results in the removal of rims from projections on grain surfaces. Inherited clay rims are absent in water-washed environments or in coastal dunes derived directly from beach sands.

In shelf environments, inherited clay rims are generated when sand grains passing through the digestive tracts of organisms, or disturbed by burrowing, become coated with clay. Later reworking may reduce the thickness of clay rims or remove them altogether. Clay rims are destroyed in high-energy nearshore environments.

Because the presence of inherited clay rims can severely limit quartz-overgrowth development, high levels of intergranular porosity may be preserved to depths where associated clay-free sandstones exhibit extremely low porosities. Inherited clay rims are not entirely beneficial. The presence of clays at grain contacts may significantly accelerate the process of pressure-solution suturing. Commonly, clays in the rims are regenerated at depth to forms having a much higher surface area and containing a large amount of microporosity.

INTRODUCTION

Inherited clay rims are defined as clay coatings that form on framework grains prior to their deposition. The surfaces of framework grains may be partially or completely covered by such rims. Although inherited clay rims on framework grains are surprisingly common in Holocene and ancient dune and shelf sandstones, they have largely gone unrecognized. The lack of recognition is primarily attributable to the fact that these clays generally occur as very thin coats and/or are covered and masked by iron oxides/hydroxides, matrix clays, or various authigenic cements.

It is the author's contention that inherited clay rims are common in a large number of, if not most, eolian sandstones in North America and in the eolian Rotliegendes sandstones of the North Sea. It is the presence of these clay coats that governs preservation of porosity in most eolian sandstones buried at moderate to great depths (3,048 to 7,620 m; 10,000 to 25,000 ft). Many of the world's largest eolian reservoirs, such as Groningen and Leman Bank (Rotliegend), Anschutz Ranch (Nugget Sandstone), Reno (Minnelusa Formation), Rangely (Weber Sandstone), and Mary Ann (Norphlet Sandstone), fit this description. These reservoirs probably owe much of their porosity to the development of inherited clay rims. Similarly, porosities of shelf-sandstone reservoirs in the Prudhoe Bay Field (Sag River Sandstone) and Kuparuk River Field (Kuparuk Formation) on the North Slope of Alaska are, in part, controlled by the occurrence of inherited clay rims. These North Slope reservoirs, which contain 4 to 5 billion barrels of oil plus oil-equivalent gas in place (Barnes, 1987; Gaynor and Scheihing, 1988), may constitute some of the world's largest shelf-sandstone reservoirs.

It is the purpose of this paper to review the distribution of these inherited clay rims, the criteria useful in their recognition, the processes by which they may form, and their influence on porosity and permeability.

PREVIOUS WORK

Inherited clay rims were not recognized as one of the known modes of occurrence of dispersed clays by Neasham (1977), or of allogenic (detrital) clays by Wilson and Pittman (1977) and Almon (1981). Recent publications dealing with Holocene and ancient eolian and shelf sandstones fail to mention the presence of inherited clay rims. Holocene dune sands have long been noted for their very low content of, or lack of, fine silt and clay (Bagnold, 1941; Kuenen, 1960; Shepard and Young, 1961; Glennie, 1970; Ahlbrandt, 1979; McKee, 1979). These conclusions are based on sieve or settling-tube analysis. Many interdune deposits, on the other hand, have been shown to contain significant amounts of clays (Ahlbrandt, 1979).

Examination of detrital-grain surfaces of Holocene desert sand using the scanning electron microscope (SEM) or transmission electron microscope (TEM) reveals the presence of scaly and flaky features (Margolis and Krinsley, 1971; Krinsley and others, 1976; Folk, 1978). In the two former articles, the authors interpreted these features to be cleavage scarps produced by grain impingements. Folk concluded the features are silica or limonitic coatings. No data bearing on the composition of these flaky projections are included in any of these papers.

Iron-stained clay rims were identified by Folk (1976, 1978) in his studies of desert-sand reddening and rounding in the Simpson Desert of central Australia. Following the terminology of soil scientists (Brewer, 1964), Folk referred to these clay rims as cutans and considered them to be detrital in origin. Walker (1979), in his study of dune-sand reddening in Holocene eolian deposits of northern Libya, concluded that inherited clay rims, which he also referred to as cutans, are common in eolian sands of this area. Walker also presented a number of SEM micrographs of red-colored clay coats along with energy-dispersive X-ray traces to document the morphology and distribution of these coats. The coats are best developed in depressions on sand-grain surfaces.

Very few studies of ancient eolian deposits have reported the presence of inherited clay rims. Schenk (1990) noted the presence of early clay-grain coatings in the Minnelusa Formation of the Powder River Basin, but did not describe them, stating that they play no role in porosity evolution. These early clay coatings were interpreted by Schenk (pers. commun., 1990) to have an infiltration origin. In his discussion of a series of North American and North Sea eolian examples, Krystinik (1990) consistently noted the presence of authigenic illite but did not mention inherited clay rims.

Dixon and others (1989) noted the development of early diagenetic iron oxide-stained clay rims on framework grains of the Norphlet Formation in southern Alabama. The clay rims were considered to have been generated by alteration of labile framework grains and to have been altered to chlorite during a later stage of diagenesis. Dixon and others attributed the abnormally high porosities in the deep (greater than 6,100 m; 20,000 ft) Norphlet of this area to inhibition of quartz-overgrowth development by these chloritized clay rims. No inherited clay rims were reported by these authors.

Stalder (1973) was the first of many authors to report the occurrence of a delicate, fibrous illite in the Rotliegend of the southern North Sea. Because of its highly intricate morphology, Stalder concluded that this illite is authigenic in origin. Seemann (1979) stated that pore-lining illite develops as a "more or less continuous layer" on detrital grains in the Rotliegend of the Dutch Sector of the North Sea. Seemann interpreted these clays to have a diagenetic origin.

Glennie and others (1978) stated that dune sands in the Rotliegend of the Leman Bank and Sole Pit areas of the southern North Sea are virtually free of detrital clay particles at the time of deposition. They also reported that clay coats, commonly with iron-oxide staining, are now present on grain surfaces. These authors interpreted the clay coats, because of their absence at points of grain contact, to have an authigenic origin and suggested that they were generated by breakdown of chemically unstable grains. The authigenic-clay coats were considered to have had a mixed-layer smectite-illite composition, which altered to a highly fibrous illite with increased thermal exposure.

Goodchild and Whitaker (1986) noted that detrital clays are generally absent in dune sands but may be common in finer grained eolian-sheet sandstones. These authors also interpreted clay rims in the Rotliegend to be authigenic in origin and to have been generated by dissolution of labile grains. Goodchild and Whitaker reported the occurrence of these authigenic clays as thin squashed films between grain contacts.

This author (Wilson, 1981) first mentioned the occurrence of inherited clay rims in ancient sandstones deposited in eolian and outer shoreface to shelf environments, but did not provide a definition for these clays or discuss a mechanism for their formation. The authigenic origin of the clay rims in the Rotliegendes dune sands was subsequently challenged (Wilson, 1982), and a detrital origin was claimed for these rims. It was also noted that, subsequent to burial, clay rims in the Rotliegend are regenerated and assume an elongate fiber to fibrous mat-type morphology. Regeneration is assumed to consist entirely of recrystallization involving potassium uptake without development of neoformed crystals or crystalline extensions.

To the author's knowledge, no workers have previously described inherited clay rims from ancient marine-shelf deposits. The term shelf as used in this paper includes both shelf and shelf-shoreface transition zone environments. The boundary between the shoreface and transition zone is taken to be normal wave base (also referred to by many workers as fairweather wave base). In articles dealing specifically with sedimentation and diagenesis of shelf sandstones, authors commonly refer to the presence of matrix clays but never to inherited clay rims. Some examples of such articles include those by Tillman and Almon (1979), Gautier (1983), Rice and Gautier (1983), and Turner and Conger (1984). On the North Slope of Alaska, an area reviewed in this paper, Barnes (1987) and Masterson and Paris (1987) did not report the presence of inherited clay rims in Jurassic (Sag River Sandstone) or Cretaceous (Kuparuk Formation) shelf-reservoir sandstones. Authigenic mixed-layer smectite-illite pore linings and pore fillings were reported to occur in the Sag River Sandstone of the Prudhoe Bay Field by Barnes (1987) and in the Kuparuk Formation of the Kuparuk Field by Eggert (1987).

Glauconitic rinds have been observed on quartz and other framework grains (Wermund, 1961; Bentor and Kastner, 1965; Triplehorn, 1966). Triplehorn (1966) noted that he and other authors have observed that glauconite selectively coats heavy minerals, whereas quartz grains are seldom coated. The coats may be very thin, or they may be so extensive that the enclosed framework grains appear to be inclusions (Triplehorn, 1966). Wermund (1961) interpreted these coats to represent replacements of the underlying silicate material.

Rude and Aller (1989) reported the occurrence of lateritic particle coatings on framework grains in Holocene continental-shelf sediments derived from the Amazon Basin river system. These coatings consist of iron hydroxide-stained clays, which tend to be concentrated in embayments and cracks in sand-grain surfaces. These authors also noted that an abundance of coated grains occurs throughout the river system but did not present data to document this statement.

TERMINOLOGY

Both Folk (1976, 1978) and Walker (1979) commonly referred to detrital clays present on framework-grain surfaces in eolian deposits as coatings and cutans. Brewer (1964) established the term cutan for concentrations or alignments of constituents along natural surfaces in soils. He stressed the fact that, by definition, cutans are pedological features and that cutans can consist of any component or components present in soil material. Cutans in soils not only include coatings on free or embedded grain surfaces but also may occur along irregular channels, planar surfaces, and walls of voids. Cutans that consist primarily of clay minerals are referred to by Brewer as argillans. Other soil scientists have referred to clay coatings on soil particles as clay skins (Dregne, 1976). Because the ultimate origin of inherited clay rims in a given sandstone is essentially impossible to determine, the term cutan, as defined by Brewer (1964), cannot be used to describe the types of clay coats discussed in this paper.

Up to this point, the terms rim, rind, coat, and coating have been used in a general sense to refer to coverings of detrital clay on framework grains. These same terms have also been used as generic terms to describe some types of authigenic clays in sandstones (Neasham, 1977; Wilson and Pittman, 1977). To avoid confusion with clay coatings having other origins, the writer here proposes the term "inherited clay rims" be used to describe coats of detrital clay on framework grains where these clays were present on the grains prior to their arrival at the site of deposition.

CHARACTERISTICS OF INHERITED CLAY RIMS

Thickness, Color, and Morphology

The thicknesses of inherited clay rims typically range from a fraction of a micrometer to 5 μm. In depressions, thicknesses as much as 10 to 15 μm have been observed. In transmitted light, inherited clay rims in both modern and ancient sandstones range in color from almost colorless to pale brown, reddish brown, olive brown, pale green, or grass green. The brown to red-brown colors are commonly produced by impurities such as pyrite, organic matter, and iron oxides/hydroxides. Grass-green rims are composed of glauconite. Other materials, which form partial to complete rims, and in some cases ooid coatings, on framework grains are phosphatic material (Notholt, 1980) and chamosite (Van Houten and Purucker, 1984). Phosphate and chamosite (a 14 Angstrom trioctahedral chlorite) rims also are beige, brown, or olive green in color and may be easily mistaken for inherited clay rims.

SEM micrographs of inherited clay rims on sand grains from Holocene dunes indicate that these rims consist of overlapping aggregates of flaky material having ragged outlines (Figs. 1B, 1D, and 1F). Rims on grains from the Simpson Desert of Australia are very well developed and relatively complete (Figs. 1A and 1C), whereas those on grains from the Rub'al Khl area of southern Saudi Arabia are present primarily in depressions on the grain surfaces (Fig. 1E). The clay rims on the Simpson Desert sand grains give these grains a waxy luster when examined with a binocular microscope.

Rims studied by this author are interpreted as clays based on microprobe elemental analysis (Figs. 1G and 1H) and on thin-section optical properties. The samples from the Simpson Desert and the Rub'al Khl are both stained by iron oxide/hydroxides. The iron present in the microprobe analyses for these samples may occur primarily as very finely granular materials coating the surfaces of the clay platelets. Clay-rim morphologies very similar to those shown in Figure 1 were documented for clay rims on Holocene eolian sands from northern Libya by Walker (1979). Folk (1978) interpreted rims having similar morphologies to those shown here as being "turtle-skin" silica coats but did not discuss the means by which the composition of these rims was determined.

Abundance

The abundances of inherited clay rims present in sandstones are difficult to estimate. Where they are very thin (less than 1 μm) or where they are heavily stained by iron oxides/hydroxides, they may be totally overlooked. Because of the white to yellow birefringence associated with smectitic or illitic clays, most clay rims tend to appear much brighter than surrounding materials and tend to be overestimated during point-count analysis. Many inherited clay rims contain significant microporosity that cannot be differentiated from the encasing clays and, therefore, is lumped with the clay materials during point-count analysis. Effective separation of inherited clay rims from the framework grains on which they occur, particularly where they are partially covered by quartz overgrowths, is very difficult and probably leads to their underestimation by X-ray diffraction analysis. In most samples, inherited clay-rim abundance probably ranges from a trace to 2%, but some samples may contain as much as 3 to 6%.

CRITERIA FOR RECOGNITION OF INHERITED CLAY RIMS

A series of criteria useful in establishing the presence of inherited clay rims follows. These criteria are based on the examination of several hundred thin sections in eolian and shelf sandstones. None of the criteria is individually definitive; thus, it is important to use a number of lines of evidence to come to a final conclusion. A chart listing the various criteria useful in differentiating the various types of detrital clays from one another and from neoformed authigenic clays and drilling mud is presented in Table 1. Because data for these clay types are lacking or very limited, the characteristics of adhesion clays and *in situ* clast-disintegration clays are primarily guesswork. Criteria for differentiation of authigenic clays from detrital clays as a group are summarized in Wilson and Pittman (1977).

Clay Distribution

Occurrence at contacts.—

Inherited clay rims occur at grain contacts (Figs. 2, 3A, 3B, and 3C). Where inherited clay rims are relatively complete, they will consistently be present at grain contacts. However, because in any given sample at least some are incomplete, they will also be absent along some contacts. Where clay rims are poorly developed, they will be missing at most grain contacts. Under conditions of deep burial, grain rotation and fracturing may cause some authigenic clays to shift to positions at grain contacts. Great caution must be exercised in using this criterion alone in such situations.

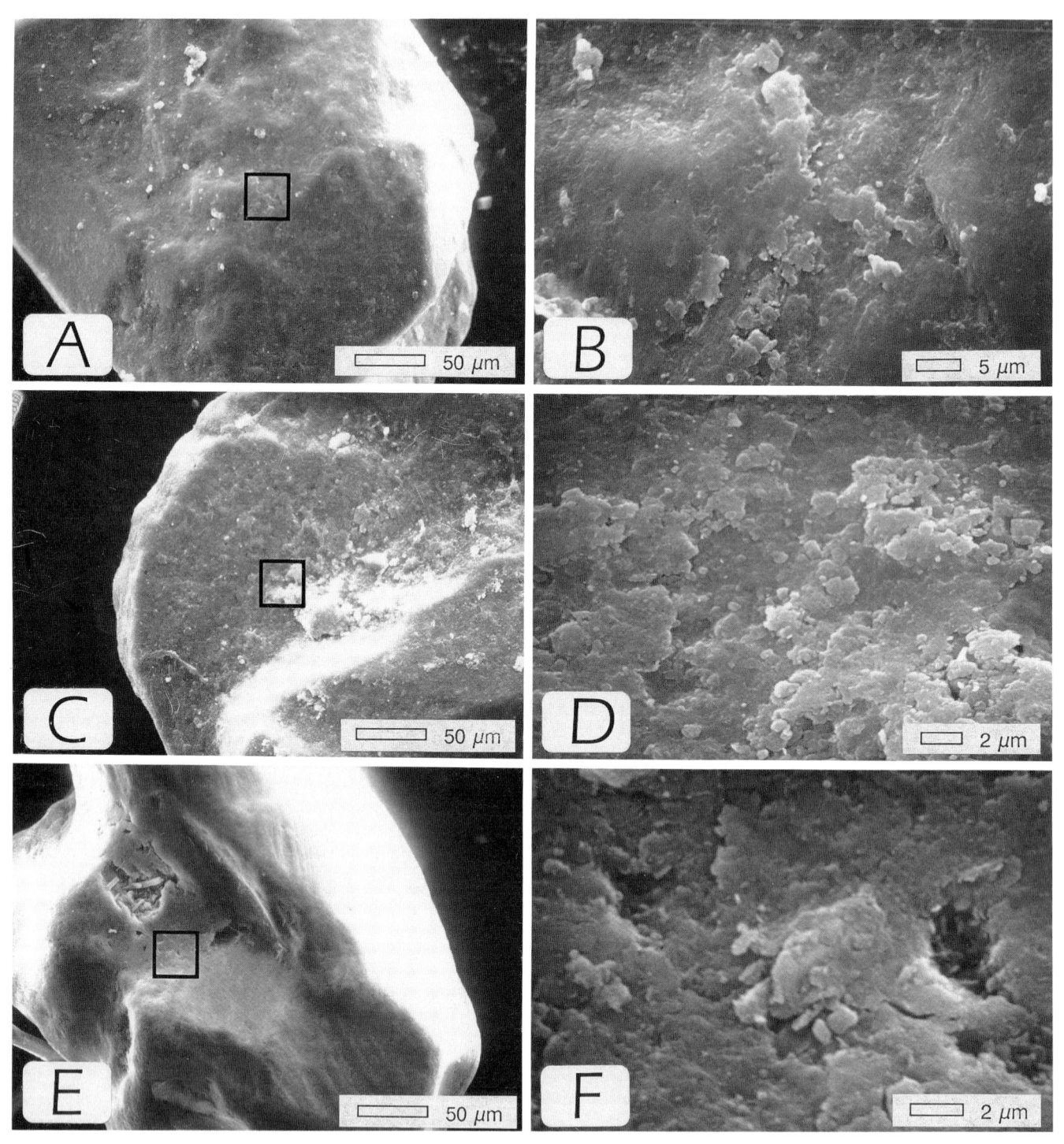

FIG. 1.—Illustration shows morphologies of inherited clay rims on Holocene dune-sand grains. (A) Low-magnification view of a sand grain from the Simpson Desert of Australia shows surface texture typical of a grain that is heavily coated by an iron-oxide/hydroxide-stained inherited clay rim. (B) High-magnification view of inset in previous micrograph. The entire field of view is characterized by overlapping flaky particles having ragged outlines. Particles are interpreted to be primarily clays based on microprobe analysis. (C) A second sand grain from the Simpson Desert has a well-developed clay rim in the depression at the center of this view. (D) Closeup of inset in previous micrograph showing well-developed clay rim. Rim exhibits overlapping flaky aggregates having ragged outlines. (E) Sand grain from a dune in the Rub'al Khl district of Saudi Arabia appears much smoother than those from the Simpson Desert. (F) Closeup of inset in depression in previous micrograph indicates localized presence of an inherited clay rim. Rim consists of overlapping irregular flaky material interpreted to be comprised primarily of clays based on microprobe analysis.

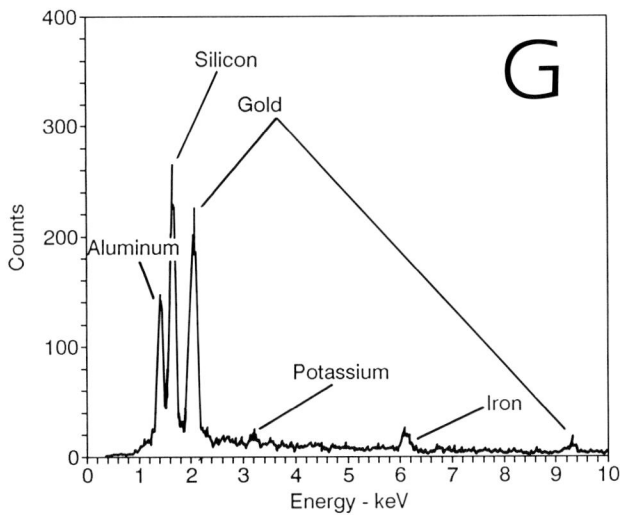

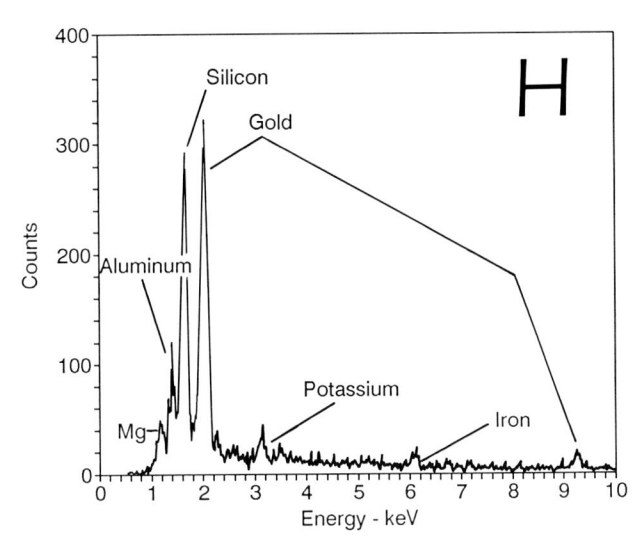

FIG. 1. (Cont.)—(G) Energy-dispersive X-ray (EDX) analysis for the clay coat shown in (B). Analysis was run on the entire area shown in the micrograph. A spectrum for the area shown in (D) is virtually identical to this spectrum. (H) EDX analysis for the entire area shown in (F). The iron in both this and the previous spectrum may occur primarily as very finely dispersed oxide/hydroxides not visible in the micrograph.

It should also be noted that pressure-solution suturing is commonly accelerated by the presence of clays at points of grain contact (Heald, 1956; Thomson, 1959; Weyl, 1959). Intense compression and shearing of clays can occur at such points, leaving only an extremely thin film. Such films may be difficult to recognize, even by means of scanning electron microscopy. In many cases, however, careful SEM analysis reveals the presence of clays in "bald" areas representing points of grain contact (Figs. 3D and 3E). Tada and Siever (1989) contend that there is no consistent relation between clay content and intergranular pressure-solution intensity.

Disturbance due to bioturbation in marine environments and root disturbance in nonmarine environments may allow clays introduced by these mechanisms to completely surround some framework grains. In general, bioturbation can

TABLE 1.—CRITERIA FOR IDENTIFICATION OF CLAY TYPE

	DETRITAL					AUTHIGENIC	
Criteria	Inherited Clay Rims	Infiltration Clays	Bioturbation Clays	Adhesion Clays	In Situ Clast Disintegration Clays	Neoformed Authigenic Clays	Drilling Mud Invasion
DISTRIBUTION							
Present at Grain Contacts	VC	R?	C	C	X	R	X
Wide Variations in Thickness	VC	VC	VC	VC?	VC?	R	VC
Thicker in Depressions in Grain Surfaces	VC	X	X	X	X?	X	X
More Extensive Development in Finer Grained Lamina or Beds	VC	R?	C	O?	X?	X	X
Meniscus Bridges	X	VC	X	X?	C?	O	VC
Geopetal Fabric	X	C	X	O?	C?	X	X
Internal layering	X	VC	X	X	X?	O	X
Cost Diagenetic Components	X	R?	X	X	X?	C	VC
Invades Late-Stage Dissolution Pores	X	R?	X	X	X	C	VC
Clay-Supported Fabric	X	O	C	O?	X	X	X
Clays Span Multiple Grains	X	O	C	VC	X?	X	X
Concentration Adjacent to Root or Bioturbation Channels	X	O	C	X	X	X	X
Abundance Related to Depositional Environment	C	C	C	VC	O?	R	O
Abundance Varies Rapidly between Lamina or Beds	VC	VC	C	VC	O?	O	C
COMPOSITION							
Clay Particle Size Varies Widely and Outlines Irregular	C	C	C	C	C	R	C
Mixed-Clay Mineral Assemblage	C	C	C	C	C	R	R
Impurities Present	C	C	C	C	C	R	VC
Well-Developed Crystal Outlines/Delicate Morphology							
Unregenerated	X	X	X	X	X	VC	X
Regenerated	C	O	O	O?	O?	VC	X

VC - Very Common, C - Common, O - Occassionally, R - Rare, X - Absent

INHERITED CLAY RIMS

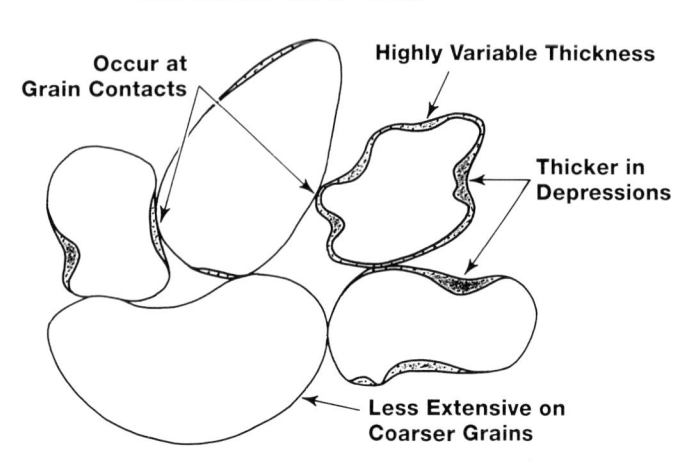

FIG. 2.—Diagram illustrating some of the spatial criteria useful in identifying inherited clay rims. Inherited clay rims commonly occur at points of framework grain contact. However, because they do not entirely coat some grains, they are absent from contacts where gaps in the rim occur. Inherited clay rims exhibit wide variations in rim thickness and are consistently thicker in depressions in framework-grain surfaces. Coarser grains are better rounded than finer grains and therefore have fewer depressions on their surfaces. As a result, inherited clay rims generally are less extensive on coarser grains.

be recognized and discounted at the megascopic level. Root disturbance may, in some cases, be similarly identified. However, because of their irregular diameter and because remnant organic material may be absent, some root-disturbance features may be very difficult to recognize. Pedogenic features useful in establishing an infiltration origin for clays in nonmarine sandstones have been discussed by Molenaar (1986) and are addressed in a later section of this paper.

Thickness variations.—

Inherited clay rims exhibit wide variations in thickness (Figs. 2, 3A, and 3B). The thickness variations are the result of irregular development in soil horizons, sabkhas, or shelf regimes, and/or of partial removal by abrasion during transport. Most authigenic pore-lining clays tend to develop as isopachous coats. However, some pore-lining authigenic clays, particularly chlorite, also may exhibit significant thickness variations (Fig. 3F).

Thickness in depressions.—

Inherited clay rims are consistently thicker in depressions on grain surfaces (Figs. 2, 4A, and 4B). Clay rims are preferentially removed from projecting portions of framework-grain surfaces as a result of grain-to-grain impingements. In contrast, where grains are not highly rounded, clays are protected within indentations and are commonly preserved.

Grain-size controls.—

Inherited clay rims are developed more extensively in finer grained laminae. As noted by Walker (1979), the coarser fraction of eolian deposits is consistently more rounded than the associated finer grained fraction. Walker cites Kuenen (1960), whose experimental data indicate that eolian abrasion decreases with grain size and is insignificant at grain sizes less than 0.15 mm. Thus, clay-filled depressions will tend to be most common in finer grained sandstones (Fig. 2). This relation has been observed in a large number of eolian deposits. Although a positive relation between grain size and rounding also exists in shelf sandstones, the presence of much larger amounts of biogenically introduced clays in many finer grained shelf sandstones commonly makes recognition of inherited clay rims in indentations very difficult.

Meniscus bridges.—

Inherited clay rims do not exhibit meniscus bridges. Meniscus bridges are ridges in clay coats that join coats on adjacent grains and that develop in areas surrounding grain contacts (Fig. 5). They form in vadose environments due to the localization of water in menisci at grain contacts. Since inherited clay rims are formed prior to transport, they exhibit no preferred relation to grain contacts. Where inherited clay rims occur at grain contacts, they may be deformed as a result of compaction and display a slight bulge or even exfoliation in the areas adjacent to grain contacts. Meniscus bridges are commonly developed by infiltration clays (Crone, 1975; Matlack and others, 1989) and by some authigenic clays (Neasham, 1977; Wilson and Pittman, 1977).

Geopetal fabric.—

Inherited clay rims do not exhibit geopetal fabric. Geopetal fabric refers to an up-down orientation displayed by a geologic feature. Infiltration clays commonly are draped over the upper surfaces of framework grains (Fig. 5) and thus indicate an up orientation (Crone, 1975; Matlack and others, 1989). Inherited clay rims have not been observed with such a fabric. In the advanced stages of infiltration, it is likely that clays also occur on the undersides of framework grains or fill entire pores.

FIG. 3.—Spatial criteria indicating an inherited clay rim origin. (A) Inherited clay rims occur at points of grain contact (arrows). Rims have a dark appearance due to the presence of iron-oxide staining. Permian Rotliegend, southern North Sea. (B) Inherited clay rims composed of glauconitic clay are present at points of grain contact (arrows). Triassic Sag River Sandstone, ARCO #1 Ugnu, North Slope, Alaska. (C) Crossed-polarizers view displaying the presence of illitic clay (thin white rims) at points of grain contact (arrows). Jurassic Nugget Sandstone, Reserve #1–16A Mohawk State, Uinta County, Wyoming. (D) Scanning electron micrograph of a grain on which smooth areas represent contacts with adjacent grains. Examination of all four smooth areas indicated by numbers reveals that each is covered by a flat fibrous mat of illitic clays (see E). Permian Rotliegend, southern North Sea. (E) Closeup of the smooth area designated 1 in (D) exhibits a compacted mass of flat illitic fibers. Thus, contact areas, which appear to be devoid of clay upon preliminary examination, commonly are coated by clays. Similar mat-like illite also coats the remainder of this grain. Permian Rotliegend, southern North Sea. (F) Chlorite pore linings (black coatings on grains) vary widely in thickness and locally are missing entirely (arrows). Most authigenic-clay pore linings are essentially isopachous. Paleocene Ft. Union Formation, Piceance Basin, Colorado.

A | 0 0.1 mm

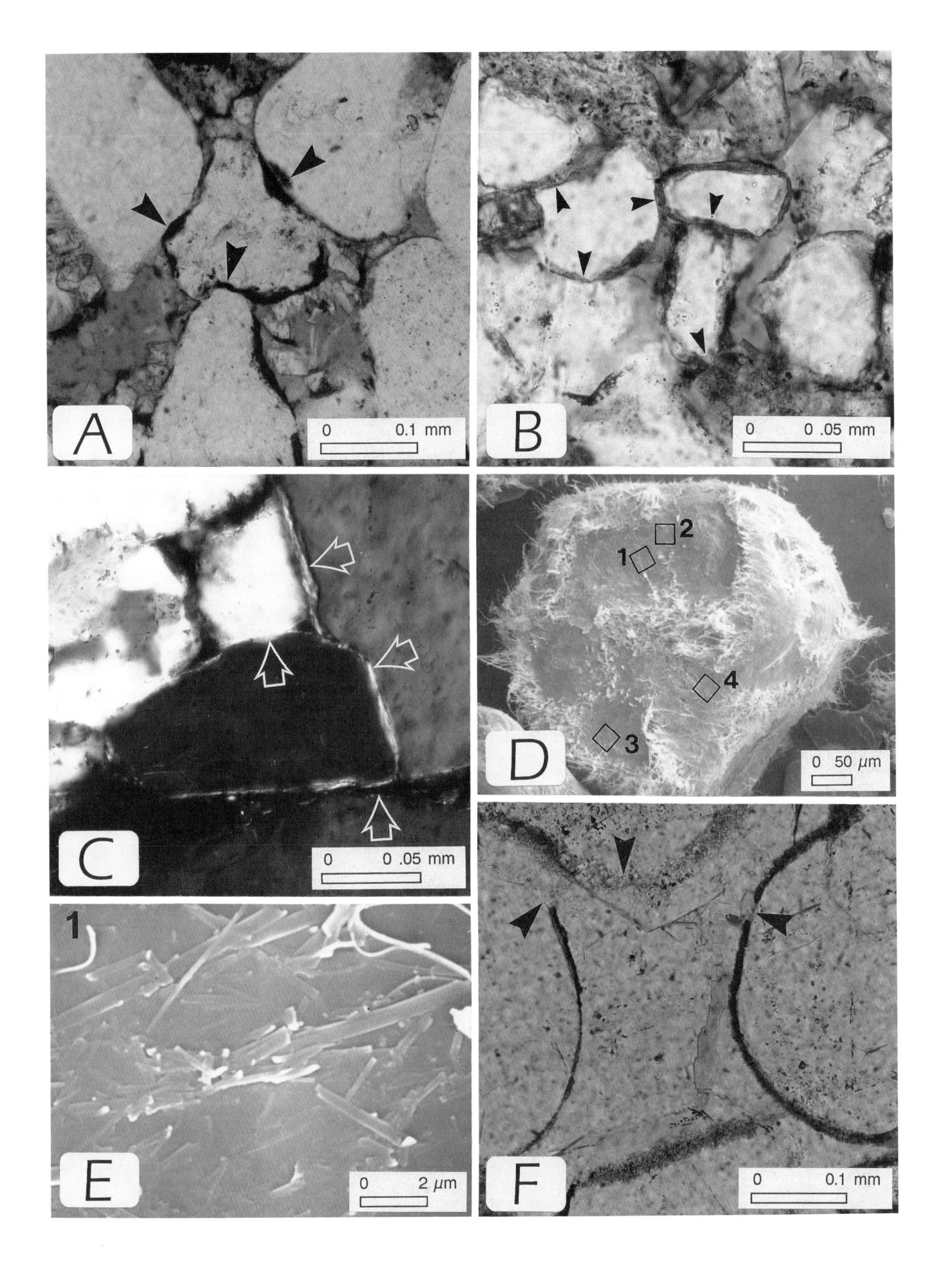

B | 0 0.05 mm

C | 0 0.05 mm

D | 0 50 μm

E | 0 2 μm

F | 0 0.1 mm

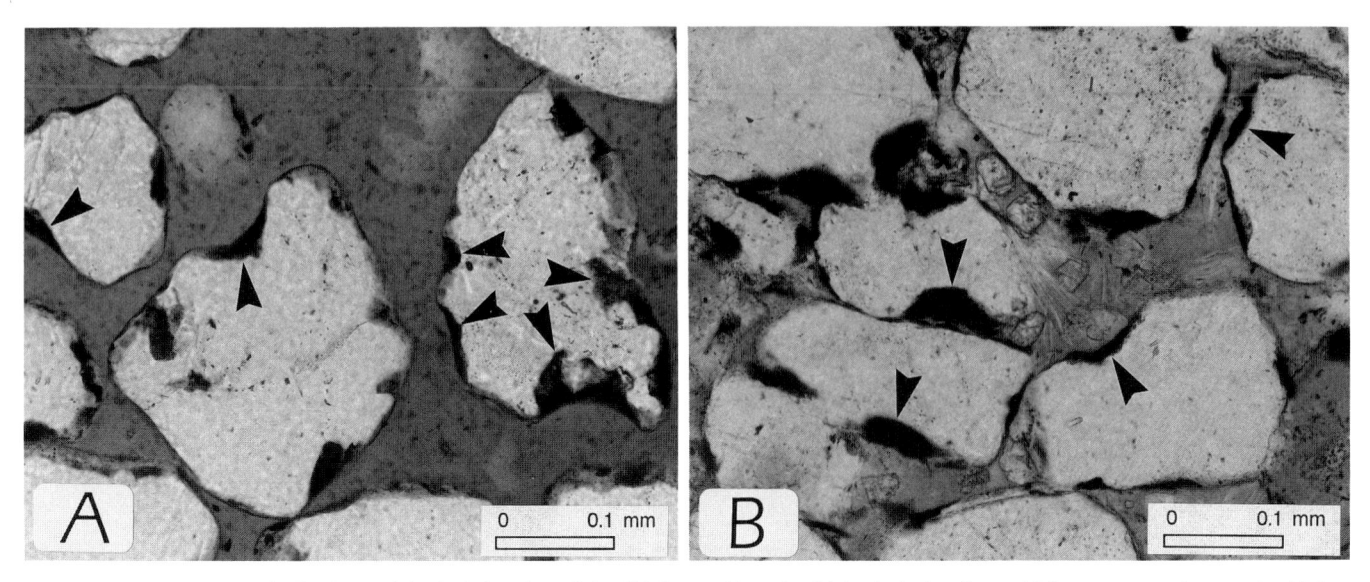

FIG. 4.—Spatial criteria indicating an inherited clay-rim origin. (A) Iron oxide-stained inherited clay rims exhibit a strong tendency to be thickest in depressions on the framework-grain surfaces (arrows). Loose sand from Holocene dune sand, Simpson Desert, central Australia. (B) Depressions on framework-grain surfaces tend to have the thickest developments of iron oxide-stained inherited clay rims (arrows). Note that not all of the thicker rims occur in depressions. Small dolomite rhombs occurring in the intergranular pores are not coated by inherited clay rims. Permian Rotliegend, southern North Sea.

Internal layering.—

Inherited clay rims do not exhibit internal layering. Infiltration clays do occasionally exhibit such layering (Fig. 5). Internal layering of infiltrated clays has been observed by this author in only a very few samples of eolian sandstone.

Relation to diagenetic components.—

Inherited clay rims do not coat diagenetic components and invade dissolution pores, even those of very early or syndepositional origin. For example, very early diagenetic dolomite is common in eolian deposits. Although framework grains adjacent to dolomite crystals exhibit well-developed inherited clay rims, the dolomite crystals are not coated by such clay (Fig. 4B). Clays that meet this criterion are not necessarily of detrital origin. They may be detrital, or they may have formed authigenically prior to any other diagenetic phase present.

Clay-supported fabric.—

Inherited clay rims do not exhibit a clay-supported fabric. Such fabrics are common in samples containing infiltration clays (Molenaar, 1986) and bioturbation clays.

Continuity.—

Inherited clay rims do not span multiple framework grains. This has been observed in argillaceous sandstones containing wavy, discontinuous, clay laminae interpreted to have been produced by adhesion in a wetted-sabkha environment. In sandstones that have been deeply buried, such features commonly represent zones of intense pressure-solution suturing.

Association with root or burrow structures.—

Inherited clay rims are not concentrated in the vicinity of root or burrow structures. Molenaar (1986) reported such concentrations for infiltration clays.

Relation to depositional environment.—

Inherited clay rims may be most extensive in, or restricted to, certain depositional environments. Inherited clay rims are rapidly eliminated in wetted environments in which grain collisions are common. Consequently, they are not present, or occur only very rarely, in fluvial and nearshore marine environments. In contrast, clay rims are common

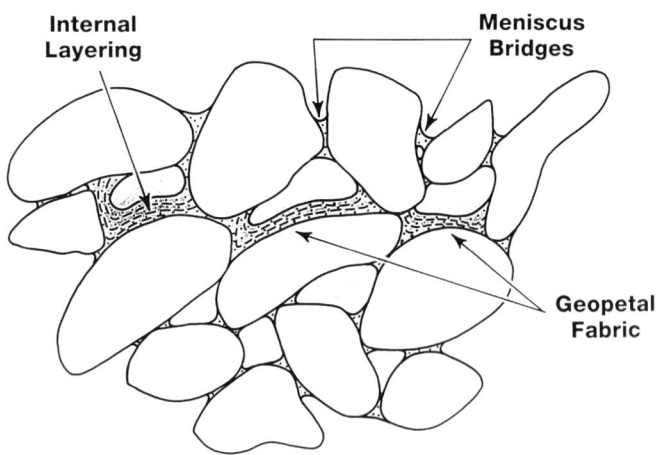

FIG. 5.—Diagram illustrating some of the common features of infiltration clays. Most commonly observed are meniscus bridges between grains at points of grain contact. Geopetal fabrics, in which clays occur primarily on the upper surfaces of grains, are observed occasionally. Internal layering appears to occur only rarely. The features illustrated are those described by Crone (1975) and Walker and others (1978).

in eolian environments, particularly dune and interdune/desert floor/sabkha (Table 1). Inherited clay rims may be absent from most coastal dunes. This absence is attributed to the derivation of these dunes from immediately adjacent foreshore sands. Clay coats cannot develop in foreshore deposits, and extensive reworking in this environment removes rims on grains introduced from environments where clay rims are formed. Walker (1979) reported a lack of clay rims on Libyan beach sands immediately adjacent to eolian deposits in which they are common.

In marine environments, inherited clay rims are best developed in shelf sands and in shelf-shoreface transition zones. Reworking in both the eolian and shelf environment can partially to completely remove clay rims. Introduction of sands from several different sources, which is quite common in some eolian regimes, may lead to the interbedding of sands that contain highly variable amounts of clay rims.

Variations in abundance.—

The abundance of inherited clay rims may vary abruptly between laminae or beds. This variation may be related to differences in grain size, provenance, or to distance of travel, all of which are controlled by shifts in wind direction and velocity. The reduced abundance of clay rims on better rounded coarser sands can lead to the interbedding of coarser sands with very limited development of rims and finer sands with very well-developed rims. Assuming an authigenic origin, it would be difficult to explain such rapid variations in clay-rim content. As noted previously, grains that are derived from coastal nearshore environments may not exhibit any inherited clay rims (Walker, 1979). Grains that experience a longer distance of travel may have most or all of their detrital rims removed by eolian abrasion. Iron-oxide coats, on the other hand, may become better developed as a result of extended transport and increased age (Walker, 1979).

Composition and Crystallinity

Particle size and crystallinity.—

Clay particles in inherited clay rims vary widely in size and have irregular outlines (Figs. 5A through 6A and B). When first deposited, all detrital clays have these characteristics (Wilson and Pittman, 1977). However, subsequent diagenesis may produce pervasive regeneration, which leads to clays having only a very limited size range and exhibiting well-developed crystalline outlines (Figs. 6C and D). The most common modifications are regeneration of detrital-clay suites to illite and/or chlorite.

Clay assemblage.—

The clay-mineral assemblage present in inherited clay rims commonly is a mix of clay types. It is similar or identical to that of associated detrital-clay laminae. The clays in many detrital-clay suites are poorly crystallized and produce relatively broad, ragged peaks on X-ray diffraction patterns (Wilson and Pittman, 1977). Most authigenic-clay suites contain only one or two relatively pure, well-crystallized components. This criterion is difficult or impossible to apply where a mix of authigenic and detrital clays occur, where low inherited clay-rim abundance prevents X-ray analysis, where a variety of detrital-clay components (matrix, clay pellets, clay rims) is present, or where regeneration or alteration of inherited clay rims homogenizes the clay assemblage. In the author's experience, regeneration of clays in many eolian sandstones is quite extensive, and clay mineral assemblages commonly consist only of highly illitic (Fig. 7) or mixed-layer illite-smectite, often in association with authigenic kaolinite. In marine environments, use of this criterion is commonly negated by the fact that inherited clay rims commonly occur in sandstones containing biogenically introduced clay matrix or are altered syndepositionally to glauconitic minerals (Figs. 6E and F).

Impurities.—

Inherited clay rims commonly contain impurities. Because they consist of detrital clays generated in low-energy depositional environments, they contain minor amounts of one or more accessory components, such as organic debris, iron oxides/hydroxides, pyrite, and very fine silt-size quartz, feldspars, or other silicates. Authigenic clays are generally free of such impurities. However, some impurities in inherited clay rims may be generated subsequent to burial, particularly pyrite and iron oxides/hydroxides.

DISTRIBUTION OF INHERITED CLAY RIMS

A large number of thin sections from ancient eolian deposits, as well as a number of samples from Holocene dune and interdune deposits, were examined to determine if inherited clay rims were present. Such rims are interpreted to be present in at least some samples from most eolian deposits (Table 2). Ancient dune-sand examples studied come primarily from North America, whereas Holocene samples represent eolian environments from a much wider geographic distribution. This author has not conducted a sufficiently extensive examination of a large number of marine-shelf and nearshore sandstones to be able to provide a detailed picture of the distribution of inherited clay rims in this regime. However, the author has encountered inherited clay rims in many samples of shelf sandstones from the North Slope of Alaska, and their occurrences are listed here to indicate their importance in such sandstones.

POSSIBLE INDICATORS OF INHERITED CLAY RIMS

Some features that may serve as indications that inherited clay rims are present are given here. However, these features can also be attributable to other mechanisms. These criteria are predicated on assumptions that may or may not be valid in a given area or unit. One such assumption is that quartz grains will develop quartz overgrowths at depths of burial greater than a few thousand feet unless overgrowth generation is prevented by development of detrital or early authigenic-clay coats or by extensive infill by other early diagenetic cements. Unless they are relatively thick, iron-oxide rims alone are generally insufficient to prevent overgrowth development (see Fig. 6–26, Pettijohn and others, 1972). A second assumption is that, under normal subsurface conditions, quartz grains will not exhibit pressure-so-

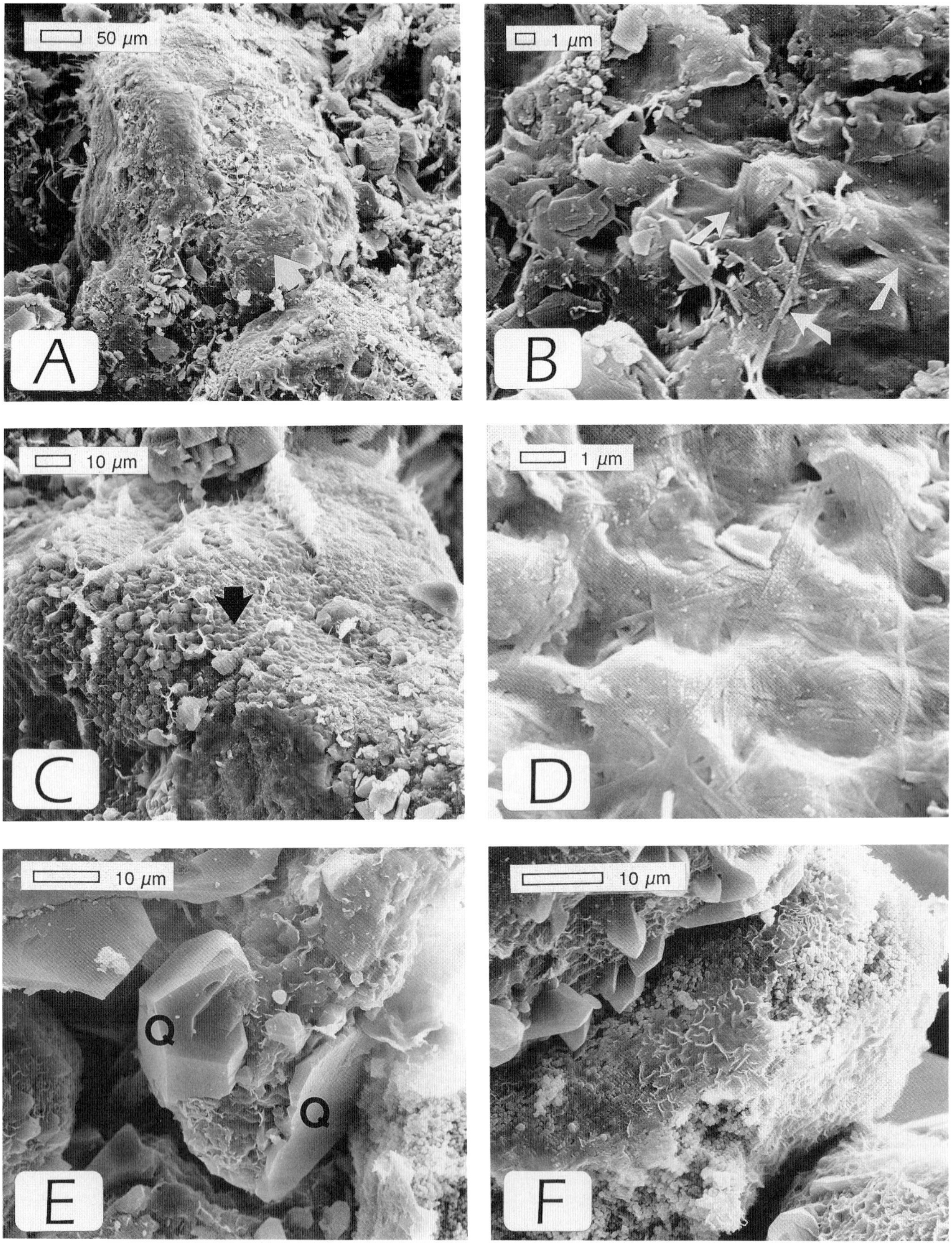
50 μm
A
1 μm
B
10 μm
C
1 μm
D
10 μm
Q
Q
E
10 μm
F

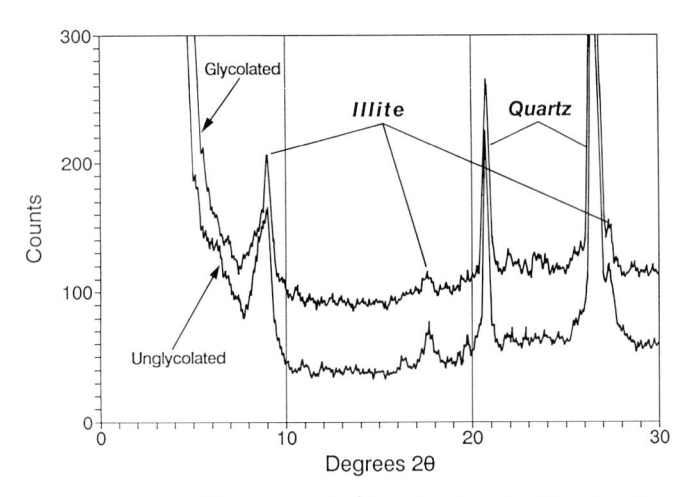

FIG. 7.—X-ray diffractogram of <5-μm fraction of a Minnelusa Formation sandstone containing inherited clay rims. Only clay present is highly illitic and contains a minor swelling component. SEM micrographs of this clay (Figs. 6C and D) suggest the clay has experienced significant regeneration. Quartz peaks represent quartz-framework grains crushed during sample preparation.

lution suturing unless a catalyst, particularly clays, occurs at points of contact. Chert grains commonly contain contaminants that can act as catalysts and will exhibit extensive suturing without the presence of a separate clay coat or clay lamina at grain contacts. Features that may serve as indications of inherited clay rims are:

(1) Older (pre-Tertiary) quartz-rich sandstones that have experienced significant burial (greater than 1,525 m; 5,000 ft) exhibit good visible porosity and very limited or no quartz-overgrowth development. No evidence for extensive removal of intergranular cements, particularly carbonates, exists. Limited quartz-overgrowth development in many eolian and shelf sandstones can commonly be attributed to the presence of inherited clay rims.
(2) Quartz grains in a sandstone buried more than several thousand feet display extensive suturing. Suturing is very commonly accelerated by the presence of detrital clays at grain contacts. Although authigenic clays have the capacity to accelerate suturing, they are seldom present at grain-to-grain contacts (Wilson and Pittman, 1977).
(3) Argillaceous marine quartzose sandstones exhibit significantly higher porosities than associated clay-free quartzose sandstones. In some instances, the clay-free sandstones may contain virtually no visible or measured porosity. Inherited clay rims present along quartz-grain surfaces prevent overgrowth development and preserve porosity. An excellent example of this relation was reported by Turner and Conger (1984), who stated that the bioturbated sandstones in the Woodbine Sandstone of the Kurten Field are porous, whereas the clay-free sandstones are nonporous.
(4) Deeply buried sandstones are relatively friable. The presence of clays at points of grain contact commonly leads to a low degree of cementation at shallow to moderate depth. Even at great depth, where these clays promote intense pressure solution, grains may still be rather weakly bonded to one another.

MODE OF FORMATION

Dispersed Detrital Clays

Introduction of dispersed detrital clays into sands can be accomplished in five very different manners: (1) biogenic reworking (plants or animals); (2) *in situ* disintegration of smectitic clay or mud clasts; (3) infiltration in surface waters; (4) adhesion on wetted grain surfaces; and (5) inheritance of previously formed clay rims.

The first three of these mechanisms may be considered as being very early diagenetic in that they constitute processes operating at or very near the depositional interface subsequent to initial deposition. These processes are described by some authors, following the terminology of Schmidt and McDonald (1979), as eogenetic. The first of these is well known and needs no discussion here. The second method was proposed by Wilson (1982), but no evidence to document this mechanism was presented. The third of these mechanisms has been investigated by a number of workers. Crone (1975), Walker (1976), and Walker and others (1978) have studied the infiltration of detrital clays into desert alluvium. Their results are based on both laboratory experiments and on studies of Holocene to Tertiary deposits from the southwestern United States and northwestern Mexico. They note that infiltration clays exhibit the following characteristics. (1) Infiltration clays form meniscus bridges at points of grain contact. (2) Clays present in coatings on framework-grain surfaces are of mixed sizes and have irregular outlines. Flakes are generally oriented parallel to grain surfaces. (3) During early stages of development, a geopetal fabric may be developed in which the clay coatings occur along the floors of intergranular pores but are absent on the undersides of framework grains.

FIG. 6.—Morphological features of inherited clay rims. (A) Low-magnification SEM micrograph of a framework grain covered by an inherited clay rim. Clays appear to be oriented parallel with the grain surface. Arrow indicates the location of (B). (B) High-magnification SEM micrograph of the inherited clay rim at arrow in (A). Note ragged appearance of clay particles. Locally, faint outlines of flat illitic fibers can be recognized (arrows). Permian Rotliegend, southern North Sea. (C) Low-magnification SEM micrograph of surface of a framework grain covered by a very thin inherited clay rim. Very small quartz overgrowths break through the clay rim along the left side of the grain. Arrow indicates location of (D). (D) High-magnification SEM micrograph of clay rim indicated by arrow in (C). Rim consists of a mat of elongate flat fibers overlapping in a random orientation but all subparallel to the underlying framework-grain surface. Clay morphology suggests significant regeneration. Permian Minnelusa Formation, Mountain Fuel #1 Bonnidee, Johnson County, Wyoming. (E) SEM micrograph displaying smooth to locally wrinkled surface of a glauconitic inherited clay rim. Morphology suggests a mixed-layer smectitic clay. Where rim is incomplete, quartz overgrowths (Q) have extended out and over it. Triassic Sag River Sandstone, Standard Alaska Prudhoe Bay Unit Term Well C, North Slope, Alaska. (F) SEM micrograph of wrinkled glauconitic clay coats on framework grains. Grains above and below exhibit small quartz overgrowths. Grain at center is a partially dissolved microporous chert. Triassic Sag River Sandstone, Standard Alaska Prudhoe Bay Unit Term Well C, North Slope, Alaska.

TABLE 2.—DISTRIBUTION OF INHERITED CLAY RIMS IN HOLOCENE AND ANCIENT SANDSTONES

AGE AND STRATIGRAPHIC UNIT	LOCATION	ENVIRONMENT	WELL DEVELOPED RIMS	WEAKLY DEVELOPED RIMS	RIMS ABSENT
ANCIENT EOLIAN SANDSTONES					
Ordovician St. Peter Sandstone	Illinois Basin	Dune/Interdune	X	X	O
Pennsylvanian/Permian Minnelusa Formation	Powder River Basin	Dune	X	X	X
		Interdune/Sabkha	X	X	X
		Beach	O	O	X
Pennsylvanian Weber Sandstone	Rangely Field	Dune	X	X	X
Permian Lyons Sandstone	Morrison, Colorado	Dune	X	O	O
Permian Lyons Sandstone	Lyons, Colorado	Dune	O	X	O
Permian Rotliegend	North Sea	Dune	X	X	X
		Interdune/Sabhka	X	X	X
Jurassic Aztec Sandstone	Southern Nevada	Dune	X	X	X
Jurassic Nuggett Sandstone	Southeastern Wyoming/ Northeastern Utah	Dune	X	X	X
Jurassic Norphlet Sandstone	Southern Alabama	Dune	X	X	X
HOLOCENE EOLIAN SANDSTONES					
Rub Al Khali	Saudi Arabia	Dune	X	O	O
Abu Simbel	Eastern Egypt	Desert Floor	X	O	O
Simpson Desert	Central Australia	Dune	X	O	O
Death Valley	Eastern California	Dune	O	X	O
Great Sand Dunes	Central Colorado	Dune	O	X	O
Monahans Erg	West Texas	Dune	X	O	O
Arcachon	Western France	Coastal Dune	O	O	X
ANCIENT SHELF AND SHELF-SHOREFACE TRANSITION SANDSTONES					
Pennsylvanian Morrow Formation	Oklahoma Panhandle	Sand Ridge ?	O	X	O
Triassic/Jurassic Sag River Sandstone	North Slope, Alaska	Shelf/Shelf-Shoreface Transition	X	O	O
Lower Cretaceous Kuparuk River Formation	North Slope, Alaska	Shelf/Shelf-Shoreface Transition	X	X	X
Lower Cretaceous Kemik Sandstone	North Slope, Alaska	Shelf/Shelf-Shoreface Transition	O	X	O
Upper Cretaceous Wall Creek Sand	Central Wyoming	Shelf/Shelf-Shoreface Transition	X	X	O

O - No samples occur in category
X - One or more samples occur in category

Matlack and others (1989) extended the work of Crone and Walker through additional laboratory experiments and by investigating infiltration in a series of Holocene deltaic and marine deposits. In infiltration clays from both laboratory sands and Holocene sands, these authors observed the same features reported by Crone (1975) and Walker (1976) and Walker and others (1978). Infiltration was found to occur in environments such as point-bar and delta-plain deposits where suspended-sediment concentrations and water-level fluctuations were extensive and where sediment reworking was limited. Essentially no infiltration was observed in beach and tidal-delta deposits where suspended-sediment concentrations were low and reworking was extensive.

Infiltration clays have been reported from the Permian Rotliegendes sandstones of the southern North Sea by Kessler (1978), but the evidence supporting his conclusions is limited. Kessler cited the presence of "textured and scattered" surfaces of clay coatings, possible meniscus bridges, and geopetal fabrics. Molenaar (1986) described what he interpreted to be infiltration clays in Devonian fluvial sands of eastern Belgium. Characteristics of this infiltration clay, as recorded by Molenaar, include: (1) lack of clay-particle orientation in some accumulations; (2) bimodal size distribution of infiltrated sandstones; (3) clay-supported fabrics in some samples or portions of samples; (4) geopetal fabrics with clay accumulations preferentially located on the upper surfaces of detrital grains; (5) clay cutans with clay particles oriented parallel to framework-grain surfaces; (6) irregular thicknesses of clay cutans; and (7) concentrations in and surrounding root or bioturbation channels.

Molenaar also observed the occurrence of clay cutans on top of well-developed quartz overgrowths, but it is likely that these accumulations of clay were formed as a result of infiltration during outcrop weathering.

The fourth method (adhesion) is discussed in general terms by a number of authors (Hunter, 1981; Kocurek and Fielder, 1982), but none mentioned the characteristics of these clays. Glennie and others (1978) noted that silt and clay are incorporated in adhesion ripples of sabkha deposits in the Rotliegend of the southern North Sea but did not give specifics regarding amounts or distribution.

An origin by inheritance implies that recycling of framework grains has occurred and that the clays present originally formed by one of the other four mechanisms. Evidence of an inherited origin for clays in sandstones, as defined by Walker (1979) and Wilson (1982), includes: (1) greater thickness in indentations; (2) presence at grain contacts; and

(3) enhanced development in finer grained sandstones, where clay-filled depressions are more abundant.

Inherited Clay Rims in Eolian Deposits

Folk (1976) attributed the inherited clay rims (cutans) in the Holocene dune sands of the Simpson Desert of central Australia to laterization during periods of higher rainfall in the Pleistocene. In Folk's interpretation, inherited clay rims form as a result of soil development. The iron oxides staining the clays form in the soil zone prior to transport.

Walker (1979) concluded that the inherited clay rims in Holocene eolian sands of northern Libya were derived from airborne dust that mechanically infiltrated dune and surface sands. The eolian dust adhered to sand-grain surfaces where moisture was present, either in the form of dew, rainfall, or fluvial-water influx. Infiltration was most active where rainfall was relatively abundant. Thus, clay-rim and iron-oxide development was promoted in areas adjacent to the Libyan coast, where rainfall was greater. Walker disagreed with Folk's interpretation that the staining of the clays was produced by prior laterization during a wetter Pleistocene climatic episode. He instead presented strong evidence that in Libya this staining was produced by *in situ* alteration of iron-rich heavy minerals and iron-bearing clays and became more extensive with transport and age. Evidence cited by Walker to support an eolian-dust infiltration origin includes: (1) clay coats are not present on beach sands that constitute the initial upstream source of sand for the eolian deposits; (2) scanning electron microscope examination reveals that the clays have "clastic" texture (irregular grain outlines and orientation parallel to sand-grain surfaces) similar to infiltration clays elsewhere; and (3) the mineralogy of the clays is similar to that expected of eolian dusts in the study area.

Based on the examination of large numbers of ancient dune sands from a variety of areas, this author is of the opinion that infiltration in dune sands is not a significant mechanism of clay-rim generation. Infiltration clays are very rare in the dune sandstones examined for this study. They were observed by this author in some interdune/desert-floor/sabkha deposits and are probably relatively common in these environments elsewhere (Glennie, 1970). Nagtegaal (1979) reported that illite, probably of detrital origin, is common in adhesion-ripple sandstones of the Rotliegendes sandstones.

Caution must be used when interpreting clays as having been introduced by infiltration. In highly porous and permeable samples, drilling-mud invasion will exhibit most of the characteristics of a true (natural) infiltration clay. Since such muds are a form of infiltrated fines, this is not surprising. The features drilling muds do not appear to share with true infiltration residues are the presence of internal layering and geopetal fabrics (Table 1).

Because framework grains with inherited clay rims represent recycled material, the specific origin or origins of the clays coating the grains is very difficult and, in most cases, impossible to ascertain. In eolian environments, clays are introduced into sandy debris primarily by adhesion on wetted surfaces and by infiltration through invasion by surface waters. Adhesion and infiltration occur primarily in low-lying sabkha, interdune, and desert-floor environments. Framework grains with inherited clay rims generated in these environments can be recycled into a variety of eolian deposits and may possibly occur within some fluvial deposits as well. Clay rims in many sabkha, interdune, and desert-floor deposits may have compound origins. Such rims may consist of inherited (recycled) rims to which additional clays have been added by adhesion or infiltration processes. Clay coats are seldom formed directly by infiltration within dune sands. This is confirmed by the absence of recognizable clay laminae or clay matrix in dune sands worldwide. An exception to this may occur in vegetated and relatively fixed (stabilized) dunes such as those described by Pye and Tsoar (1987) from the Negev desert of Israel. In the stabilized dunes studied by these authors, infiltrated dust consisting of clays and silt constitutes as much as 10% of the dune material.

Inherited Clay Rims in Shelf Deposits

Inherited clay rims in marine deposits are most abundant in shelf and shelf-shoreface transition zone sandstones. They are most obvious in sandstones that are devoid of detrital matrix and where the rims are composed of glauconitic clay. They are very difficult to recognize where abundant detrital matrix is present.

It is likely that clay coats may be generated by the passage of sand through the guts of a variety of marine organisms. Several authors have commented on the relatively high-sand content of the feces of marine sediment-feeding organisms such as polychaete worms, limpets, gastropods, and lamellibranchs, as well as the large volume of fecal matter produced by many of these organisms (Moore, 1939; Schafer, 1972). Filter feeders, on the other hand, produce pellets that consist primarily of clays (Pryor, 1975).

Burrowing activity may also constitute a major mechanism by which inherited clay rims are generated. Bioturbation is extensive to intense in much of the Sag River Sandstone (Barnes, 1987) and the Kuparuk Formation (Gaynor and Scheihing, 1988) of the North Slope, Alaska. The occurrence of complete inherited clay rims not associated with burrow structures suggests that these rims were not formed *in situ*. Despite the presence of minor to large amounts of biogenically introduced matrix, the sorting of the framework grains in sandstones that contain inherited clay rims is generally quite good. This relatively high degree of sorting is further evidence that framework grains experienced some degree of transport prior to deposition. The presence in some samples of clay peloids exhibiting differing degrees of glauconitization and/or of glauconite displaying differing degrees of oxidation suggests that glauconite development did not occur *in situ*.

Glauconite generation on modern coastlines occurs primarily in the mid- to outer shelf zone (Van Houten and Purucker, 1984). Odin and Letolle (1980) estimated the depth zone in which glauconitization generally occurs as 60 to 350 m, and also noted that glauconitization does not occur in agitated or oxidizing environments but is favored where clastic influx is low and temperatures are relatively low (7 to 15°C). Glauconite development is especially common during periods of major transgressions. However, much of the sandstone in both the Sag River Sandstone and Kuparuk Formation was deposited in the form of coarsening-upward regressive units (Barnes, 1987; Gaynor and Scheihing, 1988). Thin units of more highly glauconitic transgressive sandstone are present at the tops of the regressive units. It is quite likely that glauconite pellets and grains with glauconitic rims formed within a somewhat quieter, less agitated environment than that in which they were deposited.

Recent work by Rude and Aller (1989) on the Amazon continental shelf suggests that some inherited clay rims generated by lateritic weathering may survive transport by fluvial processes and be preserved in shelf sediments. The relative importance of fluvial sources of inherited clay rims in other areas is unknown and merits further investigation.

INFLUENCE OF INHERITED CLAY RIMS ON RESERVOIR QUALITY

Porosity

Many workers (Heald, 1965; Fuchtbauer, 1967; Cecil and Heald, 1971; Heald and Larese, 1974; Thomson, 1979) have noted the ability of clay coats to inhibit quartz-overgrowth development and, in many cases, preserve significant intergranular porosity. In most cases, the clays involved are neoformed authigenic types. In the eolian sandstones studied by this author, the bulk of the clays preserving porosity are inherited clay rims. The importance of these clay rims is commonly overlooked or is attributed to clays of authigenic origin.

It is common for porosities in the Permian Minnelusa Formation of the Powder River Basin to be 5 to 15% higher in dune sandstones containing well-developed inherited clay rims than in interbedded or laterally equivalent dune sandstones cemented by quartz overgrowths. For example, in a well where burial depths exceed 4,600 m (15,000 ft), the average and maximum porosities of Minnelusa Formation dune sandstones containing inherited clay rims are 13.0% and 16.9%, respectively. In contrast, in a series of zones immediately below in which inherited clay rims do not occur, the average and maximum porosities of quartz-overgrowth-cemented dune sandstones are 1.6% and 4.1%, respectively. Porosity histograms for these two sand types are shown in Figure 8.

A similar comparison of reservoir quality can be made for shelf sandstones with inherited clay rims from the Triassic-Jurassic Sag River Sandstone and marine shoreface and shelf-shoreface transition zone sandstones from the Triassic Eileen Sandstone on the North Slope of Alaska. A direct comparison between sandstones with inherited clay rims and those without cannot be made using sandstones from the Sag River Sandstone alone because relatively clay-free sandstones do not occur within this unit. The average and maximum porosities for Sag River Sandstone samples containing well-developed inherited clay rims in the Mobil #7-11-12 Kuparuk are 17.2% and 18.8%, respectively. The samples from the Sag River Sandstone used to calculate the average and maximum values actually were taken from a core at the base of the unit, where porosities are commonly 5 to 15% lower than in the upper portion of the unit where clay matrix and ankerite content is significantly reduced.

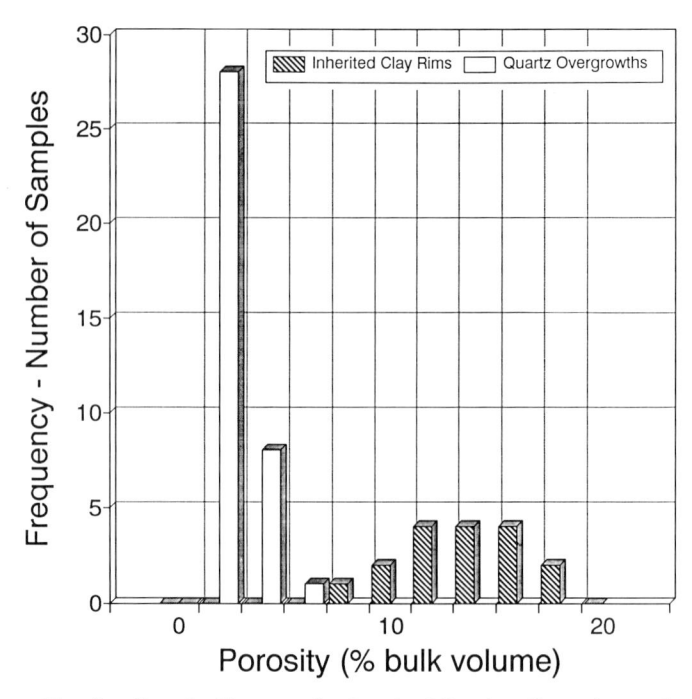

FIG. 8.—Porosity histogram for Permian Minnelusa Formation sandstones in the Mountain Fuel #1 Bonnidee, Johnson County, Wyoming. Only data for sandstones with extensive inherited clay rims and with extensive quartz-overgrowth cement are shown. All sandstones with extensive inherited clay rims have higher porosities than the most porous of those cemented by quartz overgrowths.

The quartz-overgrowth-cemented Eileen Sandstone samples from the same well, whose depths are only about 30 m (100 ft) greater than those for the Sag River Sandstone samples, have average and maximum porosities of 9.5% and 12.6%, respectively. Poorly developed inherited clay rims are present on some grains in the Eileen Sandstone and exhibit their maximum development in the most porous of the Eileen samples.

Permeability

Where inherited clay rims are thin and individual clay particles are oriented subparallel to the framework-grain surfaces, these rims may have little or no effect on permeability. Sandstones from the deeply buried Minnelusa Formation discussed in the previous section are good examples of this relation. A porosity-permeability cross-plot for these sandstones is presented in Figure 9. Note in the cross-plot that the permeabilities of the sandstones containing inherited clay rims are not abnormally low relative to those of quartz-overgrowth-cemented sandstones. This is not surprising considering the relatively thin, smooth, matlike surfaces typical of these clays (Figs. 6C and D). It is important to note that the content of inherited clay rims varies abruptly from unit to unit within the core just described and even increases from trace amounts at the base of one dune sequence to relatively complete rims within a few feet vertically.

Although critical-point drying may be necessary to prevent disturbance and collapse of fibrous illite in samples

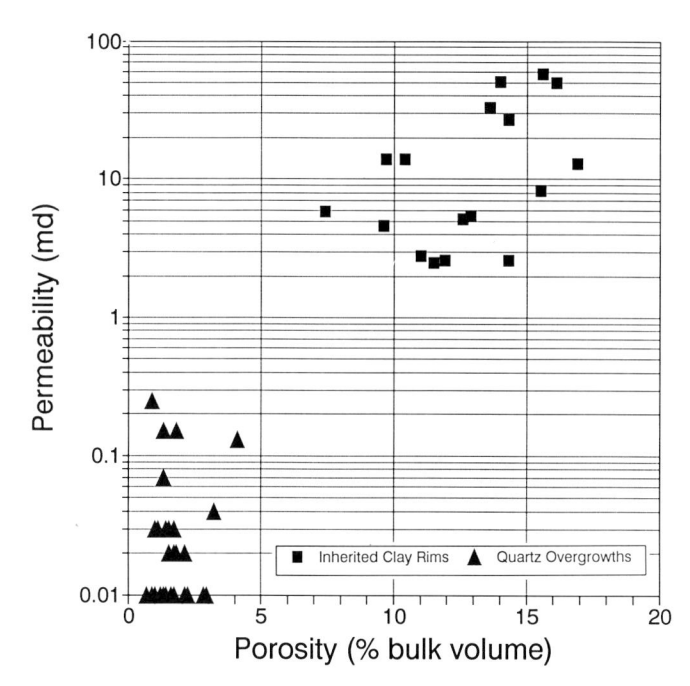

FIG. 9.—Porosity-permeability cross-plot for Permian Minnelusa Formation sandstones from the Mountain Fuel #1 Bonnidee. General trend displayed by the sandstones with inherited clay rims suggests that, despite their greater clay content, the sandstones do not have abnormally low permeability relative to quartz-overgrowth-cemented sandstones.

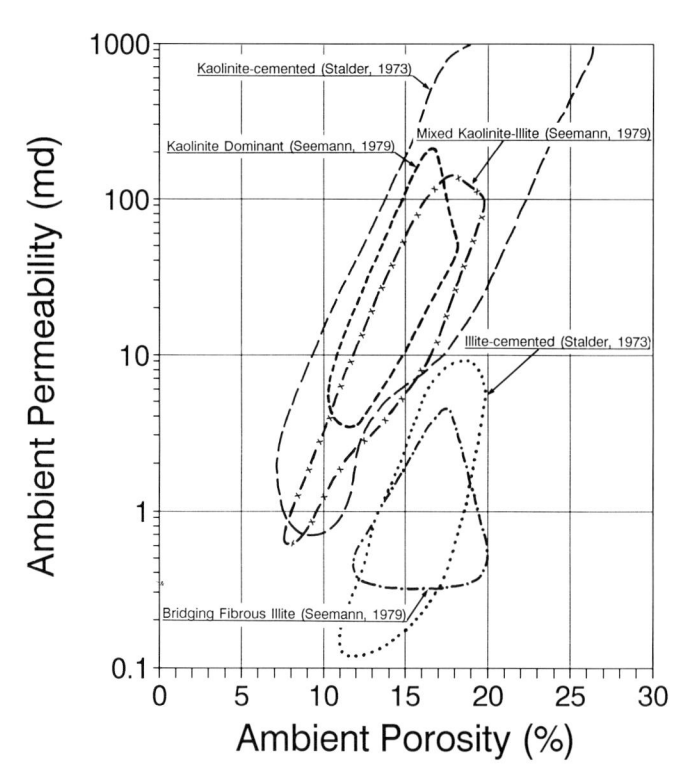

FIG. 10.—Porosity-permeability cross-plot for Permian Rotliegendes sandstones of the southern North Sea. Sandstones dominated by fibrous bridging illitic clays have permeabilities up to two magnitudes lower than those in which kaolinite is the dominant clay mineral.

analyzed using the SEM (Kantorowicz, 1990; Pallatt, 1990), the effect is only significant if the fibers of illite are quite thin (less than 60 to 80Å). Permeabilities of air-dried samples studied by Kantorowicz and Pallatt may exceed by as much as a factor of eight those measured on critical-point dried samples from the same sample site. The sizes of the illite fibers observed in the Minnelusa Formation samples shown in Figures 6C and D are comparable to the sizes of illite fibers from the Rotliegend in the Sole Pit area of the southern North Sea as reported by Kantorowicz (1990). The Rotliegendes illites studied by Kantorowicz are relatively coarse compared to illites in Jurassic reservoir units from the Viking Graben area of the North Sea and exhibit less disruption and permeability modification due to the application of standard drying techniques. Thus, though critical-point-drying techniques could not be applied in the current study (preserved core was not available for any of the wells studied), it is unlikely that morphologies observed have been significantly modified as a result.

In contrast to the Minnelusa Formation example above, some inherited clay rims may produce a significant loss of permeability. Numerous authors have noted the presence of delicate fibrous illite in Rotliegendes sandstones of the southern North Sea (Stalder, 1973; Glennie and others, 1978; Hancock, 1978; Seemann, 1979; Goodchild and Whitaker, 1986). Most have documented the dramatic reduction of permeability produced by this clay relative to other Rotliegendes sandstones containing primarily kaolinite. The latter consistently have permeabilities one to two orders of magnitude greater than sandstones having identical porosity but containing fibrous illite (Fig. 10). It is the contention of this author that many or most of these illitic clays probably originally formed as inherited clay rims. The conversion of the clays in the rims from a surface-parallel orientation to a transverse orientation may be a temperature-controlled phenomenon. It appears to occur only in wells in which maximum burial depths are relatively great.

It is interesting to note that those sandstones analyzed by Seemann (1979), which contain a mix of kaolinite and illite, have somewhat greater porosities than sandstones cemented primarily by kaolinite, yet have experienced approximately the same maximum burial depth (Fig. 10). Even more amazing is that porosities of sandstones cemented primarily by illite have even greater porosities, yet have been subjected to approximately 1,100 m greater depth of burial (Fig. 10).

CONCLUSIONS

(1) Inherited clay rims are common in many eolian and marine-shelf deposits. Such rims may be present in significant amounts in sandstones from other environments, and additional studies are needed to assess their importance in fluvial and tidal regimes.
(2) Criteria are available for the identification of inherited clay rims. The most effective criteria tend to be those relating to spatial distribution and include presence at points of grain contact, increased thickness in depressions on framework-grain surfaces, and more extensive development in finer grained laminae or beds.

(3) In eolian environments, inherited clay rims are probably formed primarily by infiltration of clay-bearing waters into, and/or adhesion of clays onto, the surficial portions of interdune/desert-floor/sabkha deposits. In marine-shelf environments, inherited clay rims are generated through the mechanisms of bioturbation and ingestion and coating of sand grains by a variety of organisms. Worms may be the most effective of these organisms.

(4) Inherited clay rims are an important, if not the most important, control on porosity in many eolian and shelf sandstones. The extensive development of inherited clay rims prevents the nucleation of quartz overgrowths on quartz-framework grains. Under conditions of very deep burial, the presence of inherited clay rims at framework-grain contacts commonly promotes intense pressure-solution suturing, resulting in the loss of most intergranular porosity. At depth, clays in detrital rims may be regenerated to form authigenic clays, which may, because of their very high surface area and bridging morphology, severely reduce permeability, possibly by as much as one to two orders of magnitude.

ACKNOWLEDGMENTS

The author thanks Robert L. Folk and Steven G. Fryberger for access to samples of Holocene eolian sands from Australia, Egypt, and Saudi Arabia. Significant improvements in the paper are credited to critical reviews by Earle F. McBride and Richard E. Larese. Sandra Wilson supplied much appreciated assistance with the typing and editing of the manuscript.

REFERENCES

AHLBRANDT, T. S., 1979, Textural parameters of eolian deposits, in McKee, E. D., ed., A Study of Global Sand Seas: U.S. Geological Survey Professional Paper 1052, p. 21–51.

ALMON, W. R., 1981, Sandstone diagenesis: applications to exploration and exploitation: American Association of Petroleum Geologists, Course Notes, Clastic Diagenesis School, p. 8–88.

BAGNOLD, R. A., 1941, The Physics of Blown Sand and Desert Dunes: London, Methuen and Co., 265 p.

BARNES, D. A., 1987, Reservoir quality in the Sag River Formation, Prudhoe Bay Field Alaska: depositional environment and diagenesis, in Tailleur, I., and Weimer, P., eds., Alaska North Slope Geology, v. 1: Pacific Section, Society of Economic Paleontologists and Mineralogists and Alaska Geological Society, p. 85–94.

BENTOR, Y. K., AND KASTNER, M., 1965, Notes on the mineralogy and origin of glauconite: Journal of Sedimentary Petrology, v. 35, p. 155–166.

BREWER, R., 1964, Fabric and Minerals Analysis of Soils: New York, Wiley and Sons, 470 p.

CECIL, C. B., AND HEALD, M. T., 1971, Experimental investigation of the effects of grain coatings on quartz growth: Journal of Sedimentary Petrology, v. 41, p. 582–584.

CRONE, A. J., 1975, Laboratory and field studies of mechanically infiltrated clay matrix in arid fluvial sediments: Unpublished Ph.D. Dissertation, University of Colorado, Boulder, 162 p.

DIXON, S. A., SUMMERS, D. M., AND SURDAM, R. C., 1989, Diagenesis and preservation of porosity in Norphlet Formation (Upper Jurassic), southern Alabama: American Association of Petroleum Geologists Bulletin, v. 73, p. 707–728.

DREGNE, H. E., 1976, Soils of Arid Regions: Developments in Soil Science, v. 6: New York, Elsevier Publishing Co., 376 p.

EGGERT, J. T., 1987, Sandstone petrology, diagenesis, and reservoir quality, Lower Cretaceous Kuparuk River Formation, Kuparuk River Field, North Slope, Alaska (abs.), in Tailleur, I., and Weimer, P., eds., Alaska North Slope Geology, v. 1: Pacific Section, Society of Economic Paleontologists and Mineralogists and Alaska Geological Society, p. 108.

FOLK, R. L., 1976, Reddening of desert sands: Simpson Desert, Northern Territory, Australia: Journal of Sedimentary Petrology, v. 46, p. 604–615.

FOLK, R. L., 1978, Angularity and silica coatings of Simpson Desert sand grains, Northern Territory, Australia: Journal of Sedimentary Petrology, v. 48, p. 611–624.

FUCHTBAUER, H., 1967, Influence of different types of diagenesis on sandstone porosity: Panel Discussion, 7th World Petroleum Congress, Mexico, Preprint, 3 p.

GAUTIER, D. L., 1983, Patterns of sedimentation, diagenesis, and hydrocarbon accumulation in Cretaceous rocks of the Rocky Mountains: Society of Economic Paleontologists and Mineralogists Short Course 11, p. 5.1–5.43.

GAYNOR, G. C., AND SCHEIHING, M. H., 1988, Shelf depositional environments and reservoir characteristics of the Kuparuk River Formation (Lower Cretaceous), Kuparuk Field, North Slope, Alaska, in Lomando, A. J., and Harris, P. M., eds., Giant Oil and Gas Fields: A Core Workshop, v. 1: Society of Economic Paleontologists and Mineralogists Core Workshop 12, Houston, March 19–20, p. 333–390.

GLENNIE, K. W., 1970, Desert Sedimentary Environments: Developments in Sedimentology, v. 14: New York, Elsevier Publishing Co., 222 p.

GLENNIE, K. W., MUDD, G. C., AND NAGTEGAAL, P. J., 1978, Depositional environment and diagenesis of Permian Rotliegendes sandstones in Leman Bank and Sole Pit areas of the UK, southern North Sea: Journal of Geological Society of London, v. 135, p. 25–34.

GOODCHILD, M. W., AND WHITAKER, J. H., 1986, A petrographic study of the Rotliegendes Sandstone reservoir (Lower Permian) in the Rough Gas Field: Clay Minerals, v. 21, p. 459–477.

HANCOCK, N. J., 1978, Possible causes of Rotliegend sandstone diagenesis in northern West Germany: Journal of Geological Society of London, v. 135, p. 35–40.

HEALD, M. T., 1956, Cementation of Simpson and St. Peter sandstones in parts of Oklahoma, Arkansas and Missouri: Journal of Geology, v. 64, p. 16–30.

HEALD, M. T., 1965, Lithification of sandstones in West Virginia: West Virginia Geological and Economic Survey Bulletin 30, 28 p.

HEALD, M. T., AND LARESE, R. E., 1974, Influence of coatings on quartz cementation: Journal of Sedimentary Petrology, v. 44, p. 1269–1274.

HUNTER, R. W., 1981, Stratification styles in eolian sandstones: some Pennsylvanian to Jurassic examples from the western interior U.S.A., in Ethridge, F. G., and Flores, R. M., eds., Recent and Ancient Nonmarine Depositional Environments: Models for Exploration: Society of Economic Paleontologists and Mineralogists Special Publication 31, p. 315–329.

KANTOROWICZ, J. D., 1990, The influence of variations in illite morphology on the permeability of Middle Jurassic Brent Group sandstones, Cormorant Field, UK, North Sea: Marine and Petroleum Geology, v. 7, p. 66–74.

KESSLER, L. G., II, 1978, Diagenetic sequence in ancient sandstones deposited under desert climatic conditions: Journal of Geology Society, v. 135, p. 41–49.

KOCUREK, G., AND FIELDER, G., 1982, Adhesion structures: Journal of Sedimentary Petrology, v. 52, p. 1229–1241.

KRINSLEY, D. H., FRIEND, P. F., AND KLIMENTIDIS, R., 1976, Eolian transport textures on the surfaces of sand grains of early Triassic age: Geological Society of America, v. 87, p. 130–132.

KRYSTINIK, L. F., 1990, Diagenesis in ancient eolian sandstone, in Fryberger, S. G., Krystinik, L. F., and Schenk, C. J., eds., Modern and Ancient Eolian Deposits: Rocky Mountain Section, Society of Economic Paleontologists and Mineralogists, p. 14.1–14.14.

KUENEN, P. H., 1960, Experimental abrasion (Part 4), eolian action: Journal of Geology, v. 68, p. 427–449.

MARGOLIS, S. V., AND KRINSLEY, D. H., 1971, Submicroscopic frosting on eolian and subaqueous quartz sands: Geological Society of America Bulletin, v. 82, p. 3395–3406.

MASTERSON, W. D., AND PARIS, C. E., 1987, Depositional history and reservoir description of the Kuparuk River Formation, North Slope, Alaska, in Tailleur, I., and Weimer, P., eds., Alaska North Slope Geology, v. 1: Pacific Section, Society of Economic Paleontologists and Mineralogists and Alaska Geological Society, p. 95–107.

MATLACK, K. S., HOUSEKNECHT, D. W., AND APPLIN, D. R., 1989, Emplacement of clay into sand by infiltration: Journal of Sedimentary Petrology, v. 59, p. 77–87.

MCKEE, E. D., 1979, Ancient sandstones considered to be eolian, in McKee, E. D., ed., A Study of Global Sand Seas: U.S. Geological Survey Professional Paper 1052, p. 187–238.

MOLENAAR, N., 1986, The interrelation between clay infiltration, quartz cementation, and compaction in Lower Givetian terrestrial sandstones, northern Ardennes, Belgium: Journal of Sedimentary Petrology, v. 56, p. 359–369.

MOORE, H. B., 1939, Faecal pellets in relation to marine deposits, in Trask, P. D., ed., Recent Marine Sediments: a Symposium: Society of Economic Paleontologists and Mineralogists Special Publication 4, p. 516–524.

NAGTEGAAL, P. J., 1979, Relationship of facies and reservoir quality in Rotliegendes desert sandstones, southern North Sea region: Journal of Petroleum Geology, v. 2, p. 145–158.

NEASHAM, J. W., 1977, The morphology of dispersed clays in sandstone reservoirs and its effect on sandstone shaliness, pore space and fluid flow properties: Society of Petroleum Engineers, 52nd Annual Technical Conference, Preprint, Society of Petroleum Engineers 6858, 8 p.

NOTHOLT, A. J., 1980, Phosphatic and glauconitic sediments: Journal of Geological Society of London, v. 137, p. 657–659.

ODIN, G. S., AND LETOLLE, R., 1980, Glauconitization and phosphatization environments: a tentative comparison: Society of Economic Paleontologists and Mineralogists Special Publication 29, p. 227–237.

PALLATT, N., 1990, Critical point drying applied to clay minerals in sandstones (abs.): Clay Minerals Society 27th Annual Meeting, Columbia, MO, Program and Abstracts, p. 100.

PETTIJOHN, F. J., POTTER, P. E., AND SIEVER, R., 1972, Sand and Sandstone: New York, Springer-Verlag, 618 p.

PRYOR, W. A., 1975, Biogenic sedimentation and alteration of argillaceous sediments in shallow marine environments: Geological Society of America Bulletin, v. 86, p. 1244–1254.

PYE, K., AND TSOAR, H., 1987, The mechanics and geological implications of dust transport and deposition in deserts with particular reference to loess formation and dune sand diagenesis in the northern Negev, Israel, in Frostick, L.E., and Reid, I., eds., Desert Sediments: Ancient and Modern: London, Blackwell Scientific Publications, p. 139–156.

RICE, D. D., AND GAUTIER, D. L., 1983, Patterns of sedimentation, diagenesis, and hydrocarbon accumulation in Cretaceous rocks of the Rocky Mountains: Society of Economic Paleontologists and Mineralogists Short Course 11, p. 7.1–7.41.

RUDE, P. D., AND ALLER, R. C., 1989, Early diagenetic alteration of lateritic particle coatings in Amazon continental shelf sediment: Journal of Sedimentary Petrology, v. 59, p. 704–716.

SCHAFER, W., 1972, Ecology and Palaeocology of Marine Environments: Chicago, University of Chicago Press, 568 p.

SCHENK, C. J., 1990, Overview of eolian sandstone diagenesis, Permian upper part of the Minnelusa Formation, Powder River Basin, Wyoming, in Fryberger, S. G., Krystinik, L. F., and Schenk, C. J., eds., Modern and Ancient Eolian Deposits: Rocky Mountain Section, Society of Economic Paleontologists and Mineralogists, p. 15.1–15.10.

SCHMIDT, V., AND MCDONALD, D. A., 1979, Texture and recognition of secondary porosity in sandstones, in Scholle, P. A., and Schluger, P. R., eds., Aspects of Diagenesis: Society of Economic Paleontologists and Mineralogists Special Publication 26, p. 209–225.

SEEMANN, U., 1979, Diagenetically formed interstitial clay minerals as a factor in Rotliegend Sandstone reservoir quality in the Dutch sector of the North Sea: Journal of Petroleum Geology, v. 1, p. 55–62.

SHEPARD, F. P., AND YOUNG, R., 1961, Distinguishing between beach and dune sands: Journal of Sedimentary Petrology, v. 31, p. 196–214.

STALDER, P. J., 1973, Influence of crystallographic habit and aggregate structure of authigenic clay minerals on sandstone permeability: Geologie en Mijnbouw, v. 52, p. 217–220.

TADA, R., AND SIEVER, R., 1989, Pressure solution during diagenesis: Annual Review of Earth and Planetary Sciences, p. 89–118.

THOMSON, A., 1959, Pressure solution and porosity, in Ireland, H. A., ed., Silica in Sediments: Society of Economic Paleontologists and Mineralogists Special Publication 7, p. 92–110.

THOMSON, A., 1979, Origin of porosity in deep Woodbine-Tuscaloosa Trend, Louisiana (abs.): American Association of Petroleum Geologists Bulletin, v. 63, p. 1611.

TILLMAN, R. W., AND ALMON, W. R., 1979, Diagenesis of Frontier Formation offshore bar sandstones, Spearhead Ranch Field, Wyoming, in Scholle, P. A., and Schluger, P. R., eds., Aspects of Diagenesis: Society of Economic Paleontologists and Mineralogists Special Publication 26, p. 337–378.

TRIPLEHORN, D. M., 1966, Morphology, internal structure, and origin of glauconite pellets: Sedimentology (Part 6), Amsterdam, Elsevier Publishing Co., p. 247–266.

TURNER, J. R., AND CONGER, S. J., 1984, Environment of deposition and reservoir properties of the Woodbine Sandstone at Kurten Field, Brazos Co., Texas, in Tillman, R. W., and Siemers, C. T., eds., Siliciclastic Shelf Sediments: Society of Economic Paleontologists and Mineralogists Special Publication 34, p. 215–249.

VAN HOUTEN, F. B., AND PURUCKER, M. E., 1984, Glauconitic peloids and chamositic ooids–favorable factors, constraints, and problems: Earth Science Reviews, v. 20, p. 211–243.

WALKER, T. R., 1976, Diagenetic origin of continental red beds, in Falke, H., ed., The Continental Permian in Central, West, and South Europe: Dordrecht, D. Reidel Publishing Co., p. 240–282.

WALKER, T. R., 1979, Red color in dune sand, in McKee, E. D., ed., A Study of Global Sand Seas: U.S. Geological Survey Professional Paper 1052, p. 61–81.

WALKER, T. R., WAUGH, B., AND CRONE, A. J., 1978, Diagenesis in first-cycle desert alluvium of Cenozoic age, southwestern United States and northwestern Mexico: Geological Society of America Bulletin, v. 89, p. 19–32.

WERMUND, E. G., 1961, Glauconite in early Tertiary sediments of Gulf Coast Province: American Association of Petroleum Geologists Bulletin, v. 45, p. 1667–1696.

WEYL, P. K., 1959, Pressure solution and the force of crystallization—a phenomenological theory: Journal of Geophysical Research, v. 64, p. 2001–2025.

WILSON, M. D., 1981, Origins of clays controlling permeability in tight gas sands: Society of Petroleum Engineers/U.S. Department of Energy, SPE/DOE Special Paper 9843, p. 157–164.

WILSON, M. D., 1982, Reservoir quality and formation damage susceptibility in sandstones: a short course: Sponsored by Poroperm Laboratories, Ltd., London, 232 p.

WILSON, M. D., AND PITTMAN, E. D., 1977, Authigenic clays in sandstones: Recognition and influence on reservoir properties and paleoenvironmental analysis: Journal of Sedimentary Petrology, v. 47, p. 1–31.

CLAY MINERALS IN ATOKAN DEEP-WATER SANDSTONE FACIES, ARKOMA BASIN: ORIGINS AND INFLUENCE ON DIAGENESIS AND RESERVOIR QUALITY

DAVID W. HOUSEKNECHT AND LOUIS M. ROSS, JR.
Department of Geological Sciences, University of Missouri, Columbia, Missouri 65211 USA

ABSTRACT: Strata of the lower and middle Atoka Formation in the Arkoma Basin comprise submarine-fan and marine-slope facies that display a variety of primary and secondary sedimentary structures, formed by sediment gravity-flow depositional processes and dewatering, respectively. Primary sedimentary structures are most common in beds deposited by unconfined sediment gravity flows on submarine-fan lobes, whereas secondary sedimentary structures are most common in beds deposited by channelized sediment gravity flows in fan channels and slope channels. Primary sedimentary structures display horizontal fabrics, whereas secondary sedimentary structures display deformed and vertical fabrics. Abundance and distribution of clay minerals in Atoka sandstones are related to sedimentary structures. Beds that display primary sedimentary structures contain little detrital clay that is sparsely disseminated through the sandstone. In contrast, beds that display secondary sedimentary structures contain more detrital clay that forms pervasive grain coatings, bridges between grains, and consolidation laminae. Other beds lack sedimentary structures and display abundant detrital clay that forms a matrix-supported fabric.

The abundance and distribution of detrital-clay minerals exerted significant influences on diagenesis and reservoir quality of Atoka sandstones. Among sandstones with grain-supported fabrics, those that display primary sedimentary structures and contain little detrital clay were pervasively cemented by quartz overgrowths and are characterized by poor reservoir quality. Those that display secondary sedimentary structures and contain more abundant detrital clay retained primary porosity because quartz-overgrowth nucleation was inhibited by clay coatings on detrital grains. Porosity was enhanced in these sandstones by dissolution of framework grains, and the sandstones are characterized by good reservoir quality. Sandstones with matrix-supported fabrics apparently had little original porosity, which was reduced by compaction of the pervasive matrix; they are characterized by poor reservoir quality. These observations suggest that channelized turbidite facies have greater potential to retain good reservoir quality than unconfined turbidite facies, because the former have detrital-clay minerals emplaced within sand during dewatering and those clay minerals inhibit destruction of porosity by quartz cementation.

INTRODUCTION

Clay minerals in sandstones may have diverse origins and may significantly influence diagenesis and reservoir quality (Wilson and Pittman, 1977). One mode of clay emplacement in sandstones that has received considerable emphasis in recent years is infiltration. Pioneering work by Crone (1975), Walker (1976), and Walker and others (1978) demonstrated that clay can accumulate in alluvial sand by infiltration of muddy water. Matlack and others (1989) documented several variables that influence the presence and abundance of infiltrated clay in experimental sand packs and in Holocene sand bodies in fluvial and coastal environments. Moraes and De Ros (1990) presented convincing evidence that much of the clay in certain Jurassic sandstones of Brazil was emplaced by infiltration in a fluvial environment. Papers by Dunn, and Moraes and De Ros in this volume amplify the importance of clay formed by infiltration in sandstones of nonmarine origin and present additional criteria by which such clays can be recognized.

Infiltrated clays are of particular importance in studies of sandstone diagenesis and reservoir quality for several reasons. They are emplaced while the sand is still in communication with depositional water and before chemical diagenesis has effectively begun. They commonly form grain coatings and meniscus-shaped bridges between detrital grains and, therefore, influence chemical interaction between grains and pore fluid. In this role, infiltrated clays influence the course of subsequent diagenesis and thereby control the presence or absence of porosity in many sandstones (Pittman and others, this volume). The presence and abundance of infiltrated clays are commonly related to depositional facies and, therefore, provide a potential key for predicting diagenetic facies through an understanding of depositional facies.

One important limitation on the occurrence of infiltrated clays in sandstones is the requirement for a mechanism to move muddy water through sand shortly following deposition. The best documentation of infiltrated clays comes from sands deposited in nonmarine environments, where vadose conditions provide that mechanism (Crone, 1975; Walker, 1976; Walker and others, 1978; Matlack and others, 1989; Moraes and De Ros, 1990). In fact, Matlack and others (1989) suggested that the absence of such a mechanism is one reason infiltrated clays are not common in Holocene beach and tidal-delta facies of the Texas Gulf Coast.

The objective of this paper is to present evidence that dewatering associated with deposition of sediment gravity flows (i.e., "turbidites") represents an additional mechanism by which "infiltrated" clays can be effectively emplaced in sand. Evidence is presented from sandstones of the Pennsylvanian Atoka Formation in the Arkoma Basin, and includes consideration of depositional facies, abundance and characteristics of clay minerals, and apparent influence of clay minerals on subsequent diagenesis and reservoir quality.

GEOLOGIC SETTING

The Arkoma Basin, which extends from east-central Arkansas to southeastern Oklahoma, is a foreland basin that developed in response to convergent tectonism that formed the Ouachita Mountains (Fig. 1). The following summary of Arkoma Basin geology is modified from Houseknecht (1986, 1987) and Houseknecht and McGilvery (1990).

Stratigraphy

Stratigraphy of the basin reflects the opening and subsequent closing of a Paleozoic ocean basin. Pre-Atokan strata

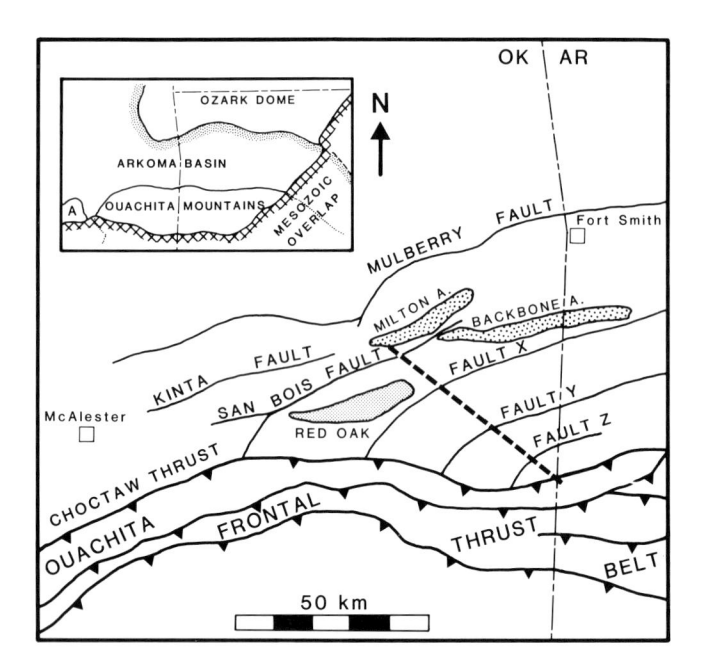

FIG. 1—Base map of part of Arkoma Basin from which samples were analyzed. Most subsurface samples were collected from conventional cores in Red Oak gas field and surface samples were collected from Atoka Formation outcrops in Ouachita frontal thrust belt and on Milton and Backbone anticlines. Named faults are south-dipping normal faults that were active during deposition of Atoka strata (see Fig. 2). Dashed line is location of cross section shown in Figure 2.

comprise mostly shelf carbonates, with subordinate sandstones and shales, that accumulated on a passive continental margin following opening of the ocean basin. These strata display gradual thickening to the south. Sediment accumulation rates were very low (average ≈7 m/Ma), sand was derived from the North American craton, and sand dispersal was mostly southward.

Most Atokan (lower middle Pennsylvanian) strata are shales and sandstones deposited during convergent tectonism that caused breakdown of the precursor shelf and development of the foreland basin. Atoka strata thicken dramatically southward, with abrupt increases in thickness across basement-rooted normal faults that were active during Atokan sedimentation (Fig. 2). Sediment accumulation rates were high (average ≈1,100 m/Ma), sand was mostly derived from the evolving, Ouachita orogenic belt to the east, and sand dispersal was mostly westward.

Post-Atokan strata are shales, sandstones, and coal beds deposited during final phases of foreland-basin sedimentation. These strata thicken to the south but do not display abrupt thickening across normal faults, indicating that syndepositional normal faulting had ceased. Sediment accumulation rates were moderate (average ≈250 m/Ma), sand was mostly derived from the evolving, Ouachita orogenic belt to the east and south, and sand dispersal was mostly westward.

Structure

The structural style of the basin reflects rheology and geometry of these stratigraphic units, as well as their age,

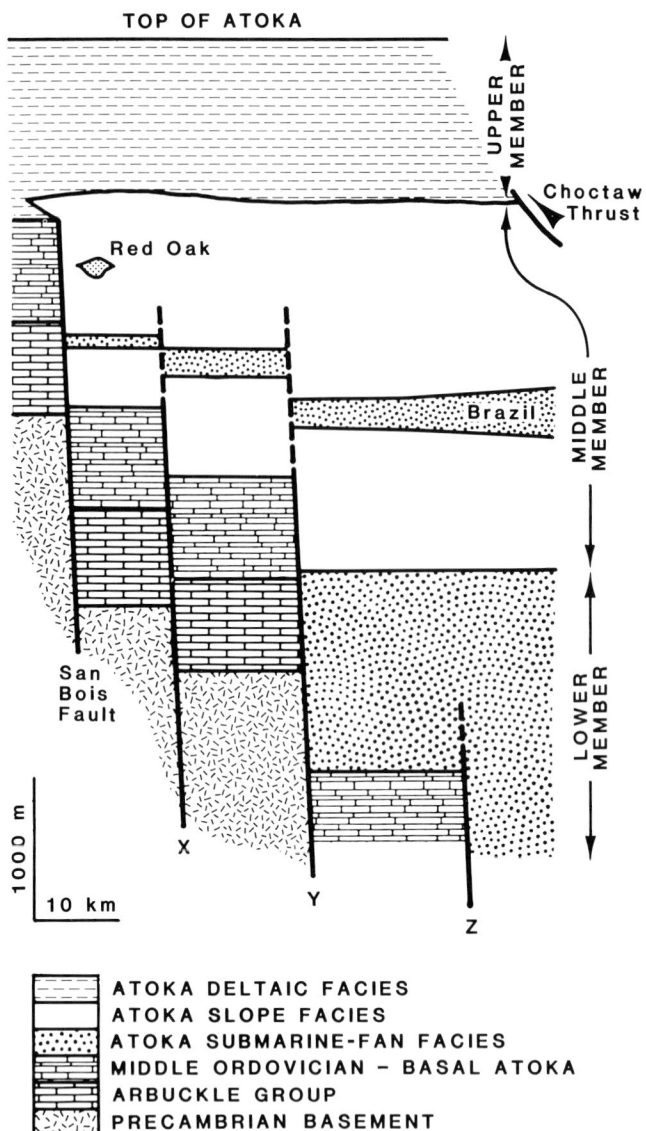

FIG. 2—Generalized stratigraphic cross section illustrating stratigraphy of Atoka Formation within study area. Named normal faults were active during deposition of lower and middle members of Atoka Formation and controlled distribution of submarine-fan and slope facies as explained in text. Figure modified from Houseknecht and others (1989).

relative to major compressional events associated with Ouachita orogenesis. Pre-Atokan shelf strata, as well as Precambrian granitic basement, are mostly deformed by Atokan-age normal faults. Most of these normal faults dip southward and display displacements of a few hundred meters to more than 4 km, thereby producing the step-like stratigraphic geometry of the Atoka Formation illustrated in Figure 2. All these Atokan-age normal faults are thought to have formed within a compressional tectonic setting as a result of lithospheric bending into a subduction zone coupled with vertical loading during obduction of the Ouachita orogenic pile onto the edge of North American continental crust (Houseknecht, 1986).

The structural style is different above the wedge of shelf strata. In the Arkoma Basin proper, Atokan and younger strata are deformed into broad synclines and narrow anticlines whose axes are parallel to the Ouachitas. Listric thrust faults underlie much of the folded section and ramp to the surface along the crests of many anticlines. Southward, in the Ouachita frontal thrust belt (Fig. 1), thrust faults increase in number and displacement. Rocks involved in this thrust faulting are mostly Atokan strata that were deposited in deeper portions of the foreland basin, and locally include the uppermost strata deposited on the precursor shelf.

Thermal Maturity

Inasmuch as many clay minerals are temperature sensitive, the thermal maturity of the Arkoma Basin is of importance to this paper. Pennsylvanian strata exposed at the surface are characterized by mean vitrinite-reflectance values that range from less than 1% in the western end of the basin in Oklahoma to more than 2% in central Arkansas (Houseknecht and others, in prep.). Thermal maturity increases with depth, and lower Atoka strata in the subsurface are characterized by mean vitrinite-reflectance values above 4% in deep parts of the basin.

Although the high values and lateral variations in thermal maturity are not completely understood, they apparently reflect the combined influences of stratigraphic burial, tectonic burial, and hydrothermal-fluid flow away from the Ouachita orogenic belt (Houseknecht and others, in prep.).

DEPOSITIONAL FACIES

The Atoka Formation represents the sedimentary fill of a foredeep developed by downfaulting of a precursor shelf during Ouachita orogenesis, as outlined earlier. The formation is composed of shale and sandstone exclusively, and can be informally divided into three members on the basis of lithology and depositional facies. The lower member is composed of submarine-fan and basin-plain facies, the middle member is a marine-slope and slope-channel facies, and the upper member is composed of shallow marine through deltaic facies. Criteria on which these interpretations are based have been presented by Houseknecht (1986, 1987) and Houseknecht and others (1989), and a brief summary of facies in the middle and lower members is provided in the following paragraphs. Strata of the upper member are not discussed in this paper.

The lower member of the Atoka Formation is known from exposures in the Ouachita frontal thrust belt (Fig. 1) and from numerous subsurface penetrations in the thrust belt and southern Arkoma Basin. It is characterized by a relatively high sandstone/shale ratio and facies indicative of submarine-fan and basin-plain deposition on the bathymetric floor of the basin (Houseknecht, 1986). Submarine-fan facies display a gradation from mostly proximal facies in the east to mostly distal facies in the west, and paleocurrent indicators that suggest a predominance of westward sediment dispersal (Fig. 3).

The middle member of the Atoka Formation is known from thousands of subsurface penetrations and a few relatively poor exposures in the Arkoma Basin and Ouachita frontal thrust belt (Fig. 1). It is characterized by a low sandstone/shale ratio and facies that indicate deposition in a marine-slope environment intermediate in both location and water depth between submarine-fan facies of the lower member and deltaic facies of the upper member. Although most of the middle member is comprised of a shale and silty shale, numerous lenticular sandstones are present (e.g., Red Oak in Fig. 2). Evidence presented by Houseknecht (1986) indicates these sandstones were deposited by sediment gravity flows within slope channels that were tectonically localized above syndepositional normal faults, although alternative interpretations have been presented (Vedros and Visher, 1978). Regional distribution of these sandstones, together with paleocurrent indicators, suggest westward sand dispersal (Fig. 3). This evidence further suggests that some slope channels curved southward and debouched onto relatively small submarine-fan lobes at the base of the slope (Fig. 3). The middle member also contains one widespread sandstone (Brazil in Fig. 2), which apparently represents submarine-fan sedimentation within a section otherwise characterized by slope facies.

As illustrated in Figure 2, distribution of these members of the Atoka Formation was controlled by syndepositional normal faulting. The lower member is thickest in the deepest part of the Arkoma Basin and in the Ouachita Mountains; toward the north, it thins abruptly across syndepositional normal faults and pinches out against the normal fault labeled "Y" in Figure 2. South of fault "Y" (Fig. 2), slope facies of the middle member lie directly on submarine-fan facies and display a fairly uniform thickness. North of fault "Y" (Fig. 2), slope facies lie directly on shelf strata, thin northward across syndepositional normal faults, and essentially pinch out against the San Bois fault (Fig. 2).

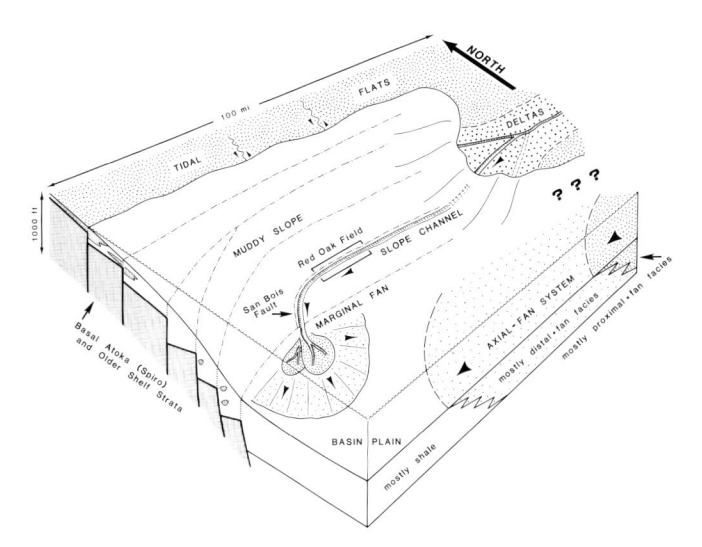

FIG. 3—Reconstruction of depositional system in which Atoka facies were deposited in the study area. Dash-dot lines represent seafloor traces of syndepositional normal faults illustrated in Figure 2. The Red Oak sandstone was deposited in the slope-channel environment depicted near the center of the figure, with approximate location of Red Oak gas field indicated. Shaftless arrowheads indicate direction of sediment transport. Vertical and lateral scales are approximate. Figure modified from Houseknecht (1986) and Houseknecht and McGilvery (1990).

SEDIMENTARY STRUCTURES

Locally, slope-channel sandstones of the middle member are prolific gas reservoirs in the Arkoma Basin (e.g., Red Oak sandstone in Red Oak field), and conventional cores are available for study. Submarine-fan facies of the lower member are not very productive in the region, so fewer subsurface samples are available. For this reason, the following descriptions of sedimentary structures are based mostly on conventional cores of slope-channel sandstones and on outcrops of submarine-fan sandstones.

More than 360 m of conventional core from the Red Oak sandstone and several hundred meters of outcrop sections have been described. The following sections present descriptions of sedimentary structures most commonly observed and provide interpretations of mechanisms responsible for those structures. These observations are divided into *primary structures*, inferred to have formed as the result of deposition from sediment gravity flows, and *secondary structures*, inferred to have formed as the result of dewatering. In most cores, the structures locally display orientations that are rotated by as much as 90° relative to similar structures elsewhere in the same cores. This observation suggests the presence of slump folds (e.g., Helwig, 1970). The occurrence of slumped intervals displaying rotated primary and secondary sedimentary structures overlain by non-slumped intervals displaying non-rotated primary and secondary sedimentary structures suggests that all structures described herein are essentially synsedimentary.

Primary Structures

Only 26% of the sandstone described within Red Oak cores displays sedimentary structures inferred to be primary. In contrast, it is estimated that more than 80% of the sandstone described in outcrops of lower Atoka submarine-fan facies displays primary sedimentary structures. It must be emphasized that the latter estimate is approximate because sedimentary structures are vague in many sandstone beds due to weathering. One sedimentary-structure sequence and four solitary sedimentary structures inferred to be primary are common in Atoka sandstones of submarine-fan and slope-channel origin.

Bouma sequence.—

Partial to complete Bouma sequences (Bouma, 1962) are the most common sedimentary structure observed in outcrops of Atoka submarine-fan facies, except in the easternmost part of the basin, where relatively proximal-facies sequences are present. Complete sequences displaying divisions A through E of the Bouma sequence are rarely observed. Instead, most beds display partial sequences of divisions B through E in thicker beds and C through E in thinner beds.

Complete and partial Bouma sequences represent waning velocity of turbidity currents (Middleton and Hampton, 1976). The pervasive absence of division A in the Atoka may be more the result of grain size than relative location within the basin of deposition. Atoka sandstones are uniform in grain size and rarely contain grains larger than fine sand. This likely is responsible for the paucity of complete Bouma sequences as well as the general lack of grading observed in Atoka "turbidites."

Massive$_H$ bedding.—

The most common primary sedimentary structure observed in Red Oak cores and in outcrops of proximal submarine-fan facies is characterized by a lack of distinct lamination, yet commonly displays a subtle horizontal fabric defined by elongate particles of humic organic material, micas, or rip-up clasts. Elongate particles are typically present in sufficient abundance to allow recognition of this bedding type in hand specimen (Fig. 4A), although local absence of such particles makes positive identification impossible. In these cases, examination of oriented thin sections reveals elongate grains, mostly foliated metamorphic lithic fragments and micas, that are oriented horizontally. This structure occurs in repetitive beds separated either by an erosional surface or by thin shale beds; internally, these beds display neither grain-size gradation nor sequences of sedimentary structures.

Massive$_H$ bedding is interpreted to represent the "massive sandstone" facies of recent literature (e.g., Walker, 1984). Massive sandstone facies have been equated to division A of the Bouma sequence, although they typically occur as amalgamated deposits that do not display an internal gradation of structures.

Diffuse$_H$ lamination.—

This structure displays a subtle horizontal fabric defined by concentration of humic organic particles, clay minerals, and micas (Fig. 4B). Close examination reveals that the horizontal fabric is actually anastomosing rather than laminar. Diffuse$_H$ bedding occurs randomly in gradational contact with massive$_H$ bedding and may, in fact, represent a more organic- or clay-rich variety of massive$_H$ bedding. This is the second most common structure, inferred to be primary, observed in Red Oak cores. It is rarely observed in outcrops of Atoka submarine-fan facies, perhaps because it is obliterated by weathering of the organic and clay particles.

Plane-parallel lamination.—

As a solitary structure, this bedding type occurs rarely in Red Oak cores and Atoka outcrops. It is characterized by horizontal, alternating, light- and dark-colored laminations (Fig. 4C) that are defined by concentrations of quartz grains (light colored) and pelitic lithic fragments (dark colored), as revealed by thin-section examination. Plane-parallel lamination occurs locally above massive$_H$ bedding and, more rarely, as the only structure present within thin beds. It is probably equivalent to division B of the Bouma sequence, and is inferred to represent deposition from traction as flow velocity waned.

Ripple cross-lamination.—

This is the rarest type of primary structure observed in Red Oak cores, although is is fairly common in Atoka outcrops. Both ripple and ripple-drift cross-lamination occur locally at the tops of relatively thick beds and as a solitary

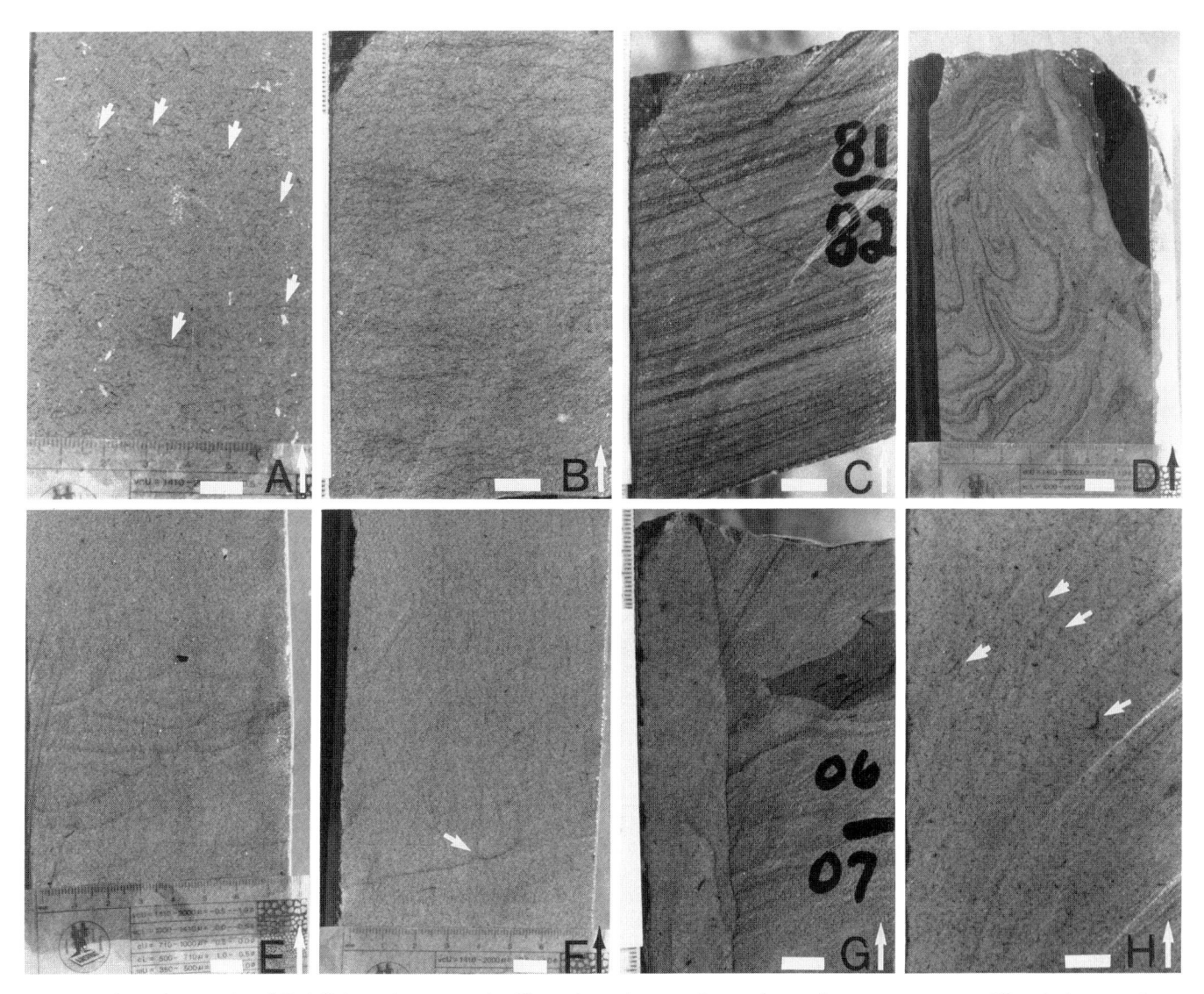

FIG. 4—Core photographs of Red Oak sandstone samples illustrating primary and secondary sedimentary structures. All scale bars are 1 cm; arrows in lower right corner of each photo point up. (A) Massive$_H$ bedding lacks clear stratification but displays a subtle horizontal fabric defined by elongate particles of humic organic material, micas, or clay rip-up clasts (black particles, some indicated by white arrows). Miscellaneous white spots on core are pieces of packing debris and are not part of rock. (B) Diffuse$_H$ lamination displays a subtle horizontal fabric defined by anastomosing concentrations of humic organic particles, clay minerals, and micas (black material in photo). (C) Plane-parallel lamination consists of horizontal, alternating light- and dark-colored laminations defined by concentrations of quartz grains and lithic fragments. Apparent inclination of laminae is structural dip. (D) Convolute lamination consists, in this case, of folded plane-parallel laminae. (E) Dish structures consist of concentrations of fine-grained, dark-colored material (see Plate 4) that accumulated as "consolidation laminations" (Lowe, 1975) during dewatering. (F) Diffuse$_V$ lamination displays a subtle vertical fabric defined by anastomosing concentrations of humic organic particles, clay minerals, and micas (black material in photo). Note that some fine-grained material can be traced downward to the upturned edges of dish structures (white arrow). (G) Pillar (vertical feature at left), with dark consolidation lamination separating it from adjacent sandstone. (H) Massive$_V$ bedding lacks clear stratification but displays a subtle vertical fabric defined by elongate particles of humic organic material, micas, or clay rip-up clasts (black particles, some indicated by white arrows).

sedimentary structure in thin beds (1 to 3 cm). The former probably represents division C of the Bouma sequence and indicates waning flow, whereas the latter may represent overbank deposition adjacent to an active channel. The paucity of ripple bedding may be the result of pervasive amalgamation (i.e., erosion of upper parts of beds prior to deposition of the overlying bed) or may reflect widespread obliteration of primary structures by dewatering (see below).

Secondary Structures

Sedimentary structures inferred to have formed by water escape occupy 74% of the sandstone measured in Red Oak cores, but less than 20% of Atoka sandstones described in outcrops of submarine-fan facies. A spectrum of structures apparently formed by hydroplastic, liquefied, and fluidized modes of deformation (terminology of Lowe, 1975) are present. These three conditions of deformation were defined by Lowe (1975, p. 163 and 165) as follows: "Hydroplastic behavior characterizes grain supported sediments with a significant yield stress and at pore fluid velocities below those required for fluidization." "Liquefied behavior involves the flowage of sediment lacking both cohesive and frictional resistance." "Fluidized deformation takes place at relative pore fluid velocities above those required for minimum fluidization of the greater part of the sediment," and it "is typically turbulent and all primary structures will be erased within fluidized sediments."

Convolute lamination.—

This structure is relatively rare in Red Oak cores, but is conspicuous in the upper few centimeters of numerous beds. In contrast, it is the most abundant secondary structure observed in outcrops of submarine-fan facies, where it can be recognized to occur within the B and C divisions of Bouma sequences. It represents hydroplastic folding of primary structures, mostly plane-parallel lamination (Fig. 4D) and ripple cross-lamination. Although the origin of convolute lamination remains somewhat equivocal (Lowe, 1975; Allen, 1982), most workers agree that it forms when relatively low-permeability laminations inhibit the upward flow of water escaping from underlying sediments, commonly while fluid drag is being exerted at the sediment-water interface (i.e., syndepositionally).

Dish structures.—

These structures are common in Red Oak cores. In hand specimen, they comprise concave-upward concentrations of fine-grained, dark-colored material (Fig. 4E) and tend to grade upward into diffuse$_V$ laminations or into massive$_V$ bedding (see later discussion). In thin section, it can be seen that individual dishes are composed mostly of elongate particles of humic organic material and relatively small proportions of clay. Dish structures form during dewatering as organic material and clay that are carried upward during liquefied or fluidized flow are concentrated beneath a relative permeability barrier (Lowe and LoPiccolo, 1974; Lowe, 1975). Their occurrence in deep-water sandstones of the North Sea has been related to reduction of reservoir quality by Hurst and Buller (1984).

Diffuse$_V$ lamination.—

This structure resembles diffuse$_H$ lamination, except that the organic material, clay minerals, and micas that define the subtle fabric are oriented vertically, or nearly so (Fig. 4F). The vertical fabric is commonly anastomosing and individual concentrations of fine material can locally be traced downward to the upturned edge of a dish structure, as illustrated in Figure 4F. Based on concepts presented by Lowe (1975), we infer that diffuse$_V$ laminations form under conditions of slightly higher fluid-escape velocity than dish structures, and probably represent conditions transitional between liquefaction and fluidization.

Pillars.—

Red Oak sandstones commonly display structures defined as "type B pillars" or "fluidized intrusions" by Lowe (1975). These occur as vertical or near-vertical concentrations of sandstone that are discordant with surrounding sandstone (Fig. 4G). Internally, pillars display fabrics similar to diffuse$_V$ lamination or massive$_V$ bedding. They represent zones "where local fluid escape velocities exceed those required for minimum sediment fluidization" (Lowe, 1975, p. 169). Organic material, clay minerals, and micas are commonly concentrated along the margins of pillars (Fig. 4G), and these represent "consolidation laminations" (Lowe, 1975), where fine and low-density material became concentrated in a fashion similar to that forming dish structures. Pillars commonly cut through sandstone that also displays evidence of liquefied or fluidized deformation (Fig. 4G), indicating that parts of the Red Oak were subjected to multiple dewatering events. Many pillars observed in Red Oak cores are different in color from the surrounding sandstone (Fig. 4G), a reflection of different contents of organic material and clay. Some pillars contain more clay and organic material than surrounding sandstone, whereas others contain less; the former are typically darker colored than surrounding sandstone, whereas the latter are typically lighter colored. Pillars are evidence of local fluidization and represent more turbulent dewatering than the secondary structures described previously.

Massive$_V$ bedding.—

The most common secondary sedimentary structure in Red Oak cores is characterized by a lack of distinct lamination, yet it commonly displays a subtle vertical fabric defined by elongate particles of humic organic material or micas. Elongate particles are commonly present in sufficient abundance to allow recognition of this bedding type in hand specimen (Fig. 4H), although local absence of such particles makes positive identification impossible. In these cases, examination of oriented thin sections usually reveals elongate grains, mostly foliated metamorphic lithic fragments and micas, that are oriented vertically. This structure is pervasive throughout the thickest sandstone accumulations and represents the most abundant bedding type observed in Red Oak cores. Massive$_V$ bedding is interpreted to represent wholesale fluidization of the bed and, as such, represents the most widespread and turbulent dewatering events inferred from the secondary structures described in this paper.

Facies-Sedimentary Structure Relation

Throughout the middle and lower members of the Atoka Formation, there appears to be a close relation between facies deposited by sediment gravity-flow mechanisms and sedimentary structures. Sandstone beds deposited as unconfined turbidites on submarine-fan lobes display a predominance of primary sedimentary structures, with partial to complete Bouma sequences most common. Secondary sedimentary structures are not as abundant and are mostly indicative of non-turbulent water escape (hydroplastic through liquefied conditions).

In contrast, sandstone beds deposited by confined (i.e., channelized) sediment gravity-flow processes, including slope-channel and fan-channel deposits, display a predominance of secondary sedimentary structures, with massive$_V$

bedding most common. Primary structures are relatively uncommon, suggesting they were obliterated by dewatering, and most secondary structures are indicative of turbulent water escape (liquefied through fluidized conditions).

TEXTURE AND FRAMEWORK-GRAIN COMPOSITION

Atoka sandstones are very fine- through fine-grained, and moderately through well sorted. Framework-grain compositions average $Q_{87}F_3L_{10}$, although this mean composition does not include pores inferred to represent dissolution of framework grains. Lithic fragments include slate, phyllite, volcanic, and sedimentary varieties, and feldspars include plagioclase and K-feldspar.

CLAY MINERALOGY

XRD analysis of the <2-μm fraction separated from Atoka sandstones reveals that a limited assemblage of clay minerals is present. Illitic clays and chlorite are pervasive, whereas vermiculite, kaolinite, and dickite are locally present.

Illitic clays are present as illite and mixed-layer illite/smectite (I/S). Some samples contain only illite, whereas others contain varying proportions of illite and I/S (Fig. 5). The relative proportion of illite and I/S present in an individual sample appears to be influenced by two factors, thermal maturity and diagenetic history. I/S tends to occur in relatively low thermal-maturity strata, whereas illite tends to occur in relatively high thermal-maturity strata. Illite also tends to occur preferentially in samples that display a relatively high volume of porosity resulting from dissolution of lithic fragments, suggesting that some of it may have precipitated as a by-product of lithic-fragment dissolution.

Chlorite is present in every sample analyzed from the subsurface (Fig. 5), although its relative abundance is variable. Petrographic and SEM observations suggest that chlorite is present as both detrital and authigenic particles, and these observations are summarized in following sections.

Vermiculite (and locally, mixed-layer vermiculite/chlorite) is present only in outcrop samples (Fig. 5B). The restricted occurrence of vermiculite in outcrop samples and the petrographic similarities of clays in samples that contain vermiculite lead us to believe that it represents weathering of chlorite, which was probably an original component of the clay fraction.

Kaolinite (Fig. 5C) is common in Atoka sandstones, although it is subordinate in volume to illitic clays and chlorite. Dickite (Fig. 5C) occurs as a fracture filling in relatively few samples. The origins of both kaolinite and dickite can be directly inferred from their modes of occurrence.

MODES OF CLAY-MINERAL OCCURRENCE

Clay minerals occur in five distinct modes in Atoka sandstones deposited in submarine-fan and slope-channel environments. These are recognized on the basis of thin-section and SEM observation.

Depositional Matrix

Depositional clay occurrence is most common in sandstones that contain relatively high volumes of clay (more

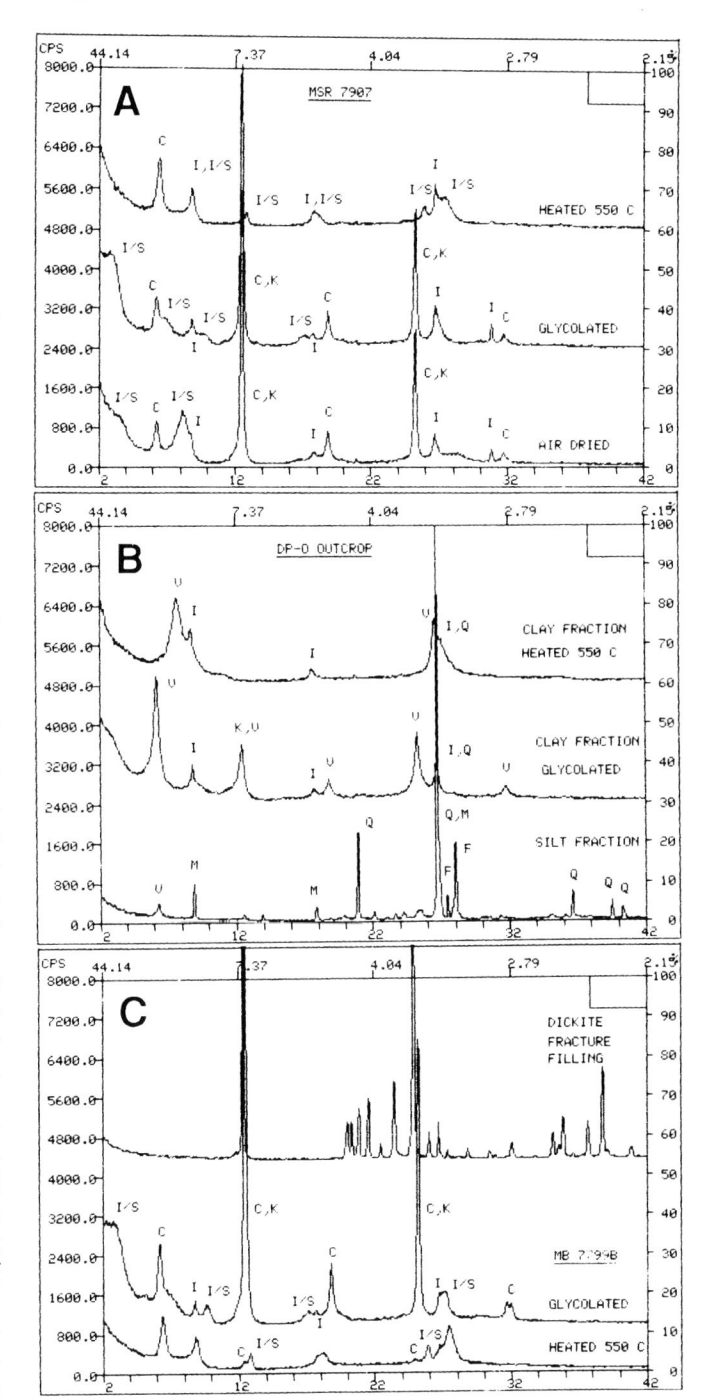

FIG. 5—XRD spectra illustrating mineralogy of fine fractions of Atoka sandstones. (A) Spectra of air-dried, glycolated, and heated samples showing mixed mineralogy of <2-μm fraction of subsurface sample from Red Oak slope-channel facies characterized by dewatering structures, abundant dispersed clay, and grain-supported fabric. (B) Spectra of <2-μm fraction (glycolated and heated) and <62.5-μm fraction (glycolated) of outcrop sample from lower Atoka submarine-fan facies characterized by lack of sedimentary structures, abundant depositional clay, and matrix-supported fabric. (C) Lower two spectra are of <2-μm fraction of glycolated and heated samples from Red Oak slope-channel facies showing mineralogy similar to that in (A). Note distinct kaolinite signature, which is typical of samples that contain grain-alteration kaolinite, as defined in text. Upper spectrum is of bulk dickite fracture fill collected from fracture in core of Red Oak sandstone. Key to labeled peaks: C = chlorite; I = illite; I/S = mixed-layer illite/smectite; K = kaolinite; V = vermiculite; Q = quartz; M = mica; F = feldspar.

than 10% by point-count estimate) and display a matrix-supported fabric characterized by sand-size grains suspended in a matrix of clay- and silt-size particles (Plate 4A). Samples that contain this type of matrix typically are poorly sorted, and it could be argued that the matrix represents the fine "tail" of a detrital-grain population. In fact, the distinction between grains and matrix is arbitrary because there appears to be a continuum of particle sizes from sand through silt and clay. In thin section, this matrix appears to be a nonporous, heterogeneous mixture of clay minerals, humic organic particles, micas, and silt-size grains of quartz (Plate 4A). However, composition is difficult to establish because the matrix typically is dark brown to opaque in plane light and birefringence is masked by the dark color. SEM observations confirm these characteristics, with the matrix appearing as a heterogeneous mixture of clay- and silt-size particles compacted between coarser grains (Fig. 6A-B). XRD analysis reveals a predominance of illite, kaolinite, and chlorite within the clay fraction (<2-μm fraction; Fig. 5B). Analysis of the clay plus silt fraction (<62.5-μm fraction) reveals the same clay minerals plus quartz, feldspar, and mica (Fig. 5B).

Sandstone beds that contain this type of matrix are common within submarine-fan facies of the lower member of the Atoka Formation. In fact, the presence of this matrix is responsible for the widespread perception of Atoka turbidites as "green-colored greywackes" where sandstone beds have been exposed to weathering in the Ouachitas. Measured sections in the Ouachita frontal thrust belt reveal that sandstone beds containing this type of matrix are commonly interbedded with sandstone beds that contain none of this matrix. The former display a general lack of internal sed-

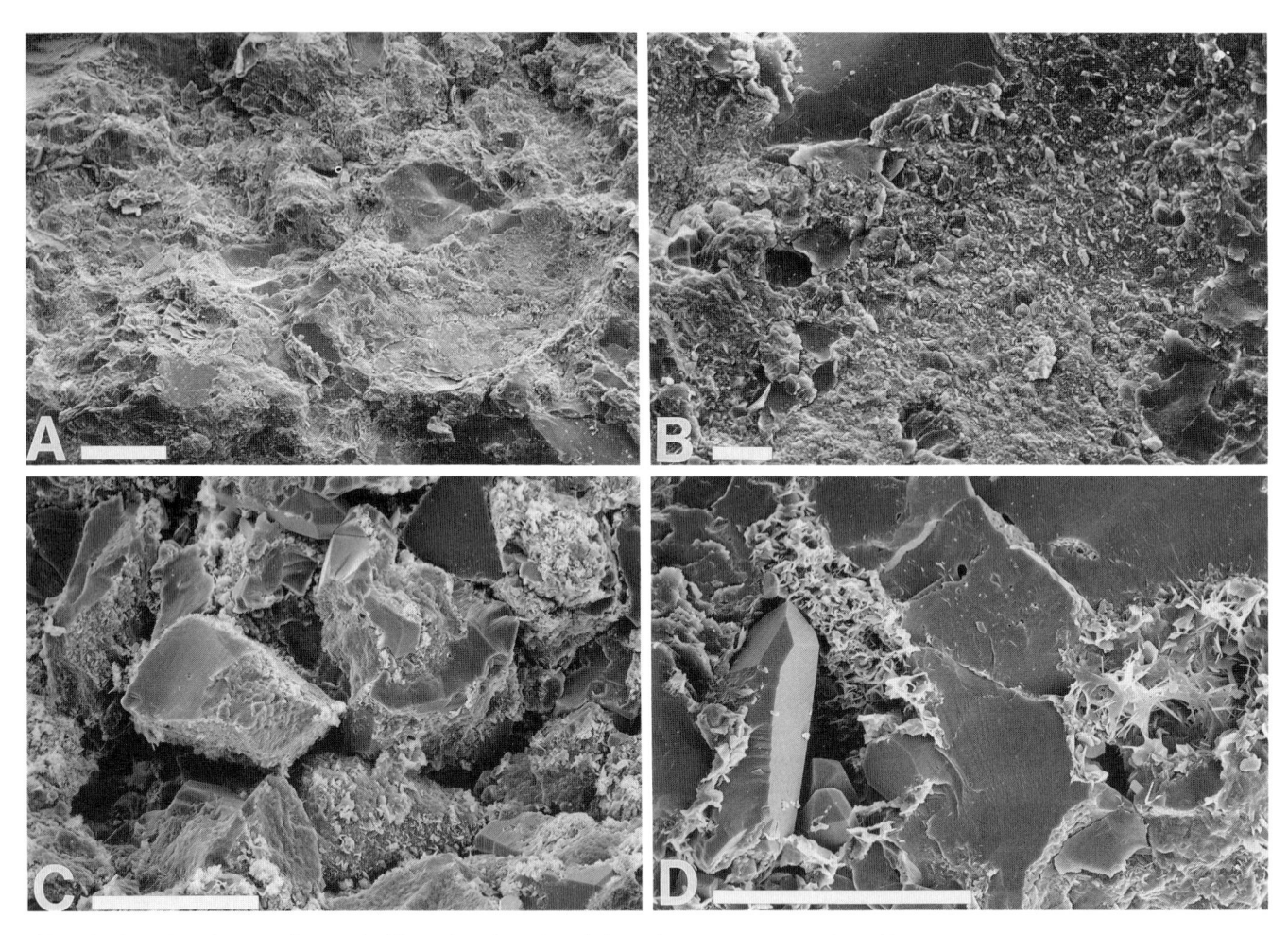

FIG. 6—Scanning electron micrographs illustrating clay minerals in Atoka sandstones. (A) Depositional clay in outcrop sample of lower Atoka submarine-fan facies illustrating matrix-supported fabric. Discrete grains are difficult to recognize and large "crater" is imprint of sand-size grain that was plucked during sample preparation. Compare to Plate 4A, a thin-section photomicrograph of the same sample. Scale bar 100 μm. (B) Higher magnification view inside crater in (A) illustrates inhomogeneity of depositional matrix. Scale bar 10 μm. (C) Dispersed clay in sample from Red Oak slope-channel facies. This sample contains approximately 5% "porous" clay by point-count estimate. Clay grain coatings have only partly inhibited nucleation of quartz overgrowths, as suggested by presence of overgrowths on parts of some grains and absence of overgrowths on other grains. Most clay crystals visible at this magnification represent facade of neoformed clays that nucleated on substrate of clay grain coatings inferred to have formed by dewatering. Scale bar 100 μm. (D) Neoformed clay minerals and authigenic quartz forming "ingrowths" in grain-dissolution pores, Red Oak slope-channel facies. Grain-dissolution pore at left has been mostly occluded by neoformed chlorite (lining pore walls) and authigenic quartz. The euhedral quartz crystals appear to have nucleated on "bald spots" where clay grain coatings were absent at grain contacts. Grain-dissolution pore at right has been mostly occluded by neoformed illite, although some chlorite crystals are also present.

imentary structures, whereas the latter typically display whole or partial Bouma sequences. Both display abundant sole marks, including flute casts and tool marks.

The characteristics described in the preceding paragraphs suggest that these beds were deposited by sandy-debris flows (Middleton and Hampton, 1976), and that the matrix is depositional in origin. That is, the matrix is composed of detrital particles that, when mixed with water, formed a cohesive "fluid" responsible for transport of sand-size grains.

It is likely that the original clay-mineral composition of this depositional matrix has been altered by diagenesis, especially considering the high thermal maturity of these strata, and weathering. Illite and chlorite appear to be the major components of the depositional matrix, although the illite likely has been transformed from a smectite or I/S precursor and the chlorite has been partly altered to vermiculite in outcrop samples. Kaolinite has been observed only as grain-size and grain-shape concentrations that probably formed by diagenetic alteration of feldspar grains (more on this later); it does not appear to have been a primary constituent of the depositional matrix.

Dispersed Clay Minerals

Dispersed clay-mineral occurrence is best developed in sandstones that contain between 4 and 10% clay (by point-count estimate), although it is also present in sandstones that contain less than 4% clay. Samples that contain this type of clay are moderately to well sorted, and there is a clear distinction between sand-size framework grains and clay particles (Plate 4B-H). These clay particles are dispersed through Atoka sandstones in the three following morphologies, as revealed by thin-section and SEM observations.

Grain coatings.—

Clay particles rest on surfaces of sand-size framework grains. Most of the clay particles within grain coatings are not euhedral and are oriented parallel to grain surfaces. Based on petrographic and SEM observations, it appears that grain coatings are mixtures of I/S or illite and chlorite. The high birefringence of the illitic clays makes grain coatings visible in most thin sections, even where they occur beneath facades of neoformed clay minerals or quartz overgrowths. In samples that contain small volumes of clay (less than 4%), framework grains are only partly coated and the partial coatings tend to be thin (less than a few μm; Plate 4B-C). In samples that contain larger volumes of clay, framework grains are completely coated, except at grain contacts, and grain coatings tend to be thicker (3 to 10 μm; Plate 4E-H).

Bridges and ridges.—

Clay particles oriented at high angles to grain surfaces form meniscus-shape bridges between adjacent grains. Bridges are commonly broken during sample preparation and appear in SEM as ridges of clay circumscribing grain-contact scars. Bridges are most pervasive in samples that contain sufficient clay to have complete clay coatings on most framework grains (Plate 4H). Petrographic and SEM observations indicate that bridges are identical in composition to grain coatings, with illite and chlorite predominant.

Consolidation laminations.—

As defined by Lowe (1975), consolidation laminations are surfaces along which hydrodynamically mobile grains have been concentrated as a result of water escape during consolidation (i.e., dewatering). In Atoka sandstones, they include dish structures (Plate 4D), anastomosing concentrations of clay particles that define the distinctive fabric in diffuse$_v$ lamination (Plate 4E), and concentrations of clay particles along the margins of pillars (Plate 4F). Humic organic material and micas are commonly associated with illite and chlorite particles in consolidation laminations. In thin section and SEM, consolidation laminations resemble laminations formed by deposition of suspended sediment (Plate 4D-F). Identification of such clay concentrations as dewatering features requires recognition of consolidation laminations in hand specimen, together with examination of oriented thin sections.

Dispersed clay minerals occurring in the three modes discussed above are absent, or present in negligible volumes, in Atoka sandstone samples collected from beds that display primary sedimentary structures (Plate 4B-C) or convolute laminations. In contrast, they are common in samples collected from beds that display secondary sedimentary structures indicative of liquefied or fluidized dewatering (Plate 4D-H). This close relation between sedimentary structures and the presence and abundance of dispersed clay minerals suggests that most dispersed clay was emplaced by elutriation during dewatering, although the micro-mechanisms by which grain coatings and bridges formed are not clear. Evaluation of these mechanisms would probably require petrographic study of experimentally dewatered sands and/or modern turbidites.

Neoformed Clay Minerals

Neoformed clay occurs in small volumes in porous Atoka sandstones; it rarely has been observed in nonporous samples. Two clay minerals occur in this mode, illite and chlorite (identification based on thin-section, SEM, and XRD criteria). Illite occurs as fibrous crystals that bridge pore spaces and are attached to pore walls at each end (Fig. 6D). Chlorite occurs as solitary, pseudohexagonal crystals that are attached edgewise to pore walls (including clay grain coatings and interior walls of grain-dissolution pores; Figs. 6D). Both minerals are commonly observed attached to the outer surface of clay grain coatings, forming a facade of euhedral crystals on a substrate of non-euhedral clay particles. In rare cases, neoformed clay minerals are so abundant that they effectively occlude most intergranular porosity. Both minerals also commonly occur as partial fillings of grain-dissolution pores; it is not unusual to observe one such pore that contains exclusively illite and a nearby pore that contains exclusively chlorite (Fig. 6D).

Atoka sandstones display abundant evidence that a variety of lithic fragments and feldspar grains dissolved during diagenesis, and it is likely that neoformed clay minerals precipitated during grain dissolution. The close proximity of grain-dissolution pores partially filled by neoformed clay minerals of different composition (illite and chlorite; Fig. 6D) supports the inference that clay precipitation was related to local availability of ions from dissolving framework grains. Neoformed clay minerals that occur as facades on grain coatings may represent recrystallization of the outer layer of an originally detrital clay, or may represent truly neoformed clays that nucleated on a favorable substrate provided by grain coatings of detrital clay.

Grain-Alteration Kaolinite

Grain-alteration kaolinite occurs in small volumes in many Atoka sandstones, although it is more common in porous sandstones than in nonporous ones. It comprises grain-size and grain-shape aggregates of kaolinite (identification based on thin-section, SEM, and XRD criteria). Within each aggregate, kaolinite occurs as euhedral, pseudohexagonal crystals organized into vermicular "books." Tattered remnants of feldspar grains have been observed within kaolinite aggregates (Fig. 7A) in a few samples, but no evidence of a precursor mineral is present in most cases. Some kaolinite aggregates display clay coatings similar to those observed on framework grains.

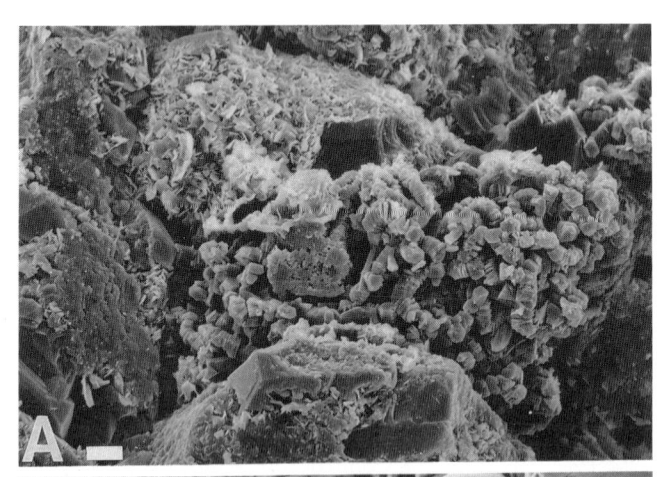

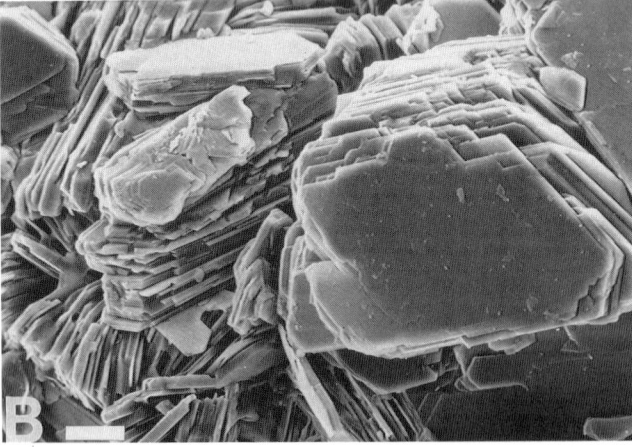

FIG. 7—Scanning electron micrographs illustrating occurrence of kaolin minerals in Red Oak slope-channel facies. (A) Grain-size and grain-shape concentration of vermicular kaolinite crystals associated with tattered remnant of a K-feldspar grain, whose composition was verified by EDS. Note local development of quartz overgrowths where grain surfaces were not completely coated by clay. Scale bar 10 μm. (B) Coarsely crystalline dickite collected from fracture. Scale bar 10 μm.

These observations suggest that kaolinite formed as a byproduct of feldspar grain dissolution during diagenesis of Atoka sandstones. The virtual absence of kaolinite occurring in other modes suggests that its precipitation was closely tied to the local availability of ions, and that it is probably a direct indicator of the former presence of detrital-feldspar grains.

Fracture-Filling Dickite

Several Red Oak cores contain fractures filled with coarsely crystalline dickite (Fig. 5C, 7B). Most of the fractures are nearly vertical, although irregularly shaped fracture openings also have been observed. The fractures and fracture-filling dickite appear to postdate all diagenetic events documented within the sandstone, and there is no mineralogical evidence of fluid interaction between the sandstone-porosity system and the fracture system.

Dickite fracture fillings most likely formed during hydrothermal activity that apparently accompanied thermal maturation of the basin (Houseknecht and others, in prep.). Dickite is considered to be a high-temperature polymorph of kaolinite and is commonly associated with hydrothermal mineralization (e.g., Marumo, 1989). Houseknecht and others (in prep.) have suggested that gas-charged Atoka sandstones were overpressured during hydrothermal activity, and this may have prevented hydrothermal fluids in fractures from entering and chemically interacting with minerals within the sandstone-porosity system.

INFLUENCE OF CLAY MINERALS ON DIAGENESIS AND RESERVOIR QUALITY

Among the five clay modes discussed in the preceding section, the two that were emplaced syndepositionally exerted important influences on subsequent diagenesis and, therefore, reservoir quality. The presence and abundance of depositional matrix and dispersed clay minerals essentially determined the diagenetic pathways that sandstone beds followed during burial diagenesis. To illustrate these influences, the following discussion is divided into end-member examples whose diagenetic histories and reservoir quality are summarized in Figures 8 and 9, respectively. Throughout this discussion, the influence of clay minerals on diagenesis is emphasized and specific diagenetic events are only briefly described. More in-depth discussion of diagenetic events is presented by Houseknecht (1987) and Houseknecht and McGilvery (1990).

Sandstones with Grain-Supported Fabrics

Atoka sandstones that contain little or no depositional matrix display grain-supported fabrics, and their diagenetic histories were significantly influenced by the amount of dispersed clay they contain.

Dispersed clay absent.—

Samples collected from beds that display primary sedimentary structures tend to contain a negligible volume of dispersed clay, less than 4% by point-count estimate. These samples display evidence of a simple diagenetic history in which most intergranular porosity was destroyed by com-

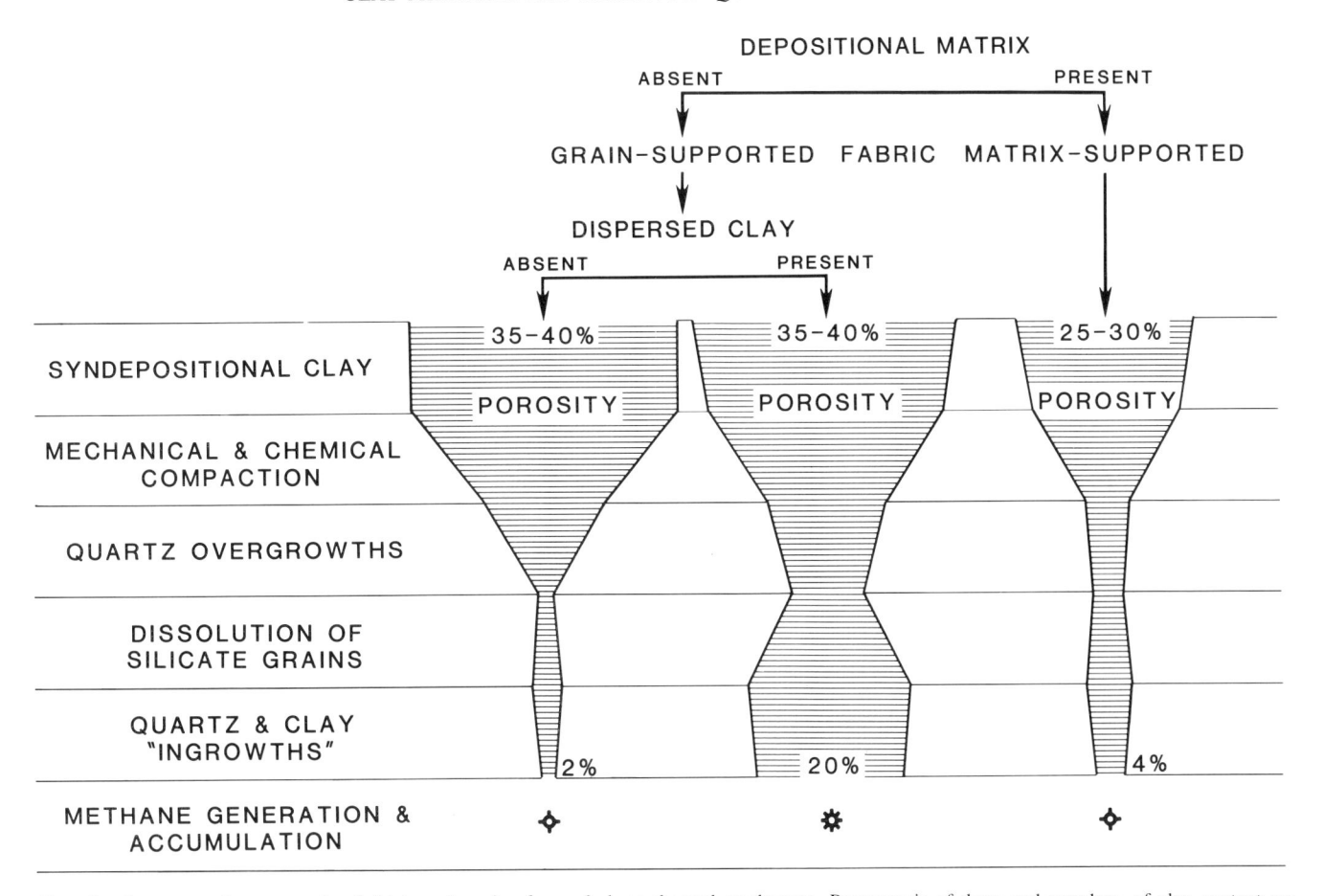

FIG. 8—Summary of paragenesis of Atoka submarine-fan and slope-channel sandstones. Paragenesis of three end members of clay content are illustrated; samples with grain-supported fabrics that contain little or no dispersed clay, samples with grain-supported fabrics that contain dispersed clay, and samples with matrix-supported fabrics that contain abundant depositional clay. Diagenetic events are listed at left in general sequence of occurrence (although there was some overlap in timing), and width of striped areas illustrates how each diagenetic event influenced average porosity.

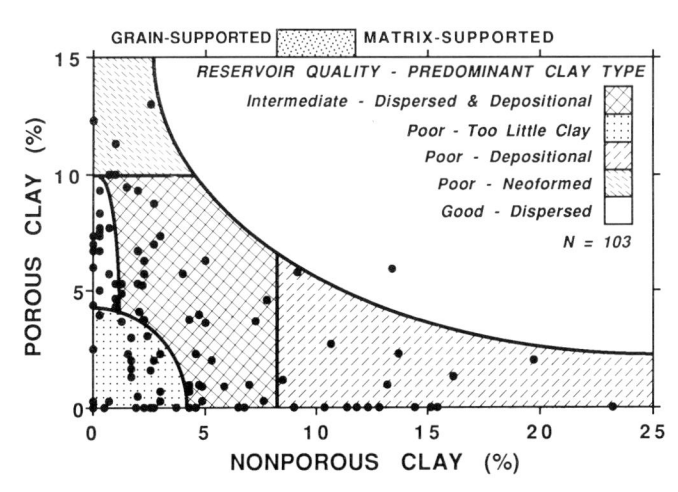

FIG. 9—Summary of relation between clay content and reservoir quality. Nonporous clay and porous clay represent arbitrary point-count categories as defined in text. Sandstones with less than 8% nonporous clay display grain-supported fabrics, those with more than 12% nonporous clay display matrix-supported fabrics, and those with 8 to 12% nonporous clay display fabrics gradational between grain- and matrix-supported. Reservoir-quality fields are subjectively defined on the basis of petrographic observations, laboratory measurements of porosity and permeability, and production history of wells from which samples were collected.

paction and quartz cementation (Fig. 8). Mechanical compaction was predominant and included rearrangement of brittle grains (mostly quartz and feldspar) and plastic deformation of ductile grains (mostly metamorphic and sedimentary lithics). Chemical compaction also occurred, as evidenced by grain contacts indicative of intergranular pressure solution, but was not as important to reduction of intergranular volume. Quartz cementation involved precipitation of syntaxial overgrowths on quartz grains, an event that destroyed virtually all intergranular porosity. Nucleation of quartz overgrowths occurred on grains that apparently have no clay grain coatings, as well as on grains that display incomplete clay grain coatings. In the latter cases, clay coatings are conspicuous in thin section because of high birefringence of the clays, which XRD analysis confirms to be mostly illite. Following precipitation of quartz overgrowths, other diagenetic events occurred in a sequence similar to that discussed in the following section, but in a subdued manner owing to the general lack of remaining porosity.

Dispersed clay present.—

Samples collected from beds that display secondary sedimentary structures, especially those indicative of fluidized dewatering, tend to contain a larger volume of dispersed clay, typically more than 4% by point-count estimate. Dispersed clay occurs as grain coatings, bridges, and consolidation laminations.

Significantly, these samples display evidence of a more complex diagenetic history (Fig. 8). Evidence of compaction is essentially identical to that discussed in the preceding section. However, syntaxial quartz overgrowths are conspicuously absent, or only are present locally and in small volumes. It appears that the presence of dispersed clay, especially as grain coatings, effectively inhibited the nucleation of syntaxial overgrowths on quartz grains. Such samples are commonly found in close association with samples that contain no clay grain coatings and abundant quartz overgrowths. For example, in a sequence of amalgamated beds, one bed may contain clay, no overgrowths, and porosity, whereas an adjacent bed may contain no clay, overgrowths, and no porosity. In other cases, the volumes of clay, overgrowths, and porosity are gradational within a single bed. These observations suggest that the presence or absence of quartz overgrowths (and therefore, primary porosity) was controlled by the presence or absence of nucleation sites, rather than the geochemistry or flow patterns of diagenetic waters.

Porosity was enhanced in these sandstones as a result of framework-grain dissolution (Plate 4H). Petrographic evidence suggests that metamorphic lithic fragments and feldspar grains were the most abundant grain types dissolved, although volcanic and sedimentary lithic fragments, originally present in smaller volumes, were also dissolved. In general, the volume of grain-dissolution porosity appears to be directly related to the volume of preserved primary porosity. In samples that have high volumes of primary porosity preserved by the presence of dispersed clay, most dissolution products appear to have been removed in solution, as dissolution pores contain little authigenic material. Some of the authigenic-clay minerals that occupy intergranular space may have precipitated as by-products of grain dissolution, but it is impossible to link those clay minerals genetically with grain dissolution on the basis of petrographic observations.

In contrast, authigenic by-products of grain dissolution are more abundant in samples that have lower volumes of preserved primary porosity. Pores generated by dissolution of lithic fragments commonly contain authigenic quartz and clays, with chlorite predominant in some pores and illite in others. Local variation in authigenic-clay mineralogy in grain-dissolution pores suggests that pore-fluid chemistry was influenced by the composition of dissolving grains on a local scale. Pores generated by dissolution of feldspar grains are commonly occupied by kaolinite or calcite, which apparently represent by-products of K-feldspar and plagioclase dissolution, respectively, as indicated by local preservation of remnants of original grains.

Grain-dissolution pores that are partly occluded by authigenic-quartz provide additional evidence as to the effectiveness of clay grain coatings in inhibiting nucleation of syntaxial-quartz overgrowths. These authigenic-quartz crystals are referred to as "ingrowths" in Figure 8 because they appear to have nucleated on the walls of grain-dissolution pores and grown inward toward the centers of those pores (Fig. 7B). Significantly, authigenic quartz that occurs within grain-dissolution pores appears to have nucleated on "bald spots" on quartz grains, which represent grain contacts where dispersed clay coatings did not form. Such sites became available for authigenic-quartz nucleation only after dissolution of a chemically unstable grain that had been in contact with a quartz grain. In the same samples, syntaxial-quartz overgrowths that occlude intergranular space are absent or rare. These observations suggest that pore fluids were supersaturated with respect to quartz, but that nucleation was only possible where "bald" surfaces of detrital-quartz grains were exposed to pore fluids.

A small number of samples also display abundant pore-filling clay that occupies intergranular space. This clay is predominantly neoformed chlorite that apparently precipitated at the same time as the clay minerals that occupy grain-dissolution pores, although unequivocal evidence of timing is lacking.

Hydrocarbon accumulation followed the events described earlier. Shales of the Atoka Formation contain a predominance of humic (type III) kerogen and it is likely that methane, which now occupies the pores, was the only hydrocarbon generated in significant volumes. Point-count data indicate that primary intergranular porosity preserved by the presence of dispersed clay accounts for 45% and pores generated by dissolution of framework grains account for 55% of the total porosity in the Red Oak sandstone. Thus, the presence of dispersed clay minerals, and specifically clay grain coatings that inhibited quartz cementation, not only preserved a significant volume of primary porosity but also allowed the development of grain-dissolution porosity in sandstones where primary porosity had been preserved.

Sandstones with Matrix-Supported Fabrics

Atoka sandstones that contain abundant depositional matrix display matrix-supported fabrics and generally poorer sorting than sandstones discussed previously. Initial porosity was probably low and most primary porosity appears to have been destroyed during shallow burial by mechanical compaction. Petrographic and SEM observations show that sand-size grains appear compressed into a pervasive matrix that is composed of a heterogeneous mixture of clay minerals and silt-size quartz grains (Fig. 6A).

Compaction of matrix-supported sandstones during shallow burial effectively destroyed most porosity and permeability, thereby inhibiting diagenetic reactions during deeper burial. These sandstones display evidence of paragenesis that was similar to that of grain-supported sandstones, but on a more subdued scale (Fig. 8).

Influence of Clay on Reservoir Quality

Figure 9 summarizes the influence of clay minerals on reservoir quality in Atoka sandstones. The diverse clay types described in preceding sections cannot always be positively identified in thin section, so clay minerals were classified as either porous or nonporous, depending on the presence or absence of visible porosity (as indicated by blue-stained epoxy) within clay masses encountered during point counting. This technique provided objectively defined categories that were further characterized by description of sedimentary structures, SEM observations, and XRD analyses. This approach allows us to infer that most nonporous clay represents depositional matrix and that most porous clay represents a combination of dispersed detrital clay and neoformed clay (Fig. 9). As indicated at the top of Figure 9, sandstones with less than 8% nonporous clay display grain-supported fabrics, those with more than 12% nonporous clay display matrix-supported fabrics, and those with 8 to 12% nonporous clay display fabrics gradational between grain and matrix supported. Reservoir-quality fields were subjectively drawn on Figure 9 by integrating petrographic observations, helium porosity and permeability data collected from conventional core samples, and production records of wells from which core samples were analyzed. We emphasize that Figure 9 is based on point-count estimates of clay abundance, and is therefore prone to errors that typically accompany attempts to estimate clay volumes from thin-section observations. Moreover, we emphasize that the reservoir-quality fields are subjectively drawn and are intended to be gradational in nature; in fact, reservoir quality varies dynamically as clay content and clay types vary among samples.

Sandstones with grain-supported fabrics and low volumes of porous clay plot in the lower left corner of Figure 9. Helium porosity and permeability determinations indicate that these samples typically have less than 5% porosity and less than 0.1 md permeability. The poor reservoir quality of these sandstones can be directly attributed to the presence of quartz overgrowths, which nucleated on surfaces of detrital-quartz grains that were not coated by clay (Fig. 9).

Sandstones with grain-supported fabrics that display more than a few percent porous clay contain little or no quartz cement and, therefore, have preserved primary porosity as well as grain-dissolution porosity. However, both porosity and permeability are widely variable. Samples that plot within a narrow "window" of clay content, 4 to 10% porous clay and 0 to 2% nonporous clay, are characterized by the highest porosities (>20%) and permeabilities (10 to 1,000 md) in the Atoka Formation. These samples have sufficient clay to have inhibited the nucleation of quartz overgrowths but not so much that permeability has been reduced to low values. Samples that contain more than 10% porous clay display intergranular space choked with dispersed and neoformed clay that renders them of poor reservoir quality; they are characterized by 5 to 10% porosity and less than 1 md permeability. Samples that have more than 2% nonporous clay and whose combined total of porous and nonporous clay is 4 to 12% are of intermediate reservoir quality, with 10 to 20% porosity and 1 to 10 md permeability.

Sandstone samples grade into matrix-supported fabrics as the content of nonporous clay exceeds 8%. These are characterized by 2 to 10% porosity and less than 0.1 md permeability. Matrix-supported sandstones have poor reservoir potential because most porosity was destroyed by compaction during shallow burial and because subsequent diagenesis did not effectively enhance porosity (Fig. 8).

CONCLUSIONS

Sandstones of the Atoka Formation display a variety of primary and secondary sedimentary structures, formed by sediment gravity-flow depositional processes and dewatering, respectively. Primary sedimentary structures include complete and partial Bouma sequences, massive$_H$ bedding, diffuse$_H$ lamination, plane-parallel lamination, and ripple cross-lamination. They are preferentially associated with facies indicative of unconfined turbidite deposition on submarine-fan lobes. Secondary sedimentary structures include convolute lamination, dish structures, diffuse$_V$ lamination, pillars, and massive$_V$ bedding. They are preferentially associated with facies indicative of channelized-turbidite deposition in fan channels and slope channels.

Oriented thin-section and SEM observations reveal that clays are present in five distinct modes in Atoka sandstones. (1) Depositional matrix is a heterogeneous mixture of illitic clays, chlorite, and silt-size quartz and feldspar grains that results in a matrix-supported fabric. It is likely detrital in origin and was probably emplaced as part of the depositional medium in sandy-debris flows. (2) Dispersed clay minerals are mixtures of illitic clays and chlorite that form grain coatings, bridges between grains and consolidation laminations. They are most abundant in samples that display secondary sedimentary structures, and are inferred to represent detrital clays emplaced during turbidite dewatering. (3) Neoformed clay minerals are pure, euhedral concentrations of chlorite and illite that line pore walls and partly fill grain-dissolution pores. Their association with grain-dissolution porosity suggests they precipitated from pore fluids during, or after, dissolution of lithic fragments and/or feldspar grains. (4) Grain-alteration kaolinite occurs as grain-size and grain-shape concentrations of vermicular aggregates that locally contain remnants of dissolved feldspar grains, indicating it represents a by-product of feldspar dissolution. (5) Dickite occurs as a local fracture fill within Atoka sandstones, and may be evidence of hydrothermal activity during late stages of diagenesis.

The distribution and abundance of depositional matrix and dispersed clay minerals exerted important influences on subsequent diagenesis of Atoka sandstones. Among sandstones with grain-supported fabrics, diagenesis was dependent on the volume of dispersed clay present. Those with little or no dispersed clay were pervasively cemented by quartz overgrowths, which nucleated on detrital-quartz grains. In contrast, those that contain dispersed clays retained primary porosity because quartz-overgrowth nucleation was inhibited by clay grain coatings on detrital-quartz grains. Having retained a primary porosity system through which fluids could migrate, these sandstones underwent porosity enhancement by dissolution of feldspar grains and lithic fragments. Sandstones with matrix-supported fabrics appear to have had little primary porosity and, as a result, no significant development of dissolution porosity.

These observations indicate that the distribution of detrital-clay minerals in Atoka sandstones was controlled by processes associated with deposition and dewatering of submarine-fan and slope-channel facies. Moreover, the presence and distribution of detrital clays influenced subsequent diagenesis, and thereby determined the amount and distribution of porosity in Atoka sandstones. These relations suggest that "infiltrated" clays emplaced by dewatering may be a key to understanding the distribution of reservoir quality in deep-water sandstone facies. Petrographic examination of experimentally dewatered sands and young turbidites that have not been subjected to high-temperature diagenesis will likely be required to evaluate this hypothesis.

REFERENCES

ALLEN, J. R. L., 1982, Sedimentary structures, their character and physical basis, v. 2, *in* Developments in Sedimentology, 30B: Amsterdam, Elsevier, 663 p.

BOUMA, A. H., 1962, Sedimentology of some flysch deposits: Amsterdam, Elsevier, 168 p.

CRONE, A. J., 1975, Laboratory and field studies of mechanically infiltrated matrix clay in arid fluvial sediments: Unpublished Ph.D. Dissertation, University of Colorado, Boulder, 162 p.

HELWIG, J., 1970, Slump folds and early structures, northeastern Newfoundland: Journal of Geology, v. 78, p. 172–187.

HOUSEKNECHT, D. W., 1986, Evolution from passive margin to foreland basin: the Atoka Formation of the Arkoma Basin, south-central U.S.A., *in* Allen, P. A., and Homewood, P., eds., Foreland Basins: International Association of Sedimentologists, Special Publication No. 8, p. 327–345.

HOUSEKNECHT, D. W., 1987, The Atoka Formation of the Arkoma Basin: Tectonics, sedimentology, thermal maturity, sandstone petrology: Tulsa Geological Society, Short Course Notes, 72 p.

HOUSEKNECHT, D. W., AND MCGILVERY, T. A., 1990, Red Oak field, *in* Beaumont, E. A., and Foster, N. H., eds., Structural Traps II. Traps Associated with Tectonic Faulting: Treatise of Petroleum Geology, Atlas of Oil and Gas Fields: American Association of Petroleum Geologists, p. 201–225.

HOUSEKNECHT, D. W., WOODS, M. O., AND KASTENS, P. H., 1989, Transition from passive margin to foreland basin sedimentation: The Atoka Formation of the Arkoma Basin, Arkansas and Oklahoma, *in* Vineyard, J. D., and Wedge, W. K., ed., Geological Society of America, 1989 Field Trip Guidebook: Missouri Department of Natural Resources, Special Publication No. 5, p. 121–137.

HURST, A., AND BULLER, A. T., 1984, Dish structures in some Paleocene deep-sea sandstones (Norwegian sector, North Sea): Origin of the dish-forming clays and their effect on reservoir quality: Journal of Sedimentary Petrology, v. 54, p. 1206–1211.

LOWE, D. R., 1975, Water escape structures in coarse-grained sediments: Sedimentology, v. 22, p. 157–204.

LOWE, D. R., AND LOPICCOLO, R. D., 1974, The characteristics and origins of dish and pillar structures: Journal of Sedimentary Petrology, v. 44, p. 484–501.

MARUMO, K., 1989, Genesis of kaolin minerals and pyrophyllite in Kuroko deposits of Japan: implications for the origins of the hydrothermal fluids from mineralogical and stable isotope data: Geochimica et Cosmochimica Acta, v. 53, p. 2915–2924.

MATLACK, K. S., HOUSEKNECHT, D. W., AND APPLIN, K. R., 1989, Emplacement of clay into sand by infiltration: Journal of Sedimentary Petrology, v. 59, p. 77–87.

MIDDLETON, G. V., AND HAMPTON, M. A., 1976, Subaqueous sediment transport and deposition by sediment gravity flow, *in* Stanley, D., and Swift, D., eds., Marine Sediment Transport and Environmental Management: New York, Wiley, p. 197–218.

MORAES, M. A. S., AND DEROS, L. F., 1990, Infiltrated clays in fluvial Jurassic sandstones of Reconcavo basin, northeastern Brazil: Journal of Sedimentary Petrology, v. 60, p. 809–819.

VEDROS, S. G., AND VISHER, G. S., 1978, The Red Oak sandstone: a hydrocarbon-producing submarine fan deposit, *in* Stanley, D., and Kelling, G., eds., Sedimentation in Submarine Canyons, Fans, and Trenches: Stroudsburg, Pennsylvania, Dowden, Hutchinson, and Ross, p. 292–308.

WALKER, R., 1984, Turbidites and associated coarse clastic deposits, *in* Walker, R., ed., Facies Models: Geoscience Canada Reprint Series 1 (2nd ed.), p. 171–188.

WALKER, T. R., 1976, Diagenetic origin of continental red beds, *in* Falke, H., ed., The Continental Permian in Central, West, and South Europe: Boston, Reidel Publishing Co., p. 240–282.

WALKER, T. R., WAUGH, B., AND CRONE, A. J., 1978, Diagenesis in first-cycle desert alluvium of Cenozoic age, southwestern United States and northwestern Mexico: Geological Society of America Bulletin, v. 89, p. 19–32.

WILSON, M. D., AND PITTMAN, E. D., 1977, Authigenic clays in sandstones: recognition and influence on reservoir properties and paleoenvironmental analysis: Journal of Sedimentary Petrology, v. 47, p. 3–31.

CLAY COATS: OCCURRENCE AND RELEVANCE TO PRESERVATION OF POROSITY IN SANDSTONES

EDWARD D. PITTMAN
Department of Geosciences, University of Tulsa, Tulsa, Oklahoma 74104
RICHARD E. LARESE
Research Center, Amoco Production Co., Box 3385, Tulsa, Oklahoma 74102
AND
MILTON T. HEALD
Geology Department, West Virginia University, Morgantown, West Virginia 26506

ABSTRACT: Clay coats, which may be continuous or discontinuous, originate from soils as cutans, from infiltration of clay in sand and sandstone, and authigenically as newly formed or regenerated clay minerals. Allogenic cutans and infiltration deposits have a laminar morphology, whereas authigenic-clay coats commonly have a radial morphology.

Thick, well-developed, continuous clay coats, regardless of origin, may retard quartz cementation by masking the surface of detrital-quartz grains and preventing the nucleation of quartz overgrowths. Chlorite is the most effective of the clay minerals in preserving intergranular porosity and appears to be important in very deep sandstone reservoirs. The most favorable amount of chlorite to preserve porosity is variable: 4 to 7 volume percent for the Berea Sandstone and 5 to 13 volume percent for the Tuscaloosa Sandstone, for example. Smaller amounts of chlorite permit quartz to nucleate and destroy porosity and greater amounts result in porosity reduction by infill of pores. Clay coats do not retard epitaxial cements (e.g., carbonates and sulfates), which may cover clay coats and occlude porosity. Clay coats may occur in highly lithic (e.g., >35% lithic material) sandstones, but are not important because physical compaction dominates diagenesis and destroys porosity.

Experimental growth of clay coats shows that clay flakes are flatly attached to detrital-sand grains and curl upward to form a radial-fibrous morphology. This attached root zone may explain why clay coats are effective at blocking nucleation of quartz cement. Experimental work also shows that mineralogy may provide an initial substrate control over the precipitation of clay coats by providing an *in situ* source of the cations needed to precipitate the clay. Later, the clay coats nucleate on other framework grains farther from the site of initial nucleation.

INTRODUCTION

Mineralogically, clay coats may be smectite, mixed-layer illite/smectite, illite, chlorite, or mixed-layer chlorite/smectite (corrensite). Clay coats, which also are commonly referred to as clay coatings, clay rims, or pore-lining clay, may be allogenic or authigenic in origin. Allogenic-clay coats, which have a laminar morphology, may form as cutans in soils and from infiltration of clay in sand or sandstone. Authigenic-clay coats originate as newly formed or regenerated clay minerals and typically have a radial morphology.

Clay coats retard nucleation of quartz overgrowths by physically blocking the nucleation sites on host detrital-quartz grains. Therefore, extensive, continuous, and thickly developed clay coats, regardless of mineralogy, origin, and morphology may be effective in retarding quartz cementation in quartz-rich sandstones. Sparse, discontinuous, and thin clay coats are ineffective at retarding quartz-overgrowth development. The importance of clay coats to the preservation of intergranular porosity has been documented by numerous workers (e.g., Heald, 1965; Horn, 1965; Pittman and Lumsden, 1968; Heald and Larese, 1974; Thomson, 1979; Smith, 1985). Most of the examples of porosity preservation involve authigenic clay; however, Molenaar (1986) and Moraes and De Ros (1990) described thick-infiltration clay coats that prevented nucleation of quartz cement. Clay coats also may serve as barriers to replacement (Bastin and others, 1931).

The purpose of this paper is to discuss the characteristics of clay coats and the nature of the clay-coat blocking mechanism, which is vital to porosity preservation in many deeply buried reservoirs. Evidence comes from the study of subsurface sandstones as well as from experimental petrology involving the precipitation of quartz overgrowths and clay coats.

We are following the recommendations of the AIPEA nomenclature committee (Bailey, 1980) and consider berthierene to be a 7Å Fe-rich 1:1 clay and chamosite to be a 14Å 2:1 clay of similar composition. In the past, there has often been confusion between these minerals and chamosite commonly was identified as a 7Å mineral. Chamosite is the Fe-rich member of the trioctahedral chlorite subgroup. Most authigenic chlorite coats are chamosite.

OCCURRENCE OF CLAY COATS

Table 1 lists occurrences of authigenic chlorite (and corrensite) coats in sandstones from a large variety of depositional environments. As discussed in detail later, experimental petrology indicates that the necessary cations for authigenic-clay coats can be derived *in situ* from alteration of lithic fragments. This can explain how chlorite coats in sublitharenites such as in the Tuscaloosa Formation formed in a variety of depositional environments.

Some occurrences of authigenic chlorite (and corrensite) coats are believed to be related to depositional environment (Table 1), although there is no universally preferred environment for coats to develop. Examples of depositional-environment control are discussed below for the Horsethief, Berea, Spiro and Belly River sandstones. In the Horsethief Sandstone, Montana, corrensite coats are developed in distributary-channel and mouth-bar facies, whereas smectite of unspecified form is the dominant clay in the various nearshore marine facies (Almon and others, 1976). This was attributed to the initial water chemistry by Almon and others (1976). Smectite developed in marine sediments where the Mg/Ca ratio was lower, whereas corrensite was favored

TABLE 1.—SANDSTONES WITH CHLORITE COATS

AGE	FORMATION	DEPOSITIONAL ENVIRONMENT	SANDSTONE COMPOSITION[2]	LOCATION	REFERENCE
Cret.	Belly River	Deltaic	Vol. Lith. Arenite	Alberta Basin, Canada	Carrigy and Mellon (1964)
Miss.	Big Injun	Deltaic	Lith. Arenite	West Virginia	Heald (1965)
Jurassic	Dogger-Beta	?	?	Holstein Trough, Germany	Horn (1965)
Penn.	Spiro	Channel	Quartz Arenite	Arkoma Basin, Oklahoma	Pittman and Lumsden (1968)
Miss.	Berea	Fluvial/Bar	Quartz Arenite-Sublith. Arenite	West Virginia	Larese (1974)
Cret.	Horsethief[1]	Distrib. Channel/Bar	Vol. Lith. Arenite	Wyoming	Almon and others (1976)
Triassic	Unnamed	Fluvial	Subarkose	UK Sector, North Sea	Taylor (1978)
Penn.	Strawn	Deltaic	Sublith. Arenite	North-Central Texas	Land and Dutton (1978)
Cret.	Tuscaloosa	Flunial/Deltaic/Bar	Lith. Arenite	Louisiana	Smith (1985); Thomson (1979)
Cret.	Frontier	Shelf Ridges	Subarkose	Powder River Basin	Winn and others (1983)
Penn.	Granite Wash	Fan Delta	Arkose	Mobeetie Field, Anadarko Basin, Texas	Dutton and Land (1985)
Miocene	Unnamed	Shallow Marine	Lith. Arenite	Matagorda, Offshore TX	Thayer (1985)
Jurassic	Cotton Valley	Marine	Mixed	Catahoula Creek Field, MS	Janks and others (1985)
Neogene	Surma Gp.	?	?	Bengal Basin, Bangladesh	Imam and Shaw (1987)
Penn.	Springer	Shallow Marine	Quartz Arenite	Anadarko Basin	McBride and others (1987)
Cret.	Parkman	Marine	Subarkose	Powder River Basin	Dogan and Brenner (1983)
Jurassic	Norphlet	Alluvial Fan/Eolian	Arkose-Subarkose	Florida, Alabama, Miss.	Dixon and others (1989)
Penn.	Red Oak	Deep Fan	Lithic Arenite	Arkoma Basin, Oklahoma	Unpublished
Penn.	Fanshawe	Deep Fan	Lithic Arenite	Artkoma Basin, Oklahoma	Pittman and Wray (1989)
Penn.	Goddard	Deltaic	Quartz Arenite	Anadarko Basin, Oklahoma	Unpublished
Penn.	Red Fork	Deltaic	Sublith. Arenite	Oklahoma	Unpublished
Penn.	Granite Wash	Channels/Bars	Arkose	Elk City Field, Anadarko Basin, Oklahoma	Unpublished
Cret.	Rijnland[1]	Offshore Bar	Subarkose	Rijn Field, Offshore Netherlands	Unpublished

[1] Corrensite coats.
[2] As published; classification varies.

by the higher Mg/Ca ratio in the other environments. The Berea Sandstone of West Virginia may be a similar situation where chlorite-coat development was controlled by water chemistry (Larese, 1974). In the Berea, chlorite coats occur in fluvial channels, but not in associated marine bars and sheet sands. Well-developed chlorite coats are restricted to the channel facies of the Spiro sand, Red Oak Field, Oklahoma (Houseknecht and McGilvery, 1990). The tidal-flat sandstone facies, into which the channels are incised, lack well-developed chlorite coats and is usually tight. The channel facies contains clay pellets, particularly near the base of the channel, which may be the source of the cations for the chlorite coats. The Belly River Sandstone, Pembina Field, Canada, has a preferential development of chlorite coats within distributary channels. This may be attributed to the mixing of fresh and saline waters (Longstaffe, 1986), because there is no apparent difference in the mineralogical composition of the sandstone to explain the chlorite coats.

Authigenic-chlorite coats commonly are Fe rich. Microprobe analyses indicate that chlorite coats in the Norphlet Sandstone, offshore Alabama and Florida, are unusually Mg rich compared with chlorite coats analyzed from the Tuscaloosa and Berea Formations (Fig. 1). The source of magnesium is believed to be waters derived from the underlying Louann Salt (McHugh, 1987). In Norphlet core studies, the Mg/Fe ratio in chlorite increases with depth over approximately a 175 m (575 ft) interval in one well (Fig. 2). Lower Mg/Fe ratios of chlorite occur in Norphlet Sandstone intervals that have been subjected to apparently reducing con-

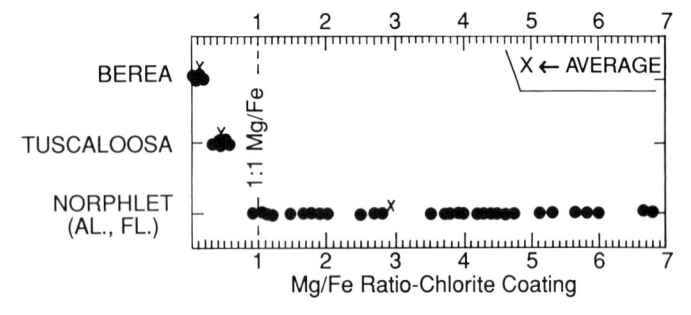

FIG. 1.—Comparison of Mg/Fe ratios, as determined by microprobe analyses, for chlorite coats in Norphlet, Tuscaloosa, and Berea sandstones. Note the high Mg/Fe ratios for the Norphlet.

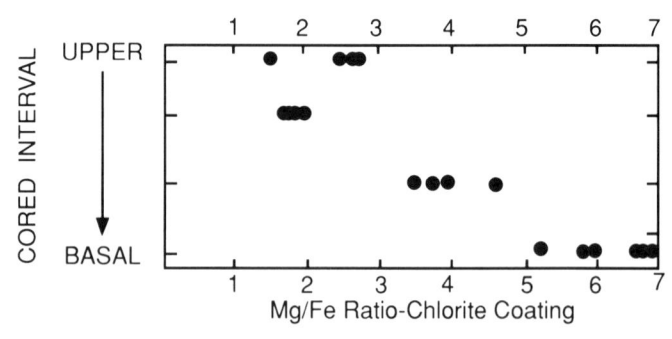

FIG. 2.—Diagram showing a downward increase in the Mg/Fe ratio for chlorite coats in the Norphlet Sandstone, which may be related to the source of the magnesium. Data are from one well with a cored interval of about 175 m (575 ft).

ditions associated with the presence of hydrocarbons, which now occur as bitumen in pores. In upper portions of the Norphlet cored intervals, near or above the oil-water contact, the reduction of precursor iron-oxide rims (Dixon and others, 1989) may have provided a source of Fe^{+2} for the replacement of Mg^{+2} within the chlorite structure.

PRESERVATION OF POROSITY

Heald and Larese (1974) noted that chlorite was more effective than illite in preserving porosity. Illite coats may retard quartz-overgrowth development, but illite also appears to catalyze pressure solution where illite coats are under stress along quartz-grain contacts (Heald, 1956; Thomson, 1959). South State Line Field, Mississippi, is interesting because the eolian Norphlet Sandstone produces from an upper and lower facies, which are characterized by illite and chlorite coats, respectively (Thomson and Stancliffe, 1990). Apparently, the pore-bridging illite coats have promoted pressure solution, whereas the chlorite coats have not promoted pressure solution and have preserved porosity. This is reflected in reservoir properties. The illitic facies has 9.5% porosity and 0.6 md permeability compared with 16.5% and 15.5 md, respectively, for the chloritic facies (Thomson and Stancliffe, 1990).

Chlorite and other related Fe-bearing clay minerals (e.g., corrensite) commonly have been reported as occurring in hydrocarbon reservoirs (Table 1). Note in Table 1 the wide range in age, depositional environment, and sandstone composition. The key to effective preservation of porosity is the blockage of nucleation sites on detrital quartz grains. This is accomplished by thick, continuous clay coats. Disruption of clay coats leading to exposure of the surface of the quartz grains to formation water results in precipitation of a prismatic-quartz overgrowth. Thin, discontinuous clay coats, even though readily visible in thin section, are generally ineffective at retarding quartz cementation. Clay coats do not affect the precipitation of epitaxial cements such as carbonates and sulfates. Moreover, clay coats are not effective at preserving porosity in lithic sandstones with a low percentage (e.g., 65% or less) of quartz. Ductile deformation of lithic fragments is the dominant porosity-destruction process in these compositionally immature sandstones, and quartz cement is usually of minor importance.

The density of the underside of a clay coat, adjacent to the surface of the detrital grain, probably is critical in blocking the nucleation of quartz overgrowths. Figures 3A and B show the underside of a critical-point-dried smectite coat. Smectite crystals form an effective coat because they nucleate flatly attached to the detrital surface and curl away from that surface. This crystal morphology leads to a dense, effective coat. The underside of a chlorite coat, although not as dense appearing as the smectite example, has a much finer crystalline appearance than the radial fabric that has grown outward into the pore (Figs. 3C–D).

The most favorable range in the volume of clay as grain coats needed to preserve porosity varies with the sandstone. For the Tuscaloosa, a sublitharenite on average, the range is 5 to 13 volume percent chlorite. This may be anomalously high because of the lithic-rich nature of the Tuscaloosa. The Berea Sandstone, a quartz arenite to sublitharenite, contains only 4 to 7 volume percent chlorite coats, which appears to be sufficient to favorably preserve porosity (Larese, 1974). There are less data on the optimum amount of illite needed as grain coats to preserve porosity. Heald and Baker (1977) showed that 3.5 to 6.5 volume percent illite was optimal for the Rose Run Sandstone in the western Appalachian Basin. Pressure solution increased as the illite content increased beyond 6.5 volume percent illite. Samples with less than 3.5 volume percent illite were cemented by quartz.

Tuscaloosa Sandstone

The subsurface Upper Cretaceous Tuscaloosa Sandstone of Louisiana will be used to illustrate the important aspects of porosity preservation by chlorite coats, although similar features occur in many formations. The Woodbine Sandstone in southwestern Arkansas is the outcrop equivalent of the Tuscaloosa Sandstone. In the shallow subsurface, oil operators use both names indiscriminately. The Woodbine Sandstone, where it crops out, is a volcanic arenite with typically less than 10% quartz. The Woodbine compositionally consists of trachytic and alkalic igneous lithic fragments, oligoclase feldspar, quartz, Ti-rich pyroxenes, and magnetite detrital grains coated with smectite (Belk and others, 1986).

Data for the following discussion came from examination of Woodbine outcrop samples and Tuscaloosa cores from seven areas or fields (Fig. 4) with depositional environments that included fluvial, deltaic, and nearshore marine. The Tuscaloosa Sandstone is a volcanic lithic/sublithic arenite throughout most of Louisiana and Arkansas. The volcanic detritus was derived from Cretaceous volcanoes and alkalic intrusives in northern Louisiana and Arkansas (Fig. 4). In Rigolets and Freeland fields, the Tuscaloosa Sandstone is quartz rich and probably was derived from an Appalachian source. Where chlorite coats are well developed and preserve porosity, the Tuscaloosa contains altered and partially dissolved volcanic lithic-framework grains, which are believed to be the *in situ* source of the cations for the chlorite (Thomson, 1979) (Fig. 5A). In sandstones lacking lithic components, for example, Rigolets and Freeland fields (Fig. 4), chlorite coats are absent or insignificant and the sandstone has significant amounts of quartz cement (Fig. 5B). Disrupted and offset chlorite coats on fractured sand grains correlate with formation of prismatic overgrowths (Fig. 5A). This suggests that pore waters were supersaturated with silica, which precipitated where nucleation sites were available. Thick, but discontinuous, clay coats allow nucleation of quartz, which may fill the pore space (Figs. 5 C–D).

Another line of evidence for silica being available is that quartz, in the form of overgrowths, has preferentially replaced calcite cement in carbonate concretions. Abundant remnants of calcite occur only in the replacing-quartz overgrowths and not in the quartz nuclei, which indicates that quartz was the replacing mineral (Figs. 6 A–B).

In some sandstones that have well-developed chlorite coats, there are pores with no chlorite because oil occupied the interstices before chlorite formation. Subsequent to oil mi-

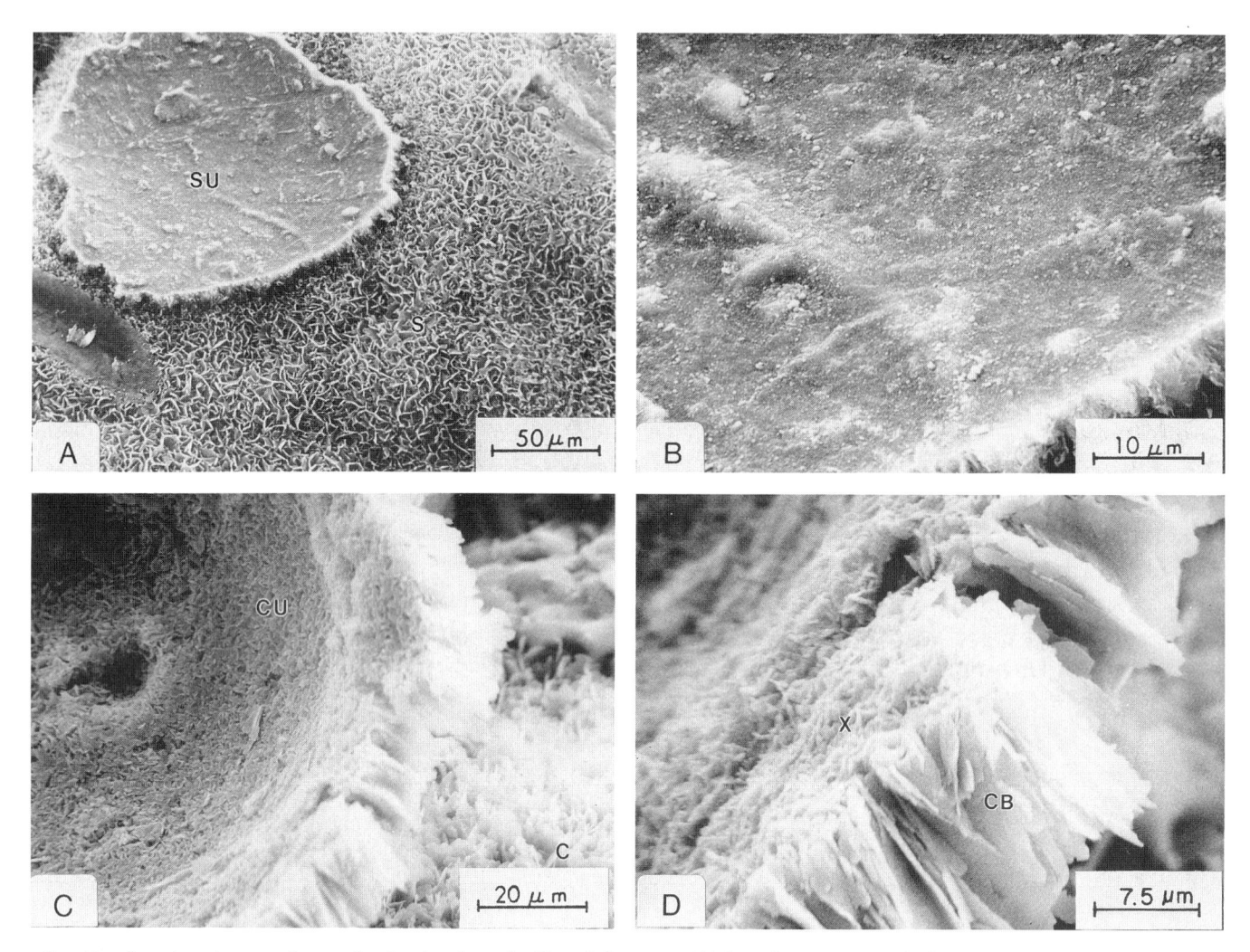

FIG. 3.—Scanning electron micrographs showing the undersides of clay coats. (A) Smectite coats grown in the laboratory and critical-point dried have the typical honeycomb morphology (S). Where the clay coats became detached from the framework grains, the undersides of the smectite coats are visible (SU). (B) Closeup of the underside of a smectite coat showing a dense and impervious barrier, which would inhibit the precipitation of quartz. (C) Chlorite coats from the Tuscaloosa Sandstone showing a well-developed, bladed habit (C) where the chlorite grows into a pore. The underside of a chlorite coat (CU) is noticeably more dense appearing than the outside. (D) At higher magnification, the underside of the chlorite coat appears to be composed of minute crystals (X), which form a relatively dense layer overlain by blades of chlorite (CB) oriented perpendicular to the detrital-grain surface.

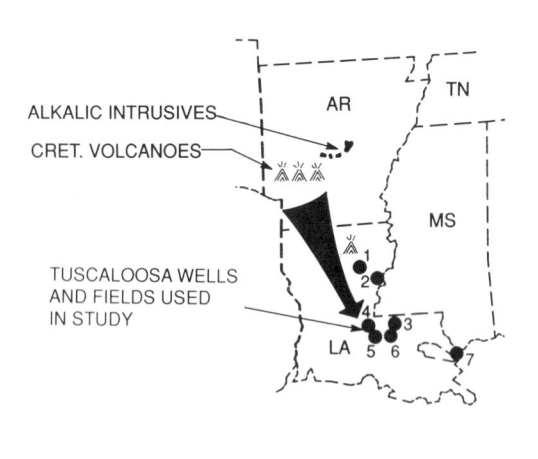

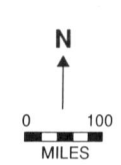

gration, silica cementation continued filling the pore space, leaving only traces of hydrocarbon to indicate its former presence. Early entry of hydrocarbon is commonly considered to be favorable for porosity preservation. However, where this fluid prevented the formation of chlorite coats and later moved out, quartz cementation could destroy po-

FIG. 4.—Location map showing Tuscaloosa wells and/or fields studied in Louisiana, as well as the location of igneous rocks in Arkansas, which are believed to be the source of the lithic material in the sandstones that supplied the cations needed to precipitate the chlorite coats. The bold arrow indicates the major direction of sediment transport. Some areas (e.g., Rigolets Field; no. 7) contain sparse or no volcanic lithic grains and little or no chlorite coats. Provenance for these areas is believed to be from an Appalachian source to the east. The numbers correspond to: (1) Big Creek Field; (2) Amoco No. 1 Lancaster 7-7 and Amoco No. 1 Strickland, Tensas Parish; (3) Freeland Field; (4) Morganza Field; (5) Moore-Sams Field; (6) Port Hudson Field; and (7) Rigolets Field.

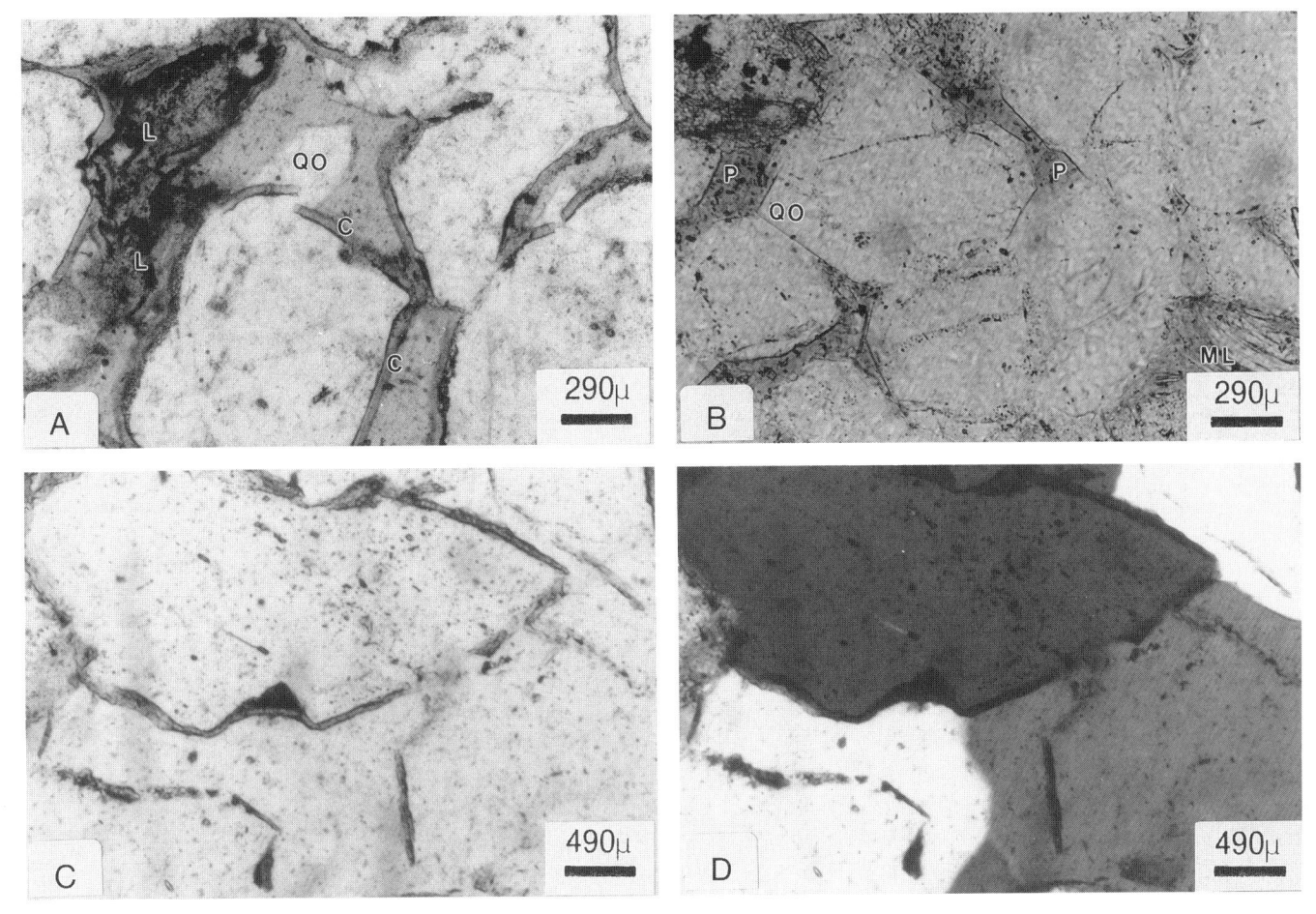

FIG. 5.—Photomicrographs of the Tuscaloosa Sandstone. (A) Example of a chlorite coat (C) offset by a fracture where a prismatic-quartz overgrowth (QO) has formed on the exposed surface of quartz in a volcanic lithic (L)-bearing sandstone. The opaque material associated with the lithic fragment is leucoxene. Plane polarized light. (B) Sample from Rigolets Field showing abundant quartz cement (QC) with minor intergranular porosity (P). Some metalithic fragments (ML), but no volcanic lithic grains or chlorite coats were observed. (C) and (D) Plane and partially crossed polarizer views, respectively, showing thick but discontinuous chlorite coats (C). Where chlorite coats were absent, extensive void-filling, eccentric-quartz overgrowths nucleated on the detrital grains.

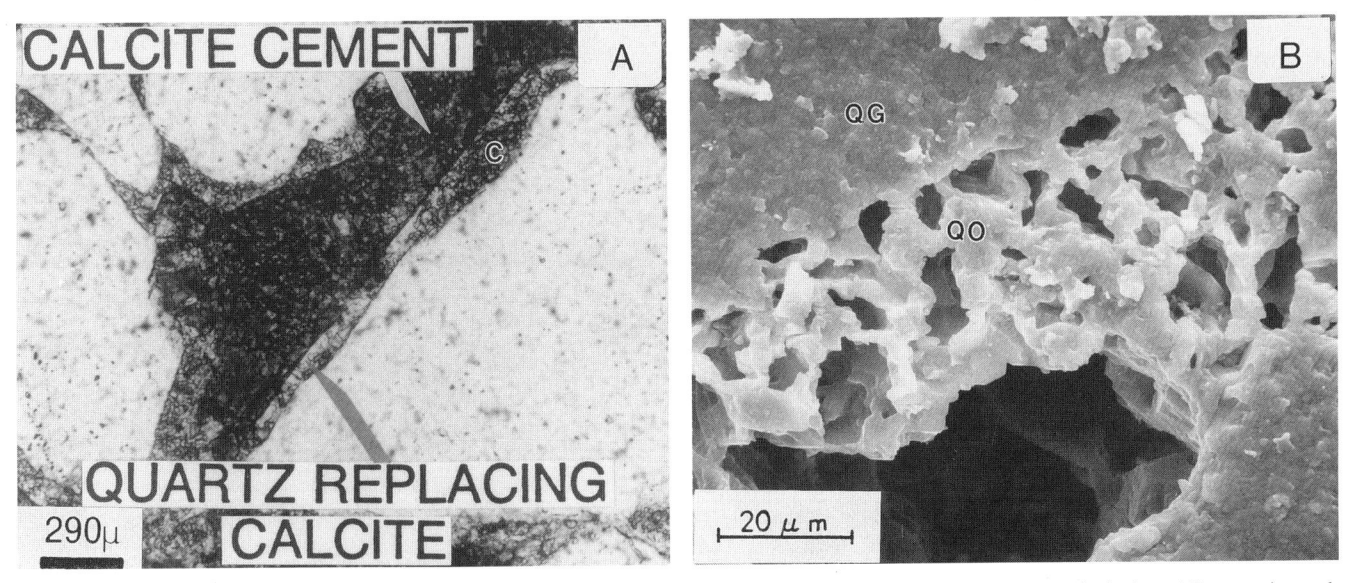

FIG. 6.—In some concretions, secondary quartz has replaced Fe-calcite. (A) Photomicrograph showing carbonate inclusions (C) occurring only in the secondary quartz and not in the nucleus, which indicates that the calcite was not replacing the quartz. Plane polarized light. (B) Scanning electron micrograph of (A) following leaching in HCl. Note the holes in the overgrowth (QO) previously occupied by calcite crystals and the absence of holes in the detrital grain (QG).

rosity and eliminate the possibility of any future accumulation of hydrocarbon in these zones.

Further support for the effectiveness of chlorite coats and the availability of silica in formation waters is furnished by the Amoco No. 1 Lancaster 7–7 core, Tensas Parish, Louisiana (Fig. 7). Comparison of two channel sandstones reveals distinct compositional differences, which affected cementation. The lower channel is lithic rich ($Q_{70.5}F_{1.5}L_{28.0}$) with 16.8 volume percent volcanic-rock fragments. The upper channel is quartz rich ($Q_{95.4}F_0L_{4.6}$) with only 0.8 volume percent volcanic-rock fragments. These compositional differences between channels are expressed in the cements. The sandstone in the lithic-rich channel has predominantly chlorite cement (88.0%) with 12% other cements and only a trace of quartz. The sandstone in the quartz-rich channel is cemented predominantly by quartz (89.4%) with 3.6% chlorite and 7.0% other cements. There is no reason to believe that silica was available only in the formation water of the upper channel. It is likely that silica precipitated in the upper, quartz-rich channel sandstone but not in the lower channel sandstone where chlorite coats blocked nucleation sites.

An additional line of evidence stems from growth of quartz into secondary pores formed by dissolution of feldspar. Consider a situation where quartz and feldspar grains are in contact. Well-developed clay coats form, but not at the grain contact. Later, the feldspar is dissolved and a quartz overgrowth nucleates at the former grain contact and grows into the secondary pore as a prismatic growth. These relations, which can be documented in some samples, show that silica was available in the pore water and that quartz nucleated if a site was available.

Some lithic grains dissolved prior to or during the growth of chlorite coats, as shown by fringing crystals of chlorite on the outer margin as well as on the remnants of the lithic material inside the partially dissolved lithic fragment (Fig. 8A). Other lithics dissolved after the precipitation of chlorite coats, as shown by the occurrence of chlorite coats only on the outer surface of the partly dissolved grain (Fig. 8B).

Figure 9 shows the effect of chlorite coats on the precipitation of quartz overgrowths and the preservation of intergranular porosity. These data are based on point counts of medium-grained, well-sorted sandstones. It is important to compare sandstones of similar textural characteristics because composition often changes with grain size. In the Tuscaloosa, finer grained sandstones have less lithic material, which is the source of the cations for forming chlorite. In Figure 9, note the high percentage of secondary quartz and low porosity where chlorite abundance is low (approximately 3 or 4 volume percent). There is a sharp decrease in the amount of quartz overgrowths with increasing amounts of chlorite due to the blocking of nucleation sites for the quartz by the clay coats. In the same thin sections, intergranular porosity is low where quartz overgrowths are abundant. Intergranular porosity increases up

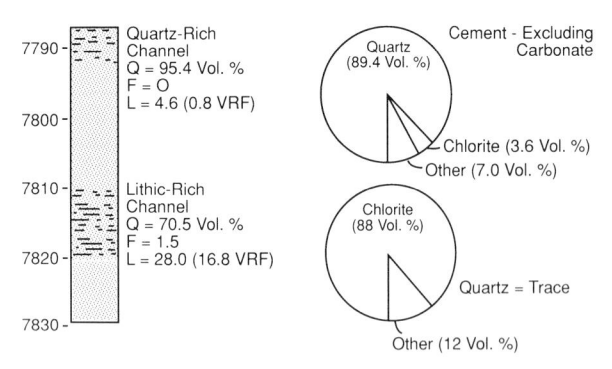

FIG. 7.—Core in the Amoco 1 Lancaster 7-7 reveals two channels. The lower channel is lithic rich and contains significant volcanic lithic fragments (16.8 volume percent). The upper channel is quartz rich with only a trace amount of volcanic lithic grains. These compositions are reflected in the cements of the two channels because the volcanic lithics are the source of the authigenic chlorite. The volcanic-rich, lower channel has extensive chlorite cement, whereas the quartz-rich, upper channel has extensive quartz cement.

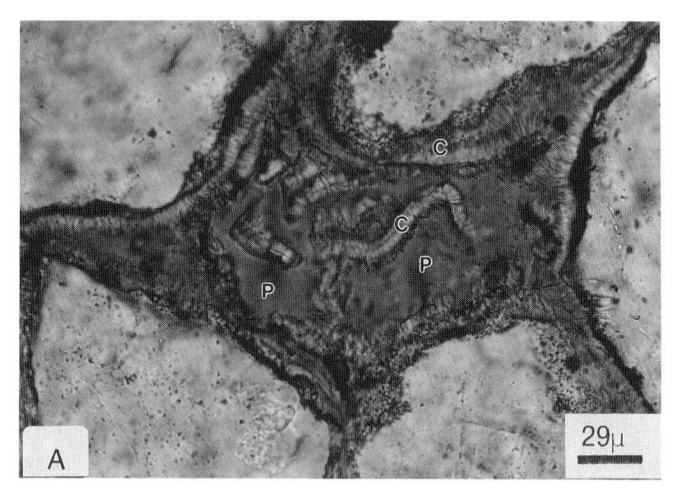

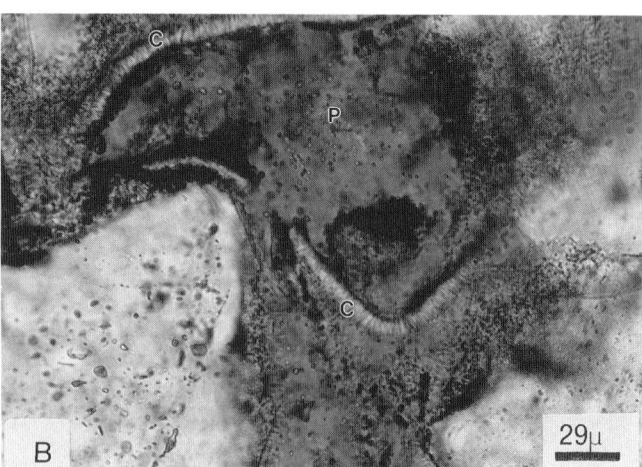

FIG. 8.—(A) Partially dissolved (P = pore) volcanic lithic grains with chlorite coats (C) on the outside as well as the inside of the lithic grain. (B) Volcanic lithic grains with authigenic chlorite (C) developed only on the exterior of the partially dissolved grain. The opaque material is leucoxene. Both photographs, plane polarized light.

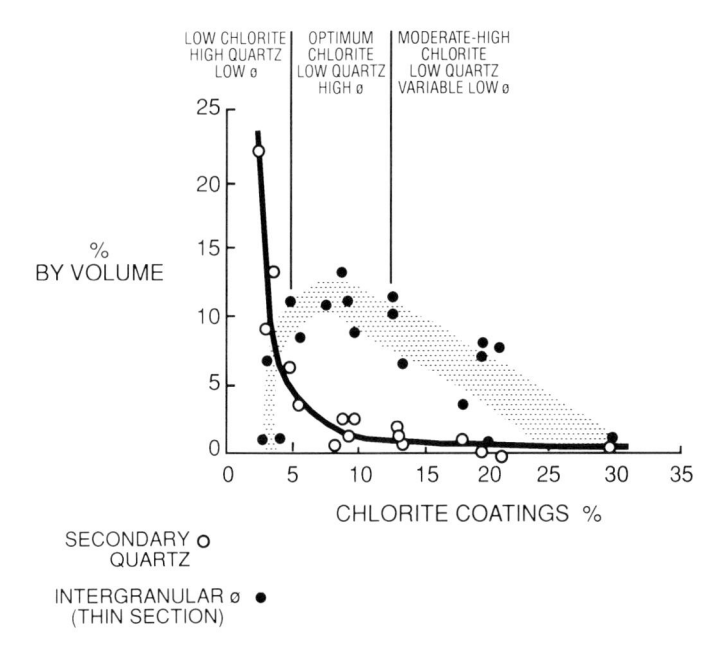

FIG. 9.—Plot showing the relation among secondary quartz, chlorite coats, and intergranular porosity for medium-grained sandstones from the Tuscaloosa. The most favorable amount of chlorite coats to preserve porosity is from 5 to 13 volume percent. Lesser amounts permit the development of extensive secondary quartz and greater amounts fill pore space to reduce reservoir-pore volume.

to about 8 volume percent chlorite and then decreases. The optimum amount of chlorite in the Tuscaloosa to preserve porosity ranges from about 5 to 13 volume percent. Above 13 volume percent, the chlorite occupies a large portion of the pores and is a detriment to porosity. The weak lithic grains, some of which are partially filled by chlorite, have deformed ductilely to further reduce porosity.

Outcrop samples of the Woodbine Sandstone and shallow subsurface samples of the Tuscaloosa Sandstone have isopachous, authigenic-smectite coats (Fig. 10). At depths as great as 1,676 m (5,500 ft) in Big Creek Field, Louisiana, the sandstone has smectite coats; whereas, at a depth of 2,371 m (7,779 ft) in Tensas Parish, Louisiana, chlorite coats occur (Fig. 4). We were unable to locate cores in the critical depth interval to evaluate the transition from smectite to chlorite, but the existence of a mixed-layer chlorite/smectite seems reasonable.

Figure 11 is a generalized diagenetic sequence for the Tuscaloosa Sandstone in the cores studied. Compaction, including mechanical processes such as grain readjustment, ductile deformation of lithics, and fracturing of grains, as well as some pressure solution, has been important. Note that there are early calcite and siderite cements, which occur in concretions, as well as late Fe-calcite and Fe-dolomite, which occur as poikilotopic crystals.

One would not expect the deep Tuscaloosa to have outstanding porosity based on shallow wells where porosity decreases at 0.66%/100 m (2%/1,000 ft) (Fig. 12). The porosity should be below economic limits at a depth of approximately 4,572 m (15,000 ft). However, the mean porosity in the gas fields from 4,877 to 6,401 m (16,000 to 21,000 ft) shown on Figure 12 ranges from 14 to 18%, although there is a wide range in porosity values. The maximum porosity in these fields is exceptional and ranges from 26 to 31%. The low-porosity rocks are commonly argillaceous or cemented by carbonates. The mean total porosity can be subdivided as follows: primary intergranular macroporosity, 38%; secondary intragranular/moldic macroporosity, 17%; and microporosity, associated with clay and lithic debris, 45%.

The Tuscaloosa Sandstone is a good deep reservoir because of a favorable pore geometry and possibly overpressure. Intragranular/moldic and intergranular porosity provide a well-interconnected macropore system. Both pore types probably are necessary to provide the porosity and permeability required of these deep reservoirs. Many of the Tuscaloosa reservoirs deeper than 18,000 feet (5,486 m) are overpressured, which, if formed early, would retard compaction and perhaps restrict fluid movement conducive to cementation.

We favor a primary origin for the intergranular porosity and will present supporting evidence from study of the sandstones using optical and scanning electron microscopy and from experimental petrology. Other workers who also

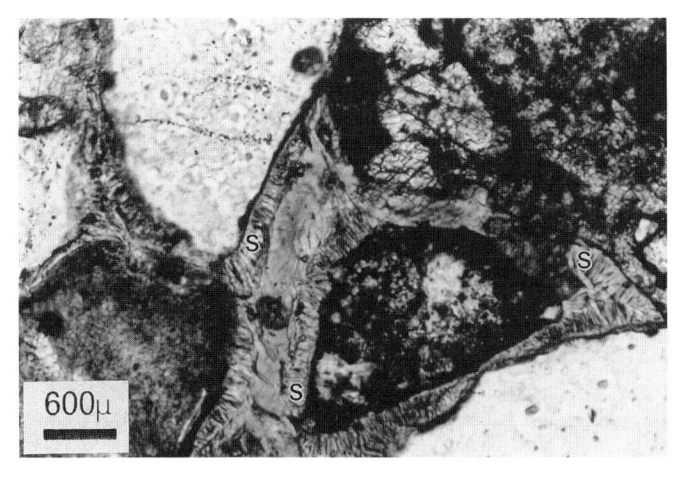

FIG. 10.—Photomicrograph showing smectite coats (S) from the lithic-rich, shallow Tuscaloosa Sandstone. Plane polarized light.

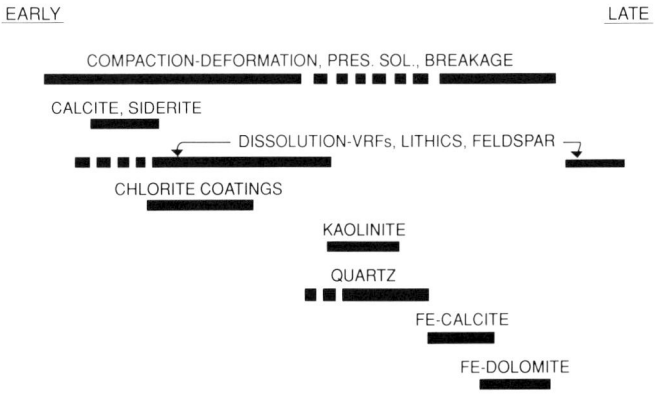

FIG. 11.—Generalized diagenetic sequence for the Tuscaloosa based on this study. There appears to have been two stages of lithic-grain dissolution based on the evidence in Figure 8.

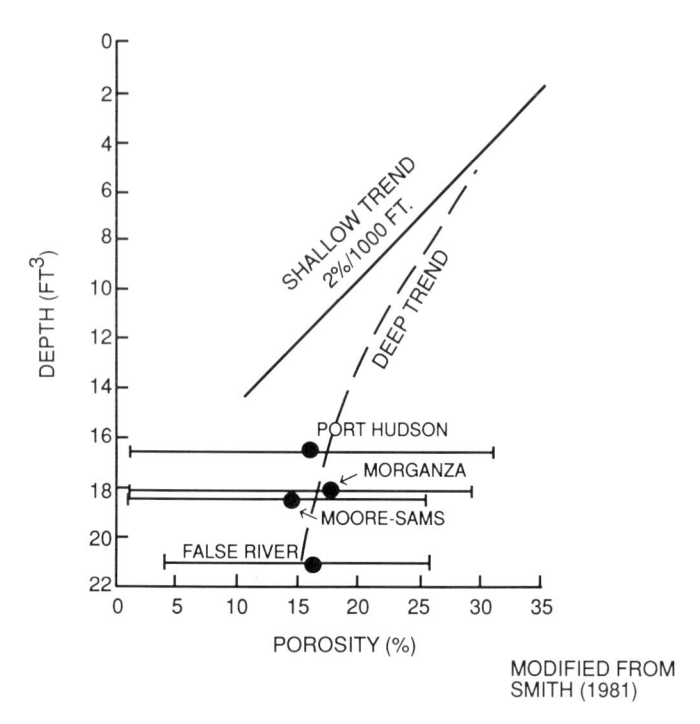

FIG. 12.—Porosity versus depth for the Tuscaloosa Sandstone. Note how the deep trend deviates from the shallow trend. The porosity ranges for the four fields are depicted by the horizontal bars with the dots indicating the average porosity. Note the exceptional porosity (>25%) at depths greater than 5,486 m (18,000 ft).

FIG. 13.—Scanning electron micrograph showing idiomorphic, late forming Fe-calcite (C) containing embedded chlorite (CH). Note that the Fe-calcite crystal faces show no evidence of being affected by dissolution. The smooth areas in the lower left and upper right are the undersides of chlorite coats exposed because the chlorite coats, which are embedded in the calcite, pulled loose from the detrital grains. Note the denser appearance of this surface.

favor a primary origin for the intergranular porosity in the Tuscaloosa are Thomson (1979), Dahl (1984), Lin (1984), Smith (1985), and Wiygul and Young (1987). Paxton and others (1990) recognized that chlorite could preserve porosity, but believed that the main reason for good primary intergranular porosity at depth was due to inadequate cement. In other words, the formation waters were not saturated with respect to quartz or a carbonate mineral for a sufficient residence time to produce significant volumes of cement. Based on textural criteria for secondary porosity developed by Schmidt and McDonald (1979) and the absence or near absence of carbonates in the high-porosity sandstones, some workers have interpreted the intergranular porosity as being secondary in origin, forming from the dissolution of carbonate cements (e.g., Franks, 1980; Hudder and Tieh, 1983; Hamlin and Cameron, 1987). Everyone recognizes that there is intragranular and moldic porosity related to dissolution of lithic fragments and feldspars. The problem is distinguishing between primary and secondary intergranular porosity.

We believe the following evidence supports a predominantly primary origin for the intergranular pores. Firstly, the porous sandstone locally contains calcareous fossils that do not appear to be undergoing dissolution. Secondly, the calcite and siderite concretions that must have formed relatively early, based on compactional drape around them, show no evidence of undergoing dissolution. Thirdly, the late forming ferroan-calcite cement, which has supposedly dissolved to form secondary porosity, has smooth pristine-appearing crystal faces (Fig. 13). These faces appear to be growth surfaces rather than dissolution surfaces. Fourthly, late forming ferroan dolomite has replaced chert grains along fractures and grain margins to create an irregular, distinctive texture that would lead to a unique pore pattern if the carbonate was removed (Fig. 14). This pore pattern was never seen on chert grains in porous sandstones.

Supporting evidence for a primary intergranular origin is derived from simple acid-leaching experiments where ferroan calcite was dissolved to exhume earlier formed authigenic minerals. The morphology of these exhumed authigenic minerals was then compared to the same authigenic

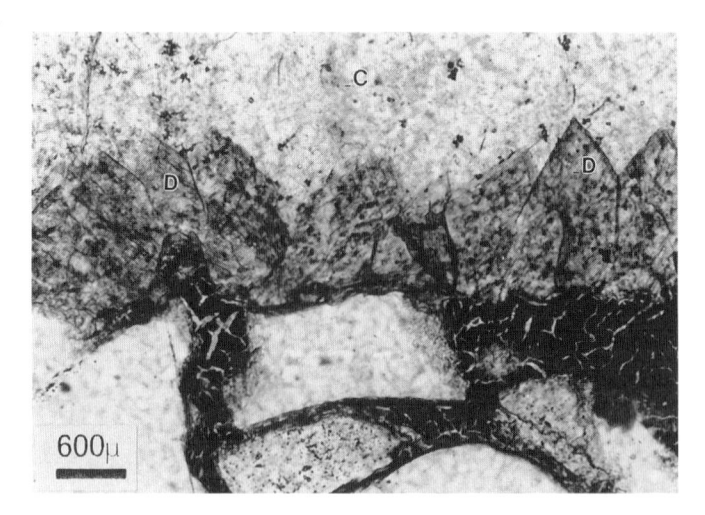

FIG. 14.—Photomicrograph showing dolomite crystals (D) replacing the margins of a chert grain (C). The dark material cementing the sand grains is calcite, stained dark red by alizarine red-S. If the dolomite was dissolved, a diagnostic replacement texture would be created for the chert-grain margins to serve as evidence that the intergranular pores were secondary in origin. This texture was never seen. Plane polarized light.

minerals in naturally porous sandstone that had been treated identically with acid. These techniques are discussed by Burley and Kantorowicz (1986) and Larese and Pittman (1987).

Two sandstone samples were selected for acid leaching: one sample had pervasive ferroan-calcite cement (Fig. 15A) and the other was porous with well-developed chlorite coats (Fig. 15C). Both samples were placed in beakers of a buffered solution of sodium acetate/acetic acid (pH ~5.5) for 2 hrs at room temperature. Samples were then washed, dried, and prepared for examination under the scanning electron microscope. There was no difference in the morphology of the chlorite coats in the pre- and post-acid samples for the naturally porous sandstone (compare Figs. 15C and D). The chlorite exhumed by leaching the ferroan calcite, however, had a decidedly different appearance from the chlorite in the naturally porous sandstone that also had been exposed to acid (compare Figs. 15B and D). The ferroan calcite appeared to have modified the morphology of the chlorite by replacing the margins of the radiating chlorite crystals. The

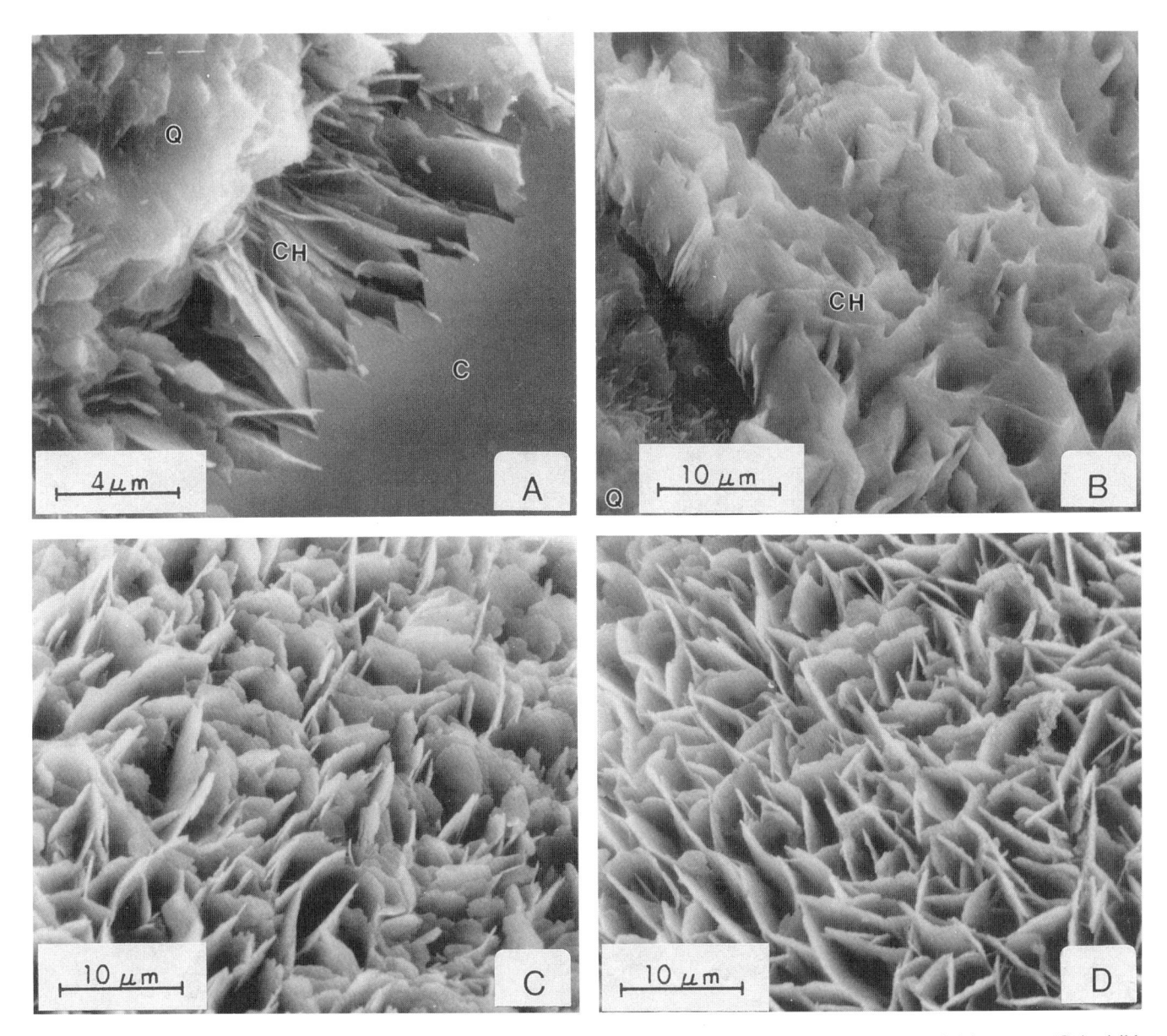

FIG. 15.—Scanning electron micrographs showing results of acid-leaching experiments on Tuscaloosa Sandstone. (A) Calcite cement (C) is visible overlying chlorite coat (CH), which is attached to a quartz-framework grain (Q). (B) Closeup of modified chlorite coat (CH) on quartz grain (Q) created by leaching sample shown in (A) in dilute acetic acid and dissolving calcite. (C) Note the typical morphology of the bladed, radial-chlorite coat from naturally porous sandstone without calcite cement. (D) Sample shown in (C) after receiving the same acid treatment as sample shown in (A). Note that the acid did not affect the morphology of the chlorite. Comparison of (B) and (D) reveals that the chlorite exhumed from under the calcite cement in (B) has a disturbed appearance because the calcite partially replaced the chlorite. This morphology is distinctive and if present would be evidence that the intergranular porosity was of secondary origin. This modified chlorite morphology was never seen in a naturally porous sandstone in the Tuscaloosa.

chlorite coats in naturally porous sandstones did not have a modified appearance (Fig. 15D), indicating that carbonate cement had not previously filled the intergranular pores.

Another leaching experiment involved examination of quartz overgrowths that precipitated before the ferroan-calcite pore fill (Fig. 16A). A hand specimen-size piece of core contained porous as well as calcite-cemented sandstone. Pieces of each rock type were treated with a buffered solution of sodium acetate/acetic acid. Comparison of quartz overgrowths in each rock type revealed a difference. The crystal faces on quartz overgrowths exhumed from under ferroan calcite were modified. They were pitted, irregular, and had rounded edges (Fig. 16B) compared with the overgrowths in the naturally porous sandstone, where the crystal faces were smooth and facet contacts sharp (Fig. 16C), as would be expected for crystals that grew unimpeded. It is obvious that the ferroan calcite partially replaced the margins of the quartz overgrowths to create a modified surface. If the Tuscaloosa Sandstone had significant secondary intergranular porosity, then the authigenic quartz in porous sandstone should have a modified appearance, but the quartz is unaltered. Authigenic chlorite also can be compared in Figures 16B and C, although not as clearly as in the earlier discussed micrographs; however, the chlorite that was under the ferroan calcite has a modified appearance. We did not find a single example of what could be interpreted as

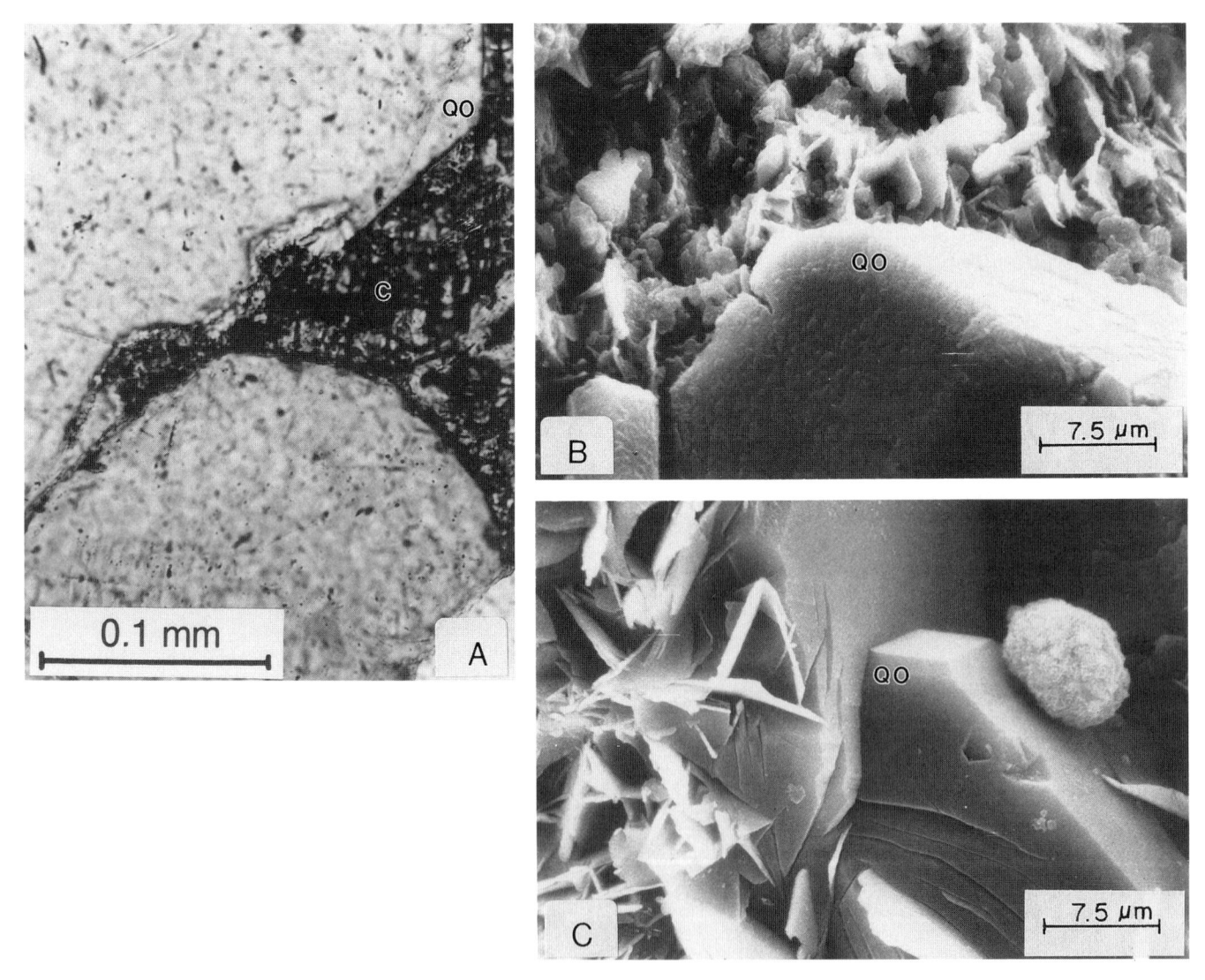

FIG. 16.—(A) Photomicrograph showing quartz overgrowths (QO) underlying dark colored Fe-calcite (C). Plane polarized light. (B) Scanning electron micrograph showing quartz overgrowth (QO) exposed after leaching of calcite in dilute acetic acid. Note the rough and pitted crystal faces and rounded edges resulting from partial replacement by the calcite that was in contact with the overgrowth. (C) Closeup showing pristine-quartz overgrowths (QO) from naturally porous sandstone, despite exposure to the same acid treatment as the sample shown in (A) and (B). The chlorite in (B) also appears to be modified when compared with the chlorite in (C). The fact that overgrowths in naturally porous sandstones have unaltered morphology is a strong argument that the intergranular porosity is primary.

modified authigenic morphology of quartz or chlorite to support the hypothesis for the secondary origin of intergranular porosity.

Applications of Experimental Petrology

The effect of artificial-clay coats on precipitation of quartz overgrowths was determined in the laboratory. A quartz crystal was cut normal to the c-axis and the surface was polished. This basal plane (0001) is not a naturally occurring face, but is used in the synthesis of quartz because growth is most rapid in the direction of the c-axis (Ballman and Laudise, 1963). One half of this artificial basal plane face was covered with a slurry of illite. This sample was then placed in an autoclave and the techniques of Cecil and Heald (1971) were used to grow quartz. After 48 hrs in a 0.56 m K_2CO_3 solvent at 318°C (604°F), a quartz overgrowth had developed on the exposed quartz surface (Figs. 17A and B). The other half of the crystal face developed small, sporadic prismatic overgrowths where the illite coat was incomplete. This experiment showed that physical blockage of nucleation sites is an effective mechanism for retarding quartz-overgrowth development.

In another experiment, highly porous Tuscaloosa Sandstone with well-developed chlorite coats was artificially compacted in a hydrothermal reactor, causing the quartz grains to fracture. This sample was then transferred to an autoclave where quartz overgrowths were experimentally precipitated. Prismatic-quartz overgrowths formed where the chlorite coats had been breached by fractures (Fig. 18A).

FIG. 17.—Scanning electron micrographs showing sample from experiment designed to test the nucleation-blocking potential of an illite slurry that covered one-half of a quartz crystal cut and polished along the basal plane (0001). (A) The right side of the quartz crystal, which was covered by illite, did not develop secondary quartz when the sample was placed in a hydrothermal reactor under conditions conducive to quartz growth. The unprotected left side did develop quartz overgrowths. (B) A higher magnification view showing the artificial illite coat and secondary quartz development. This experiment showed that clay can effectively block the nucleation of quartz by physically providing a barrier that prevents silica-saturated waters from contacting a quartz nucleus to precipitate a syntaxial overgrowth.

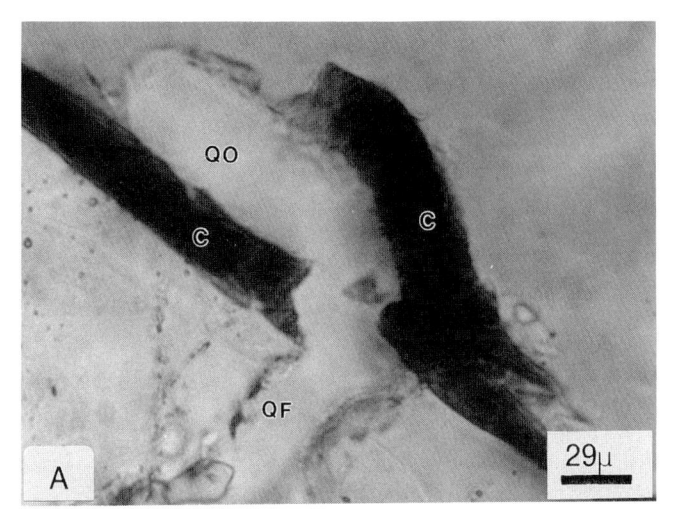

FIG. 18.—Photographs illustrating a sample of the Tuscaloosa Sandstone that was artificially compacted prior to growing quartz overgrowths in the laboratory. (A) Photomicrograph showing chlorite coats (C) offset by fractures and secondary quartz healing fracture (QF) and as an overgrowth (QO). Note how the chlorite coat (C) appears to have been dragged by the growing quartz crystal. Plane polarized light. (B) Scanning electron micrograph showing quartz overgrowths growing among chlorite crystals.

This process closely mimics the process in natural sandstones, which also leads to the precipitation of prismatic overgrowths (Fig. 18B).

A hydrothermal reactor was also used to grow smectite, illite and chlorite coats. An experiment using sand-size, rounded fragments of serpentinite and quartz, distilled water, a temperature of 355°C (671°F) and a pressure of 10,000 psi (69 MPa) for 15 days produced smectite coats that were better developed on serpentinite grains and on quartz grains near the contact with serpentinite grains. This illustrates that the development of clay coats may be influenced by a local source of cations. Figures 19 and 20 show schematically and under the scanning electron microscope, respectively, the sequence of development of a laboratory produced, critical-point-dried smectite coat. The clay developed initially as clay wisps (Fig. 19, stage 1; Fig. 20A) and progressed to clay platelets that formed a "root" zone (Fig. 19, stage 2; Fig. 20B), then to an open polygonal boxwork (Fig. 19, stage 3; Fig. 20C), and finally to a denser polygonal boxwork (Fig. 19, stage 4; Fig. 20D). This sequence was viewed on individual quartz grains in contact with serpentinite grains, with stage 4 adjacent to the serpentinite, which was the source of the cations. The flatly attached, tight root zone is probably especially effective at blocking the nucleation of quartz overgrowths. A similar clay morphology can be seen in natural sandstones (Fig. 21A), where incipient smectite grew as clay flakes that are flatly attached and curl

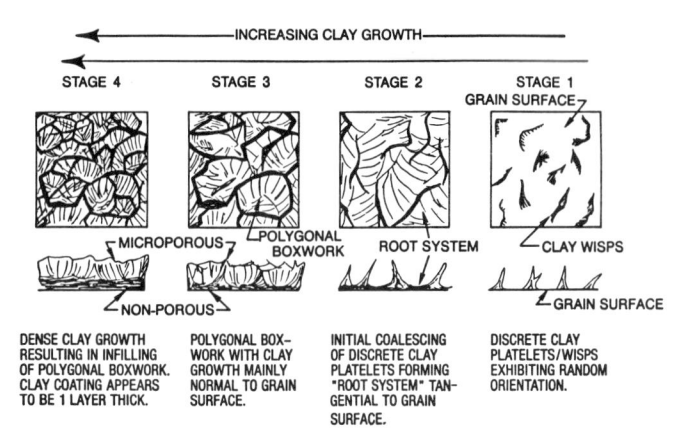

FIG. 19.—Schematic diagram showing four stages in the sequence of development of experimentally grown smectite clay coats.

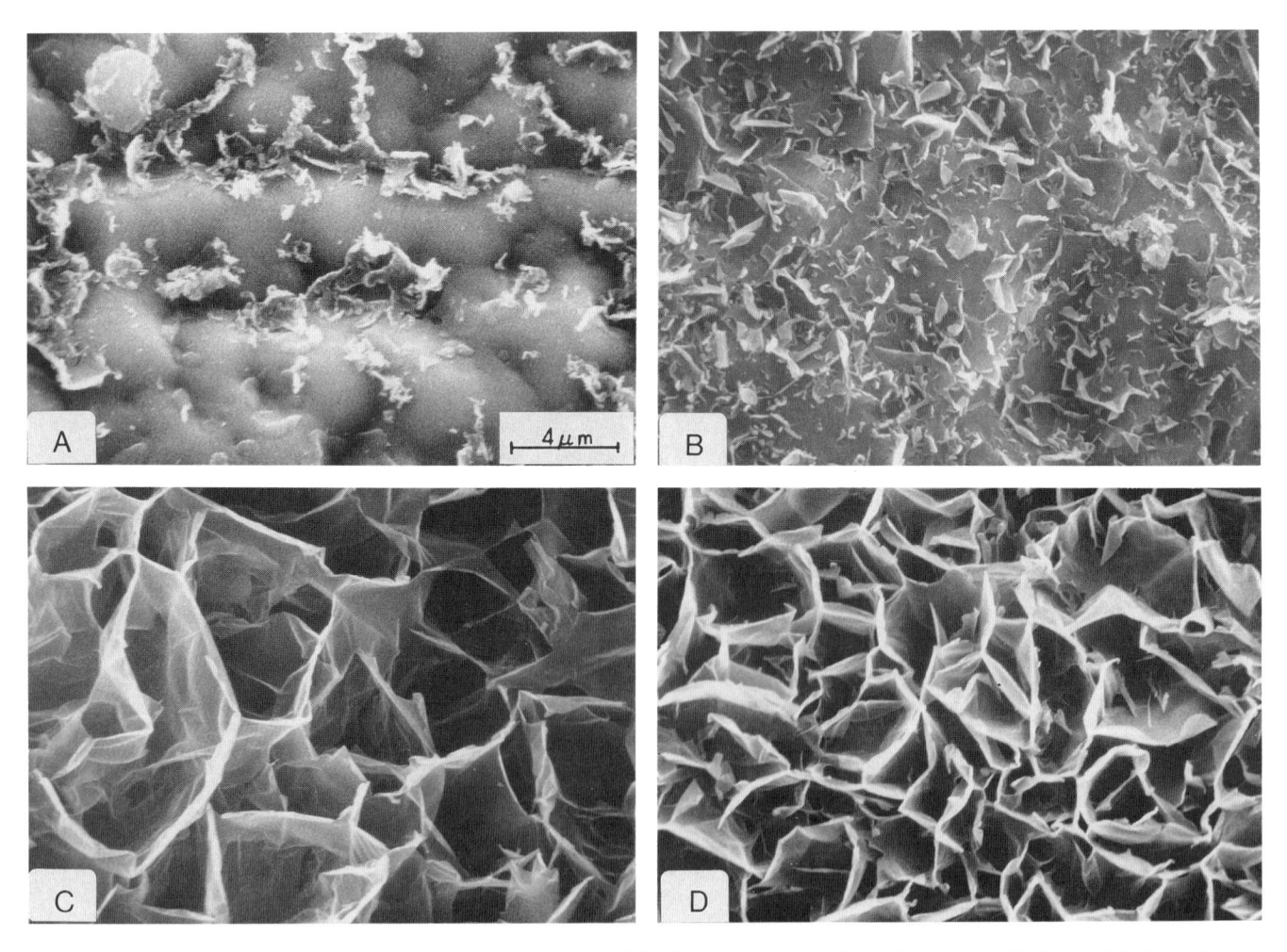

FIG. 20.—Scanning electron micrographs showing the four stages of development of an experimentally grown, critical point-dried smectite coat. (A–D) correspond to stages 1–4, respectively, as shown in Figure 19.

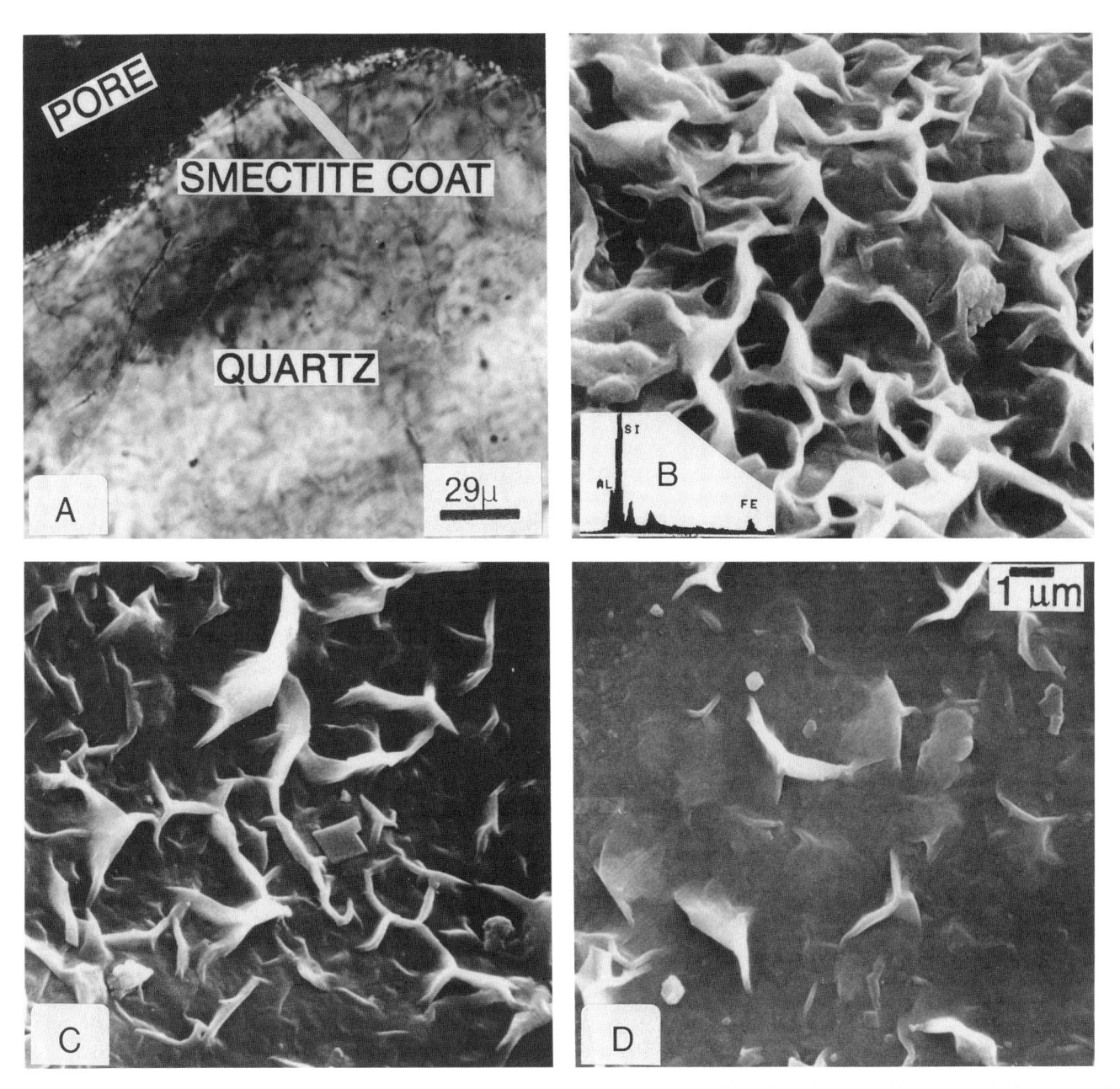

FIG. 21.—Photographs illustrating smectite coat morphology for a Cretaceous sandstone (255 m; New Jersey), which is very similar to experimentally grown smectite of Figure 20. (A) Photomicrograph showing smectite coat on quartz grain (plane polarized light). (B–D) A sequence of scanning electron micrographs showing various stages of smectite coat development from good to incipient, which are similar to growth stages in the experimentally grown sample shown in Figures 19 and 20.

upward with random orientation (Fig. 21D). As the smectite coat became thicker, a series of ridges and a polygonal boxwork formed (Fig. 21C), eventually leading to subdivision of the polygonal pattern and a dense clay growth (Fig. 21B).

SUMMARY

Isopachous, well-developed clay coats, particularly chlorite, commonly are effective physical barriers that prevent quartz overgrowths from nucleating on detrital-quartz grains. This is an effective mechanism for preserving primary intergranular porosity in quartz arenites. Many deeply buried sandstone reservoirs owe their porosity, in part at least, to clay coats. Clay coats have no effect on the precipitation of epitaxial cements such as carbonates or sulfates. Also, clay coats are not an effective porosity-preservation mechanism in lithic arenites with perhaps 35% or more lithic material because of the low-quartz content and the importance of ductile deformation in destroying porosity.

Growth of secondary quartz on quartz crystals partially covered by simulated-clay coats reveals that the clay is an effective barrier. Overgrowths form on exposed quartz surfaces, but not where the simulated-clay coat is present.

Experimental growth of clay coats reveals the existence of a root zone where the clay flakes are flatly attached and curl upward. This zone is normally hidden by later growth, with a characteristic morphology (e.g., honeycomb) that one associates with authigenic clay. This root zone, if well formed, effectively covers the detrital grain and serves as a barrier to nucleation of a syntaxial-quartz overgrowth. The effectiveness of a clay coat in preserving porosity may depend on the development of the root zone.

ACKNOWLEDGMENTS

We thank the following individuals for their contributions to the study: M.P. Smith for assistance with microprobe analyses; G.R. Powers, who supplied the X-ray diffractometry data; R.C. Adams for assistance with experimental petrology; and Sharon Clark, who made the thin sections. B.J. Clardy and M. Howard, of the Arkansas Geological Commission, helpfully showed us outcrops of the Woodbine Sandstone to facilitate the sampling program. We thank Amoco Production Co. for granting permission to publish this paper. The manuscript profited from the helpful critiques of M. Moraes and J. Welton.

REFERENCES

ALMON, W. R., FULLERTON, L. B., AND DAVIES, D. K., 1976, Pore space reduction in Cretaceous sandstones through chemical precipitation of clay minerals: Journal of Sedimentary Petrology, v. 46, p. 89–96.

BAILEY, S. W., 1980, Summary of recommendations of AIPEA nomenclature: Clay Minerals, v. 15, p. 85–93.

BALLMAN, A. A., AND LAUDISE, R. A., 1963, Hydrothermal growth, in Gilman, J., ed., The Art and Science of Growing Crystals: New York, John Wiley and Sons, p. 231–251.

BASTIN, E. S., GRATON, L. C., LINDGREN, W., NEWHOUSE, W. H., SCHWARTZ, G. M., AND SHORT, M. N., 1931, Criteria of age relations of minerals with special reference to polished sections of ores: Economic Geology, v. 26, p. 562–610.

BELK, J. K., LEDGER, E. B., AND CROCKER, M. C., 1986, Petrography of the volcaniclastic Woodbine Formation, southwest Arkansas: Transactions, Gulf Coast Association of Geological Societies, v. 36, p. 391–400.

BURLEY, S. D., AND KANTOROWICZ, J. D., 1986, Thin section and S.E.M. textural criteria for the recognition of cement-dissolution porosity in sandstones: Sedimentology, v. 33, p. 587–604.

CARRIGY, M. A., AND MELLON, G. B., 1964, Authigenic clay mineral cements in Cretaceous and Tertiary sandstones of Alberta: Journal of Sedimentary Petrology, v. 34, p. 461–472.

CECIL, C. B., AND HEALD, M. T., 1971, Experimental investigation of the effects of grain coatings on quartz growth: Journal of Sedimentary Petrology, v. 41, p. 582–584.

DAHL, W. M., 1984, Progressive burial diagenesis in Lower Tuscaloosa sandstones, Louisiana and Mississippi: Clay Minerals Society Annual Meeting, Programs with Abstracts, p. 42.

DIXON, S. A., SUMMERS, D. M., AND SURDAM, R. C., 1989, Diagenesis and preservation of porosity in Norphlet Formation (Upper Jurassic), southern Alabama: American Association of Petroleum Geologists Bulletin, v. 73, p. 707–728.

DOGAN, A. U., AND BRENNER, R. L., 1983, Effects of depositional and diagenetic history upon the reservoir properties of Parkman Sandstone (abs.): American Association of Petroleum Geologists Bulletin, v. 67, p. 451.

DUTTON, S. P., AND LAND, L. S., 1985, Meteoric burial diagenesis of Pennsylvanian arkosic sandstones, southwestern Anadarko Basin, Texas: American Association of Petroleum Geologists Bulletin, v. 69, p. 22–38.

FRANKS, S. G., 1980, Origin of porosity in deeply buried Tuscaloosa sandstones, False River Field, Louisiana (abs.): Gulf Coast Section SEPM, First Annual Research Conference, Geology of the Woodbine and Tuscaloosa Formations, p. 20.

HAMLIN, K. H., AND CAMERON, C. P., 1987, Sandstone petrology and diagenesis of Lower Tuscaloosa Formation in the McComb and Little Creek Field areas, southwest Mississippi: Transactions, Gulf Coast Association of Geological Societies, v. 37, p. 95–104.

HEALD, M. T., 1956, Cementation of Simpson and St. Peter sandstones in parts of Oklahoma, Arkansas, and Missouri: Journal of Geology, v. 64, p. 16–30.

HEALD, M. T., 1965, Lithification of sandstones in West Virginia: West Virginia Geological and Economic Survey Bulletin, v. 30, 28 p.

HEALD, M. T., AND BAKER, G. F., 1977, Diagenesis of the Mt. Simon and Rose Run sandstones in western West Virginia and southern Ohio: Journal of Sedimentary Petrology, v. 47, p. 66–77.

HEALD, M. T., AND LARESE, R. E., 1974, Influence of coatings on quartz cementation: Journal of Sedimentary Petrology, v. 44, p. 1269–1274.

HORN, D., 1965, Diagenese und porosität des Dogger-Beta Haupsandstein es den olfeldern Plön-ost und Preetz: Erdöl und Kohle-Erdgas-Petrochemie, v. 18, p. 249–255.

HOUSEKNECHT, D. W., AND MCGILVERY, T. A., 1990, Red Oak Field, in Beaumont, E. A., and Foster, N. H., eds., Structural Traps II. Atlas of Giant Oil and Gas Fields: American Association of Petroleum Geologists, p. 201–225.

HUDDER, K. G., AND TIEH, T. T., 1983, Diagenesis of deep Tuscaloosa sandstones, Profit Island Field, Louisiana: Geological Society of America, South-Central Section Annual Meeting, Abstracts with Programs, v. 15, p. 36.

IMAM, M. B., AND SHAW, H. F., 1987, Diagenetic controls on the reservoir properties of gas bearing Neogene Surma Group sandstones in the Bengal Basin, Bangladesh: Marine and Petroleum Geology, v. 4, p. 103–111.

JANKS, J. S., SANNESS, T., AND RASMUSSEN, B. A., 1985, Diagenesis of the Cotton Valley sandstones, Catahoula Creek Field, southern Mississippi: Transactions, Gulf Coast Association of Geological Societies, v. 35, p. 415–423.

LAND, L. S., AND DUTTON, S. P., 1978, Cementation of a Pennsylvanian deltaic sandstone: isotopic data: Journal of Sedimentary Petrology, v. 48, p. 1167–1176.

LARESE, R. E., 1974, Petrology and stratigraphy of the Berea Sandstone in the Cabin Creek and Gay-Fink trends, West Virginia: Unpublished Ph.D. Dissertation, University of West Virginia, Morgantown, 246 p.

LARESE, R. E., AND PITTMAN, E. D., 1987, Indirect evidence of secondary porosity in sandstones (abs.): American Association of Petroleum Geologists Bulletin, v. 71, p. 581.

LIN, F., 1984, Clays and Tuscaloosa Sandstone porosity development in Louisiana and Mississippi: Clay Minerals Society Annual Meeting, Programs with Abstracts, p. 81.

LONGSTAFFE, F. J., 1986, Oxygen isotope studies of diagenesis in the Basal Belly River Sandstone, Pembina I-Pool, Alberta: Journal of Sedimentary Petrology, v. 56, p. 77–88.

MCBRIDE, M. H., FRANKS, P. C., AND LARESE, R. E., 1987, Chlorite grain coats and preservation of primary porosity in deeply buried Springer Formation and Lower Morrow sandstones, southeastern Anadarko Basin, Oklahoma (abs.): American Association of Petroleum Geologists Bulletin, v. 71, p. 994.

MCHUGH, A., 1987, Styles of diagenesis in Norphlet sandstone (Upper Jurassic), onshore and offshore Alabama: Unpublished M.A. Thesis, University of New Orleans, New Orleans, Louisiana, 161 p.

MOLENAAR, N., 1986, The interrelation between clay infiltration, quartz cementation, and compaction in Lower Givetian terrestrial sandstones, northern Ardennes, Belgium: Journal of Sedimentary Petrology, v. 56, p. 359–369.

MORAES, M. A. S., AND DE ROS, L. F., 1990, Infiltrated clays in fluvial Jurassic sandstones of Reconcavo Basin, northeastern Brazil: Journal of Sedimentary Petrology, v. 60, p. 809–819.

PAXTON, S. T., SZABO, J. O., CALVERT, C. S., AND AJDUKIEWICZ, J. M., 1990, Preservation of primary porosity in deeply buried sandstones: a new play concept from the Cretaceous Tuscaloosa Sandstone of Louisiana (abs.): American Association of Petroleum Geologists Bulletin, v. 74, p. 737.

PITTMAN, E. D., AND LUMSDEN, D. N., 1968, Relationship between chlorite coatings on quartz grains and porosity, Spiro sand, Oklahoma: Journal of Sedimentary Petrology, v. 38, p. 668–670.

PITTMAN, E. D., AND WRAY, L. L., 1989, Sedimentation and petrology of the Fanshawe sand, Red Oak Field, Arkoma Basin, Oklahoma (abs.): American Association of Petroleum Geologists Bulletin, v. 73, p. 1049.

SCHMIDT, V., AND MCDONALD, D. A., 1979, Texture and recognition of secondary porosity in sandstone, in Scholle, P. A., and Schluger, P. R., eds., Aspects of Diagenesis: Society of Economic Paleontologists and Mineralogists Special Publication 26, p. 209–225.

SMITH, G. W., 1981, Sedimentology and reservoir quality of the "19,800" foot sandstone, False River field, Pointe Coupee and West Baton Rouge Parishes, Louisiana, in Stewart, D. B., ed., Tuscaloosa Trend of South Louisiana: New Orleans Geological Society, p. 47–81.

SMITH, G. W., 1985, Geology of the deep Tuscaloosa (Upper Cretaceous) gas trend in Louisiana, in Perkins, B. F., and Martin, G. B., eds., Habitat of Oil and Gas in Gulf Coast: Proceedings, 4th Annual Research Conference, Gulf Coast Section, Society of Economic Paleontologists and Mineralogists, p. 153–190.

TAYLOR, J. C. M., 1978, Control of diagenesis by depositional environment within a fluvial sandstone sequence in the northern North Sea Basin: Journal of the Geological Society of London, v. 135, p. 83–91.

THAYER, P. A., 1985, Diagenetic controls on reservoir quality, Matagorda Island 623 Field, offshore Texas (abs.): American Association of Petroleum Geologists Bulletin, v. 69, p. 311.

THOMSON, A., 1959, Pressure solution and porosity, in Ireland, H. A., ed., Silica in Sediments: Society of Economic Paleontologists and Mineralogists Special Publication 7, p. 92–110.

THOMSON, A., 1979, Preservation of porosity in the deep Woodbine/Tuscaloosa trend, Louisiana: Transactions, Gulf Coast Association of Geological Societies, v. 30, p. 396–403.

THOMSON, A., AND STANCLIFFE, R. J., 1990, Diagenetic controls on reservoir quality, eolian Norphlet Formation (Upper Jurassic), South State Line Field, Mississippi, U.S.A., in Barwis, J. H., McPherson, J. G., and Studlick, J. R. J., eds., Sandstone Petroleum Reservoirs: New York, Springer-Verlag, p. 205–224.

WINN, R. D., Jr., STONECIPHER, S. A., AND BISHOP, M. G., 1983, Depositional environments and diagenesis of offshore sand ridges, Frontier Formation, Spearhead Ranch Field, Wyoming: The Mountain Geologist, v. 20, p. 41–58.

WIYGUL, G. J., AND YOUNG, L. M., 1987, A subsurface study of the Lower Tuscaloosa Formation at Olive Field, Pike and Amite Counties, Mississippi: Transactions, Gulf Coast Association of Geological Societies, v. 37, p. 295–302.

INFLUENCE OF AUTHIGENIC-CLAY MINERALS ON PERMEABILITY

JAMES J. HOWARD*

Department of Geology and Geophysics, Yale University, New Haven, Connecticut 06511

ABSTRACT: Quantitative models of the reduction of permeability in reservoir sandstones due to the growth of authigenic-clay minerals in the pore space are based on the ability to estimate the permeability of the original clay-free rock. Simple physical models based on Carman-Kozeny relations are used to calculate permeability for the idealized sandstone pore space. Values for the surface-area parameter in the models are determined from proton NMR longitudinal-relaxation times and area/perimeter ratios extracted by petrographic-image analysis. Although the magnitude of the difference between measured and calculated permeabilities is model dependent, the different models characterize relative behavior for each suite of sandstones. The normalized permeability differences correlate weakly with various measures of total clay abundance. This indicates that permeability reduction is influenced more by clay distribution than by clay abundance. Cation-exchange capacity (CEC) measurements made by flow through the intact rock are lower than values determined by standard methods on powders. As the ratio of flow to bulk CEC values decreases, fewer of the clays in the pore space are accessed by the fluid. Samples with increased fractal dimensions or surface roughness have lower CEC ratios, indicating that increased roughness limits the accessibility of exchange sites. Samples with lower fractal dimensions have more authigenic kaolinite than fibrous illite, in addition to greater differences in measured and calculated permeability. This suggests that physical constrictions caused by clay growth in the throats is more important than surface-roughness effects in reducing permeability in sandstones.

INTRODUCTION

It is commonly recognized that the presence of authigenic clays in the pores of reservoir sandstones reduces permeability. Permeability, which has the dimensions of area (1 darcy = 10^{-8} cm^2), is a property of the porous medium, independent of the fluid conditions. Its measurement provides a macroscopic-level description of fluid flow inside the medium, that is, how the fluid flux through a rock of given permeability varies with the pressure gradient across the sample. The effect of clays in reducing permeability is believed to result from the constriction of pore throats and increased turbulence by roughening the pore-wall surface (Ives, 1987). The question considered here is whether certain clays reduce permeability more than others by virtue of their different morphologies and distributions within the pores, and if this difference in permeability reduction due to clays can be quantified.

There are a number of studies that link qualitative observations of authigenic clays in sandstones and their measured permeability. Often the goal is to predict the potential for formation damage during production. Perhaps the most prominent study separates authigenic clays into several morphologic groups and defines regions on a porosity-permeability cross-plot that are typical of those groups (Neasham, 1977) (Fig. 1A). Groups of pore-bridging, pore-lining and discrete particles of clay are commonly associated with illite fibers, chlorite and smectite rims, and kaolinite books, respectively. Neasham (1977) proposed that the distribution of clay morphologies significantly affected sandstone porosity-permeability relations, capillary-pressure curves and well-log shaliness indicators. The simple idea was that patchy, discrete clay particles have less effect on fluid flow than clays lining the entire pore or throat. A comparison of petrophysical properties for sandstones that contain any one type of clay, however, illustrates that these morphology groups do not fit into well-defined porosity-permeability fields. Several suites of sandstones, all with roughly 10% authigenic, pore-filling kaolinite, display a wide range of porosity and permeability (Fig. 1B). The data scatter from these different sample suites suggests that other factors besides the abundance and morphology of authigenic clays determine porosity-permeability relations. Other studies have shown that different clay types result in different porosity-permeability relations (Stadler, 1973; Nagtegaal, 1979). The key point that is missed in these porosity-permeability cross-plots, however, is how much did the presence of the authigenic clay reduce the permeability in each sample?

Several laboratory-based studies illustrate the importance of sample preparation in the measurement of permeability in sandstones that contain certain types of clay. The preservation of delicate illite fibers by critical-point drying methods results in permeabilities three to ten times less than permeability measured on the same sample that was air-dried (Pallat and others, 1984; deWaal and others, 1988). Similar permeability reduction was observed for measurements using both air and brine as the permeating fluid. The amount of permeability reduction in the various samples was not dependent upon the amount of clay. Pallat and others (1984) found similar permeability reduction for two samples containing 0.5 and 10% illite. They suggested that distribution in critical pathways is more important than amount of clay in influencing permeability.

Other studies have measured the reduction in permeability caused by fine-particle migration (Lever and Dawe, 1984; Ives, 1987). In most of these, the fine particles are authigenic-kaolinite books dislodged from pore walls by changes in ionic strength of the fluid. Fresh water (low ionic strength) greatly expands the electric double layer, causing deflocculation, and provides the material for migration and accumulation in pore throats. The constriction of pore throats by these particles is largely reversible. Permeability can be restored to 90% of original permeability by the introduction of high-salinity brines, which reduces the thickness of the electrical double layer and causes flocculation and attachment of kaolinite particles to the pore wall. Permeability also can be restored in these cases by backflushing. The amount of permeability reduction is not associated with amount of clay in the pores. Both critical-point drying and kaolinite-migration studies measure permeability reduction due to laboratory technique, but the underlying question of

---

*Present Address: Phillips Petroleum Research, 116 GB, Bartlesville, Oklahoma 74004

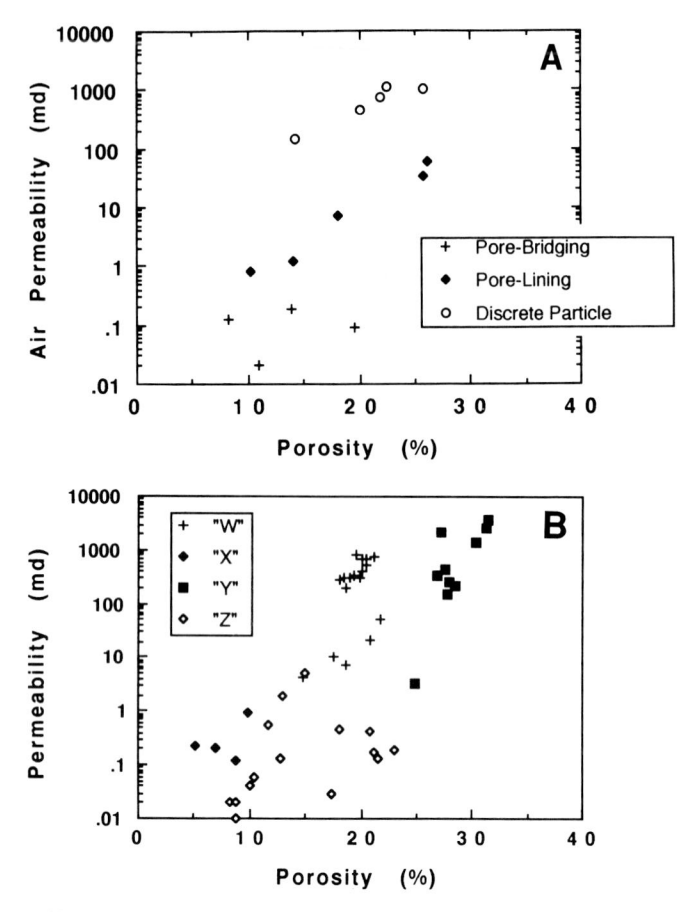

FIG. 1—(A) Porosity-permeability relations for different authigenic clay-mineral distributions (modified from Neasham, 1977). (B) Porosity-permeability relations for reservoir sandstones (designated by letter), all containing authigenic kaolinite.

how did the growth of clay contribute to the original permeability remains unanswered.

The purposes of this study are: (1) to quantify the amount of permeability reduction created by the presence of authigenic clay in the pore space; and (2) to determine the relative importance of clay abundance versus distribution in reducing permeability. A simple experiment would be to take a clean sandstone, precipitate clay in the pore space and determine the change in permeability. Lacking the ability to do this, we estimated what the permeability in a reservoir sandstone would be if all the clays were removed. Estimates of permeability for the "clean" sandstones are based on several simple physical models that utilize pore-size information. This study also proposes a clay-sensitive property measured by static and dynamic methods as a means of quantifying the importance of clay distribution on relative amounts of permeability reduction.

METHODS

Petrophysical Measurements

Petrophysical measurements were made on plugs 2 cm in diameter and 3.75 cm in length, most oriented with long axes parallel to bedding. All of the samples had been in storage and were oven dried to remove any excess water. Permeability was measured by the flow of brine through the long axes of the plugs while enclosed in a stainless steel core holder. The core holder had a rubber inner sleeve that was pressurized on the outside, forming a tight seal between the liner and core plug. Permeability was measured by the fluid flux (cm^3/s) at several different pressure gradients (atm/cm). The slope of the best fit line is k/v where k is the permeability constant in cm^2 and v is the dynamic viscosity in poise. Linear behavior between flux and pressure gradient and a y-intercept of zero define Newtonian flow, and is required for acceptable measurements. Replicate measurements indicated an error of 5% and a lower limit of measurement at 0.01 md. Porosity was measured by standard buoyancy methods. Formation factor was measured on the saturated plugs at only one salinity with a four-terminal electrode cell.

Proton NMR measurements were made on water-saturated plugs in a desk-top instrument operating on the Larmor resonance frequency of 10 MHz. The experiment measured the longitudinal relaxation or T_1 curve using a standard inversion-recovery procedure. A stretched exponential model was also fitted to the longitudinal relaxation curve

$$M(t) = Mo \exp(-t/T_{1\alpha})^\alpha \qquad (1)$$

where $M(t)$ is the magnetization at recovery time t, Mo is the initial magnetization, and $T_{1\alpha}$ is the mean T_1 value (Howard and others, 1990).

Other estimates of pore dimensions were obtained from image analysis of thin-section micrographs and mercury porosimetry (Kenyon and others, 1989). An average area/perimeter (A/P) value for each sample was determined for all the measurable pores. Mercury porosimetry curves were generated over a pressure range of 1 to 60,000 psi (413 MPa). Cylindrical samples were used instead of irregularly shaped chips to minimize intrusion into surface pores at low pressures. Mean pore-throat diameters and inflection points were extracted from the intrusion curve.

Cation-exchange measurements were made following two procedures. The standard method used NH_4^+ as the exchange cation, which in turn was replaced on the exchange sites by Na^+. The sample was prepared by grinding for only several minutes to minimize grinding effects that are common in CEC measurements. The collected NH_4^+ then was titrated in a Kjeldahl distillation apparatus (Ridge, 1983). This method gave results that are comparable to other traditional CEC techniques. The other procedure employed a fluid-flow apparatus and the original permeability plug (Crocker and others, 1983). The sample was saturated initially with 1 N $CaCl_2$, then flushed with distilled water at low-flow rates to remove excess Ca^{+2} and Cl^-. Flushing continued until Cl^- in the effluent was not detected (roughly 10 pore volumes). Immediately, a 1 N $LaCl_3$ solution was introduced at moderate-flow rates. A 10-ml aliquot of effluent was collected and the Ca^{+2} in solution was measured with a Ca-selective electrode.

Samples

The first group of samples included five suites of reservoir sandstones that cover a wide range of petrophysical properties. All had similar lithologies, distinguished by a quartz- and feldspar-rich framework and predominantly intergranular porosity, with small amounts of microporosity associated with authigenic-clay minerals filling some of the pore space (Table 1). Suite "A" sandstones had a wide range of porosity and permeability, and contained small amounts of well-formed kaolinite books in the pore space. Several "A" samples also had trace amounts of fibrous illite. The clay distribution in the "A" sandstones appeared to be irregular; many pores were clean, whereas a few were completely filled with kaolinite. The "E" group sandstones were very well-sorted quartz arenites, with large intergranular pores and small amounts (less than 5%) of kaolinite and illite in the pores and throats. Group "D" sandstones exhibited a wide range of porosity and permeability values, 6–18% and 0.01 to 100 md, respectively. The dominant clay in the pore space of group "D" samples was authigenic kaolinite. Several of the more porous and permeable samples had authigenic chlorite that lined the entire pore space. Groups "B" and "C" were classified as shaly sands, because of their greater clay content, as much as 20%. Abundant authigenic illite and some kaolinite were observed in the pores; however, there were also significant amounts of detrital clay (based on criteria of Wilson and Pittman, 1977). Most of the sandstones in this study had log-normal pore-size distributions, as determined by proton nuclear magnetic-resonance measurements (Howard and others, 1990).

A second set of six sandstones was used in the CEC-flow experiments (Table 2). These sandstones are commonly used in petrophysical studies by a number of laboratories, with their prime attraction being, with the exception of Portland sandstone, a low-clay content. Kaolinite is the dominant clay mineral in Berea, Massillon and Portland sandstones, with minor amounts of illite, whereas the Nugget and Coconino sandstones contain mostly illite. This set of "petrophysical standards" is the same as is used in a small-angle neutron scattering (SANS) study (Wong and others, 1986). In the SANS study, the scattering-intensity data taken over a length scale range of 5 to 500Å were fitted with a power-law exponential function. This exponent value is related to the fractal dimension, d, with greater surface roughness associated with higher fractal dimension values. The length scale of the SANS measurement is sensitive to variations in surface roughness on a pore scale.

TABLE 1.—DESCRIPTION OF RESERVOIR SANDSTONES

Formation	Framework[1]	% Total Clay (XRD)	Clay Minerals[2]
"A"	Qtz, Kfeld	7	Kaol
"B"	Qtz, Kfeld, Cal	15	Ill, Kaol
"C"	Qtz, Kfeld, Cal	20	Ill, Kaol
"D"	Qtz	4	Kaol
"E"	Qtz, Kfeld	9	Kaol, Chlor

[1]Qtz = Quartz, Kfeld = K-feldspar, Cal = Calcite
[2]Kaol = Kaolinite, Ill = Illite, Chlor = Chlorite

TABLE 2.—DESCRIPTION OF PETROPHYSICAL-STANDARDS SANDSTONES

Formation	Framework	Clay Minerals
Berea	Qtz, Kfeld	Kaol, Ill
Massillon 65	Qtz, Kfeld	Kaol, Ill
Massillon 85	Qtz, Kfeld	Kaol, Ill
Nugget	Qtz, Kfeld	Ill
Portland	Qtz, Kfeld, Cal	Kaol
Coconino	Qtz	Ill

Permeability Models

Permeability estimates for an idealized clay-free sandstone were generated by assuming that the pore system can be represented by an equivalent homogeneous system, in which the flow channels are identical cylinders or tubes with radius r. A number of simple physical models, generally called hydraulic-radius models, are available. The models differ primarily in how the critical-length information is obtained from measurements of the rock (Brace, 1977). The advantages of these widely used models are their theoretical simplicity and their utility in identifying the essential elements of permeability (Paterson, 1983). More complex theoretical models often fail, largely because the intricate pore geometry of the connected pore space defies quantitative characterization (Dullien, 1979).

The Carman-Kozeny parallel-tubes model is most commonly used for estimating permeability in sandstones. The model assumes a bundle of uniform capillary tubes with a total cross-sectional area equal to the sample porosity. In its simplest form, a Carman-Kozeny model includes:

$$k = \phi^3/c_1 S^2 \qquad (2)$$

where ϕ is the fractional porosity, S is the surface area to volume ratio, and c_1 is a shape factor plus tortuosity term, which equals 2 to 3 for simple packing geometries (Wyllie and Splanger, 1952). Best fits of rock data increase the value of c_1 to 5 (Wong and others, 1986). Most studies using the Carman-Kozeny model employ surface areas determined by BET gas adsorption. To emphasize the enhanced contribution to fluid flow by larger pores, this study used an area/perimeter (A/P) ratio determined by image analysis of thin sections. This A/P ratio estimates the effective pore-surface area/volume for predicting transport properties (Ehrlich and others, 1984; Doyen, 1988).

The second permeability estimate also used the Carman-Kozeny hydraulic-radius formalism, but obtained the surface area/volume information from proton NMR T_1 relaxation curves. The model predicts permeability from

$$k = \phi^{2m} T_{1\alpha}^2 \qquad (3)$$

and

$$1/T_{1\alpha} = \rho S/V \qquad (4)$$

where m is the Archie cementation exponent, $T_{1\alpha}$ is the mean longitudinal-relaxation time determined from the stretched exponent fit, and ρ is the surface-relaxation strength parameter in cm/sec. For many reservoir sandstones, empirical observations indicate that m is equal to two. For these reservoir sandstones, the ϕ^4 relation generates a NMR-based permeability estimator (Kenyon and others, 1988; Howard and others, 1990). For unimodal-size glass-bead packs, a simple analogy to clean sandstones, experimental results of $m = 3/2$ are supported by theoretical arguments based on network resistor models (Wong and others, 1984).

The pore-size information obtained from A/P and T_1 measurements focuses on the pore-body dimensions, rather than the smaller pore constrictions that ultimately have more effect on permeability. A percolation-based model of fluid flow was also evaluated in this study. This model attempts to account for more heterogeneity in the pathways and pore dimensions than the simpler Carman-Kozeny model (Thompson and others, 1987). The percolation model also suggests that the interconnectedness of the pore network is porosity independent:

$$k = c_2 \, l_c / F \tag{5}$$

where l_c is the threshold-conductance length for network percolation, F is the formation factor, which accounts for tortuosity, and c_2 is a geometrical constant. Throat diameters, or the threshold length, were obtained from the inflection point of a mercury-injection curve.

RESULTS

There was a wide range of measured permeability and porosity values for the various sandstone suites (Fig. 2). For a given porosity value, permeability varied by as much as four orders of magnitude. The "A" suite had permeabilities that ranged from 10 to 1,200 md, whereas porosities ranged from 19 to 24% (Table 3). Formation factor for these "A" sandstones was significantly less variable, 14 to 23, than measured permeabilities, even though both were percolation-based properties.

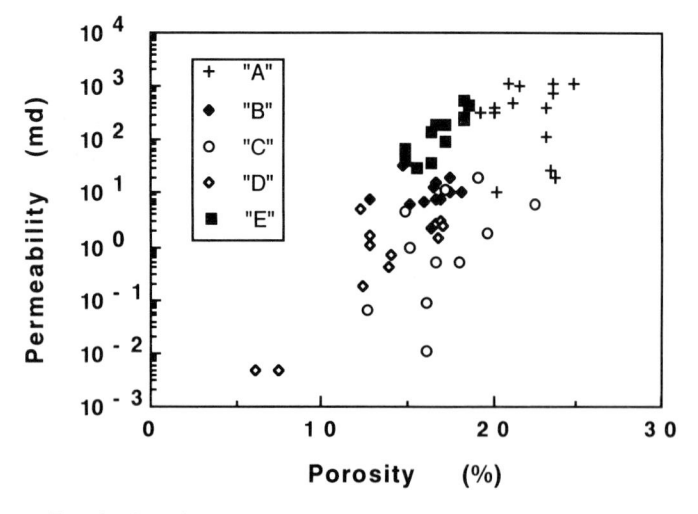

FIG. 2—Porosity-permeability relations for the five suites of reservoir sandstones (designated by letter) in this study.

TABLE 3.—PETROPHYSICAL MEASUREMENTS OF "A" SANDSTONES

Sample	Porosity (%)	Perm.[1] (md)	T_1 (ms)	l_c[2] (μm)	Form F[3]	A/P[4] (μm)
A2	20.8	1160	569	36.7	14.3	18.1
A3	23.6	1170	395		15.5	17.7
A4	24.3	1160	393	37.8	15.5	17.2
A5	20.2	10	81	2.6		8.8
A6	23.4	26	134	5.0	18.8	9.9
A7	23.2	115	146	6.1	15.8	10.0
A9	19.2	346	299	27.2	16.6	11.8
A10	23.6	769	448	35.9	14.8	16.7
A11	20.0	334	351	29.0	17.4	14.5
A13	23.7	19	181	8.8	23.4	13.9
A14	20.7	391	345	25.3	19.9	20.0
A15	23.2	408	365	34.1	17.3	21.0
A16	21.6	1030	473	18.9	16.1	17.7
A17	21.1	514	346	59.0	19.2	17.8

[1]Measured permeability.
[2]Critical throat radius, mercury porosimetry.
[3]Formation factor.
[4]Area/perimeter.

Pore dimensions determined by image analysis and mercury porosimetry also resulted in a range of values. The "A" sandstones illustrated the range of pore dimensions observed (Table 3). Mercury-porosimetry curves for the higher porosity sandstones tended to be unimodal with sharp inflection points, which characterized l_c. The "A" sandstones had mean throat diameters between 2.6 and 37.8 μm, corresponding to entry pressures of 400 and 3 psi (2.76 and 0.02 MPa), respectively. The lower permeability samples, found in suites "B" and "C," had throat dimensions determined by mercury-porosimetry as low as 0.1 μm, equivalent to an entry pressure of 1,000 psi (6.89 MPa).

Area/perimeter values for the "A" samples represented the high range for all the samples, 10–20 μm (Table 3). The "D" and "E" sandstone samples had intermediate A/P values of 8 to 13 μm. The lower porosity-permeability "B" and "C" sandstone samples had A/P values of 3 to 8 μm.

NMR $T_{1\alpha}$ values for the "A" sandstones varied between 80 and 570 ms, with a strong correlation between long relaxation times and high-permeability values (Table 3). The "E" sandstones had $T_{1\alpha}$ values between 200 and 400 ms with much less variation than the "A" sandstones. The "B" sandstones had $T_{1\alpha}$ values between 100 and 200 ms, whereas the lower permeability "C" sandstones had values between 10 and 100 ms. In addition to determining mean T_1 values, the measured NMR inversion-recovery curve was transformed into a population-density distribution of relaxation times, with shorter times corresponding to smaller pores and longer times to larger pores (Kenyon and others, 1989). The sandstone suites in this study were characterized by log-normal distributions of T_1's, which corresponded to log-normal pore-size distributions (Howard and others, 1990).

The two Carman-Kozeny models produced permeability estimates that are similar, especially for the higher porosity and permeability samples. Permeability estimates based on the simple physical models of the pore-hydraulic radius, NMR T_1 and A/P, were generally two to three times greater than the measured permeability for the "A" sandstones (Table 4). The greatest discrepancy between measured and estimated permeability occured with the percolation model that utilized pore dimensions from mercury porosimetry. Many of the estimates from the percolation model were less than or equal to the measured permeability. For the lower porosity "B," "C," and "D" sandstone samples, the permeability predicted from image-analysis A/P pore dimensions had a maximum value of 50 md. For these same low-porosity sandstones, the estimated permeability from the NMR-based model was less than 1 md.

For the reservoir-sandstone data set, the NMR-based estimated permeability was two to seven times greater than the measured value, with most of the samples being within a factor of three of the measured value (Fig. 3). The relative difference between measured and calculated permeability was the same through most of the permeability range, with greater differences at the lowest permeability values. The difference in measured and calculated permeability was evaluated as both the normalized difference between the two permeability values and as the difference between the log values (Δk and $\Delta \log k$).

The difference in permeability values is compared later to various measurements of clay content in the sandstones. Standard X-ray powder diffraction (XRD) methods of determining total-clay content were limited in this study by low levels of precision, ± 5% absolute abundance, which corresponds to as much as 100% relative error for these clay-poor samples. Because surface area and CEC measurements are not clay specific, there is greater analytical precision, if not accuracy. Newer XRD methods that improve precision were not used (Moore and Reynolds, 1989). Surface area measured by gas adsorption methods correlated poorly with the difference in log permeability (Fig. 4). Most of the samples clustered around low surface-area values. A few samples from the "C" sandstones had high surface areas (6 to 8 m^2/gm) that corresponded to large $\Delta \log k$, but several other "C" samples with large surface areas had very small differences between measured and calculated permeability. Similar relations were observed for cation-exchange capacity.

The fractal dimensions, d, of the petrophysical-standards samples obtained from SANS measurements (Wong and others, 1986) varied between 2.4 for the Berea and Massillon 85 sandstones to 2.9 for Coconino sandstone (Table 5). The samples rich in fibrous illite, Nugget and Coconino, had higher fractal-dimension values than the kaolinite-filled Berea and Massillon samples (Fig. 5). For these samples, the presence of fibrous illite resulted in a larger d value and greater surface roughness. Clay abundance did not appear to affect the fractal dimension significantly. Portland sandstone had an intermediate fractal value of 2.54, despite having more clay, mostly kaolinite, than the other samples.

The flow CEC experiments on the same petrophysical-standards samples generated CEC values less than those de-

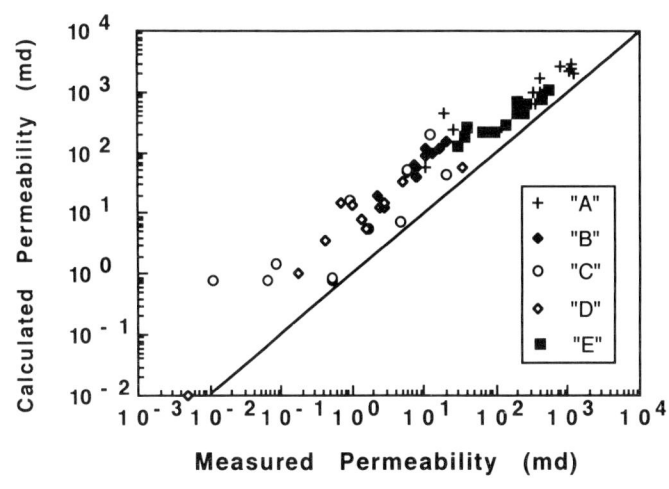

FIG. 3—Comparison of measured permeability with values calculated from the Carman-Kozeny model using NMR T_1 results as the hydraulic-radius parameter. Line represents 1:1 relation.

TABLE 4.—PERMEABILITY ESTIMATES FOR "A" SANDSTONES

Sample	NMR T$_1$ (md)	Area/Perimeter (md)	Percolation (md)
A2	2915	1477	726
A3	2059	2057	
A4	2361	2260	709
A5	55	323	
A6	231	625	10
A7	268	631	18
A9	632	490	344
A10	2640	1836	670
A11	986	846	372
A13	440	1290	25
A14	966	1621	248
A15	1666	2756	518
A16	2260	1572	171
A17	1124	1494	1398

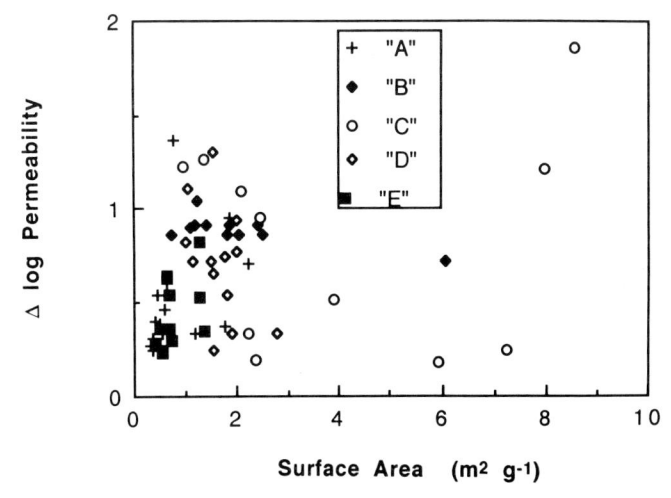

FIG. 4—Cross-plot of surface area vs. $\Delta \log k$ for suites of sandstones used in this study. See text for additional explanation.

TABLE 5.—CATION-EXCHANGE CAPACITY MEASUREMENTS FOR PETROPHYSICAL-STANDARDS SANDSTONES

Sample	Bulk CEC (meq/100 g)	Flow CEC (meq/100 g)	Fractal dimension
Berea	1.0	0.7	2.41
Massillon 65	0.4	0.2	2.62
Massillon 85	0.4	0.4	2.36
Nugget	1.8	0.8	2.68
Portland	2.9	1.5	2.54
Coconino	1.0	0.3	2.92

termined by standard bulk methods (Table 5). The ratio of flow/bulk CEC values, compared to the fractal dimension, indicates that more complete cation exchange occured in samples with lower fractal dimension (Fig. 6). As the surface roughness increased, the ability of the 10 ml of collected pore fluid to exchange cations decreased significantly.

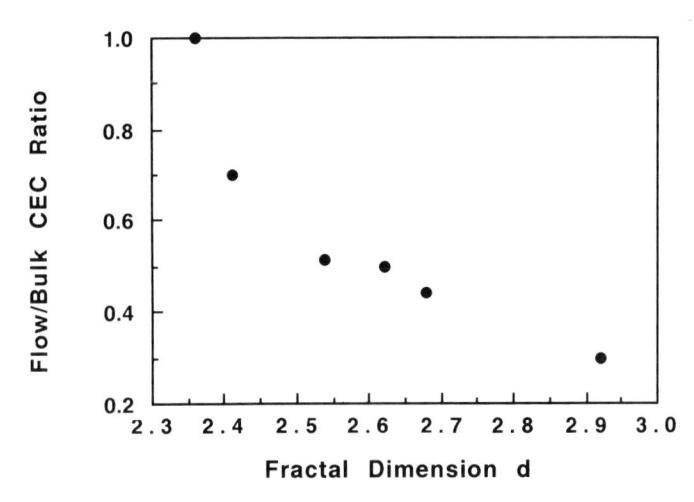

FIG. 6—Cross-plot of fractal dimension (d) vs. flow/bulk CEC ratio for petrophysical-standards sandstones.

Measured permeability for the petrophysical-standards samples varied between 0.1 and 2,233 md (Table 6). $T_{1\alpha}$ values ranged from 13 to 580 ms. The NMR-based permeability model estimated permeability values 2 to 5 times greater than measured. An increase in Δ log permeability corresponded with higher values of flow/bulk CEC ratio (Fig. 7).

DISCUSSION

The difference between measured and calculated permeability, Δk, is a quantitative estimate of how much permeability reduction is due to clays in a given sandstone. This interpretation, of course, is overly simplistic because it does not account for any other diagenetic processes, such as cementation, that reduce permeability. However, for these samples that have limited quartz and carbonate cements, this usage of Δk as a clay-sensitive term will suffice.

Physical models for estimating permeability should be simple, with estimated permeability generally being greater than the measured value. High-permeability estimates commonly result from pore-dimension measurements that are skewed toward larger pore sizes. The absolute value of the calculated permeability contains less information than the difference in measured and calculated permeability, be-

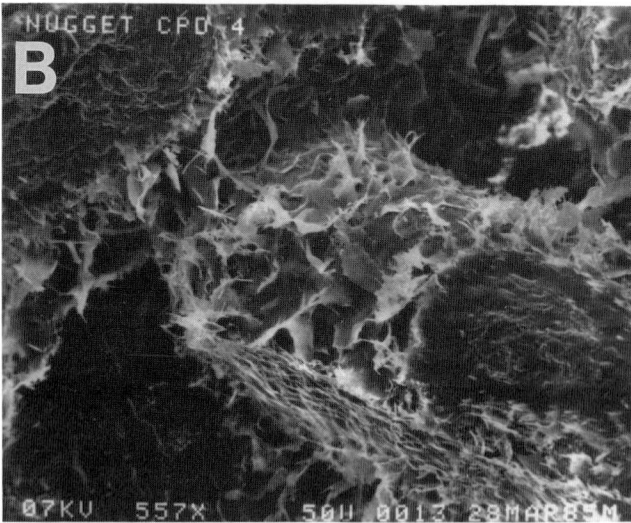

FIG. 5—(A) SEM micrograph of Berea sandstone with abundant, small particles of kaolinite (d = 2.42 and CEC ratio = 0.7). (B) SEM micrograph of Nugget sandstone; sample was prepared by critical-point drying (d = 2.68 and CEC ratio = 0.4). Width of each micrograph is 300 μm.

TABLE 6.—PERMEABILITY VALUES FOR PETROPHYSICAL-STANDARDS SANDSTONES

Sample	Porosity (%)	$T_{1\alpha}$ (ms)	Permeability (md)	Calc. Perm (md)
Berea	18.4	236	103	347
Massillon 65	23.8	407	1242	2233
Massillon 85	24.3	581	2590	4844
Nugget	6.3	21	0.1	0.1
Portland	20.0	13	0.85	1.4
Coconino	13.9	135	62.5	48.9

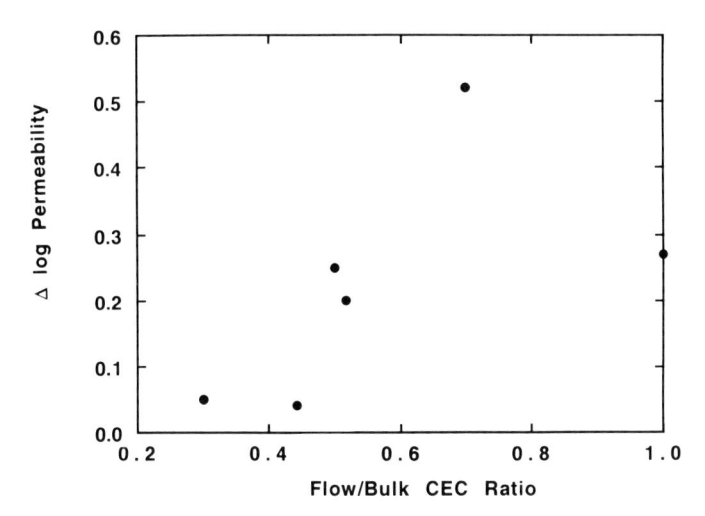

FIG. 7—Cross-plot of flow/bulk CEC ratio vs. Δ log k for the petrophysical-standards sandstones. See text for additional explanation.

cause it reflects deviations from flow behavior through an idealized pore geometry.

Is there a best model that estimates the original clay-free permeability? The answer, if one is possible, may lie more in the nature and quality of the measurements used as input parameters than in any theoretical construction of pore geometry. For the two Carman-Kozeny models, the hydraulic-radius parameter determined from A/P pore dimensions is skewed toward larger sizes than the NMR-based term due to the limited optical resolution of a microscope. For example, if the surface-relaxation term (ρ) is 0.001 cm/s (Howard and others, 1990), then 85 ms corresponds to a spherical-pore diameter of 5 μm, roughly the lower limit of optical resolution. Because several of the low porosity-permeability samples have mean T_1 values less than 85 ms, this suggests that image-analysis results for these samples utilize only a few measurable pores. Despite the limited number of pores measured by image analysis, A/P results are still useful because the largest pores contribute the most to fluid flow (Dullien, 1979). Therefore, all the smaller pores detected by NMR have no appreciable effect on permeability. Future research could be to utilize the T_1 distributions along with area/perimeter distributions to optimize these permeability models.

The percolation model gives a significantly larger range of permeability values, with greater variation in Δ log k. Many of the estimates from the percolation model are equal to or less than measured permeabilities. This results from the throat-sensitive nature of the input critical-length parameter for the percolation model that is obtained from mercury porosimetry. A mercury-injection curve is the result of percolation phenomena, and is dependent on the connectivity of the porous network. Mercury data reflect dynamic behavior and therefore contrast with the NMR and image-analysis measures of pore length that are independent of network considerations. The precipitation of small amounts of authigenic clay does little to change pore diameters, yet dramatically alters transport properties. This suggests that permeability estimates from percolation-based models are not suitable to determine original sandstone permeability.

Clay abundance does not correlate well with Δ log k. The results to date suggest that the amount of permeability reduction in clay-bearing sandstones is not dependent on the abundance of clay. One problem is estimating clay abundance in sandstones that contain low volumes of clay (less than 10%). The observation that clay abundance does not affect permeability reduction is in agreement with laboratory-based permeability-reduction studies (Pallat and others, 1984; Ives, 1987). In turn, this conclusion suggests that the distribution of clay in the pore space might be the more important parameter.

This study attempted to quantify distribution of clay by comparison of static and dynamic measurements of cation-exchange capacity. Flow or dynamic CEC measurements reflect the accessibility of flowing fluid to the exchange sites. Clays that are more directly in the flow path, at the centers of the pore spaces and throats, exchange cations more efficiently than clays along the edges of pore spaces. Less than 10 minutes was required for the collection of the 10 ml, suggesting that diffusion of Ca^{+2} from these inaccessible sites to the main flow paths (on the scale of 10 μm) should be negligible.

The relation between flow/bulk CEC ratio and fractal dimension suggests that, as the pore surface becomes rougher (e.g., Coconino and Nugget with their mostly fibrous illite), the flow experiment is less efficient in exchanging cations. The low-flow/bulk CEC value suggests that these fibrous, pore-lining clays are less accessible to the overall flow path. In a given pore, most of the flow occurs in its center, whereas the pore walls are in a no-slip boundary condition (Dullien, 1979). In contrast, the Berea sample has kaolinite books located more toward the central portions of pores and a higher flow/bulk CEC ratio. Most of the exchangeable cations in kaolinite are accessed by the flow that traverses the pores. The CEC results suggest that the kaolinite books are in the central portions of pores and throats, whereas fibrous illite is found on the edges of the pores.

The difference in measured and NMR-calculated permeability for the petrophysical-standards samples indicates that greater permeability reduction occurs in the samples with highest flow/bulk CEC ratio and lower fractal dimension. Intuition might suggest that the samples with more authigenic fibrous illite would have a greater loss of permeability, but these results indicate that authigenic kaolinite reduces permeability more. This is opposite of the proposal by Neasham (1977), who suggested that authigenic kaolinite-filled pores would have the least effect on transport properties. This suggests that the permeability-reduction mechanism is actual physical constraint and constriction in the pore throats rather than drag and increased turbulence of the laminar flow created by roughened surfaces.

CONCLUSIONS

Simple physical models are used to estimate permeability for idealized reservoir sandstones with no authigenic clays in the pore space. Comparison of these estimated values and actual measured permeabilities suggests the magnitude of permeability reduction caused by the growth of authigenic clays. The absolute magnitude of permeability reduction is model dependent; however, the models consistently yield permeability values for "clay-free" sandstones greater than measured values. The most consistent estimated permeability values are derived from Carman-Kozeny-based models, which employed surface-area terms generated from NMR T_1 and image-analysis area/perimeter measurements.

The relative difference in measured and calculated permeabilities is poorly correlated with different measures of clay abundance. The absence of a good correlation between permeability reduction and clay abundance suggests that the distribution of clays in pore space is the more important factor in controlling flow properties in the reservoir.

Quantitative measurements of clay distribution in pore space include dynamic CEC and SANS measurements. The reduction in flow/bulk CEC ratio demonstrates that clay in central pathways has more effect on reducing permeability than clays along pore edges. Rough surfaces associated with authigenic illite, for these samples, has less effect on reducing permeability than smoother kaolinite-filled pores. This suggests that the mechanism of permeability reduction by authigenic-clay minerals is associated more with physical constrictions than drag-induced turbulence.

ACKNOWLEDGMENTS

The laboratory assistance of L. McGowan, P. Dryden and C. Straley is gratefully noted. Petrographic-image analysis results were provided by E. Etris and R. Ehrlich, University of South Carolina. D. Houseknecht, J. Janks, K. Gerety and W. Kenyon are thanked for comments and discussion.

REFERENCES

BRACE, W., 1977, Permeability from resistivity and pore shape: Journal of Geophysical Research, v. 82, p. 3343–3349.

CROCKER, M., DONALDSON, E., AND MARCHIN, L., 1983, Comparison and analysis of reservoir rocks and related clays: Society of Petroleum Engineers, Paper 11973, 8 p.

DEWAAL, J., BIL, K., KANTOROWICZ, J., AND DICKER, A., 1988, Petrophysical core analysis of sandstones containing delicate illite: Log Analyst, v. 8, p. 317–330.

DOYEN, P., 1988, Permeability, conductivity, and pore geometry of sandstone: Journal of Geophysical Research, v. 93, p. 7729–7740.

DULLIEN, F. A., 1979, Porous Media: Fluid Transport and Pore Structure: New York, Academic Press, 396 p.

EHRLICH, R., KENNEDY, S., CRABTREE, S., AND CANNON, R., 1984, Petrographic image analysis, I. Analysis of reservoir pore complexes: Journal Sedimentary Petrology, v. 54, p. 1365–1378.

HOWARD, J., KENYON, W., AND STRALEY, C., 1990, Proton magnetic resonance and pore-size variations in reservoir sandstones: Society of Petroleum Engineers, Paper 20600, p. 733–742.

IVES, K., 1987, Filtration of clay suspensions through sand: Clay Minerals, v. 22, p. 49–61.

KENYON, W., DAY, P., STRALEY, C., AND WILLEMSEN, J., 1988, A three-part study of NMR longitudinal relaxation studies of water-saturated sandstones: Society of Petroleum Engineers, Formation Evaluation, v. 4, p. 622–636.

KENYON, W., HOWARD, J., SEZGINER, A., STRALEY, C., MATTESON, A., HORKOWITZ, K., AND EHRLICH, R., 1989, Pore-size distribution and NMR in microporous cherty sandstones: Transactions, Society of Professional Well Log Analysts Annual Meeting, Paper LL, 24 p.

LEVER, A., AND DAWE, R., 1984, Water sensitivity and migration of fines in the Hopeman sandstone: Journal of Petroleum Geology, v. 7, p. 97–108.

MOORE, D. M., AND REYNOLDS, R. C., 1989, X-ray Diffraction and the Identification and Analysis of Clay Minerals: New York, Oxford University Press, 332 p.

NAGTEGAAL, P., 1979, Relationship of facies and reservoir quality in Rotliegendes desert sandstones, southern North Sea region: Journal of Petroleum Geology, v. 2, p. 145–158.

NEASHAM, J. W., 1977, The morphology of dispersed clay in sandstone reservoirs and its effect on sandstone shaliness, pore space, and fluid flow properties: Society of Professional Engineers, Paper 6858, 8 p.

PALLAT, N., WILSON, J., AND MCHARDY, W., 1984, The relationship between permeability and the morphology of diagenetic illite in reservoir rocks: Journal of Petroleum Technology, v. 25, p. 2225–2227.

PATERSON, M., 1983, The equivalent channel model for permeability and resistivity in fluid-saturated rock—a re-appraisal: Mechanics Materials, v. 2, p. 345–352.

RIDGE, M., 1983, A combustion method for measuring the cation exchange capacity of clay minerals: Log Analyst, v. 3, p. 6–11.

STADLER, P., 1973, Influence of crystallographic habit and aggregate structure of authigenic clay minerals on sandstone permeability: Geologie en Mijnbouw, v. 52, p. 217–220.

THOMPSON, A., KATZ, A., AND KHRON, C., 1987, The microgeometry and transport properties of sedimentary rock: Advances in Physics, v. 36, p. 625–694.

WILSON, M., AND PITTMAN, E., 1977, Authigenic clays in sandstones: Recognition and influence on reservoir properties and paleoenvironmental analysis: Journal of Sedimentary Petrology, v. 47, p. 1–31.

WONG, P., HOWARD, J., AND LIN, J., 1986, Surface roughening and the fractal nature of rocks: Physics Review Letters, v. 57, p. 637–640.

WONG, P., KOPLIK, J., AND TOMANIC, J., 1984, Conductivity and permeability of rocks: Physics Review B, v. 30, p. 6606–6614.

WYLLIE, M., AND SPANGLER, M., 1952, Application of electrical resistivity measurements to the problem of fluid flow in porous media: American Association of Petroleum Geologists Bulletin, v. 36, p. 359–403.

…

FORMATION CLAYS: ARE THEY REALLY A PROBLEM IN PRODUCTION?

GEORGE E. KING

Amoco Production, Research, Tulsa, Oklahoma 74102

ABSTRACT: Clay minerals in reservoir sandstones may react with drilling and completion fluids. Reactions are influenced also by clay mineralogy, clay form, and location of clay minerals. Smectite and mixed-layer illite/smectite clay minerals may swell to produce blockage, or occur as loose particles that migrate. Illite has a diverse morphology, but commonly occurs as a fibrous "spiderweb" form that may migrate when subjected to high-velocity flow, trap fine particles or water, or be virtually inert. Loose particles of kaolinite may migrate, but when water chemistry is controlled, this problem is minimized. Chlorite may serve as a trap for fine particles, as a habitat for microporosity, and has the potential to react with acids to create a ferric-hydroxide precipitate. Clays in high-permeability "channels," through which the fluid is flowing, are particularly prone to produce formation damage.

Laboratory flow-test results show that there may not be a correlation of permeability damage from the flowing fluid with the quantity of clay, even authigenic clay, reported to occur in the rock. Scanning electron microscopy and X-ray diffractometry are not independent methods for predicting clay-reaction problems. However, these techniques are helpful, when combined with flow testing and interpretation experience, for predicting clay reactions.

INTRODUCTION

The paper is written from a completion engineer's point of view and thus lacks some of the wonder of the microscopic arena and all of the SEMs. No apology is made for the tongue-in-cheek writing style because no offense is intended. Clays have a unique place in the minds of geologists and more than a few well-completion engineers; stored alongside memories of vivid nightmares, ex-spouses and the IRS. Clays have been blamed for all manners of damage when the well did not produce sufficient quantities of the types of fluids that were the target of the zone. Excuses and delusions aside, clays are not a universal cause of formation damage. The real effect of clay depends on factors as easily identified as type and location, and as difficult as exact completion fluid and stimulation and production history. This paper will examine the more common clay types for potential problems indicated by flowing-fluid tests on core plugs and suggest practical tests to determine the possibility of damage. In some cases of migrating fines, clays may not be the culprit at all; rather, feldspars may be the source of migrating fines.

PHYSICAL AND CHEMICAL EFFECTS OF CLAY MINERALS RELATED TO FORMATION DAMAGE

Smectite, illite, chlorite and kaolinite are among the most commonly known reservoir clays (Wilson and Pittman, 1977). The presence or absence of these clays in a reservoir rock, as determined by bulk-analysis techniques, is a virtually meaningless piece of information. For clays to cause a problem, the type, form, location, or surface charge must be affected in some manner (physical and/or chemical reactivity) by the flowing fluid. When fluid does not interfere with one or more of these factors, there are no problems.

It must also be remembered that several of the permeability reductions blamed on clay minerals may involve clay- or silt-size particles of other minerals (e.g., feldspar), microporosity, capillary effects, wetting activity, surface and interfacial tensions, and a host of other factors that do not even involve clay minerals. These reactions are well known, although they do not lend themselves to explanation by that most overworked and abused tool, the SEM. This paper focuses on additional and sometimes better methods of spotting problems.

Clay Type and Form

Clay swelling is the absorption of water into the interlayer portion of the clay's crystal structure. Swelling is commonly associated with, but not limited to, smectite and clay mixtures containing smectite. The reason for swelling is also one of the keys to its prevention. On contact with water, smectite matrix (matrix is used in the engineering context to include clay in the pore network of the rock) may behave in one of four ways: (1) the clay structure may admit the water and swell to accommodate the extra mass (Hower, 1974); (2) it may admit the water and disintegrate into fine particles (Sharma and Yortsos, 1986; Azari and Leimkuhler, 1988); (3) the clay may give up water and shrink (Azari and Leimkuhler, 1988); or (4) the water may not interact with the structure. Admission of water depends on the type and concentration of ions in the water, and the ion-charge type and strength initially on the surface of the clay (Jones, 1964). This is the ion-exchange capacity that has been adequately described by several authors (Hewitt, 1963; Jones, 1964; Khilar and Fogler, 1981; Sharma and others, 1985; Kia and others, 1987b; Vaidya and Fogler, 1990). Fluids that interrupt the charge balance on the clay may cause severe problems of swelling and particle dispersion. Fluids that are similar in salinity type, ion concentration and pH to the initial connate water (used in the engineering context; formation water) usually react less with the clay than waters with different chemical character. The pH of the liquids is important due to effects of mineral solubility, dispersion, and flocculation. Minor changes in pH may trigger chemical reactions where none was occurring (Simon and others, 1976; Coulter and others, 1983; Priisholm and others, 1987; Kia and others, 1987a). The fluids, such as drilling-mud filtrate, cement filtrate, completion fluids, stimulation fluids, kill fluids and secondary- and tertiary-reservoir flood liquids, are the keys to clay reactions. The problem, of course, is describing the chemical state of the clay and getting an uncontaminated sample of formation fluid and a representative formation-rock sample from an early test. The samples are best obtained from DST (drill stem test) flows, produced fluids on offset wells and RFT (repeat formation tester) tools. Where good fluid samples are available and sensitive clays are known to exist, clay problems can usually be minimized by addition of certain salts to the drilling, completion and stimulation fluids.

The test for chemical reactions involving clay minerals is the standard sensitivity test, whereby the initial permeability is established with a formation water and the effect of other fluids is evaluated by flowing several pore volumes through the rock and determining the permeability trend. The test concludes with flow of the initial fluid; the behavior trends are a function of the change from the initial permeability. Although recreated formation water (from an analysis) is frequently used for the test, the actual formation water may produce slightly different results. Actual formation water should be used when available (have it shipped in plastic containers and use it as rapidly as possible after sampling). The concept of using actual formation water is not universally accepted, especially where the formation brine is unstable due to outgassing of hydrocarbon or other gas. Field-by-field evaluation of water use is needed. The sensitivity test will be amended for other factors as they are discussed. Use of a 0.45-μm filter in front of the core is common, but the filter can mask some problems with solids carryover in injection fluids and particles in brines, such as calcium carbonate. By removing the filter from the flowing stream after the core has reached stable-flow behavior, the effect of any entrained solids on permeability can be evaluated. If precipitates such as iron and carbonate form, they usually may be dissolved by achieving original pH and pressure levels. Precipitates must not be allowed to influence the tests.

As hinted previously, smectite and its mixtures are not the only types of clay that will swell on contact with water of different chemistry than the connate fluid. There are other clay minerals, not usually thought to be water sensitive, that will react with water.

Illitic clays (mixed-layer illite/smectite and illite) are chameleons because of their diverse morphologies and compositions. Mixed-layer varieties may swell, thereby producing blockage or loose particles that may migrate. All varieties, especially the fibrous or "spiderweb" forms, may migrate when subjected to high-velocity flow, may trap mineral particles or water, or may be virtually inert.

Kaolinite is the occupant of the well-completion engineer's anxiety closet. Its precarious-looking arrangement in pore throats makes it appear more vulnerable to migration than it commonly is. Proper selection of fluid chemistry can often control it (Leone and Scott, 1987). From personal experience, when water chemistry is intelligently controlled, problems with kaolinites are minimized. In the literature investigation for this paper and in years of core-test experience, few actual production-affecting problems have been proven to be associated with kaolinites. The key to kaolinite problems is their chemical charge (the reactivity control), their location, and the strength of attachment to the pore wall.

Chlorites have a Dr. Jeckel and Mr. Hyde personality. They are usually tightly attached to the pore wall, do not swell in the most common form, and have a slow reaction with even strong acids at moderate temperatures (Simon and Anderson, 1990). Reaction time depends on surface area and temperature. Their undesirable characteristics include their capacity to trap particles, their contribution to microporosity, and their ferric iron content. The slow reaction of HCl acid with chlorite is a positive benefit in limiting the iron in solution.

Differentiating between movable fines and swollen clays can sometimes be accomplished with the sensitivity test. This operation involves backflowing the core after damage has occurred. If the permeability in the reversed direction increases rapidly and then decreases, the problem is movable fines. If there is no change in permeability, the damage may be either swollen clays or very tightly packed fines. If, after completion of flowing saturation with connate water, there is no decrease in the permeability when the test fluids are flowed, the core may not be sensitive. Lower permeability than expected should not be necessarily interpreted as damage. Minor permeability difference between flow directions can be misleading; stable permeabilities in opposite directions on undamaged cores can be different by as much as 10 to 15% or more. Sequential core plugs of 1 in. length from a single 3-in.-long, horizontal core plug have been tested with results differing by as much as 50%.

Many of the clay minerals discussed earlier occur as clay coats on detrital-sand grains. Dissolution of certain grain types (e.g., feldspars) results in unsupported clay rims that are particularly susceptible to migration during fluid flow. Such clay rims have high surface areas and tend to be associated with large pores, characteristics that promote chemical interaction with pore fluids.

By now, you may be forming an idea that form and location of clay are as almost as important as presence of clay. That idea is wrong; form and location are more important.

Clay Location

Whereas a speeding, on-coming dump truck in its own lane is a commonplace occurrence, a speeding, on-coming dump truck in your lane is a more immediate concern. The same is true for clays on a more mundane scale (unless it's your money in the well). Remember, mere presence of clay is almost meaningless. What can be important is when reactive clays are in the pore throats through which most of the fluid passes.

Formations are not homogeneous, as many reservoir engineers want to believe (hence, some of the difference between grandiose production estimates from computer modeling and the true production figures). At any given pressure, fluid will flow through only part of the pores in the rock. The amount of pores contributing to flow is too complex to address here, but it hinges on the wetting of the pore surfaces (Honarpour and Mahmood, 1988), capillary effects (Jones and Owens, 1980), gas slippage, saturation of the different phases (Jennings, 1975; Anderson, 1986; Crocker and Marchin, 1988), flowing-fluid viscosity (Honarpour, 1986) and a host of other, equally interconnected factors.

Among the easier items to explain in a relatively few words is the variation in pore size. Pore size varies widely in many reservoir rocks. A plot of pore-size distribution in even a "homogeneous" Berea Sandstone core confirms this fact (Fig. 1). The pore size-distribution test measures the size of the pore throats by forcing a non-wetting viscous liquid (mercury) through the pores at increasing pressure and measuring the pressure response (the pressure drop or leakoff). The liquid goes through the large pore throats easiest and will enter the smaller pores only as the driving pressure is increased. Figure 1 shows the percent of pores that have a certain size pore throat. Granting that mercury-producing formations are a bit rare, hydrocarbons and water behave in much the same manner, although wetting character and the associated capillary effects can play a big role. The majority of flow in any formation is in high-permeability "channels" that are defined by location of the larger pores (and larger pore throats). These are the same channels (natural fractures constitute many of these channels) that deplete first in primary and secondary production. In the tests of Figure 1, the graphic difference of the presence of large pores on permeability is easily seen in the comparison of the pore apertures for the 700-md and 150-md rocks.

Parts of the formation with smaller pores can contribute to flow, but only when the pressure drawdown is high or the flow effects of a high-permeability zone are removed from the producing environment. Draining the lower permeability zones is a serious problem in efficiently producing a reservoir. A visual test that illustrates the effects of the relative size of pores uses metal casts of the pores to demonstrate the pore size and the parts of the rock that are being swept by a fluid with certain wetting character (Swanson, 1979).

This rambling discourse eventually leads to a conclusion: If the clays are only in the smaller pores, they will probably have little effect on production of fluids regardless of type, quantity, reactivity, form, or fluids.

In closing this section, a minor tirade is in order. The measurement of pore size and water saturation uses, as one of its preparative steps, core cleaning by solvent extraction. Cleaning removes the wetting films and most bitumen from the pores. Whereas cleaning may be required in these experiments, it ruins the core for reliable permeability and sensitivity measurements. Oils, connate water, and natural surfactants coat the pores and clays, and inhibit contact of the clay with many completion or stimulation fluids. Although the coatings may be thin, their effectiveness on protecting clays against short-term exposure to completion fluids has been documented (Barkman and others, 1975). Removal of the coatings by cleaning or use of other solvents will make the rock seem more reactive than it truly is. Removal of coatings and bitumen will also make the core appear to be more permeable, sometimes much more permeable.

The best cores for sensitivity work are native state, carefully preserved, and as fresh as possible. If drilling-mud effects can be separated, the sensitivity should match that of the formation. Restored-state core is next best and, if restored correctly, is almost identical to native state. Restored core can be the core of choice when drilling-mud filtrate has swept the core. Dried core, resaturated and restored to pressure and temperature by use of a well-designed test vessel, can provide usable results, but caution is suggested and field pilots of treatments based on these cores are highly recommended.

Clay Form and Reactivity

With all the subtle hints thus far, it should come as no surprise that the form of the clay is an important control on reactivity. The reactivity of clay results from its high surface area-to-mass ratio. Almost all reactions of interest are controlled by the amount of surface area on which to react. The more surface area, the faster the reaction. Rough and arguable estimates of the clay-surface areas quoted in the literature are reproduced in Table 1 (Davies, 1975). Comparing the surface-area difference between sand grains and clays is somewhat like comparing the surface-area difference between a 7.6-cm (3 in.) cube of wood and roughly the same amount of wood in the form of a newspaper with a surface area of ≈ 92.9 m^2 (1,000 ft^2). Obviously, detrital clays in matrix, or covered by quartz or other low-reactivity cement, are of limited concern. Only the authigenic forms are of importance, and morphology plays an important role in the reactivity.

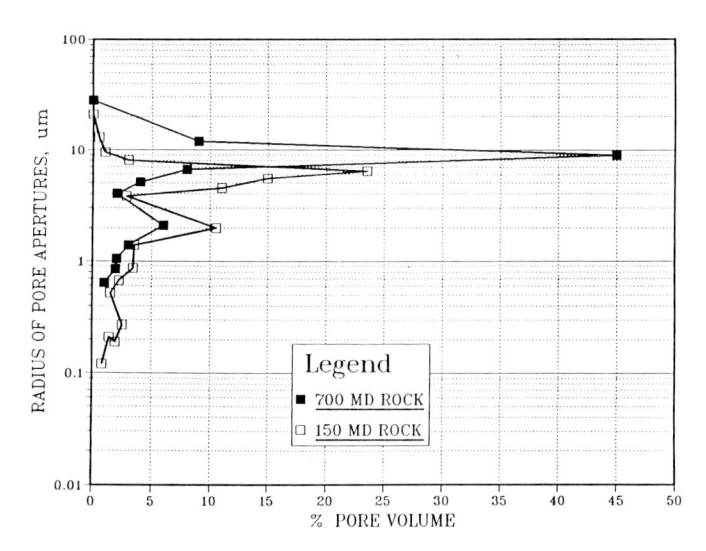

FIG. 1.—Percentage of pore volume versus pore-aperture radius for high- and moderate-permeability (to air) Berea Sandstone.

TABLE 1.—SURFACE AREA OF SOME COMMON MINERALS AS MEASURED BY NITROGEN ADSORBTION (DAVIES, 1975)

MINERAL	SURFACE AREA (m^2/g)
Quartz	0.000015
Kaolinite	22
Smectite	82
Illite	113

The reactivity of clay, contrary to consultants' opinions, can only be hinted at with the SEM. To really determine reactivity requires, at minimum, flowing tests with representative core samples, reservoir fluids and treating fluids at the reservoir temperature and pressure. Condition of representative core and fluid samples is of special importance. Almost any clay can be made to react if treated in the wrong manner. The classic "horror story" core-test example is flowing a 1 N solution of sodium chloride followed by fresh water. The "damage" produced where authigenic clay exists is usually spectacular and also usually misleading. The object of this test is to determine, in the absence of other analytical equipment, if reactable clays are present in the pore passages. However, unless injection of sodium chloride brine followed by fresh water is planned, the test reveals nothing about the reactivity of clay under operating conditions. In many cases, field tests and successful experience with stimulation and production in an area are better indicators than any laboratory tests. Field tests do have their own set of limitations; too many reactions may occur in a well to distinguish a specific cause of a response.

Clay reactivity can take many forms. Although swelling has already been addressed, an additional comment is needed. Swelling clays react on the basis of ion charge and ion type on the clay surface. Often, the trigger for water absorption into a clay structure is a change in the ion strength, or the rate of change in ion strength, of the flowing water. Small changes in the ion strength of the water usually produce small changes in the clay, as evidenced by the permeability behavior in the test of Figure 2. Large changes can produce severe permeability changes (Fig. 3), whereas stepwise changes to the same ionic-strength end point produce minor permeability changes (Khilar and others, 1983; Leone and Scott, 1987; Priisholm and others, 1987; Vaidya and Fogler, 1990).

Clays also react to changes in ion type at the same relative ion-strength level (Jones, 1964; Azari and Leimkuhler, 1988). This behavior, known as salinity shock, is also a variation in the quantity of water in the clay structure but, as shown in Figure 4, the permeability usually returns to near initial levels when the initial fluid resaturates the core.

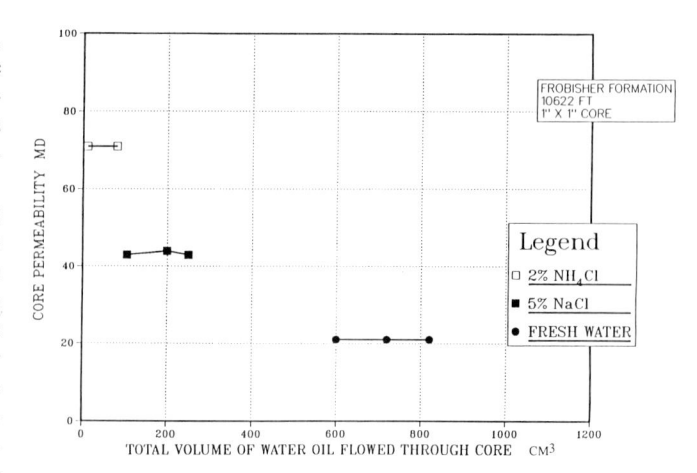

FIG. 3.—Brine sensitivity test, Frobisher Formation.

Particle migration is a reaction that may involve clays. An examination of migrating particles collected from a 0.45-μm downstream filter during a core test on a Gulf Coast sand has led to the conclusion that many of these are detrital particles. Moreover, most known migration problems occur in young sands, where bonding is poor and production flow is usually radial (unfractured). Highly altered forms of feldspar are reportedly more commonly involved in migrating reactions than are clays (Conway, pers. commun., 1991). Use of a downstream filter in laboratory experiments is frequently useful, but when particles are trapped, the downstream "cut" face of the core plug contributes to the debris, and the results are compromised.

Migration reactions involving clays may be initiated by swelling, fluid velocity, chemicals and physical processes such as compaction. During absorption of water by swelling clays, mobile-clay particles may be produced. Matching ionic type and concentration between indigenous and introduced waters will minimize swelling and should solve the problem. Velocity effects related to dynamic-drag forces produced by liquids flowing past the clays are most obvious during high-rate drawdown. If fluid flows too rapidly, clay

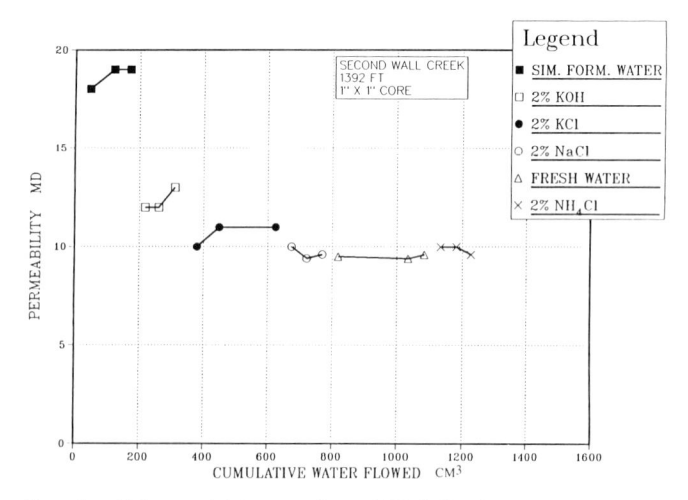

FIG. 2.—Brine sensitivity test, Second Wall Creek Sandstone.

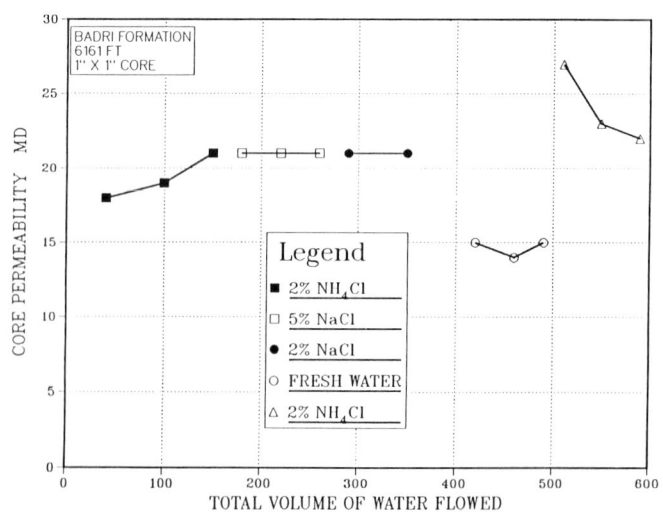

FIG. 4.—Brine sensitivity test, Badri Formation.

may be sheared away from the pore wall and moved until it lodges in a pore throat or against clays or other fine particles. Although fines movement is possible due solely to velocity, swelling may exacerbate the problem (Zaitoun and Berton, 1990). Other effects, such as chemical disruption of silt and clay, are also known (Muecke, 1979). The effect of velocity is most apparent in an unfractured well, where flow is totally radial and velocity increases as the fluid converges near the wellbore (Gabriel and Inamdar, 1983; Gruesbeck and Collins, 1982; Leone and Scott, 1987). The result is a series of blocked pores, usually the larger pores where velocity is highest. Formation damage may be characterized by a "critical velocity." At velocities greater than the critical velocity, fines (clay and non-clay) will move in sufficient quantities to alter permeability.

A test for this type of damage uses single-phase linear flow through a core plug at increasing rates until the critical velocity is reached and damage is produced, or until the pore will not flow any more fluid at the maximum differential pressure involved in the test. If no damage is produced, the critical velocity has not been defined by the test. A critical velocity may or may not exist at other flowing conditions. Example plots of ideal cases (Figs. 5 and 6; flowing brine), with flow rate plotted against permeability, show permeability response to increased flow rate (at increasing differential flow pressure) as a relatively flat line until the flow rate or pressure limit of the test equipment is reached or until damage occurs. The permeabilities measured at decreasing flow rates are plotted on the same axis. Plots of permeabilities measured during increasing and decreasing flow periods that overlay and reach the same end points are indications of no change in the pore, hence no damage (Fig. 5). Plots that widely diverge indicate a change in pore geometry that simulates formation damage (Fig. 6). Interpretation of the results is critical, and the test is not totally diagnostic. SEM analysis of the core and of particles collected on a downstream 0.45-μm filter are needed for confirmation. Possible indications of this type of velocity damage from field tests are development of a skin (wellbore damage zone) during flow testing, a common occurrence in unstable formations.

The core-test results discussed are for ideal cases. In the real world, the plots will commonly have some slope. Equipment leaks, multiphase flow, turbulence, capillary effects, and other effects cause the test to be an indicator and not a conclusive proof. In single-phase laminar flow of an ideal incompressible fluid through a homogeneous sample, the plotted values should have no slope. D'Arcy (1856) made his observations using homogeneous sand-pack filters, whereas reservoir sandstones are anything but homogeneous. In the author's experience, the amount of pressure differential used in a permeability test does affect the value of permeability recorded for cores with a wide range of pore sizes.

In fractured wells (naturally fractured or wells stimulated by hydraulic fractures), the flow toward the well is not strictly radial. Less damage from velocity is expected in these cases.

CHEMICAL EFFECTS ON CLAY

Clay- and silt-size particles may be released by contact with fluids that dissolve oil films, scale, wetting surfactants or heavy hydrocarbon deposits (Muecke, 1979). This may occur in single- or multiphase flow, and the fluids may include alcohols, mutual solvents, acids and alkalis. The amount of fines liberated will depend on the amount of fine particles in the pores and what is binding them. This type of fines movement is commonly observed in laboratory tests. It accounts, along with carbon dioxide-gas liberation from hydrochloric or hydrochloric/hydrofluoric-acid reactions, for the rapid, usually temporary, decrease of permeability following acid injection into a core plug. The evolved gas is rapidly dissolved in the liquid, but the particles can be swept into traps or drop out harmlessly in a large pore, depending on sizes of particles, pores, and pore throats. Small dolomite crystals, feldspar grains, and kaolinite are commonly identified as acid-loosened particles. Use of a back-pressure regulator can remove the evolved-gas prob-

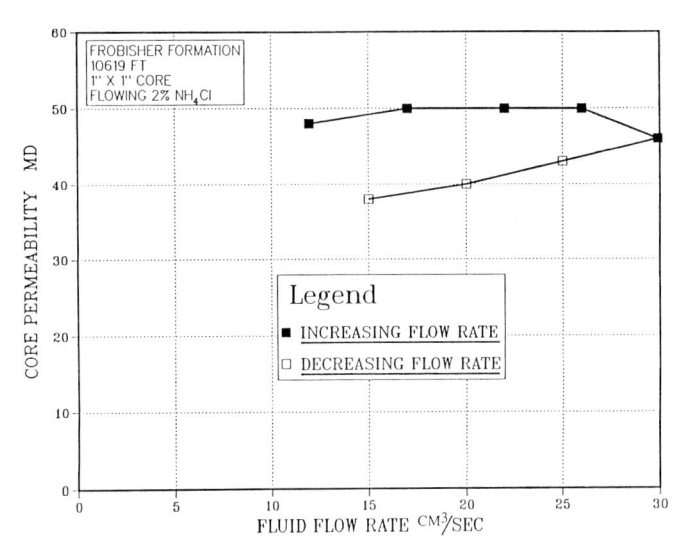

FIG. 6.—Fines migration test, velocity effect, Frobisher Formation.

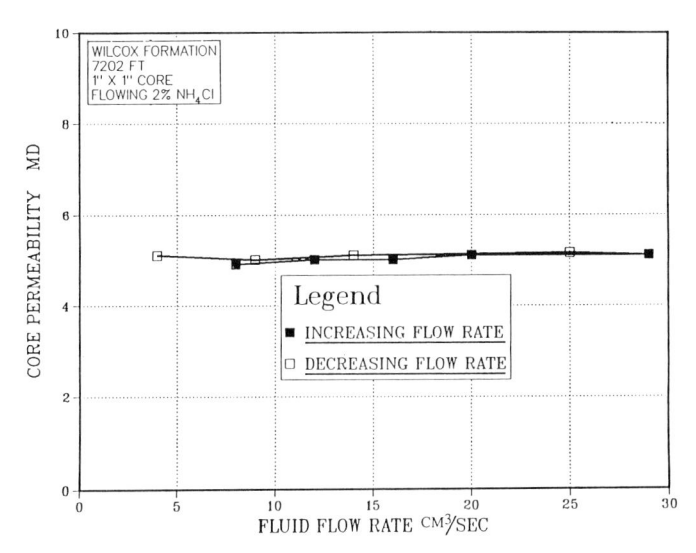

FIG. 5.—Fines migration test, velocity effect, Wilcox Formation.

lem, but commonly masks the core response because of operational surges.

Although most drillers understand the importance of water chemistry in control of heaving and sticky shales, a few forget the real purpose of the well when drilling through a pay zone. Although reactions of mud filtrates with clay minerals are well known, freshwater muds continue to be widely used. In addition, formulation of a mud that will form a low-permeability, thin, hard mudcake to limit leakoff is at least as important as use of a salt in the water that will render the filtrate non-damaging to the clay and minerals in the rock.

In cased wells, the cement circulated behind the pipe to bond the casing to the wellbore contributes cement filtrate, a pH-12 fluid that can be damaging to both clays and some minerals found in the matrix of the formation. Drilling mud and cement filtrates have shallow invasion in a well-designed completion. This filtrate-invasion distance frequently can be exceeded by commonly used perforation and breakdown procedures.

OTHER CLAY AND NON-CLAY PROBLEMS

This section addresses factors that can contribute to well-completion problems, without swelling, moving or reacting clays. A few of the problems do not involve clays at all.

Highly microporous clay minerals trap water. These clays, along with zeolites and a few other minerals, may produce high, apparent water saturations, while contributing little, if any, free water to the produced fluids. Examples are known where calculated water saturations were 50% or more and yet the formation produced little or no water (Zemanek, 1989). In some instances, the oil industry has probably abandoned producible hydrocarbon accumulations, assuming from log analysis that the high-water saturation of the formations would preclude successful production.

There are also formations with low-water saturations that produce large amounts of water, usually as the result of water being a non-wetting and mobile component of the formation fluids (Desbrandes, 1989).

In this section, a geologist will probably be surprised to find that many completion engineers consider clay to be only one of the factors affecting potential formation damage. The most serious problems involved in preventing skin damage in a new completion are plugging by, and fluid reactions with, non-clay particles. From the first time that a drill bit penetrates the formation, the potential for formation damage exists. Drilling-mud filtrate, which has to pass into the formation for the mudcake to form, displaces the connate fluid and changes the salinity. The filtrate fluids, however, are only part of the problem.

The pressure differential of the mud to the pore fluids in the formation is of critical importance. It is known that when pressure differentials exceed 1,000 psi (6.9 MPa), the success of a DST, the most common wildcat evaluation tool, is usually a "no show" or a dry test, even in a productive zone (Paccaloni and others, 1988). When high-pressure differentials are involved, the well must be perforated (even open-hole wells) prior to obtaining a successful DST. The damage from this type of high pressure-differential operation is virtually never related to formation clay. Shallow invasion of drilling-mud particles into the pores of the formation forms blockage to flow. Drilling-mud particles ideally are sized to stop at the formation face. In the higher pressure application, tight bridges form and mud particles are commonly forced into permeable formations by high-pressure differentials, with little chance of removal by backflow.

The core test to distinguish this surface blockage response from "true clay" damage involves trimming a very thin layer from the injection face of a core that has been damaged by drilling mud. If the permeability is restored to near initial levels after removal of the injection face, then the damage is concluded to be caused by drilling-mud particles and is confined to the injection face. Reverse flow with a clean fluid prior to trimming the injection face is also a diagnostic test, but should be used after the results of the injection-face plugging test are known. This test addresses the issue of whether backflow can remove drilling-mud damage in open-hole completions without resorting to acidizing. Longer cores (0.2–2.4 m; 0.5–8.0 ft), with sampling taps at regular intervals (King and Lee, 1988), are also used for this and many other tests for formation-damage potential. The longer cores with sample taps afford the same core flow with fewer effects of cut or broken core ends. The disadvantage is that most long cores are quarried and few truly represent reservoir rock.

Perforating uses a shaped-explosive charge to punch a hole into a formation. The small, shaped charges focus immense energy into making a hole through the casing, cement and formation. The process is best described as a focused pressure pulse. The characteristics of the highspeed (6,400 m/s; 21,000 ft/s) punch process dictate that all the rock material in the path of the jet is thrust to the sides and forms a crushed zone at the edges of the perforation tunnel (Saucier and Lands, 1978; McLeod, 1982). This type of damage has little to do with clays or other minerals in the formation. However, the crushed zone can act as a significant barrier to the flow of fluids from an otherwise undamaged reservoir into the perforations. The final step in some natural completions is often a "perforation breakdown step." This operation involves injection of acid or water into the perforations to establish communication between the perforation and the reservoir. Although it is commonly thought that the acid removes the permeability damage by dissolving the crushed zone, it has been shown that the crushed zone has a limited solubility in acid, and the actual permeability linkage between the perforation and the reservoir is generally a small fracture. This has been documented in an Amoco field test in the Wamsutter Area, Wyoming, a tight-gas reservoir requiring hydraulic-fracture stimulation for economic completions. The 15% HCl used in the perforation breakdown operations was replaced with a 2% KCl solution; the wells performed the same after cleanup, regardless of breakdown fluids.

In this author's opinion, more formation damage is caused by solids introduced with injection fluids than by migration of indigenous fines in the reservoir. The existence of a perfectly clean fluid is unknown, and even a small amount of particles can reduce permeability. Minimizing these particles is a primary task in assuring a low-damage completion. In many operations, such as gravel packing, the filtration of injected fluids has become a common practice. In stimulations and kill fluids, however, injected fluids are generally not filtered. As long as fluids continue to be injected in an unfiltered state, formation damage from entrained particles will continue to cause skin damage in the wellbore, even after stimulation.

Studies of contamination of injection fluids have revealed that tanks used to haul cement, drilling mud, sewage, and other debris-laden fluids are often used on the next load to haul water for completion and stimulation fluids without cleaning the tank (Maly, 1976; McCune, 1982; Patton, 1990). The conclusion of this observation is simple; clean the tanks and filter the fluids, or expect permeability damage in a multi-million dollar well.

SUMMARY

This paper has attempted to describe a few tests that can be used to identify formation damage in the laboratory and in the reservoir. The following engineering-based observations may contribute to a geological understanding of formation-damage problems.

1. The potential for clay reaction within formation pores is directly related to the presence of a reactive clay located in pore channels through which the liquid is flowing, and the reactivity of that clay with the liquid.
2. SEMs and X-ray analysis are not "stand-alone" methods for predicting clay reaction problems. They are very useful, however, when combined with flow testing and interpretation experience, for predicting clay reactions.

A few rules of thumb are valid for clay form and reactivity:

1. Ragged, irregularly shaped clay and non-clay minerals are commonly the most fragile and reactive, although exceptions are known.
2. Unsupported rims of clay that were part of a clay coating on a grain that has been dissolved are especially prone to causing formation damage.
3. The spiderweb or hairy form of illite, in the main pore throats, indicates a need for very clean fluids to avoid "brush piles" of pore-blocking, micron-size debris.
4. Rapidly changing the salinity of the flowing fluid can produce damage where gradual changes would not.
5. Formation damage due to high fluid-velocity effects are more noticeable in unfractured wells, but are rare in actual operations.

REFERENCES

ANDERSON, W. G., 1986, Wettability literature survey—part 1: rock/oil/brine interactions and the effects of core handling on wettability: Journal of Petroleum Technology, October, p. 1125–1144.
AZARI, M., AND LEIMKUHLER, J., 1988, Permeability changes due to invasion of sodium and potassium-based completion brines in Berea and Casper sandstones: Society of Petroleum Engineers preprint 17149, 12 p.
BARKMAN, J. H., ABRAMS, A., DARLEY, H. C., AND HILL, H. S., 1975, An oil coating process to stabilize clays in fresh water flooding operations: Journal of Petroleum Technology, September, p. 1053–1058.
COULTER, A. W., Jr., FRICK, E. K., AND SAMUELSON, M. L., 1983, Effect of fracturing fluid pH on formation permeability: Society of Petroleum Engineers preprint 12150, 11 p.
CROCKER, M. E., AND MARCHIN, L. M., 1988, Wettability and adsorption characteristics of crude-oil asphaltene and polar fractions: Journal of Petroleum Technology, April, p. 470–482.
D'ARCY, H., 1856, Determination des lois d'ecoulement de l'eau a travers le sable, in Hubert, M. K., 1969, The Theory of Ground Water Motion and Related Papers: New York, Hafner Publishing Company, p. 303–311.
DAVIES, D. K., 1975, Clay technology and well stimulation: Southwestern Petroleum Short Course, 7 p.
DESBRANDES, R., 1989, In situ wettability determination improves formation evaluation: part 1—wettability concept: Petroleum Engineer International, August, p. 38–46.
GABRIEL, G. A., AND INAMDAR, G. R., 1983, An experimental investigation of fines migration in porous media: Society of Petroleum Engineers preprint 12168, 12 p.
GRUESBECK, C., AND COLLINS, R. E., 1982, Entrainment and deposition of fine particles in porous media: Society of Petroleum Engineers Journal, December, p. 847–856.
HEWITT, C. H., 1963, Analytical techniques for recognizing water-sensitive reservoir rocks: Journal of Petroleum Technology, August, p. 813–818.
HONARPOUR, M., 1986, How temperature affects relative permeability measurement: World Oil, May, p. 116–126.
HONARPOUR, M., AND MAHMOOD, S. M., 1988, Relative-permeability measurements: an overview: Journal of Petroleum Technology, August, p. 963–966.
HOWER, W. F., 1974, Influence of clays on the production of hydrocarbons: Society of Petroleum Engineers preprint 4785, 11 p.
JENNINGS, A. R., 1975, The effect of surfactant-bearing fluids on permeability behavior in oil-producing formations: Society of Petroleum Engineers preprint 5635, 10 p.
JONES, F. O., Jr., 1964, Influence of chemical composition of water on clay blocking of permeability: Journal of Petroleum Technology, April, p. 441–446.
JONES, F. O., AND OWENS, W. W., 1980, A laboratory study of low-permeability gas sands: Journal of Petroleum Technology, September, p. 1631–1640.
KHILAR, K. C., AND FOGLER, H. S., 1981, Water sensitivity of sandstones: Society of Petroleum Engineers preprint 10103, 15 p.
KHILAR, K. C., FOGLER, H. S., AND AHLUWALIA, J. S., 1983, Sandstone water sensitivity: existence of a critical rate salinity decrease for particle capture: Chemical Engineering Science, v. 38, p. 789–800.
KIA, S. F., FOGLER, H. S., AND REED, M. G., 1987a, Effect of pH on colloidally induced fines migration: Journal of Colloid and Interface Science, v. 118, p. 158–168.
KIA, S. F., FOGLER, H. S., REED, M. G., AND VAIDYA, R. N., 1987b, Effect of salt composition on clay release in Berea sandstones: Society of Petroleum Engineers, Production Engineering, November, p. 277–283.
KING, G. E., AND LEE, R. M., 1988, Adsorption and chlorination of mutual solvents used in acidizing: Society of Petroleum Engineers, Production Engineering, May, p. 205–209.
LEONE, J. A., AND SCOTT, E. M., 1987, Characterization and control of formation damage during waterflooding of a high-clay-content reservoir: Society of Petroleum Engineers Preprint 16234, 10 p.
MALY, G. P., 1976, Close attention to the smallest job details vital for minimizing formation damage: Society of Petroleum Engineers preprint 5702, 9 p.
McCUNE, C. C., 1982, Seawater injection experience—an overview: Journal of Petroleum Technology, October, p. 2265–2270.
McLEOD, H. O., 1982, The effect of perforating conditions on well performance: Society of Petroleum Engineers preprint 10649, 11 p.
MUECKE, T. W., 1979, Formation fines and factors controlling their movement in porous media: Journal of Petroleum Technology, February, p. 144–150.
PACCALONI, G., TAMBINI, M., AND GALOPPINI, M., 1988, Key factors for enhanced results of matrix stimulation treatments: Society of Petroleum Engineers preprint 17154, 12 p.

PATTON, C. C., 1990, Injection water quality: Society of Petroleum Engineers, October, p. 1238–1240.

PRIISHOLM, S., NIELSEN, B. L., AND HASLUND, O., 1987, Fines migration, blocking, and clay swelling of potential geothermal sandstone reservoirs, Denmark: Society of Petroleum Engineers, Formation Evaluation, June, p. 168–178.

SAUCIER, R. J., AND LANDS, J. F., 1978, A laboratory study in stressed formation rocks: Journal of Petroleum Technology, September, p. 1347–1353.

SHARMA, M. M., AND YORTSOS, Y. C., 1986, Permeability impairment due to fines migration in sandstones: Society of Petroleum Engineers preprint 14819, 11 p.

SHARMA, M. M., YORTSOS, Y. C., AND HANDY, L. L., 1985, Release and deposition of clays in sandstones: Society of Petroleum Engineers preprint 13562, p. 125–138.

SIMON, D. E., AND ANDERSON, M. S., 1990, Stability of clay minerals in acid: Society of Petroleum Engineers preprint 19422, 12 p.

SIMON, D. E., McDANIEL, B. W., AND COON, R. M., 1976, Evaluation of fluid pH effects on low permeability sandstones: Society of Petroleum Engineers preprint 6010, 12 p.

SWANSON, B. F., 1979, Visualizing pores and nonwetting phase in porous rock: Journal of Petroleum Technology, January, p. 10–18.

VAIDYA, R. N., AND FOGLER, H. S., 1990, Fines migration and formation damage: influence of pH and ion exchange: Society of Petroleum Engineers preprint 19413, 8 p.

WILSON, M. D., AND PITTMAN, E. D., 1977, Authigenic clays in sandstones: recognition and influence on reservoir properties and paleoenvironmental analysis: Journal of Sedimentary Petrology, v. 47, p. 1–31.

ZAITOUN, A., AND BERTON, N., 1990, Stabilization of montmorillonite clay in porous media by high-molecular-weight polymers: Society of Petroleum Engineers 19416, 11 p.

ZEMANEK, J., 1989, Low-resistivity hydrocarbon-bearing sand reservoirs: Society of Petroleum Engineers, Formation Evaluation, December, p. 515–521.

COLOR PLATES

David A. Barnes, Jean-Pierre Girard, and James L. Aronson
K-Ar Dating of Illite Diagenesis in the Middle Ordovician St. Peter Sandstone, Central Michigan Basin, USA.
PLATE 1.—(A) Scour contact (arrow) between quartz-cemented, bioturbated, slightly coarse-grained sand (right of arrow) and fine-grained, friable, clay-cemented, poorly sorted and bioturbated sand (left of arrow) in the offshore-shelf facies (lithofacies 4). Stratigraphic "up" to right. Hunt Martin, 3,447 m (11,302 ft). Illite age is 343 ± 6 Ma (see Table 1). Scale bar in cm. (B) Comparison of authigenic-clay-cemented and friable sandstone (left) and detrital-clay-rich bioturbated sandstone (right) in the offshore-shelf facies (lithofacies 4). Stratigraphic "up" at top. Federated Kitchenhoff, 3,447 m (11,025 ft)/3,447 m (10,982 ft). Scale bar in cm. (C) Loosely consolidated, medium- to fine-grained, authigenic-clay-cemented sandstone with low-angle cross-bedding (possibly swaley or hummocky cross-sets) at the base of the slab (left of arrow), overlain at the sharp scour surface (arrow) by medium-grained, quartz-cemented sandstone (right of arrow). Stratigraphic "up" to right. Hunt Martin, 3,422 m (11,408 ft). Illite age is 339 ± 7 Ma (see Table 1). Scale bar in cm. (D) Interfingering contact between open-packed and dolomite-cemented sandstone (dark area, d) adjacent to light-colored, loosely consolidated, clay-cemented sandstone (c). Petrographic textures support the replacement of carbonate cement in intergranular pores by clay in this sample. Stratigraphic "up" to right. Hunt Martin, 3,446 m (11,297 ft). Illite age is 353 ± 6 Ma (see Table 1). Scale bar in cm. (E) Photomicrograph from thin section centered on a skeletal K-feldspar grain and intragranular secondary porosity with partial, authigenic-clay infill adjacent to an oversize, authigenic-clay-lined, secondary intergranular pore. Sun Mentor, 3,078 m (10,092 ft). Illite age is 342 ± 6 Ma (see Table 1). Scale bar 0.1 mm. (F) Photomicrograph of remnant, intergranular dolomite (d) in a loosely consolidated, clay-rich (c) sandstone. Note extensive replacement of detrital quartz by the dolomite (filled arrow) and clay-filled (open arrow) embayments in some quartz grains. The latter is interpreted as precipitation of clay in secondary pores formed after dissolution of dolomite. Hunt Martin, 3,446 m (11,298.5 ft). Illite age is 353 ± 6 Ma (see Table 1). Scale bar 0.1 mm.

Stuart D. Burley and Joe H. S. MacQuaker
Authigenic Clays, Diagenetic Sequences and Conceptual Diagenetic Models in Contrasting Basin Margin and Basin Center North Sea Jurassic Sandstones and Mudstones
PLATE 2.—Thin-section photomicrographs showing evidence of diagenetic reactions in sandstones. (A) Detrital-feldspar grain showing initial stages of dissolution and replacement by illite. Relic feldspar (F) is still present and the original grain outline is clearly visible. Replacement illite (il) is oil stained. Note absence of authigenic kaolinite within or adjacent to the dissolving feldspar. Field of view 220 μm, 3.2 km depth, Witch Ground Graben. (B) An oversized pore, the end product of feldspar dissolution, within an otherwise quartz-overgrowth-cemented sandstone. Small remnants of the detrital feldspar are still visible (F, arrowed). Field of view 400 μm, 4.3 km depth, Witch Ground Graben. (C) Dense aggregate of blocky kaolinite filling oversized pore and adjacent intergranular pore space. Field of view 180 μm, 2.8 km depth, Brent Province. (D) Illitization of kaolinite aggregates. Field of view 400 μm, 4.2 km depth, Brent Province. (E) Kaolinite (K, oil stained, hence brown color) enclosed within zoned-quartz overgrowth. The enclosed kaolinite is restricted to a growth zone (z2) that is preceded and followed by kaolinite-free quartz-overgrowth zones (z1 and z3). Boundary (B) between zone 1 and zone 2 overgrowth is distinct and abrupt, indicating kaolinite precipitation was not contemporary with zone 1 quartz, but may have been so with zone 2 quartz. Detrital grain/overgrowth boundary annotated b and partially highlighted with dashed line. Field of view 60 μm, 3.9 km depth, Witch Ground Graben. (F) Coarsely crystalline, pore-filling kaolinite (identified by powder XRD). Field of view 80 μm, 4.1 km depth, Witch Ground Graben. (G) Authigenic illite associated with oversized pore after feldspar dissolution. Field of view 120 μm, 4.2 km depth, South Viking Graben.

Thomas L. Dunn
Infiltrated Materials in Cretaceous Volcanogenic Sandstones, San Jorge Basin, Argentina
PLATE 3.—(A) Geopetal structures (arrows) of light tan, laminated mixture of quartz, clay and oxides, which have accumulated along the bottoms of pores; partially crossed polars, PE-815, 2,262.35 m (7,423 ft). (B) Massive, quartz-rich pore filling above geopetal laminae (arrows). Irregular bleb within the massive portion is an artifact of thin-section preparation. (C) Cutans of the tan-colored, submicron-size quartz, clay and oxides rimming the sand grains. Faint laminations parallel to grain surfaces are locally visible. Interior of the remaining pore space is sparsely lined with minute quartz crystals and stained with hydrocarbons; PE-1, 2,200 m (7,218 ft). (D) Geopetal structures of quartz-rich material, which is internally laminated (arrows) with the pore space above filled with quartz prisms, albite laths and chlorite rosettes; PZ-823, 2,054.35 m (6,740 ft). (E) Oxide segregation, within this quartz-rich geopetal (arrows) and massive pore fill, is an arcuate band of Fe-oxides that is crudely concentric with respect to the pore outline, but cuts across the infiltration laminations. (F) Within the clay-rich material, such segregations are less common and typically reversed, with the oxide concentrations adjacent to the pore surface; ICD-6, 1,577.8 m (5,177 ft).

David W. Houseknecht and Louis M. Ross, Jr.
Clay Minerals in Atokan Deep-water Sandstone Facies, Arkoma Basin: Origins and Influence on Diagenesis and Reservoir Quality
PLATE 4.—Thin-section photomicrographs of Atoka sandstones illustrating microscopic aspects of primary vs. secondary sedimentary structures and typical clay fabrics. All samples are from conventional cores of Red Oak sandstone except (A). All thin sections were oriented and all black arrows point to stratigraphic up. Samples were vaccum impregnated with blue epoxy; all photos were taken in plane polarized light. (A) Outcrop sample from unconfined turbidite facies of lower Atoka Formation illustrating poor sorting, abundant depositional clay, and matrix-supported fabric. Scale bar 0.2 mm. (B) Sample displays plane parallel lamination in core. Quartz grains display incomplete clay coatings and pervasive overgrowths. Note horizontal orientation of dark, elongate grains (mostly slate lithic fragments, labeled S). Scale bar 0.2 mm. (C) Sample displays massive$_H$ bedding in core. Clay and overgrowth content is similar to that in (B). Note horizontal orientation of elongate grains of mica (M) and slate lithic fragment (S). Scale bar 0.2 mm. (D) View of dish structure, defined by dark concentration of clay and organic material. Note finer grained sand above dish, concave-upward geometry of dish, and coarser grained and more clay-rich nature of sand below dish. Scale bar 0.2 mm. (E) Sample displays diffuse$_V$ laminations in core. Intergranular clay is abundant, quartz overgrowths are present in only small volumes, and elongate particles of organic material (dark color) are vertically oriented. Scale bar 0.5 mm. (F) View of dewatering pillar (clean area at right), consolidation laminations (dark, vertically oriented organics), and sandstone adjacent to pillar (left half of photo). The pillar (right) contains little clay and adjacent sandstone displays massive$_V$ lamination, suggesting at least two episodes of fluidized dewatering. Scale bar 0.5 mm. (G) Sample displays massive$_V$ lamination in core. Dispersed clay is pervasively distributed through intergranular space, quartz overgrowths are rare, and slate lithic fragments (brown, foliated grains) are vertically oriented. Note grain-size and grain-shape aggregate of microporous kaolinite just above scale bar. Scale bar 0.2 mm. (H) Field of view is an example of optimum reservoir quality developed in a sample that has 7% porous clay, <1% nonporous clay, and 19% total porosity by point-count estimate. Not only is primary porosity preserved by pervasive clay-grain coatings, but framework-grain dissolution has enhanced porosity. Scale bar 0.2 mm.

PLATE 1

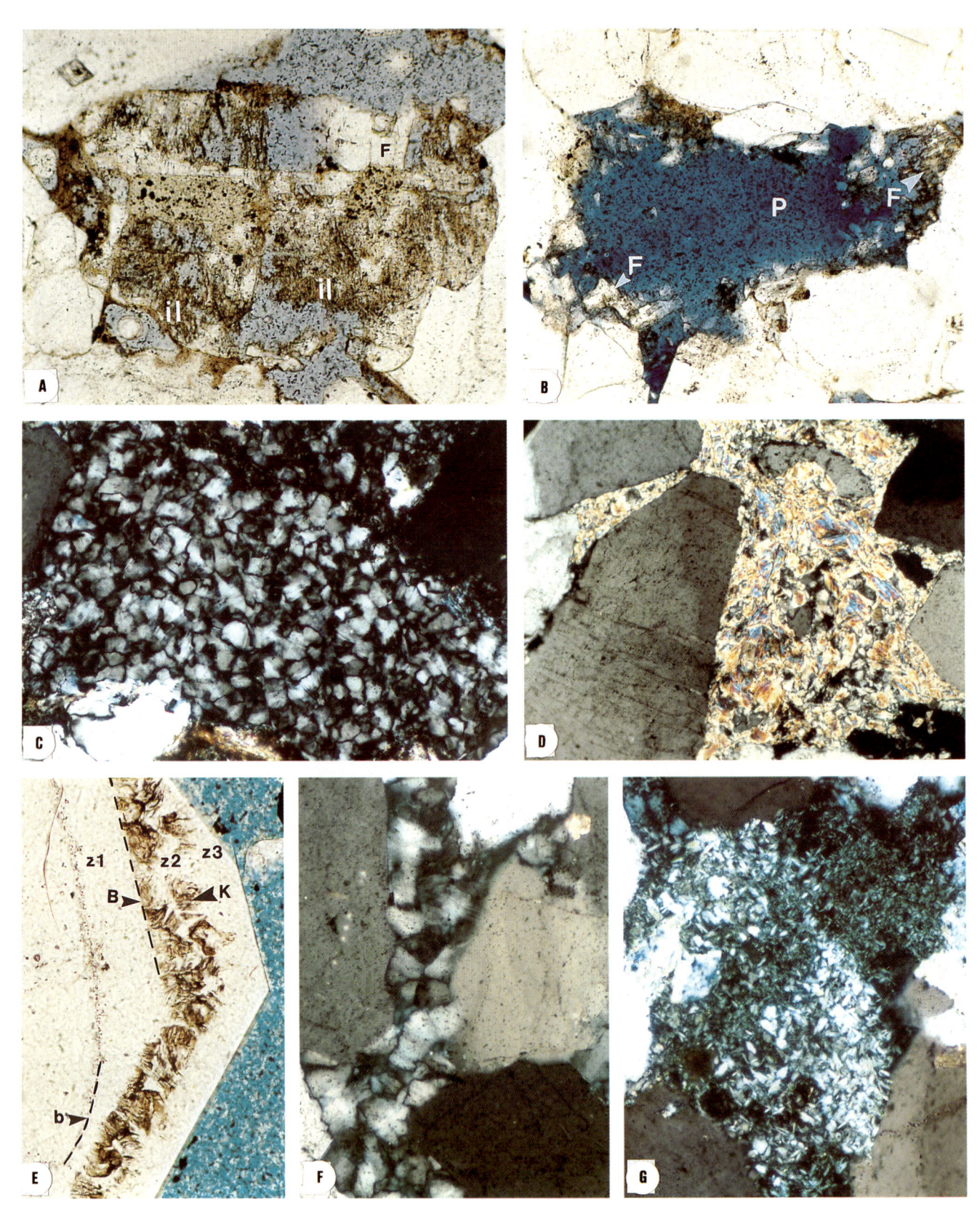

PLATE 2

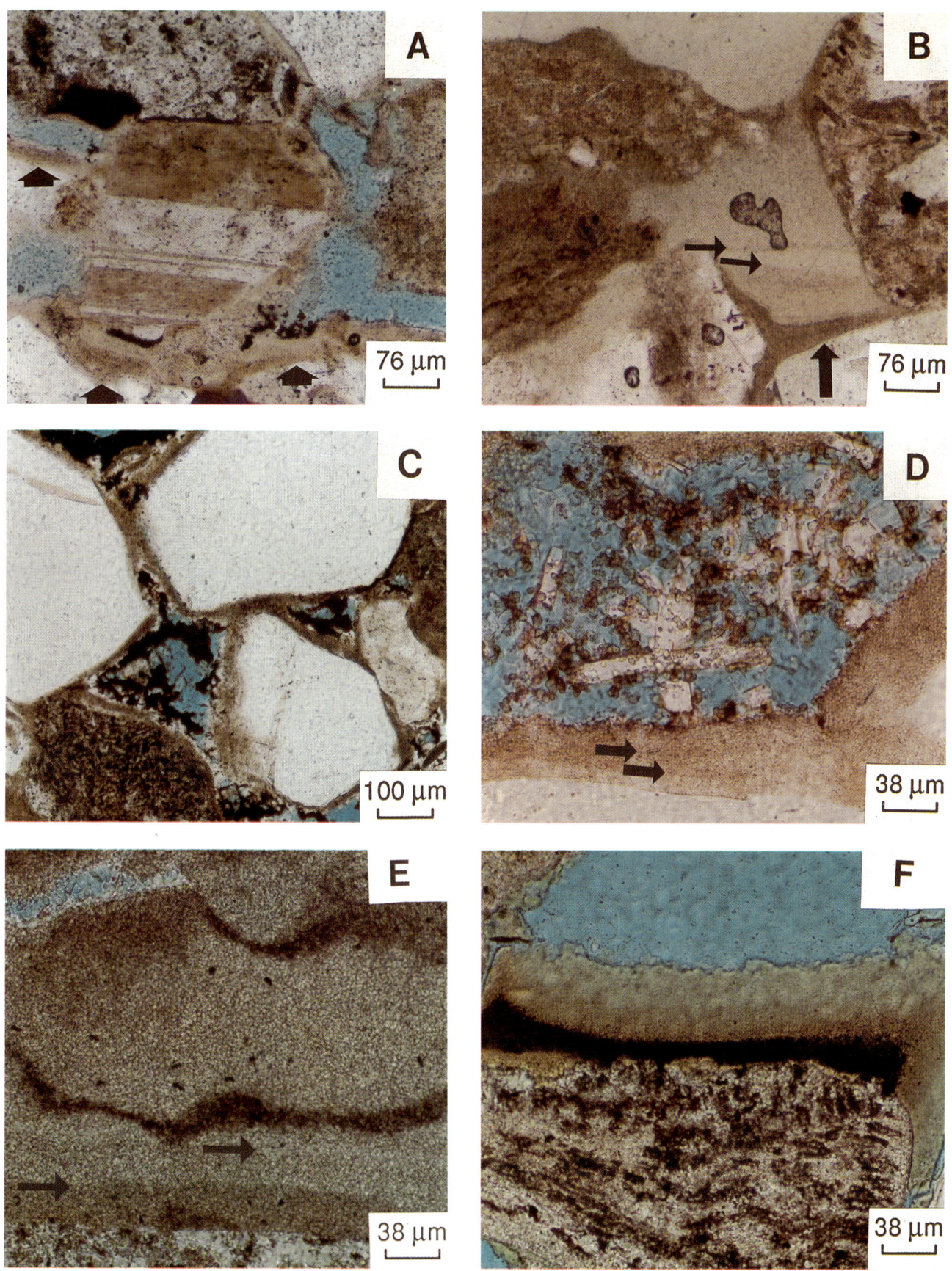

PLATE 3

277

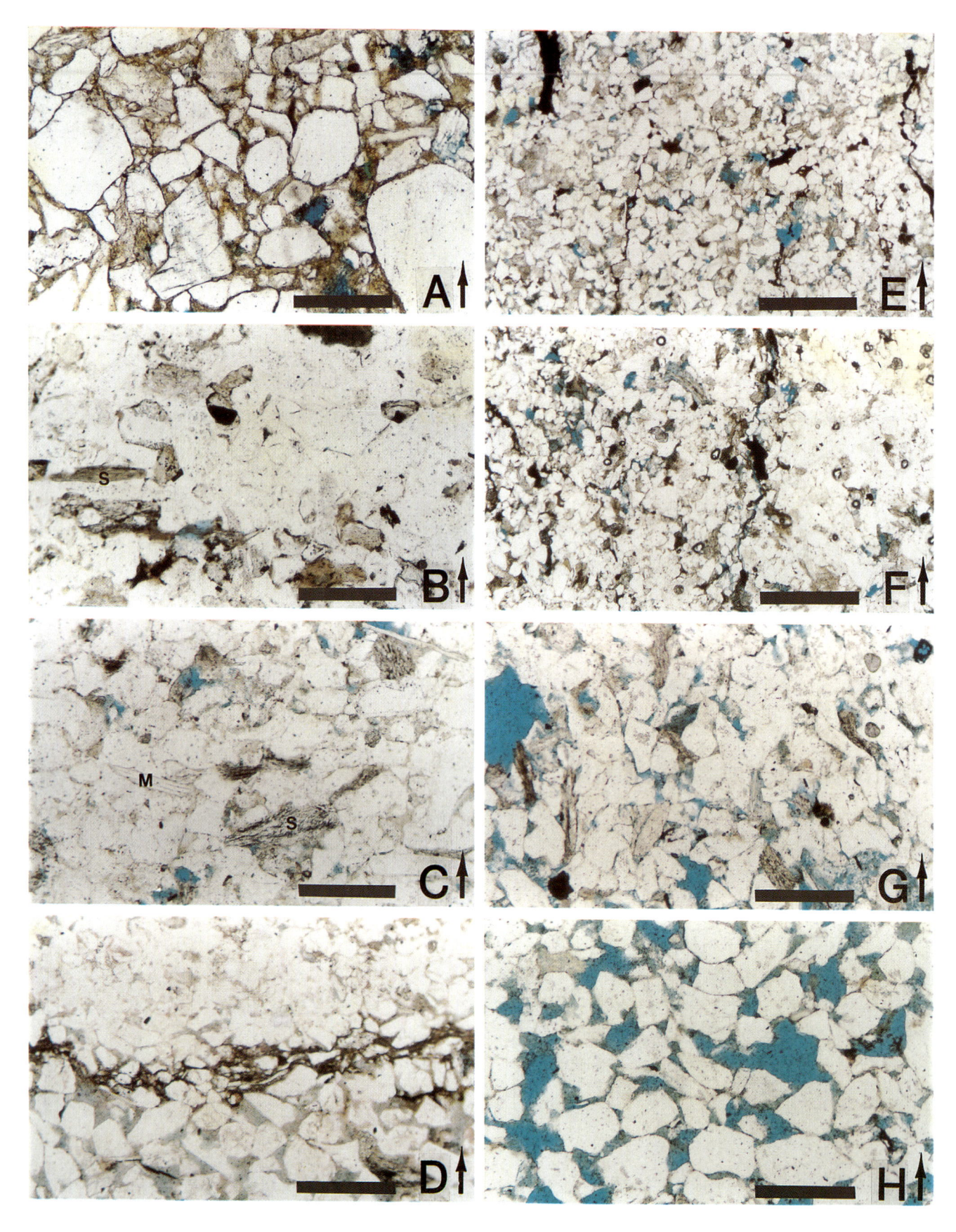

PLATE 4

SUBJECT INDEX

SUBJECT INDEX

A
Acidizing 270
Alberta 13, 14, 17, 21, 23, 25, 29, 30, 31, 32
Aluminum 60, 61, 76, 77, 111, 113, 114, 115, 116, 117, 118, 119, 120, 121, 137, 142
Argentina 159, 172, 173
Arkoma Basin 194, 227, 229, 230
Atoka 227, 228, 229, 230, 231, 232, 233, 234, 235, 236, 238, 239, 240
Authigenic clay 1, 5, 6, 9, 11, 40, 41, 47, 61, 65, 66, 67, 72, 73, 81, 86, 88, 92, 94, 95, 113, 121, 125, 127, 147, 154, 155, 162, 163, 185, 187, 190, 192, 194, 197, 198, 200, 201, 203, 206, 207, 210, 211, 214, 217, 219, 224, 238, 241, 254, 257, 258, 259, 263, 264, 265, 268

B
Brazil 172, 197, 227, 229
Burial history 3, 31, 35, 46, 47, 65, 94, 103, 106, 111, 137, 138, 140, 143, 147, 153, 160, 161, 186, 207

C
Chlorite 9, 10, 17, 19, 29, 35, 38, 41, 42, 43, 47, 60, 61, 67, 71, 73, 74, 78, 81, 82, 88, 92, 111, 113, 125, 128, 132, 133, 134, 135, 137, 141, 142, 145, 146, 147, 151, 152, 155, 159, 161, 163, 164, 165, 167, 168, 170, 171, 172, 173, 178, 179, 185, 187, 188, 190, 191, 192, 193, 194, 197, 198, 200, 201, 203, 204, 206, 207, 210, 211, 214, 217, 233, 234, 235, 238, 239, 241, 242, 243, 244, 246, 247, 248, 249, 250, 251, 252, 253, 257, 259, 265, 266
Clay coats 113, 116, 134, 209, 210, 211, 214, 217, 221, 222, 241, 243, 246, 251, 252, 253, 254, 266
Clay mineral diagenesis 74, 76, 185, 194
Clay mineralogy 1, 66, 67, 75, 86, 130, 132, 133, 145, 151, 161, 175, 178, 190, 192, 194, 209, 233, 238, 265
Clay minerals 1, 2, 3, 5, 6, 8, 9, 10, 11, 31, 46, 65, 66, 67, 69, 72, 73, 74, 76, 77, 78, 81, 86, 87, 88, 90, 92, 103, 125, 127, 133, 134, 135, 140, 141, 145, 147, 151, 152, 154, 155, 159, 172, 175, 178, 179, 180, 185, 187, 188, 190, 192, 193, 194, 197, 200, 211, 217, 227, 229, 230, 232, 233, 234, 235, 236, 238, 239, 241, 243, 257, 259, 264, 265, 266, 270
Clay rims 1, 67, 162, 200, 209, 210, 211, 214, 216, 217, 219, 221, 222, 223, 224, 241, 266
Clays 1, 9, 10, 19, 25, 29, 30, 40, 49, 50, 51, 52, 53, 55, 56, 57, 58, 59, 60, 61, 62, 70, 73, 74, 77, 78, 81, 84, 86, 88, 89, 90, 91, 92, 94, 95, 96, 105, 113, 121, 125, 127, 132, 133, 134, 135, 137, 141, 145, 147, 151, 152, 153, 154, 159, 162, 163, 167, 170, 172, 175, 178, 185, 187, 190, 192, 194, 197, 198, 199, 200, 201, 203, 204, 205, 206, 207, 209, 210, 211, 213, 214, 216, 217, 219, 220, 221, 222, 223, 224, 227, 233, 235, 236, 237, 238, 239, 240, 257, 258, 262, 263, 264, 265, 266, 267, 268, 269, 270
Comodoro Rivadavia Formation 159, 160, 161, 163, 171, 172, 173
Completion fluids 265, 267
Corrensite 74, 132, 134, 145, 147, 151, 153, 154, 155, 160, 192, 41, 243
Cretaceous 6, 13, 14, 15, 17, 19, 26, 27, 28, 29, 30, 31, 32, 65, 67, 69, 71, 78, 81, 82, 83, 84, 86, 88, 94, 96, 101, 102, 103, 126, 128, 159, 161, 172, 175, 185, 197, 207, 210, 243

D
Depositional 1, 5, 13, 14, 23, 29, 37, 38, 40, 45, 47, 61, 65, 67, 72, 73, 74, 75, 78, 81, 84, 87, 92, 99, 100, 101, 102, 105, 106, 111, 120, 121, 125, 126, 127, 132, 133, 134, 141, 142, 145, 153, 154, 172, 175, 185, 188, 190, 192, 194, 197, 198, 207, 216, 217, 219, 227, 228, 229, 233, 235, 236, 238, 239, 241, 243
Depositional matrix 185, 188, 190, 192, 194, 233, 235, 236, 238, 239
Detrital clays 40, 145, 147, 152, 203, 209, 210, 211, 217, 219, 239, 267
Dewatering 77, 227, 230, 231, 232, 233, 235, 238, 239, 240
Diagenesis 1, 3, 9, 13, 14, 17, 19, 21, 23, 25, 26, 28, 29, 30, 31, 32, 35, 37, 40, 47, 49, 56, 60, 61, 65, 67, 69, 70, 71, 74, 75, 76, 77, 78, 81, 82, 83, 86, 88, 89, 92, 99, 103, 104, 105, 106, 111, 113, 114, 120, 125, 126, 137, 139, 141, 142, 143, 147, 153, 154, 155, 161, 171, 172, 175, 178, 179, 182, 185, 188, 190, 192, 194, 198, 199, 200, 203, 207, 210, 217, 227, 235, 236, 239, 240, 241
Diagenetic minerals 13, 14, 17, 19, 21, 23, 30, 32, 61, 74, 86, 93, 141, 161
Dispersed clays 209, 239
Dissolved feldspar 88, 120, 239

E
Early diagenesis 13, 25, 29, 31, 32, 67, 69, 74, 75, 86, 88, 92, 141, 142, 179, 198, 203
Eolian sandstones 67, 154, 209, 217, 222

F
Fibrous illite 51, 52, 53, 57, 61, 91, 210, 223, 257, 259, 261, 263
Fluvial sandstone reservoirs 197
Formation damage 3, 257, 265, 269, 270, 271

G
Geochemistry 1, 3, 15, 65, 77, 125, 131, 137, 145, 238

H
Historical perspective 1

I
Illite 5, 6, 7, 8, 9, 10, 11, 13, 17, 19, 23, 25, 26, 27, 30, 31, 32, 35, 38, 41, 42, 43, 44, 45, 46, 47, 49, 50, 51, 52, 53, 55, 56, 57, 58, 59, 60, 61, 62, 65, 67, 69, 70, 71, 72, 73, 74, 75, 77, 78, 81, 82, 88, 89, 90, 91, 92, 93, 95, 96, 97, 98, 99, 114, 121, 132, 133, 134, 137, 140, 141, 142, 145, 151, 152, 154, 155, 175, 178, 179, 181, 182, 185, 190, 192, 200, 210, 217, 221, 223, 233, 234, 235, 237, 238, 239, 241, 243, 251, 252, 257, 259, 261, 263, 264, 265, 266, 271
Illite/smectite 30, 31, 49, 50, 55, 56, 57, 58, 65, 86, 125, 152, 175, 178, 179, 181, 182, 185, 190, 192, 194, 233, 241, 265, 266
Infiltrated 31, 111, 113, 118, 120, 121, 159, 161, 162, 163, 164, 165, 167, 168, 169, 170, 171, 172, 173, 185, 188, 190, 192, 194, 197, 199, 203, 205, 207, 216, 220, 221, 227, 240
Infiltrated clay 1, 162, 185, 190, 192, 194, 203, 216, 227

J
Jurassic 8, 15, 29, 30, 31, 46, 61, 62, 65, 67, 69, 70, 71, 72, 73, 74, 75, 76, 77, 78, 81, 82, 83, 84, 86, 88, 89, 90, 92, 94, 95, 96, 97, 98, 99, 100, 101, 102, 103, 105, 106, 121, 159, 172, 173, 185, 197, 210, 223, 227

K
Kaolinite 9, 13, 17, 19, 21, 25, 26, 29, 49, 51, 52, 53, 55, 56, 57, 58, 59, 60, 61, 65, 67, 69, 70, 71, 72, 73, 74, 75, 76, 77, 78, 81, 86, 87, 88, 89, 90, 91, 92, 93, 94, 95, 98, 101, 102, 103, 111, 113, 114, 115, 116, 117, 118, 119, 120, 121, 125, 128, 132, 133, 134, 136, 141, 142, 159, 161, 162, 163, 167, 175, 178, 179, 180, 181, 182, 185, 187, 190, 192, 193, 194, 197, 198, 210, 203, 206, 207, 217, 223, 233, 234, 235, 236, 238, 239, 265, 266, 269
Kolo Creek 175, 179, 180

M
Mass transfer 65, 104, 111, 113, 114, 118, 120, 121
Mesaverde Group 125, 126, 142
Meteoric water 13, 14, 23, 25, 26, 28, 29, 30, 31, 32, 59, 62, 65, 67, 69, 70, 71, 72, 73, 75, 76, 77, 78, 81, 82, 83, 92, 94, 95, 99, 100, 101, 102, 103, 106, 111, 112, 116, 118, 120, 121, 143, 180, 193
Michigan Basin 35, 36, 37, 40, 41, 46, 47
Middle Miocene 175

N
Neoformed clays 49, 50, 53, 60, 209, 227, 236
Niger delta 175, 179, 180, 181, 182
North Sea 5, 6, 8, 45, 61, 65, 66, 67, 69, 71, 72, 73, 75, 76, 77, 78, 81, 82, 83, 84, 86, 88, 89, 90, 91, 92, 94, 95, 96, 97, 98, 99, 101, 102, 103, 105, 106, 114, 116, 121, 185, 209, 210, 220, 223, 232

O

Organic maturation 125
Oxygen isotopes 94, 95, 154

P

Permeability models 259, 263
Petrography 3, 15, 95, 125, 126, 127, 133, 145, 147, 194
Piceance basin 125, 126, 137, 142
Porewater evolution 21, 23, 29
Porosity 2, 17, 19, 21, 23, 32, 35, 38, 40, 41, 46, 47, 65, 70, 73, 75, 76, 77, 78, 102, 103, 104, 105, 111, 113, 114, 115, 116, 117, 118, 119, 120, 121, 125, 127, 128, 135, 136, 137, 141, 142, 145, 147, 158, 160, 188, 197, 198, 199, 201, 203, 204, 205, 206, 209, 210, 219, 222, 223, 224, 227, 233, 235, 236, 237, 238, 239, 240, 241, 243, 244, 246, 247, 248, 250, 251, 253, 254, 257, 258, 259, 260, 261
Preservation of porosity 209, 243
Primary sedimentary structures 227, 230, 232, 235, 236, 239

R

Reservoir 3, 5, 8, 35, 41, 45, 49, 51, 56, 60, 61, 65, 66, 67, 70, 71, 72, 73, 75, 76, 77, 78, 81, 82, 83, 89, 90, 92, 93, 97, 99, 101, 102, 103, 105, 106, 111, 114, 119, 120, 121, 125, 145, 155, 172, 175, 178, 179, 180, 181, 182, 185, 188, 197, 198, 203, 204, 205, 206, 207, 208, 209, 210, 222, 223, 227, 232, 236, 238, 239, 240, 243, 247, 257, 258, 260, 263, 264, 265, 266, 267, 268, 269, 270, 271
Reservoir quality 1, 3, 35, 77, 78, 105, 111, 114, 185, 188, 198, 204, 205, 207, 209, 222, 227, 232, 236, 238, 239, 240
Rotated fault blocks 65, 71, 81, 82

S

San Jorge basin 159, 170
Sandstone 1, 2, 5, 6, 8, 9, 10, 11, 13, 14, 15, 16, 17, 19, 21, 23, 28, 29, 30, 31, 35, 36, 37, 38, 40, 41, 43, 44, 45, 46, 47, 49, 51, 57, 59, 60, 61, 62, 65, 67, 69, 70, 71, 72, 73, 75, 76, 77, 82, 83, 84, 86, 88, 90, 91, 92, 99, 102, 103, 105, 111, 112, 113, 116, 117, 118, 119, 120, 121, 125, 126, 127, 130, 131, 133, 134, 136, 137, 145, 147, 151, 153, 154, 155, 161, 172, 173, 175, 178, 181, 185, 187, 192, 194, 197, 200, 205, 206, 207, 209, 210, 211, 216, 219, 221, 222, 227, 229, 230, 231, 232, 234, 235, 236, 238, 239, 240, 241, 242, 243, 246, 247, 248, 249, 250, 251, 253, 254, 257, 258, 259, 260, 261, 262, 263, 267
Sandstone diagenesis 30, 98, 125, 185, 227
Sandstone reservoirs 2, 5, 6, 8, 9, 10, 12, 35, 77, 91, 99, 103, 120, 125, 126, 172, 197, 205, 207, 209, 241, 253
Secondary sedimentary structures 227, 230, 232, 235, 237, 239
Sedimentary basins 76, 77, 95, 111, 120, 121, 125, 126
Sedimentary petrology 1
Sespe 185, 186, 187, 188, 190, 192, 194
Shale 14, 15, 16, 27, 30, 31, 36, 40, 41, 42, 43, 45, 46, 47, 72, 113, 128, 133, 145, 147, 152, 159, 161, 167, 175, 187, 192, 197, 229, 230
Shelf sandstones 40, 41, 75, 111, 120, 121, 209, 210, 211, 214, 217, 219, 222, 224
Smectite 13, 29, 30, 31, 32, 51, 55, 57, 60, 61, 65, 67, 69, 71, 73, 74, 75, 77, 78, 81, 111, 113, 115, 120, 140, 141, 142, 145, 146, 151, 154, 155, 159, 161, 163, 165, 167, 170, 171, 172, 173, 178, 179, 182, 185, 190, 192, 194, 200, 206, 235, 241, 243, 247, 252, 253, 257, 265, 266
Spore coloration 175, 179, 182
St. Peter 35, 36, 37, 38, 40, 41, 45, 46, 47
Stable isotope 15, 17, 21, 65, 145, 146, 152, 154, 155
Stimulation 207, 265, 266, 267, 268, 270, 271

T

Thermal history 35, 142, 143, 152, 159
Thermal maturity 3, 143, 180, 181, 194, 229, 233, 235
Turbidites 73, 75, 197, 227, 230, 232, 234, 235, 240
Tuscaloosa Sandstone 241, 243, 247, 250, 251